T0317447

FUNDAMENTALS OF LIQUID CRYSTAL DEVICES

Wiley-SID Series in Display Technology

Series Editors:
Anthony C. Lowe and Ian Sage

Display Systems: Design and Applications
Lindsay W. MacDonald and Anthony C. Lowe (Eds.)

Electronic Display Measurement: Concepts, Techniques, and Instrumentation
Peter A. Keller

Reflective Liquid Crystal Displays
Shin-Tson Wu and Deng-Ke Yang

Colour Engineering: Achieving Device Independent Colour
Phil Green and Lindsay MacDonald (Eds.)

Display Interfaces: Fundamentals and Standards
Robert L. Myers

Digital Image Display: Algorithms and Implementation
Gheorghe Berbecel

Flexible Flat Panel Displays
Gregory Crawford (Ed.)

Polarization Engineering for LCD Projection
Michael G. Robinson, Jianmin Chen, and Gary D. Sharp

Fundamentals of Liquid Crystal Devices
Deng-Ke Yang and Shin-Tson Wu

Introduction to Microdisplays
David Armitage, Ian Underwood, and Shin-Tson Wu

Mobile Displays: Technology and Applications
Achintya K. Bhowmik, Zili Li, and Philip Bos (Eds.)

Photoalignment of Liquid Crystalline Materials: Physics and Applications
Vladimir G. Chigrinov, Vladimir M. Kozenkov and Hoi-Sing Kwok

Projection Displays, Second Edition
Matthew S. Brennesholtz and Edward H. Stupp

Introduction to Flat Panel Displays
Jiun-Haw Lee, David N. Liu and Shin-Tson Wu

LCD Backlights
Shunsuke Kobayashi, Shigeo Mikoshiba and Sungkyoo Lim (Eds.)

Liquid Crystal Displays: Addressing Schemes and Electro-Optical Effects, Second Edition
Ernst Lueder

Transflective Liquid Crystal Displays
Zhibing Ge and Shin-Tson Wu

Liquid Crystal Displays: Fundamental Physics and Technology
Robert H. Chen

3D Displays
Ernst Lueder

OLED Display Fundamentals and Applications
Takatoshi Tsujimura

Illumination, Colour and Imaging: Evaluation and Optimization of Visual Displays
Tran Quoc Khanh and Peter Bodrogi

Interactive Displays: Natural Human-Interface Technologies
Achintya K. Bhowmik (Ed.)

Modeling and Optimization of LCD Optical Performance
Dmitry A. Yakovlev, Vladimir G. Chigrinov, Hoi-Sing Kwok

Addressing Techniques of Liquid Crystal Displays
Temkar N. Ruckmongathan

FUNDAMENTALS OF LIQUID CRYSTAL DEVICES

Second Edition

Deng-Ke Yang
Liquid Crystal Institute, Kent State University, USA

Shin-Tson Wu
College of Optics and Photonics, University of Central Florida, USA

This edition first published 2015

First Edition published in 2006

Registered Office
John Wiley & Sons, Ltd, The Atrium, Southern Gate, Chichester, West Sussex, PO19 8SQ, United Kingdom

For details of our global editorial offices, for customer services and for information about how to apply for permission to reuse the copyright material in this book please see our website at www.wiley.com.

Library of Congress Cataloging-in-Publication Data

Yang, Deng-Ke.
Fundamentals of liquid crystal devices / Deng-Ke Yang and Shin-Tson Wu. – Second edition.
pages cm – (Wiley series in display technology)
Includes bibliographical references and index.
ISBN 978-1-118-75200-5 (hardback)
1. Liquid crystal displays. 2. Liquid crystal devices. 3. Liquid crystals. I. Wu, Shin-Tson. II. Title.
TK7872.L56Y36 2014
621.3815′422–dc23

2014027707

A catalogue record for this book is available from the British Library.

Set in 10/12pt Times by SPi Publisher Services, Pondicherry, India

1 2015

Contents

Series Editor's Foreword

The first edition of this book marked a new departure for the Wiley-SID Series in Display Technology because it had been written primarily as a text for postgraduate and senior undergraduate students. It fulfilled that objective admirably, but the continuing advances in liquid crystal display technology over the intervening eight years have made necessary some additions to keep the book current.

Accordingly, the following new sections have been added: elastic deformation of liquid crystals in Chapter 1, polarisation conversion with narrow and broadband quarter-wave plates in Chapter 3, and measurement of anchoring strength and pretilt angle in Chapter 5.

With each chapter is designed to be self-contained, the first chapters cover the basic physics of liquid crystals, their interaction with light and electric fields, and the means by which they can be modelled. Next are described the majority of the ways in which liquid crystals can be used in displays, and Chapter 12, the final chapter of the first edition, deals with photonic devices such as beam steerers, tunable-focus lenses and polarisation-independent devices. In this second edition, four new chapters have been added: two on blue phase and polymer stabilised blue phase liquid crystals, which are emerging from the realm of academic research to show promise for very fast response display and photonic devices, a chapter which discusses LCD componentry, and a final chapter on the use of LCDs in 3D display systems.

As with the first edition, and following a standard textbook format, each chapter concludes with a set of problems, the answers to which may be found on the Wiley web site.

New electro-optic technologies continue to be developed, and some of them make inroads into the LCD market. Nevertheless, liquid crystal technology – the first other than the CRT to make a significant breakthrough into the mass market and which made possible flat displays and transformed projection display technology – continues to hold a dominant position. This second edition of Fundamentals of Liquid Crystal Devices, with its additions which include references to some very recent work, will ensure that this volume will continue to provide students and other readers at the professional level with a most useful introduction to the subject.

Anthony C Lowe
Braishfield, UK 2014

Preface to the First Edition

Liquid crystal displays have become the leading technology in the information display industry. They are used in small-sized displays such as calculators, cellular phones, digital cameras, and head-mounted displays; in medium-sized displays such as laptop and desktop computers; and in large-sized displays such as direct-view TVs and projection TVs. They have the advantages of high resolution and high brightness, and, being flat paneled, are lightweight, energy saving, and even flexible in some cases. They can be operated in transmissive and reflective modes. Liquid crystals have also been used in photonic devices such as laser beam steering, variable optical attenuators, and tunable-focus lenses. There is no doubt that liquid crystals will continue to play an important role in the era of information technology.

There are many books on the physics and chemistry of liquid crystals and on liquid crystal devices. There are, however, few books covering both the basics and applications of liquid crystals. Our main goal, therefore, is to provide a textbook for senior undergraduate and graduate students. The book can be used for a one- or two-semester course. The instructors can selectively choose the chapters and sections according to the length of the course and the interest of the students. The book can also be used as a reference book by scientists and engineers who are interested in liquid crystal displays and photonics.

The book is organized in such a way that the first few chapters cover the basics of liquid crystals and the necessary techniques to study and design liquid crystal devices. The later chapters cover the principles, design, operation, and performance of liquid crystal devices. Because of limited space, we cannot cover every aspect of liquid crystal chemistry and physics and all liquid crystal devices, but we hope this book will introduce readers to liquid crystals and provide them with the basic knowledge and techniques for their careers in liquid crystals.

We are greatly indebted to Dr A. Lowe for his encouragement. We are also grateful to the reviewers of our book proposal for their useful suggestions and comments. Deng-Ke Yang would like to thank Ms E. Landry and Prof. J. Kelly for patiently proofreading his manuscript. He would also like to thank Dr Q. Li for providing drawings. Shin-Tson Wu would like to thank his research group members for generating the new knowledge included in this book, especially Drs Xinyu Zhu, Hongwen Ren, Yun-Hsing Fan, and Yi-Hsin Lin, and Mr Zhibing Ge for kind help during manuscript preparation. He is also indebted to Dr Terry Dorschner of Raytheon, Dr Paul McManamon of the Air Force Research Lab, and Dr Hiroyuki Mori of Fuji Photo Film for sharing their latest results. We would like to thank our colleagues and friends for useful discussions and drawings and our funding agencies (DARPA, AFOSR, AFRL, and Toppoly) for providing financial support. Finally, we also would like to thank our families (Xiaojiang Li, Kevin Yang, Steven Yang, Cho-Yan Wu, Janet Wu, and Benjamin Wu) for their spiritual support, understanding, and constant encouragement.

Deng-Ke Yang
Shin-Tson Wu

Preface to the Second Edition

Liquid crystal displays have become the leading technology in the information display industry. They are used in small-sized displays such as calculators, smart phones, digital cameras, and wearable displays; medium-sized displays such as laptop and desktop computers; and large-sized displays such as direct view TVs and data projectors. They have the advantages of having high resolution and high brightness, and being flat paneled, lightweight, energy saving, and even flexible in some cases. They can be operated in transmissive and reflective modes. Liquid crystals have also been used in photonic devices such as switching windows, laser beam steering, variable optical attenuators, and tunable-focus lenses. There is no doubt that liquid crystals will continue to play an important role in information technology.

There are many books on the physics and chemistry of liquid crystals and on liquid crystal devices. There are, however, few books covering both the basics and the applications of liquid crystals. The main goal of this book is to provide a textbook for senior undergraduate and graduate students. This book can be used for a one- or two-semester course. The instructors can selectively choose the chapters and sections according to the length of the course and the interest of the students. It can also be used as a reference book for scientists and engineers who are interested in liquid crystal displays and photonics.

The book is organized in such a way that the first few chapters cover the basics of liquid crystals and the necessary techniques to study and design liquid crystal devices. The later chapters cover the principles, design, operation, and performance of liquid crystal devices. Because of limited space, we cannot cover every aspect of liquid crystal chemistry and physics and all the different liquid crystal devices. We hope that this book can introduce readers to liquid crystals and provide them with the basic knowledge and techniques for their career in liquid crystals.

Since the publication of the first edition, we have received a lot of feedback, suggestions, corrections, and encouragements. We appreciate them very much and have put them into the second edition. Also there are many new advances in liquid crystal technologies. We have added new chapters and sections to cover them.

We are greatly indebted to Dr A. Lowe for his encouragement. We are also grateful to the reviewers of our book proposal for their useful suggestions and comments. Deng-Ke Yang would like to thank Ms E. Landry, Prof. P. Crooker, his research group, and coworkers for patiently proofreading and preparing his sections of the book. He would also like to thanks Dr Q. Li for providing drawings. Shin-Tson Wu would like to thank his research group members for generating new knowledge included in this book, especially Drs Xinyu Zhu, Hongwen Ren, Yun-Hsing Fan, Yi-Hsin Lin, Zhibing Ge, Meizi Jiao, Linghui Rao, Hui-Chuan Cheng, Yan Li, and Jin Yan for their kind help during manuscript preparation. He is also indebted to Dr Terry Dorschner of Raytheon, Dr Paul McManamon of Air Force Research Lab, and Dr Hiroyuki Mori of Fuji Photo Film for sharing their latest results. We would like to thank our colleagues and friends for useful discussions and drawings and our funding agencies (DARPA, AFOSR, AFRL, ITRI, AUO, and Innolux) for providing financial support. We would also like to thank our family members (Xiaojiang Li, Kevin Yang, Steven Yang, Cho-Yan Wu, Janet Wu, and Benjamin Wu) for their spiritual support, understanding, and constant encouragement.

Deng-Ke Yang
Shin-Tson Wu

1

Liquid Crystal Physics

1.1 Introduction

Liquid crystals are mesophases between crystalline solid and isotropic liquid [1–3]. The constituents are elongated rod-like (calamitic) or disk-like (discotic) organic molecules as shown in Figure 1.1. The size of the molecules is typically a few nanometers (nm). The ratio between the length and the diameter of the rod-like molecules or the ratio between the diameter and the thickness of disk-like molecules is about 5 or larger. Because the molecules are non-spherical, besides positional order, they may possess orientational order.

Figure 1.1(a) shows a typical *calamitic* liquid crystal molecule. Its chemical name is 4′-n-Pentyl-4-cyano-biphenyl and is abbreviated as 5CB [4,5]. It consists of a biphenyl, which is the rigid core, and a hydrocarbon chain which is the flexible tail. The space-filling model of the molecule is shown in Figure 1.1(c). Although the molecule itself is not cylindrical, it can be regarded as a cylinder, as shown in Figure 1.1(e), in considering its physical behavior, because of the fast rotation (on the order of 10^{-9} s) around the long molecule axis due to thermal motion. The distance between two carbon atoms is about 1.5 Å; therefore the length and the diameter of the molecule are about 2 nm and 0.5 nm, respectively. The molecule shown has a permanent dipole moment (from the CN head), but it can still be represented by the cylinder whose head and tail are the same, because in non-ferroelectric liquid crystal phases, the dipole has equal probability of pointing up or down. It is necessary for a liquid crystal molecule to have a rigid core(s) and flexible tail(s). If the molecule is completely flexible, it will not have orientational order. If it is completely rigid, it will transform directly from isotropic liquid phase at high temperature to crystalline solid phase at low temperature. The rigid part favors both orientational and positional order while the flexible part disfavors them. With balanced rigid and flexible parts, the molecule exhibits liquid crystal phases.

Fundamentals of Liquid Crystal Devices, Second Edition. Deng-Ke Yang and Shin-Tson Wu.

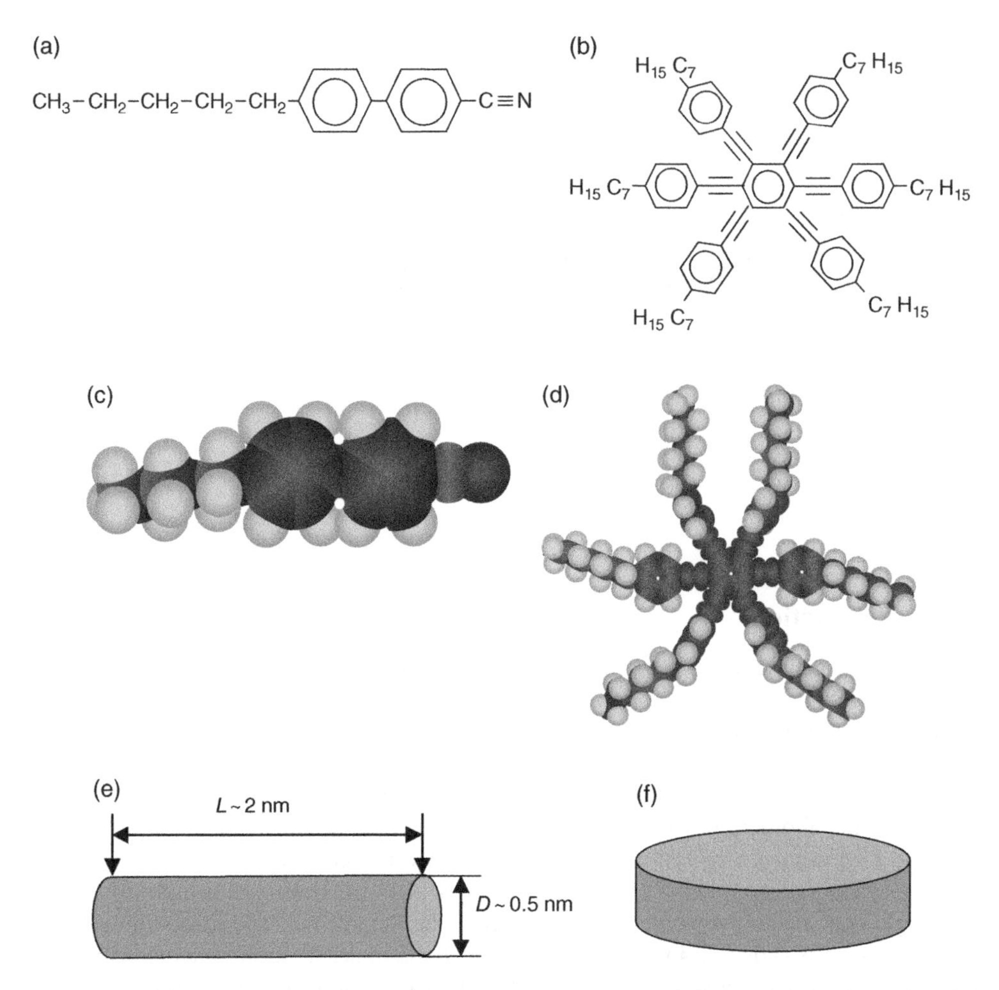

Figure 1.1 Calamitic liquid crystal: (a) chemical structure, (c) space-filling model, (e) physical model. Discostic liquid crystal: (b) chemical structure, (d) space-filling mode, (f) physical model.

Figure 1.1(b) shows a typical *discotic* liquid crystal molecule [6]. It also has a rigid core and flexible tails. The branches are approximately on one plane. The space-filling model of the molecule is shown in Figure 1.1(d). If there is no permanent dipole moment perpendicular to the plane of the molecule, it can be regarded as a disk in considering its physical behavior as shown in Figure 1.1(f) because of the fast rotation around the axis which is at the center of the molecule and perpendicular to the plane of the molecule. If there is a permanent dipole moment perpendicular to the plane of the molecule, it is better to visualize the molecule as a bowl, because the reflection symmetry is broken and all the permanent dipoles may point in the same direction and spontaneous polarization occurs. The flexible tails are also necessary, otherwise the molecules form a crystal phase where there is positional order.

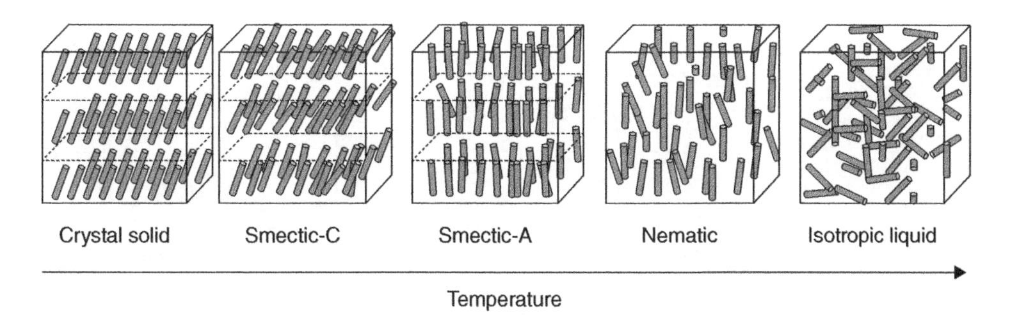

Figure 1.2 Schematic representation of the phases of rod-like molecules.

The variety of phases that may be exhibited by rod-like molecules are shown in Figure 1.2. At high temperature, the molecules are in the isotropic liquid state where they do not have either positional or orientational order. The molecules can easily move around, and the material can flow like water. The translational viscosity is comparable to that of water. Both the long and short axes of the molecules can point in any direction.

When the temperature is decreased, the material transforms into the *nematic* phase, which is the most common and simplest liquid crystal phase, where the molecules have orientational order but still no positional order. The molecules can still diffuse around, and the translational viscosity does not change much from that of the isotropic liquid state. The long axis of the molecules has a preferred direction. Although the molecules still swivel due to thermal motion, the time-averaged direction of the long axis of a molecule is well defined and is the same for all the molecules at macroscopic scale. The average direction of the long molecular axis is denoted by $\vec{n}$ which is a unit vector called the liquid crystal director. The short axes of the molecules have no orientational order in a uniaxial nematic liquid crystal.

When the temperature is decreased further, the material may transform into the *Smectic-A* phase where, besides the orientational order, the molecules have partial positional order, i.e., the molecules form a layered structure. The liquid crystal director is perpendicular to the layers. Smectic-A is a one-dimensional crystal where the molecules have positional order in the layer normal direction. The cartoon shown in Figure 1.2 is schematic. In reality, the separation between neighboring layers is not as well defined as that shown by the cartoon. The molecule number density exhibits an undulation with the wavelength about the molecular length. Within a layer, it is a two-dimensional liquid crystal in which there is no positional order, and the molecules can move around. For a material in poly-domain smectic-A, the translational viscosity is significantly higher, and it behaves like a grease. When the temperature is decreased further, the material may transform into the *smectic-C* phase, where the liquid crystal director is no longer perpendicular to the layer but tilted.

At low temperature, the material is in the crystal solid phase where there are both positional and orientational orders. The translational viscosity becomes infinitely high and the molecules (almost) do not diffuse anymore.

Liquid crystals get the 'crystal' part of their name because they exhibit optical birefringence as crystalline solids. They get the 'liquid' part of their name because they can flow and do not support shearing as regular liquids. Liquid crystal molecules are elongated and have different

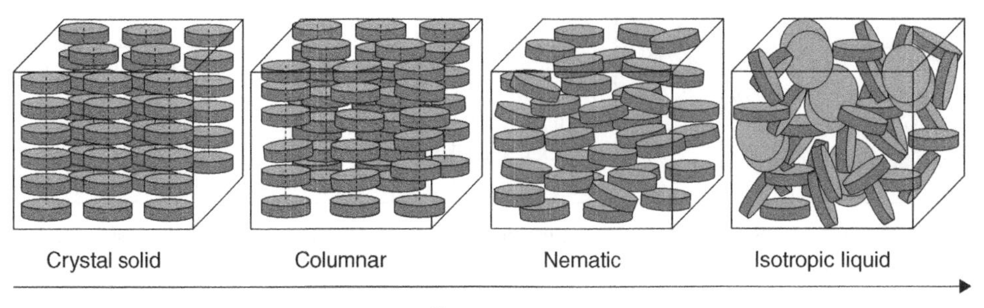

Figure 1.3 Schematic representation of the phases of disk-like molecules.

molecular polarizabilities along their long and short axes. Once the long axes of the molecules orient along a common direction, the refractive indices along and perpendicular to the common direction are different. It should be noted that not all rod-like molecules exhibit all the liquid crystal phases. They may exhibit some of the liquid crystal phases.

Some of the liquid crystal phases of disk-like molecules are shown in Figure 1.3. At high temperature, they are in the isotropic liquid state where there are no positional and orientational orders. The material behaves in the same way as a regular liquid. When the temperature is decreased, the material transforms into the nematic phase, which has orientational order but not positional order. The average direction of the short axis perpendicular to the disk is oriented along a preferred direction, which is also called the liquid crystal director and denoted by a unit vector $\vec{n}$. The molecules have different polarizabilities along a direction in the plane of the disk and along the short axis. Thus the discotic nematic phase also exhibits birefringence as crystals.

When the temperature is decreased further, the material transforms into the columnar phase where, besides orientational order, there is partial positional order. The molecules stack up to form columns. Within a column, it is a liquid where the molecules have no positional order. The columns, however, are arranged periodically in the plane perpendicular to the columns. Hence it is a two-dimensional crystal. At low temperature, the material transforms into the crystalline solid phase where the positional order along the columns is developed.

The liquid crystal phases discussed so far are called thermotropic liquid crystals and the transitions from one phase to another are driven by varying temperature. There is another type of liquid crystals, called lyotropic liquid crystals, exhibited by molecules when they are mixed with a solvent of some kind. The phase transitions from one phase to another phase are driven by varying the solvent concentration. Lyotropic liquid crystals usually consist of amphiphilic molecules that have a hydrophobic group at one end and a hydrophilic group at the other end and the water is the solvent. The common lyotropic liquid crystal phases are micelle phase and lamellar phase. Lyotropic liquid crystals are important in biology. They will not be discussed in this book because the scope of this book is on displays and photonic devices.

Liquid crystals have a history of more than 100 years. It is believed that the person who discovered liquid crystals is Friedrich Reinitzer, an Austrian botanist [7]. The liquid crystal phase observed by him in 1888 was a cholesteric phase. Since then, liquid crystals have come a long way and become a major branch of interdisciplinary sciences. Scientifically, liquid crystals are important because of the richness of structures and transitions. Technologically, they have won tremendous success in display and photonic applications [8–10].

1.2 Thermodynamics and Statistical Physics

Liquid crystal physics is an interdisciplinary science: thermodynamics, statistical physics, electrodynamics, and optics are involved. Here we give a brief introduction to thermodynamics and statistical physics.

1.2.1 Thermodynamic laws

One of the important quantities in thermodynamics is entropy. From the microscopic point of view, entropy is a measurement of the number of quantum states accessible to a system. In order to define entropy quantitatively, we first consider the fundamental logical assumption that *for a closed system (no energy and particles exchange with other systems), quantum states are either accessible or inaccessible to the system, and the system is equally likely to be in any one of the accessible states as in any other accessible state* [11]. For a macroscopic system, the number of accessible quantum states g is a huge number ($\sim 10^{23}$). It is easier to deal with $\ln g$, which is defined as the *entropy* σ:

$$\sigma = \ln g \tag{1.1}$$

If a closed system consists of subsystem 1 and subsystem 2, the numbers of accessible states of the subsystems are g_1 and g_2, respectively. The number of accessible quantum states of the whole system is $g = g_1 g_2$ and the entropy is $\sigma = \ln g = \ln(g_1 g_2) = \ln g_1 + \ln g_2 = \sigma_1 + \sigma_2$.

Entropy is a function of the energy u of the system $\sigma = \sigma(u)$. The second law of thermodynamics states that *for a closed system, the equilibrium state has the maximum entropy.* Let us consider a closed system which contains two subsystems. When two subsystems are brought into thermal contact (energy exchange between them is allowed), the energy is allocated to maximize the number of accessible states, that is, the entropy is maximized. Subsystem 1 has the energy u_1 and entropy σ_1; subsystem 2 has the energy u_2 and entropy σ_2. For the whole system, $u = u_1 + u_2$ and $\sigma = \sigma_1 + \sigma_2$. The first law of thermodynamics states that *energy is conserved*, that is, $u = u_1 + u_2 = \text{constant}$. For any process inside the closed system, $\delta u = \delta u_1 + \delta u_2 = 0$. From the second law of thermodynamics, for any process, we have $\delta\sigma = \delta\sigma_1 + \delta\sigma_2 \geq 0$. When the two subsystems are brought into thermal contact, at the beginning, energy flows. For example, an amount of energy $|\delta u_1|$ flows from subsystem 1 to subsystem 2, $\delta u_1 < 0$ and $\delta u_2 = -\delta u_1 > 0$, and $\frac{\partial\sigma}{\partial u_2} = \frac{\partial\sigma_1}{\partial u_2} + \frac{\partial\sigma_2}{\partial u_2} = \frac{\partial\sigma_1}{\partial u_1}\frac{\partial u_1}{\partial u_2} + \frac{\partial\sigma_2}{\partial u_2} = -\frac{\partial\sigma_1}{\partial u_1} + \frac{\partial\sigma_2}{\partial u_2} \geq 0$. When equilibrium is reached, the entropy is maximized and $\frac{\partial\sigma_1}{\partial u_1} - \frac{\partial\sigma_2}{\partial u_2} = 0$, that is, $\frac{\partial\sigma_1}{\partial u_1} = \frac{\partial\sigma_2}{\partial u_2}$. We know that when two systems reach equilibrium, they have the same temperature. Accordingly the *fundamental temperature* τ is defined by

$$1/\tau = \left(\frac{\partial\sigma}{\partial u}\right)_{N,V}. \tag{1.2}$$

Energy flows from a high temperature system to a low temperature system. The *conventional temperature* (Kelvin temperature) is defined by

$$T = \tau/k_B, \tag{1.3}$$

where $k_B = 1.381 \times 10^{-23}$ *Joule/Kelvin* is the Boltzmann constant. *Conventional entropy S* is defined by

$$1/T = \partial S/\partial u. \tag{1.4}$$

Hence

$$S = k_B \sigma. \tag{1.5}$$

1.2.2 Boltzmann Distribution

Now we consider the thermodynamics of a system at a constant temperature, that is, in thermal contact with a thermal reservoir. The temperature of the thermal reservoir (named B) is τ. The system under consideration (named A) has two states with energy 0 and ε, respectively. A and B form a closed system, and its total energy $u = u_A + u_B = u_o =$ constant. When A is in the state with energy 0, B has the energy u_o, the number of accessible states: $g_1 = g_A \times g_B = 1 \times g_B(u_o) = g_B(u_o)$. When A has the energy ε, B has the energy $u_o - \varepsilon$, the number of accessible states is $g_2 = g_A \times g_B = 1 \times g_B(u_o - \varepsilon) = g_B(u_o - \varepsilon)$. For the whole system, the total number of accessible states is

$$G = g_1 + g_2 = g_B(u_o) + g_B(u_o - \varepsilon). \tag{1.6}$$

(A + B) is a closed system, and the probability in any of the G states is the same. When the whole system is in one of the g_1 states, A has the energy 0. When the whole system is in one of the g_2 states, A has the energy ε. Therefore the probability for A in the state with energy 0 is $P(0) = \dfrac{g_1}{g_1+g_2} = \dfrac{g_B(u_o)}{g_B(u_o)+g_B(u_o-\varepsilon)}$. The probability for A in the state with energy ε is $P(\varepsilon) = \dfrac{g_2}{g_1+g_2} = \dfrac{g_B(u_o-\varepsilon)}{g_B(u_o)+g_B(u_o-\varepsilon)}$. From the definition of entropy, we have $g_B(u_o) = e^{\sigma_B(u_o)}$ and $g_B(u_o - \varepsilon) = e^{\sigma_B(u_o-\varepsilon)}$. Because $\varepsilon \ll u_o$, $\sigma_B(u_o-\varepsilon) \approx \sigma_B(u_o) - \dfrac{\partial \sigma_B}{\partial u_B}\varepsilon = \sigma_B(u_o) - \frac{1}{\tau}\varepsilon$. Therefore we have

$$P(0) = \frac{e^{\sigma_B(u_o)}}{e^{\sigma_B(u_o)} + e^{\sigma_B(u_o)-\varepsilon/\tau}} = \frac{1}{1+e^{-\varepsilon/\tau}} = \frac{1}{1+e^{-\varepsilon/k_BT}} \tag{1.7}$$

$$P(\varepsilon) = \frac{e^{\sigma_B(u_o)-\varepsilon/\tau}}{e^{\sigma_B(u_o)} + e^{\sigma_B(u_o)-\varepsilon/\tau}} = \frac{e^{-\varepsilon/\tau}}{1+e^{-\varepsilon/\tau}} = \frac{e^{-\varepsilon/k_BT}}{1+e^{-\varepsilon/k_BT}} \tag{1.8}$$

$$\frac{p(\varepsilon)}{P(0)} = e^{-\varepsilon/k_BT}. \tag{1.9}$$

For a system having N states with energies $\varepsilon_1, \varepsilon_2, \ldots\ldots, \varepsilon_i, \varepsilon_{i+1}, \ldots\ldots, \varepsilon_N$, the probability for the system in the state with energy ε_i is

$$P(\varepsilon_i) = e^{-\varepsilon_i/\tau} \Big/ \sum_{j=1}^{N} e^{-\varepsilon_j/k_BT}. \tag{1.10}$$

The *partition function* of the system is defined as

$$Z = \sum_{i=1}^{N} e^{-\varepsilon_i/k_B T}. \tag{1.11}$$

The *internal energy* (average energy) of the system is given by

$$U = <\varepsilon> = \sum_i \varepsilon_i P(\varepsilon_i) = \frac{1}{Z}\sum_i \varepsilon_i e^{-\varepsilon_i/k_B T}. \tag{1.12}$$

Because $\frac{\partial Z}{\partial T} = \sum_i \left(\frac{\varepsilon_i}{k_B T^2}\right) e^{-\varepsilon_i/k_B T} = \frac{1}{k_B T^2}\sum_i \varepsilon_i e^{-\varepsilon_i/k_B T}$,

$$U = \frac{k_B T^2}{Z}\frac{\partial Z}{\partial T} = k_B T^2 \frac{\partial(\ln Z)}{\partial T}. \tag{1.13}$$

1.2.3 Thermodynamic quantities

As energy is conserved, the change of the internal energy U of a system equals the heat dQ absorbed and the mechanical work dW done to the system, $dU = dQ + dW$. When the volume of the system changes by dV under the pressure P, the mechanical work done to the system is given by

$$dW = -PdV. \tag{1.14}$$

When there is no mechanical work, the heat absorbed equals the change of internal energy. From the definition of temperature $1/T = \left(\frac{\partial S}{\partial U}\right)_V$, the heat absorbed in a reversible process at constant volume is

$$dU = dQ = TdS. \tag{1.15}$$

When the volume is not constant, then

$$dU = TdS - PdV. \tag{1.16}$$

The derivatives are

$$T = \left(\frac{\partial U}{\partial S}\right)_V, \tag{1.17}$$

$$P = -\left(\frac{\partial U}{\partial V}\right)_S. \tag{1.18}$$

The internal energy U, entropy S, and volume V are extensive quantities, while temperature T and pressure P are intensive quantities. The *enthalpy* H of the system is defined by

$$H = U + PV. \tag{1.19}$$

Its variation in a reversible process is given by

$$dH = dU + d(PV) = (TdS - PdV) + (PdV + VdP) = TdS + VdP. \tag{1.20}$$

From this equation, it can be seen that the physical meaning of enthalpy is that in a process at constant pressure ($dP = 0$), the change of enthalpy dH is equal to the heat absorbed dQ ($=TdS$)). The derivatives of the enthalpy are

$$T = \left(\frac{\partial H}{\partial S}\right)_P, \tag{1.21}$$

$$V = \left(\frac{\partial H}{\partial P}\right)_S. \tag{1.22}$$

The *Helmholtz free energy* F of the system is defined by

$$F = U - TS. \tag{1.23}$$

Its variation in a reversible process is given by

$$dF = dU - d(TS) = (TdS - PdV) - (TdS + SdT) = -SdT - PdV. \tag{1.24}$$

The physical meaning of Helmholtz free energy is that in a process at constant temperature ($dT = 0$), the change of Helmholtz free energy is equal to the work done to the system.

The derivatives are

$$S = -\left(\frac{\partial F}{\partial T}\right)_V, \tag{1.25}$$

$$P = -\left(\frac{\partial F}{\partial V}\right)_T. \tag{1.26}$$

The *Gibbs free energy* G of the system is defined by

$$G = U - TS + PV. \tag{1.27}$$

The variation in a reversible process is given by

$$dG = dU - d(TS) - d(PV) = -SdT + VdP. \tag{1.28}$$

In a process at constant temperature and pressure, Gibbs free energy does not change. The derivatives are

$$S = -\left(\frac{\partial G}{\partial T}\right)_P, \tag{1.29}$$

$$V = \left(\frac{\partial G}{\partial P}\right)_T. \tag{1.30}$$

The Helmholtz free energy can be derived from the partition function. Because of Equations (1.13) and (1.25),

$$F = U - TS = k_B T^2 \frac{\partial(\ln Z)}{\partial T} + T\left(\frac{\partial F}{\partial T}\right)_V$$

$$F - T\left(\frac{\partial F}{\partial T}\right)_V = -T^2\left\{\frac{1}{T}\left(\frac{\partial F}{\partial T}\right)_V + F\left[\frac{\partial\left(\frac{1}{T}\right)}{\partial T}\right]_V\right\} = -T^2\left[\frac{\partial\left(\frac{F}{T}\right)}{\partial T}\right]_V = k_B T^2 \frac{\partial(\ln Z)}{\partial T}.$$

Hence

$$F = -k_B T \ln Z = -k_B T \ln\left(\sum_i e^{-\varepsilon_i/k_B T}\right). \tag{1.31}$$

From Equations (1.11), (1.25) and (1.31), the entropy of a system at a constant temperature can be calculated:

$$S = -k_B < \ln \rho > = -k_B \sum_i \rho_i \ln \rho_i \tag{1.32}$$

1.2.4 *Criteria for thermodynamical equilibrium*

Now we consider the criteria which can used to judge whether a system is in its equilibrium state under given conditions. We already know that for a closed system, as it changes from a non-equilibrium state to the equilibrium state, the entropy increases,

$$\delta S \geq 0. \tag{1.33}$$

It can be stated in a different way that for a closed system the entropy is maximized in the equilibrium state.

In considering the equilibrium state of a system at constant temperature and volume, we construct a closed system which consists of the system (subsystem 1) under consideration and a thermal reservoir (subsystem 2) with the temperature T. When the two systems are brought into thermal contact, energy is exchanged between subsystem 1 and subsystem 2. Because the whole system is a closed system, $\delta S = \delta S_1 + \delta S_2 \geq 0$. For system 2, $1/T = \left(\frac{\partial S_2}{\partial U_2}\right)_V$, and therefore $\delta S_2 = \delta U_2/T$ (this is true when the volume of subsystem is fixed, which also means that the volume of subsystem 1 is fixed). Because of energy conservation, $\delta U_2 = -\delta U_1$. Hence $\delta S = \delta S_1 + \delta S_2 = \delta S_1 + \delta U_2/T = \delta S_1 - \delta U_1/T \geq 0$. Because the temperature and volume are constant for subsystem 1, $\delta S_1 - \delta U_1/T = (1/T)\delta(TS_1 - U_1) \geq 0$, and therefore

$$\delta(U_1 - TS_1) = \delta F_1 \leq 0. \tag{1.34}$$

At constant temperature and volume, the equilibrium state has the minimum Helmholtz free energy.

In considering the equilibrium state of a system at constant temperature and pressure, we construct a closed system which consists of the system (subsystem 1) under consideration and a thermal reservoir (subsystem 2) with the temperature T. When the two systems are brought into thermal contact, energy is exchanged between subsystem 1 and subsystem 2. Because the whole system is a closed system, $\delta S = \delta S_1 + \delta S_2 \geq 0$. For system 2, because the volume is not fixed, and mechanical work is involved. $\delta U_2 = T\delta S_2 - P\delta V_2$, that is, $\delta S_2 = (\delta U_2 + P\delta V_2)/T$. Because $\delta U_2 = -\delta U_1$ and $\delta V_2 = -\delta V_1$, $\delta S = \delta S_1 + (\delta U_2 + P\delta V_2)/T = \delta S_1 - (\delta U_1 + P\delta V_1)/T = (1/T)\delta(TS_1 - U_1 - PV_1) \geq 0$. Therefore

$$\delta(U_1 + PV_1 - TS_1) = \delta G_1 \leq 0. \tag{1.35}$$

At constant temperature and pressure, the equilibrium state has the minimum Gibbs free energy. If electric energy is involved, then we have to consider the electric work done to the system by external sources such as a battery. In a thermodynamic process, if the electric work done to the system is dW_e, $\delta S \geq \frac{dQ}{T} = \frac{dU - dW_m - dW_e}{T} = \frac{dU + PdV - dW_e}{T}$. Therefore at constant temperature and pressure

$$\delta(U - W_e + PV - TS) = \delta(G - W_e) \leq 0. \tag{1.36}$$

In the equilibrium state, $G - W_e$ is minimized.

1.3 Orientational Order

Orientational order is the most important feature of liquid crystals. The average directions of the long axes of the rod-like molecules are parallel to each other. Because of the orientational order, liquid crystals possess anisotropic physical properties, that is, in different directions they have different responses to external fields such as electric field, magnetic field and shear. In this section, we will discuss how to specify quantitatively orientational order and why rod-like molecules tend to parallel each other.

For a rigid elongated liquid crystal molecule, three axes can be attached to it to describe its orientation. One is the long molecular axis and the other two axes are perpendicular to the long molecular axis. Usually the molecule rotates fast around the long molecular axis. Although the molecule is not cylindrical, if there is no hindrance in the rotation in nematic phase, the fast rotation around the long molecular axis makes it behave as a cylinder. There is no preferred direction for the short axes and thus the nematic liquid crystal is usually uniaxial. If there is hindrance in the rotation, the liquid crystal is biaxial. Biaxial nematic liquid crystal is a long-sought material. A lyotropic biaxial nematic phase has been observed [12]. A thermotropic biaxial nematic phase is still debatable, and it may exist in systems consisting of bent-core molecules [13,14]. Also the rotation symmetry around the long molecular axis can be broken by confinements. In this book, we deal with uniaxial liquid crystals consisting of rod-like molecules unless otherwise specified.

1.3.1 Orientational order parameter

In uniaxial liquid crystals, we have only to consider the orientation of the long molecular axis. The orientation of a rod-like molecule can be represented by a unit vector $\hat{a}$ which is attached to the molecule and parallel to the long molecular axis. In the nematic phase, the average directions of the long molecular axes are along a common direction: the liquid crystal director denoted by the unit vector $\vec{n}$. The orientation of $\hat{a}$ in 3-D can be specified by the polar angle θ and the azimuthal angle ϕ where the z axis is chosen parallel to $\vec{n}$ as shown in Figure 1.4. In general, the orientational order of $\hat{a}$ is specified by an orientational distribution function $f(\theta, \phi)$. $f(\theta, \phi)d\Omega$ ($d\Omega = \sin\theta d\theta d\phi$) is the probability that $\hat{a}$ orients along the direction specified by θ and ϕ within the solid angle $d\Omega$. In isotropic phase, $\hat{a}$ has equal probability of pointing any direction and therefore $f(\theta, \phi) =$ costant. For uniaxial liquid crystals, there is no preferred orientation in the azimuthal direction, and then $f = f(\theta)$, which depends only on the polar angle θ.

Rod-like liquid crystal molecules may have permanent dipole moments. If the dipole moment is perpendicular to the long molecule axis, the dipole has equal probability of pointing along any direction because of the fast rotation around the long molecular axis in uniaxial liquid crystal phases. The dipoles of the molecules cannot generate spontaneous polarization. If the permanent dipole moment is along the long molecular axis, the flip of the long molecular axis is much slower (of the order of 10^{-5} s); the above argument does not hold. In order to see the orientation of the dipoles in this case, we consider the interaction between two dipoles [15]. When one dipole is on top of the other dipole, if they are parallel, the interaction energy is low and thus parallel orientation is preferred. When two dipoles are side by side, if they are anti-parallel, the interaction energy is low and thus anti-parallel orientation is preferred. As we know, the molecules cannot penetrate each other. For elongated molecules, the distance between the two dipoles when on top of each other is farther than that when the two dipoles

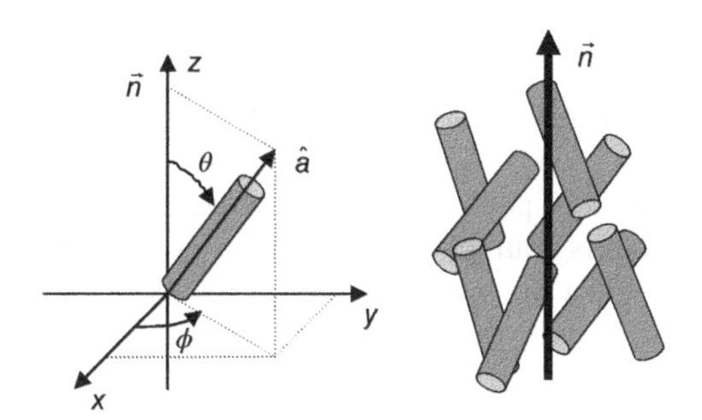

Figure 1.4 Schematic diagram showing the orientation of the rod-like molecule.

are side by side. The interaction energy between two dipoles is inversely proportional to the cubic power of the distance between them. Therefore anti-parallel orientation of dipoles is dominant in rod-like molecules. There are the same number of dipoles aligned parallel to the liquid crystal director $\vec{n}$ as the number of diploes aligned anti-parallel to $\vec{n}$. The permanent dipole along the long molecular axis cannot generate spontaneous polarization. Thus even when the molecules have permanent dipole moment along the long molecule axes, they can be regarded as cylinders whose top and end are the same. It can also be concluded that $\vec{n}$ and $-\vec{n}$ are equivalent.

An order parameter must be defined in order to specify quantitatively the orientational order. The order parameter is usually defined in such a way that it is zero in the high temperature unordered phase and non-zero in the low temperature ordered phase. By analogy with ferromagnetism, we may consider the average value of the projection of $\hat{a}$ along the director $\vec{n}$:

$$<\cos\theta> = \int_0^\pi \cos\theta f(\theta)\sin\theta d\theta \bigg/ \int_0^\pi f(\theta)\sin\theta d\theta, \tag{1.37}$$

where $< \ >$ indicates the average (temporal and spatial averages are the same), and $\cos\theta$ is the first Legendre polynomial. In isotropic phase, the molecules are randomly oriented, $<\cos\theta>$ is zero. We also know that in nematic phase the probabilities that the molecule orients at the angles θ and $\pi-\theta$ are the same, that is, $f(\theta)=f(\pi-\theta)$, therefore $<\cos\theta>=0$, and is not a good choice for the orientational order parameter. Next let us try the average value of the second Legendre polynomial for the order parameter:

$$S = <P_2(\cos\theta)> = <\frac{1}{2}(3\cos^2\theta-1)> = \int_0^\pi \frac{1}{2}(3\cos^2\theta-1)f(\theta)\sin\theta d\theta \bigg/ \int_0^\pi f(\theta)\sin\theta d\theta \tag{1.38}$$

In the isotropic phase, as shown in Figure 1.5(b), $f(\theta)=c$, a constant. $\int_0^\pi \frac{1}{2}(3\cos^2\theta-1)f(\theta)\sin\theta d\theta = \int_0^\pi \frac{1}{2}(3\cos^2\theta-1)c\sin\theta d\theta = 0$. In nematic phase, $f(\theta)$ depends on θ. For a perfectly ordered nematic phase as shown in Figure 1.5(d), $f(\theta)=\delta(\theta)$, where $\sin\theta\delta(\theta)=\infty$ when $\theta=0$, $\sin\theta\delta(\theta)=0$ when $\theta\neq 0$ and $\int_0^\pi \delta(\theta)\sin\theta d\theta = 1$, the order parameter is $S=(1/2)(3\cos^2 0-1)=1$. It should be pointed out that the order parameter can be positive or negative. For two order parameters with the same absolute value but different signs, they correspond to different states. When the molecules all lie in a plane but randomly orient in the plane, as shown in Figure 1.5(a), the distribution function is $f(\theta)=\delta(\theta-\pi/2)$, where $\delta(\theta-\pi/2)=\infty$ when $\theta=\pi/2$, $\delta(\theta-\pi/2)=0$ when $\theta\neq\pi/2$ and $\int_0^\pi \delta(\theta-\pi/2)\sin\theta d\theta = 1$, the order

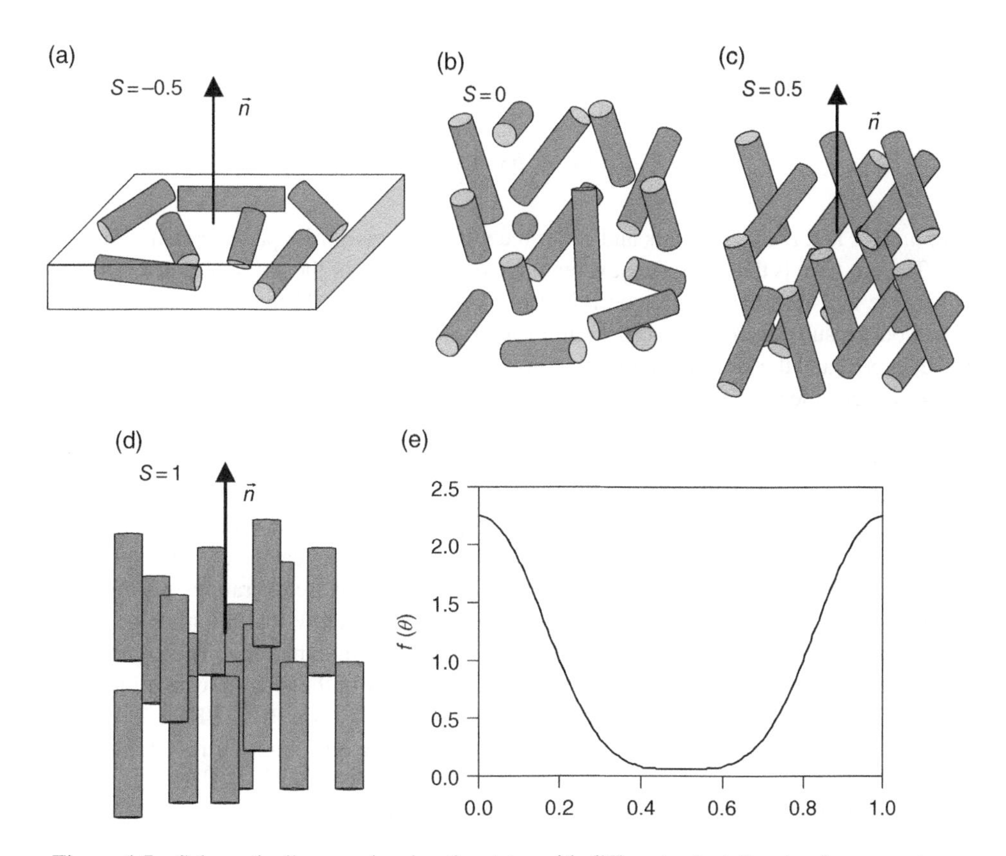

Figure 1.5 Schematic diagram showing the states with different orientational order parameters.

parameter is $S = (1/2)[3\cos^2(\pi/2) - 1)/1 = -0.5$. In this case, the average direction of the molecules is not well defined. The director $\vec{n}$ is defined by the direction of the uniaxial axis of the material. Figure 1.5(c) shows the state with the distribution function $f(\theta) = (35/16)[\cos^4\theta + (1/35)]$, which is plotted vs. θ in Figure 1.5(e). The order parameter is $S = 0.5$. Many anisotropies of physical properties are related to the order parameter and will be discussed later.

1.3.2 Landau–de Gennes theory of orientational order in nematic phase

Landau developed a theory for second-order phase transition [16], such as from diamagnetic phase to ferromagnetic phase, in which the order parameter increases continuously from zero as the temperature is decreased across the transition temperature T_c from the high temperature disordered phase to the low temperature ordered phase. For a temperature near T_c, the order is very small. The free energy of the system can be expanded in terms of the order parameter.

The transition from water to ice at 1 atmosphere pressure is a first-order transition, and the latent heat is about 100 J/g. The isotropic–nematic transition is a weak first-order transition because the order parameter changes discontinuously across the transition but the latent heat is only about 10 J/g. De Gennes extended Landau's theory into isotropic–nematic transition because it is a weak first-order transition [1,17]. The free energy density f of the material can be expressed in terms of the order parameter S,

$$f=\frac{1}{2}a(T-T^*)S^2-\frac{1}{3}bS^3+\frac{1}{4}cS^4+\frac{1}{2}L(\nabla S)^2, \tag{1.39}$$

where a, b, c and L are constants and T^* is the virtual second-order phase transition temperature. The last term is the energy cost when there is a variation of the order parameter in space, and here we will consider only the uniform order parameter case. There is no linear term of S, which would result in a non-zero order parameter at any temperature; a is positive, otherwise S will never be 0 and the isotropic phase will not be stable at any temperature. A significant difference between the free energy here and that of a magnetic system is the cubic term. In a magnetic system, the magnetization m is the order parameter. For a given value of $|m|$, there is only one state, and the sign of m is decided by the choice of the coordinate. The free energy must be the same for a positive m and a negative m, and therefore the coefficient of the cubic term must be zero. For nematic liquid crystal, positive and negative values of the order parameter S correspond to two different states, and the corresponding free energies can be different, and therefore b is not zero. b must be positive because at sufficiently low temperatures positive-order parameters have the global minimum free energies. We also know that the maximum value of S is 1. The quadratic term with a positive c prevents S from exploding. The values of the coefficients can be estimated in the following way: the energy of the intermolecular interaction between molecules associated with orientation is about $k_BT = 1.38\times10^{-23}$(J/K) $\times$ 300 K $\approx 4\times10^{-21}$ J and the molecular size is about 1 nm, f is the energy per unit volume, and therefore Ta (or b or c) ~ k_BT/*volume of one molecule* ~ 4×10^{-21}joule/$(10^{-9}$m$)^3$ ~ 10^6J/m^3.

For a given temperature, the order parameter S is found by minimizing f,

$$\frac{\partial f}{\partial S}=a\left(T-T^*\right)S-bS^2+cS^3=\left[a\left(T-T^*\right)-bS+cS^2\right]S=0. \tag{1.40}$$

There are three solutions:

$$S_1=0,$$

$$S_2=\frac{1}{2c}\left[b+\sqrt{b^2-4ac(T-T^*)}\right],$$

$$S_3=\frac{1}{2c}\left[b-\sqrt{b^2-4ac(T-T^*)}\right].$$

$S_1=0$ corresponds to the isotropic phase and the free energy is $f_1=0$. The isotropic phase has the global minimum free energy at high temperature. It will be shown that at low temperature S_2 has the global minimum free energy $f_2=\frac{1}{2}a(T-T^*)S_2^2-\frac{1}{3}bS_2^3+\frac{1}{4}cS_2^4$. And S_3 has a local maximum free energy. At the isotropic–nematic phase transition temperature T_{NI}, the order parameter is $S_c=S_{2c}$, and $f_2(S_2=S_c)=f_1=0$, that is,

$$\frac{1}{2}a\left(T_{NI}-T^*\right)S_c^{\,2}-\frac{1}{3}bS_c^{\,3}+\frac{1}{4}cS_c^{\,4}=0. \tag{1.41}$$

From Equation (1.40), at this temperature, we also have

$$a\left(T_{NI}-T^{*}\right)-bS_c+cS_c{}^2=0. \tag{1.42}$$

From the two equations above, we can obtain

$$a\left(T_{NI}-T^{*}\right)-\frac{1}{3}bS_c=0.$$

Therefore

$$S_c=\frac{3a}{b}\left(T_{NI}-T^{*}\right). \tag{1.43}$$

Substitute Equation (1.43) into Equation (1.42), we will get the transition temperature

$$T_{NI}=T*+\frac{2b^2}{9ac}, \tag{1.44}$$

and the order parameter at the transition temperature

$$S_c=\frac{2b}{3c}. \tag{1.45}$$

For liquid crystal 5CB, the experimentally measured order parameter is shown by the solid circles in Figure 1.6(a) [6]. In fitting the data, the following parameters are used: $a=0.023\sigma$ J/K · m^3, $b=1.2\sigma$ J/m^3 and $c=2.2\sigma$ J/m^3, where σ is a constant which has to be determined by latent heat of the isotropic–nematic transition.

Because S is a real number in the region from –0.5 to 1.0, when $T-T*>b^2/4ac$, that is, when $T-T_{NI}>b^2/4ac-2b^2/9ac=b^2/36ac$, S_2 and S_3 are not real. The only real solution is $S=S_1=0$, corresponding to the isotropic phase. When $T-T_{NI}<b^2/36ac$, there are three solutions. However, when $0<T-T_{NI}\leq b^2/36ac$, the isotropic phase is the stable state because its free energy is still the global minimum, as shown in Figure 1.6(b). When $T-T_{NI}\leq 0$, the nematic phase with the order parameter $S=S_2=\left(b+\sqrt{b^2-4ac(T-T^{*})}\right)/2c$ is the stable state because its free energy is the global minimum.

In order to see clearly the physical meaning, let us plot f vs. S at various temperatures as shown in Figure 1.7. First we consider what occurs with decreasing temperature. At temperature $T_1=T_{NI}+b^2/36ac+1.0°\text{C}$, the curve has only one minimum at $S=0$, which means that $S_1=0$ is the only solution, and the corresponding isotropic phase is the stable state. At temperature $T_3=T_{NI}+b^2/36ac-0.5°\text{C}$, there are two local minima and one local maximum, where there are three solutions: $S_1=0$, $S_2>0$, and $S_3>0$. Here, $S_1=0$ corresponds to the global minimum and the isotropic phase is still the stable state. At $T_4=T_{NI}$, the free energies of the isotropic phase with the order parameter S_1 and the nematic phase with the order parameter S_2 become the same; phase transition takes place and the order parameter changes discontinuously from 0 to $S_c=2b/3c$. This is a first-order transition. It can be seen from the figure that at this

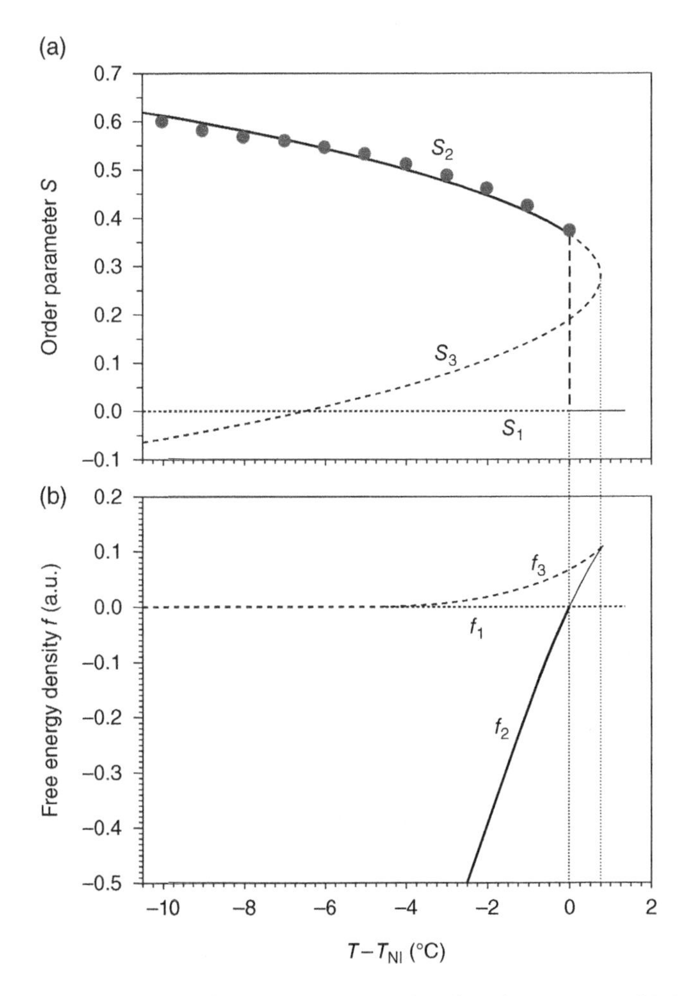

Figure 1.6 (a) The three solutions of order parameter as a function of temperature, (b) the corresponding free energies as a function of temperature, in Landau–de Gennes theory.

temperature there is an energy barrier between S_1 and S_2. The height of the energy barrier is $b^4/81c^3$. If the system is initially in the isotropic phase and there are no means to overcome the energy barrier, it will remain in the isotropic phase at this temperature. As the temperature is decreased, the energy barrier is lowered. At $T_5 = T_{NI} - 3°\text{C}$, the energy barrier is lower. At $T_6 = T^*$, the second-order derivative of f with respect to S at $S_1 = 0$ is

$$\left.\frac{\partial^2 f}{\partial S^2}\right|_{S=0} = a\left(T - T^*\right) - 2bS + 3cS^2\Big|_{S=0} = a\left(T - T^*\right) = 0.$$

S_1 is no longer a local minimum and the energy barrier disappears. T^* is therefore the supercooling temperature below which the isotropic phase becomes absolutely unstable. At this temperature, $S_1 = S_3$. At $T_7 = T^* - 2°\text{C}$, there are two minima located at $S_2(>0)$ and $S_3(<0)$ (the minimum value is slightly below 0), and a maximum at $S_1 = 0$.

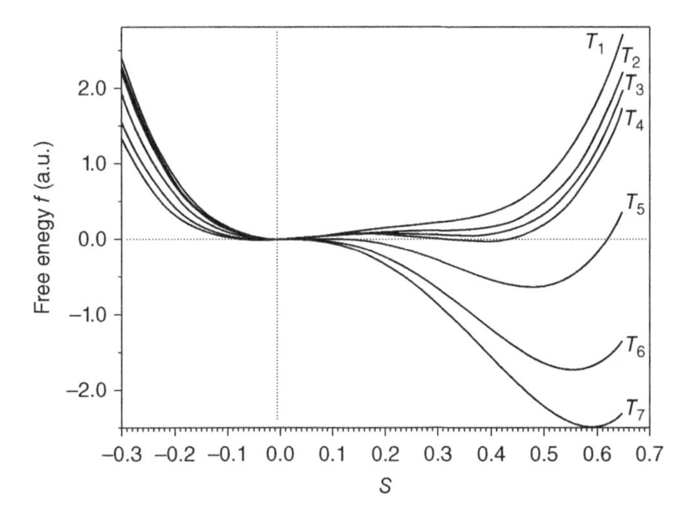

Figure 1.7 Free energy vs. order parameter at various temperatures in Landau–de Gennes theory.

Now we consider what occurs with increasing temperature. If initially the system is in the nematic phase, it will remain in this phase even at temperatures higher than T_{NI} and its free energy is higher than that of the isotropic phase, because there is an energy barrier preventing the system to transform from the nematic phase to the isotropic phase. The temperature T_2 (superheating temperature) at which the nematic phase becomes absolutely unstable can be found by

$$\left.\frac{\partial^2 f}{\partial S^2}\right|_{S_2} = a\left(T_2 - T^*\right) - 2bS_2 + 3cS_2^2 = 0. \tag{1.46}$$

Using $S_2 = \frac{1}{2c}\left[b + \sqrt{b^2 - 4ac(T_2 - T^*)}\right]$, we can get $T_2 = T_{NI} + b^2/36ac$.

In reality, there are usually irregularities, such as impurities and defects, which can reduce the energy barrier against the isotropic–nematic transition. The phase transition takes place before the thermodynamic instability limits (supercooling or superheating temperature). Under an optical microscope, it is usually observed that with decreasing temperature nematic 'islands' are initiated by irregularities and growing out the isotropic 'sea' and with increasing temperature isotropic 'lakes' are produced by irregularities and grow on the nematic 'land.' The irregularities are called nucleation seeds and the transition is a nucleation process. In summary, the nematic–isotropic transition is a first-order transition; the order parameter changes discontinuously; there is an energy barrier against the transition and the transition is a nucleation process; there are superheating and supercooling. In second-order transition, there is no energy barrier and the transition occurs simultaneously everywhere at the transition temperature (the critical temperature).

There are a few points worth mentioning in Landau–de Gennes theory. It works well at temperatures near the transition. At temperatures far below the transition temperature, the order parameter increases without limit with decreasing temperature, and the theory does not work well because we know that the maximum order parameter should be 1. In Figure 1.6, the parameters are chosen in such a way that the fitting is good for a relatively wide range of temperatures. $T_{NI} - T^* = 2b^2/9ac = 6.3°C$, which is much larger than the value (~1 °C) measured by light scattering experiments in isotropic phase [18]. There are fluctuations in orientational order in the isotropic phase, which results in a variation of refractive index in space and causes light scattering. The intensity of the scattering light is proportional to $1/(T - T^*)$.

1.3.3 *Maier–Saupe theory*

In the nematic phase, there are interactions, such as the Van der Waals interaction, between the liquid crystal molecules. Because the molecular polarizability along the long molecular axis is larger than that along the short transverse molecular axis, the interaction is anisotropic and results in the parallel alignment of the rod-like molecules. In the spirit of the mean field approximation, Maier and Saupe introduced an effective single molecule potential V to describe the intermolecular interaction [19,20]. The potential has the following properties. (1) It must be a minimum when the molecule orients along the liquid crystal director (the average direction of the long molecular axis of the molecules). (2) Its strength is proportional to the order parameter $S = \langle P_2(\cos\theta) \rangle$ because the potential well is deep when the molecules are highly orientationally ordered and vanishes when the molecules are disordered. (3) It ensures that the probabilities for the molecules pointing up and down are the same. The potential in Maier–Saupe theory is given by

$$V(\theta) = -\nu S\left(\frac{3}{2}\cos^2\theta - \frac{1}{2}\right), \tag{1.47}$$

where ν is a interaction constant of the order of $k_B T$ and θ is the angle between the long molecular axis and the liquid crystal director as shown in Figure 1.4. The probability f for the molecule orienting along the direction with the polar angle θ is governed by the Boltzmann distribution:

$$f(\theta) = e^{-V(\theta)/k_B T} \Big/ \int_0^\pi e^{-V(\theta)/k_B T} \sin\theta d\theta \tag{1.48}$$

The single molecule partition function is

$$Z = \int_0^\pi e^{-V(\theta)/k_B T} \sin\theta d\theta. \tag{1.49}$$

From the orientational distribution function we can calculate the order parameter:

$$S = \frac{1}{Z}\int_0^\pi P_2(\cos\theta) e^{-V(\theta)/k_B T} \sin\theta d\theta = \frac{1}{Z}\int_0^\pi P_2(\cos\theta) e^{\nu S P_2(\theta)/k_B T} \sin\theta d\theta \tag{1.50}$$

Introduce a normalized temperature $\tau = k_B T/v$. For a given value of τ, the order parameter S can be found by numerically solving Equation (1.50). An iteration method can be used in the numerical calculation of the order parameter: (1) use an initial value for the order parameter, (2) substitute into the right side of Equation (1.50), and (3) calculate the order parameter. Use the newly obtained order parameter to repeat the above process until a stable value is obtained. As shown in Figure 1.8(a), there are three solutions: S_1, S_2, and S_3. In order to determine which is the actual solution, we have to examine the corresponding free energies. The free energy F has two parts: $F = U - TE_n$, where U is the intermolecular interaction energy and E_n is the

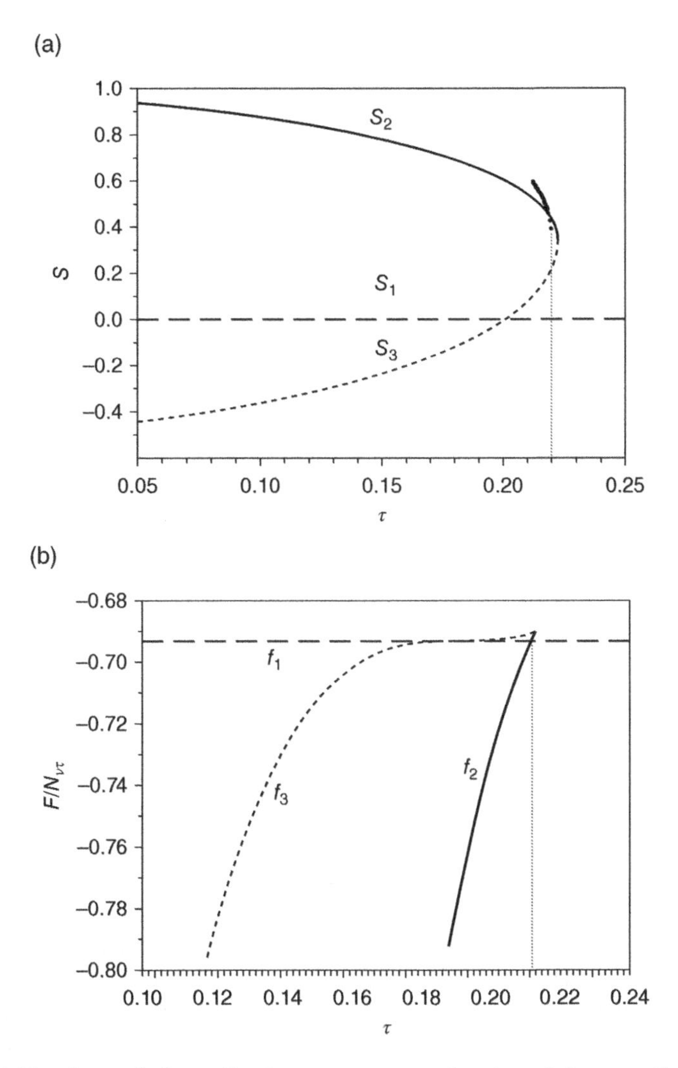

Figure 1.8 (a) The three solutions of order parameter as a function of the normalized temperature in Maier–Saupe theory. The solid circles represent the experimental data. (b) The normalized free energies of the three solutions of the order parameter.

entropy. The single molecular potential describes the interaction energy between one liquid crystal molecule and the rest of the molecules of the system. The interaction energy of the system with N molecules is given by

$$U=\frac{1}{2}N< V> =\frac{N}{2Z}\int_0^\pi V(\theta)\mathrm{e}^{-V(\theta)/k_BT}\sin\theta d\theta \tag{1.51}$$

where the factor 1/2 avoids counting the intermolecular interaction twice. The entropy is calculated by using Equation (1.32):

$$En=-Nk_B<\ln f> =-\frac{Nk_B}{Z}\int_0^\pi \ln[f(\theta)]\mathrm{e}^{-V(\theta)/k_BT}\sin\theta d\theta \tag{1.52}$$

From Equation (1.48) we have $\ln[f(\theta)]=-V(\theta)/k_BT-\ln Z$, and therefore $En=\frac{N}{T}< V> +$ $Nk_B\ln Z$ and the free energy is

$$F=U-TEn=-Nk_BT\ln Z-\frac{1}{2}N< V>. \tag{1.53}$$

From Equations (1.47) we have $< V>=-\nu S^2$ and therefore

$$F=U-TEn=-Nk_BT\ln Z+\frac{1}{2}N\nu S^2 \tag{1.54}$$

Although the second term of the above equation looks abnormal, this equation is correct and can be checked by calculating the derivative of F with respect to S:

$$\frac{\partial F}{\partial S}=-Nk_BT\frac{\partial \ln Z}{\partial S}-\frac{1}{2}N\frac{\partial< V>}{\partial S}=-\frac{Nk_BT}{Z}\frac{\partial Z}{\partial S}+N\nu S$$

Letting $\frac{\partial F}{\partial S}=0$, we have

$$S=\frac{k_BT}{\nu Z}\frac{\partial Z}{\partial S}=\frac{k_BT}{\nu Z}\int_0^\pi\frac{-1}{k_BT}\frac{\partial V}{\partial S}e^{-V(\theta)/k_BT}\sin\theta d\theta=\frac{1}{Z}\int_0^\pi P_2(\cos\theta)e^{\nu SP_2(\theta)/k_BT}\sin\theta d\theta,$$

which is consistent with Equation (1.50). The free energies corresponding to the solutions are shown in Figure 1.8(b). The nematic–isotropic phase transition temperature is $\tau_{NI}=0.22019$. For temperature higher than τ_{NI}, the isotropic phase with the order parameter $S=S_1=0$ has lower free energy and thus is stable. For temperature lower than τ_{NI}, the nematic phase with the order parameter $S=S_2$ has lower free energy and thus is stable. The order parameter jumps from 0 to $S_c=0.4289$ at the transition.

In the Maier–Saupe theory there are no fitting parameters. The predicted order parameter as a function of temperature is universal, and agrees qualitatively – but not quantitatively – with experimental data. This indicates that higher-order terms are needed in the single molecule potential, that is,

$$V(\theta)=\sum_{i}[-v_i<P_i(\cos\theta)>P_i(\cos\theta)] \tag{1.55}$$

where $P_i(\cos\theta)$ ($i=2,\ 4,\ 6,......$) are the ith -order Legendre polynomial. The fitting parameters are v_i. With higher-order terms, better agreement with experimental results can be achieved.

The Maier–Saupe theory is very useful in considering liquid crystal systems consisting of more than one type of molecules, such as mixtures of nematic liquid crystals and dichroic dyes. The interactions between different molecules are different and the constituent molecules have different order parameters.

None the theories discussed above predicts well the orientational order parameter for temperatures far below T_{NI}. The order parameter as a function of temperature is better described by the empirical formula [21]

$$S=\left(1-\frac{0.98TV^2}{T_{NI}V_{NI}^2}\right)^{0.22}, \tag{1.56}$$

where V and V_{NI} are the molar volumes at T and T_{NI}, respectively.

1.4 Elastic Properties of Liquid Crystals

In nematic phase, the liquid crystal director $\vec{n}$ is uniform in space in the ground state. In reality, the liquid crystal director $\vec{n}$ may vary spatially because of confinements or external fields. This spatial variation of the director, called the deformation of the director, costs energy. When the variation occurs over a distance much larger than the molecular size, the orientational order parameter does not change, and the deformation can be described by a continuum theory in analogue to the classic elastic theory of a solid. The elastic energy is proportional to the square of the spatial variation rate.

1.4.1 Elastic properties of nematic liquid crystals

There are three possible deformation modes of the liquid crystal director as shown in Figure 1.9. Choose the cylindrical coordinate such that the z axis is parallel to the director at the origin of the coordinate: $\vec{n}(0)=\hat{z}$. Consider the variation of the director at an infinite small distance away from the origin. When moving along the radial direction, there are two possible modes of variation: (1) the director tilts toward the radial direction $\hat{\rho}$, as shown in Figure 1.9(a), and (2) the director tilts toward the azimuthal direction $\hat{\phi}$, as shown in Figure 1.9(b). The first mode is called splay, where the director at ($\delta\rho,\ \phi,\ z=0$) is

$$\vec{n}(\delta\rho,\varphi,z=0)=\delta n_\rho(\delta\rho)\hat{\rho}+[1+\delta n_z(\delta\rho)]\hat{z}, \tag{1.57}$$

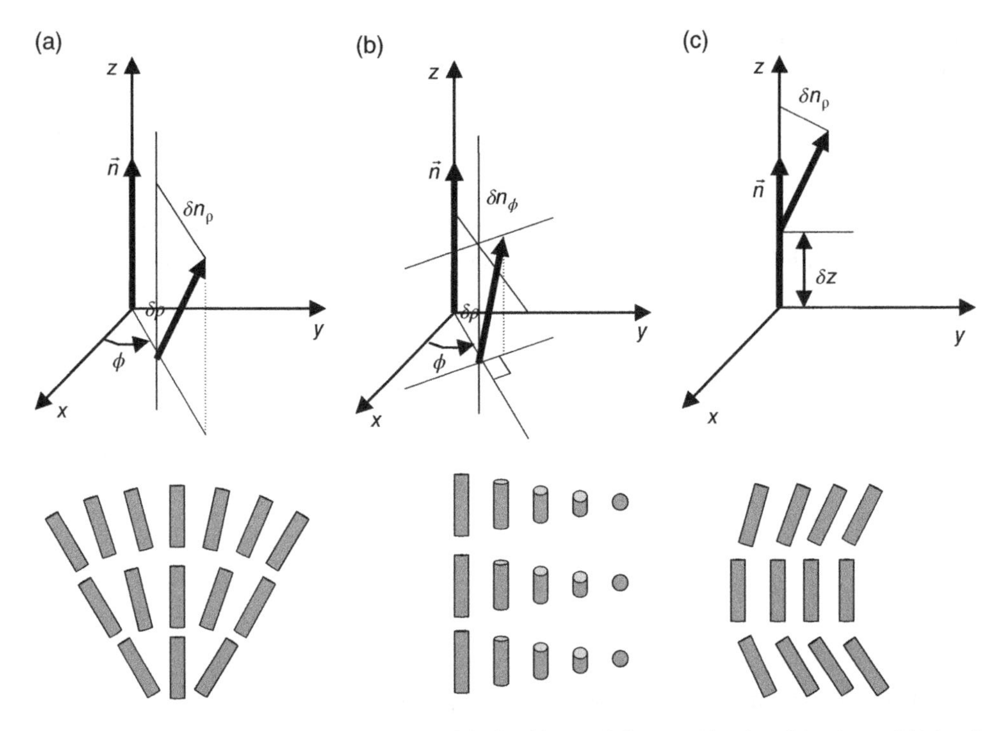

Figure 1.9 The three possible deformations of the liquid crystal director: (a) splay, (b) twist, and (c) bend.

where $\delta n_\rho \ll 1$ and $\delta n_z \ll 1$. Because $|\vec{n}|^2 = n_\rho^2 + n_\phi^2 + n_z^2 = (\delta n_\rho)^2 + (1 + \delta n_z)^2 = 1$, therefore $\delta n_z = -(\delta n_\rho)^2/2$, δn_z is a higher-order term and can be neglected. The spatial variation rate is $\partial n_\rho/\partial \rho$ and the corresponding elastic energy is

$$f_{splay} = (1/2)K_{11}(\partial n_\rho/\partial \rho)^2, \tag{1.58}$$

where K_{11} is the splay elastic constant. The second mode is called twist, where the director at $(\delta\rho, \phi, z=0)$ is

$$\vec{n}(\delta\rho, \phi, z=0) = \delta n_\phi(\delta\rho)\hat{\phi} + [1 + \delta n_z(\delta\rho)]\hat{z}, \tag{1.59}$$

where $\delta n_\phi \ll 1$ and $\delta n_z = -(\delta n_\phi)^2/2$, a higher-order term which can be neglected. The spatial variation rate is $\partial n_\phi/\partial \rho$ and the corresponding elastic energy is

$$f_{twist} = (1/2)K_{22}(\partial n_\phi/\partial \rho)^2 \tag{1.60}$$

where K_{22} is the twist elastic constant.

When moving along the z direction, there is only one possible mode of variation, as shown in Figure 1.9(c), which is called bend. The director at $(\rho = 0, \phi, \delta z)$ is

$$\vec{n}(\rho=0, \phi, \delta z) = \delta n_\rho(\delta z)\hat{\rho} + [1 + \delta n_z(\delta z)]\hat{z} \tag{1.61}$$

where $\delta n_\rho \ll 1$ and $\delta n_z = -(\delta n_\phi)^2/2$, a higher-order term which can be neglected. Note that when $\rho = 0$, the azimuthal angle is not well defined and we can choose the coordinate such that the director tilts toward the radial direction. The corresponding elastic energy is

$$f_{bend} = (1/2)K_{33}\left(\partial n_\rho/\partial z\right)^2, \tag{1.62}$$

where K_{33} is the bend elastic constant. Because δn_z is a higher-order term, $\partial n_z/\partial z \approx 0$ and $\partial n_z/\partial \rho \approx 0$. Recall $\nabla\cdot\vec{n}\big|_{\rho=0,\ z=0} = (1/\rho)\partial(\rho n_\rho)/\partial\rho + (1/\rho)\partial n_\phi/\partial\phi + \partial n_z/\partial z = \partial n_\rho/\partial\rho + \delta n_\rho$. Because $\partial n_\rho/\partial\rho$ is finite and $\delta n_\rho \ll 1$, then $\nabla\cdot\vec{n}\big|_{\rho=0,\ z=0} = \partial n_\rho/\partial\rho$. The splay elastic energy can be expressed as $f_{splay} = (1/2)K_{11}\left(\nabla\cdot\vec{n}\right)^2$. Because at the origin $\vec{n} = \hat{z}$, then $\vec{n}\cdot\nabla\times\vec{n}\big|_{\rho=0,\ z=0} = \left(\nabla\times\vec{n}\right)_z = \partial n_\phi/\partial\rho$. The twist elastic energy can be expressed as $f_{twist} = (1/2)K_{22}\left(\vec{n}\cdot\nabla\times\vec{n}\right)^2$. Because $\vec{n}\times\nabla\times\vec{n}\big|_{\rho=0,\ z=0} = \left(\nabla\times\vec{n}\right)_\rho - \left(\nabla\times\vec{n}\right)_\phi = \partial n_\rho/\partial z$, the bend elastic energy can be expressed as $f_{bend} = (1/2)K_{33}\left(\vec{n}\times\nabla\times\vec{n}\right)^2$. Putting all the three terms together, we the elastic energy density:

$$f_{ela} = \frac{1}{2}K_{11}\left(\nabla\cdot\vec{n}\right)^2 + \frac{1}{2}K_{22}\left(\vec{n}\cdot\nabla\times\vec{n}\right)^2 + \frac{1}{2}K_{33}\left(\vec{n}\times\nabla\times\vec{n}\right)^2 \tag{1.63}$$

This elastic energy is often referred to as the *Oseen–Frank energy,* and K_{11}, K_{22}, and K_{33} are referred to as the Frank elastic constants, because of his pioneering work on the elastic continuum theory of liquid crystals [22]. The value of the elastic constants can be estimated in the following way. When a significant variation of the director occurs in a length L, the angle between the average directions of the long molecules axes of two neighboring molecules is a/L, where a is the molecular size. When the average directions of the long molecular axes of two neighboring molecules are parallel, the intermolecular interaction energy between them is a minimum. When the angle between the average directions of the long molecular axes of two neighboring molecules makes the angle of a/L, the intermolecular interaction energy increases $(a/L)^2u$, where u is the intermolecular interaction energy associated with orientation and is about k_BT. The increase of the interaction energy is the elastic energy, that is, $\left(\frac{a}{L}\right)^2 u = K_{ii}\left(\nabla\vec{n}\right)^2 \times \textit{molecular voume} = K_{ii}\left(\frac{1}{L}\right)^2 a^3$. Therefore $K_{ii} = \frac{u}{a} \sim 1.38\times10^{-23}(\mathrm{J/K})\times300\mathrm{K}/\left(10^{-9}\mathrm{m}\right) = 4\times10^{-12}\mathrm{N}$. Experiments show that usually the bend elastic constant K_{33} is the largest and twist elastic constant K_{22} is the smallest. As an example, at room temperature the liquid crystal 5CB has these elastic constants: $K_{11} = 0.64\times10^{-11}$N, $K_{22} = 0.3\times10^{-11}$N, and $K_{33} = 1\times10^{-11}$N.

The elastic constants depend on the product of the order parameters of two neighboring molecules. If one of the molecules had the order of 0, the second molecule can orient along any direction with the same inter-molecular interaction energy even if it has non-zero order parameter. Therefore the elastic constants are proportional to S^2. When the temperature changes, the order parameter will change and so will the elastic constants.

It is usually adequate to consider the splay, twist, and bend deformations of the liquid crystal director in determining the configuration of the director, except in some cases where the surface-to-volume ratio is high, and a further two terms, called divergence terms (or surface terms), may have to be considered. The elastic energy densities of these terms are given by $f_{13} = K_{13}\nabla\cdot\left(\vec{n}\nabla\cdot\vec{n}\right)$ and $f_{24} = -K_{24}\nabla\cdot\left(\vec{n}\nabla\cdot\vec{n} + \vec{n}\times\nabla\times\vec{n}\right)$, respectively [23]. The volume integral of these two terms can be changed to surface integral because of the Gauss theorem.

1.4.2 Elastic properties of cholesteric liquid crystals

So far we have considered liquid crystals consisting of molecules with reflection symmetry. The molecules are the same as their mirror images, and are called *achiral* molecules. 5CB shown in Figure 1.1(a) is an example of an achiral molecule. Now we consider liquid crystals consisting of molecules without reflection symmetry. The molecules are different from their mirror images and are called *chiral* molecules. One such example is CB15 shown in Figure 1.10(a). It can be regarded as a screw, instead of a rod, considering its physical properties. After considering the symmetry that $\vec{n}$ and $-\vec{n}$ are equivalent, the generalized elastic energy density is

$$f_{ela}=\frac{1}{2}K_{11}(\nabla\cdot\vec{n})^2+\frac{1}{2}K_{22}(\vec{n}\cdot\nabla\times\vec{n}+q_o)^2+\frac{1}{2}K_{33}(\vec{n}\times\nabla\times\vec{n})^2, \tag{1.64}$$

where q_o is the *chirality,* and its physical meaning will be discussed in a moment. Note that $\nabla\times\vec{n}$ is a pseudo-vector which does not change sign upon reflection symmetry operation, and $\vec{n}\cdot\nabla\times\vec{n}$ is a pseudo-scalar which changes sign upon reflection symmetry operation. Upon reflection symmetry operation, the elastic energy changes to

$$f'_{ela}=\frac{1}{2}K_{11}(\nabla\cdot\vec{n})^2+\frac{1}{2}K_{22}(-\vec{n}\cdot\nabla\times\vec{n}+q_o)^2+\frac{1}{2}K_{33}(-\vec{n}\times\nabla\times\vec{n})^2 \tag{1.65}$$

If the liquid crystal molecule is achiral, and thus has reflection symmetry, the system does not change and the elastic energy does not change upon reflection symmetry operation. It is required that $f_{ela}=f'_{ela}$, then $q_o=0$. When the liquid crystal is in the ground state with the minimum free energy, $f_{ela}=0$, and this requires $\nabla\cdot\vec{n}=0$, $\vec{n}\cdot\nabla\times\vec{n}=0$, and $\vec{n}\times\nabla\times\vec{n}=0$. This means that in the ground state, the liquid crystal director $\vec{n}$ is uniformly aligned along one direction.

If the liquid crystal molecule is chiral, and thus has no reflection symmetry, the system changes upon a reflection symmetry operation. The elastic energy may change. It is no longer required that $f_{ela}=f'_{ela}$, and thus q_o may not be zero. When the liquid crystal is in the ground state with the minimum free energy, $f_{ela}=0$, and this requires $\nabla\cdot\vec{n}=0$, $\vec{n}\cdot\nabla\times\vec{n}=-q_o$, and $\vec{n}\times\nabla\times\vec{n}=0$. A director configuration which satisfies the above conditions is

$$n_x=\cos(q_oz),\quad n_y=\sin(q_oz),\quad n_z=0; \tag{1.66}$$

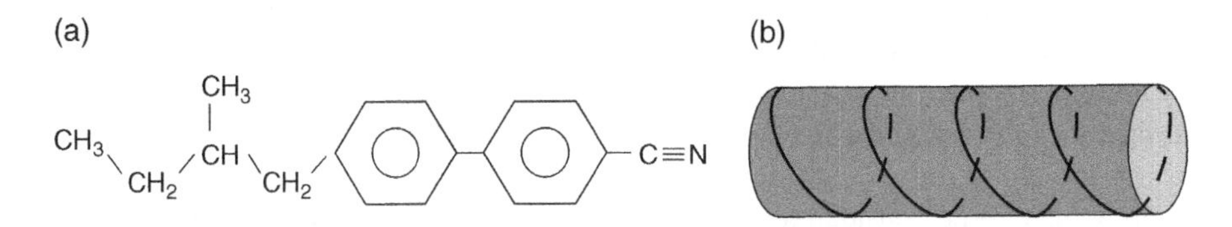

Figure 1.10 (a) Chemical structure of a typical chiral liquid crystal molecule; (b) physical model of a chiral liquid crystal molecule.

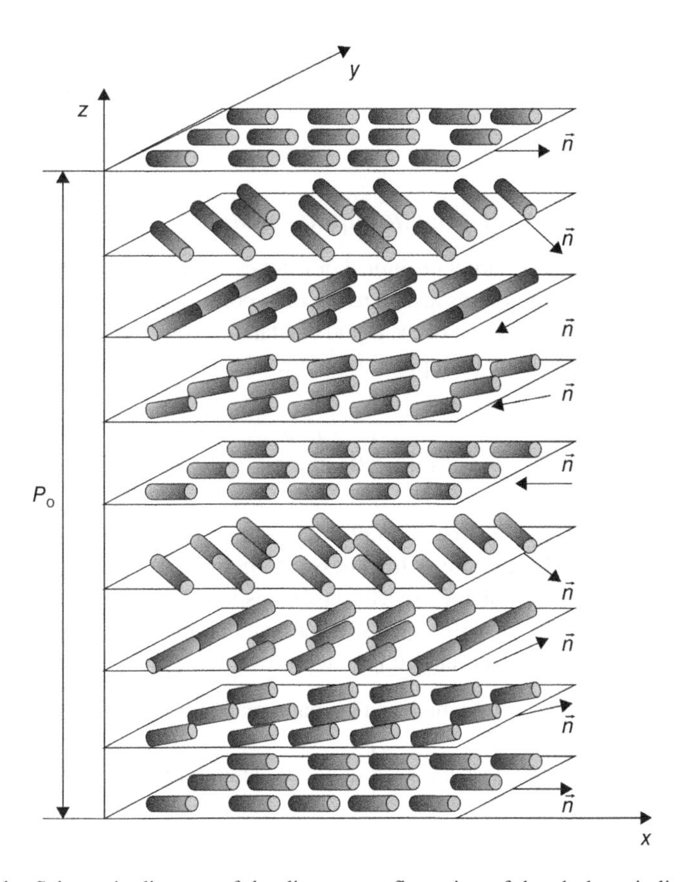

Figure 1.11 Schematic diagram of the director configuration of the cholesteric liquid crystal.

this is schematically shown in Figure 1.11 The liquid crystal director twists in space. This type of liquid crystal is called a *cholesteric* liquid crystal. The axis around which the director twists is called the *helical axis* and is chosen to be parallel to the z axis here. The distance P_o over which the director twists 360° is called the *pitch* and is related to the chirality by

$$P_o = \frac{2\pi}{q_o}. \tag{1.67}$$

Depending on its chemical structure, the pitch of a cholesteric liquid crystal could take any value in the region from a few tenths of a micron to infinitely long. The periodicity of a cholesteric liquid crystal with the pitch P_o is $P_o/2$, because $\vec{n}$ and $-\vec{n}$ are equivalent. Cholesteric liquid crystals are also called *chiral nematic* liquid crystals and denoted as N*. Nematic liquid crystals can be considered as a special case of cholesteric liquid crystals with an infinitely long pitch.

In practice, a cholesteric liquid crystal is usually obtained by mixing a nematic host and a chiral dopant. The pitch of the mixture is given by

$$P = \frac{1}{(HTP)\cdot x} \tag{1.68}$$

where x is the concentration of the chiral dopant and (HTP) is the *helical twisting power* of the chiral dopant, which is mainly determined by the chemical structure of the chiral dopant and depends slightly on the nematic host.

1.4.3 Elastic properties of smectic liquid crystals

Smectic liquid crystals possess partial positional orders besides the orientational order exhibited in nematic and cholesteric liquid crystals. Here we only consider the simplest case: smectic-A. The elastic energy of the deformation of the liquid crystal director in smectic-A is the same as in nematic. In addition, the dilatation (compression) of the smectic layer also costs energy, which is given by [23]

$$f_{layer} = \frac{1}{2}B\left(\frac{d-d_o}{d_o}\right)^2, \tag{1.69}$$

where B is elastic constant for the dilatation of the layer and is referred to as the Young modulus, d_o and d are the equilibrium layer thickness (the periodicity of the density undulation) and the actual layer thickness of the smectic layer, respectively. The typical value of B is about $10^6 - 10^7$ joule/m^3, which is $10^3 - 10^4$ smaller than that in a solid. In a slightly deformed smectic-A liquid crystal, consider a closed loop as shown in Figure 1.12. The total number of layers traversed by the loop is zero, which can be mathematically expressed as $\oint \vec{n}\cdot dl = 0$. Using the Stokes theorem, we have $\int \nabla\times\vec{n}\cdot d\vec{s} = \oint \vec{n}\cdot d\vec{l} = 0$. Therefore in smectic-A we have

$$\nabla\times\vec{n} = 0, \tag{1.70}$$

which ensures that $\vec{n}\cdot\nabla\times\vec{n} = 0$ and $\vec{n}\times\nabla\times\vec{n} = 0$. The consequence is that twist and bend deformation of the director are not allowed (because they significantly change the layer thickness and cost too much energy). The elastic energy in a smectic-A liquid crystal is

$$f_{elas} = \frac{1}{2}K_{11}(\nabla\cdot\vec{n})^2 + \frac{1}{2}B\left(\frac{d-d_o}{d_o}\right)^2. \tag{1.71}$$

Some chiral liquid crystals, as temperature is decreased, exhibit the mesophases: isotropic → cholesteric → smectic-A. Because of the property shown by Equation (1.70), there is no

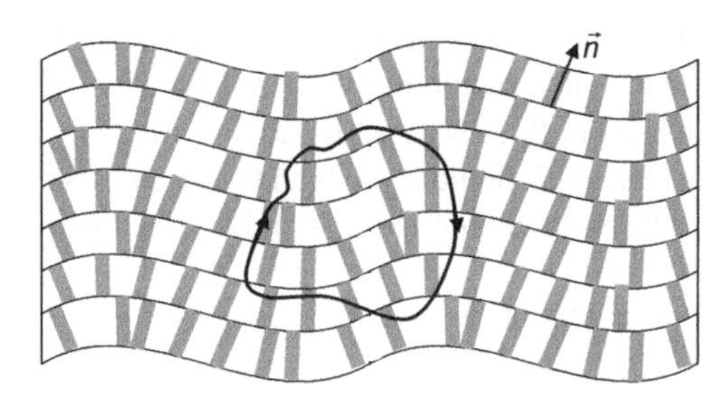

Figure 1.12 Schematic diagram showing the deformation of the liquid crystal director and the smectic layer in the smectic-A liquid crystal.

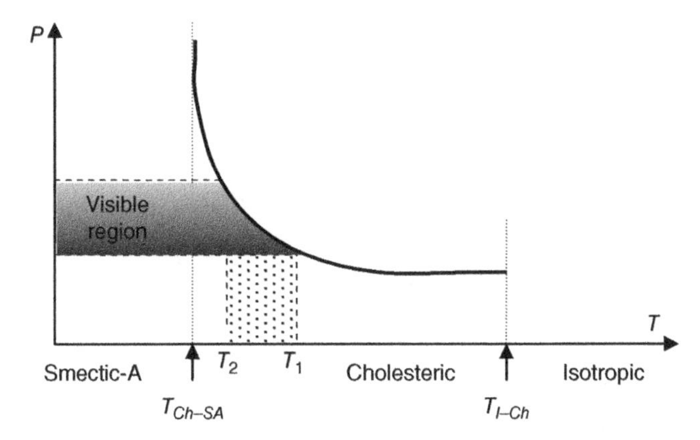

Figure 1.13 Schematic diagram showing how the pitch change of a thermochromic cholesteric liquid crystal.

spontaneous twist in smectic-A. To express this in another way, the pitch in smectic-A is infinitely long. In the cholesteric phase, as the temperature is decreased toward the cholesteric–smectic-A transition, there is a pretransitional phenomenon that smectic-A order forms in short space-scale and time scale due to thermal fluctuation. This effect causes the pitch of the cholesteric liquid crystal to increase with decreasing temperature and diverges at the transition temperature as shown in Figure 1.13. As will be discussed later, a cholesteric liquid crystal with the pitch P exhibits Bragg reflection at the wavelength $\lambda = \bar{n}P$ where $\bar{n}$ is the average refractive index of the material. If $\lambda = \bar{n}P$ is in the visible light region, the liquid crystal reflects colored light. When the temperature is varied, the color of the liquid crystal changes. Such cholesteric liquid crystals are known as thermochromic cholesteric liquid crystals [24]. As shown in Figure 1.13, the reflected light is in the visible region for temperature in the region T_1 to T_2. There are liquid crystals with $\Delta T = T_1 - T_2$ about 1 degree. If there are two thermochromic cholesteric liquid crystals with different cholesteric–smectic-A transition temperatures, mixtures with different concentrations of the two components will exhibit color reflection at different temperatures. This is how thermochromic cholesteric liquid crystals are used to make thermometers.

1.5 Response of Liquid Crystals to Electromagnetic Fields

Liquid crystals are anisotropic dielectric and diamagnetic media [1,25]. Their resistivities are very high ($\sim 10^{10}\Omega \cdot \text{cm}$). Dipole moments are induced in them by external fields. They have different dielectric permittivities and magnetic susceptibilities along the directions parallel to and perpendicular to the liquid crystal director.

1.5.1 Magnetic susceptibility

We first consider magnetic susceptibility. Because the magnetic interaction between the molecules is weak, the local magnetic field on the molecules is approximately the same as the externally applied magnetic field. For a uniaxial liquid crystal, the molecule can be regarded as a cylinder. When a magnetic field $\vec{H}$ is applied to the liquid crystal, it has different responses

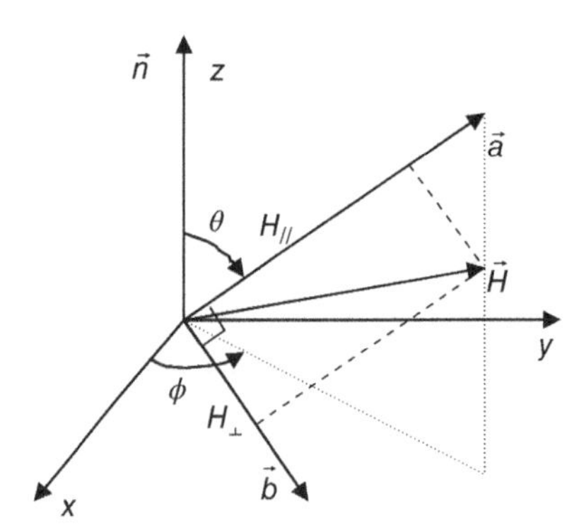

Figure 1.14 Schematic diagram showing the field decomposed into the components parallel to and perpendicular to the long molecular axis. $\vec{a}$: unit vector parallel to the long molecular axis, $\vec{b}$: unit vector perpendicular to the long molecular axis.

to the applied field, depending on the angle between the long molecular axis $\vec{a}$ and the field $\vec{H}$. The magnetic field can be decomposed into a parallel component and a perpendicular component, as shown in Figure 1.14. The magnetization $\vec{M}$ is given by

$$\begin{aligned}\vec{M} &= N\kappa_{//}\left(\vec{a}\cdot\vec{H}\right)\vec{a} + N\kappa_{\perp}\left[\vec{H} - \left(\vec{a}\cdot\vec{H}\right)\vec{a}\right] \\ &= N\kappa_{\perp}\vec{\mathrm{H}} + N\Delta\kappa\left(\vec{a}\cdot\vec{H}\right)\vec{\mathrm{a}} \\ &= N\kappa_{\perp}\vec{\mathrm{H}} + N\Delta\kappa\left(\vec{a}\vec{a}\right)\cdot\vec{\mathrm{H}},\end{aligned} \tag{1.72}$$

where N is the molecular number density, $\kappa_{//}$ and $\kappa_{\perp}$ are molecular magnetic polarizabilities along and perpendicular to the long molecular axis, respectively, and $\Delta\kappa = \kappa_{//} - \kappa_{\perp}$. Expressed in matrix form, Equation (1.72) changes to

$$\vec{M} = N\begin{pmatrix} \kappa_{\perp} + \Delta\kappa a_x a_x & \Delta\kappa a_x a_y & \Delta\kappa a_x a_z \\ \Delta\kappa a_y a_x & \kappa_{\perp} + \Delta\kappa a_y a_y & \Delta\kappa a_y a_z \\ \Delta\kappa a_z a_x & \Delta\kappa a_z a_y & \kappa_{\perp} + \Delta\kappa a_z a_z \end{pmatrix}\cdot\vec{H} = N\overleftrightarrow{\kappa}\cdot\vec{H}, \tag{1.73}$$

where a_i ($i = x,\ y,\ z$) are the projection of $\vec{a}$ in the x, y, and z directions in the lab frame whose z axis is parallel to the liquid crystal director, and $a_z = \cos\theta$, $a_x = \sin\theta\cos\phi$, and $a_y = \sin\theta\sin\phi$. The molecule swivels because of thermal motion. The averaged magnetization is $\vec{M} = N < \overleftrightarrow{\kappa} > \cdot\vec{H}$. For a uniaxial liquid crystal, recall $<\cos^2\theta> = (2S+1)/3$, $<\sin^2\theta> = (2-2S)/3$, $<\sin^2\phi> = <\cos^2\phi> = 1/2$, and $<\sin\phi\cos\phi> = 0$. Therefore

$$<\overleftrightarrow{\kappa}> = \begin{pmatrix} \kappa_{\perp} + \frac{1}{3}(1-S)\Delta\kappa & 0 & 0 \\ 0 & \kappa_{\perp} + \frac{1}{3}(1-S)\Delta\kappa & 0 \\ 0 & 0 & \kappa_{\perp} + \frac{1}{3}(2S+1)\Delta\kappa \end{pmatrix}. \tag{1.74}$$

Because $\vec{M} = \overleftrightarrow{\chi} \cdot \vec{H}$, the magnetic susceptibility tensor is

$$\overleftrightarrow{\chi} = \begin{pmatrix} \chi_\perp & 0 & 0 \\ 0 & \chi_\perp & 0 \\ 0 & 0 & \chi_{//} \end{pmatrix} = N \begin{pmatrix} \kappa_\perp + \frac{1}{3}(1-S)\Delta\kappa & 0 & 0 \\ 0 & \kappa_\perp + \frac{1}{3}(1-S)\Delta\kappa & 0 \\ 0 & 0 & \kappa_\perp + \frac{1}{3}(2S+1)\Delta\kappa \end{pmatrix}. \quad (1.75)$$

The anisotropy is

$$\Delta\chi = \chi_{//} - \chi_\perp = N\Delta\kappa S. \quad (1.76)$$

For most liquid crystals, $\chi_{//}$ and $\chi_\perp$ are negative and small ($\sim 10^{-5}$ in SI units). $\Delta\chi$ is usually positive. From Equation (1.75) it can be seen that $(2\chi_\perp + \chi_{//})/3 = N(3\kappa_\perp + \Delta\kappa)/3 = N(2\kappa_\perp + \kappa_{//})/3$, which is independent of the order parameter. The quantity $(2\chi_\perp + \chi_{//})/3N$ does not change discontinuously when crossing the isotropic–nematic transition.

1.5.2 Dielectric permittivity and refractive index

When an electric field is applied to a liquid crystal, it will induce dipole moments in the liquid crystal. For a uniaxial liquid crystal, the molecule can be regarded as a cylinder, and it has different molecular polarizabilities along and perpendicular to the long molecular axis $\vec{a}$. Similar to the magnetic case, when a local electric field $\vec{E}_{loc}$ (also called internal field) is applied to the liquid crystal, the polarization (dipole moment per unit volume) is given by

$$\begin{aligned} \vec{P} &= N\alpha_{//}\left(\vec{a}\cdot\vec{E}_{loc}\right)\vec{a} + N\alpha_\perp\left[\vec{E}_{loc} - \left(\vec{a}\cdot\vec{E}_{loc}\right)\vec{a}\right] \\ &= N\alpha_\perp\vec{E}_{loc} + N\Delta\alpha\left(\vec{a}\cdot\vec{E}_{loc}\right)\vec{a} \\ &= N\alpha_\perp\vec{E}_{loc} + N\Delta\alpha\left(\vec{a}\,\vec{a}\right)\cdot\vec{E}_{loc}, \end{aligned} \quad (1.77)$$

where N is the molecular density, $\alpha_{//}$ and $\alpha_\perp$ are the molecular polarizabilities along and perpendicular to the long molecular axis, respectively, and $\Delta\alpha = \alpha_{//} - \alpha_\perp$. Different from the magnetic case, the dipole–dipole interaction between the molecules are strong or, stated in another way, the local electric field on a molecule is the sum of the externally applied electric field and the electric field produced by the dipole moment of other molecules. We can approach this problem in the following way. Imagine a cavity created by removing the molecule under consideration, as shown in Figure 1.15. The macroscopic field $\vec{E}$ is the sum of the field $\vec{E}_{self}$ produced by the molecule itself and the field $\vec{E}_{else}$, which is the local field $\vec{E}_{local}$, produced by the external source and the rest of the molecules of the system:

$$\vec{E} = \vec{E}_{self} + \vec{E}_{else} = \vec{E}_{self} + \vec{E}_{local} \quad (1.78)$$

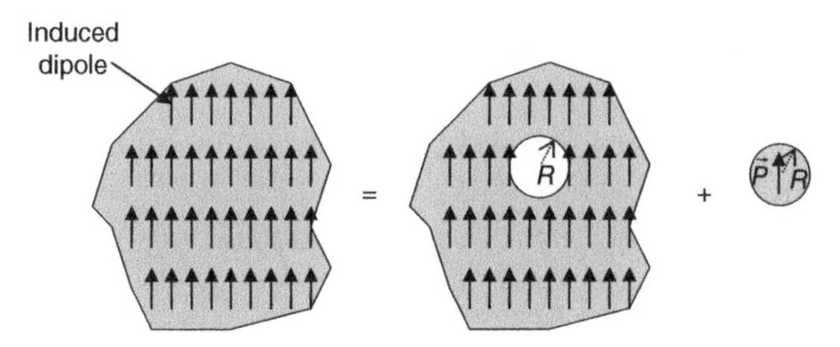

Figure 1.15 Schematic diagram showing how the macroscopic field is produced in the medium.

In order to illustrate the principle, let us first consider an isotropic medium. The cavity can be regarded as a sphere. The field $\vec{E}_{self}$ is produced by the dipole moment inside the sphere, which can be calculated in the following way. In the calculation of the field, the dipole moment can be replaced by the surface charge produced by the dipole moment on the surface of the sphere. The surface charge density is $\sigma = \vec{P} \cdot \vec{m}$. The field produced by the surface charge is $\vec{E}_{self} = -\vec{P}/3\varepsilon_o$. The local field is $\vec{E}_{local} = \vec{E} + \vec{P}/3\varepsilon_o$. Hence the polarizability is

$$\vec{P} = N\alpha\vec{E}_{loc} = N\alpha\left(\vec{E} + \vec{P}/3\varepsilon_o\right), \tag{1.79}$$

$$\vec{P} = \frac{N\alpha\vec{E}}{1 - N\alpha/3\varepsilon_o}. \tag{1.80}$$

The electric displacement $\vec{D} = \varepsilon_o \varepsilon \vec{E} = \varepsilon_o \vec{E} + \vec{P}$, where $\varepsilon_o = 8.85 \times 10^{12} N/V^2$ is the permittivity of vacuum, and ε is the (relative) dielectric constant, which is given by

$$\varepsilon = 1 + \frac{\vec{P}}{\varepsilon_o \vec{E}} = 1 + \frac{N\alpha/\varepsilon_o}{1 - N\alpha/3\varepsilon_o}, \tag{1.81}$$

$$\frac{\varepsilon - 1}{\varepsilon + 2} = \frac{1}{3\varepsilon_o} N\alpha, \tag{1.82}$$

which is called the Clausius–Mossotti relation. At optical frequency, the refractive index n is given by $n^2 = \varepsilon$, and therefore

$$\frac{n^2 - 1}{n^2 + 2} = \frac{1}{3\varepsilon_o} N\alpha, \tag{1.83}$$

which is called the Lorentz–Lorenz relation. The local field is related to the macroscopic field by

$$\vec{E}_{local} = \vec{E} + \vec{P}/3\varepsilon_o = \vec{E} + \frac{N\alpha/3\varepsilon_o}{1 - N\alpha/3\varepsilon_o}\vec{E} = \frac{1}{1 - N\alpha/3\varepsilon_o}\vec{E} = K\vec{E}, \tag{1.84}$$

where the defined $K = 1/(1 - N\alpha/3\varepsilon_o)$ is called the internal field constant.

Liquid crystals are anisotropic. In them the local field $\vec{E}_{local}$ depends on the macroscopic field $\vec{E}$ as well as the angles between $\vec{E}$ and the long molecular axis $\vec{a}$ and the liquid crystal director $\vec{n}$. They are related to each other by

$$\vec{E}_{local} = \overleftrightarrow{K} \cdot \vec{E}, \tag{1.85}$$

where $\overleftrightarrow{K}$ is the internal field tensor which is a second-rank tensor. After taking account of the internal field tensor and the thermal motion of the molecules, the polarization is

$$\vec{P} = N\alpha_{\perp} < \vec{K} > \cdot \vec{E} + N\Delta\alpha < \left[\vec{K} \cdot (\vec{a}\vec{a})\right] > \cdot \vec{E}. \tag{1.86}$$

The macroscopic dielectric tensor is

$$\overleftrightarrow{\varepsilon} = \overleftrightarrow{I} + \frac{N}{\varepsilon_o}\left[\alpha_{\perp} < \overleftrightarrow{K} > + \Delta\alpha < \overleftrightarrow{K} \cdot (\vec{a}\vec{a}) > \right]. \tag{1.87}$$

In a material consisting of non-polar molecules, the induced polarization is contributed by two parts: (1) electronic polarization, $P_{electronic}$, which comes from the deformation of the electron clouds of the constituting atoms of the molecule, (2) ionic polarization, P_{ionic}, which comes from the relative displacement of the atoms constituting the molecule. For a material consisting of polar molecules, there is a third contribution: dipolar polarization, $P_{dipolar}$, which comes from the reorientation of the dipole. These contributions to the molecular polarizability depend on the frequency of the applied field. The rotation of the molecule is slow and therefore the dipole-orientation polarization can only contribute up to a frequency of the order of MHz. The vibration of atoms in molecules is faster and the ionic polarization can contribute up to the frequency of infrared light. The motion of electrons is the fastest and the electronic polarization can contribute up to the frequency of UV light. In relation to their magnitudes, the order is $P_{electronic} < P_{ionic} < P_{dipolar}$.

At optical frequencies, only the electronic polarization contributes to the molecular polarizability, which is small, and the electric field is usually low. De Jeu and Bordewijk experimentally showed that: (1) $(2\varepsilon_{\perp} + \varepsilon_{//})/3\rho$ is a constant through the nematic and isotropic phases [25,26], where ρ is the mass density, (2) the dielectric anisotropy $\Delta\varepsilon = \varepsilon_{//} - \varepsilon_{\perp}$ is directly proportional to the anisotropy of the magnetic susceptibility. Based on these facts, it was concluded that $\overleftrightarrow{K}$ is a molecular tensor independent of the macroscopic dielectric anisotropy. In the molecular principal frame $\eta\varsigma\xi$ with the ξ axis parallel to the long molecular axis $\vec{a}$, $\vec{K}$ has the form

$$\vec{K} = \begin{pmatrix} K_{\perp} & 0 & 0 \\ 0 & K_{\perp} & 0 \\ 0 & 0 & K_{//} \end{pmatrix}. \tag{1.88}$$

Next we need to find the form of $\overleftrightarrow{K}$ in the lab frame xyz with the z axis parallel to the liquid crystal director $\vec{n}$. Because of the axial symmetry around $\vec{a}$, we only need to consider the transformation of the matrix between the two frames as shown in Figure 1.16. The frame $\eta\varsigma\xi$ is

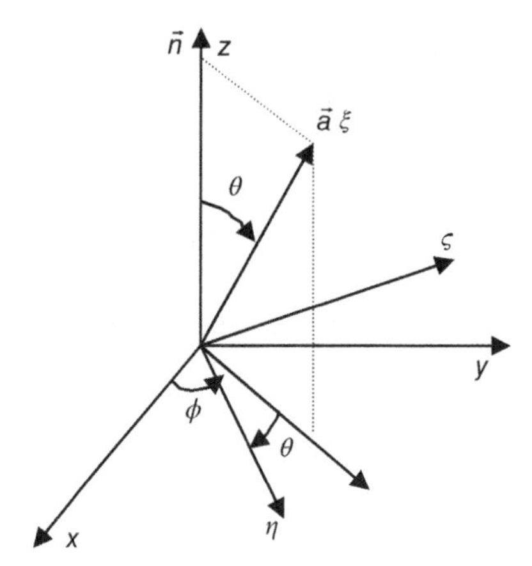

Figure 1.16 Schematic diagram showing the transformation between the molecular principal frame $\eta\varsigma\xi$ and the lab frame xyz.

achieved by first rotating the frame xyz around the z axis by the angle ϕ and then rotating the frame around the ς axis by the angle θ. The rotation matrix is

$$\overleftrightarrow{R} = \begin{pmatrix} \cos\theta\cos\phi & -\sin\phi & \sin\theta\cos\phi \\ \cos\theta\sin\phi & \cos\phi & \sin\theta\sin\phi \\ -\sin\theta & 0 & \cos\theta \end{pmatrix}. \tag{1.89}$$

The reserve rotation matrix is

$$\overleftrightarrow{R}^{-1} = \begin{pmatrix} \cos\theta\cos\phi & \cos\theta\sin\phi & -\sin\theta \\ -\sin\phi & \cos\phi & 0 \\ \sin\theta\cos\phi & \sin\theta\sin\phi & \cos\theta \end{pmatrix}. \tag{1.90}$$

In the lab frame $\overleftrightarrow{K}$ has the form

$$\vec{K} = \overleftrightarrow{R} \cdot \begin{pmatrix} K_\perp & 0 & 0 \\ 0 & K_\perp & 0 \\ 0 & 0 & K_{//} \end{pmatrix} \cdot \overleftrightarrow{R}^{-1}$$

$$= \begin{pmatrix} K_\perp + \Delta K \sin^2\theta\cos^2\phi & \Delta K \sin^2\theta\sin\phi\cos\phi & \Delta K \sin\theta\cos\theta\cos\phi \\ \Delta K \sin^2\theta\sin\phi\cos\phi & K_\perp + \Delta K \sin^2\theta\sin^2\phi & \Delta K \sin\theta\cos\theta\sin\phi \\ \Delta K \sin\theta\cos\theta\cos\phi & \Delta K \sin\theta\cos\theta\sin\phi & K_\perp + \Delta K \cos^2\theta \end{pmatrix}, \tag{1.91}$$

where $\Delta K = K_{//} - K_\perp$, and $\overline{aa}$ has the form

$$\overrightarrow{a}\overrightarrow{a} = \begin{pmatrix} \sin\theta\cos\phi \\ \sin\theta\sin\phi \\ \cos\theta \end{pmatrix} (\sin\theta\cos\phi \sin\theta\sin\phi\cos\theta)$$

$$= \begin{pmatrix} \sin^2\theta\cos^2\phi & \sin^2\theta\sin\phi\cos\phi & \sin\theta\cos\theta\cos\phi \\ \sin^2\theta\sin\phi\cos\phi & \sin^2\theta\sin^2\phi & \sin\theta\cos\theta\sin\phi \\ \sin\theta\cos\theta\cos\phi & \sin\theta\cos\theta\sin\phi & \cos^2\theta \end{pmatrix}, \tag{1.92}$$

and $\overrightarrow{a}\overrightarrow{a}\cdot\overleftrightarrow{K}$ has the form

$$\overrightarrow{a}\overrightarrow{a}\cdot\overleftrightarrow{K} = \begin{pmatrix} K_{//}\sin^2\theta\cos^2\phi & K_{//}\sin^2\theta\sin\phi\cos\phi & K_{//}\sin\theta\cos\theta\cos\phi \\ K_{//}\sin^2\theta\sin\phi\cos\phi & K_{//}\sin^2\theta\sin^2\phi & K_{//}\sin\theta\cos\theta\sin\phi \\ K_{//}\sin\theta\cos\theta\cos\phi & K_{//}\sin\theta\cos\theta\sin\phi & K_{//}\cos^2\theta \end{pmatrix}. \tag{1.93}$$

Recall that $<\cos^2\theta> = (2S+1)/3$, $<\sin^2\theta> = (2-2S)/3$, $<\sin^2\phi> = <\cos^2\phi> = 1/2$, $<\cos\theta> = <\sin\phi> = <\cos\phi> = <\sin\phi\cos\phi> = 0$, therefore their averaged values are

$$<\overleftrightarrow{K}> = \begin{pmatrix} K_\perp + \Delta K(1-S)/3 & 0 & 0 \\ 0 & K_\perp + \Delta K(1-S)/3 & 0 \\ 0 & 0 & K_\perp + \Delta K(2S+1)/3 \end{pmatrix}, \tag{1.94}$$

$$<\overrightarrow{a}\overrightarrow{a}\cdot\overleftrightarrow{K}> = \begin{pmatrix} K_{//}(1-S)/3 & 0 & 0 \\ 0 & K_{//}(1-S)/3 & 0 \\ 0 & 0 & K_{//}(2S+1)/3 \end{pmatrix}, \tag{1.95}$$

$$\overleftrightarrow{\varepsilon} = \overleftrightarrow{I} + \frac{N}{\varepsilon_o}\left[\alpha_\perp <\overleftrightarrow{K}> + \Delta\alpha <\overleftrightarrow{K}\cdot(\vec{a}\vec{a})>\right].$$

Therefore

$$\overleftrightarrow{\varepsilon} = \begin{pmatrix} 1+\frac{N}{3\varepsilon_o}[\alpha_\perp K_\perp(2+S)+\alpha_{//}K_{//}(1-S)] & 0 & 0 \\ 0 & 1+\frac{N}{3\varepsilon_o}[\alpha_\perp K_\perp(2+S)+\alpha_{//}K_{//}(1-S)] & 0 \\ 0 & 0 & 1+\frac{N}{3\varepsilon_o}[\alpha_\perp K_\perp(2-2S)+\alpha_{//}K_{//}(1+2S)] \end{pmatrix}. \tag{1.96}$$

The anisotropy is

$$\Delta\varepsilon = \varepsilon_{//} - \varepsilon_\perp = \frac{N}{\varepsilon_o}(\alpha_{//}K_{//} - \alpha_\perp K_\perp)S, \tag{1.97}$$

which is linearly proportional to the order parameter S. In terms of the refractive indices, Equation (1.97) becomes

$$n_{//}^2 - n_{\perp}^2 = 2\bar{n}\Delta n = \frac{N}{\varepsilon_o}\left(\alpha_{//}K_{//} - \alpha_{\perp}K_{\perp}\right)S,$$

where $\bar{n} = \left(n_{//} + n_{\perp}\right)/2$ and $\Delta n = (n_{//} - n_{\perp})$. Approximately, the birefringence Δn is linearly proportional to the order parameter. For most liquid crystals, $\bar{n} \sim 1.5$–2.0 and $\Delta n \sim 0.05$–0.3.

The electronic polarization may be treated by using classical mechanics, where the system is regarded as a simple harmonic oscillator. There are three forces acting on the electron: (1) elastic restoring force $-Kx$, where K is the elastic constant and x is the displacement of the electron from its equilibrium position, (2) viscosity force $-\gamma\partial x/\partial t$, and (3) the electric force $-eE_o e^{i\omega t}$, where E_o and ω are the amplitude and frequency of the applied electric field, respectively. The dynamic equation is

$$m\frac{d^2x}{dt^2} = -kx - eE_o e^{i\omega t} - \gamma\frac{\partial x}{\partial t}. \tag{1.98}$$

The solution is $x = x_o e^{i\omega t}$ and the amplitude of the oscillation is

$$x_o = \frac{-eE_o}{m\left(\omega^2 - \omega_o^2\right) + i\gamma\omega}, \tag{1.99}$$

where $\omega_o = \sqrt{k/m}$ is the frequency of the oscillator (the frequency of the transition dipole moment in quantum mechanics). The induced dipole moment is $p = -ex_o$. The molecule polarizability is

$$\alpha = p/E_{loc} = \frac{e^2}{m}\frac{\left(\omega_o^2 - \omega^2\right)}{\left(\omega_o^2 - \omega^2\right)^2 + (\gamma\omega/m)^2} - i\frac{e^2}{m}\frac{\gamma\omega/m}{\left(\omega_o^2 - \omega^2\right)^2 + (\gamma\omega/m)^2}, \tag{1.100}$$

which is a complex number, and the imaginary part corresponds to absorption. When the frequency of the light is far away from the absorption frequency ω_o or the viscosity is small, the absorption is negligible, $\alpha = p/E_{loc} = (e^2/m)/(\omega_o^2 - \omega^2)$. The refractive index is $n^2 \propto \alpha \propto \frac{1}{(\omega_o^2 - \omega^2)} = \frac{1}{\left[(2\pi/C\lambda_o)^2 - (2\pi/C\lambda)^2\right]} = \frac{C^2\lambda_o^2}{4\pi^2}\frac{\lambda^2}{\lambda^2 - \lambda_o^2}$, as expressed in Sellmeier's equation

$$n^2 = 1 + \frac{H\lambda^2}{\lambda^2 - \lambda_o^2}, \tag{1.101}$$

where H is a constant. When λ is much longer than λ_o, expanding the above equation, we have

$$n \approx A + \frac{B}{\lambda^2} + \frac{C}{\lambda^4}. \tag{1.102}$$

This is Cauchy's Equation. The refractive index increases with decreasing wavelength. For liquid crystals, along different directions with respect to the long molecular axis, the molecular polarizabilities are different. Also along different directions, the frequencies of the transition dipole moments are different, which results in *dichroic absorption*: when the electric field of light is parallel to the transition dipole moment, light is absorbed, but when the electric field is perpendicular to the transition moment, light is not absorbed. Positive dichroic dyes have transition dipole moments parallel to the long molecular axis while negative dichroic dyes have transition dipole moments perpendicular to the long molecular axis.

At DC or at low frequency applied electric fields, for liquid crystals of polar molecules, the dipolar polarization is dominant. For a liquid crystal with a permanent dipole moment $\vec{p}$, the polarization is now given by

$$\vec{P} = N\alpha_{\perp} < \vec{K} > \cdot \vec{E} + N\Delta\alpha < \left[\vec{K} \cdot (\vec{a}\vec{a})\right] > \cdot \vec{E} + N < \vec{p} > . \quad (1.103)$$

The macroscopic dielectric tensor is

$$\overleftrightarrow{\varepsilon} = \overleftrightarrow{I} + \frac{N}{\varepsilon_o}\left[\alpha_{\perp} < \overleftrightarrow{K} > + \Delta\alpha < \overleftrightarrow{K} \cdot (\vec{a}\vec{a}) > + < \vec{p} > \vec{E} / E^2\right]. \quad (1.104)$$

The energy of the dipole in the directing electric field $\vec{E}_d$ is $u = -\vec{p} \cdot \vec{E}_d$. The directing field $\vec{E}_d$ is different from the local field $\vec{E}_{loc}$ because the dipole polarizes its surroundings, which in turn results in a reaction field $\vec{E}_r$ at the position of the dipole. As $\vec{E}_r$ is always parallel to the dipole, it cannot affect the orientation of the dipole. As an approximation, it is assumed that $\vec{E}_d = d \cdot \vec{E}$, where d is a constant. Usually the dipole moment p is about $1e \times 1$ Å $= 1.6 \times 10^{-19}$ C $\times 10^{-10}$ m $= 1.6 \times 10^{-29}$ m · C. At room temperature ($T \sim 300$ K) and under the normal strength field $E \sim 1$ V/μm $= 10^6$ V/m, $pE/3k_BT \ll 1$. Consider a liquid crystal molecule with a permanent dipole moment making the angle β with the long molecular axis. In the molecular frame $\eta\varsigma\xi$, the components of $\vec{p}$ are $(p \sin\beta \cos\psi, p \sin\beta \sin\psi, p \cos\beta)$, as shown in Figure 1.17. Using the

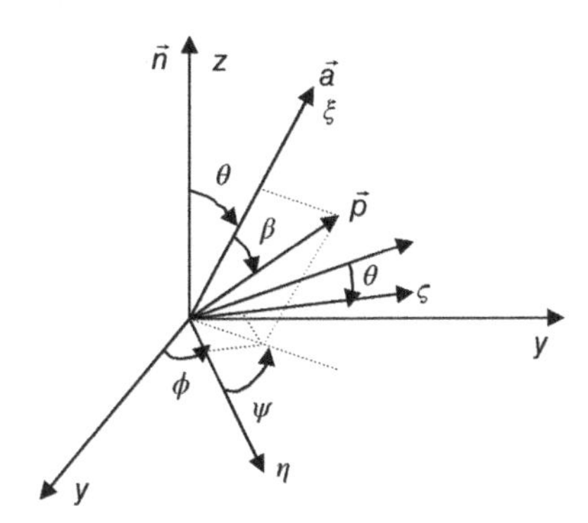

Figure 1.17 Schematic diagram showing the orientation of the dipole $\vec{p}$ in the molecular principal frame $\eta\varsigma\xi$ and the lab frame xyz.

rotation matrix given by Equation (1.90), we can calculate the components of $\vec{p}$ in the lab frame xyz:

$$\vec{p} = \begin{pmatrix} \cos\phi & -\cos\theta\sin\phi & -\sin\theta\sin\phi \\ \sin\phi & \cos\theta\cos\phi & \sin\theta\cos\phi \\ 0 & -\sin\theta & \cos\theta \end{pmatrix} \cdot p \begin{pmatrix} \sin\beta\cos\psi \\ \sin\beta\sin\psi \\ \cos\beta \end{pmatrix}$$

$$= p \begin{pmatrix} \sin\beta\cos\psi\cos\phi - \sin\beta\sin\psi\cos\theta\sin\phi - \cos\beta\sin\theta\sin\phi \\ \sin\beta\sin\psi\sin\phi + \sin\beta\sin\psi\cos\theta\cos\phi + \cos\beta\sin\theta\cos\phi \\ -\sin\theta\sin\beta\sin\psi + \cos\theta\cos\beta \end{pmatrix}. \tag{1.105}$$

When the applied field is parallel to $\vec{n}$, $\vec{E} = E_{//}\vec{z}$, the projection of the dipole along the applied field is

$$p_{//} = -p(\sin\beta\sin\psi\sin\theta + \cos\beta\cos\theta), \tag{1.106}$$

and the energy is

$$u = -dp(\cos\beta\cos\theta - \sin\beta\sin\psi\sin\theta)E_{//}. \tag{1.107}$$

The average value of the projection is

$$<p_{//}> = \frac{\int (p\cos\beta\cos\theta - p\sin\beta\sin\psi\sin\theta)e^{-u/k_BT - V(\theta)/k_BT}\sin\theta d\theta d\phi d\psi}{\int e^{-u/k_BT - V(\theta)/k_BT}\sin\theta d\theta d\phi d\psi}. \tag{1.108}$$

Because $-u \ll k_BT$, $e^{-u/k_BT} \approx (1 - u/k_BT)$, then

$$<p_{//}> = \frac{dE_{//}}{k_BT}\frac{\int (p\cos\beta\cos\theta - p\sin\beta\sin\psi\sin\theta)^2 e^{-V(\theta)/k_BT}\sin\theta d\theta d\phi d\psi}{\int e^{-u/k_BT - V(\theta)/k_BT}\sin\theta d\theta d\phi d\psi}$$

$$= \frac{dE_{//}p^2}{k_BT} < \left(\cos^2\beta\cos^2\theta + \sin^2\beta\sin^2\theta\sin^2\psi - \sin\beta\cos\beta\sin\theta\cos\theta\sin\psi\right) >$$

Because $<\sin^2\psi> = 1/2$, $<\sin\psi> = 0$, $<\cos^2\theta> = (2S+1)/3$ and $<\sin^2\theta> = (2-2S)/3$,

$$<p_{//}> = \frac{dE_{//}p^2}{3k_BT}\left[\cos^2\beta(2S+1) + \sin^2\beta(1-S)\right] = \frac{dE_{//}p^2}{3k_BT}\left[1 - (1 - 3\cos^2\beta)S\right] \tag{1.109}$$

From Equations (1.96), (1.104), and (1.109), we have

$$\varepsilon_{//} = 1 + + \frac{N}{3\varepsilon_o}\left\{\alpha_\perp K_\perp(2-2S) + \alpha_{//}K_{//}(1+2S) + \frac{dp^2}{k_BT}\left[1 - (1 - 3\cos^2\beta)S\right]\right\}. \tag{1.110}$$

Note that $\alpha_{//}$ and $\alpha_{\perp}$ are the molecular polarizabilities contributed by the electronic and ionic polarizations.

When the applied field is perpendicular to $\vec{n}$, say $\vec{E} = E_{\perp}\hat{x}$, the projection of the dipole along the applied field is

$$p_{\perp} = p(\sin\beta\cos\psi\cos\phi - \sin\beta\sin\psi\cos\theta\sin\phi - \cos\beta\sin\theta\sin\phi), \tag{1.111}$$

and the energy is

$$u = -dp(\sin\beta\cos\psi\cos\phi - \sin\beta\sin\psi\cos\theta\sin\phi - \cos\beta\sin\theta\sin\phi)E_{\perp}. \tag{1.112}$$

The average value of the projection is

$$<p_{\perp}> = \frac{dE_{\perp}p^2}{k_BT}\left[\frac{1}{4}\sin^2\beta + \frac{1}{4}\sin^2\beta\frac{(2S+1)}{3} + \cos^2\beta\frac{(1-S)}{3}\right] = \frac{dE_{\perp}p^2}{3k_BT}\left[1 + \frac{1}{2}(1-3\cos^2\beta)S\right].$$

From Equations (1.96), (1.104) and (1.112), we have

$$\varepsilon_{\perp} = 1 + \frac{N}{3\varepsilon_o}\left\{\alpha_{\perp}K_{\perp}(2+S) + \alpha_{//}K_{//}(1-S) + \frac{dp^2}{k_BT}\left[1 + \frac{1}{2}(1-3\cos^2\beta)S\right]\right\}. \tag{1.113}$$

The dielectric anisotropy is

$$\Delta\varepsilon = \varepsilon_{//} - \varepsilon_{\perp} = \frac{N}{\varepsilon_o}\left[(\alpha_{//}K_{//} - \alpha_{\perp}K_{\perp}) - \frac{dp^2}{2k_BT}(1-3\cos^2\beta)\right]S, \tag{1.114}$$

which is proportional to the order parameter S. The contribution of induced polarization (electronic and ionic polarizations) changes with temperature as S, while the contribution of the orientation polarization changes with temperature as S/T. When the angle between the permanent dipole and the long molecular axis is $\beta = 55\,°$, $(1 - 3\cos^2\beta) = 0$, the orientation polarization of the permanent dipole does not contribute to $\Delta\varepsilon$.

The permanent dipole moment is fixed on the molecule. Thus the molecule has to reorient in order to contribute to the dielectric constants. Qualitatively speaking, only when the frequency of the applied field is lower than a characteristic frequency ω_c, can the molecule rotate to follow the oscillation of the applied field and therefore to contribute to the dielectric constants. For rod-like liquid crystal molecules, it is easier to spin around the long molecular axis than to rotate around a short molecular axis. Therefore the characteristic frequency $\omega_{\perp c}$ for $\varepsilon_{\perp}$ is higher than the characteristic frequency $\omega_{//c}$ for $\varepsilon_{//}$. For molecules on which the angle β between the permanent dipole and the long molecular axis is very small, $\Delta\varepsilon$ is always positive at all frequencies. For molecules with large permanent dipole moment p and large β, $\Delta\varepsilon$ is negative at low frequencies. For molecules with large permanent dipole moment p and intermediate β, $\Delta\varepsilon$ is positive at low frequencies, then changes to negative when the frequency is increased above a crossover frequency ω_o. The crossover frequency is in the region from a few kHz to a few tens of kHz. At infrared light or higher frequencies, the dipolar polarization does not contribute anymore, and $\Delta\varepsilon$ is always positive.

1.6 Anchoring Effects of Nematic Liquid Crystal at Surfaces

In most liquid crystal devices, the liquid crystals are sandwiched between two substrates coated with alignment layers. In the absence of externally applied fields, the orientation of the liquid crystal in the cell is determined by the anchoring condition of the alignment layer [26–28].

1.6.1 Anchoring energy

Consider an interface between a liquid crystal ($z > 0$) and an alignment layer ($z < 0$), as shown in Figure 1.18. For a liquid crystal molecule on the interface, some of the surrounding molecules are liquid crystal molecules and the other surrounding molecules are the alignment layer molecules. The potential for the molecule's orientation is different from that of the liquid crystal in the bulk, where all the surrounding molecules are liquid crystal molecules. At the interface, the orientational and positional orders may be different from those in the bulk. Here we only discuss the anisotropic part of the interaction between the liquid crystal molecule and the alignment layer molecule. The liquid crystal is anisotropic. If the alignment layer is also anisotropic, then there is a preferred direction – referred to as the easy axis – for the liquid crystal director at the interface, as shown in Figure 1.18. The interaction energy is a minimum when the liquid crystal director is along the easy axis. The z axis is perpendicular to the interface and pointing toward the liquid crystal side. The polar angle and azimuthal angle of the easy axis are θ_o and ϕ_o, respectively. If $\theta_o = 0°$, the anchoring is referred to as homeotropic. If $\theta_o = 90°$ and ϕ_o is well defined, the anchoring is termed homogeneous. If $\theta_o = 90°$ and there is no preferred azimuthal angle, the anchoring is called planar. If $0 < \theta_o < 90°$, the anchoring is referred to as tilted.

When the liquid crystal director $\vec{n}$ is aligned along the direction specified by the polar angle θ and azimuthal angle ϕ, the anisotropic part of the surface energy – referred to as the anchoring energy function – of the liquid crystal is $f_s = f_s(\theta, \phi)$. When $\theta = \theta_o$ and $\phi = \phi_o$, f_s has the minimum value of 0, and thus $\partial f_s/\partial\theta|_{\theta=\theta_o} = 0$ and $\partial f_s/\partial\phi|_{\phi=\phi_o} = 0$.

The materials above and below the interface are different, and there is no reflection symmetry about the interface. If $\theta_o \neq 0$, the anchoring energy does not have azimuthal rotational

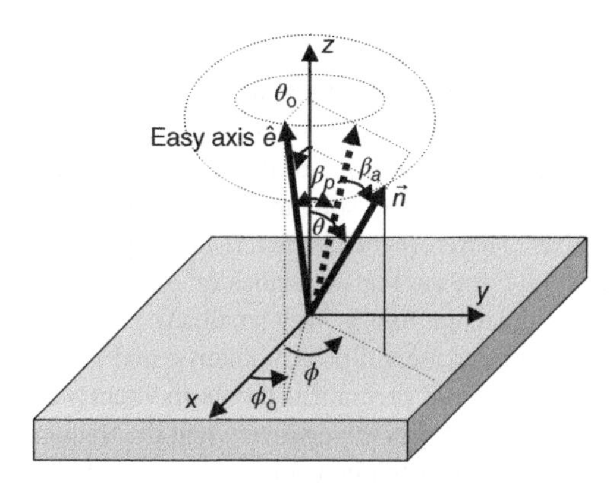

Figure 1.18 Schematic diagram showing the easy direction of the surface anchoring and the deviation of the liquid crystal director.

symmetry around the easy direction. Therefore the anchoring energies are different for deviation in polar angle and azimuthal angle. For small deviations, in the Rapini–Papoular model [29], the anchoring energy function can be expressed as

$$f_s = \frac{1}{2} W_p \sin^2 \beta_p + \frac{1}{2} W_a \sin^2 \beta_a, \tag{1.115}$$

where β_p and β_a are the angles between $\vec{n}$ and the easy axis when $\vec{n}$ deviates from the easy axis in the polar angle direction and the azimuthal angle direction, respectively. W_p and W_a are the polar and azimuthal anchoring strengths, respectively. For small values of $\theta - \theta_o$ and $\phi - \phi_o$, we have the approximations $\sin^2\beta_p = \sin^2(\theta - \theta_o)$ and $\sin^2\beta_a = \sin^2(\phi - \phi_o)\sin^2\theta_o$. Therefore the anchoring energy function is

$$f_s = \frac{1}{2} W_p \sin^2(\theta - \theta_o) + \frac{1}{2} W_a \sin^2\theta_o \sin^2(\phi - \phi_o). \tag{1.116}$$

For a homogeneous anchoring, $\theta_o \sim 90°$, If we define $\alpha = \pi/2 - \theta$, which is the polar angle defined with respect to the cell surface, Equation (1.116) becomes

$$f_s = \frac{1}{2} W_p \sin^2(\alpha - \alpha_o) + \frac{1}{2} W_a \sin^2(\phi - \phi_o). \tag{1.117}$$

Equations (1.116) and (1.117) are valid only for small deviations. For large deviations, quartic terms must be included. The anchoring strengths can be determined experimentally, which will be discussed in Chapter 5.

1.6.2 Alignment layers

Homogeneous anchoring can be achieved by mechanically rubbing the surface of the substrate, such as glass, of the liquid crystal cell with a cotton ball or cloth. The rubbing creates micro-grooves along the rubbing direction in the form of ridges and troughs, as shown in Figure 1.19 (a). When the liquid crystal is aligned parallel to the grooves, there is no orientation deformation. If the liquid crystal were perpendicular to the groves, there would be orientation deformation which costs elastic energy. Therefore the liquid crystal will be homogeneously aligned along the groves (the rubbing direction). The problem with alignment created in this way is that the anchoring strength ($\sim 10^{-5}$ J/m^2) is weak. The widely used homogeneous alignment layers are rubbed polyimides. The rubbing not only creates the micro-grooves but also aligns the polymer chains. The intermolecular interaction between the liquid crystal and the aligned polymer chains also favors the parallel alignment and thus increases the anchoring energy. The anchoring strength can become as high as 10^{-3} J/m^2. Furthermore, pretilt angle of a few degrees can be generated. Homogeneous anchoring can also be achieved by using obliquely evaporated SiO film.

Homeotropic anchoring can be achieved using monolayer surfactants such as lecithin and silane. The polar head of the surfactant is chemically attached to the glass substrate, and the hydrocarbon tail points out and perpendicular to the surface, as shown in Figure 1.19(b).

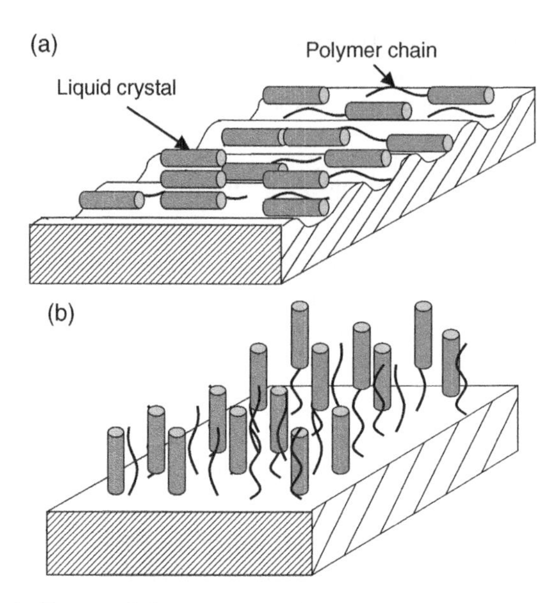

Figure 1.19 Schematic diagram of alignment layers: (a) homogeneous alignment layer, (b) homeotropic alignment layer.

The intermolecular interaction between the surfactant and the liquid crystal promotes the homeotropic alignment.

1.7 Liquid crystal director elastic deformation

When a nematic liquid crystal is in the ground state the direction of the liquid crystal director $\vec{n}$ is uniform in space. When $\vec{n}$ is deformed, there will be elastic energy. Although liquid crystal director deformations cost energy, they do occur in reality because of surface anchoring, spatial confinements, impurities, irregularities, and externally applied fields [1,23]. In this section, we consider possible director deformations, associated elastic energies, and transformation between deformations

1.7.1 Elastic deformation and disclination

We consider the possible deformations of a nematic liquid crystal confined between two parallel substrates with tangential anchoring condition (parallel to the substrates but no preferred direction on the plane of the substrate). We use the one elastic constant approximation ($K_{11} = K_{22} = K_{33} = K$), the elastic energy is given by

$$f = \frac{1}{2}K(\nabla\cdot\vec{n})^2 + \frac{1}{2}K(\vec{n}\cdot\nabla\times\vec{n})^2 + \frac{1}{2}K(\vec{n}\times\nabla\times\vec{n})^2 = \frac{1}{2}K(\nabla\cdot\vec{n})^2 + \frac{1}{2}K(\nabla\times\vec{n})^2. \quad (1.117)$$

Because of the anchoring condition, the liquid crystal director is parallel to the plane of the substrate and depends on the coordinates x and y on the plane. The liquid crystal director is described by

$$\vec{n} = \cos\theta(x,y)\hat{x} + \sin\theta(x,y)\hat{y}, \tag{1.118}$$

where θ is the angle between the director and the x axis. Substituting Equation (1.118) into Equation (1.117) we get

$$f = \frac{1}{2}K\left(\frac{\partial(\cos\theta)}{\partial x} + \frac{\partial(\sin\theta)}{\partial y}\right)^2 + \frac{1}{2}K\left[\left(\frac{\partial(\sin\theta)}{\partial x} - \frac{\partial(\cos\theta)}{\partial y}\right)\hat{z}\right]^2 = \frac{1}{2}K\left[\left(\frac{\partial\theta}{\partial x}\right)^2 + \left(\frac{\partial\theta}{\partial y}\right)^2\right]. \tag{1.119}$$

Using the Euler–Lagrange method (which will be presented in Chapter 5) to minimize the free energy we get

$$\frac{\partial^2\theta}{\partial x^2} + \frac{\partial^2\theta}{\partial y^2} = 0. \tag{1.120}$$

When we switch from Cartesian coordinates to cylindrical coordinates, the above equation becomes

$$\frac{1}{r}\frac{\partial}{\partial r}\left(r\frac{\partial\theta}{\partial r}\right) + \frac{1}{r^2}\frac{\partial^2\theta}{\partial\phi^2} = 0. \tag{1.121}$$

We consider the case where θ only depends on the azimuthal angle ϕ, but not the radius r. The solution to the above equation is

$$\theta = S\phi + \theta_o = S\tan^{-1}(x/y) + \theta_o. \tag{1.122}$$

When we go around one complete circle, the azimuthal angle changes by 2π, the liquid crystal director must be in the same direction. Furthermore $\vec{n}$ and $-\vec{n}$ are equivalent. Therefore it is required that

$$\theta(\phi = 2\pi) - \theta(\phi = 0) = (S\cdot 2\pi + \theta_o) - (S\cdot 0 + \theta_o) = 2\pi S = \pm m\cdot\pi, \quad S = \pm m/2, \tag{1.123}$$

where $m = 0,\ 1,\ 2,$..... The liquid crystal director configurations for the deformations with variety of S and θ_o values are shown in Figure 1.20. In the center there is a singularity. In the liquid crystal cell the singularity goes from one surface to the other surface of the cell. Thus it is a line singularity and is called disclination. S is called the strength of the disclination. For a positive S, the liquid crystal director rotates counterclockwise when the azimuthal angle changes counterclockwise. For a negative S, the liquid crystal director rotates clockwise when the azimuthal angle changes counterclockwise. The elastic energy density of the deformation is given by

$$f = \frac{1}{2}K\left[\left(\frac{\partial[S\tan^{-1}(x/y)]}{\partial x}\right)^2 + \left(\frac{\partial[S\tan^{-1}(x/y)]}{\partial y}\right)^2\right] = \frac{KS^2}{2r^2}. \tag{1.124}$$

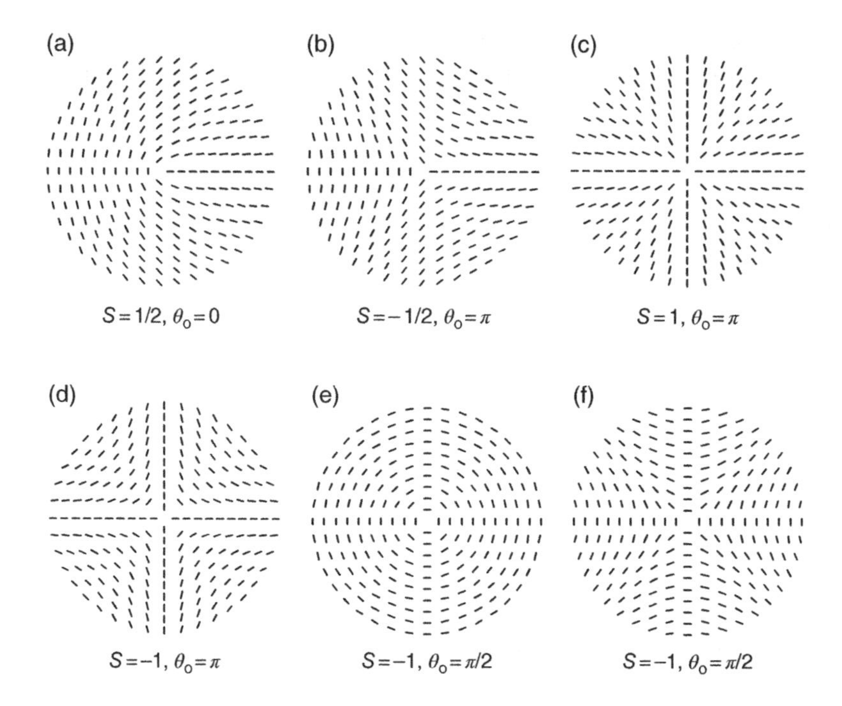

Figure 1.20 The liquid crystal director configurations of the disclinations with variety of strengths.

The elastic energy is proportional to the square of the strength of the disclination and increases when approaching the singularity.

1.7.2 Escape of liquid crystal director in disclinations

For a disclination with the strength S, as one approaches the center of the disclination, the elastic energy diverges, as shown by Equation (1.124). In reality this will not occur. The liquid crystal will transform either into isotropic phase at the center of the disclination or a different deformation where there is no singularity. Here we only discuss the cases of cylindrical confinements (two-dimensional confinement) where it is possible to obtain analytical solutions. The mechanism of liquid crystal director escape in spherical confinement (three-dimensional confinement) is similar to that of two-dimensional.

1.7.2.1 Escape to isotropic phase

We consider a nematic liquid crystal confined in a cylinder with a radius of R. The anchoring condition on the surface of the cylinder is perpendicular, as shown in Figure 1.21. The liquid crystal director aligns along the radial axis direction, as shown in Figure 1.21(a), and is described by $\vec{n} = \hat{r}$. The elastic deformation of the liquid crystal director is splay with the strength of $S = 1$. The elastic energy is

$$f = \frac{1}{2}K_{11}\frac{1}{r^2}. \tag{1.125}$$

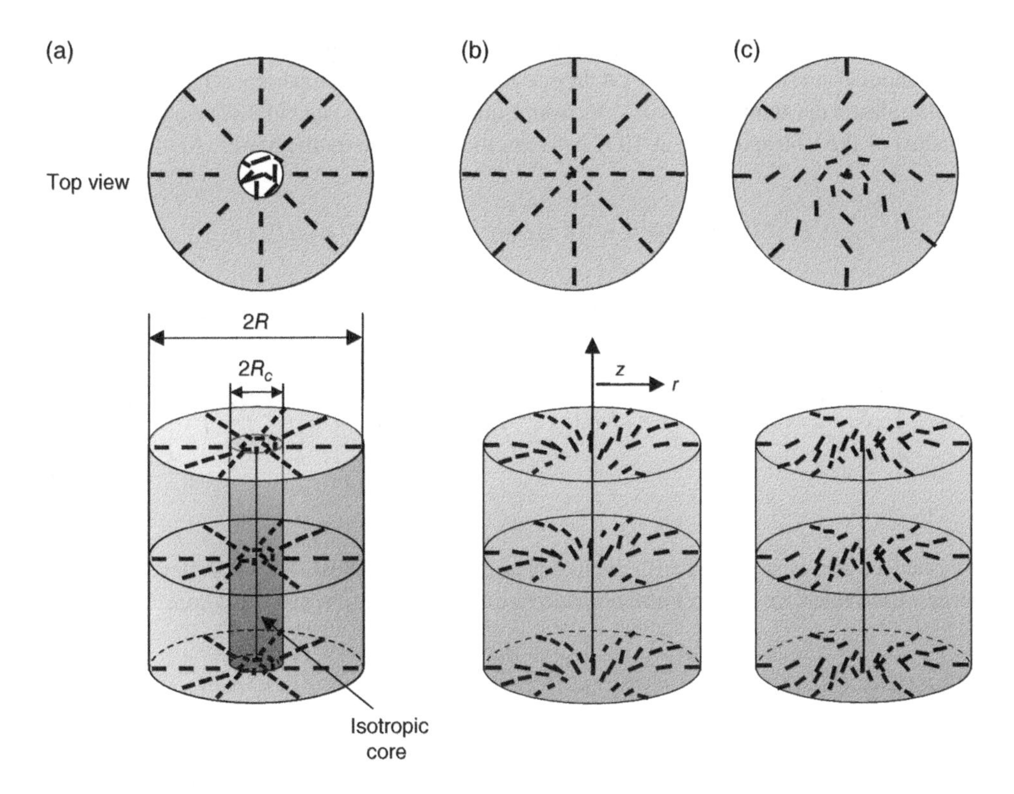

Figure 1.21 Liquid crystal director configurations in the cylinder: (a) escape from splay to isotropic phase, (b) escape from splay to bend, and (c) escape from bend to twist.

When $r \to 0, f \to \infty$. In order to avoid the divergence of the elastic energy, the liquid crystal transforms into the isotropic phase inside a core with the radius R_c, as shown in Figure 1.21(a) [24]. The radius of the isotropic core is given by

$$\frac{1}{2}K_{11}\frac{1}{R_c^2} = \Delta f = f_I - f_N,$$

$$R_c = \sqrt{K_{11}/(2\Delta f)}, \tag{1.126}$$

where Δf is the free energy difference between the isotropic phase and the nematic phase. At temperature T, the free energy difference can be estimated by

$$\Delta f = f_I - f_N \approx \left[\left(\frac{\partial f_I}{\partial T}\right)_{T_{NI}} - \left(\frac{\partial f_N}{\partial T}\right)_{T_{NI}}\right](T - T_{NI}) = \left[-\frac{S_I}{T} + \frac{S_N}{T}\right](T - T_{NI}) = \frac{L}{T_{NI}}(T_{NI} - T), \tag{1.127}$$

where S_I and S_N are the entropies of the isotropic and nematic phases, respectively, and L is the latent heat per unit volume of the nematic–isotropic transition. A typical latent heat is 100 cal/mole = 100 × 4.18 J/mole. The density of the liquid crystal is about 1 g/cm^3. One mole of liquid crystal weighs about 300 g and occupies the volume 300 g/(1 g/cm^3) = 300 cm^3. The latent heat is about $L = 418$ J/300 cm^3 = 1.4 J/cm^3. If the transition temperature T_{NI} is 300 K and the splay elastic constant K_{11} is 10^{-11} N, from Equations (1.126) and (1.127) we can calculate the radius of the isotropic core at 10 K between the transition temperature:

$$R_c = \sqrt{10^{-11}\ \mathrm{N}/[2\times 1.4\ (\mathrm{J/cm^3})\times 10\ \mathrm{K}/300\ \mathrm{K}]} \approx 10\ \mathrm{nm} \tag{1.128}$$

The total free energy per unit length of the configuration with the isotropic core is

$$F_{IC} = \int_{R_c}^{R} \frac{1}{2}K_{11}\frac{1}{r^2}2\pi r dr + \pi R_c^2 \Delta f = \pi K_{11}\ln\left(\frac{R}{R_c}\right) + \frac{1}{2}\pi K_{11} = K_{11}\pi \ln\left[R\left(\frac{2\Delta f}{K_{11}}\right)^{1/2}\right] + \frac{1}{2}\pi K_{11}. \tag{1.129}$$

The total free energy increases with the radius of the cylinder. Note that we neglect the surface energy of the interface between the nematic liquid crystal and the isotropic core.

1.7.2.2 Escape to bend

The second possibility to avoid the singularity at the center of the cylinder is to escape from the splay deformation to the bend deformation, as shown in Figure 1.21(b) [30–32]. The liquid crystal director tilts to the z direction and is given by

$$\vec{n} = \sin\theta(r)\hat{r} + \cos\theta(r)\hat{z}, \tag{1.130}$$

where θ is the angle between the liquid crystal director and the z axis. Under the approximation $K_{33} = K_{11}$, the elastic energy density is

$$f = \frac{1}{2}K_{11}\left[\frac{\sin^2\theta}{r^2} + 2\sin\theta\cos\theta\frac{1}{r}\frac{d\theta}{dr} + \left(\frac{d\theta}{dr}\right)^2\right]. \tag{1.131}$$

The total elastic energy per unit length of the cylinder is

$$F_B = 2\pi\int_0^R f r dr = \pi K_{11}\int_0^R K_{11}\left[\frac{\sin^2\theta}{r} + 2\sin\theta\cos\theta\frac{d\theta}{dr} + r\left(\frac{d\theta}{dr}\right)^2\right]dr. \tag{1.132}$$

Using Euler–Lagrange method to minimize the elastic energy, we obtain

$$\frac{\delta(rf)}{\delta\theta} = \frac{\partial(rf)}{\partial\theta} - \frac{d}{dr}\left(\frac{\partial(rf)}{\partial(\partial\theta/\partial r)}\right) = \frac{2\sin\theta\cos\theta}{r} - 2\frac{d\theta}{dr} - 2r\frac{d^2\theta}{dr^2} = 0, \tag{1.133}$$

$$\frac{d\theta}{dr} + r\frac{d^2\theta}{dr^2} = \frac{\sin\theta\cos\theta}{r},$$

$$\frac{d}{dr}\left(r\frac{d\theta}{dr}\right) = \frac{1}{r}\sin\theta\cos\theta. \tag{1.134}$$

Multiplying both sides of the above equation by $rd\theta/dr$, we get

$$\left(r\frac{d\theta}{dr}\right)\frac{d}{dr}\left(r\frac{d\theta}{dr}\right) = \frac{1}{2}\frac{d}{dr}\left(r\frac{d\theta}{dr}\right)^2 = \frac{1}{r}\sin\theta\cos\theta\left(r\frac{d\theta}{dr}\right) = \frac{1}{2}\frac{d}{dr}\left(\sin^2\theta\right). \tag{1.135}$$

Integrating the above equation from 0 to r, and using the boundary condition $\theta(r=0)=0$, we obtain

$$r\frac{d\theta}{dr} = \sin\theta,$$

$$\frac{dr}{r} = \frac{d\theta}{\sin\theta}. \tag{1.136}$$

Integrating the above equation from r to R, and using the boundary condition $\theta(r=R)=\pi/2$, we obtain

$$\theta = 2\tan^{-1}\left(\frac{r}{R}\right). \tag{1.137}$$

Substituting Equations (1.136) and (1.137) into Equation (1.132) we can get the total free energy,

$$F_B = 3\pi K_{11}, \tag{1.138}$$

which is independent of the radius R.

In order to determine which escape will occur, we compare the total free energies of the two escapes. From Equations (1.129) and (1.138), we get

$$\Delta F = F_B - F_{IC} = 3\pi K_{11} - \left[K_{11}\pi\ln\left(\frac{R}{R_c}\right) + \frac{1}{2}\pi K_{11}\right] = \frac{5}{2}\pi K_{11} - K_{11}\pi\ln\left(\frac{R}{R_c}\right). \tag{1.139}$$

We can see that if

$$\frac{5}{2}\pi K_{11} \le K_{11}\pi\ln\left(\frac{R}{R_c}\right),$$

$$R \ge e^{5/2}R_c \approx 12R_c. \tag{1.140}$$

The free energy of the state with the bend escape is lower than that of the state with the isotropic core. When the radius of the cylinder is small, it costs too much elastic energy for the liquid crystal to change from 0° at the center to 90° at the surface, and the liquid crystal system will have an isotropic core to avoid the divergence of the splay elastic energy. When the radius of the cylinder is sufficiently large, the liquid crystal will escape from the splay state to the bend state to further reduce the free energy.

1.7.2.3 Escape to twist

For most liquid crystals, the twist elastic constant is smaller than the bend elastic constant. Therefore it is possible to reduce the total elastic energy by escaping from the bend deformation to the twist deformation as shown in Figure 1.21(c). The liquid crystal director is no longer on the r–z plane but twists out of the plane and is given by

$$\vec{n} = \sin\theta(r)\hat{r} + \cos\theta(r)\sin\alpha(r)\hat{\phi} + \cos\theta(r)\cos\alpha(r)\hat{z}, \tag{1.141}$$

where θ is the angle of the liquid crystal director defined with respect to the ϕ–z plane, and the twist angle α is the angle of the projection of the liquid crystal director on the ϕ–z plane defined with respect to the z axis. The divergence and curl of $\vec{n}$ are

$$\nabla\cdot\vec{n} = \frac{1}{r}\frac{\partial}{\partial r}(r\sin\theta) = \frac{\sin\theta}{r} + \cos\theta\frac{\partial\theta}{\partial r} \tag{1.142}$$

$$\begin{aligned}\nabla\times\vec{n} &= \frac{1}{r}\frac{\partial}{\partial r}(r\cos\theta\sin\alpha)\hat{z} - \frac{\partial}{\partial r}(\cos\theta\cos\alpha)\hat{\phi} \\ &= \left(\frac{1}{r}\cos\theta\sin\alpha - \sin\theta\sin\alpha\frac{\partial\theta}{\partial r} + \cos\theta\cos\alpha\frac{\partial\alpha}{\partial r}\right)\hat{z} + \left(\sin\theta\cos\alpha\frac{\partial\theta}{\partial r} + \cos\theta\sin\alpha\frac{\partial\alpha}{\partial r}\right)\hat{\phi}\end{aligned} \tag{1.143}$$

$$\begin{aligned}(\nabla\times\vec{n})^2 = &\left(\frac{1}{r}\cos\theta\sin\alpha\right)^2 + \left(\sin\theta\frac{\partial\theta}{\partial r}\right)^2 + \left(\cos\theta\frac{\partial\alpha}{\partial r}\right)^2 \\ &+ 2\frac{1}{r}\cos\theta\sin\alpha\left(-\sin\theta\sin\alpha\frac{\partial\theta}{\partial r} + \cos\theta\cos\alpha\frac{\partial\alpha}{\partial r}\right)\end{aligned} \tag{1.144}$$

$$\begin{aligned}\vec{n}\cdot\nabla\times\vec{n} = &\left(\frac{1}{r}\cos\theta\sin\alpha - \sin\theta\sin\alpha\frac{\partial\theta}{\partial r} + \cos\theta\cos\alpha\frac{\partial\alpha}{\partial r}\right)\cos\theta\cos\alpha \\ &+ \left(\sin\theta\cos\alpha\frac{\partial\theta}{\partial r} + \cos\theta\sin\alpha\frac{\partial\alpha}{\partial r}\right)\cos\theta\sin\alpha.\end{aligned} \tag{1.145}$$

The elastic energy density is

$$\begin{aligned}f &= \frac{1}{2}K_{11}(\nabla\cdot\vec{n})^2 + \frac{1}{2}K_{22}(\vec{n}\cdot\nabla\times\vec{n})^2 + \frac{1}{2}K_{33}\left[(\nabla\times\vec{n})^2 - (\vec{n}\cdot\nabla\times\vec{n})^2\right] \\ f &= \frac{1}{2}K_{11}(\nabla\cdot\vec{n})^2 + \frac{1}{2}(K_{22} - K_{33})(\vec{n}\cdot\nabla\times\vec{n})^2 + \frac{1}{2}K_{33}(\nabla\times\vec{n})^2.\end{aligned} \tag{1.146}$$

By introducing the twist, the bend elastic can be reduced, but the trade-off is that the twist elastic energy is increased. The change of the elastic energy density [Eq. (1.146) subtracted by Eq. (1.131)] is

$$\Delta f = \frac{1}{2}(K_{22} - K_{33})f_1 + \frac{1}{2}K_{33}f_2, \tag{1.147}$$

where

$$f_1 = \left[\left(\frac{1}{r}\cos\theta\sin\alpha - \sin\theta\sin\alpha\frac{\partial\theta}{\partial r} + \cos\theta\cos\alpha\frac{\partial\alpha}{\partial r}\right)\cos\alpha \right.$$
$$\left. + \left(\sin\theta\cos\alpha\frac{\partial\theta}{\partial r} + \cos\theta\sin\alpha\frac{\partial\alpha}{\partial r}\right)\sin\alpha\right]^2\cos^2\theta, \tag{1.148}$$

$$f_2 = \left(\frac{1}{r}\cos\theta\sin\alpha\right)^2 + 2\frac{1}{r}\cos\theta\sin\alpha\left(-\sin\theta\sin\alpha\frac{\partial\theta}{\partial r} + \cos\theta\cos\alpha\frac{\partial\alpha}{\partial r}\right) + \left(\cos\theta\frac{\partial\alpha}{\partial r}\right)^2. \tag{1.149}$$

The total change of the elastic energy per unit length of the cylinder is

$$\Delta F = 2\pi\int_0^R \Delta f r dr = \pi\int_0^R [(K_{22} - K_{33})f_1 + K_{33}f_2]rdr = \pi K_{22}\Delta F_1 - \pi K_{33}(\Delta F_1 - \Delta F_2). \tag{1.150}$$

The total free energy is decreased if

$$K_{22}/K_{33} \leq (\Delta F_1 - \Delta F_2)/\Delta F_1. \tag{1.151}$$

Instead of using the Euler–Lagrange method to find the exact solutions for θ and α, we use approximations for them. For $\theta(r)$, we use the same solution found in the last section. For $\alpha(r)$, because the boundary conditions are $\alpha(r=0)=0$, we use the approximation

$$\alpha = Ar^b. \tag{1.152}$$

We minimize ΔF with respect to the amplitude A and power b of the twist angle. It is numerically found that when $A=0.34\pi$ and $b=0.3$, $\Delta F_1 = 1.14$ and $\Delta F_2 = 0.75$. From Equation (1.151) we have

$$K_{22}/K_{33} \leq (1.14 - 0.75)/1.14 = 0.34. \tag{1.153}$$

If the ratio between the twist and bend elastic constants is smaller than 0.34, the total elastic energy is reduced by escaping from the bend deformation to the twist deformation.

Homework Problems

1.1. Consider a nematic liquid crystal. The molecule can be regarded as a cylinder with the length of 2 nm and diameter of 0.5 nm. The molecule has a permanent dipole moment of 10^{-29} m · C at the center of the molecule. The interaction between the molecules comes from the interactions between the permanent dipoles. Calculate the interaction between two molecules in the following cases: (1) one molecule is on top of the other molecule and the dipoles are parallel, (2) one molecule is on top of the other molecule and the dipoles are anti-parallel, (3) the molecules are side by side and the dipoles are parallel, and (4) the molecules are side by side and the dipoles are anti-parallel.

1.2. Using Equations (1.11), (1.25), and (1.31), prove that the entropy of a system at a constant temperature is $S = -k_B < \ln \rho > = -k_B \sum_i \rho_i \ln \rho_i$.

1.3. Calculate the orientational order parameter in the following two cases. (1) The orientational distribution function is $f(\theta) = \cos^2 \theta$. (2) The orientational distribution function is $f(\theta) = \sin^2 \theta$. θ is the angle between the long molecular axis and the liquid crystal director.

1.4. *Landau–de Gennes theory.* For a liquid crystal with the parameters $a = 0.1319 \times 10^5$ J/K · m^3, $b = -1.836 \times 10^5$ J/m^3, and $c = 4.05 \times 10^5$ J/m^3. Numerically calculate the free energy as a function of the order parameter, and identify the order parameters corresponding to the maximum and minimum free energy at the following temperatures. (1) $T - T^* = 4.0$°C, (2) $T - T^* = 3.0$°C, (3) $T - T^* = 2.0$°C, (4) $T - T^* = 1.0$°C, (5) $T - T^* = 0.0$°C, (6) $T - T^* = -10.0$°C.

1.5. *Maier–Saupe theory.* Use Equation (1.50) to numerically calculate all the possible order parameters as a function of the normalized temperature $\tau = k_B T / v$, and use Equation (1.54) to calculate the corresponding free energy.

1.6. Use Maier–Saupe theory to study isotropic–nematic phase transition of a binary mixture consisting of two components A and B. For molecule A, when its long molecular axis makes the angle θ_A with respect to the liquid crystal director, the single molecular potential is $V_A(\theta) = -v_{AA}(1-x)S_A\left(\frac{3}{2}\cos^2\theta_A - \frac{1}{2}\right) - v_{AB}xS_B\left(\frac{3}{2}\cos^2\theta_A - \frac{1}{2}\right)$. For molecule B, when its long molecular axis makes the angle θ_B with respect to the liquid crystal director, the single molecular potential is $V_B(\theta) = -v_{AB}(1-x)S_A\left(\frac{3}{2}\cos^2\theta_B - \frac{1}{2}\right) - v_{BB}xS_B\left(\frac{3}{2}\cos^2\theta_B - \frac{1}{2}\right)$. x is the molar fraction of component B. The interaction constants are $v_{BB} = 1.05 v_{AA}$ and $v_{AB} = 0.95 v_{AA}$. Express the normalized temperature by $\tau = k_B T / v_{AA}$. Assume that the two components are miscible at any fraction. Numerically calculate the transition temperature as a function of the molar fraction x.

1.7. Consider a nematic liquid crystal cell with the thickness of 10 μm. On the bottom surface the liquid crystal is aligned parallel to the cell surface, and on top of the top surface the liquid crystal is aligned perpendicular to the cell surface. Assume the tilt angle of the liquid crystal director changes linearly with the coordinate z, which is in the cell normal direction. Calculate the total elastic energy per unit area. The elastic constants of the liquid crystal are $K_{11} = 6 \times 10^{-12}$N, $K_{22} = 3 \times 10^{-12}$N and $K_{11} = 10 \times 10^{-12}$N.

1.8. The Cano-wedge method is an experimental technique to measure the pitch of cholesteric liquid crystals. It consists of a flat substrate and a hemisphere with a cholesteric liquid crystal sandwiched between them as shown in Figure 1.22(a). At the center, the spherical surface touches the flat surface. On both the flat and spherical surfaces there is a homogeneous alignment layer. The intrinsic pitch of the liquid crystal is P_o. Because of the boundary

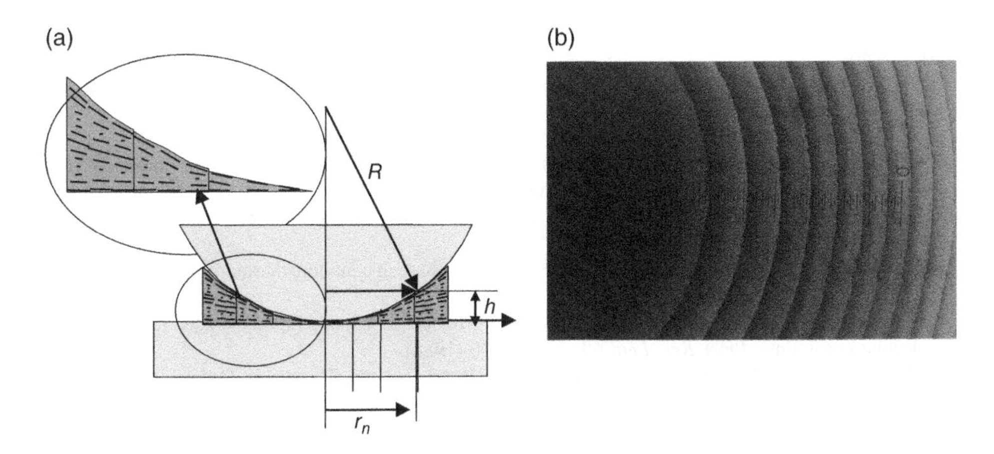

Figure 1.22 Figure for homework #1.8

condition, the pitch of the liquid crystal is quantized to match the boundary condition. In region n, $h = n(P/2)$. In each region, on the inner side, the pitch is compressed, that is, $P < P_o$, while on the outer side, the pitch is stretched, that is, $P > P_o$. Between region $(n-1)$ and region n, there is a disclination ring as shown in Figure 1.22(b). Find the square of the radius of the nth disclination ring r_n^2 as a function of the intrinsic pitch P_o, the radius R of the hemisphere and the ring number n. $R \gg P_o$ and for small r only twist elastic energy has to be considered. Hint, r_n^2 vs. n is a straight line with a slope dependent on P_o and R.

1.9. Consider a sphere of radius R. The polarization inside the sphere is $\vec{P}$. Calculate the electric field at the center of the sphere produced by the polarization. Hint, the polarization can be replaced by a surface charge whose density is given by $\vec{P} \cdot \vec{n}$, where $\vec{n}$ is the unit vector along the surface normal direction.

1.10. Using Equations (1.87), (1.91), and (1.93), calculate the dielectric tensor $\overleftrightarrow{\varepsilon}$ in terms of the order parameter S.

References

1. P. G. de Gennes and J. Prost, *The physics of liquid crystals* (Oxford University Press, New York, 1993).
2. S. Chandrasekhar *Liquid crystals*, 2nd edn (Cambridge University Press, New York, 1997).
3. L. M. Blinov and V. G. Chigrinov, *Electro-optical effects in liquid crystal materials* (Springer-Verlag, New York, 1994).
4. G. W. Gray, K. J. Harrison and J. A. Nash, *Electron. Lett.*, **9**, 130 (1973).
5. K. J. Toyne, Liquid crystal behavior in relation to molecular structure, in *Thermotropic liquid crystals*, ed. G. W. Gray (John Wiley & Son, Chichester, 1987).
6. P. J. Collings and M. Hird, *Introduction to liquid crystals, chemistry and physics*, (Taylor and Francis, London, 1997).
7. F. Reinitzer, *Monatsh Chem*, **9**, 421 (1898).

8. B. Bahadur, (ed.) *Liquid crystals: applications and uses*, Vol. 1, 2, and 3, (Singapore, World Scientific, 1990).
9. E. Lueder, *Liquid crystal displays: addressing schemes and electro-optical effects* (John Wiley & Sons, Chichester, 2001).
10. S.-T. Wu and D.-K. Yang, *Reflective liquid crystal displays* (John Wiley & Sons, Ltd., 2001).
11. C. Kittel and H. Kroemer, *Thermal physics*, 2nd ed. (W. H. Freeman and Company, San Francisco, 1980).
12. L. J. Lu and A. Saupe, *Phys. Rev. Lett.*, **45**, 1000 (1980).
13. L. A. Madsen, T. J. Dingemans, M. Nakata, and E. T. Samulski, Thermotropic biaxial nematic liquid crystals, *Phys. Rev. Lett.*, **92,** 145505 (2004).
14. B. R. Acharya, A. Primak, and S. Kumar, Biaxial nematic phase in bent-core thermotropic mesogens, *Phys. Rev. Lett.*, **92**, 145505 (2004).
15. P. Palffy-Muhoray, M. A. Lee and R. G. Petschek, Ferroelectric nematic liquid crystals: realizability and molecular constraints, *Phys. Rev. Lett.* **60**, 2303–2306 (1988).
16. L. D. Landau and E. M. Lifshitz, *Statistical Physics*, Part I, 3rd edn (Pergamon, Oxford, 1980).
17. P. G. de Gennes, *Mol. Cryst. Liq. Cryst.*, **12**, 193 (1971).
18. T. W. Stinson and J. D. Lister, *Phys. Rev. Lett.*, **25**, 503 (1970).
19. W. Maier and A. Saupe, *Z. Naturforsch.*, **13a**, 564 (1958).
20. E. B. Priestley, P. J. Wojtoicz and P. Sheng, *Introduction to liquid crystals* (Plenum, New York 1979).
21. I. C. Khoo, *Liquid crystals, physical properties and non-linear optical phenomena*, (John Wiley & Sons, New York, 1995).
22. W. H. de Jeu, Physical properties of liquid crystal materials, in *Liquid crystal monographs*, vol. 1, ed. G. W. Gray (Gordon Breach, London, 1980).
23. F. C. Frank, *Disc. Faraday*, **25**, 19 (1958).
24. M. Kleman and O. D. Lavrentovich, *Soft matter physics, Introduction* (Springer-Verlag, New York, 2003).
25. I. Sage, *Thermochromic liquid crystal devices in liquid crystals-applications and uses*, Vol. 3, ed. B. Bahadur (World Scientific, New Jersey, 1990).
26. W. H. de Jeu and P. Bordewijk, *J. Chem. Phys.*, **68**, 109 (1978).
27. A. A. Sonin, *The surface physics of liquid crystals*, (Gordon and Breach), Luxembourg, 1995).
28. T. Uchida, Surface alignment of liquid crystals, in *Liquid crystals-applications and uses*, Vol. 3, ed. B. Bahadur (World Scientific, New Jersey, 1990).
29. A. Rapini and M. Papoular, *J. Phys. (Paris) Colloq.* **30**, C-4 (1969).
30. P.E. Cladis and M. Kléman, Non-singular disclinations of strength S = +1 in nematics, *J. Phys.* (Paris) **33**, 591 (1972).
31. I. Vilfan, M. Vilfan, and S. Zumer, Defect structures of nematic liquid crystals in cylindrical cavities, *Phys. Rev. A*, **43**, 6873 (1991).
32. R. J. Ondris-Crawford, G. P. Crawford, S. Zumer, and J. W. Doane, Curvature-induced configuration transition in confined nematic liquid crystals, *Phys. Rev. Lett.*, **70**, 195 (1993).

2

Propagation of Light in Anisotropic Optical Media

2.1 Electromagnetic Wave

In wave theory, light is electromagnetic waves propagating in space [1–3]. There are four fundamental quantities in electromagnetic wave: *electric field* $\vec{E}$, *electric displacement* $\vec{D}$, *magnetic field* $\vec{H}$, and *magnetic induction* $\vec{B}$. These quantities are vectors. In the SI system, the unit of electric field is *volt/meter*; the unit of electric displacement is *coulomb/meter*^{2}, which equals *newton/volt · meter*; the unit of magnetic field is *ampere/meter*, which equals *newton/volt · second*, and the unit of magnetic induction is *tesla*, which equals *volt · second/meter*^{2}. In a medium, the electric displacement is related to the electric field by

$$\vec{D} = \varepsilon_o \overleftrightarrow{\varepsilon} \cdot \vec{E}, \tag{2.1}$$

where $\varepsilon_o = 8.85 \times 10^{-12}$ farad/meter $= 8.85 \times 10^{-12}$ newton/volt^2 and $\overleftrightarrow{\varepsilon}$ is the (relative) dielectric tensor of the medium. The magnetic induction is related to the magnetic field by

$$\vec{B} = \mu_o \overleftrightarrow{\mu} \cdot \vec{H}, \tag{2.2}$$

where $\mu_o = 4\pi \times 10^{-7}$ henry/meter $= 4\pi \times 10^{-7}$ volt^2 · second^2/newton · meter^2 is the permeability of vacuum and $\overleftrightarrow{\mu}$ is the (relative) permeability tensor of the medium. Liquid crystals are non-magnetic media, and the permeability is close to 1 and approximately we have $\overleftrightarrow{\mu} = \overleftrightarrow{I}$, where $\overleftrightarrow{I}$ is the identity tensor. In a medium without free charge, the electromagnetic wave is governed by the Maxwell equations:

Fundamentals of Liquid Crystal Devices, Second Edition. Deng-Ke Yang and Shin-Tson Wu.

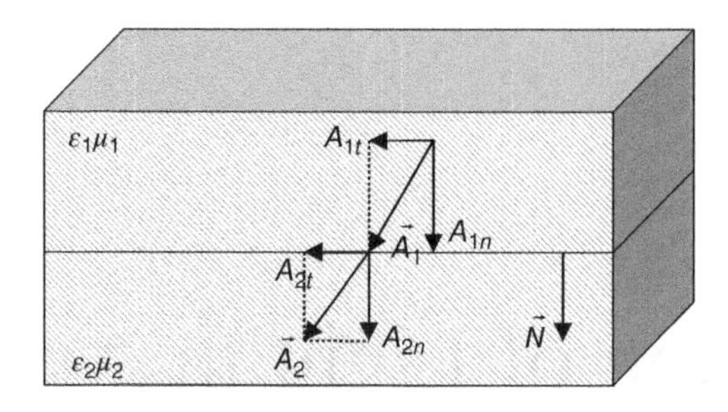

Figure 2.1 Schematic diagram showing the electromagnetic fields at the boundary between two media; $\vec{A}$ represents $\vec{E}$, $\vec{D}$, $\vec{H}$, and $\vec{B}$; $\vec{N}$ is a unit vector along the normal direction of the interface; n: normal component and t: tangential component.

$$\nabla\cdot\vec{D}=0 \tag{2.3}$$

$$\nabla\cdot\vec{B}=0 \tag{2.4}$$

$$\nabla\times\vec{E}=-\frac{\partial\vec{B}}{\partial t} \tag{2.5}$$

$$\nabla\times\vec{H}=\frac{\partial\vec{D}}{\partial t} \tag{2.6}$$

When light propagates through more than one medium, at the boundary between two media, there are boundary conditions:

$$D_{2n}-D_{1n}=0 \tag{2.7}$$

$$B_{2n}-B_{1n}=0 \tag{2.8}$$

$$E_{2t}-E_{1t}=0 \tag{2.9}$$

$$H_{2t}-H_{1t}=0 \tag{2.10}$$

At the boundary, the normal components of $\vec{D}$ and $\vec{B}$ and the tangential components of $\vec{E}$ and $\vec{H}$ are continuous. These boundary condition equations are derived from the Maxwell equations.

We first consider light propagating in an isotropic uniform medium where $\vec{D}=\varepsilon_o\varepsilon\vec{E}$ and $\vec{B}=\mu_o\mu\vec{H}$. The Maxwell equations become

$$\nabla\cdot\vec{D}=\nabla\cdot\left(\varepsilon_o\varepsilon\vec{E}\right)=\varepsilon_o\varepsilon\nabla\cdot\vec{E}=0, \tag{2.11}$$

$$\nabla\cdot\vec{B}=\nabla\cdot\left(\mu_o\mu\vec{H}\right)=\mu_o\mu\nabla\cdot\vec{H}=0, \tag{2.12}$$

$$\nabla \times \vec{E} = -\frac{\partial \vec{B}}{\partial t} = -\mu_o \mu \frac{\partial \vec{H}}{\partial t}, \tag{2.13}$$

$$\nabla \times \vec{H} = \frac{\partial \vec{D}}{\partial t} = \varepsilon_o \varepsilon \frac{\partial \vec{E}}{\partial t}. \tag{2.14}$$

From Equation (2.13) and (2.14), we have

$$\nabla \times \left(\nabla \times \vec{E}\right) = \nabla\left(\nabla \cdot \vec{E}\right) - \nabla^2 \vec{E} = -\nabla^2 \vec{E} = -\mu_o \mu \nabla \times \left(\frac{\partial \vec{H}}{\partial t}\right) = -\mu_o \mu \frac{\partial\left(\nabla \times \vec{H}\right)}{\partial t} = -\varepsilon_o \varepsilon \mu_o \mu \frac{\partial^2 \vec{E}}{\partial t^2}$$

that is,

$$\nabla^2 \vec{E} = \varepsilon_o \varepsilon \mu_o \mu \frac{\partial^2 \vec{E}}{\partial t^2}. \tag{2.15}$$

This is a wave equation. In the complex-function formulism, the solution for a monochromatic wave is

$$\vec{E}\left(\vec{r}, t\right) = \vec{E}_o e^{i\left(\omega t - \vec{k} \cdot \vec{r}\right)} \tag{2.16}$$

where ω is the angular frequency and $k = 2\pi/\lambda$ (λ is the wavelength in the medium) is the wavevector. The real part of the electric field vector in Equation (2.16) is the actual electric field of the light. Substituting Equation (2.16) into Equation (2.15), we have

$$\omega^2 / k^2 = 1/\varepsilon_o \mu_o \varepsilon \mu \tag{2.17}$$

The propagation velocity of the wave is

$$V = \frac{\omega}{k} = \sqrt{\frac{1}{\varepsilon_o \mu_o \varepsilon \mu}}. \tag{2.18}$$

In vacuum, $\varepsilon = 1$ and $\mu = 1$, $V = c = [1/(8.85 \times 10^{-12} \times 4\pi \times 10^{-7})]^{1/2} = 3 \times 10^8$ m/s. In the non-magnetic medium, $V = c/\sqrt{\varepsilon} = c/n$, where $n = \sqrt{\varepsilon}$ is the refractive index. Here ε is the dielectric constant that is usually frequency-dependent. The wavevector is

$$k = \frac{\omega}{V} = \frac{\omega}{c/n} = \frac{2\pi n}{\lambda_o} \tag{2.19}$$

where λ_o is the wavelength in vacuum. From Equation (2.3), we have

$$\nabla \cdot \vec{D} = -i \vec{k} \cdot \vec{D}_o e^{i\left(\omega t - \vec{k} \cdot \vec{r}\right)} = 0. \tag{2.20}$$

The electric displacement vector is perpendicular to the propagation direction (the direction of the wavevector), which is true even in anisotropic media. Therefore light is a transverse wave.

The electric field vector is perpendicular to the wavevector in isotropic media, but not in anisotropic media.

When light propagates in a homogeneous isotropic medium, all the fields have the form as Equation (2.16). The amplitudes do not change with time and position. Because of the wave form of the fields, as shown by Equation (2.16), we have

$$\frac{\partial}{\partial t} = i\omega, \tag{2.21}$$

$$\nabla = -i\vec{k}. \tag{2.22}$$

From Equation (2.13), we can get

$$\vec{B} = \frac{\vec{k} \times \vec{E}}{\omega}. \tag{2.23}$$

Therefore $\vec{B}$ and $\vec{E}$ are orthogonal to each other. Their magnitudes are related by $\left|\vec{B}\right| = (n/c)\left|\vec{E}\right|$. The energy density of the EM wave is

$$u = \frac{1}{2}\left(\vec{E}\cdot\vec{D} + \vec{H}\cdot\vec{B}\right). \tag{2.24}$$

The *Poynting vector* (energy flux) is

$$\vec{S} = \vec{E} \times \vec{H}. \tag{2.25}$$

The magnitude of $\vec{S}$ is

$$S = \frac{n}{\mu_o C}E^2. \tag{2.26}$$

2.2 Polarization

2.2.1 *Monochromatic plane waves and their polarization states*

When a monochromatic plane light wave is propagating in a homogeneous isotropic medium, only the electric field is needed to characterize it, because the other quantities can be calculated from the electric field. The electric field is a vector and generally has the form

$$\vec{E} = \vec{A}e^{i\left(\omega t - \vec{k}\cdot\vec{r}\right)}, \tag{2.27}$$

where $\vec{A}$ is a constant. It is understood that the real part of above equation represents the actual electric field. This representation is called an *analytic representation*. The polarization state of a light beam is specified by the electric field vector. In many liquid crystal devices, the liquid crystal is used to manipulate the polarization state of the light.

When the propagation direction is along the z axis, the real electric field has two components (along the x and y axes) [3–4]:

$$E_x = A_x \cos(\omega t - kz + \delta_x) \tag{2.28}$$

$$E_y = A_y \cos(\omega t - kz + \delta_y) \tag{2.29}$$

A_x and A_y are positive numbers representing the amplitudes. δ_x and δ_y are the phases and are defined in the range $-\pi < \delta_i \le \pi$ $(i = 1, 2)$. The important quantity is the phase difference defined by

$$\delta = \delta_y - \delta_x. \tag{2.30}$$

δ is also defined in the range $-\pi < \delta \le \pi$. We will show that only two parameters are needed to specify the polarization state of a beam. One of the ways to specify the polarization state is the ratio A_y/A_x and phase difference δ.

2.2.2 *Linear polarization state*

Let us consider the time evolution of the electric field vector at a given position (z is fixed). If the electric field vibrates in a constant direction (in the x-y plane), the light is said to be *linearly polarized*. This occurs when $\delta = 0$ or $\delta = \pi$. The angle ϕ of the electric field with respect to the x axis is given by $\tan\phi = A_y/A_x$ for $\delta = 0$ or $\tan\phi = -A_y/A_x$ for $\delta = \pi$. If we examine the spatial evolution of the electric field vector at a fixed time (say, $t = 0$), for a linearly polarized light, the curve traced by the electric field in space is confined in a plane. For this reason linearly polarized light is also called *plane polarized* light.

2.2.3 *Circular polarization states*

If amplitudes in the x and y directions are the same and the phase difference is $\delta = \pi/2$, then

$$E_x = A\cos(\omega t - kz), \tag{2.31}$$

$$E_y = A\cos\left(\omega t - kz + \frac{\pi}{2}\right) = -A\sin(\omega t - kz). \tag{2.32}$$

At a fixed position, say $z = 0$, $E_x = A\cos(\omega t)$ and $E_y = -A\sin(\omega t)$. The endpoint of the electric field vector will clockwise trace out a circle on the xy plane (the light is coming toward the observer). At a given time, say $t = 0$, $E_x = A\cos(kz)$ and $E_y = A\sin(kz)$. The endpoint of the electric vector along a line in the propagation direction traces out a right-handed helix in space. The polarization is referred to as *right-handed circular polarization*. If $\delta = -\pi/2$, at given time, the endpoint of the electric vector will trace out a left-handed helix in space, and is referred to as *left-handed circular polarization*.

2.2.4 Elliptical polarization state

Generally, the amplitudes in the x and y directions are not the same, and the phase difference is neither 0 nor π. For the purpose of simplicity, let $\delta_x = 0$. From Equations (2.28) and (2.29), we have

$$E_x/A_x = \cos(\omega t - kz)$$

$$E_y/A_y = \cos(\omega t - kz + \delta) = \cos(\omega t - kz)\cos\delta - \sin(\omega t - kz)\sin\delta.$$

By eliminating $\sin(\omega t - kz)$ and $\cos(\omega t - kz)$, we get

$$\left(\frac{E_x}{A_x}\right)^2 + \left(\frac{E_y}{A_y}\right)^2 - 2\left(\frac{E_x}{A_x}\right)\left(\frac{E_y}{A_y}\right)\cos\delta = \sin^2\delta. \tag{2.33}$$

This is an elliptical equation. At a given position, the endpoint of the electric vector traces out an ellipse on the xy plane, as shown in Figure 2.2. For this reason, the light is said to be *elliptically polarized*. In the $x'y'$ frame where the coordinate axes are along the major axes of the ellipse, the components, E'_x and E'_y, of the electric vector satisfy the equation

$$\left(\frac{E'_x}{a}\right)^2 + \left(\frac{E'_y}{b}\right)^2 = 1, \tag{2.34}$$

where a and b are the lengths of the principal semi-axes of the ellipse. ϕ is the *azimuthal angle* of the major axis x' with respect to the x axis. The transformation of the components of the electric vector between the two frames are given by

$$E_x = E'_x\cos\phi - E'_y\sin\phi, \tag{2.35}$$

$$E_y = E'_x\sin\phi + E'_y\cos\phi. \tag{2.36}$$

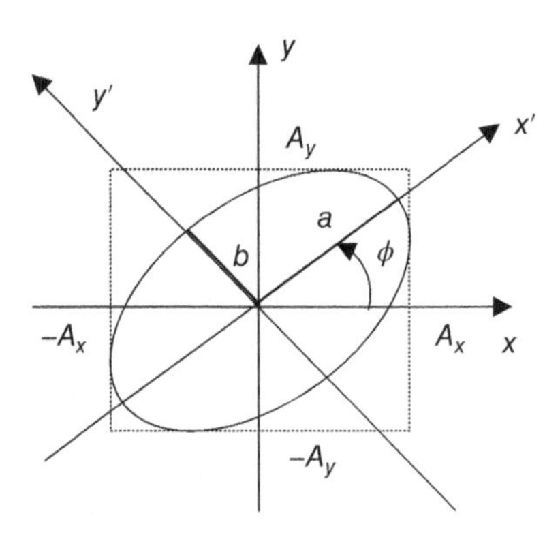

Figure 2.2 The polarization ellipse.

Substituting Equations (2.35) and (2.36) into Equation (2.33), we have

$$\left[\left(\frac{\cos\phi}{A_x}\right)^2+\left(\frac{\sin\phi}{A_y}\right)^2-\frac{\sin 2\phi\cos\delta}{A_xA_y}\right]E_x'^2+\left[\left(\frac{\sin\phi}{A_x}\right)^2+\left(\frac{\cos\phi}{A_y}\right)^2+\frac{\sin 2\phi\cos\delta}{A_xA_y}\right]E_y'^2$$

$$-\left[\frac{\sin 2\phi}{A_x^2}-\frac{\sin 2\phi}{A_y^2}+\frac{2\cos 2\phi}{A_xA_y}\cos\delta\right]E_x'E_y'=\sin^2\delta. \tag{2.37}$$

Comparing Equation (2.37) with Equation (2.34), we have

$$\left(\frac{\cos\phi}{A_x}\right)^2+\left(\frac{\sin\phi}{A_y}\right)^2-\frac{\sin 2\phi\cos\delta}{A_xA_y}=\frac{\sin^2\delta}{a^2}, \tag{2.38}$$

$$\left(\frac{\sin\phi}{A_x}\right)^2+\left(\frac{\cos\phi}{A_y}\right)^2+\frac{\sin 2\phi\cos\delta}{A_xA_y}=\frac{\sin^2\delta}{b^2}, \tag{2.39}$$

$$\frac{\sin 2\phi}{A_x^2}-\frac{\sin 2\phi}{A_y^2}+\frac{2\cos 2\phi}{A_xA_y}\cos\delta=0. \tag{2.40}$$

The azimuthal angle, ϕ, can be calculated from Equation (2.40):

$$\tan 2\phi=\frac{A_xA_y\cos\delta}{\left(A_x^2-A_y^2\right)} \tag{2.41}$$

Note that if ϕ is a solution, then $\phi+\pi/2$ is also a solution. From Equations (2.38) and (2.39) we have

$$\left(\frac{A_xA_y\sin\delta}{a}\right)^2=A_x^2\cos^2\phi+A_y^2\sin^2\phi-A_xA_y\sin 2\phi\cos\delta, \tag{2.42}$$

$$\left(\frac{A_xA_y\sin\delta}{b}\right)^2=A_x^2\sin^2\phi+A_y^2\cos^2\phi+A_xA_y\sin 2\phi\cos\delta. \tag{2.43}$$

Adding the above two equations together we have

$$\left(a^2+b^2\right)\left(\frac{\sin\delta A_xA_y}{ab}\right)^2=A_x^2+A_y^2.$$

Because the light intensity does not change upon a transformation between two frames, $a^2+b^2=A_x^2+A_y^2$, and therefore

$$\left(A_xA_y\sin\delta\right)^2=(ab)^2. \tag{2.44}$$

From Equations (2.43) and (2.44) we have

$$a^2 = A_x^2 \cos^2\phi + A_y^2 \sin^2\phi + A_x A_y \sin 2\phi \cos\delta. \tag{2.45}$$

From Equations (2.42) and (2.44) we have

$$b^2 = A_x^2 \sin^2\phi + A_y^2 \cos^2\phi - A_x A_y \sin 2\phi \cos\delta. \tag{2.46}$$

Table 2.1 Polarization states in the three representations.

Polarization Ellipse	$(\phi,\ \nu)$	Jones vector	Stokes vector
	(0, 0)	$\begin{pmatrix}1\\0\end{pmatrix}$	$\begin{pmatrix}1\\1\\0\\0\end{pmatrix}$
	$(\pi/2,\ 0)$	$\begin{pmatrix}0\\1\end{pmatrix}$	$\begin{pmatrix}1\\-1\\0\\0\end{pmatrix}$
	$(\pi/4,\ 0)$	$\frac{1}{\sqrt{2}}\begin{pmatrix}1\\1\end{pmatrix}$	$\begin{pmatrix}1\\0\\1\\0\end{pmatrix}$
	$(-\pi/4,\ 0)$	$\frac{1}{\sqrt{2}}\begin{pmatrix}1\\-1\end{pmatrix}$	$\begin{pmatrix}1\\0\\-1\\0\end{pmatrix}$
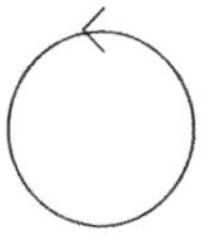	$(0,\ \pi/4)$	$\frac{1}{\sqrt{2}}\begin{pmatrix}1\\i\end{pmatrix}$	$\begin{pmatrix}1\\0\\0\\1\end{pmatrix}$
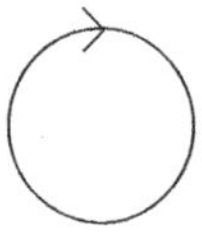	$(0,\ -\pi/4)$	$\frac{1}{\sqrt{2}}\begin{pmatrix}1\\-i\end{pmatrix}$	$\begin{pmatrix}1\\0\\0\\-1\end{pmatrix}$
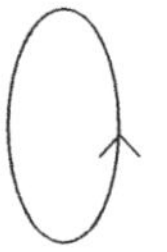	$\left[\pi/2,\ \tan^{-1}\left(\frac{1}{2}\right)\right]$	$\frac{1}{\sqrt{5}}\begin{pmatrix}1\\2i\end{pmatrix}$	$\begin{pmatrix}1\\-3/5\\0\\4/5\end{pmatrix}$
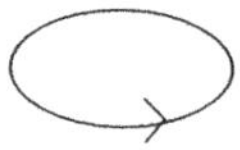	$\left[0,\ \tan^{-1}\left(\frac{1}{2}\right)\right]$	$\frac{1}{\sqrt{5}}\begin{pmatrix}2\\i\end{pmatrix}$	$\begin{pmatrix}1\\3/5\\0\\4/5\end{pmatrix}$
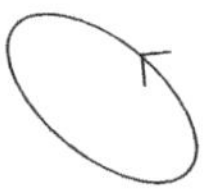	$\left[\pi/4,\ \tan^{-1}\left(\frac{1}{2}\right)\right]$	$\frac{1}{\sqrt{10}}\begin{pmatrix}2+i\\2-i\end{pmatrix}$	$\begin{pmatrix}1\\0\\3/5\\4/5\end{pmatrix}$

The lengths of the principal semi-major axes can be calculated from the above two equations. The sense of the revolution of an elliptical polarization is determined by the sign of $\sin\delta$. If $\sin\delta > 0$, the endpoint of the electric vector revolves clockwise (the light is coming toward the observer). If $\sin\delta < 0$, the endpoint of the electric vector revolves counterclockwise. The *ellipticity* of the polarization ellipse is define by

$$e = \pm\frac{b}{a}. \tag{2.47}$$

The positive sign is used for right-handed circular polarization while the negative sign is used for left-handed circular polarization. The *ellipticity angle* ν is defined by

$$\tan\nu = e. \tag{2.48}$$

We can also use the azimuthal angle, ϕ, of the major axis and the ellipticity angle, ν, to represent the polarization state. The values of ϕ and ν of various polarization states are listed in Table 2.1.

2.3 Propagation of Light in Uniform Anisotropic Optical Media

Now we consider the propagation of light in uniform non-magnetic anisotropic media [3, 5]. The speed of light in the medium, and thus the phase variation in space, depends on the direction of the electric field with respect to the optical axis of the medium. The optical properties of an anisotropic medium are described by the dielectric tensor $\overleftrightarrow{\varepsilon}$.

$$\overleftrightarrow{\varepsilon} = \begin{pmatrix} \varepsilon_{11} & \varepsilon_{12} & \varepsilon_{13} \\ \varepsilon_{21} & \varepsilon_{22} & \varepsilon_{23} \\ \varepsilon_{31} & \varepsilon_{32} & \varepsilon_{33} \end{pmatrix} \tag{2.49}$$

If the medium is non-absorbing, the dielectric tensor is real and symmetric ($\varepsilon_{ij} = \varepsilon_{ji}$). The values of the elements depend on the choice of the coordinate axes. Because the tensor is symmetric, it is always possible to choose a frame (the principal frame) with three orthogonal axes such that only the diagonal elements of the dielectric tensor are not zero. If the xyz frame is the *principal frame*, the dielectric tensor has the form

$$\overleftrightarrow{\varepsilon} = \begin{pmatrix} \varepsilon_x & 0 & 0 \\ 0 & \varepsilon_y & 0 \\ 0 & 0 & \varepsilon_z \end{pmatrix}. \tag{2.50}$$

In the following discussion in this section, the reference frame used is the principal frame. From the Maxwell Equations we will get

$$\nabla\times\left(\nabla\times\vec{E}\right) = \nabla\left(\nabla\cdot\vec{E}\right) - \nabla^2\vec{E} = -\mu_o\frac{\partial^2\vec{D}}{\partial t^2} = -\varepsilon_o\mu_o\overleftrightarrow{\varepsilon}\cdot\frac{\partial^2\vec{E}}{\partial t^2}. \tag{2.51}$$

Note that $\nabla \cdot \vec{D} = \nabla \cdot \left(\varepsilon_o \overleftrightarrow{\varepsilon} \cdot \vec{E} \right) = 0$ only ensures that $\vec{D}$, but not $\vec{E}$, is perpendicular to the propagation direction.

2.3.1 Eigenmodes

Generally speaking, when light is propagating in a uniform anisotropic medium, the direction of the electric field, and therefore the polarization state, will vary in space. Only when the electric field is in some special directions, known as the *eigenmode*, will its direction remain invariant in space, which will be proved to be true in this section. In the eigenmode, the electric field has the form

$$\vec{E} = \vec{E}_o e^{i\left(\omega t - \vec{k} \cdot \vec{r}\right)}, \tag{2.52}$$

where $\vec{E}_o$ is a constant vector known as the *eigenvector*; the corresponding refractive index is called the *eigenvalue*. The wavevector is

$$\vec{k} = \frac{2\pi}{\lambda_o} n\hat{s} = k_o n\hat{s} = k_o n\left(s_x\hat{x} + s_y\hat{y} + s_z\hat{z}\right) = k_x\hat{x} + k_y\hat{y} + k_z\hat{z}, \tag{2.53}$$

where $\hat{s} = s_x\hat{x} + s_y\hat{y} + s_z\hat{z}$ is a unit vector along the propagation direction, n is the refractive index which depends on the directions of the electric field and the propagation. $k_o = 2\pi/\lambda_o$ is the wavevector in vacuum. Because of the form of the electric field shown in Equation (2.52), for the monochromatic plane wave, we have

$$\frac{\partial}{\partial t} = i\omega, \tag{2.54}$$

$$\nabla = -i\vec{k}. \tag{2.55}$$

The wave equation (2.51) becomes

$$\left(\vec{k} \cdot \vec{E}\right)\vec{k} - k^2\vec{E} = -\varepsilon_o \mu_o \omega^2 \overleftrightarrow{\varepsilon} \cdot \vec{E} = -k_o^2 \overleftrightarrow{\varepsilon} \cdot \vec{E}. \tag{2.56}$$

In the principal frame, in component form, Equation (2.56) becomes

$$\begin{pmatrix} k_o^2\varepsilon_x - k_y^2 - k_z^2 & k_x k_y & k_x k_z \\ k_y k_x & k_o^2\varepsilon_y - k_x^2 - k_z^2 & k_y k_z \\ k_z kx & k_z k_y & k_o^2\varepsilon_z - k_x^2 - k_y^2 \end{pmatrix} \cdot \begin{pmatrix} E_x \\ E_y \\ E_z \end{pmatrix} = 0. \tag{2.57}$$

In order to have non-zero solution, the determinant must be zero:

$$\det = \begin{vmatrix} k_o^2\varepsilon_x - k_y^2 - k_z^2 & k_x k_y & k_x k_z \\ k_y k_x & k_o^2\varepsilon_y - k_x^2 - k_z^2 & k_y k_z \\ k_z k_x & k_z k_y & k_o^2\varepsilon_z - k_x^2 - k_y^2 \end{vmatrix} = 0 \tag{2.58}$$

This equation is also called the eigen equation. Define

$$n_x^2=\varepsilon_x, n_y^2=\varepsilon_y, n_z^2=\varepsilon_y, \tag{2.59}$$

$$a_x=(k_x/k_o)^2=(ns_x)^2, a_y=(k_y/k_o)^2=(ns_y)^2, a_z=(k_z/k_o)^2=(ns_z)^2, \tag{2.60}$$

$$a=a_x+a_y+a_z=n^2. \tag{2.61}$$

Equation (2.58) becomes

$$\det=\begin{vmatrix} n_x^2-a_y-a_z & \sqrt{a_xa_y} & \sqrt{a_xa_z} \\ \sqrt{a_xa_y} & n_y^2-a_x-a_z & \sqrt{a_ya_z} \\ \sqrt{a_xa_z} & \sqrt{a_ya_z} & n_z^2-a_x-a_y \end{vmatrix}=0.$$

After some manipulation we obtain

$$\begin{aligned}&\left(n^2-n_x^2\right)\left(n^2-n_y^2\right)\left(n^2-n_z^2\right)\\&=s_x^2n^2\left(n^2-n_y^2\right)\left(n^2-n_z^2\right)+s_y^2n^2\left(n^2-n_x^2\right)\left(n^2-n_z^2\right)+s_z^2n^2\left(n^2-n_x^2\right)\left(n^2-n_y^2\right).\end{aligned} \tag{2.62}$$

If $\left(n^2-n_x^2\right)\neq 0$, $\left(n^2-n_y^2\right)\neq 0$, and $\left(n^2-n_z^2\right)\neq 0$, then the above equation can be put into the form

$$\frac{s_x^2}{\left(n^2-n_x^2\right)}+\frac{s_y^2}{\left(n^2-n_y^2\right)}+\frac{s_z^2}{\left(n^2-n_z^2\right)}=\frac{1}{n^2}. \tag{2.63}$$

Equation (2.63) looks simpler than Equation (2.62), and is popularly used; it is referred to as *Fresnel's equation of wavenormals*. However, one must be careful in using Equation (2.63) to calculate the refractive index; an erroneous value may be obtained if $\left(n^2-n_i^2\right)=0 \ (i=x,y,z)$. For a given propagation direction, the eigenvalue refractive index of the eigenmode can be calculated by using Equation (2.62). On the right side of this equation, the coefficient of the term containing $(n^2)^3$ is 1; on the left side, the coefficient of the term containing $(n^2)^3$ is $\left(s_x^2+s_y^2+s_z^2\right)$, which is also 1. Therefore Equation (2.62) is a quadratic in n^2. For a given propagation direction, there are two solutions of n^2, and thus there are also two solutions of n, because $n>0$.

Now we consider the eigenmodes, also referred to as normal modes. In component form, Equation (2.57) can be rewritten as three equations:

$$\left(k_o^2n_x^2-k_y^2-k_z^2\right)E_x+k_xk_yE_y+k_xk_zE_z=0 \tag{2.64}$$

$$k_xk_yE_x+\left(k_o^2n_y^2-k_x^2-k_z^2\right)E_y+k_yk_zE_z=0 \tag{2.65}$$

$$k_x k_z E_x + k_y k_z E_z + \left(k_o^2 n_z^2 - k_x^2 - k_y^2\right) E_y = 0 \tag{2.66}$$

By eliminating E_z from Equations (2.64) and (2.65), we get

$$\left(k_o^2 n_x^2 - k^2\right) k_y E_x = \left(k_o^2 n_y^2 - k^2\right) k_x E_y. \tag{2.67}$$

If $s_x \neq 0$ and $s_y \neq 0$, it can put into the form

$$\frac{\left(n_x^2 - n^2\right) E_x}{s_x} = \frac{\left(n_x^2 - n^2\right) E_y}{s_y}. \tag{2.68}$$

In the same way we can get

$$\left(k_o^2 n_x^2 - k^2\right) k_z E_x = \left(k_o^2 n_z^2 - k^2\right) k_x E_z. \tag{2.69}$$

If $s_x \neq 0$ and $s_z \neq 0$, it can be put into the form

$$\frac{\left(n_x^2 - n^2\right) E_x}{s_x} = \frac{\left(n_z^2 - n^2\right) E_z}{s_z}. \tag{2.70}$$

Therefore the eigen field is

$$\vec{E} = \begin{pmatrix} \dfrac{s_x}{\left(n_x^2 - n^2\right)} \\ \dfrac{s_y}{\left(n_y^2 - n^2\right)} \\ \dfrac{s_z}{\left(n_z^2 - n^2\right)} \end{pmatrix}, \tag{2.71}$$

which is linearly polarized in uniform anisotropic media. The physical meaning of the eigenmode is that for a given propagation direction, if the initial polarization of the light corresponds to an eigenmode, when it propagates through the medium, its polarization remains *invariant* and it propagates at the speed c/n, where the refractive index n is the corresponding eigenvalue. If the initial polarization is not along the eigenmodes, its electric field can be decomposed into two components along the two eigenmodes, respectively. These two components retain their directions, but propagate with different speeds. The resultant polarization changes in space.

It may be easier to visualize the eigen refractive indices and the eigen electric field vectors using the *refractive index ellipsoid* [5]. The major axes of the refractive index ellipsoid are parallel to the x, y, and z axes of the principal frame and have the lengths $2n_x$, $2n_y$, and $2n_z$, respectively, as shown in Figure 2.3. The ellipsoid is described by the equation

$$\frac{x^2}{n_x^2} + \frac{y^2}{n_y^2} + \frac{z^2}{n_z^2} = 1. \tag{2.72}$$

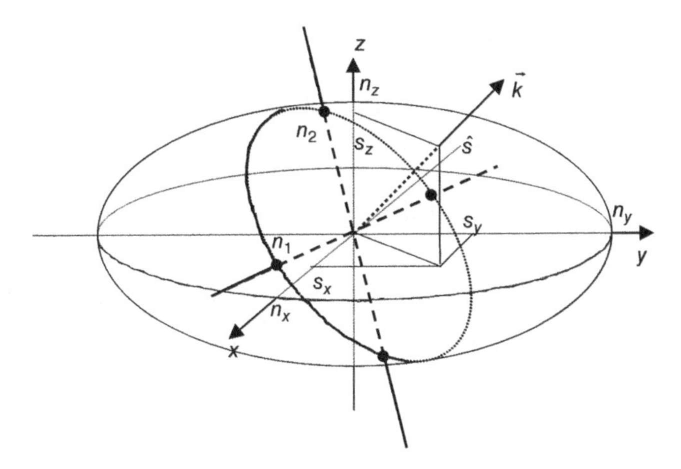

Figure 2.3 The refractive index ellipsoid.

Draw a straight line which is through the origin and parallel to $\hat{s}$. Cut a plane through the origin, which is perpendicular to $\hat{s}$. This is plane is described by

$$\hat{s}\cdot(x\hat{x}+y\hat{y}+z\hat{z}) = s_x x + s_y y + s_z z = 0. \tag{2.73}$$

The intersection of this plane and the ellipsoid is an ellipse. Any point (x, y, z) on the ellipse must satisfy both Equations (2.72) and (2.73), and its distance n from the origin is given by

$$n = \left(x^2 + y^2 + z^2\right)^{1/2}. \tag{2.74}$$

The maximum and minimum value of n are, respectively, half the length of the major axes of the ellipse. Now we maximize (or minimize) n^2 under the constraints given by Equations (2.72) and (2.73). Using the Lagrange multipliers, we maximize (or minimize)

$$g = \left(x^2 + y^2 + z^2\right) - \lambda_1\left(\frac{x^2}{n_x^2} + \frac{y^2}{n_y^2} + \frac{z^2}{n_z^2}\right) - \lambda_2\left(s_x x + s_y y + s_z z\right), \tag{2.75}$$

where λ_1 and λ_2 are the Lagrange multipliers. From $\partial g/\partial x = 0$, $\partial g/\partial y = 0$, and $\partial g/\partial z = 0$, we get

$$x_m = \frac{\lambda_2 s_x n_x^2}{2\left(n_x^2 - \lambda_1\right)}, y_m = \frac{\lambda_2 s_y n_y^2}{2\left(n_y^2 - \lambda_1\right)}, z_m = \frac{\lambda_2 s_z n_z^2}{2\left(n_z^2 - \lambda_1\right)}, \tag{2.76}$$

$$n_m^2 = x_m^2 + y_m^2 + z_m^2 = \frac{\lambda_2^2}{4}\left[\left(\frac{s_x n_x^2}{n_x^2 - \lambda_1}\right)^2 + \left(\frac{s_y n_y^2}{n_y^2 - \lambda_1}\right)^2 + \left(\frac{s_z n_z^2}{n_z^2 - \lambda_1}\right)^2\right], \tag{2.77}$$

$$\frac{\lambda_2^2}{4}\left[\frac{n_x^2}{n_m^2}\left(\frac{s_x n_x}{n_x^2 - \lambda_1}\right)^2 + \frac{n_y^2}{n_m^2}\left(\frac{s_y n_y}{n_y^2 - \lambda_1}\right)^2 + \frac{n_z^2}{n_m^2}\left(\frac{s_z n_z}{n_z^2 - \lambda_1}\right)^2\right] = 1. \tag{2.78}$$

From Equations (2.72) and (2.76) we have

$$\frac{\lambda_2^2}{4}\left[\left(\frac{s_x n_x}{n_x^2-\lambda_1}\right)^2+\left(\frac{s_y n_y}{n_y^2-\lambda_1}\right)^2+\left(\frac{s_z n_z}{n_z^2-\lambda_1}\right)^2\right]=1. \tag{2.79}$$

From Equations (2.78) and (2.79) we have

$$\frac{s_x^2 n_x^2}{\left(n_x^2-\lambda_1\right)}\frac{\left(n_x^2-n_m^2\right)}{\left(n_x^2-\lambda_1\right)}+\frac{s_y^2 n_y^2}{\left(n_y^2-\lambda_1\right)}\frac{\left(n_y^2-n_m^2\right)}{\left(n_y^2-\lambda_1\right)}+\frac{s_z^2 n_z^2}{\left(n_z^2-\lambda_1\right)^2}\frac{\left(n_z^2-n_m^2\right)}{\left(n_z^2-\lambda_1\right)}=0. \tag{2.80}$$

From Equations (2.74) and (2.76), we have

$$\frac{s_x^2 n_x^2}{n_x^2-\lambda_1}+\frac{s_y^2 n_y^2}{n_y^2-\lambda_1}+\frac{s_z^2 n_z^2}{n_z^2-\lambda_1}=0. \tag{2.81}$$

Comparing above two equations we have

$$\lambda_1=n_m^2. \tag{2.82}$$

From Equations (2.81) and (2.82), we have

$$n_m^2\left[\left(\frac{s_x^2}{n_x^2-n_m^2}+\frac{s_x^2}{n_m^2}\right)+\left(\frac{s_y^2}{n_y^2-n_m^2}+\frac{s_y^2}{n_m^2}\right)+\left(\frac{s_z^2}{n_z^2-n_m^2}+\frac{s_z^2}{n_m^2}\right)\right]=0,$$

that is,

$$\frac{s_x^2}{\left(n_m^2-n_x^2\right)}+\frac{s_y^2}{\left(n_m^2-n_y^2\right)}+\frac{s_z^2}{\left(n_m^2-n_z^2\right)}=\frac{s_x^2}{n_m^2}+\frac{s_y^2}{n_m^2}+\frac{s_z^2}{n_m^2}=\frac{1}{n_m^2}. \tag{2.83}$$

This equation is the same as Equation (2.63). Therefore half the lengths of the major axes of the ellipse are the two eigen refractive indices. The vectors along the major axes of the ellipse are

$$\vec{r}=\begin{bmatrix}\dfrac{s_x n_x^2}{\left(n_x^2-n_m^2\right)}\\[2ex] \dfrac{s_y n_y^2}{\left(n_y^2-n_m^2\right)}\\[2ex] \dfrac{s_z n_z^2}{\left(n_z^2-n_m^2\right)}\end{bmatrix}. \tag{2.84}$$

Comparing Equation (2.84) with Equation (2.71), it can be seen that the eigen electric displacements are along the major axes of the ellipse.

2.3.2 *Orthogonality of eigenmodes*

When light propagates in a uniform anisotropic medium, there are two eigenmodes, represented by $\vec{E}_1$ and $\vec{E}_2$, which are linearly polarized and invariant in space. The corresponding eigen refractive indices are n_1 and n_2. Here we discuss some of the basic properties of the eigenmodes.

1. *The electric displacement of the eigenmodes, $\vec{D}_i$ $(i=1,2)$, is perpendicular to the propagation direction.*
   From Equations (2.3) and (2.55) we have

$$\hat{s}\cdot\vec{D}_i=0. \tag{2.85}$$

   Therefore the propagation direction $\hat{s}$ and the electric displacement $\vec{D}_i$ are orthogonal to each other.
2. *$\hat{s}$, $\vec{E}_i$ and $\vec{D}_i$ are on the same plane.*
   From Equations (2.51), (2.54), and (2.55) we have the electric displacement of the eigenmodes:

$$\vec{D}_i=\frac{n^2}{c^2\mu_o}\left[\vec{E}_i-\left(\hat{s}\cdot\vec{E}_i\right)\hat{s}\right] \quad i=1,2. \tag{2.86}$$

   Therefore $\hat{s}$, $\vec{E}_i$ and $\vec{D}_i$ lie in the same plane.
3. *$\vec{D}_1\perp\vec{D}_2$, $\vec{D}_1\perp\vec{E}_2$ and $\vec{D}_2\perp\vec{E}_1$*
   Equation (2.56) can be rewritten as

$$k^2\left(\hat{s}\hat{s}-\overleftrightarrow{I}\right)\cdot\vec{E}=-k_o^2\overleftrightarrow{\varepsilon}\cdot\vec{E}, \tag{2.87}$$

   where $\overleftrightarrow{I}$ is the identity matrix and

$$\hat{s}\hat{s}=\begin{pmatrix} s_x \\ s_y \\ s_z \end{pmatrix}\left(s_x s_y s_z\right)=\begin{pmatrix} s_x^2 & s_x s_y & s_x s_z \\ s_x s_y & s_y^2 & s_y s_z \\ s_x s_z & s_y s_z & s_z^2 \end{pmatrix}.$$

   Because $\vec{E}_1$ is the eigenmode 1 with the eigen wavevector k_1, and $\vec{E}_2$ is the eigenmode 2 with the eigen wavevector k_2,

$$k_1^2\left(\hat{s}\hat{s}-\overleftrightarrow{I}\right)\cdot\vec{E}_1=-k_o^2\overleftrightarrow{\varepsilon}\cdot\vec{E}_1, \tag{2.88}$$

$$k_2^2\left(\hat{s}\hat{s}-\overleftrightarrow{I}\right)\cdot\vec{E}_2=-k_o^2\overleftrightarrow{\varepsilon}\cdot\vec{E}_2. \tag{2.89}$$

   From above two equations, we can get

$$k_1^2\vec{E}_2\cdot\left(\hat{s}\hat{s}-\overleftrightarrow{I}\right)\cdot\vec{E}_1-k_2^2\vec{E}_1\cdot\left(\hat{s}\hat{s}-\overleftrightarrow{I}\right)\cdot\vec{E}_2=-k_o^2\left(\vec{E}_2\cdot\overleftrightarrow{\varepsilon}\cdot\vec{E}_1-\vec{E}_1\cdot\overleftrightarrow{\varepsilon}\cdot\vec{E}_2\right)=0. \tag{2.90}$$

   Because $\hat{s}\hat{s}$ is a symmetric matrix, $\vec{E}_1\cdot(\hat{s}\hat{s})\cdot\vec{E}_2=\left(\hat{s}\cdot\vec{E}_1\right)\left(\hat{s}\cdot\vec{E}_2\right)$. The above equation is

$$\left(k_1^2-k_2^2\right)\left[\vec{E}_1\cdot\vec{E}_2-\left(\hat{s}\cdot\vec{E}_1\right)\left(\hat{s}\cdot\vec{E}_2\right)\right]=0.$$

Because $k_1 \neq k_2$, we must have

$$\vec{E}_1\cdot\vec{E}_2-\left(\hat{s}\cdot\vec{E}_1\right)\left(\hat{s}\cdot\vec{E}_2\right)=0. \tag{2.91}$$

On the other hand, $\vec{D}_1=\frac{n_1^2}{C^2\mu_o}\left[\vec{E}_1-\left(\hat{s}\cdot\vec{E}_1\right)\hat{s}\right]$ and $\vec{D}_2=\frac{n_2^2}{C^2\mu_o}\left[\vec{E}_2-\left(\hat{s}\cdot\vec{E}_2\right)\hat{s}\right]$, therefore

$$\vec{D}_1\cdot\vec{D}_2=\frac{n_1^2n_2^2}{\left(C^2\mu_o\right)^2}\left[\vec{E}_1\cdot\vec{E}_2-\left(\hat{s}\cdot\vec{E}_1\right)\left(\hat{s}\cdot\vec{E}_2\right)\right]=0.$$

Namely, $\vec{D}_1$ and $\vec{D}_2$ are orthogonal to each other. We also have

$$\vec{D}_1\cdot\vec{E}_2=\frac{n_1^2}{C^2\mu_o}\left[\vec{E}_1\cdot\vec{E}_2-\left(\hat{s}\cdot\vec{E}_1\right)\left(\hat{s}\cdot\vec{E}_2\right)\right]=0,$$

$$\vec{D}_2\cdot\vec{E}_1=\frac{n_2^2}{C^2\mu_o}\left[\vec{E}_2\cdot\vec{E}_1-\left(\hat{s}\cdot\vec{E}_2\right)\left(\hat{s}\cdot\vec{E}_1\right)\right]=0.$$

Generally $\vec{E}_1$ and $\vec{E}_2$ are not perpendicular to $\hat{s}$. From Equation (2.91) we have $\vec{E}_1\cdot\vec{E}_2=\left(\hat{s}\cdot\vec{E}_1\right)\left(\hat{s}\cdot\vec{E}_2\right)$. If both $\vec{E}_1$ and $\vec{E}_2$ are not perpendicular to $\hat{s}$, then $\vec{E}_1\cdot\vec{E}_2\neq 0$. If at least one of the eigen electric field is perpendicular to $\hat{s}$, then $\vec{E}_1\cdot\vec{E}_2=0$.

2.3.3 Energy flux

The energy flux in a uniform anisotropic medium is still given by the Poynting vector $\vec{S}=\vec{E}\times\vec{H}$. From Equations (2.13), (2.54) and (2.55), for an eigenmode $\vec{E}_i$ $(i=1,2)$, we have the magnetic field

$$\vec{H}_i=\frac{n_i}{\mu_o C}\hat{s}\times\vec{E}_i, \tag{2.92}$$

$$\vec{S}_i=\frac{n_i}{\mu_o C}\vec{E}_i\times\left(\hat{s}\times\vec{E}_i\right)=\frac{n_i}{\mu_o C}\left[E_i^2\hat{s}-\left(\hat{s}\cdot\vec{E}_i\right)\vec{E}_i\right]. \tag{2.93}$$

Because in general $\vec{E}_i$ is not perpendicular to $\hat{s}$, $\vec{S}_i$ is not parallel to $\hat{s}$. For a light beam, if the initial polarization is not an eigenmode, then the electric field can be decomposed into the two eigenmodes:

$$\vec{E}=c_1\vec{E}_1+c_2\vec{E}_2, \tag{2.94}$$

where c_1 and c_2 are constants. The magnetic field is given by

$$\vec{H}=c_1\vec{H}_1+c_2\vec{H}_2=c_1\frac{n_1}{\mu_o C}\hat{s}\times\vec{E}_1+c_2\frac{n_2}{\mu_o C}\hat{s}\times\vec{E}_2. \tag{2.95}$$

The Poynting vector is

$$\vec{S}=\left(c_1\vec{E}_1+c_2\vec{E}_2\right)\times\left(c_1\vec{H}_1+c_2\vec{H}_2\right)=\left(c_1\vec{E}_1+c_2\vec{E}_2\right)\times\left(\frac{n_1}{\mu_o C}\hat{s}\times\vec{E}_1+\frac{n_2}{\mu_o C}\hat{s}\times\vec{E}_2\right), \quad (2.96)$$

which is not equal to $c_1^2\vec{S}_1+c_2^2\vec{S}_2$, namely, the total energy flux is not equal to the sum of the energy fluxes of the two eigenmodes because of the cross terms between the two eigenmodes. Now let us consider the projections of the cross terms in the propagation direction. Because of Equation (2.91), for $i\neq j$ $(i,j=1,2)$,

$$\hat{s}\cdot\left(\vec{E}_i\times\vec{H}_j\right)=\vec{H}_i=\frac{n_j}{\mu_o C}\hat{s}\cdot\left[\vec{E}_i\times\hat{s}\times\vec{E}_j\right]=\frac{n_j}{\mu_o C}\left[\vec{E}_i\cdot\vec{E}_j-\left(\hat{s}\cdot\vec{E}_i\right)\left(\hat{s}\cdot\vec{E}_j\right)\right]=0. \quad (2.97)$$

Hence $\hat{s}\cdot\vec{S}=\hat{s}\cdot\left(c_1^2\vec{S}_1\right)+\hat{s}\cdot\left(c_2^2\vec{S}_2\right)$, indicating that the total energy flux in the propagation direction is equal to the sum of the energy fluxes of the two eigenmodes. This is known as the *power orthogonal theorem.*

2.3.4 Special cases

First we consider a uniaxial medium whose dielectric constants along the principal frame axes are $\varepsilon_x=\varepsilon_y=n_o^2\neq\varepsilon_z=n_e^2$, and n_o and n_e are the ordinary and extraordinary refractive index, respectively. When the propagation direction is along $\hat{s}=\sin\theta\cos\psi\hat{x}+\sin\theta\sin\psi\hat{y}+\cos\theta\hat{z}$, as shown in Figure 2.4, Equation (2.62) becomes

$$\left(n^2-n_o^2\right)^2\left(n^2-n_e^2\right)=\sin^2\theta n^2\left(n^2-n_o^2\right)\left(n^2-n_e^2\right)+\cos^2\theta n^2\left(n^2-n_o^2\right)^2.$$

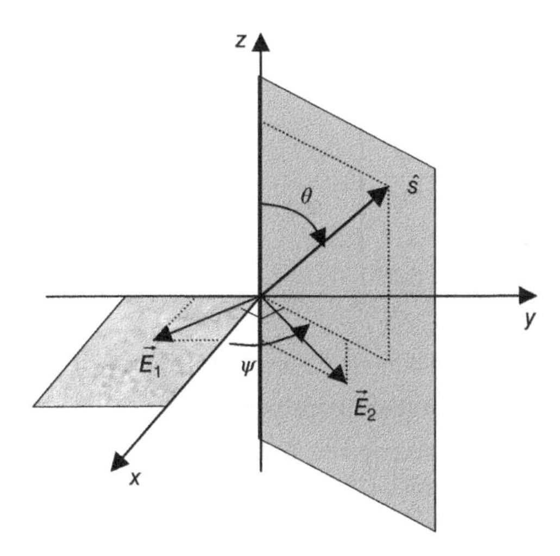

Figure 2.4 Diagram showing the propagation direction and the corresponding eigen modes.

Solution 1: $\left(n^2-n_o^2\right)=0$, namely,

$$n=n_1=n_o. \tag{2.98}$$

From Equation (2.69) we have $k_o^2\left(n_e^2-n_o^2\right)k_xE_z=0$, and therefore $E_z=0$. In this case, Equation (2.67) does not provide any information on the eigenvector. In order to get the eigenvector corresponding to the refractive index n_o, we use the condition $\vec{k}\cdot\overleftrightarrow{\varepsilon}\cdot\vec{E}=0$ because of $\nabla\cdot\vec{D}=\nabla\cdot\left(\overleftrightarrow{\varepsilon}\cdot\vec{E}\right)=0$ and $\nabla=i\vec{k}$.

$$\vec{k}\cdot\overleftrightarrow{\varepsilon}\cdot\vec{E}=k(\sin\theta\cos\psi,\sin\theta\sin\psi,\cos\theta)\begin{pmatrix} n_o^2 & 0 & 0\\ 0 & n_o^2 & 0\\ 0 & 0 & n_e^2\end{pmatrix}\begin{pmatrix}E_x\\ E_y\\ 0\end{pmatrix}=0,$$

which gives $\cos\psi E_x+\sin\psi E_y=0$. Hence the eigenmode is

$$\vec{E}_1=\begin{pmatrix}\sin\psi\\ -\cos\psi\\ 0\end{pmatrix}, \tag{2.99}$$

which is in the xy plane and perpendicular the projection direction $\hat{k}_{xy}$ of $\vec{k}$ on the xy plane. This eigenmode is sometimes referred to as the *ordinary wave* or simply *O wave*.

Solution 2: $\left(n^2-n_o^2\right)\left(n^2-n_e^2\right)=\sin^2\theta n^2\left(n^2-n_e^2\right)+\cos^2\theta n^2\left(n^2-n_o^2\right)$, namely

$$n=n_2=\frac{n_on_e}{\left(n_e^2\cos^2\theta+n_o^2\sin^2\theta\right)^{1/2}}. \tag{2.100}$$

From Equation (2.71) we have the eigenmode

$$\vec{E}_2=\begin{pmatrix}\dfrac{s_x}{\left(n_x^2-n^2\right)}\\ \dfrac{s_y}{\left(n_y^2-n^2\right)}\\ \dfrac{s_z}{\left(n_z^2-n^2\right)}\end{pmatrix}=\begin{pmatrix}\dfrac{\cos\psi}{n_o^2\sin\theta}\\ \dfrac{\sin\psi}{n_o^2\sin\theta}\\ \dfrac{-1}{n_e^2\cos\theta}\end{pmatrix}=\frac{1}{n_o^2n_e^2\sin\theta\cos\theta}\begin{pmatrix}n_e^2\cos\psi\cos\theta\\ n_e^2\sin\psi\cos\theta\\ -n_o^2\sin\theta\end{pmatrix}, \tag{2.101}$$

which is on the zk_{xy} plane. This eigenmode is sometimes referred to as the *extraordinary wave* or simply *E wave*. It can be shown that $\hat{s}$ is perpendicular to $\vec{E}_1$ but not to $\vec{E}_2$.

When $\psi=\pi/2$, the light wave propagates in the yz plane. Eigenmode 1 is $\vec{E}_1^T=(1,0,0)$, where T stands for the transpose, which is a long the x axis, and the corresponding eigenvalue is n_o. Eigenmode 2 is $\vec{E}_2^T=\left(0,n_e^2\cos\theta,-n_o^2\sin\theta\right)$, which is in the yz plane, and the corresponding eigenvalue is $n = n_e n_o / \left(n_e^2 \cos^2\theta + n_o^2 \sin^2\theta\right)^{1/2}$. Note that the angle between $\vec{E}_2$ and the z axis is $\arctan\left(n_e^2 \cos\theta / n_o^2 \sin\theta\right) \neq \pi/2 - \theta$. $\vec{D}_2^T = \overleftrightarrow{\varepsilon} \cdot \vec{E}_2 = n_e^2 n_o^2 (0,\ \cos\theta,\ -\sin\theta)$. The angle between $\vec{D}_2$ and the z axis is $\pi/2 - \theta$.

When $\theta = \pi/2$, the light is propagating in the xy plane. Eigenmode 1 is $\vec{E}_1^T = (\sin\psi, -\cos\psi, 0)$, and the corresponding eigenvalue is n_o. Eigenmode 2 is $\vec{E}_2^T = (0, 0, 1)$, which is along the z axis, and the corresponding eigenvalue is n_e.

2.3.5 *Polarizers*

Polarizers are an essential component in many liquid crystal devices. Most sheet polarizers are uniaxially anisotropic in their absorption. One way to make a sheet polarizer is by embedding elongated absorbing molecules (or tiny rod-like crystals), which exhibit strong absorption for light polarized along their long axis, in a polymer film, and stretching the polymer, which produces a unidirectional alignment of the embedded molecules. Small needle-like crystals of herapathite in polyvinyl alcohol is one example. The refractive indices of a uniaxial polarizer can be written as

$$n_o' = n_o - i a_o \tag{2.102}$$

$$n_e' = n_e - i a_e \tag{2.103}$$

where the imaginary parts a_o, a_e are referred to as the extinction coefficients and are responsible for the absorption, and n'_e *and* n'_o are the refractive indices, respectively, parallel and perpendicular to the uniaxial axis. Sometimes polarizers are divided into two types: *O-type polarizer* where $a_e \gg a_o \approx 0$ and *E-type polarizer* where $a_o \gg a_e \approx 0$. The transmittances of the polarizer for light polarized parallel and perpendicular to its transmission axis are, respectively,

$$T_1 = e^{-2(2\pi a_{\min}/\lambda)h}, \tag{2.104}$$

$$T_2 = e^{-2(2\pi a_{\max}/\lambda)h}, \tag{2.105}$$

where h is the optical path inside the polarizer, $a_{\max}$ is the bigger of (a_o, a_e), and $a_{\min}$ is the smaller of (a_o, a_e). An ideal polarizer would have $T_1 = 1$ and $T_2 = 0$. For a real polarizer $T_1 < 1$ and $T_2 > 0$. The extinction ratio of a polarizer is defined as T_1/T_2. When an unpolarized light is incident on one polarizer, the transmittance is $(T_1 + T_2)/2$. When an unpolarized light beam is incident on two parallel polarizers, the transmittance is $\left(T_1^2 + T_2^2\right)/2$. When an unpolarized light beam is incident on two crossed polarizers, the transmittance is $T_1 T_2/2$.

For obliquely incident light, some light will leak through a set of two crossed polarizers even if the polarizers are ideal [4]. We consider the leakage of two crossed ideal O-type polarizers as shown in Figure 2.5. The normal of the polarizer films is in the z direction in the lab frame. The propagation direction is specified by the polar angle θ and the azimuthal

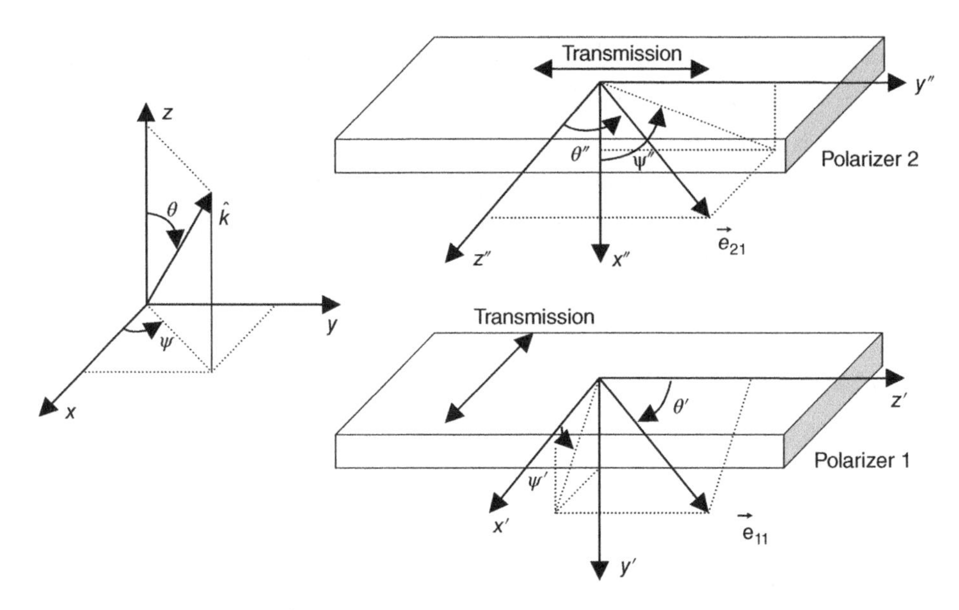

Figure 2.5 Schematic diagram of the two crossed polarizers.

angle ψ. For polarizer 1, the transmission axis is parallel to the x axis in the lab frame. In the local frame $x'y'z'$, the transmission axis is parallel to the x' axis and the uniaxial axis is parallel to the z' axis. In this local frame, the formulas derived in the last section can be used. Eigenmode 1 has the non-absorbing refractive index n_o and thus can pass polarizer 1. The direction of the eigen electric field vector is along the direction $\vec{e}'_{11}$ which in the local frame is given by

$$\vec{e}'_{11} = \begin{pmatrix} \sin\psi' \\ -\cos\psi' \\ 0 \end{pmatrix}. \tag{2.106}$$

In the lab frame xyz, this vector is given by

$$\vec{e}_{11} = \begin{pmatrix} \sin\psi' \\ 0 \\ \cos\psi' \end{pmatrix}. \tag{2.107}$$

The relations between the propagation angles in these frames are $\cos\theta = -\sin\theta'\sin\psi'$, $\sin\theta\cos\psi = \sin\theta'\cos\psi'$, and $\sin\theta\sin\psi = \cos\theta'$. Therefore we have

$$\vec{e}_{11} = \frac{1}{\sqrt{1-\sin^2\theta\sin^2\psi}} \begin{pmatrix} -\cos\theta \\ 0 \\ \sin\theta\cos\psi \end{pmatrix} \tag{2.108}$$

For polarizer 2, the transmission axis is parallel to the y axis in the lab frame. In the local frame $x''y''z''$, the transmission axis is parallel to the y'' axis and the uniaxial axis is parallel to the z'' axis. For the eigenmode that has the non-absorbing refractive index n_o and can pass polarizer 2, the direction of the eigen electric field vector is along the direction $\vec{e}''_{21}$, which in the local frame is given by

$$\vec{e}''_{21} = \begin{pmatrix} \sin\psi'' \\ -\cos\psi'' \\ 0 \end{pmatrix}. \tag{2.109}$$

In the lab frame, this vector is given by

$$\vec{e}_{21} = \begin{pmatrix} 0 \\ -\cos\psi'' \\ \sin\psi'' \end{pmatrix}. \tag{2.110}$$

The relations between the propagation angles in the lab frame xyz and the local frame $x''y''z''$ are $\cos\theta = -\sin\theta''\cos\psi''$, $\sin\theta\cos\psi = \cos\theta''$ and $\sin\theta\sin\psi = \sin\theta''\sin\psi''$. Therefore we have

$$\vec{e}_{21} = \frac{1}{\sqrt{1-\sin^2\theta\cos^2\psi}} \begin{pmatrix} 0 \\ \cos\theta \\ \sin\theta\sin\psi \end{pmatrix} \tag{2.111}$$

The light (eigenmode $\vec{e}_{11}$) coming out of polarizer 1 can be decomposed into two components in the the eigenvector directions of polarizer 2. The component along the eigenmode $\vec{e}_{21}$ passes polarizer 2 without absorption. The leakage of unpolarized light through the two crossed polarizers is given by

$$T_{leakage} = \frac{1}{2}\left(\vec{e}_{11}\cdot\vec{e}_{21}\right)^2 = \frac{\sin^4\theta\sin^2\psi\cos^2\psi}{2(1-\sin^2\theta\sin^2\psi)(1-\sin^2\theta\cos^2\psi)} \tag{2.112}$$

Note that the angle θ here is the polar angle of the propagation direction inside the polarizers. This leakage, if not compensated, will limit the viewing angle of liquid crystal displays. The iso-transmission (leakage) diagram of the crossed polarizers as a function of the polar and azimuthal angles is shown in Figure 2.6. It resembles the appearance of the crossed polarizers under isotropic incident light when viewed at various polar and azimuthal angles. The black indicates low transmittance (leakage) and the white indicates high transmittance (leakage). At the azimuthal angle of 45°, when the polar angle is 30°, 60°, and 90°, the transmittance is 0.01, 0.18, and 0.5, respectively. The leakage of the crossed polarizers can be reduced by using compensation films [6].

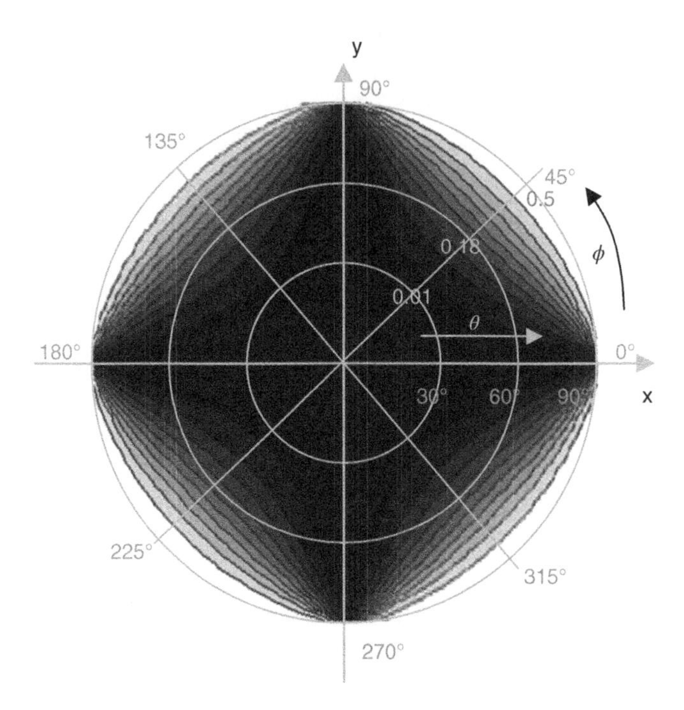

Figure 2.6 Iso-transmittance diagram of the crossed polarizers. Black: low transmittance; white: high transmittance.

2.4 Propagation of Light in Cholesteric Liquid Crystals

2.4.1 Eigenmodes

We showed in last section that in a uniform anisotropic medium, for each propagation direction, there are two eigenmodes which are linearly polarized. The polarization state of the eigenmodes is invariant in space. In this section, we discuss the propagation of light in a special case of a non-uniform anisotropic medium: a cholesteric liquid crystal which locally is optically uniaxial, but the optic axis twists uniformly in space [6,7]. Choose the z axis of the lab frame to be parallel to the helical axis of the cholesteric liquid crystal. The pitch P of the liquid crystal is the distance over which the liquid crystal director twists 2π. The components of the liquid crystal director of a right-handed cholesteric liquid crystal ($q > 0$) are given by

$$n_x = \cos(qz), \quad n_y = \sin(qz), \quad n_z = 0, \tag{2.113}$$

where the twisting rate (chirality) q is related to the pitch by $q = 2\pi/P$. We consider light propagating in the z direction, $\vec{E}(z,t) = \vec{A}(z)e^{i\omega t}$, and therefore $\nabla = \hat{z}\frac{\partial}{\partial z}$ and $\frac{\partial}{\partial t} = i\omega$. From Equation (2.51) we have

$$\frac{\partial^2 \vec{A}(z)}{\partial z^2} = -k_o^2 \overleftrightarrow{\varepsilon}(z) \cdot \vec{A}(z), \tag{2.114}$$

where $k_o = \omega/c = 2\pi/\lambda$ (λ is the wavelength in vacuum). The dielectric constants of the liquid crystal for light polarized parallel and perpendicular to the liquid crystal director are $\varepsilon_{//} = n_o^2$ and $\varepsilon_{\perp} = n_e^2$, respectively. The dielectric tensor in the xy plane in the lab frame is

$$\overleftrightarrow{\varepsilon}(z) = \varepsilon_{\perp}\overleftrightarrow{I} + 2\delta\vec{n}\vec{n} = \begin{pmatrix} \varepsilon_{\perp} + 2\delta n_x^2 & 2\delta n_x n_y \\ 2\delta n_y n_x & \varepsilon_{\perp} + 2\delta n_y^2 \end{pmatrix} = \begin{pmatrix} \bar{\varepsilon} + \delta\cos(2qz) & \delta\sin(2qz) \\ \delta\sin(2qz) & \bar{\varepsilon} - \delta\cos(2qz) \end{pmatrix}, \quad (2.115)$$

where $\delta = (\varepsilon_{//} - \varepsilon_{\perp})/2$ and $\bar{\varepsilon} = (\varepsilon_{//} + \varepsilon_{\perp})/2$. As we will show that there is no mode whose polarization state is invariant in space in the lab frame; consequently we employ the local frame whose x' axis is parallel to the liquid crystal director. The angle between the x' axis and the x axis is $\phi = qz$. The relation between the two frames is

$$\hat{x}' = \cos(qz)\hat{x} + \sin(qz)\hat{y}, \quad (2.116)$$

$$\hat{y}' = -\sin(qz)\hat{x} + \cos(qz)\hat{y}. \quad (2.117)$$

In the $x'y'$ frame, the electric field is

$$\vec{A}' = \begin{pmatrix} A'_x \\ A'_y \end{pmatrix} = \begin{pmatrix} \cos\phi & \sin\phi \\ -\sin\phi & \cos\phi \end{pmatrix} \begin{pmatrix} A_x \\ A_y \end{pmatrix} \equiv \overleftrightarrow{S}^{-1}(\phi)\vec{A}, \quad (2.118)$$

where $\overleftrightarrow{S}$ is the transformation matrix. The dielectric tensor in the local frame is

$$\overleftrightarrow{\varepsilon}' = \overleftrightarrow{S}^{-1}\overleftrightarrow{\varepsilon}\,\overleftrightarrow{S} = \begin{pmatrix} \cos\phi & \sin\phi \\ -\sin\phi & \cos\phi \end{pmatrix} \begin{pmatrix} \bar{\varepsilon} + \delta\cos(2\phi) & \delta\sin(2\phi) \\ \delta\sin(2\phi) & \bar{\varepsilon} - \delta\cos(2\phi) \end{pmatrix} \begin{pmatrix} \cos\phi & -\sin\phi \\ \sin\phi & \cos\phi \end{pmatrix}$$
$$= \begin{pmatrix} \varepsilon_{//} & 0 \\ 0 & \varepsilon_{\perp} \end{pmatrix}. \quad (2.119)$$

Because the dielectric tensor in the local frame is a constant tensor, we presume that the polarization of the eigenmodes is invariant in space in this frame [4, 8], which will be proved true,

$$\vec{A}'(z) = \vec{A}'_o e^{-ikz} = \left(A'_{ox}\hat{x}' + A'_{oy}\hat{y}'\right)e^{-ikz}, \quad (2.120)$$

where A'_{ox} and A'_{oy} are constants. In the lab frame, the electric field is

$$\begin{aligned} \vec{A}(z) &= \overleftrightarrow{S}(qz)\cdot\vec{A}'(z) \\ &= \left[A'_{ox}\cos(qz) - A'_{oy}\sin(qz)\right]e^{-ikz}\hat{x} + \left[A'_{ox}\sin(qz) + A'_{oy}\cos(qz)\right]e^{-ikz}\hat{y} \\ &= \left(A'_{ox}\hat{x} + A'_{oy}\hat{y}\right)\cos(qz)e^{-ikz} + \left(-A'_{oy}\hat{x} + A'_{ox}\hat{y}\right)\sin(qz)e^{-ikz} \end{aligned} \quad (2.121)$$

$$\frac{\partial \vec{A}}{\partial z} = (-ik)\left(A'_{ox}\hat{x} + A'_{oy}\hat{y}\right)\cos(qz)e^{-ikz} + (-ik)\left(-A'_{oy}\hat{x} + A'_{ox}\hat{y}\right)\sin(qz)e^{-ikz}$$

$$+(-q)\left(A'_{ox}\hat{x} + A'_{oy}\hat{y}\right)\sin(qz)e^{-ikz} + (q)\left(-A'_{oy}\hat{x} + A'_{ox}\hat{y}\right)\cos(qz)e^{-ikz}$$

$$\frac{\partial^2 \vec{A}}{\partial z^2} = \left(-k^2 - q^2\right)\vec{A} + (i2kq)\vec{B} \tag{2.122}$$

where

$$\vec{B} = \left\{\left[A'_{ox}\sin(qz) + A'_{oy}\cos(qz)\right]\hat{x} - \left[A'_{ox}\cos(qz) - A'_{oy}\sin(qz)\right]\hat{y}\right\}e^{-ikz} \tag{2.123}$$

Equation (2.114) becomes

$$-\left(k^2 + q^2\right)\vec{A}(z) + (i2kq)\vec{B} = -k_o^2\overleftrightarrow{\varepsilon}(z)\cdot\vec{A}(z). \tag{2.124}$$

Multiplying both sides by the transformation matrix, we get

$$-(k^2+q^2)\overleftrightarrow{S}^{-1}\cdot\vec{A}(z) + (i2kq)\overleftrightarrow{S}^{-1}\cdot\vec{B} = -k_o^2\overleftrightarrow{S}^{-1}\overleftrightarrow{\varepsilon}(z)\cdot\overleftrightarrow{S}\cdot\overleftrightarrow{S}^{-1}\cdot\vec{A}(z)$$

$$-(k^2+q^2)\vec{A}'(z) + (i2kq)\overleftrightarrow{S}^{-1}\cdot\vec{B} = -k_o^2\overleftrightarrow{\varepsilon}'(z)\cdot\vec{A}'(z).$$

Because $\overleftrightarrow{S}^{-1}\cdot\vec{B} = \begin{pmatrix}\cos\phi & \sin\phi\\ -\sin\phi & \cos\phi\end{pmatrix}\begin{pmatrix}A'_{ox}\sin\phi + A'_{oy}\cos\phi\\ -A'_{ox}\cos\phi + A'_{oy}\sin\phi\end{pmatrix} = \begin{pmatrix}A'_{oy}\\ -A'_{ox}\end{pmatrix}$,

Equation (2.124) can be put into the form

$$\begin{pmatrix} n_e^2k_o^2 - k^2 - q^2 & i2qk\\ -i2qk & n_o^2k_o^2 - k^2 - q^2\end{pmatrix}\begin{pmatrix}A'_{ox}\\ A'_{oy}\end{pmatrix} = 0. \tag{2.125}$$

For non-zero solutions, it is required that

$$\begin{vmatrix} n_e^2k_o^2 - k^2 - q^2 & i2qk\\ -i2qk & n_o^2k_o^2 - k^2 - q^2\end{vmatrix} = 0. \tag{2.126}$$

Define $k = nk_o$ and $\alpha = q/k_o = \lambda/P$. Equation (2.126) becomes

$$n^4 - \left(2\alpha^2 + n_e^2 + e_o^2\right)n^2 + \left(\alpha^2 - n_e^2\right)\left(\alpha^2 - n_o^2\right) = 0,$$

$$n_1^2 = \alpha^2 + \bar{\varepsilon} + \left(4\alpha^2\bar{\varepsilon} + \delta^2\right)^{1/2}, \tag{2.127}$$

$$n_2^2 = \alpha^2 + \bar{\varepsilon} - \left(4\alpha^2\bar{\varepsilon} + \delta^2\right)^{1/2}. \tag{2.128}$$

n_1^2 is always positive. n_2^2 can be either positive or negative, depending on the ratio between the wavelength and the pitch, as shown in Figure 2.7.

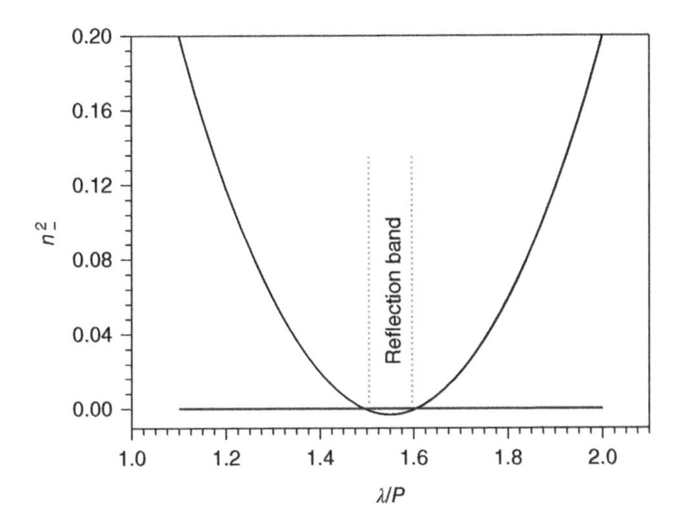

Figure 2.7 n_-^2 vs. λ/P curve. $n_e = 1.6$ and $n_o = 1.5$ are used in the calculation.

$$n_{1\pm} = \pm\left[\alpha^2 + \bar{\varepsilon} + \left(4\alpha^2\bar{\varepsilon} + \delta^2\right)^{1/2}\right]^{1/2} = \pm\left\{(\lambda/P)^2 + \bar{\varepsilon} + \left[4(\lambda/P)^2\bar{\varepsilon} + \delta^2\right]^{1/2}\right\}^{1/2}, \tag{2.129}$$

$$n_{2\pm} = \pm\left[\alpha^2 + \bar{\varepsilon} - \left(4\alpha^2\bar{\varepsilon} + \delta^2\right)^{1/2}\right]^{1/2} = \pm\left\{(\lambda/P)^2 + \bar{\varepsilon} - \left[4(\lambda/P)^2\bar{\varepsilon} + \delta^2\right]^{1/2}\right\}^{1/2} \tag{2.130}$$

are the eigenvalues. For each eigenvalue, there is an eigenmode. Altogether there are four eigenmodes. Two of the eigenmodes propagate in the $+z$ direction, and the other two eigenmodes propagate in the $-z$ direction. When '+' is used from the sign '±', the corresponding eigenmode is not necessarily propagating in the $+z$ direction. From Equation (2.125) we can calculate the polarization of the eigenmodes,

$$\vec{A}'_{o1\pm} = a\begin{pmatrix} 1 \\ i\left[n_e^2 - n_{1\pm}^2 - \alpha^2\right]/2\alpha n_{1\pm} \end{pmatrix}, \tag{2.131}$$

$$\vec{A}'_{o2\pm} = b\begin{pmatrix} 1 \\ i\left[n_e^2 - n_{2\pm}^2 - \alpha^2\right]/2\alpha n_{2\pm} \end{pmatrix}, \tag{2.132}$$

where a and b are the normalization constant. Generally they are elliptically polarized because the $\pi/2$ phase difference between A'_{ox} and A'_{oy}. $n_{1\pm}$ is always real for any frequency ω. $n_{2\pm}$ can be real or imaginary depending on the frequency ω.

We consider some special cases.

2.4.1.1 $P \gg \lambda$

In this case $\alpha = \lambda/P \ll 1$. Using the approximation $\bar{\varepsilon} \approx \bar{n}^2$ and the assumption $\Delta n \ll 1$, from Equation (2.129) we have

$$n_1 = \left[\alpha^2 + \bar{n}^2 + 2\bar{n}\left(\alpha^2 + (\Delta n/2)^2\right)^{1/2}\right]^{1/2} = \bar{n} + \sqrt{\alpha^2 + (\Delta n/2)^2} = \bar{n} + \alpha\sqrt{1 + u^2}, \tag{2.133}$$

where $u = \Delta n/2\alpha = \Delta nP/2\lambda$. The corresponding eigenmode has the polarization

$$\vec{A}'_{o1} \approx a\begin{pmatrix} 1 \\ i\left(n_e^2 - \bar{n}^2 - (2+u^2)\alpha^2 - 2\bar{n}\alpha\sqrt{1+u^2}\right)/(2\alpha\bar{n}) \end{pmatrix} \approx \begin{pmatrix} 1 \\ i\left(u - \sqrt{1+u^2}\right) \end{pmatrix}, \tag{2.134}$$

which is elliptically polarized in the local frame. From Equation (2.130) we have

$$n_2 = \left[\alpha^2 + \bar{n}^2 - 2\bar{n}\left(\alpha^2 + (\Delta n/2)^2\right)^{1/2}\right]^{1/2} = \bar{n} - \sqrt{\alpha^2 + (\Delta n/2)^2} = \bar{n} - \alpha\sqrt{1+u^2}. \tag{2.135}$$

The corresponding eigenmode has the polarization

$$\vec{A}'_{o2} \approx a\begin{pmatrix} 1 \\ i\left(n_e^2 - \bar{n}^2 - (2+u^2)\alpha^2 + 2\bar{n}\alpha\sqrt{1+u^2}\right)/(2\alpha\bar{n}) \end{pmatrix} \approx \begin{pmatrix} 1 \\ i\left(u + \sqrt{1+u^2}\right) \end{pmatrix}, \tag{2.136}$$

which is also elliptically in the local frame. Consider light propagating in the $+z$ direction. At the entrance plane the electric field vector is $\vec{E}_{in}^{T} = (E_{xi}, E_{yi})$; the local frame is the same as the lab frame. If the amplitudes of the two eigenmodes are u_i and v_i, we have

$$\begin{pmatrix} E_{xi} \\ E_{yi} \end{pmatrix} = u_i\begin{pmatrix} 1 \\ i\left(u - \sqrt{1+u^2}\right) \end{pmatrix} + v_i\begin{pmatrix} 1 \\ i\left(u + \sqrt{1+u^2}\right) \end{pmatrix} = \begin{pmatrix} 1 & 1 \\ i\left(u - \sqrt{1+u^2}\right) & i\left(u + \sqrt{1+u^2}\right) \end{pmatrix}\begin{pmatrix} u_i \\ v_i \end{pmatrix}. \tag{2.137}$$

From the above equation we can get

$$\begin{pmatrix} u_i \\ v_i \end{pmatrix} = \frac{1}{2\sqrt{1+u^2}}\begin{pmatrix} u + \sqrt{1+u^2} & i \\ -u + \sqrt{1+u^2} & -i \end{pmatrix}\begin{pmatrix} E_{xi} \\ E_{yi} \end{pmatrix}. \tag{2.138}$$

After the light propagates the distance h along the $+z$ direction, in the local frame the amplitude of the eigenmodes becomes

$$\begin{pmatrix} u_o \\ v_o \end{pmatrix} = \begin{pmatrix} e^{-i2\pi h\left(\bar{n} + \alpha\sqrt{1+u^2}\right)/\lambda} & 0 \\ 0 & e^{-i2\pi h\left(\bar{n} - \alpha\sqrt{1+u^2}\right)/\lambda} \end{pmatrix}\begin{pmatrix} u_{in} \\ v_{in} \end{pmatrix} = e^{-i2\pi h\bar{n}/\lambda}\begin{pmatrix} e^{-i\Theta} & 0 \\ 0 & e^{i\Theta} \end{pmatrix}\begin{pmatrix} u_{in} \\ v_{in} \end{pmatrix}, \tag{2.139}$$

where $\Theta = 2\pi h\alpha\sqrt{1+u^2}/\lambda$. The electric field in the local frame becomes

$$\begin{pmatrix} E'_{xo} \\ E'_{yo} \end{pmatrix} = u_o\begin{pmatrix} 1 \\ i\left(u - \sqrt{1+u^2}\right) \end{pmatrix} + v_o\begin{pmatrix} 1 \\ i\left(u + \sqrt{1+u^2}\right) \end{pmatrix} = \begin{pmatrix} 1 & 1 \\ i\left(u - \sqrt{1+u^2}\right) & i\left(u + \sqrt{1+u^2}\right) \end{pmatrix}\begin{pmatrix} u_o \\ v_o \end{pmatrix}. \tag{2.140}$$

Substituting Equations (2.138) and (2.139) into Equation (2.140) we get

$$\begin{aligned}
\begin{pmatrix} E'_{xo} \\ E'_{yo} \end{pmatrix} &= \frac{e^{-i2\pi h\bar{n}/\lambda}}{2\sqrt{1+u^2}} \begin{pmatrix} 1 & 1 \\ i\left(u-\sqrt{1+u^2}\right) & i\left(u+\sqrt{1+u^2}\right) \end{pmatrix} \begin{pmatrix} e^{-i\Theta} & 0 \\ 0 & e^{i\Theta} \end{pmatrix} \begin{pmatrix} u+\sqrt{1+u^2} & i \\ -u+\sqrt{1+u^2} & -i \end{pmatrix} \begin{pmatrix} E_{xi} \\ E_{yi} \end{pmatrix} \\
&= \frac{e^{-i2\pi h\bar{n}/\lambda}}{2\sqrt{1+u^2}} \begin{pmatrix} 2\sqrt{1+u^2}\cos\Theta - i2u\sin\Theta & 2\sin\Theta \\ -2\sin\Theta & 2\sqrt{1+u^2}\cos\Theta + i2u\sin\Theta \end{pmatrix} \begin{pmatrix} E_{xi} \\ E_{yi} \end{pmatrix} \\
&= e^{-i2\pi h\bar{n}/\lambda} \begin{pmatrix} \cos\Theta - i\dfrac{u}{\sqrt{1+u^2}}\sin\Theta & \dfrac{1}{\sqrt{1+u^2}}\sin\Theta \\ -\dfrac{1}{\sqrt{1+u^2}}\sin\Theta & \cos\Theta + i\dfrac{u}{\sqrt{1+u^2}}\sin\Theta \end{pmatrix} \begin{pmatrix} E_{xi} \\ E_{yi} \end{pmatrix}.
\end{aligned} \tag{2.141}$$

Defining the total twist angle $\Phi = 2\pi h/P$ and total retardation angle $\Gamma = 2\pi\Delta nh/\lambda$, then $\Theta = 2\pi h\frac{\lambda}{P}\sqrt{1+(\Delta nP/2\lambda)^2}/\lambda = \left[\left(\frac{2\pi h}{P}\right)^2 + \left(\frac{2\pi h}{2\lambda}\right)^2\right]^{1/2} = \left[\Phi^2 + (\Gamma/2)^2\right]^{1/2}$, $\frac{u}{\sqrt{1+u^2}} = \frac{(\Gamma/2)}{\Theta}$, and $\frac{1}{\sqrt{1+u^2}} = \frac{\Phi}{\Theta}$. Equation (2.141) becomes

$$\begin{pmatrix} E'_{xo} \\ E'_{yo} \end{pmatrix} = \mathrm{e}^{-i2\pi h\bar{n}/\lambda} \begin{pmatrix} \cos\Theta - i\dfrac{(\Gamma/2)}{\Theta}\sin\Theta & \dfrac{\Phi}{\Theta}\sin\Theta \\ -\dfrac{\Phi}{\Theta}\sin\Theta & \cos\Theta + i\dfrac{(\Gamma/2)}{\Theta}\sin\Theta \end{pmatrix} \begin{pmatrix} E_{xi} \\ E_{yi} \end{pmatrix}. \tag{2.142}$$

The factor $\mathrm{e}^{-i2\pi h\bar{n}/\lambda}$ can be omitted. In the lab frame we have

$$\begin{pmatrix} E_{xo} \\ E_{yo} \end{pmatrix} = \begin{pmatrix} \cos\Phi & -\sin\Phi \\ \sin\Phi & \cos\Phi \end{pmatrix} \begin{pmatrix} \cos\Theta - i\dfrac{(\Gamma/2)}{\Theta}\sin\Theta & \dfrac{\Phi}{\Theta}\sin\Theta \\ -\dfrac{\Phi}{\Theta}\sin\Theta & \cos\Theta + i\dfrac{(\Gamma/2)}{\Theta}\sin\Theta \end{pmatrix} \begin{pmatrix} E_{xi} \\ E_{yi} \end{pmatrix}. \tag{2.143}$$

Under the Mauguin condition $\Delta nP \gg \lambda$ [9], $u = \Delta n/2\alpha = \Delta nP/2\lambda \gg 1$, Equation (2.133) becomes

$$n_1 = \bar{n} + \alpha\sqrt{1+u^2} \approx \bar{n} + \alpha u = \bar{n} + \frac{\lambda}{P}\frac{\Delta nP}{2\lambda} = \bar{n} + \frac{\Delta n}{2} = n_e. \tag{2.144}$$

The corresponding eigenmode has the polarization

$$\vec{A}'_{o1} = \begin{pmatrix} 1 \\ i\left(u-\sqrt{1+u^2}\right) \end{pmatrix} = \begin{pmatrix} 1 \\ -i/2u \end{pmatrix} \approx \begin{pmatrix} 1 \\ 0 \end{pmatrix}, \tag{2.145}$$

which is linearly polarized along the liquid crystal director. Equation (2.135) becomes

$$n_2=\bar{n}-\alpha\sqrt{1+u^2}=\bar{n}-\alpha u=\bar{n}-\frac{\lambda}{P}\frac{\Delta nP}{2\lambda}=\bar{n}-\frac{\Delta n}{2}=n_o. \tag{2.146}$$

The corresponding eigenmode has the polarization

$$\vec{A}'_{o2}=b\begin{pmatrix}1\\ i\left(u+\sqrt{1+u^2}\right)\end{pmatrix}=b\begin{pmatrix}1\\ 2iu\end{pmatrix}=\begin{pmatrix}0\\ 1\end{pmatrix}, \tag{2.147}$$

which is linearly polarized perpendicular to the liquid crystal director. Note that in the above equation is the normalization constant and, in this regime, for the eigenmodes, the polarization twists in phase with the liquid crystal director in space. This is the 'waveguide' regime.

2.4.1.2 $\Delta nP \ll \lambda$

In this case, $4\alpha^2\bar{\varepsilon} \gg \delta^2$. From Equation (2.129) we have

$$n_{1\pm}\approx\pm\left(\alpha^2+\bar{\varepsilon}+2\alpha\sqrt{\bar{\varepsilon}}+\frac{\delta^2}{4\alpha\sqrt{\bar{\varepsilon}}}\right)^{1/2}\approx\pm\left[\alpha+\sqrt{\bar{\varepsilon}}+\frac{\delta^2}{8\alpha\left(\alpha+\sqrt{\bar{\varepsilon}}\right)\sqrt{\bar{\varepsilon}}}\right] \tag{2.148}$$

The corresponding eigenmodes have the polarization

$$\vec{A}'_{o1\pm}\approx a\begin{pmatrix}1\\ \dfrac{i\left[n_e^2-\left(\alpha^2+\bar{\varepsilon}+2\alpha\sqrt{\bar{\varepsilon}}+\dfrac{\delta^2}{4\alpha\sqrt{\bar{\varepsilon}}}\right)-\alpha^2\right]}{2\alpha\left[\pm\left(\alpha+\sqrt{\bar{\varepsilon}}+\dfrac{\delta^2}{8\alpha\left(\alpha+\sqrt{\bar{\varepsilon}}\right)\sqrt{\bar{\varepsilon}}}\right)\right]}\end{pmatrix}\approx a\begin{pmatrix}1\\ \dfrac{-i\left[2\alpha^2+2\alpha\sqrt{\bar{\varepsilon}}\right]}{2\alpha\left[\pm\left(\alpha+\sqrt{\bar{\varepsilon}}\right)\right]}\end{pmatrix}\approx\frac{1}{\sqrt{2}}\begin{pmatrix}1\\ \mp i\end{pmatrix}.$$

In the lab frame, the polarization is

$$A_{o1\pm x}=\left[\cos(qz)\pm i\sin(qz)\right]e^{-ik_on_{1\pm}z}=e^{i(\pm q-k_on_{1\pm})z},$$
$$A_{o1\pm y}=\left[\sin(qz)\mp i\cos(qz)\right]e^{-ik_on_{1\pm}z}=(\mp i)e^{i(\pm q-k_on_{1\pm})z}.$$

Because $\pm q-k_on_{1\pm}=\pm k_o\alpha\mp k_o\left[\alpha+\sqrt{\bar{\varepsilon}}+\frac{\delta^2}{8\alpha\left(\alpha+\sqrt{\bar{\varepsilon}}\right)\sqrt{\bar{\varepsilon}}}\right]=\mp k_o\left[\sqrt{\bar{\varepsilon}}+\frac{\delta^2}{8\alpha\left(\alpha+\sqrt{\bar{\varepsilon}}\right)\sqrt{\bar{\varepsilon}}}\right]$, we have

$$\vec{A}_{1\pm}=\frac{1}{\sqrt{2}}\begin{pmatrix}1\\ \mp i\end{pmatrix}e^{-i(\pm k_o)\left[\sqrt{\bar{\varepsilon}}+\frac{\delta^2}{8\alpha\left(\alpha+\sqrt{\bar{\varepsilon}}\right)\sqrt{\bar{\varepsilon}}}\right]z}. \tag{2.149}$$

Eigenmode 1 is left-handed circularly polarized and propagates in the $+z$ direction with the refractive index

$$n_1 = \sqrt{\bar{\varepsilon}} + \delta^2 / \left[8\alpha\left(\alpha + \sqrt{\bar{\varepsilon}}\right)\sqrt{\bar{\varepsilon}} \right]. \tag{2.150}$$

Eigenmode 2 is also left-handed circularly polarized but propagates in the $-z$ direction with the same refractive index.

From Equation (2.130) we have

$$n_{2\pm} \approx \pm \left[\alpha^2 + \bar{\varepsilon} - 2\alpha\sqrt{\bar{\varepsilon}} - \frac{\delta^2}{4\alpha\sqrt{\bar{\varepsilon}}} \right]^{1/2} \approx \pm \left[\alpha - \sqrt{\bar{\varepsilon}} - \frac{\delta^2}{8\alpha\left(\alpha - \sqrt{\bar{\varepsilon}}\right)\sqrt{\bar{\varepsilon}}} \right]. \tag{2.151}$$

The corresponding eigenmodes have the polarization

$$\vec{A}'_{o2\pm} \approx b \left(\begin{array}{c} 1 \\ \dfrac{ i\left[n_e^2 - \left(\alpha^2 + \bar{\varepsilon} - 2\alpha\sqrt{\bar{\varepsilon}} - \dfrac{\delta^2}{4\alpha\sqrt{\bar{\varepsilon}}} \right) - \alpha^2 \right] }{ 2\alpha\left[\pm\left(\alpha - \sqrt{\bar{\varepsilon}} - \dfrac{\delta^2}{8\alpha\left(\alpha - \sqrt{\bar{\varepsilon}}\right)\sqrt{\bar{\varepsilon}}} \right) \right] } \end{array} \right) \approx b \left(\begin{array}{c} 1 \\ \dfrac{-2i\left[\alpha^2 - \alpha\sqrt{\bar{\varepsilon}}\right]}{2\alpha\left[\pm\left(\alpha - \sqrt{\bar{\varepsilon}}\right)\right]} \end{array} \right) = \frac{1}{\sqrt{2}} \begin{pmatrix} 1 \\ \mp i \end{pmatrix}.$$

In the lab frame, the polarization is

$$A_{2\pm x} = \left[\cos(qz) \pm i\sin(qz) \right] e^{-ik_o n_{2\pm} z} = e^{i(\pm q - k_o n_{2\pm})z},$$

$$A_{2\pm y} = \left[\sin(qz) \mp i\cos(qz) \right] e^{-ik_o n_{2\pm} z} = (\mp i) e^{i(\pm q - k_o n_{2\pm})z}.$$

Because $\pm q - k_o n_{2\pm} = \pm k_o \alpha \mp k_o \left[\alpha - \sqrt{\bar{\varepsilon}} - \frac{\delta^2}{8\alpha\left(\alpha - \sqrt{\bar{\varepsilon}}\right)\sqrt{\bar{\varepsilon}}} \right] = \pm k_o \left[\sqrt{\bar{\varepsilon}} + \frac{\delta^2}{8\alpha\left(\alpha - \sqrt{\bar{\varepsilon}}\right)\sqrt{\bar{\varepsilon}}} \right]$, we have

$$\vec{A}_{2\pm} = \frac{1}{\sqrt{2}} \begin{pmatrix} 1 \\ \mp i \end{pmatrix} e^{-i(\mp k_o)\left[\sqrt{\bar{\varepsilon}} + \frac{\delta^2}{8\alpha\left(\alpha - \sqrt{\bar{\varepsilon}}\right)\sqrt{\bar{\varepsilon}}} \right] z}. \tag{2.152}$$

Eigenmode 3 is right-handed circularly polarized and propagates in the $-z$ direction with the refractive index

$$n_2 = \sqrt{\bar{\varepsilon}} + \delta^2 / \left[8\alpha\left(\alpha - \sqrt{\bar{\varepsilon}}\right)\sqrt{\bar{\varepsilon}} \right]. \tag{2.153}$$

Eigenmode 4 is also right-handed circularly polarized but propagates in the $+z$ direction with the same refractive index. In the above calculation, the higher-order terms $\delta^2 / \left[8\alpha\left(\alpha + \sqrt{\bar{\varepsilon}}\right)\sqrt{\bar{\varepsilon}} \right]$ and $\delta^2 / \left[8\alpha\left(\alpha - \sqrt{\bar{\varepsilon}}\right)\sqrt{\bar{\varepsilon}} \right]$ are kept because they are important in calculating the optical rotatary power of the cholesteric liquid crystal.

2.4.1.3 $\sqrt{\bar{\varepsilon}}P \sim \lambda$ and $\delta/\bar{\varepsilon} \ll 1$

In this case $\alpha = q/k_o = \lambda/P \sim \sqrt{\bar{\varepsilon}}$. From Equation (2.129) we have

$$n_{1\pm} = \pm\left[\bar{\varepsilon} + \bar{\varepsilon} + 2\bar{\varepsilon}\right]^{1/2} = \pm 2\sqrt{\bar{\varepsilon}}. \tag{2.154}$$

The corresponding eigenmodes have the polarization

$$\vec{A}'_{o1\pm} = a\begin{pmatrix} 1 \\ i\left[n_e^2 - 4\bar{\varepsilon} - \bar{\varepsilon}\right]/2\bar{\varepsilon}(\pm 2\bar{\varepsilon}) \end{pmatrix} = \frac{1}{\sqrt{2}}\begin{pmatrix} 1 \\ \mp i \end{pmatrix}. \tag{2.155}$$

In the lab frame, the polarization is

$$A_{1\pm x} = \left[\cos(qz) \pm i\sin(qz)\right]e^{-ik_o n_{1\pm} z} = e^{i(\pm q - k_o n_{1\pm})z},$$
$$A_{1\pm y} = \left[\sin(qz) \mp i\cos(qz)\right]e^{-ik_o n_{1\pm} z} = (\mp i)e^{i(\pm q - k_o n_{1\pm})z}.$$

Because $\pm q - k_o n_{1\pm} = \pm k_o\sqrt{\bar{\varepsilon}} \mp k_o 2\sqrt{\bar{\varepsilon}} = \mp k_o\sqrt{\bar{\varepsilon}}$, we have

$$\vec{A}_{1\pm} = \frac{1}{\sqrt{2}}\begin{pmatrix} 1 \\ \mp i \end{pmatrix} e^{-i(\pm k_o)\sqrt{\bar{\varepsilon}} z}. \tag{2.156}$$

Eigenmode 1 is left-handed circularly polarized and propagates in the + z direction with the speed of $c/\sqrt{\bar{\varepsilon}}$. Eigenmode 2 is also left-handed circularly polarized but propagates in the $-z$ direction with the speed of $c/\sqrt{\bar{\varepsilon}}$. The instantaneous electric field pattern is of opposite sense to the cholesteric helix which is right-handed.

From Equation (2.130) we have

$$n_{2\pm} = \pm\left[\bar{\varepsilon} + \bar{\varepsilon} - \left(4\bar{\varepsilon}^2 + \delta^2\right)^{1/2}\right]^{1/2} = \pm i\frac{\delta}{2\sqrt{\bar{\varepsilon}}}, \tag{2.157}$$

which is imaginary. The corresponding polarization is

$$\vec{A}'_{o2\pm} \approx a\begin{pmatrix} 1 \\ i\left[n_e^2 + \dfrac{\delta^2}{4\bar{\varepsilon}} - \bar{\varepsilon}\right]/[2\sqrt{\bar{\varepsilon}}(\pm i\delta/2\sqrt{\bar{\varepsilon}}) \end{pmatrix} = a\begin{pmatrix} 1 \\ \pm\left[1 + \dfrac{\delta}{4\bar{\varepsilon}}\right] \end{pmatrix} \approx \frac{1}{\sqrt{2}}\begin{pmatrix} 1 \\ \pm 1 \end{pmatrix}, \tag{2.158}$$

which is linear polarization making ±45° with the x' axis. In the lab frame, the polarization is

$$A_{2\pm x} = \frac{1}{\sqrt{2}}[\cos(qz) \mp \sin(qz)]e^{-ik_o n_{2\pm} z} = [\cos(qz) \mp \sin(qz)]e^{\pm k_o \delta z/(2\sqrt{\bar{\varepsilon}})},$$
$$A_{2\pm y} = \frac{1}{\sqrt{2}}[\sin(qz) \pm \cos(qz)]e^{-ik_o n_{2\pm} z} = [\sin(qz) \pm \cos(qz)]e^{\pm k_o \delta z/(2\sqrt{\bar{\varepsilon}})}, \tag{2.159}$$
$$\vec{A}_{2\pm} = \begin{pmatrix} \cos(qz \pm \pi/4) \\ \sin(qz \pm \pi/4) \end{pmatrix} e^{\pm k_o \delta z/(2\sqrt{\bar{\varepsilon}})}.$$

Because the refractive index is imaginary, these eigenmodes are non-propagating waves. The instantaneous electric field pattern of these eigenmodes varies in space in the same way as the cholesteric helix. The light intensity decays as these eigenmodes propagate into the liquid crystal. This means that the cholesteric liquid crystal (ChLC) reflects circularly polarized light with the same handedness and the same periodicity. The reflection band can be calculated from the equation $\alpha^2+\bar{\varepsilon}-\left(4\alpha^2\bar{\varepsilon}+\delta^2\right)^{1/2}=0$, which gives

$$\lambda_1=\sqrt{\varepsilon_{//}}P=n_eP, \tag{2.160}$$

$$\lambda_2=\sqrt{\varepsilon_{\perp}}P=n_oP. \tag{2.161}$$

When $n_oP<\lambda<n_eP$, the refractive index is imaginary. The width of this region is

$$\Delta\lambda=\lambda_1-\lambda_2=(n_e-n_o)P=\Delta nP. \tag{2.162}$$

At λ_2, $\alpha=n_o$, $n_2=0$, the polarization of the eigenmodes is $\vec{A}_2^T=(0,\ 1)$, (i.e. linearly polarized perpendicular to the liquid crystal director). At λ_1, $\alpha=n_e$, $n_2=0$, the polarization of the eigenmodes is $\vec{A}_2^T=(1,0)$, (i.e. linearly polarized parallel to the liquid crystal director). When the wavelength changes from λ_1 to λ_2, the angle χ between the electric vector and the liquid crystal director changes from 0° to 90°. For the light having the wavelength in the region from λ_2 to λ_1 (in vacuum), χ varies in such a way that the wavelength of the light inside the ChLC is equal to the helical pitch.

2.4.2 *Reflection of cholesteric liquid crystals*

Now we consider the reflection of the cholesteric liquid crystal [10]. In the wavelength region from $\lambda_2(=n_oP)$ to $\lambda_1(=n_eP)$, for light which is circularly polarized with the same helical sense as the helix of the liquid crystal, the angle between the electric vector of the light and the liquid crystal director is fixed because the light propagates along the helical axis. Light reflected from different positions is always in phase, and interferes constructively and results in strong reflection. We analytically calculate the reflection from a cholesteric (Ch) film in a simple case in which media below and above the Ch film are isotropic and have the refractive index $\bar{n}=(n_e+n_o)/2$, as shown in Figure 2.8. The incident light is right-handed circularly polarized and has the field amplitude u. The electric field is

$$\vec{E}_i=\frac{u}{\sqrt{2}}\begin{pmatrix}1\\ i\end{pmatrix}e^{-ik_o\bar{n}z}. \tag{2.163}$$

The reflected light is also right-handed circularly polarized and has the amplitude r. The fields is

$$\vec{E}_r=\frac{r}{\sqrt{2}}\begin{pmatrix}1\\ -i\end{pmatrix}e^{ik_o\bar{n}z}. \tag{2.164}$$

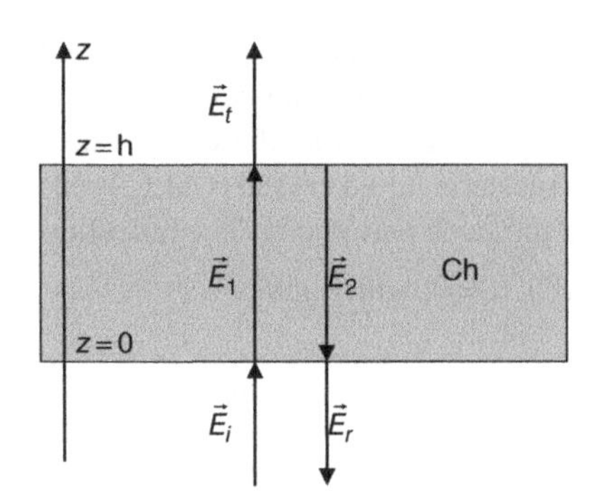

Figure 2.8 Schematic diagram showing the reflection of the Ch film.

Above the Ch film, there is only light propagating in the $+z$ direction, which is the transmitted light that is right-handed circularly polarized and has the field

$$\vec{E}_t = \frac{t}{\sqrt{2}}\begin{pmatrix}1\\ i\end{pmatrix}e^{-ik_o\bar{n}z}, \tag{2.165}$$

where t is the amplitude of the wave. Outside the Ch film, the magnetic field is related to the electric field by

$$\vec{H} = \frac{1}{i\omega\mu_o}\nabla\times\vec{E} = \frac{-k_o\bar{n}}{i\omega\mu_o}\hat{z}\times\vec{E}.$$

Generally speaking, there are four eigenmodes inside the Ch film. Two of the eigenmodes (eigenmodes 1 and 2) are left-handed circularly polarized (one mode propagating in the $+z$ direction and the other mode propagating in the $-z$ direction). Then two other eigenmodes (eigenmodes 3 and 4) are linearly polarized (one mode propagating in the $+z$ direction and the other mode propagating in the $-z$ direction). Inside the Ch film, the amplitudes of eigenmodes 1 and 2 are much smaller than those of eigenmodes 3 and 4. As an approximation, we neglect eigenmodes 1 and 2. From Equation (2.132) we know that the electric field in the Ch film is

$$\vec{E}_{ch} = \frac{v_1}{\sqrt{2}}\begin{pmatrix}1\\ w\end{pmatrix}e^{-ik_on_2z} + \frac{v_2}{\sqrt{2}}\begin{pmatrix}1\\ -w\end{pmatrix}e^{+ik_on_2z}, \tag{2.166}$$

where

$$w = i\left(n_e^2 - n_2^2 - \alpha^2\right)/2\alpha n_2, \tag{2.167}$$

and v_1 and v_2 are electric field amplitudes of eigenmodes 3 and 4. Inside the Ch film, the magnetic field is related to the electric field by

$$\vec{H} = \frac{1}{i\omega\mu_o}\nabla\times\vec{E} = \frac{-k_on}{i\omega\mu_o}\hat{z}\times\vec{E}.$$

The refractive index n depends on the eigenmode. The eigenmodes with left-handed circular polarization are n_1 and the eigenmodes with linear polarization are n_2. From Equations (2.127) and (2.128), we know that within and near the reflection band $|n_2| \ll n_1$.

The relations between u, r, v_1, v_2, and t can be found by using the boundary conditions at the surface of the Ch film. At the interfaces, the tangential components of the electric and magnetic fields are continuous. Although the amplitudes of eigenmodes 1 and 2 are small, their amplitudes of magnetic field are not small, compared with those of eigenmodes 3 and 4, and will make the boundary conditions of the magnetic field satisfied. As an approximation, it is only necessary to consider the boundary conditions of the electric field of eigenmodes 3 and 4. We consider a Ch film with $h/P =$ integer. The boundary conditions at $z = h$ are

$$t = v_1 e^{-ik_o n_2 h} + v_2 e^{ik_o n_2 h}, \tag{2.168}$$

$$it = v_1 w e^{-ik_o n_2 h} - v_2 w e^{ik_o n_2 h}. \tag{2.169}$$

Note that the local frame is the same as the lab frame at the bottom and top surface of the Ch film because the film has m pitch. From these two equations we can get

$$v_1 = \frac{t}{2}(w + i) e^{ik_o n_2 h}, \tag{2.170}$$

$$v_2 = \frac{t}{2}(w - i) e^{-ik_o n_2 h}. \tag{2.171}$$

The boundary conditions at $z = 0$ are

$$u + r = v_1 + v_2, \tag{2.172}$$

$$u - r = -iwv_1 + iwv_2. \tag{2.173}$$

From these two equations we get

$$u = \frac{1}{2}(1 - iw)v_1 + \frac{1}{2}(1 + iw)v_2, \tag{2.174}$$

$$r = \frac{1}{2}(1 + iw)v_1 + \frac{1}{2}(1 - iw)v_2. \tag{2.175}$$

The reflectance is given by [11]

$$R = \left|\frac{r}{u}\right|^2 = \left|\frac{(1 + wi) + (1 - wi)(v_2/v_1)}{(1 - wi) + (1 + wi)(v_2/v_1)}\right|^2 = \left|\frac{(w^2 + 1)(1 - e^{-i2k_o n_2 h})}{2w(1 + e^{-i2k_o n_2 h}) - i(w^2 - 1)(1 - e^{-i2k_o n_2 h})}\right|^2. \tag{2.176}$$

The calculated reflection spectra of Ch films with a few film thicknesses are shown in Figure 2.9 [12,13]. For a sufficiently thick Ch film, within the reflection band, $e^{-i2k_o n_2 h} \approx 0$, $R \approx \left|\frac{(w^2 + 1)}{2w - i(w^2 - 1)}\right|^2 = 1$.

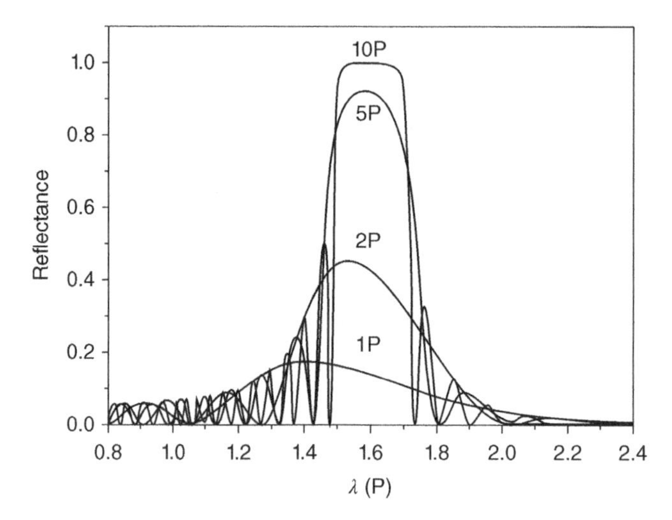

Figure 2.9 Reflection spectra of Ch films with various film thicknesses. The refractive indices used are 1.7 and 1.5. Reproduced with permission from the American Institute of Physics.

The thickness-dependence of the reflectance can be estimated in the following way. At the center of the reflection band, $\lambda = \sqrt{\bar{\varepsilon}}P$, $w = 1$, and $n_2 = i\delta/\left(2\sqrt{\bar{\varepsilon}}\right) \approx i\Delta n/2$. The reflectance is given by

$$R = \left(\frac{\exp(2\Delta n\pi h/\bar{n}P) - 1}{\exp(2\Delta n\pi h/\bar{n}P) + 1}\right)^2. \tag{2.177}$$

2.4.3 *Lasing in cholesteric liquid crystals*

Cholesteric liquid crystals are periodic optical media. When doped with fluorescent dyes, they can be used to make cavity-free lasers [14,15]. In lasers, one of the important properties is spontaneous emission rate W, which is proportional to the density of states ρ, as pointed out by Purcell [16]. The density of states function is given by

$$\rho = \frac{dk}{d\omega} = \frac{d(nk_o)}{d(2\pi C/\lambda)} = \frac{d(2\pi n/\lambda)}{d(2\pi C/\lambda)} = \frac{1}{C}\left(n - \lambda\frac{dn}{d\lambda}\right). \tag{2.178}$$

For the eigenmodes corresponding to the eigen refractive index $n_2 = \left\{(\lambda/P)^2 + \bar{\varepsilon} - \left[4(\lambda/P)^2\bar{\varepsilon} + \delta^2\right]^{1/2}\right\}^{1/2}$, the density of states function has the term $\frac{dn_2}{d\lambda} = \frac{1}{2n_2}\frac{dg}{d\lambda}$, where $g = (\lambda/P)^2 + \bar{\varepsilon} - \left[4(\lambda/P)^2\bar{\varepsilon} + \delta^2\right]^{1/2}$. At the edges of the reflection band of the cholesteric liquid crystal, the density of states is very large because $n_2 = 0$. Therefore lasing can occur at the edges of the reflection band.

Homework Problems

2.1 Calculate and draw the end point of the electric field vector at a fixed position in space as a function time for all the polarization states listed in Table 2.1.

2.2 Consider a homogeneously aligned nematic film with the thickness h shown in Figure 2.10. The ordinary and extraordinary refractive indices of the liquid crystal are n_o and n_e. Light with wavelength λ is incident on the film at the angle θ. The refractive angles of the ordinary and extraordinary rays are θ_e and θ_o, respectively. Prove that the phase difference between the extraordinary ray and ordinary ray when they come out of the film is $\Gamma = \frac{2\pi h}{\lambda}\left(n_{eff}\cos\theta_e - n_o\cos\theta_o\right) = h(k_{ez} - k_{oz})$, where $\sin\theta = n_o\sin\theta_o$, $\sin\theta = n_{eff}\sin\theta_e = n_o n_e \sin\theta_e / \left(n_o^2\cos^2\theta_e + n_e^2\sin^2\theta_e\right)^{1/2}$, k_{ez} and k_{oz} are the projections in the film normal direction of the wavevectors of the extraordinary and ordinary rays, respectively.

2.3 What is the transmittance of a stack of three ideal polarizers? The angle between the transmission axes of the first and third polarizers is 90°. The transmission axis of the second polarizer is 45° with respect the transmission axes of the other polarizers.

2.4 Linearly polarized light is normally incident on a uniformly aligned nematic liquid crystal cell that has a pretilt angle of 45°. The refractive indices of the liquid crystal are $n_o = 1.5$ and $n_e = 1.7$. If the polarization is in the plane defined by the director and the wavevector, determine the angle that the Poynting vector makes with the wavevector.

2.5 A wedge cell is filled with a homogeneously aligned nematic liquid crystal whose director is aligned along the wedge direction. The angle of the wedge is 3°. The wedge is sandwiched between two crossed polarizers with the entrance polarizer placed at 45° to the director. When the cell is illuminated at normal incidence with light at the wavelength of 620 nm and viewed in transmission with a microscope, dark fringes are observed at intervals of 100 microns along the wedge. What is the birefringence of the liquid crystal.

2.6 *Crossed polarizer with compensation films.* Consider two crossed O-type polarizers. A uniaxial a plate and a uniaxial c plate are sandwiched between the two polarizers. The a plate has its optical axis parallel to the transmission axis of the first polarizer and has the retardation of $\lambda/4$ [$(\Delta nd)_a = \lambda/4$]. The c plate has the retardation of $2\lambda/9$ [$(\Delta nd)_c = 2\lambda/9$]. Calculate the transmission of the system at the azimuthal angle of 45° as a function of the polar angle θ.

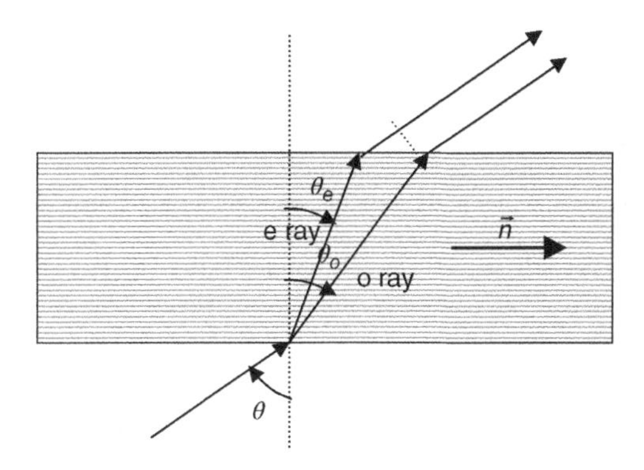

Figure 2.10 Figure for Problem 2.2.

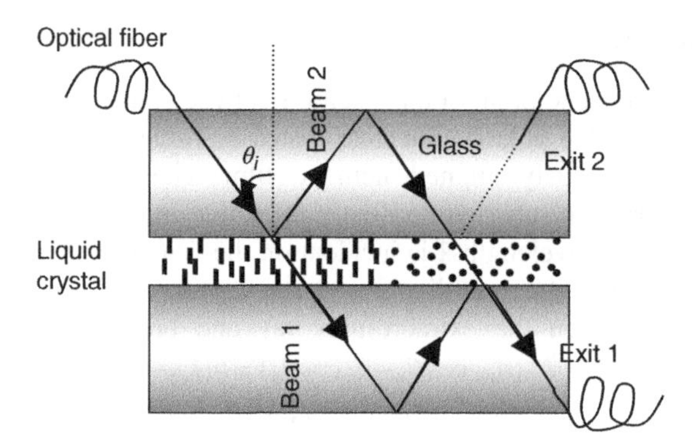

Figure 2.11 Figure for Problem 2.7.

2.7 A liquid crystal optical switch, based on total internal reflection, is shown in Figure 2.11. It consists of two thick glass plates with a thin layer of liquid crystal sandwiched between them. On the left side, the liquid crystal is aligned homeotropically by a homeotropic alignment layer and serves as a beam split. On the right side, the liquid crystal is aligned homogeneously (in the direction perpendicular to the paper plane) by a homogeneous alignment layer, and serves as the switch. The refractive index of the glass is 1.7. The refractive indices of the liquid crystal are $n_o = 1.5$ and $n_e = 1.7$. The incident light is unpolarized. When the liquid crystal on the right side is in the field-off state, light is switched to Exit 1. When the liquid crystal on the right side is switched into the homeotropic state by an external field applied across the cell, light is switched to Exit 2. What are the polarization states of Beam 1 and 2? Explain how the switch works.

References

1. J. D. Jackson, *Classic electrodynamics* (John Wiley and Sons, third edn 1998).
2. M. Born and E. Wolf, *Principle of optics: electromagnetic theory of propagation, interference and diffraction of light* (Cambridge University Press, 7th edn 1999).
3. R. D. Guenther, *Modern optics* (John Wiley and Sons, New York, 1990).
4. P. Yeh and C. Gu, *Optics of liquid crystal displays* (John Wiley and Sons, New York, 1999).
5. A. Yariv and P. Yeh, *Optical waves in crystals* (John Wiley and Sons, New York, 1984).
6. J. Chen, K-H. Kim, J.-J. Jyu, et al., Optimum film compensation modes for TN and VA LCDs, *SID Symposium Digest of Tech. Papers*, **29**, 315 (1998).
7. E. B. Priestley, P. J. Wojtoicz, and P. Sheng, *Introduction to liquid crystals* (Plenum, New York 1979).
8. P. Palffy-Mohuray's lecture note.
9. C. Mauguin, Sur les cristaux liquids de Lehman, *Bull. Soc. Franc. Mineral,* **34**, 71–117 (1911).
10. W. Cao, Fluorescence and lasing in liquid crystalline photonic bandgap materials, dissertation, (Kent State University, 2005).
11. S. Chandrasekhar *Liquid crystals*, second edn (Cambridge University Press, New York, 1997).
12. S.-T. Wu and D.-K. Yang, *Reflective liquid crystal displays* (John Wiley & Sons, Ltd., 2001).
13. M. Xu, F.D. Xu and D.-K. Yang, Effects of cell structure on the reflection of cholesteric liquid crystal display, *J. Appl. Phys.*, **83**, 1938 (1998).
14. V. I. Kopp, B. Fan, H. K. M. Vithana, and A. Z. Genack, *Opt. Lett.*, **23**, 1707–1709 (1998).
15. A. Muñoz, P. Palffy-Muhoray, and B. Taheri, *Opt. Lett.*, **26**, 804 (2001).
16. E. M. Purcell, *Phys. Rev.*, **69**, 181 (1946).

3

Optical Modeling Methods

For many liquid crystal devices, their optical properties cannot be calculated analytically because their refractive indices vary in space. In this chapter we will discuss methods which can be used to numerically calculate the optical properties of liquid crystal devices.

3.1 Jones Matrix Method

3.1.1 Jones vector

For a light beam with frequency ω propagating in a uniform medium, the electric field vector is sufficient to specify the beam. In this chapter, the coordinate frame is always chosen in such a way that the propagation direction is the z direction except when otherwise specified. In this section, we only consider the case where the light propagation direction is parallel to the normal direction of the optical film, i.e. normal incident light. As discussed in Chapter 2, light is a transverse wave. If the medium under consideration is isotropic, the electric field vector lies in the xy plane. If the medium is uniaxial and its optic axis is in the xy plane or parallel to the z axis, the electric field vector is also in the xy plane. In these cases, the only quantities needed to specify a light beam are its electric field components, E_x and E_y, in the x and y directions. Thus the wave can be represented by the Jones vector defined by [1,2]

$$\vec{E} = \begin{pmatrix} E_x \\ E_y \end{pmatrix}. \tag{3.1}$$

If we are interested only in the polarization state of the wave, it is convenient to use the normalized *Jones vector* which satisfies the condition

Fundamentals of Liquid Crystal Devices, Second Edition. Deng-Ke Yang and Shin-Tson Wu.

$$\vec{E}^{*} \cdot \vec{E} = 1, \tag{3.2}$$

where $\vec{E}^{*}$ is the complex conjugate. For light linearly polarized along a direction making an angle ϕ with respect to the x axis, the Jones vector is

$$\vec{L}(\phi) = \begin{pmatrix} \cos\phi \\ \sin\phi \end{pmatrix}. \tag{3.3}$$

Jones vectors for right- and left-handed circularly polarized light are

$$\vec{C}_R = \frac{1}{\sqrt{2}} \begin{pmatrix} 1 \\ i \end{pmatrix}, \tag{3.4}$$

$$\vec{C}_L = \frac{1}{\sqrt{2}} \begin{pmatrix} 1 \\ -i \end{pmatrix}, \tag{3.5}$$

respectively. The Jones vector of various polarization states is listed in Table 2.1.

3.1.2 Jones matrix

In the Jones representation, the effect of an optical element can be represented by a 2×2 matrix known as the Jones matrix. We first consider the Jones matrix of a uniaxial birefringent film with the ordinary and extraordinary refractive indices n_o and n_e, respectively. A uniformly aligned nematic is one such example. As discussed in Chapter 2, in the uniaxial birefringence film there are two eigenmodes whose eigen electric field vectors do not change in space. If the optic axis (the uniaxis, also called the *c axis*) is along the x axis of the lab frame, one of the eigenmodes has the eigenvector along the x axis and propagates at speed c/n_e; the other eigenmode has the eigenvector along the y axis and propagates at the speed c/n_o. If $n_e > n_o$, the x axis is called the slow axis and the y axis is called the fast axis. If the incident light on the film has the Jones vector $\vec{E}_i^T = (E_{xi}, E_{yi})$, when the component of the electric vector along the x axis propagates through the film, its amplitude remains as E_{xi} and its phase changes according to $\exp(-2\pi n_e z/\lambda)$. When the component of the electric vector along the y axis propagates through the film, its amplitude remains as E_{yi} and its phase changes according to $\exp(-2\pi n_o z/\lambda)$. Therefore the Jones vector $\vec{E}_o^T = (E_{xo}, E_{yo})$ of outgoing light will be

$$\vec{E}_o = \begin{pmatrix} E_{xo} \\ E_{yo} \end{pmatrix} = e^{-i[\pi(n_e+n_o)h/\lambda]} \begin{pmatrix} e^{-i\Gamma/2} & 0 \\ 0 & e^{i\Gamma/2} \end{pmatrix} \begin{pmatrix} E_{xi} \\ E_{yi} \end{pmatrix}, \tag{3.6}$$

where Γ is the phase retardation and is given by $\Gamma = 2\pi(n_e - n_o)h/\lambda$, where h is the thickness of the film, λ is the wavelength of the light in vacuum. Uniform birefringent films are also called retardation films or wave plates. If $\Gamma = \pi/2$, the film is called a quarter-wave plate. If $\Gamma = \pi$, the film is called a half-wave plate. The phase factor $e^{-i[\pi(n_e+n_o)h/\lambda]}$ can be neglected when the absolute phase is not important. Defining the Jones matrix of retardation

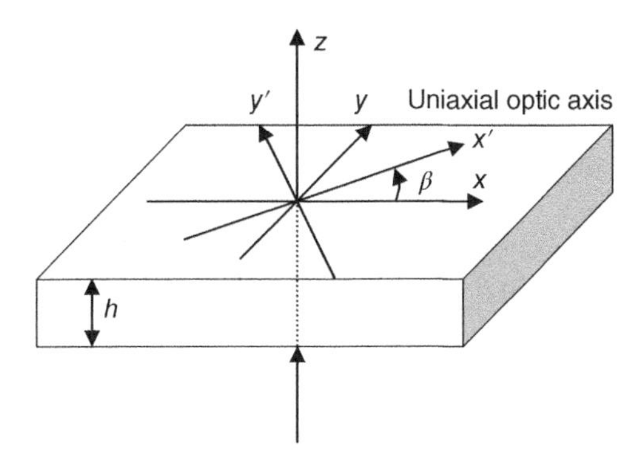

Figure 3.1 Schematic diagram showing a light beam propagating through a uniaxial birefringent film.

$$\overleftrightarrow{G}(\Gamma) = \begin{pmatrix} e^{-i\Gamma/2} & 0 \\ 0 & e^{i\Gamma/2} \end{pmatrix}, \tag{3.7}$$

Equation (3.6) becomes

$$\vec{E}_o = \overleftrightarrow{G} \cdot \vec{E}_i. \tag{3.8}$$

If the optic axis makes an angle β with the x axis of the lab frame, as shown in Figure 3.1, the eigenvectors are no longer in the x and y directions. For incident light not polarized along the optic axis, its polarization will vary in space when propagating through the film. In this case, we have to use the principal frame $x'y'$ whose x' is parallel to the optic axis of the film. In the principal frame, Equation (3.6) is valid. The Jones vector $\vec{E}'_i$ of the incident light in the principal frame is related to the Jones vector $\vec{E}_i$ in the lab frame by

$$\vec{E}'_i = \begin{pmatrix} E'_{xi} \\ E'_{yi} \end{pmatrix} = \begin{pmatrix} \cos\beta & \sin\beta \\ -\sin\beta & \cos\beta \end{pmatrix} \begin{pmatrix} E_x \\ E_y \end{pmatrix} = \begin{pmatrix} \cos\beta & \sin\beta \\ -\sin\beta & \cos\beta \end{pmatrix} \vec{E}_i. \tag{3.9}$$

Define the matrix for the rotation of the frame

$$\overleftrightarrow{R}(\beta) = \begin{pmatrix} \cos\beta & -\sin\beta \\ \sin\beta & \cos\beta \end{pmatrix}. \tag{3.10}$$

Note $\overleftrightarrow{R}^{-1}(\beta) = \overleftrightarrow{R}(-\beta)$. Then

$$\vec{E}'_i = \overleftrightarrow{R}^{-1}(\beta) \cdot \vec{E}_i. \tag{3.11}$$

In the principal frame, the amplitudes of the electric fields in the x' and y' directions do not change with position z, but they propagate with different speeds given by c/n_e and c/n_o, respectively, and thus their phase delays are different. The outgoing light in the principal frame is given by

$$\vec{E}'_o = \begin{pmatrix} E'_{xo} \\ E'_{yo} \end{pmatrix} = \begin{pmatrix} e^{-i\Gamma/2} & 0 \\ 0 & e^{i\Gamma/2} \end{pmatrix} \begin{pmatrix} E'_{xi} \\ E'_{yi} \end{pmatrix} = \overset{\leftrightarrow}{G}(\Gamma)\cdot \vec{E}'_i. \tag{3.12}$$

The Jones vector of the outgoing light in the lab frame is

$$\vec{E}_o = \begin{pmatrix} E_{xo} \\ E_{yo} \end{pmatrix} = \begin{pmatrix} \cos\beta & -\sin\beta \\ \sin\beta & \cos\beta \end{pmatrix} \begin{pmatrix} E'_{xo} \\ E'_{yo} \end{pmatrix} = \overset{\leftrightarrow}{R}(\beta)\cdot \overset{\leftrightarrow}{G}(\Gamma)\cdot \overset{\leftrightarrow}{R}^{-1}(\beta)\vec{E}_i. \tag{3.13}$$

From Equations (3.10) and (3.12) we have

$$\vec{E}_o = \begin{pmatrix} \cos^2\beta e^{-i\Gamma/2} + \sin^2\beta e^{i\Gamma/2} & \sin\beta\cos\beta\left(e^{-i\Gamma/2} - e^{i\Gamma/2}\right) \\ \sin\beta\cos\beta\left(e^{-i\Gamma/2} - e^{i\Gamma/2}\right) & \sin^2\beta e^{-i\Gamma/2} + \cos^2\beta e^{i\Gamma/2} \end{pmatrix} \vec{E}_i. \tag{3.14}$$

The polarization of a light beam can be changed into any other polarization state by using a proper birefringent film. If the incident light is linearly polarized along the x axis and the uniaxial birefringent film is a quarter plate with its slow axis (the uniaxial axis) at 45° with respect to the x axis, $\vec{E}_o^T = (1/\sqrt{2})(1, -i)$, and the outgoing light is left-handed circularly polarized. If the slow axis is at –45° with respect to the x axis, $\vec{E}_o^T = (1/\sqrt{2})(1, i)$, and the outgoing light is right-handed circularly polarized. If the film is a half-wave plate and the slow axis is at 45°, $\vec{E}_o^T = (-i)(0,1)$, and the outgoing light is linearly polarized along the y axis.

If the birefringent film is sandwiched between two polarizers with the transmission axis of the (bottom) polarizer along the x axis, then $\vec{E}_i^T = (1,0)$. The polarization of the outgoing light is

$$\begin{aligned} \vec{E}_o &= \begin{pmatrix} \cos^2\beta e^{-i\Gamma/2} + \sin^2\beta e^{i\Gamma/2} & \sin\beta\cos\beta\left(e^{-i\Gamma/2} - e^{i\Gamma/2}\right) \\ \sin\beta\cos\beta\left(e^{-i\Gamma/2} - e^{i\Gamma/2}\right) & \cos^2\beta e^{-i\Gamma/2} + \sin^2\beta e^{i\Gamma/2} \end{pmatrix} \begin{pmatrix} 1 \\ 0 \end{pmatrix} \\ &= \begin{pmatrix} \cos^2\beta e^{-i\Gamma/2} + \sin^2\beta e^{i\Gamma/2} \\ \sin\beta\cos\beta\left(e^{-i\Gamma/2} - e^{i\Gamma/2}\right) \end{pmatrix}. \end{aligned} \tag{3.15}$$

If the transmission axis of the analyzer (top polarizer) is along the y axis, only the y component of the outgoing light can pass the analyzer, and the transmittance is

$$T = \left|E_{yo}\right|^2 = \left|\sin\beta\cos\beta\left(e^{-i\Gamma/2} - e^{i\Gamma/2}\right)\right|^2 = \sin^2(2\beta)\sin^2(\Gamma/2). \tag{3.16}$$

When $\beta = \pi/4$ and $\Gamma = \pi$, the maximum transmittance $T = 1$ is obtained.

Birefringent films are used as compensation films to improve the viewing angle of liquid crystal displays. They can be divided into three groups according to the orientation of the uniaxis (c axis) with respect to the normal of the film.

c plate: the *c* axis is perpendicular to the film normal
a plate: the *c* axis is parallel to the film normal
o plate: the *c* axis makes an angle α ($\neq 0°$, $\neq 90°$) with respect to the film normal

3.1.3 Jones matrix of non-uniform birefringent film

When light propagates through films in which the slow and fast axes as well as the refractive indices are a function of position z, the Jones matrix method can still be used as an approximation method as long as the the refractive indices do not change much in one wavelength. We divide the film into N slabs as shown in Figure 3.2. When the thickness $\Delta h = h/N$ of the slabs is sufficiently small, then within each slab, the slow axis can be considered fixed.

For layer j, the angle of the slow axis with respect to the x axis is β_j and the phase retardation is $\Gamma_j = 2\pi[n_e(z=j\Delta h) - n_o(z=j\Delta h)]\Delta h/\lambda$. In the lab frame, the Jones vector of the incident light on the layer is $\vec{E}_{ji}$, which is the same as the Jones vector, $\vec{E}_{(j-1)o}$, of the light exiting layer $(j-1)$, and the Jones vector of the outgoing light is $\vec{E}_{jo}$ [3]:

$$\vec{E}_{jo} = \overleftrightarrow{R}(\beta_j)\cdot\overleftrightarrow{G}(\Gamma_j)\cdot\overleftrightarrow{R}^{-1}(\beta_j)\vec{E}_{ji} = \overleftrightarrow{R}(\beta_j)\cdot\overleftrightarrow{G}(\Gamma_j)\cdot\overleftrightarrow{R}^{-1}(\beta_j)\cdot\vec{E}_{(j-1)o} \tag{3.17}$$

The Jones vector, $\vec{E}_o$, of the outgoing light is related to the Jones vector, $\vec{E}_i$ of the incident light by

$$\begin{aligned}\vec{E}_o = &\left[\overleftrightarrow{R}(\beta_N)\cdot\overleftrightarrow{G}(\Gamma_N)\cdot\overleftrightarrow{R}^{-1}(\beta_N)\right]\cdot\\ &\left[\overleftrightarrow{R}(\beta_{N-1})\cdot\overleftrightarrow{G}(\Gamma_{N-1})\cdot\overleftrightarrow{R}^{-1}(\beta_{N-1})\right]\cdot\\ &\ldots\ldots\cdot\\ &\left[\overleftrightarrow{R}(\beta_1)\cdot\overleftrightarrow{G}(\Gamma_1)\cdot\overleftrightarrow{R}^{-1}(\beta_1)\right]\cdot\vec{E}_i\\ = &\prod_{j=1}^{N}\left[\overleftrightarrow{R}(\beta_j)\cdot\overleftrightarrow{G}(\Gamma_j)\cdot\overleftrightarrow{R}^{-1}(\beta_j)\right]\cdot\vec{E}_i.\end{aligned} \tag{3.18}$$

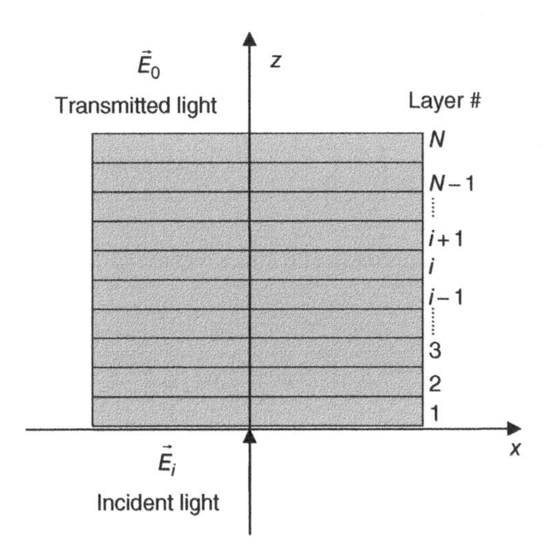

Figure 3.2 Schematic diagram showing the propagation of light through a birefringent film with varying slow axis.

Usually the multiplication of the matrices is carried out numerically. Analytical solutions can be obtained in some special cases.

The Jones matrix method has the limitation that it only works well for normally incident and paraxial rays. It neglects the reflection and refraction from the interface between two optic media whose refractive indices are different. The extended Jones matrix method takes account of the reflection and refraction, but still neglect multiple reflection, and can be used to calculate the optical properties of media for obliquely incident light [4–7].

3.1.4 Optical properties of twisted nematic

Nematic liquid crystals are usually uniaxial. Twisted nematic (TN) liquid crystals have been used in many applications, especially in flat panel displays [8]. A twisted nematic cell consists of two parallel substrates with a nematic liquid crystal sandwiched between them. The inner surfaces of the cell are coated with homogeneous anchoring alignment layers. At the surfaces, the liquid crystal director is aligned along the alignment direction. The alignment directions of the bottom and top alignment layers are different. The angle between the alignment directions is Φ, which is referred to as the total twist angle. The liquid crystal director twists at a constant rate from the bottom to the top to match the boundary condition. The twist rate is $\tau = \Phi/h$, where h is the thickness.

In the calculation of the optical properties, the twisted nematic film is divided into N thin slabs with the thickness $\Delta h = h/N$. Within each slab, the liquid crystal director can be approximately considered uniform. If the alignment direction of the liquid crystal director at the entrance plane is along the x axis, the rotation matrix of the ith layer is

$$\overleftrightarrow{S}_i(\beta_i) = \begin{pmatrix} \cos\beta_i & -\sin\beta_i \\ \sin\beta_i & \cos\beta_i \end{pmatrix} = \begin{pmatrix} \cos[i\Delta\beta] & -\sin[i\Delta\beta] \\ \sin[i\Delta\beta] & \cos[i\Delta\beta] \end{pmatrix} \equiv \vec{S}(i\Delta\beta), \tag{3.19}$$

where $\Delta\beta = (\Delta h/h)\Phi = \Phi/N$. The Jones matrix of the ith layer in the principal frame is

$$\overleftrightarrow{G}_i(\Gamma_i) = \begin{pmatrix} e^{-i\Delta\Gamma/2} & 0 \\ 0 & e^{i\Delta\Gamma/2} \end{pmatrix} \equiv \overleftrightarrow{G}(\Delta\Gamma), \tag{3.20}$$

where $\Delta\Gamma = 2\pi(n_e - n_o)\Delta h/\lambda = 2\pi\Delta n\Delta h/\lambda$. The Jones vector of the outgoing light is related to that of the incident light by

$$\vec{E}_o = \prod_{j=1}^{N}\left[\overleftrightarrow{S}_j(\beta_j)\cdot\overleftrightarrow{G}_j(\Gamma_j)\cdot\overleftrightarrow{S}_j^{-1}(\beta_j)\right]\cdot\vec{E}_i = \prod_{j=1}^{N}\left[\overleftrightarrow{S}(j\Delta\beta)\cdot\overleftrightarrow{G}(\Delta\Gamma)\cdot\overleftrightarrow{S}^{-1}(j\Delta\beta)\right]\cdot\vec{E}_i.$$

Note $\overleftrightarrow{S}_i^{-1}(\beta_i)\overleftrightarrow{S}_{(i-1)}\left(\beta_{(i-1)}\right) = \begin{pmatrix} \cos[\Delta\beta] & \sin[\Delta\beta] \\ -\sin[\Delta\beta] & \cos[\Delta\beta] \end{pmatrix} \equiv \vec{S}(\Delta\beta)$. Therefore

$$\vec{E}_o = \overleftrightarrow{S}(N\Delta\beta)\cdot\left[\overleftrightarrow{G}(\gamma)\cdot\overleftrightarrow{S}^{-1}(\Delta\beta)\right]^N\cdot\vec{E}_i. \tag{3.21}$$

Define a new matrix $\overleftrightarrow{A}$

$$\overleftrightarrow{A} = \overleftrightarrow{G}(\Delta\Gamma)\cdot\overleftrightarrow{S}^{-1}(\Delta\beta) = \begin{pmatrix} e^{-i\Gamma/2}\cos\Delta\beta & -e^{-i\Gamma/2}\sin\Delta\beta \\ e^{i\Gamma/2}\sin\Delta\beta & e^{i\Gamma/2}\cos\Delta\beta \end{pmatrix}. \tag{3.22}$$

Because $\overleftrightarrow{A}$ is a 2×2 matrix, from Cayley–Hamilton theory [9], $\overleftrightarrow{A}^N$ can be expanded as

$$\overleftrightarrow{A}^N = \lambda_1 \overleftrightarrow{I} + \lambda_2 \overleftrightarrow{A}, \tag{3.23}$$

where λ_1 and λ_2 are found from the equations

$$q_1^N = \lambda_1 + \lambda_2 A q_1, \tag{3.24}$$

$$q_2^N = \lambda_1 + \lambda_2 A q_2, \tag{3.25}$$

where q_1 and q_2 are the eigenvalues of $\overleftrightarrow{A}$ and can be calculated from

$$\begin{vmatrix} e^{-i\Gamma/2}\cos\Delta\beta - q & -e^{-i\Gamma/2}\sin\Delta\beta \\ e^{i\Gamma/2}\sin\Delta\beta & e^{i\Gamma/2}\cos\Delta\beta - q \end{vmatrix} = 0,$$

which is

$$1 - 2q\cos(\Delta\beta)\cos(\Delta\Gamma/2) + q^2 = 0. \tag{3.26}$$

Defining angle θ by

$$\cos\theta = \cos(\Delta\beta)\cos(\Delta\Gamma/2), \tag{3.27}$$

Equation (3.26) becomes $(q - \cos\theta)^2 = -\sin^2\theta$, and therefore the solutions are

$$q = \cos\theta \pm i\sin\theta = e^{\pm i\theta}. \tag{3.28}$$

Equations (3.24) and (3.25) become $e^{iN\theta} = \lambda_1 + \lambda_2 e^{i\theta}$ and $e^{-iN\theta} = \lambda_1 + \lambda_2 e^{-i\theta}$, and the solutions are $\lambda_1 = -\sin(N-1)\theta/\sin\theta$ and $\lambda_2 = \sin N\theta/\sin\theta$. From Equation (3.23) we have

$$\overleftrightarrow{A}^N = -\frac{\sin(N-1)\theta}{\sin\theta}\overleftrightarrow{I} + \frac{\sin N\theta}{\sin\theta}\overleftrightarrow{A}$$

$$= \begin{pmatrix} \frac{\sin N\theta}{\sin\theta}\cos\Delta\beta e^{-i\Delta\Gamma/2} - \frac{\sin(N-1)\theta}{\sin\theta} & \frac{\sin N\theta}{\sin\theta}\sin\Delta\beta e^{-i\Gamma/2} \\ -\frac{\sin N\theta}{\sin\theta}\sin\Delta\beta e^{i\Gamma/2} & \frac{\sin N\theta}{\sin\theta}\cos\Delta\beta e^{i\Gamma/2} - \frac{\sin(N-1)\theta}{\sin\theta} \end{pmatrix}. \tag{3.29}$$

We also have

$$\overleftrightarrow{S}(N\Delta\beta)=\begin{pmatrix}\cos[N\Delta\beta] & -\sin[N\Delta\beta]\\ \sin[N\Delta\beta] & \cos[N\Delta\beta]\end{pmatrix}=\begin{pmatrix}\cos\Phi & -\sin\Phi\\ \sin\Phi & \cos\Phi\end{pmatrix}. \tag{3.30}$$

The total phase retardation is

$$\Gamma=N\Delta\Gamma=\frac{2\pi}{\lambda}(n_e-n_o)h. \tag{3.31}$$

When $N\to\infty$, $\Delta\beta\to 0$, and $\Gamma\to 0$, we have $\sin\Delta\beta=\Delta\beta$, $\cos\Delta\beta=1$, and $\sin(\Gamma/2)=\Gamma/2$, $\cos(\Gamma/2)=1$. Also from Equation (3.27) we have the following:

$$\theta=\left[(\Delta\beta)^2+(\Delta\Gamma/2)^2\right]^{1/2} \tag{3.32}$$

$$N\theta=\left[\Phi^2+(\Gamma/2)^2\right]^{1/2}\equiv\Theta \tag{3.33}$$

$$\left(\overleftrightarrow{A}^N\right)_{11}=\frac{\sin N\theta}{\sin\theta}\cos\Delta\beta\left[\cos\left(\frac{\Delta\Gamma}{2}\right)-i\sin\left(\frac{\Delta\Gamma}{2}\right)\right]-\frac{\sin N\theta\cos\theta-\sin\theta\cos N\theta}{\sin\theta}$$

$$=\frac{\sin N\theta}{\sin\theta}\cos\Delta\beta\cos\left(\frac{\Delta\Gamma}{2}\right)-i\frac{\sin N\theta}{\sin\theta}\cos\Delta\beta\sin\left(\frac{\Delta\Gamma}{2}\right)-\frac{\sin N\theta\cos\theta-\sin\theta\cos N\theta}{\sin\theta}$$

$$\approx\frac{\sin N\theta}{\sin\theta}\cos\theta-i\frac{\sin N\theta}{\theta}\cdot 1\cdot\left(\frac{\Delta\Gamma}{2}\right)-\frac{\sin N\theta\cos\theta-\sin\theta\cos N\theta}{\sin\theta}$$

$$=\cos\Theta-i\frac{\Delta\Gamma}{2\theta}\sin\Theta$$

$$=\cos\Theta-i\frac{\Gamma}{2\Theta}\sin\Theta$$

$$\left(\overleftrightarrow{A}^N\right)_{12}=\frac{\sin N\theta}{\sin\theta}\sin\Delta\beta e^{-i\Gamma/2}=\frac{\Delta\beta}{\theta}\sin\Theta=\frac{\Phi}{\Theta}\sin\Theta$$

$$\left(\overleftrightarrow{A}^N\right)_{21}=-\frac{\Phi}{\Theta}\sin\Theta$$

$$\left(\vec{A}^N\right)_{22}=\cos\Theta+i\frac{\Gamma}{2\Theta}\sin\Theta$$

Equation (3.21) becomes

$$\vec{E}_o = \begin{pmatrix} \cos\Phi & -\sin\Phi \\ \sin\Phi & \cos\Phi \end{pmatrix} \begin{pmatrix} \cos\Theta - i\dfrac{(\Gamma/2)}{\Theta}\sin\Theta & \dfrac{\Phi}{\Theta}\sin\Theta \\ -\dfrac{\Phi}{\Theta}\sin\Theta & \cos\Theta + i\dfrac{(\Gamma/2)}{\Theta}\sin\Theta \end{pmatrix} \cdot \vec{E}_i. \quad (3.34)$$

A twisted nematic liquid crystal is the same as a cholesteric liquid crystal. The pitch P is related to the film thickness h and the total twist angle Φ by $P = h/(\Phi/2\pi)$. It should be noted that in the derivation of Equation (3.34), reflection and interference effects have not been considered, which are important when the wavelength is comparable with the pitch.

We consider the optical properties of a twisted nematic liquid crystal film in a few special cases.

3.1.4.1 $\Delta nP \gg \lambda$

The twisting rate is $\tau = \Phi/h$. The retardation angle per unit length is $2\pi\Delta n/\lambda$. When $\Delta nP \gg \lambda$ (the *Mauguin* condition) [10], $\tau/(2\pi\Delta n/\lambda) = \lambda/[\Delta n \cdot h/(\Phi/2\pi)] = \lambda/\Delta nP \ll 1$. In this case, the twisting rate is low. $\Delta\beta/(\Delta\Gamma/2) = (\Phi/N)/(\pi\Delta nh/N\lambda) = (\Phi/\pi)(\lambda/\Delta nh) \ll 1$. Therefore $\Phi/\Theta \ll 1$, $\Gamma/2\Theta \approx 1$, and Equation (3.34) becomes

$$\vec{E}_o = \begin{pmatrix} \cos\Phi & -\sin\Phi \\ \sin\Phi & \cos\Phi \end{pmatrix} \begin{pmatrix} e^{-i\Gamma/2} & 0 \\ 0 & e^{i\Gamma/2} \end{pmatrix} \vec{E}_i = \begin{pmatrix} \cos\Phi e^{-i\Gamma/2} & -\sin\Phi e^{i\Gamma/2} \\ \sin\Phi e^{-i\Gamma/2} & \cos\Phi e^{i\Gamma/2} \end{pmatrix} \vec{E}_i. \quad (3.35)$$

When the incident light is linearly polarized along the liquid crystal director at the entrance plane (the *e-mode*), namely, $\vec{E}_i^T = (1,\ 0)$, then $\vec{E}_o^T = (\cos\Phi,\ \sin\Phi)e^{-i\Gamma/2}$. This indicates that the polarization remains parallel to the liquid crystal director as the light propagates through the TN and the propagating speed is c/n_e. When the incident light is linearly polarized perpendicular to the liquid crystal director at the entrance plane (the *o-mode*), namely, $\vec{E}_i^T = (0,\ 1)$, then $\vec{E}_o^T = (-\sin\Phi,\ \cos\Phi)e^{i\Gamma/2}$. This indicates that the polarization remains perpendicular to the liquid crystal director as the light propagates through the TN, and the propagating speed is c/n_o. This result is the same as that obtained by solving the Maxwell equation in Section 2.4.

3.1.4.2 $\Delta nP \ll \lambda$

The twisting rate is very high (much larger than the retardation angle per unit length $2\pi\Delta n/\lambda$). $\Gamma/\Phi = (2\pi\Delta nh/\lambda)/\Phi = \Delta nP/\lambda = \Delta n/\alpha \ll 1$, where $\alpha = \lambda/P$:

$$\Theta = \left[\Phi^2 + \left(\frac{\Gamma}{2}\right)^2\right]^{1/2} \approx \Phi + \frac{\Delta n^2}{8\alpha^2}\Phi = \Phi + \frac{\Delta n^2}{8\alpha}\left(\frac{2\pi h}{\lambda}\right) \quad (3.36)$$

Equation (3.34) becomes

$$\vec{E}_o = \begin{pmatrix} \cos\Phi & -\sin\Phi \\ \sin\Phi & \cos\Phi \end{pmatrix} \begin{pmatrix} \cos\Theta & \sin\Theta \\ -\sin\Theta & \cos\Theta \end{pmatrix} \vec{E}_i. \quad (3.37)$$

If the incident light is right-handed circularly polarized, namely, $\vec{E}_i = \frac{1}{\sqrt{2}}\begin{pmatrix} 1 \\ i \end{pmatrix}$, the outgoing light is

$$\vec{E}_o = \frac{1}{\sqrt{2}}\begin{pmatrix} \cos\Phi & -\sin\Phi \\ \sin\Phi & \cos\Phi \end{pmatrix}\begin{pmatrix} \cos\Theta & \sin\Theta \\ -\sin\Theta & \cos\Theta \end{pmatrix}\begin{pmatrix} 1 \\ i \end{pmatrix} = \frac{1}{\sqrt{2}}\begin{pmatrix} 1 \\ i \end{pmatrix}\mathrm{e}^{i(\Theta-\Phi)}, \tag{3.38}$$

which is still right-handed circularly polarized. Therefore right-handed circular polarization is an eigenmode. Recall the omitted factor $\mathrm{e}^{-i2\pi\bar{n}h/\lambda}$. The total phase angle is

$$\Theta-\Phi-\frac{2\pi\bar{n}h}{\lambda} = -\frac{2\pi\bar{n}h}{\lambda} + \frac{\Delta n^2}{8\alpha}\left(\frac{2\pi h}{\lambda}\right) \equiv -\frac{2\pi n_1 h}{\lambda}. \tag{3.39}$$

The corresponding refractive index is

$$n_1 = \bar{n} - \frac{\Delta n^2}{8\alpha}. \tag{3.40}$$

This result is the same as Equation (2.141) when $\alpha \ll \bar{n}$.

If the incident light is left-handed circularly polarized, namely, $\vec{E}_i = \frac{1}{\sqrt{2}}\begin{pmatrix} 1 \\ -i \end{pmatrix}$, the outgoing light is

$$\vec{E}_o = \frac{1}{\sqrt{2}}\begin{pmatrix} \cos\Phi & -\sin\Phi \\ \sin\Phi & \cos\Phi \end{pmatrix}\begin{pmatrix} \cos\Theta & \sin\Theta \\ -\sin\Theta & \cos\Theta \end{pmatrix}\begin{pmatrix} 1 \\ -i \end{pmatrix} = \frac{1}{\sqrt{2}}\begin{pmatrix} 1 \\ -i \end{pmatrix}\mathrm{e}^{i(\Phi-\Theta)}, \tag{3.41}$$

which is still left-handed circularly polarized. Therefore left-handed circular polarization is another eigenmode. The total phase angle is

$$\Phi-\Theta-\frac{2\pi\bar{n}h}{\lambda} = -\frac{2\pi\bar{n}h}{\lambda} - \frac{\Delta n^2}{8\alpha}\left(\frac{2\pi h}{\lambda}\right) \equiv -\frac{2\pi n_2 h}{\lambda}. \tag{3.42}$$

The corresponding refractive index is

$$n_2 = \bar{n} + \frac{\Delta n^2}{8\alpha}. \tag{3.43}$$

This result is the same as Equation (2.138) when $\alpha \ll \bar{n}$.

Now we consider a TN display whose geometry is shown in Figure 3.3 [11,12]. The TN liquid crystal is sandwiched between two polarizers. The x axis of the lab frame is chosen parallel to the liquid crystal director at the entrance plane. The angles of the entrance and exit polarizers are α_i and α_o, respectively. The Jones vector of the incident light is $\vec{E}_i^T = (\cos\alpha_i,\ \sin\alpha_i)$. The Jones vector of the existing light is given by

$$\vec{E}_o = \begin{pmatrix} \cos\Phi & -\sin\Phi \\ \sin\Phi & \cos\Phi \end{pmatrix}\begin{pmatrix} \cos\Theta - i\frac{(\Gamma/2)}{\Theta}\sin\Theta & \frac{\Phi}{\Theta}\sin\Theta \\ -\frac{\Phi}{\Theta}\sin\Theta & \cos\Theta + i\frac{(\Gamma/2)}{\Theta}\sin\Theta \end{pmatrix}\begin{pmatrix} \cos\alpha_i \\ \sin\alpha_i \end{pmatrix}.$$

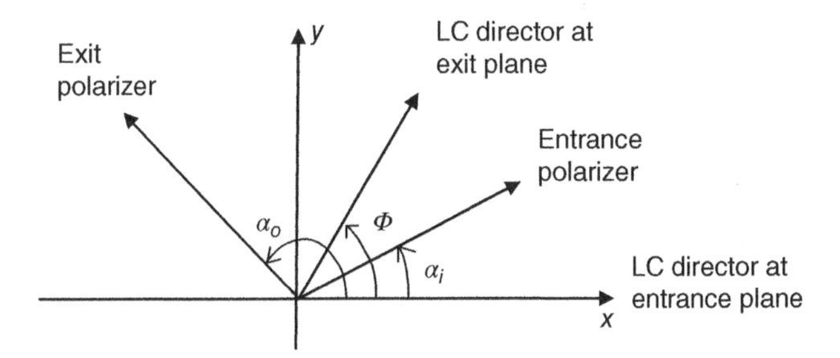

Figure 3.3 Geometry of the TN display.

The exit polarizer is along the direction represented by the unit vector $\vec{P}^T_{exit} = (\cos\alpha_o,\ \sin\alpha_o)$. The electric field of the light which can pass the exit polarizer is given by

$$
\begin{aligned}
E_{exit} &= \vec{P}_{exit} \cdot \vec{E}_o \\
&= (\cos\alpha_o,\ \sin\alpha_o)\begin{pmatrix} \cos\Phi & -\sin\Phi \\ \sin\Phi & \cos\Phi \end{pmatrix}\begin{pmatrix} \cos\Theta - i\dfrac{(\Gamma/2)}{\Theta}\sin\Theta & \dfrac{\Phi}{\Theta}\sin\Theta \\ -\dfrac{\Phi}{\Theta}\sin\Theta & \cos\Theta + i\dfrac{(\Gamma/2)}{\Theta}\sin\Theta \end{pmatrix}\begin{pmatrix} \cos\alpha_i \\ \sin\alpha_i \end{pmatrix} \\
&= \cos\Theta\cos(\alpha_o-\alpha_i-\Phi) - \frac{\Phi}{\Theta}\sin\Theta\sin(\alpha_o-\alpha_i-\Phi) - i\frac{(\Gamma/2)}{\Theta}\sin\Theta\cos(\alpha_o+\alpha_i-\Phi).
\end{aligned}
\tag{3.44}
$$

The intensity of the light is

$$
\begin{aligned}
I_o = |E_{exit}|^2 &= \cos^2(\alpha_o-\alpha_i-\Phi) - \sin^2\Theta\sin[2(\alpha_o-\Phi)]\sin(2\alpha_i) \\
&\quad - \frac{\Phi^2}{\Theta^2}\sin^2\Theta\cos[2(\alpha_o-\Phi)]\cos(2\alpha_i) - \frac{\Phi}{2\Theta}\sin(2\Theta)\sin[2(\alpha_o-\alpha_i-\Phi)].
\end{aligned}
\tag{3.45}
$$

For a normal-black 90° TN display where the transmission axes of the two polarizers are parallel to the liquid crystal director, $\Phi = \pi/2$, $\alpha_i = \alpha_o = 0$, $\Theta = [(\pi/2)^2 + (\Gamma/2)^2]^{1/2} = (\pi/2)[1 + (2\Delta nh/\lambda)^2]^{1/2}$ and the transmittance of the TN cell (normalized to the light intensity before the entrance polarizer) is given by

$$
T = \frac{1}{2}\frac{\Phi^2}{\Theta^2}\sin^2\Theta = \frac{\sin^2\left[\dfrac{\pi}{2}\sqrt{1+(2\Delta nh/\lambda)^2}\right]}{2\left[1+(2\Delta nh/\lambda)^2\right]}
\tag{3.46}
$$

The transmittance T vs. the retardation $u = 2\Delta nh/\lambda$ is shown in Figure 3.4. Generally the transmittance of the display is not 0, except when $\sqrt{1+u^2} = 2i$ $(i = 1,2,3,\ldots)$, namely

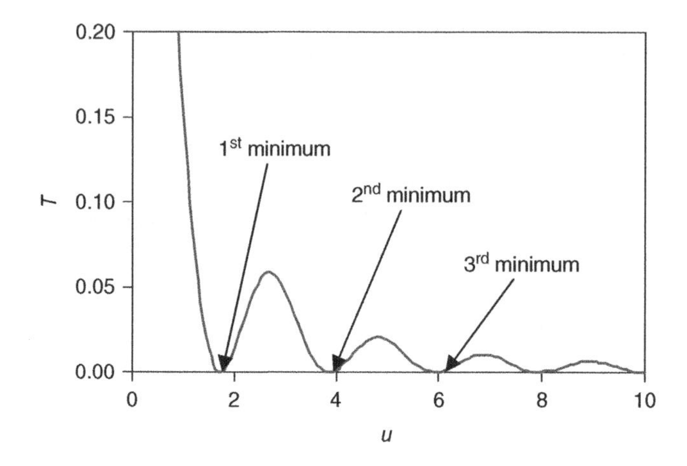

Figure 3.4 Transmittance of normal-black 90° TN vs. retardation.

$u = \sqrt{3}, \sqrt{15}, \ldots$. These values are known as the first, second, etc. minimum conditions, respectively. When $u \gg 1$, the case where the polarization rotates with the liquid crystal director, the denominator becomes very large and the transmittance becomes very small.

3.2 Mueller Matrix Method

3.2.1 *Partially polarized and unpolarized light*

If light is not absolutely monochromatic, the amplitudes and relative phase between the x and y components can both vary with time. As a result, the polarization state of a polychromatic plane wave may be constantly changing. If the polarization state changes more rapidly than the speed of observation, the light is *partially polarized* or *unpolarized* depending on the time-averaged behavior of the polarization state. In optics, we often deal with light with oscillation frequencies of about $10^{14}\ \mathrm{s}^{-1}$, whereas the polarization state may change in a time period of about 10^{-8} s. In order to describe unpolarized and partially polarized light, the *Stokes vector* is introduced.

A quasimonochromatic wave, whose frequency spectrum is confined to a narrow bandwidth $\Delta\lambda$ ($\Delta\lambda \ll \lambda$), can still be described by

$$\vec{E} = \left[A_x(t)\hat{x} + A_y(t)e^{i\delta}\hat{y}\right]e^{i(\omega t - kz)}, \tag{3.47}$$

where the wave is propagating in the z direction; A_x and A_y are positive numbers which may be time-dependent. At a given position, the components of the Stokes vector are defined by [13]

$$S_o = \langle E_x E_x^* + E_y E_y^* \rangle = \langle A_x^2 + A_y^2 \rangle, \tag{3.48}$$

which describes the light intensity,

$$S_1 = \langle E_x E_x^* - E_y E_y^* \rangle = \langle A_x^2 - A_y^2 \rangle, \tag{3.49}$$

which describes the difference in intensity between components along x and y axes;

$$S_2 = \langle E_x E_y{}^* + E_y E_x{}^* \rangle = 2\langle A_x A_y \cos\delta \rangle, \tag{3.50}$$

which describes the component along the direction at ±45°,

$$S_3 = \langle i\left(E_x E_y{}^* - E_y E_x{}^*\right) \rangle = 2\langle A_x A_y \sin\delta \rangle, \tag{3.51}$$

which describes the circular polarization. The $\langle\ \rangle$ denotes the average performed over a time interval τ_D, that is the characteristic time constant of the detection process. S_o specifies the intensity of the light beam. By considering the following cases, the rationale of defining the parameters will be shown.

3.2.1.1 Unpolarized Light

The average amplitudes of the electric field components in the x and y directions are the same, but the phase difference between them is completely random, i.e., $\langle A_x^2 \rangle = \langle A_y^2 \rangle, \langle \cos\delta \rangle = \langle \sin\delta \rangle = 0$, and therefore the normalized Stokes vector is

$$\vec{S}^T = (1,0,0,0). \tag{3.52}$$

3.2.1.2 Linearly polarized light along a direction which makes the angle ψ with respect to the x axis

The amplitudes of the electric field components in the x and y directions are $A_x = \cos\psi$ and $A_y = \sin\psi$, respectively. The phase difference is $\delta = 0$. The Stokes vector is

$$\vec{S}^T = [1, \cos(2\psi), \sin(2\psi), 0]. \tag{3.53}$$

When the light is linearly polarized along the x axis, $\vec{S}^T = (1,1,0,0)$. When the light is linearly polarized along the y axis, $\vec{S}^T = (1,-1,0,0)$. When it is linearly polarized along the direction at 45°, $\vec{S}^T = (1,0,1,0)$. When it is linearly polarized along the direction at –45°, $\vec{S}^T = (1,0,-1,0)$.

3.2.1.3 Circularly polarized

The amplitudes of the electric field components in the x and y directions are the same, i.e., $A_x = A_y = 1/\sqrt{2}$ and the phase difference is δ. The Stokes vector is

$$\vec{S}^T = (1,0,0,\sin\delta). \tag{3.54}$$

For right-handed circular polarization, the phase difference is $\delta = \pi/2$, $\vec{S}^T = (1,0,0,1)$; for left-handed circular polarization, $\delta = -\pi/2$, $\vec{S}^T = (1,0,0,-1)$.

If there were only two parameters S_o and S_1, when $S_1 = 0$, there are three possibilities: (1) unpolarized, (2) linearly polarized along the direction at ±45°, and (3) circularly polarized. Therefore more parameters are needed to differentiate them. If there were only three parameters S_o, S_1, and S_3, when $S_1 = 0$ and $S_3 = 0$, there are two possibilities: (1) unpolarized and (2) linearly polarized along the direction at ±45°. Therefore one more parameter is needed to differentiate them. The four parameters are necessary and also sufficient to describe the polarization of a light beam. The Stokes vectors of various polarizations are listed in Table 2.1.

When a light beam is completely polarized, $\sqrt{S_1^2 + S_2^2 + S_3^2} = S_o = 1$. When a light beam is unpolarized $S_1 = S_2 = S_3 = 0$. The degree of polarization can be described by

$$\gamma = \frac{\sqrt{S_1^2 + S_2^2 + S_3^2}}{S_o}. \tag{3.55}$$

For partially polarized light, $0 < \gamma < 1$.

3.2.2 *Measurement of the Stokes parameters*

The light beam to be studied is incident on a combination of a quarter-wave plate and a polarizer as shown in Figure 3.5. The slow optical axis of the wave plate is along the x axis and the retardation angle is $\beta = 90°$. The transmission axis of the polarizer is at the angle α.

The electric field (in Jones vector form) of the incident light (before the quarter-wave plate) is

$$\vec{E} = \begin{pmatrix} A_x \\ A_y e^{i\delta} \end{pmatrix} \tag{3.56}$$

where A_x and A_y are positive numbers which may be time-dependent. After the quarter-wave plate, the field is

$$\vec{E}_1 = \begin{pmatrix} A_x \\ A_y e^{i(\delta+\beta)} \end{pmatrix}. \tag{3.57}$$

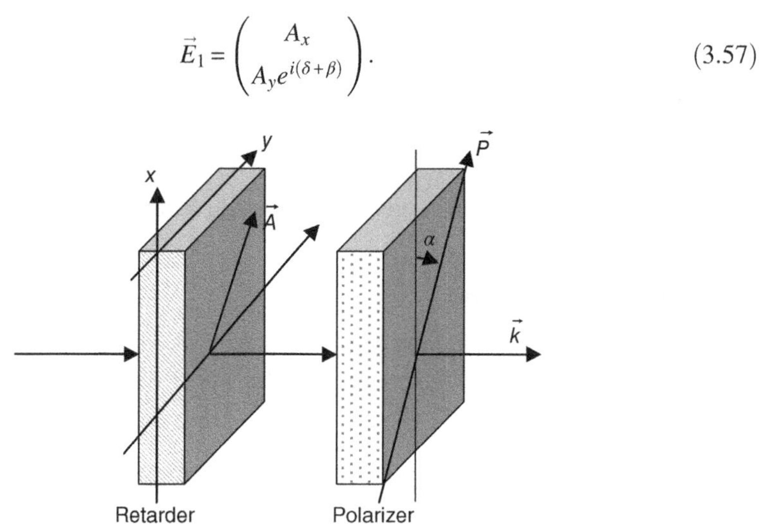

Figure 3.5 Schematic diagram of the setup which is used to measure the Stokes parameters.

At the polarizer, the electric field along the transmission axis is

$$E = A_x \cos\alpha + A_y e^{i(\delta+\beta)} \sin\alpha \tag{3.58}$$

The intensity of the outgoing light is

$$\begin{aligned}
I &= \langle |E|^2 \rangle = \langle \left(A_x \cos\alpha + A_y e^{i(\delta+\beta)} \sin\alpha\right)\left(A_x \cos\alpha + A_y e^{-i(\delta+\beta)} \sin\alpha\right) \rangle \\
&= \langle A_x^2 \rangle \cos^2\alpha + \langle A_y^2 \rangle \sin^2\alpha + \sin(2\alpha)\left(\langle A_x A_y \cos\delta \rangle \cos\beta + \langle A_x A_y \sin\delta \rangle \sin\beta\right) \\
&= \frac{1}{2}(S_o + S_1)\cos^2\alpha + \frac{1}{2}(S_o - S_1)\sin^2\alpha + \sin(2\alpha)\left(\frac{1}{2}S_2 \cos\beta + \frac{1}{2}S_3 \sin\beta\right) \\
&= \frac{1}{2}[S_o + S_1\cos(2\alpha) + \sin(2\alpha)(S_2\cos\beta + S_3\sin\beta)].
\end{aligned} \tag{3.59}$$

In the first step of the measurement, the quarter-wave plate is removed, which is equivalent to $\beta = 0$. The light intensity is measured when the polarizer is at the following positions:

$$I(\beta=0,\ \alpha=0) = \frac{1}{2}(S_o + S_1) \tag{3.60}$$

$$I(\beta=0,\ \alpha=45^\circ) = \frac{1}{2}(S_o + S_2) \tag{3.61}$$

$$I(\beta=0,\ \alpha=90^\circ) = \frac{1}{2}(S_o - S_1) \tag{3.62}$$

When the quarter-wave plate is inserted and the polarizer is at 45°, the measured light intensity will be

$$I(\beta=90^\circ,\ \alpha=45^\circ) = \frac{1}{2}(S_o + S_3). \tag{3.63}$$

From these four equations, the Stokes parameters can be calculated.

It is impossible by means of any instrument to distinguish between various incoherent superpositions of wave fields, having the same frequency, that may together form a beam with the same Stokes parameters. This is known as the *principle of optical equivalence.*

The Stokes vectors of incoherent beams can be composed and decomposed. For example, an unpolarized beam can be decomposed into two oppositely elliptically polarized light (with the same ellipticity but opposite handedness and orthogonal major axes):

$$I_o\begin{pmatrix}1\\0\\0\\0\end{pmatrix} = \frac{I_o}{2}\begin{pmatrix}1\\ \cos(2\nu)\cos(2\phi)\\ \cos(2\nu)\sin(2\phi)\\ \sin(2\nu)\end{pmatrix} + \frac{I_o}{2}\begin{pmatrix}1\\ \cos(-2\nu)\cos[2(\phi+\pi/2)]\\ \cos(-2\nu)\sin[2(\phi+\pi/2)]\\ \sin(-2\nu)\end{pmatrix}, \tag{3.64}$$

where ν and ϕ are ellipticity angle and azimuthal angle of the polarization ellipse, respectively (see Section 2.2.4 for details).

A partially polarized beam can be decomposed into a completely polarized beam and an unpolarized beam

$$\begin{pmatrix} S_o \\ S_1 \\ S_2 \\ S_3 \end{pmatrix} = (1-\gamma)\begin{pmatrix} S_o \\ 0 \\ 0 \\ 0 \end{pmatrix} + \begin{pmatrix} \gamma S_o \\ S_1 \\ S_2 \\ S_1 \end{pmatrix}, \tag{3.65}$$

where $\gamma = \frac{\sqrt{S_1^2+S_2^2+S_3^2}}{S_o}$. A partially polarized beam can also be decomposed into two oppositely polarized beams:

$$\begin{pmatrix} S_o \\ S_1 \\ S_2 \\ S_3 \end{pmatrix} = \frac{(1+\gamma)}{2\gamma}\begin{pmatrix} \gamma S_o \\ S_1 \\ S_2 \\ S_3 \end{pmatrix} + \frac{(1-\gamma)}{2\gamma}\begin{pmatrix} \gamma S_o \\ -S_1 \\ -S_2 \\ -S_1 \end{pmatrix} \tag{3.66}$$

3.2.3 *The Mueller matrix*

When the polarization state of a light beam is represented by the Stokes vector, the effect of an optical element can be represented by the Mueller matrix $\overleftrightarrow{M}$ which operates on the Stokes vector, $\vec{S}_i$, of the incident light to generate the Stokes vector, $\vec{S}_o$, of the outgoing light:

$$\vec{S}_o = \overleftrightarrow{M} \cdot \vec{S}_i \tag{3.67}$$

The Mueller matrix $\overleftrightarrow{M}$ has the form

$$\overleftrightarrow{M} = \begin{pmatrix} m_{00} & m_{01} & m_{02} & m_{03} \\ m_{10} & m_{11} & m_{12} & m_{13} \\ m_{20} & m_{21} & m_{22} & m_{23} \\ m_{30} & m_{31} & m_{32} & m_{33} \end{pmatrix}. \tag{3.68}$$

If the Jones matrix of the optical element is

$$\overleftrightarrow{G} = \begin{pmatrix} g_{11} & g_{12} \\ g_{21} & g_{22} \end{pmatrix}, \tag{3.69}$$

the Mueller matrix of the element is

$$\overleftrightarrow{M} = \frac{1}{2}\begin{pmatrix} 1 & 0 & 0 & 1 \\ 1 & 0 & 0 & -1 \\ 0 & 1 & 1 & 0 \\ 0 & i & -i & 0 \end{pmatrix}\begin{pmatrix} g_{11}g_{11}^* & g_{11}g_{12}^* & g_{12}g_{11}^* & g_{12}g_{12}^* \\ g_{11}g_{21}^* & g_{11}g_{22}^* & g_{12}g_{21}^* & g_{12}g_{22}^* \\ g_{21}g_{11}^* & g_{21}g_{12}^* & g_{22}g_{11}^* & g_{22}g_{12}^* \\ g_{21}g_{21}^* & g_{21}g_{22}^* & g_{22}g_{21}^* & g_{22}g_{22}^* \end{pmatrix}\begin{pmatrix} 1 & 1 & 0 & 0 \\ 0 & 0 & 1 & -i \\ 0 & 0 & 1 & i \\ 1 & -1 & 0 & 0 \end{pmatrix}. \tag{3.70}$$

We first consider the Mueller matrix of an absorber. The transmission coefficients along the x and y axes are p_x and p_y, respectively:

$$\begin{pmatrix} E'_x \\ E'_y \end{pmatrix} = \begin{pmatrix} p_x & 0 \\ 0 & p_y \end{pmatrix} \begin{pmatrix} E_x \\ E_y \end{pmatrix} \tag{3.71}$$

If $p_x = p_y = p$, the absorber is a neutral density filter. If $p_x = 0$ and $p_y = p = 1$, it is a vertical polarizer (transmission axis parallel to the y axis). If $p_x = p = 1$ and $p_y = 0$, it is a horizontal polarizer. The Mueller matrix of the absorber is

$$\overleftrightarrow{M}_P = \frac{1}{2} \begin{pmatrix} p_x^2 + p_y^2 & p_x^2 - p_y^2 & 0 & 0 \\ p_x^2 - p_y^2 & p_x^2 + p_y^2 & 0 & 0 \\ 0 & 0 & 2p_x p_y & 0 \\ 0 & 0 & 0 & 2p_x p_y \end{pmatrix}. \tag{3.72}$$

We now consider the Mueller matrix of a rotator. In the xy frame, the electric field vector is $\vec{E} = E_x\hat{x} + E_y\hat{y}$. In another frame $x'y'$, which is in the same plane but the x' axis makes the angle ϕ with the x axis, the electric vector is $\vec{E}' = E'_x\hat{x}' + E'_y\hat{y}'$. The components of the electric field in the two frames are transformed according to Equation (3.9). The Mueller matrix that transfers the Stokes vector in the xy frame into the Stokes vector in the $x'y'$ frame is

$$\overleftrightarrow{M}_R = \begin{pmatrix} 1 & 0 & 0 & 0 \\ 0 & \cos(2\phi) & \sin(2\phi) & 0 \\ 0 & -\sin(2\phi) & \cos(2\phi) & 0 \\ 0 & 0 & 0 & 1 \end{pmatrix}. \tag{3.73}$$

The Mueller matrix of a ideal polarizer at the angle ϕ is

$$\overleftrightarrow{M}_P(\phi) = \overleftrightarrow{M}_R(-\phi)\cdot\overleftrightarrow{M}_P(0)\cdot\overleftrightarrow{M}_R(\phi) = \frac{1}{2} \begin{pmatrix} 1 & \cos(2\phi) & \sin(2\phi) & 0 \\ \cos(2\phi) & \cos^2(2\phi) & \sin(2\phi)\cos(2\phi) & 0 \\ \sin(2\phi) & \sin(2\phi)\cos(2\phi) & \sin^2(2\phi) & 0 \\ 0 & 0 & 0 & 0 \end{pmatrix}. \tag{3.74}$$

For a retarder with retardation angle $\Gamma = 2\pi\Delta nh/\lambda$ and the slow axis along x axis

$$\begin{pmatrix} E'_x \\ E'_y \end{pmatrix} = \begin{pmatrix} e^{i\Gamma/2} & 0 \\ 0 & e^{-i\Gamma/2} \end{pmatrix} \begin{pmatrix} E_x \\ E_y \end{pmatrix}, \tag{3.75}$$

the corresponding Mueller matrix is

$$\overleftrightarrow{M}_{Retarder}(\Gamma, 0) = \begin{pmatrix} 1 & 0 & 0 & 0 \\ 0 & 1 & 0 & 0 \\ 0 & 0 & \cos\Gamma & -\sin\Gamma \\ 0 & 0 & \sin\Gamma & \cos\Gamma \end{pmatrix}. \tag{3.76}$$

The Mueller matrix of a retarder, whose slow axis makes the angle ϕ with the x axis is

$$\overleftrightarrow{M}_{\mathrm{Retarder}}(\Gamma,\phi)=\overleftrightarrow{M}_R(-\phi)\cdot\overleftrightarrow{M}_{retarder}(\Gamma,0)\cdot\overleftrightarrow{M}_R(\phi)$$
$$=\begin{pmatrix} 1 & 0 & 0 & 0 \\ 0 & \cos^2(2\phi)+\sin^2(2\phi)\cos\Gamma & \sin(2\phi)\cos(2\phi)(1-\cos\Gamma) & \sin(2\phi)\sin\Gamma \\ 0 & \sin(2\phi)\cos(2\phi)(1-\cos\Gamma) & \sin^2(2\phi)+\cos^2(2\phi)\cos\Gamma & -\cos(2\phi)\sin\Gamma \\ 0 & -\sin(2\phi)\sin\Gamma & \cos(2\phi)\sin\Gamma & \cos\Gamma \end{pmatrix}. \tag{3.77}$$

If there is no absorption element involved, we need to consider the three-component vector $\vec{S}^T=(S_1,S_2,S_3)$. The function of optical elements is described by 3×3 matrices:

$$\overleftrightarrow{M}=\begin{pmatrix} m_{11} & m_{12} & m_{13} \\ m_{21} & m_{22} & m_{23} \\ m_{31} & m_{32} & m_{33} \end{pmatrix} \tag{3.78}$$

3.2.4 *Poincaré sphere*

For completely polarized light, the normalized Stokes parameters satisfy the condition $S_1^2+S_2^2+S_3^2=S_o^2=1$. Therefore the point with the coordinates (S_1, S_2, S_3) is on the surface of a unit sphere in 3-D space. This sphere is known as the *Poincaré sphere* and is shown in Figure 3.6.

For completely polarized light, the Jones vector is

$$\vec{E}=\begin{pmatrix} A_x \\ A_y e^{i\delta} \end{pmatrix}, \tag{3.79}$$

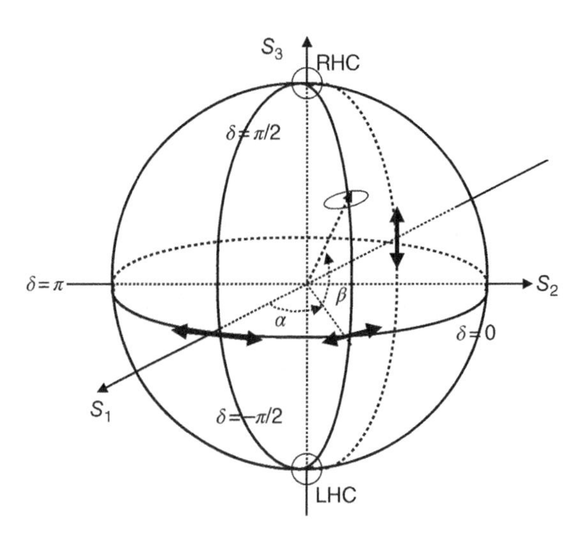

Figure 3.6 Poincaré sphere.

where A_x and A_y are time-independent positive numbers and δ is a time-independent number. Generally, it is elliptically polarized. The azimuthal angle ϕ of the polarization ellipse is given by Equation (2.41):

$$\tan 2\phi = \frac{A_x A_y \cos\delta}{\left(A_x^2 - A_y^2\right)} \tag{3.80}$$

The ellipticity angle ν is given by Equation (2.48)

$$\sin 2\nu = \frac{2\tan\nu}{1+\tan^2\nu} = \frac{2b/a}{1+(b/a)^2} = \frac{2ab}{a^2+b^2}, \tag{3.81}$$

where a and b are the lengths of the principal semi-axes of the polarization ellipse. From Equations (2.45) and (2.46) we have

$$\sin 2\nu = \frac{2A_x A_y \sin\delta}{A_x^2 + A_y^2}. \tag{3.82}$$

The stokes parameters are given by

$$S_1 = \left(A_x^2 - A_y^2\right)/\left(A_x^2 + A_y^2\right), \tag{3.83}$$

$$S_2 = 2A_x A_y \cos\delta/\left(A_x^2 + A_y^2\right), \tag{3.84}$$

$$S_3 = 2A_x A_y \sin\delta/\left(A_x^2 + A_y^2\right). \tag{3.85}$$

The longitudinal angle α of the point representing the polarization on the Poincaré sphere is given by

$$\tan\alpha = \frac{S_2}{S_1} = \frac{2A_x A_y \cos\delta}{A_x^2 - A_y^2} = \tan 2\phi. \tag{3.86}$$

Therefore $\alpha = 2\phi$. The latitude angle β is given by

$$\sin\beta = S_3 = 2A_x A_y \sin\delta/\left(A_x^2 + A_y^2\right) = \sin 2\nu. \tag{3.87}$$

Therefore $\beta = 2\nu$. If we know the azimuthal angle ϕ and the ellipticity angle ν of the polarization ellipse, the Stokes vector is

$$\vec{S} = \begin{pmatrix} \cos(2\nu)\cos(2\phi) \\ \cos(2\nu)\sin(2\phi) \\ \sin(2\nu) \end{pmatrix}. \tag{3.88}$$

If we know the angle $\chi = \arctan(A_x/A_y)$ and the phase difference δ, then the Stokes vector is

$$\vec{S} = \begin{pmatrix} \cos(2\chi) \\ \sin(2\chi)\cos\delta \\ \sin(2\chi)\cos\delta \end{pmatrix}. \tag{3.89}$$

The points corresponding to some special polarization states are as follows:

north pole (0,0,1):	right-handed polarized
south pole (0,0,–1):	left-handed polarized
a point on the equator:	linearly polarized
(1,0,0):	linearly polarized along the x axis
(–1,0,0):	linearly polarized along the y axis
(0,1,0):	linearly polarized along 45°

3.2.4.1 Features of Poincaré Sphere

1. Two diametrically opposed points on the sphere correspond to states with orthogonal polarization.
2. For any point on a half circle connecting the north and south poles (fixed longitude), the inclination angle ϕ of the polarization ellipse is the same, because S_2/S_1 = constant.
3. For any point on a circle with fixed S_3 (fixed latitude) on the sphere, the ellipticity is the same.

3.2.5 Evolution of the polarization states on the Poincaré sphere

We consider how the three-component Stokes vector $\vec{S}$ evolves on the Poincaré sphere under the action of retardation films. The Mueller matrix of a uniform uniaxial retarder with the retardation angle Γ and the slow axis making the angle ϕ with the x axis is given by (from Equations (3.77) and (3.78))

$$\overleftrightarrow{M}(\Gamma,\psi) = \begin{pmatrix} \cos^2(2\phi)+\sin^2(2\phi)\cos\Gamma & \sin(2\phi)\cos(2\phi)(1-\cos\Gamma) & \sin(2\phi)\sin\Gamma \\ \sin(2\phi)\cos(2\phi)(1-\cos\Gamma) & \sin^2(2\phi)+\cos^2(2\phi)\cos\Gamma & -\cos(2\phi)\sin\Gamma \\ -\sin(2\phi)\sin\Gamma & \cos(2\phi)\sin\Gamma & \cos\Gamma \end{pmatrix}. \tag{3.90}$$

For a thin retardation film with the thickness $dz \to 0$, the retardation angle $d\Gamma = (2\pi\Delta n/\lambda)dz = k_o\Delta n dz \to 0$, we have the approximations that $\cos d\Gamma = 1$ and $\sin d\Gamma = d\Gamma$, and the Mueller matrix becomes

$$\overleftrightarrow{M}(d\Gamma,\phi) = \begin{pmatrix} 1 & 0 & \sin(2\phi)d\Gamma \\ 0 & 1 & -\cos(2\phi)d\Gamma \\ -\sin(2\phi)d\Gamma & \cos(2\phi)d\Gamma & 1 \end{pmatrix}. \tag{3.91}$$

If the Stokes vector of the incident light is $\vec{S}_i$, the Stokes vector of the outgoing light is $\vec{S}_o = \vec{M}(\Gamma,\psi)\cdot\vec{S}_i$. The change of the Stokes vector caused by the retardation film is

$$d\vec{S} = S_o - \vec{S}_i = \overleftrightarrow{M}(d\Gamma,\psi)\cdot\vec{S}_i - \vec{S}_i = d\Gamma\begin{pmatrix} 0 & 0 & \sin(2\phi) \\ 0 & 0 & -\cos(2\phi) \\ -\sin(2\phi) & \cos(2\phi) & 0 \end{pmatrix}\cdot\vec{S}_i. \tag{3.92}$$

This can be rewritten as

$$d\vec{S} = d\Gamma\,\vec{\Omega}\times\vec{S}_i, \tag{3.93}$$

where $\vec{\Omega}$ is a unit vector and has the form

$$\vec{\Omega} = \begin{pmatrix} \cos(2\phi) \\ \sin(2\phi) \\ 0 \end{pmatrix}. \tag{3.94}$$

From Equation (3.93), we can see that the effect of the retarder is to rotate $\vec{S}$ around the axis represented by $\vec{\Omega}$ with the rotation angle of $d\Gamma$, as shown in Figure 3.7. From Equation (3.94), it can be seen that the rotation axis is on the equator and makes an angle of 2ϕ with the S_1 axis, which is twice the angle of the slow axis (with respect to the x axis) of the retarder (in the xy frame).

For a uniform retardation film with the retardation angle Γ, even if its thickness is not small, the Stokes vector of the outgoing light can be derived from Equation (3.93) and is

$$\vec{S}_o = \vec{S}_i + \Gamma\,\vec{\Omega}\times\vec{S}_i. \tag{3.95}$$

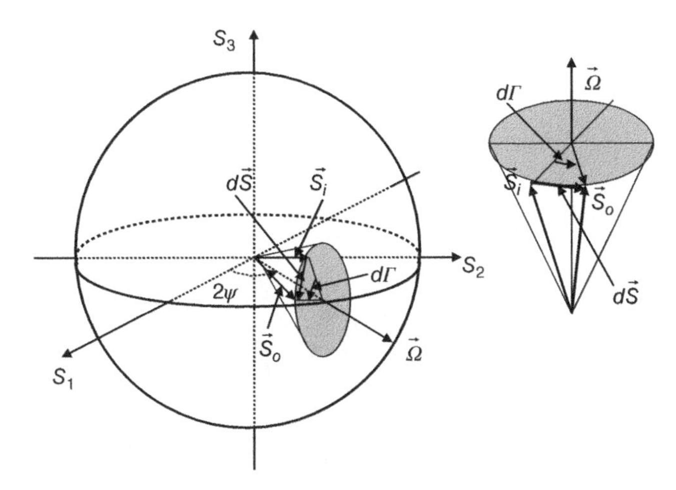

Figure 3.7 Schematic diagram showing the rotation of the Stokes vector under the action of the retardation film.

One of the reasons to use the Poincaré sphere is that the effect of retardation films and the evolution of the polarization state can be easily visualized. We consider polarization conversion in the following few special cases.

3.2.5.1 Polarization conversion using quarter-wave plates

An elliptical polarization state (with inclination angle ϕ and ellipticity angle ν), represented by $\vec{S}_1$ in Figure 3.8(a), can be converted into a circular polarization state by using two quarter-wave plates [4]. This can be done in two steps:

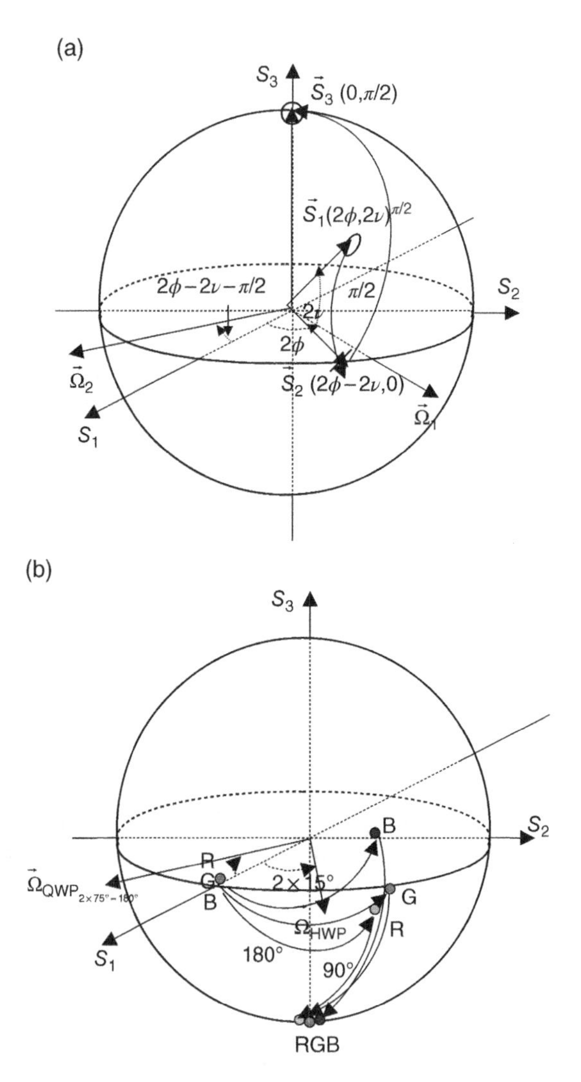

Figure 3.8 (a) Converting an elliptically polarized light by two quarter-wave plates. (b) Broadband quarter-wave plate.

i. Using the first quarter-wave plate to convert the elliptical polarization to linear polarization.
   The c axis of the quarter-wave plate is parallel to the major axis of the polarization ellipse. The rotation axis Ω_1 is at 2ϕ and the cone angle of the rotation cone is $2\times 2\nu$. The rotation angle is $\pi/2$. After this quarter-wave plate, the Stokes vector, represented by $\vec{S}_2$, is on the equator and the longitudinal angle is $2\phi - 2\nu$ on the Poincaré sphere, i.e., the light is linearly polarized at the angle $(2\phi - 2\nu)/2 = (\phi - \nu)$ with respect to the x axis in the xy frame.
ii. Using the second quarter-wave plate to convert the linear polarization into right-handed circular polarization.
   The c axis of the quarter-wave plate is at an angle of $\pi/4$ with respect to the linear polarization, which is at an angle of $(\phi - \nu - \pi/4)$ with respect to the x axis in the xy frame. On the Poincaré sphere, the rotation axis $\vec{\Omega}_2$ is at an angle of $2(\phi - \nu - \pi/4)$. The rotation angle is $\pi/2$. After this quarter-wave plate, the Stokes vector, represented by $\vec{S}_3$, is on the north pole on the Poincaré sphere, i.e., the light is circularly polarized.

By reversing this procedure, circularly polarized light can be converted into elliptically polarized light by using two quarter-wave plates. By combining the above two procedures, any elliptically polarized light can be converted into any other elliptically polarized light by using four properly oriented quarter-wave plates.

3.2.5.2 Broadband quarter-wave plate to convert a linear polarization to circular polarization

The optical retardation angle, given by $2\pi\Delta nd/\lambda$, of a uniform birefringent film is wavelength dependent. For a quarter-wave plate from a single uniform film, the retardation angle is $\pi/2$ at only one wavelength, which is typically chosen to be in the middle of visible light region. For green (G) light, the optical retardation angle is $\pi/2$. For blue (B) light the optical retardation angle is more than $\pi/2$. For red (R) light the optical retardation angle is less than $\pi/2$. A broadband quarter-wave plate can be achieved by using a combination of a quarter-wave plate and a half-wave plate.

i. Using a half-wave plate at 15°
   The incident light is linearly polarized along the x axis. The half-wave plate is at 15° with respect to the x axis in the xy frame. On the Poincaré sphere, the rotation axis of the half-wave plate $\vec{\Omega}_{HWP}$ is at $2\times 15°$, as shown in Figure 3.8(b). The green (G) light is rotated 180° to a point on the equator. The blue (B) light is rotated more than 180° to a point on the north hemisphere. The red (R) light is rotated less than 180° to a point on the south hemisphere.
ii. Using a quarter-wave plate at 75°
   The half-wave plate is at 75° with respect to the x axis in the xy frame. On the Poincaré sphere, the rotation axis of the quarter-wave plate $\vec{\Omega}_{QWP}$ is at $2\times 75°$, which is equivalent to −30°. The green (G) light is rotated 90° to the south pole. The blue (B) light is rotated more than 90° to the south pole. The red (R) light is rotated less than 90° to the south pole.

3.2.6 Mueller matrix of twisted nematic liquid crystals

We now consider the Mueller matrix of a uniformly twisted nematic (or cholesteric) liquid crystal.,The problem can be simplified if we consider the Stokes vector and the Mueller matrix in the local frame $x'y'$, in which the liquid crystal director lies along the x' axis. Divide the liquid crystal film into N thin slabs. The thickness of each slab is $dz = h/N$, where h is the thickness of the liquid crystal film. The angle between the liquid crystal director of two neighboring slabs is $d\psi = qdz$, where q is the twisting rate. The retardation angle of a slab is $d\Gamma = k_o\Delta ndz$. If the Stokes vector (in the local frame) of the light incident on a slab is $\vec{S}'$, then the Stokes vector of the light incident on the next slab is (from Equations (3.73) and (3.76))

$$\vec{S}' + d\vec{S}' = \begin{pmatrix} 1 & 2qdz & 0 \\ -2qdz & 1 & 0 \\ 0 & 0 & 1 \end{pmatrix} \begin{pmatrix} 1 & 0 & 0 \\ 0 & 1 & -k_o\Delta ndz \\ 0 & k_o\Delta ndz & 1 \end{pmatrix} \vec{S}' = \begin{pmatrix} 1 & 2qdz & 0 \\ -2qdz & 1 & -k_o\Delta ndz \\ 0 & k_o\Delta ndz & 1 \end{pmatrix} \vec{S}'. \tag{3.96}$$

In deriving the above equation, only first-order terms are kept when $dz \to 0$. In component form we have

$$\frac{dS_1'}{dz} = 2qS_2', \tag{3.97}$$

$$\frac{dS_2'}{dz} = -2qS_1' - k_o\Delta nS_3', \tag{3.98}$$

$$\frac{dS_3'}{dz} = K_o\Delta nS_2'. \tag{3.99}$$

From Equation (3.98) we can get

$$\frac{d^2S_2'}{dz^2} = -2q\frac{dS_1'}{dz} - K_o\Delta n\frac{dS_3'}{dz} = -\left[(2q)^2 + (K_o\Delta n)^2\right]S_2'. \tag{3.100}$$

Define

$$\chi = \left[(2q)^2 + (K_o\Delta n)^2\right]^{1/2} = 2\left[\Phi^2 + (\Gamma/2)^2\right]^{1/2}/h, \tag{3.101}$$

where Φ and Γ are the total twist angle and retardation angle of the liquid crystal film, respectively. The solution of Equation (3.100) is

$$S_2' = A_{21}\sin(\chi z) + A_{22}\cos(\chi z). \tag{3.102}$$

From Equation (3.97) we get

$$S_1' = \frac{2q}{\chi}\left[-A_{21}\cos(\chi z) + A_{22}\sin(\chi z)\right] + A_{11}. \tag{3.103}$$

From Equation (3.99) we get

$$S_3' = \frac{k_o \Delta n}{\chi}[-A_{21}\cos(\chi z) + A_{22}\sin(\chi z)] + A_{33}. \quad (3.104)$$

If the Stokes vector of the light incident on the liquid crystal film is $\vec{S}'^T_i = (S'_{10} S'_{20} S'_{30})$, then we have the boundary condition equations ($z = 0$):

$$A_{22} = S'_{20} \quad (3.105)$$

$$-\frac{2q}{\chi}A_{21} + A_{11} = S'_{10} \quad (3.106)$$

$$-\frac{k_o \Delta n}{\chi}A_{21} + A_{33} = S_{30}' \quad (3.107)$$

Also from Equation (3.98) we get

$$-2qA_{11} - k_o \Delta n A_{33} = 0. \quad (3.108)$$

From the above four equations we can find the four coefficients. The final results are

$$S_1' = \left[1 - 2\left(\frac{2q}{\chi}\right)^2 \sin^2\left(\frac{\chi}{2}z\right)\right] S'_{10} + \frac{2q}{\chi}\sin(\chi z) S'_{20} - \frac{4qk_o\Delta n}{\chi^2}\sin^2\left(\frac{\chi}{2}z\right) S'_{30}, \quad (3.109)$$

$$S_2' = -\frac{2q}{\chi}S'_{10}\sin(\chi z) + S'_{20}\cos(\chi z) - \frac{k_o\Delta n}{\chi}S'_{30}\sin(\chi z), \quad (3.110)$$

$$S_3' = -\frac{4qk_o\Delta n}{\chi^2}\sin^2\left(\frac{\chi}{2}z\right) S'_{10} + \frac{k_o\Delta n}{\chi}\sin(\chi z) S'_{20} + \left[1 - 2\left(\frac{2q}{\chi}\right)^2 \sin^2\left(\frac{\chi}{2}z\right)\right] S'_{30}. \quad (3.111)$$

Therefore the Stokes vector $\vec{S}'_o$ after the TN film is related to the Stokes vector $\vec{S}'_i$ before the film by

$$\vec{S}'_o = \begin{pmatrix} 1 - 2\frac{\Phi^2}{X^2}\sin^2 X & \frac{\Phi}{X}\sin(2X) & -2\frac{\Phi(\Gamma/2)}{X^2}\sin^2 X \\ -\frac{\Phi}{X}\sin(2X) & \cos(2X) & -\frac{(\Gamma/2)}{X}\sin(2X) \\ -2\frac{\Phi(\Gamma/2)}{X^2}\sin^2 X & \frac{(\Gamma/2)}{X}\sin(2X) & 1 - 2\frac{(\Gamma/2)^2}{X^2}\sin^2 X \end{pmatrix} \cdot \vec{S}'_i, \quad (3.112)$$

where

$$X = \left[\Phi^2 + (\Gamma/2)^2\right]^{1/2}. \quad (3.113)$$

We know that at the exit plane the local frame makes an angle Φ with the lab frame, and at the entrance plane the x' axis is parallel to the x axis and therefore $\vec{S}_i' = \vec{S}_i$, therefore in the lab frame we have

$$\vec{S}_o = \begin{pmatrix} \cos(2\Phi) & -\sin(2\Phi) & 0 \\ \sin(2\Phi) & \cos(2\Phi) & 0 \\ 0 & 0 & 1 \end{pmatrix} \cdot \begin{pmatrix} 1-2\frac{\Phi^2}{X^2}\sin^2 X & \frac{\Phi}{X}\sin(2X) & -2\frac{\Phi(\Gamma/2)}{X^2}\sin^2 X \\ -\frac{\Phi}{X}\sin(2X) & \cos(2X) & -\frac{(\Gamma/2)}{X}\sin(2X) \\ -2\frac{\Phi(\Gamma/2)}{X^2}\sin^2 X & \frac{(\Gamma/2)}{X}\sin(2X) & 1-2\frac{(\Gamma/2)^2}{X^2}\sin^2 X \end{pmatrix} \cdot \vec{S}_i. \tag{3.114}$$

This equation can also be obtained from Equations (3.34) and (3.70).

For example, for the normal black 90° TN where the two polarizers are parallel to each other, $\Phi = \pi/2$ and $\vec{S}_i^T = (1,0,0)$, so

$$\vec{S}_o = \begin{pmatrix} -1 & 0 & 0 \\ 0 & -1 & 0 \\ 0 & 0 & 1 \end{pmatrix} \begin{pmatrix} 1-2\frac{\Phi^2}{X^2}\sin^2 X & \frac{\Phi}{X}\sin(2X) & -2\frac{\Phi(\Gamma/2)}{X^2}\sin^2 X \\ -\frac{\Phi}{X}\sin(2X) & \cos(2X) & -\frac{(\Gamma/2)}{X}\sin(2X) \\ -2\frac{\Phi(\Gamma/2)}{X^2}\sin^2 X & \frac{(\Gamma/2)}{X}\sin(2X) & 1-2\frac{(\Gamma/2)^2}{X^2}\sin^2 X \end{pmatrix} \begin{pmatrix} 1 \\ 0 \\ 0 \end{pmatrix}$$

$$= \begin{pmatrix} -1+2\frac{\Phi^2}{X^2}\sin^2 X \\ \frac{\Phi}{X}\sin(2X) \\ 2\frac{\Phi(\Gamma/2)}{X^2}\sin^2 X \end{pmatrix}.$$

The polarizer after the TN is also along the x axis. From the definition of Stokes vector we can get the transmittance $T = (1 + S_{o1})/2 = (\Phi/X)^2 \sin^2 X$, which is the same as that given by Equation (3.46).

3.2.7 *Mueller matrix of non-uniform birefringence film*

In the same way that the Jones matrix can be used to numerically calculate the optical properties of non-uniform birefringence film, the Meuller matrix can be used to numerically calculate the optical properties of a non-uniform birefringence film. We divide the film into N slabs as shown in Figure 3.2. When the thickness $\Delta h = h/N$ of the slabs is sufficiently small so that within each slab, the slow axis can be considered fixed. For layer i, the angle of the slow axis with respect the x axis is β_i and the phase retardation is $\Gamma_i = 2\pi[n_e(z = i\Delta h) - n_o(z = i\Delta h)]\Delta h/\lambda$. In the lab frame, the Stokes vector of the incident light on the layer is $\vec{S}_{ii}$, which is the same as the Stokes vector, $\vec{S}_{(i-1)o}$, of the light exiting the layer $(i-1)$, and the Stokes vector of the light coming out of layer i is $\vec{S}_{io}$ [9]:

$$\vec{S}_{io} = \overleftrightarrow{M}_{rotator}(\beta_i)\cdot\overleftrightarrow{M}_{retardar}(\Gamma_i)\cdot\overleftrightarrow{M}^{-1}_{rotator}(\beta_i)\cdot\vec{S}_{ii} = \overleftrightarrow{M}_{rotator}(\beta_i)\cdot\overleftrightarrow{M}_{retardar}(\Gamma_i)\cdot\overleftrightarrow{M}^{-1}_{rotator}(\beta_i)\cdot\vec{S}_{(i-1)o} \tag{3.115}$$

The Stokes vector, $\vec{S}_o$, of the outgoing light is related to the Stokes vector, $\vec{S}_i$ of the incident light by

$$\begin{aligned}\vec{S}_o = & \left[\overleftrightarrow{M}_{rotator}(\beta_N)\cdot\overleftrightarrow{M}_{retarder}(\Gamma_N)\cdot\overleftrightarrow{M}^{-1}_{rotator}(\beta_N)\right]\cdot \\ & \left[\overleftrightarrow{M}_{rotator}(\beta_{N-1})\cdot\overleftrightarrow{M}_{retarder}(\Gamma_{N-1})\cdot\overleftrightarrow{M}^{-1}_{rotator}(\beta_{N-1})\right]\cdot \\ & \ldots\ldots\cdot \\ & \left[\overleftrightarrow{M}_{rotator}(\beta_1)\cdot\overleftrightarrow{M}_{retarder}(\Gamma_1)\cdot\overleftrightarrow{M}^{-1}_{rotator}(\beta_1)\right]\cdot\vec{S}_i \\ = & \prod_{j=1}^{N}\left[\overleftrightarrow{M}_{rotator}(\beta_j)\cdot\overleftrightarrow{M}_{retarder}(\Gamma_j)\cdot\overleftrightarrow{M}^{-1}_{rotator}(\beta_j)\right]\cdot\vec{S}_i\end{aligned} \tag{3.116}$$

Usually the multiplication of the matrices is carried out numerically.

3.3 Berreman 4 × 4 Method

For stratified optical media (whose refractive indices are only a function of the coordinate normal to the film), Berreman introduced a 4 × 4 matrix method (now known as the Berreman 4 × 4 method) [14–18], in which the electric field and magnetic field (the sum of the fields of the light beam propagating in forward and backward directions) are considered. When the film is divided into slabs, the reflection at the interface between the slabs is taken into account. The Berreman 4 × 4 method works well for both normal and obliquely incident light. Consider an optical film, such as a cholesteric liquid crystal in the planar texture (twisted nematic), whose dielectric tensor is only a function of the coordinate z, which is perpendicular to the film:

$$\overleftrightarrow{\varepsilon}(z) = \begin{pmatrix} \varepsilon_{11}(z) & \varepsilon_{12}(z) & \varepsilon_{13}(z) \\ \varepsilon_{21}(z) & \varepsilon_{22}(z) & \varepsilon_{23}(z) \\ \varepsilon_{31}(z) & \varepsilon_{32}(z) & \varepsilon_{33}(z) \end{pmatrix} \tag{3.117}$$

For a light incident in the xz plane with incident angle α with respect to the z axis (see Figure 3.9), the fields of the optical wave are

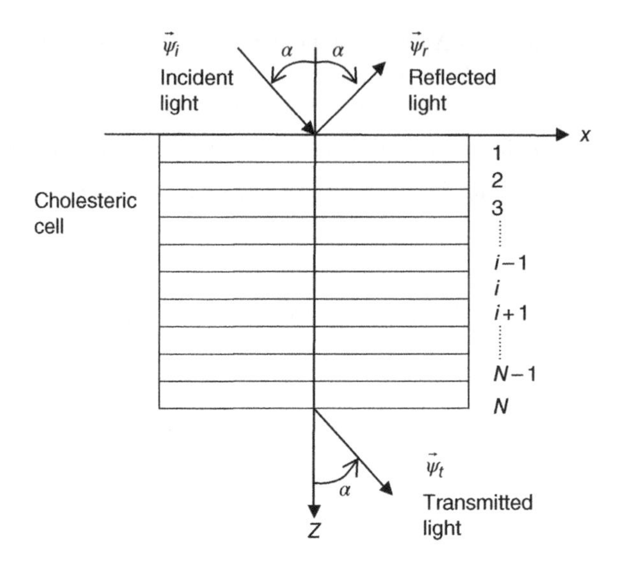

Figure 3.9 The coordinate system used to describe the light propagation in the Berreman 4×4 method.

$$\vec{E} = \vec{E}(z)e^{-ik_x x + i\omega t}, \tag{3.118}$$

$$\vec{H} = \vec{H}(z)e^{-ik_x x + i\omega t}. \tag{3.119}$$

The Maxwell's equations for the optical wave are

$$\nabla\cdot\vec{D} = \nabla\cdot\left(\varepsilon_o \overleftrightarrow{\varepsilon}\cdot\vec{E}\right) = 0, \tag{3.120}$$

$$\nabla\cdot\vec{B} = \nabla\cdot\left(\mu_o \vec{H}\right) = 0, \tag{3.121}$$

$$\nabla\times\vec{E} = -\frac{\partial\vec{B}}{\partial t} = -i\mu_o\omega\vec{H}, \tag{3.122}$$

$$\nabla\times\vec{H} = \frac{\partial\vec{D}}{\partial t} = i\varepsilon_o\omega\overleftrightarrow{\varepsilon}\cdot\vec{E}. \tag{3.123}$$

Because of Equation (3.120), it is required that

$$k_x = k_o \sin\alpha = \frac{2\pi}{\lambda}\sin\alpha = \text{const.} \tag{3.124}$$

Therefore

$$\frac{\partial}{\partial x} = -ik_x. \tag{3.125}$$

Because the light is propagating in the xz plane,

$$\frac{\partial}{\partial y}=0. \tag{3.126}$$

From Equation (3.122) we have

$$\begin{aligned}\nabla\times\vec{E}&=\left(\frac{\partial E_z}{\partial y}-\frac{\partial E_y}{\partial z}\right)\hat{x}+\left(\frac{\partial E_x}{\partial z}-\frac{\partial E_z}{\partial x}\right)\hat{y}+\left(\frac{\partial E_y}{\partial x}-\frac{\partial E_x}{\partial y}\right)\hat{z}\\&=\left(-\frac{\partial E_y}{\partial z}\right)\hat{x}+\left(\frac{\partial E_x}{\partial z}+ik_xE_z\right)\hat{y}+\left(-ik_xE_y\right)\hat{z}\\&=-i\mu_o\omega\vec{H}=-i\mu_o\omega\left(H_x\hat{x}+H_y\hat{y}+H_z\hat{z}\right)\end{aligned} \tag{3.127}$$

In terms of components, we have

$$\frac{\partial E_y}{\partial z}=i\mu_o\omega H_x, \tag{3.128}$$

$$\frac{\partial E_x}{\partial z}=-ik_xE_z-i\mu_o\omega H_y, \tag{3.129}$$

$$E_y=\frac{\mu_o\omega}{k_x}H_z. \tag{3.130}$$

From Equation (3.123) we can get

$$\begin{aligned}\nabla\times\vec{H}&=\left(\frac{\partial H_z}{\partial y}-\frac{\partial H_y}{\partial z}\right)\hat{x}+\left(\frac{\partial H_x}{\partial z}-\frac{\partial H_z}{\partial x}\right)\hat{y}+\left(\frac{\partial H_y}{\partial x}-\frac{\partial H_x}{\partial y}\right)\hat{z}\\&=\left(-\frac{\partial H_y}{\partial z}\right)\hat{x}+\left(\frac{\partial H_x}{\partial z}+ik_xH_z\right)\hat{y}+\left(-ik_xH_y\right)\hat{z}\\&=i\varepsilon_o\omega\overleftrightarrow{\varepsilon}\cdot\vec{E}\end{aligned} \tag{3.131}$$

In components we have

$$\frac{\partial H_y}{\partial z}=-i\varepsilon_o\omega\left(\varepsilon_{11}E_x+\varepsilon_{12}E_y+\varepsilon_{13}E_z\right) \tag{3.132}$$

$$\frac{\partial H_x}{\partial z}=-ik_xH_z+i\varepsilon_o\omega\left(\varepsilon_{21}E_x+\varepsilon_{22}E_y+\varepsilon_{23}E_z\right) \tag{3.133}$$

$$H_y=-\frac{\mu_o\omega}{k_x}\left(\varepsilon_{31}E_x+\varepsilon_{32}E_y+\varepsilon_{33}E_z\right) \tag{3.134}$$

From Equation (3.130) we have

$$H_z = \frac{k_x}{\mu_o \omega} E_y = \frac{k_x}{k_o} \frac{k_o}{\mu_o \omega} E_y = \frac{k_x}{k_o} \frac{\sqrt{\varepsilon_o \mu_o}}{\mu_o} E_y = \frac{k_x}{k_o} \frac{1}{\sqrt{\mu_o / \varepsilon_o}} E_y = \frac{k_x}{k_o} \frac{1}{\eta_o} E_y, \tag{3.135}$$

where $\eta_o = \sqrt{\mu_o / \varepsilon_o} = 376.98\Omega$ is known as the resistance of vacuum. Equation (3.135) can be used to replace H_z in Equation (3.133). From Equation (3.134) we can find

$$E_z = \frac{-1}{\varepsilon_{33}} \left[(k_x / \mu_o \omega) H_y + \varepsilon_{31} E_x + \varepsilon_{32} E_y \right) = \frac{-1}{\varepsilon_{33}} \left[(\eta_o k_x / k_o) H_y + \varepsilon_{31} E_x + \varepsilon_{32} E_y \right], \tag{3.136}$$

which can be used to replace E_z in Equation (3.129). Therefore only four components of the electric and magnetic fields are needed to specify the light. Define the Berreman vector

$$\vec{\psi}^T = (E_x \quad \eta_o H_y \quad E_y \quad -\eta_o H_x). \tag{3.137}$$

Then we have

$$\frac{\partial \vec{\psi}}{\partial z} = -ik_o \begin{pmatrix} -\chi \frac{\varepsilon_{31}}{\varepsilon_{33}} & -\chi^2 \frac{1}{\varepsilon_{33}} + 1 & -\chi \frac{\varepsilon_{32}}{\varepsilon_{33}} & 0 \\ -\frac{\varepsilon_{13}\varepsilon_{31}}{\varepsilon_{33}} + \varepsilon_{11} & -\chi \frac{\varepsilon_{13}}{\varepsilon_{33}} & -\frac{\varepsilon_{13}\varepsilon_{32}}{\varepsilon_{33}} + \varepsilon_{12} & 0 \\ 0 & 0 & 0 & 1 \\ -\frac{\varepsilon_{23}\varepsilon_{31}}{\varepsilon_{33}} + \varepsilon_{21} & -\chi \frac{\varepsilon_{23}}{\varepsilon_{33}} & -\chi^2 - \frac{\varepsilon_{23}\varepsilon_{32}}{\varepsilon_{33}} + \varepsilon_{22} & 0 \end{pmatrix} \vec{\psi} \equiv -ik_o \overleftrightarrow{Q} \cdot \vec{\psi}, \tag{3.138}$$

where $\chi = k_x / k_o$. Note that $Q_{13} = Q_{42}$, $Q_{11} = Q_{22}$, and $Q_{41} = Q_{23}$ because the dielectric tensor is symmetric. Equation (3.138) is known as the Berreman equation. If the dielectric tensor does not change in the region from z to $z + \Delta z$, then $\overleftrightarrow{Q}(z)$ does not change in this region and the solution to Equation (3.138) is

$$\vec{\psi}(z + \Delta z) = \mathrm{e}^{-ik_o \overleftrightarrow{Q}(z) \Delta z} \cdot \vec{\psi}(z) \equiv \overleftrightarrow{P}(z) \cdot \vec{\psi}(z). \tag{3.139}$$

This equation can be used to calculate the Berreman vector in the optical film. In the calculation, the film is divided into N slabs as shown in Figure 3.9. If the Berreman vector of the incident light is $\vec{\psi}_i$, the Berreman vector $\vec{\psi}_t$ of the outgoing light can be numerically calculated by

$$\vec{\psi}_t = \prod_{i=1}^{N} \overleftrightarrow{P}(i\Delta z) \cdot (\vec{\psi}_i + \vec{\psi}_r). \tag{3.140}$$

In this method, in order to obtain accurate results, the thickness of the slabs must be much thinner than $\lambda / 2\pi$ and thus the computation time is long. The number of slabs can be dramatically reduced and therefore the computation is much faster if the fast Berreman method is used [19, 20]. The fast Berreman method utilizes the Cayley–Hamilton theory that states that $\overleftrightarrow{P}$ can be expanded as

$$\overleftrightarrow{P} = e^{-ik_o\overleftrightarrow{Q}\Delta z} = \gamma_o \overleftrightarrow{I} + \gamma_1\left(-ik_o \overleftrightarrow{Q}\Delta z\right) + \gamma_2\left(-ik_o \overleftrightarrow{Q}\Delta z\right)^2 + \gamma_3\left(-ik_o \overleftrightarrow{Q}\Delta z\right)^3, \qquad (3.141)$$

where $\overleftrightarrow{I}$ is the identity matrix and γ_i ($i = 0, 1, 2, 3$) are the solution of the following equations:

$$\gamma_o + \gamma_1(-ik_o\Delta z q_i) + \gamma_2(-ik_o\Delta z q_i)^2 + \gamma_3(-ik_o\Delta z q_i)^3 = e^{(-ik_o\Delta z q_i)} \quad i = 1,2,3,4, \qquad (3.142)$$

where q_i ($i = 1, 2, 3, 4$) are the eigenvalues of Berreman matrix $\overleftrightarrow{Q}$; namely, they are solutions to the equation

$$\left|\overleftrightarrow{Q} - q\overleftrightarrow{I}\right| = q^4 - 2Q_{11}q^3 - (Q_{43} - Q_{11}^2 + Q_{12}Q_{21})q^2 - 2(Q_{13}Q_{23} - Q_{43}Q_{11})q$$
$$-\left(Q_{11}^2Q_{43} + Q_{21}Q_{13}^2 + Q_{12}Q_{23}^2 - 2Q_{11}Q_{13}Q_{23} - Q_{12}Q_{21}Q_{43}\right) = 0. \qquad (3.143)$$

For a uniaxial liquid crystal with the ordinary and extraordinary refractive indices n_e and n_o, respectively, when the liquid crystal director is $\vec{n} = (n_x, n_y, n_z)$, the dielectric tensor is

$$\overleftrightarrow{\varepsilon} = \begin{pmatrix} \varepsilon_\perp + \Delta\varepsilon n_x^2 & \Delta\varepsilon n_x n_y & \Delta\varepsilon n_x n_z \\ \Delta\varepsilon n_x n_y & \varepsilon_\perp + \Delta\varepsilon n_y^2 & \Delta\varepsilon n_y n_z \\ \Delta\varepsilon n_x n_z & \Delta\varepsilon n_y n_z & \varepsilon_\perp + \Delta\varepsilon n_z^2 \end{pmatrix}, \qquad (3.144)$$

where $\varepsilon_\perp = n_o^2$ and $\Delta\varepsilon = \varepsilon_{//} - \varepsilon_\perp = n_e^2 - n_o^2$. The Berreman matrix is given by

$$\overleftrightarrow{Q} = \frac{1}{(\varepsilon_\perp + \Delta\varepsilon n_z^2)} \begin{pmatrix} -\chi\Delta\varepsilon n_x n_z & (\varepsilon_\perp + \Delta\varepsilon n_z^2) - \chi^2 & -\chi\Delta\varepsilon n_y n_z & 0 \\ \varepsilon_\perp\left[\varepsilon_\perp + \Delta\varepsilon(n_x^2 + n_z^2)\right] & -\chi\Delta\varepsilon n_x n_z & \varepsilon_\perp\Delta\varepsilon n_x n_y & 0 \\ 0 & 0 & 0 & 1 \\ \varepsilon_\perp\Delta\varepsilon n_x n_y & -\chi\Delta\varepsilon n_y n_z & (\varepsilon_\perp - \chi^2)(\varepsilon_\perp + \Delta\varepsilon n_z^2) + \varepsilon_\perp\Delta\varepsilon n_y^2 & 0 \end{pmatrix}. \qquad (3.145)$$

The solutions to Equation (3.143) are

$$q_{1/2} = \pm(Q_{43} - Q_{13}Q_{23}/Q_{11})^{1/2} = \pm\left[\varepsilon_\perp - \chi^2\right]^{1/2}, \qquad (3.146)$$

$$q_{3/4} = Q_{11} \pm (Q_{12}Q_{21} + Q_{13}Q_{23}/Q_{11})^{1/2} = \frac{-\chi\Delta\varepsilon n_x n_z}{\varepsilon_\perp + \Delta\varepsilon n_z^2} \pm \frac{(\varepsilon_{//}\varepsilon_\perp)^{1/2}}{\varepsilon_\perp + \Delta\varepsilon n_z^2}\left[\varepsilon_{33} - \chi^2\left(1 - \frac{\Delta\varepsilon}{\varepsilon_{//}} n_y^2\right)\right]^{1/2}. \qquad (3.147)$$

Note when $n_z \to 0$, $Q_{11} \to 0$ and $Q_{13} \to 0$, but not Q_{13}/Q_{11}. It can be derived as follows: [16]

$$(-ik_o\Delta z)^0\gamma_o = -\frac{q_2q_3q_4\mathrm{e}^{-ik_o\Delta zq_1}}{(q_1-q_2)(q_1-q_3)(q_1-q_4)} - \frac{q_1q_3q_4\mathrm{e}^{-ik_o\Delta zq_2}}{(q_2-q_1)(q_2-q_3)(q_2-q_4)} - \frac{q_1q_2q_4\mathrm{e}^{-ik_o\Delta zq_3}}{(q_3-q_1)(q_3-q_2)(q_3-q_4)} - \frac{q_1q_2q_3\mathrm{e}^{-ik_o\Delta zq_4}}{(q_4-q_1)(q_4-q_2)(q_4-q_3)} \tag{3.148}$$

$$(-ik_o\Delta z)\gamma_1 = \frac{(q_2q_3+q_2q_4+q_3q_4)\mathrm{e}^{-ik_o\Delta zq_1}}{(q_1-q_2)(q_1-q_3)(q_1-q_4)} + \frac{(q_1q_3+q_1q_4+q_3q_4)\mathrm{e}^{-ik_o\Delta zq_2}}{(q_2-q_1)(q_2-q_3)(q_2-q_4)} + \frac{(q_1q_2+q_1q_4+q_2q_4)\mathrm{e}^{-ik_o\Delta zq_3}}{(q_3-q_1)(q_3-q_2)(q_3-q_4)} + \frac{(q_1q_2+q_1q_3+q_2q_3)\mathrm{e}^{-ik_o\Delta zq_4}}{(q_4-q_1)(q_4-q_2)(q_4-q_3)} \tag{3.149}$$

$$(-ik_o\Delta z)^2\gamma_2 = -\frac{(q_2+q_3+q_4)\mathrm{e}^{-ik_o\Delta zq_1}}{(q_1-q_2)(q_1-q_3)(q_1-q_4)} - \frac{(q_1+q_3+q_4)\mathrm{e}^{-ik_o\Delta zq_2}}{(q_2-q_1)(q_2-q_3)(q_2-q_4)} - \frac{(q_1+q_2+q_4)\mathrm{e}^{-ik_o\Delta zq_3}}{(q_3-q_1)(q_3-q_2)(q_3-q_4)} - \frac{(q_1+q_2+q_3)\mathrm{e}^{-ik_o\Delta zq_4}}{(q_4-q_1)(q_4-q_2)(q_4-q_3)} \tag{3.150}$$

$$(-ik_o\Delta z)^3\gamma_3 = \frac{q_1\mathrm{e}^{-ik_o\Delta zq_1}}{(q_1-q_2)(q_1-q_3)(q_1-q_4)} + \frac{q_2\mathrm{e}^{-ik_o\Delta zq_2}}{(q_2-q_1)(q_2-q_3)(q_2-q_4)} + \frac{q_3\mathrm{e}^{-ik_o\Delta zq_3}}{(q_3-q_1)(q_3-q_2)(q_3-q_4)} + \frac{q_4\mathrm{e}^{-ik_o\Delta zq_4}}{(q_4-q_1)(q_4-q_2)(q_4-q_3)} \tag{3.151}$$

We consider some special cases below.

3.3.0.1 Isotropic medium

Consider an isotropic medium with the dielectric constant $\varepsilon = n^2$, where n is the refractive index. From Equation (3. 138) we have

$$\overleftrightarrow{Q} = \begin{pmatrix} 0 & \frac{-\chi^2}{\varepsilon}+1 & 0 & 0 \\ \varepsilon & 0 & 0 & 0 \\ 0 & 0 & 0 & 1 \\ 0 & 0 & -\chi^2+\varepsilon & 0 \end{pmatrix} = \begin{pmatrix} 0 & \frac{k_z^2}{(nk_o)^2} & 0 & 0 \\ n^2 & 0 & 0 & 0 \\ 0 & 0 & 0 & 1 \\ 0 & 0 & \frac{k_z^2}{k_o^2} & 0 \end{pmatrix}. \tag{3.152}$$

In the medium the wave vector is $k = nk_o$. From Equations (3.146) and (3.147) we have eigenvalues of the Berreman matrix

$$q_{1,3/2,4} = \pm\left(n^2-\chi^2\right)^{1/2} = \frac{k_z}{k_o} = \pm n\cos\alpha. \tag{3.153}$$

The eigenvalues are degenerate and in this case $e^{-ik_o\overleftrightarrow{Q}\Delta z}$ can be expanded as

$$e^{-ik_o\overleftrightarrow{Q}\Delta z} = \gamma_o \overleftrightarrow{I} + \gamma_1\left(-ik_o \overleftrightarrow{Q}\,\Delta z\right), \tag{3.154}$$

where γ_i ($i = 0, 1$) are the solution of the following equations:

$$\gamma_o + \gamma_1(-ik_z\Delta z) = e^{-ik_z\Delta z} \tag{3.155}$$

$$\gamma_o + \gamma_1(+ik_z\Delta z) = e^{+ik_z\Delta z} \tag{3.156}$$

From these two equations we find

$$\gamma_o = \cos(k_z\Delta z), \tag{3.157}$$

$$\gamma_1 = \frac{1}{k_z\Delta z}\sin(k_z\Delta z), \tag{3.158}$$

$$e^{-ik_o\overleftrightarrow{Q}\Delta z} = \cos(k_z\Delta z)\begin{pmatrix}1&0&0&0\\0&1&0&0\\0&0&1&0\\0&0&0&1\end{pmatrix} - \frac{ik_o}{k_z}\sin(k_z\Delta z)\begin{pmatrix}0&\frac{k_z^2}{k^2}&0&0\\n^2&0&0&0\\0&0&0&1\\0&0&\frac{n^2k_z^2}{k^2}&0\end{pmatrix}$$

$$= \begin{pmatrix}\cos(k_z\Delta z) & -\frac{ik_z}{nk}\sin(k_z\Delta z) & 0 & 0\\ -\frac{ink}{k_z}\sin(k_z\Delta z) & \cos(k_z\Delta z) & 0 & 0\\ 0 & 0 & \cos(k_z\Delta z) & -\frac{ik}{nk_z}\sin(k_z\Delta z)\\ 0 & 0 & -\frac{ink_z}{k}\sin(k_z\Delta z) & \cos(k_z\Delta z)\end{pmatrix}. \tag{3.159}$$

We also know that $E_i(z+\Delta z) = e^{-ik_z\Delta z}E_i(z)$ and $H_i(z+\Delta z) = e^{-ik_z\Delta z}H_i(z)$ $(i = x, y)$. From Equation (3.139), in components we have

$$E_x(z+\Delta z) = e^{-ik_z\Delta z}E_x(z) = \cos(k_z\Delta z)E_x(z) - \frac{ik_z}{nk}\sin(k_z\Delta z)\eta_o H_y(z), \tag{3.160}$$

$$\eta_o H_y(z+\Delta z) = e^{-ik_z\Delta z}\eta_o H_y(z) = \cos(k_z\Delta z)\eta_o H_y(z) - \frac{ink}{k_z}\sin(k_z\Delta z)E_x(z), \tag{3.161}$$

$$E_y(z+\Delta z) = e^{-ik_z\Delta z}E_y(z) = \cos(k_z\Delta z)E_y(z) + \frac{ik_z}{nk}\sin(k_z\Delta z)\eta_o H_x(z), \tag{3.162}$$

$$\eta_o H_x(z+\Delta z) = e^{-ik_z\Delta z}\eta_o H_x(z) = \cos(k_z\Delta z)\eta_o H_x(z) + \frac{ink}{k_z}\sin(k_z\Delta z)E_y(z). \tag{3.163}$$

Therefore

$$\eta_o H_y(z) = \frac{nk}{k_z}E_x(z) = \frac{n}{\cos\alpha}E_x(z), \tag{3.164}$$

$$-\eta_o H_x(z) = \frac{nk}{k_z}E_y(z) = n\cos\alpha E_y(z). \tag{3.165}$$

The Berreman vector in the isotropic medium is

$$\vec{\psi}^T = \begin{pmatrix} E_x & \frac{n}{\cos\alpha}E_x & E_y & n\cos\alpha E_y \end{pmatrix}. \tag{3.166}$$

From Equation (3.136) we have

$$E_z = \frac{-1}{\varepsilon_{33}}\left[(\eta_o k_x/k_o)H_y + \varepsilon_{31}E_x + \varepsilon_{32}E_y\right] = \frac{-1}{\varepsilon}(k_x/k_o)\frac{n}{\cos\alpha}E_x(z) = -\frac{\sin\alpha}{\cos\alpha}E_x(z). \tag{3.167}$$

The intensity of light is

$$I = n\left(|E_p|^2 + |E_s|^2\right) = n\left[\left(|E_x|^2 + |E_z|^2\right) + |E_y|^2\right] = n\left(|E_x|^2/\cos^2\alpha + n|E_y|^2\right), \tag{3.168}$$

where E_p and E_s are the components of the electric vector in and perpendicular to the incident plane, respectively.

3.3.0.2 Cholesteric liquid crystal

For a uniaxial cholesteric liquid crystal that has the chirality q_o and the refractive indices n_o and n_e, when it is in the planar state, $n_x = \cos(q_o z)$, $n_y = \sin(q_o z)$, and $n_z = 0$. The dielectric tensor is

$$\overleftrightarrow{\varepsilon} = \begin{pmatrix} \varepsilon_{11} & \varepsilon_{12} & 0 \\ \varepsilon_{12} & \varepsilon_{22} & 0 \\ 0 & 0 & \varepsilon_{33} \end{pmatrix} = \begin{pmatrix} \varepsilon_\perp + \Delta\varepsilon\cos^2(q_o z) & \Delta\varepsilon\sin(q_o z)\cos(q_o z) & 0 \\ \Delta\varepsilon\sin(q_o z)\cos(q_o z) & \varepsilon_\perp - \Delta\varepsilon\cos^2(q_o z) & 0 \\ 0 & 0 & \varepsilon_\perp \end{pmatrix}. \tag{3.169}$$

The Berreman matrix is

$$\overleftrightarrow{Q} = \begin{pmatrix} 0 & 1-\chi^2/\varepsilon_\perp & 0 & 0 \\ \varepsilon_\perp + \Delta\varepsilon n_x^2 & 0 & \Delta\varepsilon n_x n_y & 0 \\ 0 & 0 & 0 & 1 \\ \Delta\varepsilon n_x n_y & 0 & (\varepsilon_\perp - \chi^2) + \Delta\varepsilon n_y^2 & 0 \end{pmatrix}. \tag{3.170}$$

From Equations (3.146) and (3.147) we have eigenvalues of the Berreman matrix

$$q_{1/2} = \pm \left[\varepsilon_{\perp} - \chi^2\right]^{1/2}, \tag{3.171}$$

$$q_{3/4} = \pm \left(\frac{\varepsilon_{//}}{\varepsilon_{\perp}}\right)^{1/2} \left[\varepsilon_{\perp} - \chi^2\left(1 - \frac{\Delta\varepsilon}{\varepsilon_{//}} n_y^2\right)\right]^{1/2} = \pm \left[\varepsilon_{//} - \chi^2 \frac{1}{\varepsilon_{\perp}}\left(\varepsilon_{//} - \Delta\varepsilon n_y^2\right)\right]^{1/2}. \tag{3.172}$$

Outside the cholesteric cell, the medium is an isotropic medium with the refractive index n_g. On top of the Ch film (incident side), there is incident light and reflected light, and the actual Berreman vector is the sum of the Berreman vectors of the incident light and reflected light. From Equation (3.166) we know that for the incident light, the Berreman vector is

$$\vec{\psi}_i^T = \begin{pmatrix} E_{xi} & \frac{n_g}{\cos\alpha} E_{xi} & E_{yi} & n_g \cos\alpha E_{yi} \end{pmatrix}. \tag{3.173}$$

For the reflected light, because it propagates in the reverse direction, the Berreman vector is

$$\vec{\psi}_i^T = \begin{pmatrix} E_{xr} & \frac{-n_g}{\cos\alpha} E_{xr} & E_{yr} & -n_g \cos\alpha E_{yr} \end{pmatrix}. \tag{3.174}$$

At the bottom of the Ch film, there is only the transmitted light whose Berreman vector is

$$\vec{\psi}_i^T = \begin{pmatrix} E_{xt} & \frac{n_g}{\cos\alpha} E_{xt} & E_{yt} & n_g \cos\alpha E_{yt} \end{pmatrix}. \tag{3.175}$$

We divide the cholesteric film into N slabs with thicknesses Δz. The Berreman vectors at the boundaries between the slabs are:

$$\vec{\psi}(0) = \vec{\psi}_i + \vec{\psi}_r,$$

$$\vec{\psi}(1) = \overleftrightarrow{P}(z_1) \cdot \vec{\psi}(0),$$

$$\vec{\psi}(2) = \overleftrightarrow{P}(z_2) \cdot \vec{\psi}(1) = \overleftrightarrow{P}(z_2) \cdot \overleftrightarrow{P}(z_1) \cdot \vec{\psi}(0),$$

$$\vdots$$

$$\vec{\psi}_t = \vec{\psi}(N) = \prod_{i=1}^{N} \overleftrightarrow{P}(z_i) \cdot \vec{\psi}(0) = \prod_{i=1}^{N} \overleftrightarrow{P}(z_i) \cdot (\vec{\psi}_i + \vec{\psi}_r) \equiv \overleftrightarrow{B} \cdot (\vec{\psi}_i + \vec{\psi}_r), \tag{3.176}$$

where the $\overleftrightarrow{P}$ for each slab can be numerically calculated by using the fast Berreman method. In components, Equation (3.176) contains four equations. $\vec{\psi}_i$ is given. $\vec{\psi}_r$ and $\vec{\psi}_t$ have two unknown variables each and can be found by solving Equation (3.176).

Define a new vector $\vec{\psi}_{t/r}$

$$\vec{\psi}^{\,T}_{t/r} = (E_{tx} \;\; E_{ty} \;\; E_{rx} \;\; E_{ry}). \tag{3.177}$$

The Berreman vectors of the transmitted and reflected light are related to $\vec{\psi}_{t/r}$ by

$$\vec{\psi}_t = \begin{pmatrix} 1 & 0 & 0 & 0 \\ n_g/\cos\alpha & 0 & 0 & 0 \\ 0 & 1 & 0 & 0 \\ 0 & n_g \cos\alpha & 0 & 0 \end{pmatrix} \cdot \vec{\psi}_{t/r} \equiv \overleftrightarrow{A}_t \cdot \vec{\psi}_{t/r}, \tag{3.178}$$

$$\vec{\psi}_t = \begin{pmatrix} 0 & 0 & 1 & 0 \\ 0 & 0 & -n_g/\cos\alpha & 0 \\ 0 & 0 & 0 & 1 \\ 0 & 0 & 0 & -n_g \cos\alpha \end{pmatrix} \cdot \vec{\psi}_{t/r} \equiv \overleftrightarrow{A}_r \cdot \vec{\psi}_{t/r}. \tag{3.179}$$

From Equation (3.176) we have

$$\overleftrightarrow{A}_t \cdot \vec{\psi}_{t/r} \equiv \overleftrightarrow{B} \cdot \left(\vec{\psi}_i + \overleftrightarrow{A}_r \cdot \vec{\psi}_{t/r} \right), \tag{3.180}$$

$$\vec{\psi}_{t/r} \equiv \left(\overleftrightarrow{A}_t + \overleftrightarrow{B} \cdot \overleftrightarrow{A}_r \right)^{-1} \cdot \overleftrightarrow{B} \cdot \vec{\psi}_i, \tag{3.181}$$

The electric field components of the transmitted and reflected light can be calculated from the above equation.

$$R = \left[(E_{xr}/\cos\alpha)^2 + E_{yr}^2 \right] \Big/ \left[(E_{xi}/\cos\alpha)^2 + E_{yi}^2 \right]. \tag{3.182}$$

Therefore the transmittance can be calculated by

$$T = \left[(E_{xt}/\cos\alpha)^2 + E_{yt}^2 \right] \Big/ \left[(E_{xi}/\cos\alpha)^2 + E_{yi}^2 \right]. \tag{3.183}$$

The reflection of a cholesteric liquid crystal in the planar texture depends on the polarization state of the incident light and how the reflected light is measured. As an example, we consider a cholesteric liquid crystal with the following parameters: $P_o = 338$ nm, cell thickness $h = 5070$ nm, $n_o = 1.494$ and $n_e = 1.616$. The incident angle is 22.5°. The refractive index of the glass substrates is 1.5. On the surface of the glass substrates there are an ITO conducting film and a polyimide alignment layer. The thickness of the ITO is 25 nm and its refractive index is $n(\lambda) = 2.525 - 0.001271\lambda$, where λ is the wavelength of light with the unit of nm. The thickness of the polyimide is 98 nm and its refractive index is 1.7. The reflection spectra are shown in Figure 3.10 [21]. In Figure 3.10(a) and (b), the incident light is linearly polarized perpendicular to the incident plane and the component of the reflected light in the same direction is measured. In Figure 3.10(a), the polarization of the incident light is parallel to the liquid crystal director in the entrance plane, while in Figure 3.10(b), the polarization is perpendicular to the liquid crystal director in the entrance plane. The spectra are very different because of the interference

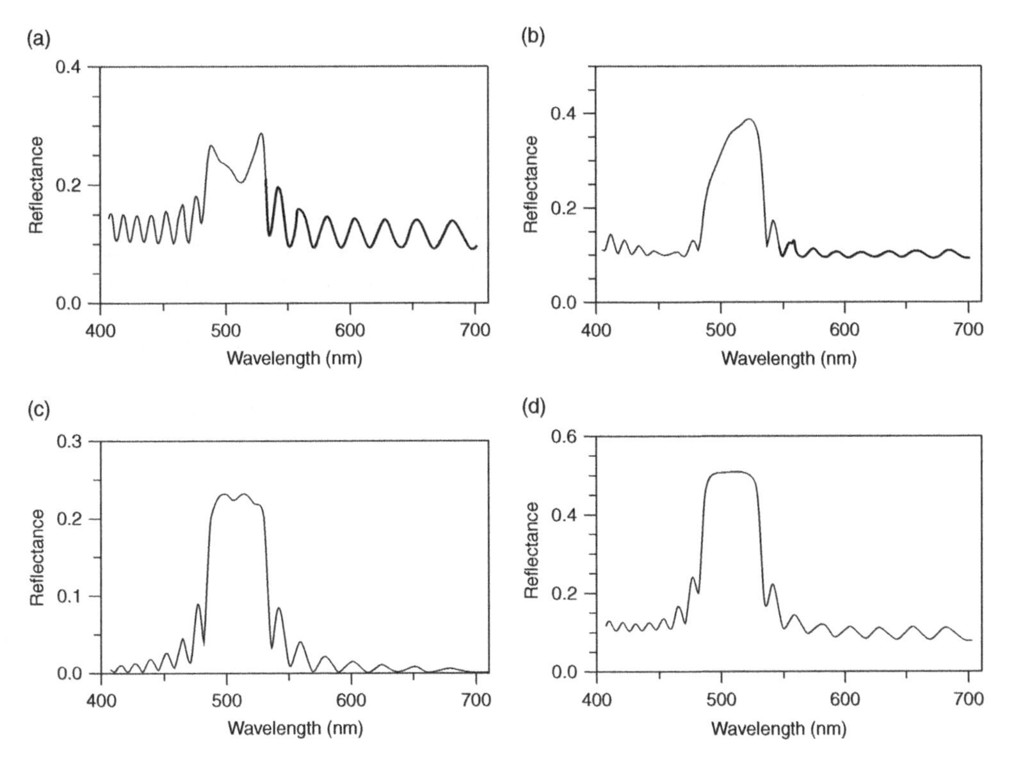

Figure 3.10 The reflection spectra of the cholesteric liquid crystal. (a) Incident light: σ-polarization and parallel to the liquid crystal director on the entrance plane; detection: σ-polarization. (b) Incident light: σ-polarization and perpendicular to the liquid crystal director on the entrance plane; detection: σ-polarization. (c) Incident light: σ-polarization; detection: π-polarization. (d) Incident light: unpolarized; detection: unpolarized. [21]. Reproduced with permission from the American Institute of Physics.

between the light reflected from the liquid crystal and the light reflected from the interfaces between the glass substrate, the ITO electrode, the alignment layer, and the liquid crystal. In Figure 3.10(a) the components interfere destructively and therefore there is a dip in the middle of the reflection band. In Figure 3.10(b) they interfere constructively and therefore the reflection is higher in the middle of the reflection band. In Figure 3.10(c), crossed polarizers are used. The light reflected from the interfaces cannot go through the analyzer and is not detected. The linearly polarized incident light can be decomposed into two circularly polarized components, and one of them is reflected. The reflected circularly polarized light can be decomposed into two linearly polarized components, and one of them passes the analyzer. Therefore the maximum reflection is 25%. In Figure 3.10(d), the incident light is unpolarized and all the reflected light is detected. The reflection in the band is slightly higher than 50% because of the light reflected from the interfaces. The fringes are due to the finite thickness of the liquid crystal. It exists even when the substrates have a refractive index match to that of the liquid crystal, but disappears for infinitely thick samples. These simulated results agree very well with experimental results.

Homework Problems

3.1 Use the Jones matrix method to numerically calculate the transmittance of a 90° twisted nematic display in the field-off state as a function of the retardation $u = 2\Delta nh/\lambda$. The polarizers are parallel to each other and are also parallel to the liquid crystal director at the entrance plane. Compare your result with Figure 3.4.

3.2 Consider a 90° twisted nematic cell sandwiched between two polarizers. Use Equation (3.45) to calculate and plot the transmittance as a function of $u = 2\Delta nh/\lambda$ in the following case. The polarizers are parallel to each other and the transmission axis of the polarizer at the entrance plane is parallel to the liquid crystal director.

3.3 Using the definition of Stokes vector, derive the Mueller matrix given by Equation (3.70) of an optical element whose Jones matrix given by Equation (3.69).

3.4 Linearly polarized light is normally incident on a homogeneously aligned nematic liquid crystal that acts as a half wave plate. The polarization is along the x direction. The liquid crystal director is at an angle of 22.5^o with respect to the x axis. Sketch the polarization trajectory on the Poincare sphere when the light propagates through the liquid crystal.

3.5 Derive the Mueller matrix given by Equation (3.77) of a retarder whose retardation is Γ and its slow axis makes the angle ϕ with respect to the x axis.

3.6 Use Equation Equations (3.34) and (3.70) to derive Equation (3.114)

3.7 In the Berreman 4 × 4 method, using $\nabla \cdot \vec{D} = 0$, prove k_x = constant.

3.8 Using the $\overleftrightarrow{Q}$ given by Equation (3.145), prove that the q given by Equations (3.146) and (3.147) are solution to Equation (3.143).

3.9 *Cell thickness-dependence of the reflection of a cholesteric liquid crystal in the planar state*. The pitch of the liquid crystal is $P = 350\ nm$. The refractive indices of the liquid crystal are $n_e = 1.7$ and $n_o = 1.5$. The liquid crystal is sandwiched between two glass plates with the refractive index $n_g = 1.6$. The incident light is circularly polarized with the same helical handedness as the liquid crystal. Neglect the reflection from the glass-air interface. Use two methods to calculate the reflection spectrum of the liquid crystal with the following cell thicknesses: P, $2P$, $5P$ and $10P$. The first method is the Berreman 4 × 4 method and the second method is using Equation (2.186). Compare the results from the two methods.

3.10 (a) A system consists of a polarizer, a quarter-wave plate and a mirror. The quarter-wave plate is sandwiched between the polarizer and the mirror. Use Jones matrix method to calculate the reflection spectrum in visible light region. (b) A systems consists of a polarizer, a broadband quarter-wave plate, described in Figure 3.8(b), and a mirror. The quarter-wave plate is sandwiched between the polarizer and the mirror. Use Jones matrix method to calculate the reflection spectrum in visible light region.

3.11 Use Berreman 4 × 4 method for Problem 3.10

3.12 Use the Berreman 4 × 4 method to calculate the reflection spectra of the cholesteric film under the polarization conditions specified in Figure 3.10. The parameters of the cholesteric liquid crystal are also given in Figure 3.10.

3.13 Use the Jones matrix method and the Berreman matrix method separately to calculate the transmittance pattern in the following two cases as a function the polar and azimuthal angles θ and ϕ of the incident light. A uniaxial birefringent film is sandwiched between two crossed polarizers. The transmission axis of the polarizer at the entrance plane is along the x axis. (1) an a plate has the retardation $\Delta nd = \lambda$ and its slow axis makes the angle of 45° with respect to the x axis. (2) a c plate has the retardation $\Delta nd = \lambda$. If the results obtained by the two methods are different, explain the difference. Neglect reflection in the Berreman method.

3.14 A reflection system consists of a linear polarizer on the top, a quarter-wave plate in the middle, and a mirror at the bottom. Calculate the reflection spectrum of the system in the wavelength region from 400 nm to 700 nm. (a) The quarter-wave plate is a regular quarter-wave plate with its principal axis at 45° with respect to the transmission axis of the polarizer. (b) The quarter-wave plate is the broadband quarter-wave plate as shown in Figure 3.8(b).

References

1. R. C. Jones, *J. Opt. Soc. Am.*, **31**, 488 (1941).
2. I. J. Hodgkinson and Q. H. Wu, *Birefringent thin films and polarizing elements* (World Scientific, Singapore, 1997).
3. S. Chandrasekhar *Liquid crystals*, 2nd edn (Cambridge University Press, New York, 1997).
4. P. Yeh and C. Gu, *Optics of liquid crystal displays* (John Wiley and Sons, New York, 1999).
5. P. Yeh, Extended Jones matrix method, *J. Opt. Soc. Am.*, **72**, 507 (1982)
6. C. Gu and P. Yeh, Extended Jones matrix method II, *J. Opt. Soc. Am. A*, **10**, 966 (1993)
7. A. Lien, The general and simplified Jones matrix representations for the high pretilt twisted nematic cell, *J. Appl. Phys.* **67**, 2853 (1990).
8. T. Scheffer and J. Nehring, Twisted nematic and super-twisted nematic mode LCDs, in *Liquid crystals – applications and uses*, Vol. 1, ed. by B. Bahadur (World Scientific, New Jersey, 1990).
9. R. A. Horn and C. R. Johnson, *Matrix analysis* (Cambridge University Press, 1985)
10. C. Mauguin, Sur les cristaux liquids de Lehman, *Bull. Soc. Franc. Mineral,* **34**, 71–117 (1911).
11. H. L. Ong, Optical properties of general twisted nematic liquid-crystal displays, *Appl. Phys. Lett.*, **51**, 1398 (1987).
12. H. L. Ong, Origin and characteristics of the optical properties of general twisted nematic liquid crystals, *J. Appl. Phys.*, **64**, 614 (1988).
13. H. C. van de Hulst, *Light scattering by small particles* (Dover Publications, New York, 1957).
14. D. W. Berreman, Optics in stratified and anisotropic media: 4 × 4-matrix formulation, *J. Opt. Soc. Am.*, **62**, 502 (1972).
15. D. W. Berreman, Optics in smoothly varying anisotropic planar structures: application to liquid-crystal twist cell*, *J. Opt. Soc. Am.*, **63**, 1374 (1973).
16. H. Wöhler and M. E. Becker, The optics of liquid crystals, Seminar Lecture Notes, *EuroDisplay '93.*
17. D. W. Berreman and T. J. Scheffer, Bragg reflection of light from single-domain cholesteric liquid crystal films, *Phys. Rev. Lett.*, **25**, 577 (1970).
18. D. W. Berreman and T. J. Scheffer, Reflection and transmission by single-domain cholesteric liquid crystal films: theory and verification, *Mol. Cryst. Liq. Cryst.*, **11**, 395 (1970).
19. D. W. Berreman, Ultrafast 4x4 matrix optics with averaged interference fringes, *SID International Symposium, Seminar and Exhibition,* 101–104, Seattle, USA 16–21. May (1993).
20. H. Wöhler, G. Haas, M. Fritsch, and D. A. Mlynski, Faster 4 × 4 method, for uniaxial inhomogeneous media, *J. Opt. Soc. Am. A*, **5**, 1554 (1988).
21. M. Xu, F.D. Xu and D.-K. Yang, Effects of cell structure on the reflection of cholesteric liquid crystal display, *J. Appl. Phys.*, **83**, 1938 (1998).

4

Effects of Electric Field on Liquid Crystals

One of the main reasons, if not the only reason, that liquid crystals are of great importance in display applications is their ready response to externally applied electric fields [1,2]. Their direction can be easily changed by electric fields produced by the application of a few volts across the liquid crystal cells. They are either dielectric or ferroelectric materials with high resistivities and thus consume little energy. When the liquid crystals reorient, their optical properties change dramatically because of their large birefringences. In this chapter, we will first discuss how liquid crystals interact with externally applied electric fields, and then consider their applications.

4.1 Dielectric Interaction

Uniformly oriented uniaxial nematic liquid crystals of rod-like molecules are non-polar because of the intermolecular interaction and the resulting symmetry of $D_{\infty h}$ (in Schoenflies notation) [3]. The continuous rotational symmetry axis is parallel to the liquid crystal director $\vec{n}$. A uniformly oriented nematic liquid crystal is invariant for a rotation of any angle around $\vec{n}$. It is also invariant for the reflection symmetry operation about the plane perpendicular to $\vec{n}$. In the absence of an external electric field, they have non-polar cylindrical symmetry. If the liquid crystal molecules have a permanent dipole along the long molecular axis, the dipole has the same probability of pointing up and pointing down with respect to the liquid crystal director $\vec{n}$. If the permanent dipole is perpendicular to the long molecular axis, it has the same probability of pointing in any direction perpendicular to the director. There is no spontaneous polarization and therefore uniformly aligned nematic liquid crystals are dielectrics.

Fundamentals of Liquid Crystal Devices, Second Edition. Deng-Ke Yang and Shin-Tson Wu.

4.1.1 Reorientation under dielectric interaction

When an electric field is applied to a nematic liquid crystal, it induces polarization. As discussed in Chapter 1, the induced polarization depends on the orientation of the liquid crystal director with respect to the applied field because the permittivity in the direction parallel to $\vec{n}$ is different from that in the direction perpendicular to $\vec{n}$. When the applied field is parallel to $\vec{n}$, the permittivity is $\chi_{//}$; when the applied field is perpendicular to $\vec{n}$, the permittivity is $\chi_{\perp}$. When the applied field is neither parallel to nor perpendicular to $\vec{n}$, as shown in Figure 4.1, the applied electric field can be decomposed into a component parallel to $\vec{n}$ and another component perpendicular to $\vec{n}$. The induced polarization is given by

$$\vec{P} = \varepsilon_o\chi_{//}\left(\vec{E}\cdot\vec{n}\right)\vec{n} + \varepsilon_o\chi_{\perp}\left[\vec{E}-\left(\vec{E}\cdot\vec{n}\right)\vec{n}\right] = \varepsilon_o\left[\chi_{\perp}\vec{E}+\Delta\chi\left(\vec{E}\cdot\vec{n}\right)\vec{n}\right]. \tag{4.1}$$

The dielectric constants $\varepsilon_{//}$ and $\varepsilon_{\perp}$ are related to the permittivities by $\varepsilon_{//} = 1 + \chi_{//}$ and $\varepsilon_{\perp} = 1 + \chi_{\perp}$. Therefore $\Delta\chi = \chi_{//} - \chi_{\perp} = \varepsilon_{//} - \varepsilon_{\perp} = \Delta\varepsilon$. The electric energy of the liquid crystal per unit volume is approximately given by (detailed discussion will be presented in Chapter 7)

$$f_{electric} = -\frac{1}{2}\vec{P}\cdot\vec{E} = -\frac{1}{2}\varepsilon_o\left[\chi_{\perp}\vec{E}+\Delta\chi\left(\vec{E}\cdot\vec{n}\right)\vec{n}\right]\cdot\vec{E} = -\frac{1}{2}\varepsilon_o\chi_{\perp}E^2 - \frac{1}{2}\varepsilon_o\Delta\varepsilon\left(\vec{E}\cdot\vec{n}\right)^2 \tag{4.2}$$

When the applied field is low, $\Delta\varepsilon$ can be approximately considered as a constant, independent of the field. The first term on the right side of Equation (4.2) is independent of the orientation of the director with respect to the applied field, and thus can be neglected in considering the reorientation of liquid crystals in electric fields. The second term depends on the orientation of the director with respect to to the applied field. When $\vec{n}$ is perpendicular to $\vec{E}$, $\left(\vec{E}\cdot\vec{n}\right)^2 = 0$. When $\vec{n}$ is parallel or anti-parallel to $\vec{E}$, $\left(\vec{E}\cdot\vec{n}\right)^2 = E^2$. If the liquid crystal has a positive dielectric anisotropy ($\Delta\varepsilon > 0$), the electric energy is minimized when the liquid crystal director is parallel or anti-parallel to the applied field; therefore the liquid crystal tends to align parallel (or anti-parallel) to the applied field. Conversely, if the dielectric anisotropy is negative ($\Delta\varepsilon < 0$), then the electric energy is low when the liquid crystal director is perpendicular to the applied field; therefore the liquid crystal tends to align perpendicular to the applied field. The dielectric responses of liquid crystals to DC and AC electric fields are same (except that the dielectric

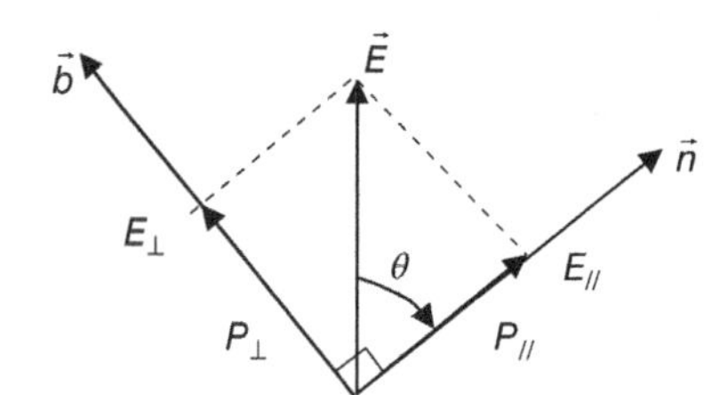

Figure 4.1 Schematic diagram showing the field decomposed into the components parallel and perpendicular to the liquid crystal director.

constants may be frequency dependent). For most nematic liquid crystals, the dielectric anisotropy is in the region from −5 to +30. For example when $\Delta\varepsilon = 10$ and the applied electric field is 1 V/μm = 10^6 V/m, the electric energy density is $\frac{1}{2}\varepsilon_o\Delta\varepsilon E^2 = 44.2\ \mathrm{J/m^3}$. The reorientation of liquid crystals under dielectric interaction will be discussed more in detail in Chapters 5 and 7.

4.1.2 *Field-induced orientational order*

Besides aligning liquid crystals, external electric fields can also change the orientational order and thus the electro-optical properties of liquid crystals. When the long molecular axis of a liquid crystal molecule, whose anisotropy of polarizability is positive, is parallel to the applied field, the potential of the molecule is low. Thus the applied field suppresses the thermal fluctuation and increases the order parameter. Now we discuss how the orientational order of a nematic liquid crystal changes with applied fields. Using the Landau–de Gennes theory, the free energy density of a liquid crystal in an electric field (when the liquid crystal director is parallel to the field) is [4]

$$f = -\frac{1}{2}\varepsilon_o\varepsilon_{//}E^2 + \frac{1}{2}a(T - T^*)S^2 - \frac{1}{3}bS^3 + \frac{1}{4}cS^4. \tag{4.3}$$

As discussed in Section 1.5.2, $\varepsilon_{//} + 2\varepsilon_\perp = \text{constant}$, namely, $3\varepsilon_{//} - 2\Delta\varepsilon = \text{constant}$ and therefore $\varepsilon_{//} = (2/3)(\Delta\varepsilon + \text{constant})$. Equation (1.114) shows that the dielectric anisotropy $\Delta\varepsilon$ is a linear function of the order parameter S. At a temperature below the isotropic–nematic transition temperature and under zero applied field, when the order parameter is S_o, the dielectric anisotropy is $(\Delta\varepsilon)_o$. Approximately we have

$$\varepsilon_{//} = \frac{2(\Delta\varepsilon)_o}{3S_o}S + \frac{2}{3}\times\text{constant}. \tag{4.4}$$

In the calculation of the order parameter by minimizing the free energy, the constant term can be neglected. The free energy density becomes

$$f = -\frac{1}{2}\varepsilon_o\alpha E^2 S + \frac{1}{2}a(T - T^*)S^2 - \frac{1}{3}bS^3 + \frac{1}{4}cS^4 \tag{4.5}$$

where $\alpha = 2(\Delta\varepsilon)_o/3S_o$. The first term on the right side of Equation (4.5) is negative and decreases with increasing order parameter, provided the dielectric anisotropy is positive. Therefore the applied field tends to increase the order parameter. Even in the isotropic phase at temperatures above the nematic–isotropic phase transition temperature, nematic order is induced by the applied electric field. This phase with field-induced order parameter is referred to as the *paranematic* phase.

Define the normalized field $e = \varepsilon_o\alpha E^2/2c$, the normalized temperature $t = a(T - T^*)/c$, and $\beta = b/c$. The normalized free energy density becomes

$$\frac{f}{c} = -eS + \frac{1}{2}tS^2 - \frac{1}{3}\beta S^3 + \frac{1}{4}S^4. \tag{4.6}$$

The order parameter S as a function of the applied field E can be found by minimizing the free energy:

$$\frac{\partial(f/c)}{\partial S} = -e + tS - \beta S^2 + S^3 \equiv 0 \tag{4.7}$$

There are three solutions to Equation (4.7). At a given temperature, the real order parameter is the one that minimizes the free energy. When the applied field is low, the induced order parameter in the paranematic phase is small. When the temperature is lowered, there is a paranematic–nematic phase transition. At the phase transition, the order parameter changes discontinuously. As the applied field is increased, the paranematic–nematic phase transition temperature t_{PN} increases, as shown in Figure 4.2(a), and the jump of the order parameter at the transition becomes smaller, as shown in Figure 4.2(b). When the applied field is increased above a critical field E_c, the jump of the order parameter becomes zero. The phase transition temperature, the jump of the order parameter ΔS at the transition, and the critical field E_c can be derived without explicitly calculating the order parameter. Let the order parameter in the paranematic phase be S_1 and the order parameter in the nematic phase be S_2. At the transition temperature t_{PN}, we have

$$-e + t_{PN}S_1 - \beta S_1^2 + S_1^3 \equiv 0, \tag{4.8}$$

$$-e + t_{PN}S_2 - \beta S_2^2 + S_2^3 \equiv 0. \tag{4.9}$$

Also at the transition temperature t_{PN}, the free energies corresponding to these two solutions are the same, namely,

$$-eS_1 + \frac{1}{2}t_{PN}S_1^2 - \frac{1}{3}\beta S_1^3 + \frac{1}{4}S_1^4 = -\frac{1}{2}eS_2 + \frac{1}{2}t_{PN}S_2^2 - \frac{1}{3}\beta S_2^3 + \frac{1}{4}S_2^4,$$

which gives

$$-4e + 2t_{PN}(S_2 + S_2) - \frac{4}{3}\beta\left(S_2^2 + S_1S_2 + S_1^2\right) + \left(S_2^2 + S_2^2\right)(S_2 + S_2) = 0. \tag{4.10}$$

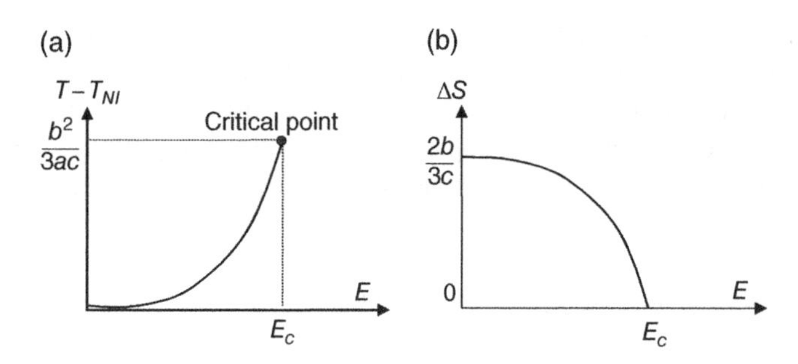

Figure 4.2 Schematic diagram showing how the transition temperature changes with the applied field.

In Equations (4.8), (4.9), and (4.10), there are three variables S_1, S_2, and t_{PN}. Solving these three equations we can find these three variables. For example, we can take the following approach to find them. Equation (4.9) – (4.8) gives

$$t_{PN}(S_2-S_1)-\beta(S_2^2-S_1^2)+(S_2^3-S_1^3)=0,$$

which gives

$$t_{PN}-\beta(S_2+S_1)+(S_2^2+S_2S_1+S_1^2)=0. \tag{4.11}$$

Then $S_2\times(4.9)-S_1\times(4.8)$ gives

$$-e(S_2-S_1)+t_{PN}(S_2^2-S_1^2)-\beta(S_2^3-S_1^3)+(S_2^4-S_1^4)=0,$$

which gives

$$-e+t_{PN}(S_2+S_1)-\beta(S_2^2+S_1S_2+S_1^2)+(S_2^2+S_2^2)(S_2+S_2)=0. \tag{4.12}$$

Equation (4.12) – (4.10) gives

$$3e-t_{PN}(S_2+S_1)+\frac{1}{3}\beta(S_2^2+S_1S_2+S_1^2)=0. \tag{4.13}$$

Then $4\times(4.12)-3\times(4.10)$ gives

$$8e-2t_{PN}(S_2+S_1)+(S_2^2+S_1^2)(S_2+S_1)=0. \tag{4.14}$$

Using $\beta\times(4.11)-3\times(4.13)$ gives

$$S_2+S_1=\frac{\beta t_{PN}-9e}{\beta^2-3t_{PN}}\equiv h. \tag{4.15}$$

Substituting (4.15) into (4.11) we have

$$S_2^2+S_1S_2+S_1^2=\beta h-t_{PN}=\frac{3(t_{PN}^2-3e\beta)}{(\beta^2-3t_{PN})}. \tag{4.16}$$

Substituting (4.15) into (4.14) we have

$$S_2^2+S_1^2=2t_{PN}-\frac{8e}{h}=\frac{2(\beta t_{PN}^2+3et_{PN}-4e\beta^2)}{(\beta t_{PN}-9e)}. \tag{4.17}$$

Because $2(S_2^2+S_2S_1+S_1^2)-(S_2^2+S_1^2)=(S_2+S_1)^2$, from (4.15), (4.16), and (4.17) we have

$$\frac{6(t_{PN}^2-3e\beta)}{(\beta^2-3t_{PN})}-\frac{2(\beta t_{PN}^2+3et_{PN}-4e\beta^2)}{(\beta t_{PN}-9e)}=\frac{(\beta t_{PN}-9e)^2}{(\beta^2-3t_{PN})^2}. \tag{4.18}$$

From (4.18) the transition temperature is found to be

$$t_{NP} = \frac{2\beta^2}{9} + \frac{3e}{\beta}. \tag{4.19}$$

The un-normalized transition temperature is

$$T_{PN} = T^* + \frac{2b^2}{9ac} + \frac{3c\varepsilon_o \alpha E^2}{2ab} = T_{NI} + \frac{3c\varepsilon_o \alpha E^2}{2ab}, \tag{4.20}$$

where T_{NI} is the nematic–isotropic phase transition temperature under zero field. At the paranematic–nematic transition

$$S_2 + S_1 = \frac{2}{3}\beta = \frac{2b}{3c}, \tag{4.21}$$

which is a constant independent of the applied field. The jump of the order parameter at the transition is given by

$$\Delta S = S_2 - S_1 = \left[2\left(S_2^2 + S_1^2\right) - (S_2 + S_1)^2\right]^{1/2} = \left(\frac{4\beta^2}{9} - \frac{12e}{\beta}\right)^{1/2}. \tag{4.22}$$

At the critical field e_c, the jump of the order parameter becomes 0, namely $\Delta S = 0$. Therefore the critical field is

$$e_c = \frac{\beta^3}{27}. \tag{4.23}$$

The un-normalized critical field is

$$E_c = \left(\frac{2b^3}{27c^2\varepsilon_o \alpha}\right)^{1/2}. \tag{4.24}$$

For example, for a liquid crystal with $b = 1.6 \times 10^6$ J/m^3, $c = 3.9 \times 10^6$ J/m^3 and $\alpha = 10$ (the dielectric anisotropy in the nematic phase at zero field is about 10), the critical field is $E_c = 15$ V/μm. The induced order parameter at the critical point is about 0.15. This has been experimentally confirmed [5]. At such a high field, attention must be paid to avoid the heating effect of the field on the liquid crystal cell and electrical breakdown of the material.

4.2 Flexoelectric Effect

4.2.1 Flexoelectric effect in nematic liquid crystals

Uniformly oriented uniaxial nematic liquid crystals have non-polar cylindrical symmetry in the absence of an external electric field. If the liquid crystal molecules have a permanent dipole along the long molecular axis, the potential for the orientation of the dipole has reflection symmetry about the plane perpendicular to the director $\vec{n}$, and the dipole has the same probability to be parallel and anti-parallel to the director, as shown in Figure 4.3(a). If the permanent dipole is perpendicular to the long molecular axis, the potential for the orientation of the dipole is

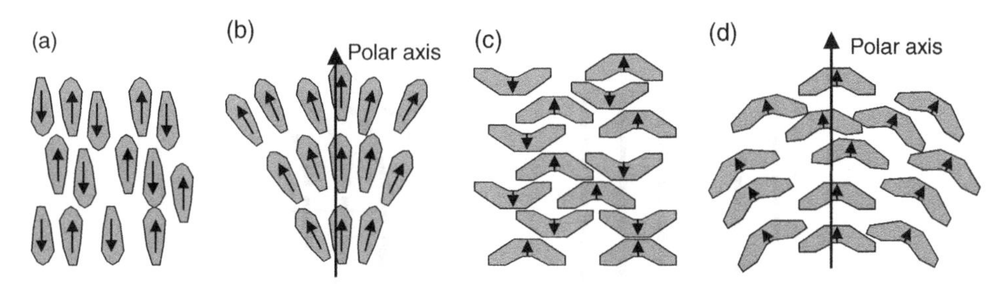

Figure 4.3 Schematic diagram showing the deformation of the liquid crystal director and induce spontaneous polarization.

cylindrically symmetric around the director and the dipole has the same probability of pointing in any direction perpendicular to the director, as shown in Figure 4.3(c).

If the orientation of the liquid crystal is not uniform and the constituent molecules are not cylindrical, the properties discussed in the previous paragraph are no longer true [1,2]. For pear-shaped molecules, because of the stero-interaction, splay deformation of the liquid crystal director will destroy the reflection symmetry about the plane perpendicular to the director, as shown in Figure 4.3(b). The permanent dipole along the long molecular axis has a higher probability of pointing in one direction than the opposite direction, and therefore spontaneous polarization along $\vec{n}$ becomes possible. For banana-shaped molecules, bend deformation will destroy the rotational symmetry around the director, as shown in Figure 4.3(d). The permanent dipole perpendicular to the long molecular axis has a higher (or lower) probability of pointing in the direction of $\vec{n} \times \nabla \times \vec{n}$ than in other directions perpendicular to the director $\vec{n}$, and spontaneous polarization along the direction of $\vec{n} \times \nabla \times \vec{n}$ may occur. This director deformation induced polarization was first pointed out by Meyer and was called the 'piezoelectric' effect in analogue to induced polarization in solid crystals by strain under externally applied pressure [6]. Because the direction deformations in nematics are usually not produced by pressure, 'flexoelectric' is more popularly used in order to avoid confusion.

In the case of pear-shaped molecules, the value of the induced polarization is proportional to the splay deformation $\nabla \cdot \vec{n}$, and its direction is along $\vec{n}$. In the case of the banana-shaped molecules, the induced polarization is proportional to the bend deformation $\vec{n} \times \nabla \times \vec{n}$. Including both cases, the induced polarization is given by

$$\vec{P}_f = e_s\left(\vec{n}\nabla \cdot \vec{n}\right) + e_b\left(\vec{n} \times \nabla \times \vec{n}\right) \tag{4.25}$$

where e_s and e_b are the flexoelectric coefficients and have the dimension of electric potential (volt). The magnitude of the flexoelectric coefficients depends on the asymmetry of the molecule shape and the permanent dipole moment and their sign could be either positive or negative. The energy of the induced polarization in an electric field $\vec{E}$ is $-\vec{P}_f \cdot \vec{E}$.

In liquid crystals with the capability of flexoelectric effect, in the absence of external electric fields, the state with uniform director configuration, which has no induced polarization, is the ground state and is stable. When an electric field is applied to the liquid crystal, the uniform orientation becomes unstable, because any small orientation deformation produced by thermal fluctuation or boundary condition will induce a polarization which will interact with the electric field and results in a lower free energy. The torque on the molecules due to the applied field and

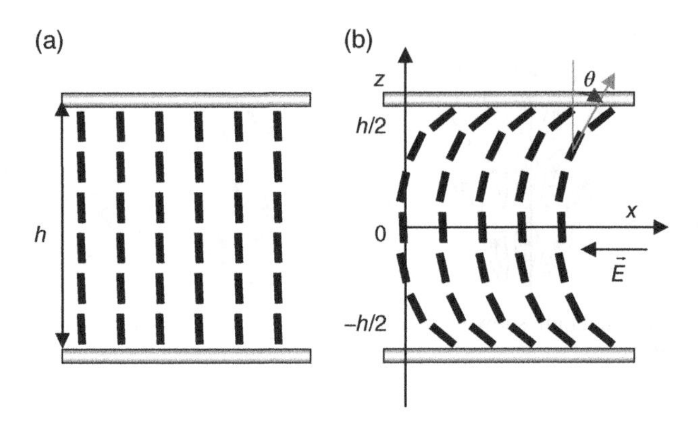

Figure 4.4 Schematic diagram showing the flexoelectric effect in the bend deformation.

the induced polarization tends to make the deformation grow. Of course, the deformation costs elastic energy which is against the deformation. The electric energy of the induced polarization, which is linearly proportional to the deformation, dominates in the beginning where the deformation is small. The elastic energy, which is proportional to the square of the deformation, dominates when the deformation is large. In the end, the system reaches the equilibrium state where the electric torque and the elastic torque balance each other.

Now we consider the experiments which can be used to study the flexoelectric effect and can also be used in electro-optical applications. Figure 4.4 shows the geometry for studying the flexoelectric effect in bend geometry [2,7]. The substrates are coated with a homeotropic alignment layer with very weak anchoring strength. The liquid crystal used has a small negative dielectric anisotropy $\Delta\varepsilon$. In the absence of external electric fields, the liquid crystal is in the uniform homeotropic state, as shown in Figure 4.4(a). When an electric field is applied along the $-x$ direction, a bend deformation occurs, as shown in Figure 4.4(b), because of the flexoelectric effect. The dielectric energy in the field is neglected as an approximation because $\Delta\varepsilon$ is very small. The free energy density is

$$f=\frac{1}{2}K_{33}\left(\vec{n}\times\vec{n}\times\nabla\vec{n}\right)^2-\vec{P}_f\cdot\vec{E}=\frac{1}{2}K_{33}\left(\vec{n}\times\vec{n}\times\nabla\vec{n}\right)^2-e_b\left(\vec{n}\times\vec{n}\times\nabla\vec{n}\right)\cdot\vec{E}. \tag{4.26}$$

The components of the liquid crystal director are

$$n_x=\sin\theta(z),\ n_y=0,\ n_z=\cos\theta(z). \tag{4.27}$$

When the applied voltage is low, the deformation is very small and θ is very small. The divergence $\nabla\cdot\vec{n}=-\sin\theta\partial\theta/\partial z$ is very small and the splay elastic energy is very small and can be neglected. The curl of $\vec{n}$ is

$$\nabla\times\vec{n}=\cos\theta\frac{\partial\theta}{\partial z}\hat{y}, \tag{4.28}$$

and the bend is

$$\vec{n}\times\nabla\times\vec{n}=-\cos^2\theta\frac{\partial\theta}{\partial z}\hat{x}+\sin\theta\cos\theta\frac{\partial\theta}{\partial z}\hat{z}\approx-\frac{\partial\theta}{\partial z}\hat{x}. \tag{4.29}$$

Equation (4.26) becomes

$$f = \frac{1}{2}K_{33}\left(\frac{\partial\theta}{\partial z}\right)^2 - e_b E\frac{\partial\theta}{\partial z}. \tag{4.30}$$

Minimizing the free energy

$$\frac{\delta f}{\delta\theta} = -\frac{d}{dz}\left(\frac{\partial f}{\partial\theta}\right) = -\frac{d}{dz}\left[K_{33}\left(\frac{\partial\theta}{\partial z}\right) - e_b E\right] = 0, \tag{4.31}$$

because when $E = 0$, $\partial\theta/\partial z = 0$,

$$\frac{\partial\theta}{\partial z} = \frac{e_b E}{K_{33}}. \tag{4.32}$$

If the anchoring of the liquid crystal at the cell surface is very weak, the solution to Equation (4.32) is

$$\theta = \frac{e_b E}{K_{33}} z. \tag{4.33}$$

It is worthwhile to point out two characteristics of the flexoelectric effect. First, there is no threshold for the applied field, which is different from Fréedericksz transition, where there is a threshold below which no deformation occurs. Deformation of the director configuration occurs under any field. Second, the direction of the bend depends on the polarity of the applied field, which is also different from Fréedericksz transition where the deformation is independent of the polarity of the applied field.

One of the experimental methods for studying the orientation of the liquid crystal in the flexoelectric effect is to measure the retardation of the liquid crystal cell, which is given by

$$\Delta nd = \int_{-h/2}^{h/2} \left[n_{eff}(z) - n_o\right] dz = 2\int_{0}^{h/2} \left[n_{eff}(z) - n_o\right] dz, \tag{4.34}$$

where n_{eff} is the effective refractive index of the liquid crystal and is given (for normal incident light with linear polarization along the x direction) by

$$n_{eff} = \frac{n_e n_o}{\sqrt{n_e^2\cos^2\theta + n_o^2\sin^2\theta}} = \frac{n_e n_o}{\sqrt{n_e^2 - \left(n_e^2 - n_o^2\right)\sin^2\theta}}. \tag{4.35}$$

For small θ we have the approximation $n_{eff} = n_o\left[1 + (1/2)\left(1 - n_o^2/n_e^2\right)\theta^2\right]$. Thus

$$\Delta nd = \int_0^{h/2} n_o\left(1 - \frac{n_o^2}{n_e^2}\right)\theta^2 dz = n_o\left(1 - \frac{n_o^2}{n_e^2}\right)\int_0^{\theta(z=h/2)} \theta^2\left(1/\frac{d\theta}{dz}\right)d\theta = n_o\left(1 - \frac{n_o^2}{n_e^2}\right)\frac{1}{24}\left(\frac{e_b}{K_{33}}\right)^2 E^2 h^3. \tag{4.36}$$

In Fréedericksz transition, when the applied field is slightly above the threshold E_c, the tilt is proportional to $\sqrt{E-E_c}$ and the retardation is proportional to $(E-E_c)$ (see Chapter 5 for details).

In the experiment performed by Schmidt, Schadt, and Helfrich on the liquid crystal MBBA [8], for the field of 0.3 V/μm, the variation rate $\partial\theta/\partial z$ was about $1\times10^{-4}\mu m^{-1}$. The elastic constant was $K_{33}=7.5\times10^{-12}N$. The flexoelectric coefficient e_b has the value about $2.5\times10^{-15}V$.

If $\Delta\varepsilon$ were negative but not very small, the dielectric interaction would prevent the deformation of the director configuration. We estimate the dielectric anisotropy $\Delta\varepsilon$, which will make the flexoelectric effect disappear, in the following way. The dielectric energy is $-(1/2)\varepsilon_o\Delta\varepsilon\left(\vec{E}\cdot\vec{n}\right)^2=-(1/2)\varepsilon_o\Delta\varepsilon E^2\sin^2\theta$, which is against the deformation. The average tilt angle is $(\partial\theta/\partial z)(h/4)$. The average dielectric energy is approximately $-(1/2)\varepsilon_o\Delta\varepsilon E^2[(\partial\theta/\partial z)(h/4)]^2$. The flexoelectric energy is $-e_bE(\partial\theta/\partial z)$, which favors the deformation. If both energies are the same, the deformation would be hindered. Thus $-(1/2)\varepsilon_o\Delta\varepsilon E^2[(\partial\theta/\partial z)(h/4)]^2=e_bE(\partial\theta/\partial z)$, which gives $\Delta\varepsilon=-8e_b/[\varepsilon_oE(\partial\theta/\partial z)h^2]$. Using the values in the previous paragraph, we get $\Delta\varepsilon=-0.75$. Moreover because the flexoelectric energy is linearly proportional to E while the dielectric energy is proportional to E^2, dielectric effect will become dominant at high fields. Therefore only when $|\Delta\varepsilon|\ll0.75$ and at low fields, do we have a pure flexoelectric effect. If $\Delta\varepsilon$ were positive and not very small, the liquid crystal would be aligned parallel to the applied field under the weak anchoring condition. Therefore in order to have the flexoelectric effect described in the previous paragraph under the field of 0.3 V/μm, the dielectric anisotropy should be negative and its absolute value should be much smaller than 0.75.

The geometry shown in Figure 4.4 can be used for an electrically controlled birefringence device. For normal incident light with polarization parallel to the x axis, it encounters the refractive index n_o in the field-off state. When an electric field is applied in the x direction, the liquid crystal molecules tilt and the light will encounter the refractive index n_{eff}. Therefore the retardation of the liquid crystal cell is changed. The flexoelectric effect can also be used in the switching bistable nematic display with asymmetrical anchoring conditions [9].

Figure 4.5 shows the geometry to study the flexoelectric effect in splay geometry [2]. The substrates are coated with a homogeneous alignment layer with very weak anchoring strength. The liquid crystal used has a small positive dielectric anisotropy $\Delta\varepsilon$. In the absence of external electric fields, the liquid crystal is in the uniform homogeneous state as shown in Figure 4.5(a). When an electric field is applied along the x direction, a splay deformation occurs because of the flexoelectric effect, as shown in Figure 4.5(b). The director configuration in the equilibrium state can be calculated in a similar way as in the bend case.

4.2.2 *Flexoelectric effect in cholesteric liquid crystals*

The flexoelectric effect also exists in cholesteric liquid crystals because the orientational order is locally the same as in nematic liquid crystals. Here we consider a cell geometry shown in Figure 4.6 [2,10–12]. The Ch liquid crystal is sandwiched between two parallel substrates with transparent electrode. A homogeneous alignment layer is coated on the inner surface of the substrates. When the liquid crystal is cooled down from the isotropic phase under an external electric field, the helical axis $\vec{h}$ of the liquid crystal is parallel to the substrate and uniformly aligned along the x axis by the alignment layer, as shown in Figure 4.6(a) and (b). The liquid

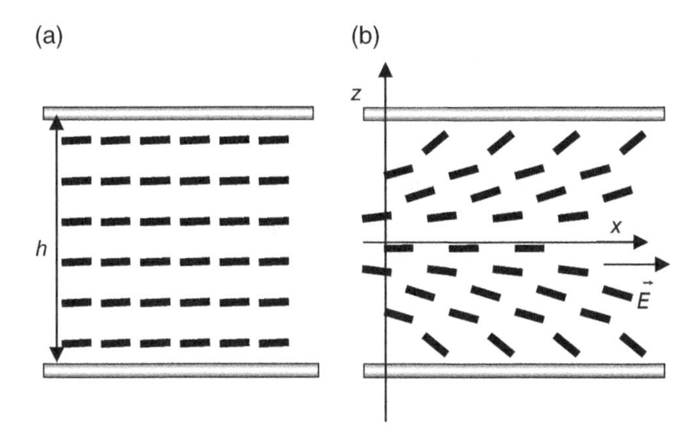

Figure 4.5 Schematic diagram showing the flexoelectric effect in the splay deformation.

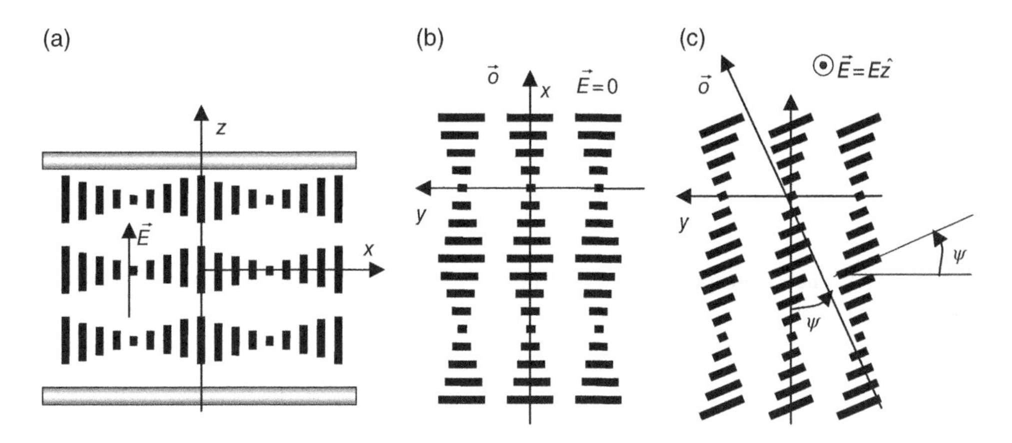

Figure 4.6 Schematic diagram showing the flexoelectric effect in the cholesteric liquid crystal.

crystal with undistorted helical structure behaves like an optically uniaxial medium (when the pitch is much smaller than the wavelength of light) with uniaxis $\vec{o}$ coincident with the helical axis. The dielectric anisotropy $\Delta\varepsilon$ is very small, and the dielectric interaction is negligible. When an electric field is applied across the cell along the z axis, the helical structure is preserved and the pitch is unchanged and the helical axis remains parallel to the substrate. Because the helical structure is incompatible with the planar boundary condition, there are director deformations near the surface and thus there is induced polarization. The applied field interacts with the induced polarization and makes the in-plane component of the liquid crystal director tilt, as shown in Figure 4.6(c). This tilting will produce more distortion and therefore induce more polarization.

The components of the liquid crystal director in the distorted state are

$$n_x = \sin\psi \sin(q_o x). \quad n_y = -\cos\psi \sin(q_o x), \quad n_z = \cos(q_o x), \tag{4.37}$$

where ψ is the tilt angle of the in-plane component of the director. Now the uniaxial optical axis $\vec{o}$ is also tilted by the same angle. The induced polarization is

$$\begin{aligned}\vec{P}_f &= e_s(\vec{n}\nabla\cdot\vec{n}) + e_b(\vec{n}\times\nabla\times\vec{n}) \\ &= e_s q_o \sin\psi \sin(q_o x)\cos(q_o x)\ [\sin.......\psi\hat{x} - \cos\psi\hat{y}] + e_s q_o \sin\psi\cos^2(q_o x)\hat{z} \\ &\quad + e_b q_o \sin\psi\sin(q_o x)\cos(q_o x)[-\sin\psi\hat{x} + \cos\psi\hat{y}] + e_b q_o \sin\psi\sin^2(q_o x)\hat{z} \\ &= e_s q_o \sin\psi\hat{z},\end{aligned} \tag{4.38}$$

where the approximation $e_s = e_b$ is used. The free energy density is

$$f = \frac{1}{2}K_{11}q_o^2\sin^2\psi + \frac{1}{2}K_{22}(q_o - q_o\cos\psi)^2 - e_s q_o E\sin\psi, \tag{4.39}$$

where the approximation $K_{11} = K_{33}$ is used. For small ψ, $\sin\psi \approx \psi$, $(1 - \cos\psi) \approx 0$, and then we have the approximation

$$f = \frac{1}{2}K_{11}q_o^2\psi^2 - e_s q_o E\psi. \tag{4.40}$$

Minimizing f with respect to ψ, we get

$$\psi = e_s E / K_{11} q_o. \tag{4.41}$$

The tilt angle is linearly proportional to E. When the polarity of the applied voltage is reversed, the optical axis will be tilted in the opposite direction. In the rotation, the torque due to the elastic and electric energies is balanced by the viscosity torque,

$$\gamma\frac{\partial\psi}{\partial t} = -\frac{\partial f}{\partial\psi} = -K_{11}q_o^2\psi + e q_o E \tag{4.42}$$

where γ is an effective viscosity coefficient. If the applied field is turned off from a distorted state with the tilt angle ψ_o, the solution to Equation (4.42) is $\psi = \psi_o e^{-t/\tau}$. The relaxation time is $\tau = \gamma/K_{11}q_o^2$. For short pitch Ch liquid crystals, τ can be as small as 100 μs [13]. This flexoelectric effect of the Ch liquid crystal can be used to modulate light intensity when the liquid crystal cell is placed between two crossed polarizers because the optical axis can be tilted by applying an electric field.

4.3 Ferroelectric Liquid Crystal

4.3.1 Symmetry and polarization

We have mentioned that it is impossible for uniformly oriented nematic liquid crystals to have spontaneous polarization because of their $D_{h\infty}$ symmetry. Now let us consider the possibility of spontaneous polarization in other liquid crystal phases. For rod-like molecules, it is impossible in any liquid crystal phase to have spontaneous polarization along the liquid crystal director because $\vec{n}$ and $-\vec{n}$ are equivalent.

Cholesteric liquid crystals consist of chiral molecules and therefore do not have reflection symmetry. The symmetry group of cholesteric liquid crystals is D_2 [1,3]. A cholesteric liquid crystal is invariant for the two-fold (180°) rotation around $\vec{n}$, which rules out the possibility of spontaneous polarization perpendicular to $\vec{n}$. It is also invariant for the two-fold rotation around an axis that is perpendicular to the $\vec{n} - \vec{h}$ (the helical axis) plane, which rules out the possibility of spontaneous polarization parallel to $\vec{n}$. Therefore there is no ferroelectricity in the cholesteric phase.

Smectic-A liquid crystals, besides the orientational order as nematics, possess one-dimensional positional order. They have a layered structure. The liquid crystal director $\vec{n}$ is perpendicular to the smectic layers. The symmetry of smectic-A is $D_{\infty h}$ if the constituent molecule is achiral or D_{∞} if the constituent molecule is chiral. It is invariant for any rotation around $\vec{n}$. It is also invariant for the two-fold rotation around any axis perpendicular to $\vec{n}$. The continuous rotational symmetry is around $\vec{n}$ and therefore there is no spontaneous polarization in any direction perpendicular to $\vec{n}$. Hence it is impossible to have spontaneous polarization in smectic-A, even when the constituent molecule is chiral.

Smectic-C liquid crystals are similar to smectic-A liquid crystals except that the liquid crystal director is no longer perpendicular to the layer but tilted. For the convenience of symmetry discussion, let us introduce a unit vector $\vec{a}$ which is perpendicular to the layer. The symmetry group is C_{2h}. The two-fold rotational symmetry is around the axis that is perpendicular to the $\vec{n}\vec{a}$ plane (which contains both $\vec{n}$ and $\vec{a}$). This implies that there is no spontaneous polarization in the $\vec{n}\vec{a}$ plane. The reflection symmetry is about the $\vec{n}\vec{a}$ plane, and therefore there is no spontaneous polarization perpendicular to the $\vec{n}\vec{a}$ plane either. This rules out the possibility of spontaneous polarization in smectic-C liquid crystals.

As pointed out by Meyer [14], the reflection symmetry of smectic-C liquid crystals can be removed if the constituent molecules are chiral, and thus it becomes possible to have spontaneous polarization. This phase is called the chiral smectic-C or smectic-C*, and its structure is shown in Figure 4.7. Within a layer, the structure is the same as in smectic-C. The liquid crystal director $\vec{n}$ is, however, no longer oriented unidirectionally in space but twists from layer to layer as in the cholesteric phase [15]. The symmetry group is C_2. The two-fold rotational symmetry axis is perpendicular to both the layer normal $\vec{a}$ and the director $\vec{n}$. Now it is possible to have spontaneous polarization along the two-fold rotational symmetry axis.

In order to illustrate the spontaneous polarization in smectic-C* liquid crystals, the liquid crystal molecule can be regarded as a parallelepiped with an attached arrow, as shown in Figure 4.8. The parallelepiped does not have reflection symmetry: its top differs from its bottom, its front differs from its back and its left differs from its right. The arrow represents the lateral permanent dipole and is perpendicular to the director $\vec{n}$. Because of the C_2 symmetry, the parallelepiped has equal probability of pointing up (parallel to $\vec{n}$) and down (anti-parallel to $\vec{n}$), but in both cases the dipole points out of the paper. When the dipole points into the paper, it does not belong to the same domain because the tilt angle is $-\theta$. If the tilt angle θ is 0, as in smectic-A liquid crystals, then the dipole has equal probability of pointing out and pointing in. This explains that it is impossible to have spontaneous polarization in smectic-A liquid crystals even if the constituent molecule is chiral. In smectic-C liquid crystals, when the constituent molecules are a racemic mixture (equal amounts of left- and right-handed molecules), if the left-handed molecule has its permanent dipole pointing out, then the right-handed molecule has its dipole pointing in. They cancel each other and thus there is no spontaneous polarization.

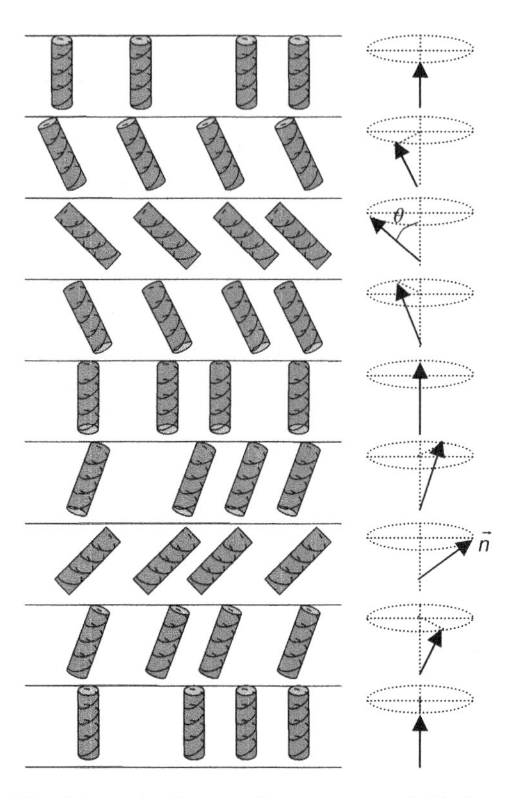

Figure 4.7 Schematic diagram of the structure of chiral smectic-C.

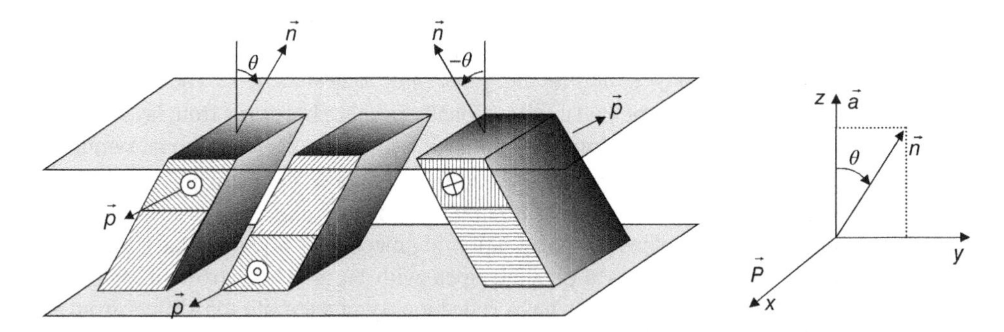

Figure 4.8 Schematic diagram showing the spontaneous polarization in smectic-C*.

4.3.2 Tilt angle and polarization

We first consider the temperature dependence of the tilt angle θ of a liquid crystal that exhibits smectic-A and smectic-C. In smectic-A, the tilt angle is zero. The transition from smectic-A to smectic-C is a second-order transition. Below the transition, the tilt angle increases gradually with increasing temperature. The tilt angle, θ, can be used as the order parameter. Near the transition, the free energy density of the system can be expressed in terms of Landau expansion in the powers of θ [15],

$$f = f_o + \frac{1}{2}a(T - T_c)\theta^2 + \frac{1}{4}b\theta^4, \tag{4.43}$$

where a and b are temperature-independent positive coefficients, and T_c is the A–C transition temperature. The equilibrium value of θ can be obtained by minimizing the free energy:

$$\frac{\partial f}{\partial \theta} = a(T - T_c)\theta + b\theta^3 = 0$$

The stable solution for temperature below T_c is

$$\theta = (a/b)^{1/2}(T_c - T)^{1/2}. \tag{4.44}$$

This result is also valid for smectic-C*.

We now consider the relation between the tilt angle and the spontaneous polarization. In smectic-A, the director $\vec{n}$ is normal to the smectic layers, and the rotation around the long molecular axis is not biased at zero applied field. If the constituent molecule has a lateral dipole, the dipole has equal probability of pointing any direction perpendicular to the director, independent of whether the constituent molecule is chiral or not. The average direction of the long molecular axis is along $\vec{n}$, and therefore there is no spontaneous polarization perpendicular to $\vec{n}$. In smectic-C, the tilt angle is no longer zero, and the rotational symmetry around the long molecular axis is broken. The rotation along the long molecular axis is biased. The molecule, however, has equal probability of pointing up and down with respect to $\vec{n}$ because of the symmetry that $\vec{n}$ and $-\vec{n}$ are equivalent. This rules out the possibility that net polarization adds up in the $\vec{a}\vec{n}$ plane. If the constituent molecule is achiral, the lateral dipole has equal probability of pointing out of and into the plane of the paper (the $\vec{a}\vec{n}$ plane), therefore no net polarization can add up in the direction perpendicular to the $\vec{a}\vec{n}$ plane. When the constituent molecule is chiral, the reflection symmetry about the $\vec{a}\vec{n}$ plane is broken; it becomes possible for net polarization to add up in the direction perpendicular to the $\vec{a}\vec{n}$ plane. Whether the spontaneous polarization is out of the plane or into the plane is determined by the molecular structure.

When the tilt angle changes from θ to $-\theta$, which is the same as the rotation of the system around the layer normal by 180°, $\vec{P}_s$ points in the opposite direction and the polarization changes sign. Therefore the spontaneous polarization must be an odd function of the tilt angle. For small θ, we must have

$$P_s = c \cdot \theta, \tag{4.45}$$

where c is a constant. The larger the tilt angle, the more biased the rotation around the long molecular axis becomes, and therefore the larger the spontaneous polarization.

4.3.3 *Surface stabilized ferroelectric liquid crystals*

In electro-optical devices, it is usually required that the liquid crystal director is unidirectionally oriented. In the smectic-C* phase, however, the liquid crystal director twists from layer to layer.

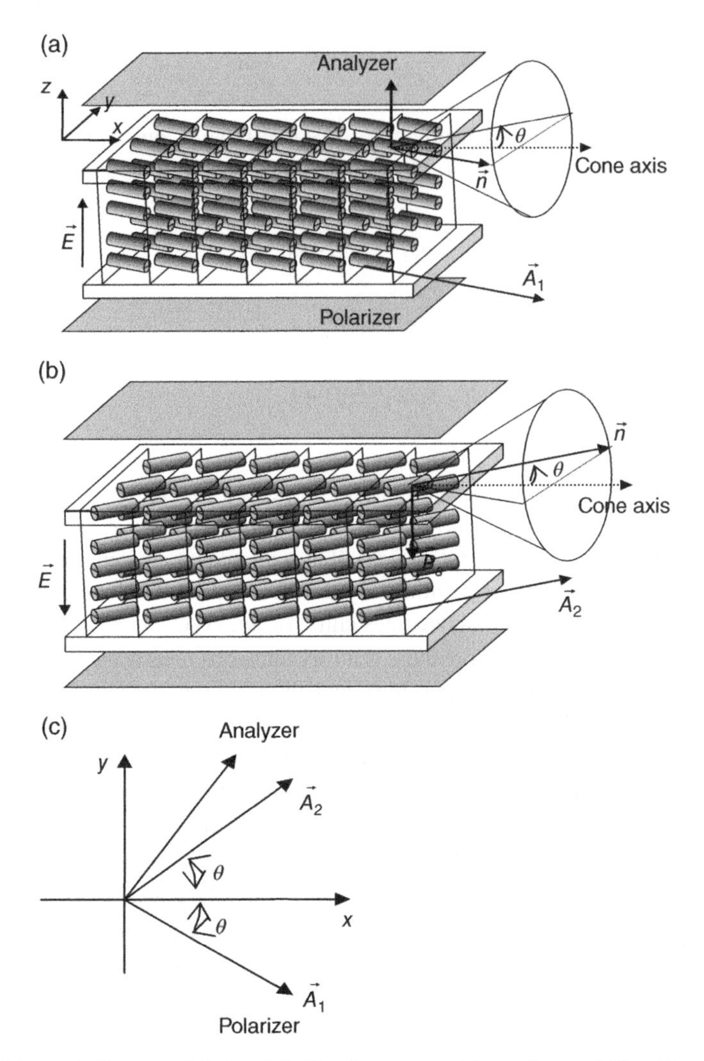

Figure 4.9 Schematic diagram of the bookshelf cell structure of the surface-stabilized ferroelectric liquid crystal display. (a) The director is along the direction $\vec{A}_1$ when the applied field is up. (b) The director is along the direction $\vec{A}_2$ when the applied field is down. (c) Directions of the polarizer and analyzer.

This problem is overcome by Clark and Lagerwall in their invention of the surface-stabilized ferroelectric liquid crystal (SSFLC) device [16], shown in Figure 4.9. The liquid crystal is sandwiched between two parallel substrates with the cell gap, h, thinner than the helical pitch, P, of the liquid crystal. The inner surface of the substrates is coated with alignment layers which promote parallel (to the substrate) anchoring of the liquid crystal on the surface of the substrate. The smectic layers are perpendicular to the substrate of the cell, while the helical axis is parallel to the substrate. Now the helical twist is suppressed and unwound by the anchoring. There are only two directions, $\vec{A}_1$ and $\vec{A}_2$, on the helical cone, and these are compatible with the boundary condition at the substrate surface. The cone angle is 2θ, and therefore the angle between $\vec{A}_1$ and $\vec{A}_2$ is 2θ. The two orientational states can be further selected by applying a DC electric field across the cell. For example, an electric field in the $+z$ direction selects the orientational state $\vec{A}_1$ where the spontaneous polarization, $\vec{P}_s$, is pointing up and parallels the field. Then an electric field in the $-z$ direction will select the orientational state $\vec{A}_2$ where the spontaneous polarization, $\vec{P}_s$, is pointing down. Once the liquid crystal is switched into the state $\vec{A}_1$ (or $\vec{A}_2$) by an externally applied electric field, it will remain in that state after the field is removed, because there is an energy barrier between these two states. Therefore, SSFLC is ideally bistable, which is a very useful property in multiplexed display applications. In reality, it is difficult to make a large-area truly bistable SSFLC display because of surface irregularities.

In the SSFLC display, the polarizer is chosen to be parallel to one of the stable orientational states, say, $\vec{A}_1$, and the analyzer is perpendicular to the polarizer. When the liquid crystal is in state $\vec{A}_1$, the polarization of the incident light is parallel to the liquid crystal director, and remains in this direction when propagating through the cell. When the light is coming out the cell, its polarization is perpendicular to the analyzer and therefore the display is black. When the liquid crystal is switched into the state $\vec{A}_2$, the polarization of the incident light makes the angle 2θ with respect to the liquid crystal director, and therefore when the light propagates through the liquid crystal, its polarization is rotated. The transmittance of the display is

$$T = \sin^2(4\theta)\sin^2\left(\frac{1}{2}\frac{2\pi\Delta nh}{\lambda}\right), \tag{4.46}$$

where $2\pi\Delta nh/\lambda$ is the retardation angle. The maximum transmittance of 100% can be achieved if $2\theta = \pi/4$ and $2\pi\Delta nh/\lambda = \pi$.

For a ferroelectric liquid crystal with the spontaneous polarization $\vec{P}_s$, the electric energy density in an electric field $\vec{E}$ is $-\vec{P}_s\cdot\vec{E}$. The typical value of the spontaneous polarization of ferroelectric liquid crystals is $100\ \mathrm{nC/cm^2} = 10^{-3}\ \mathrm{C/m^2}$. When the strength of the applied field is $1\ \mathrm{V/\mu m} = 10^6\ \mathrm{V/m}$ and $\vec{P}_s$ is parallel to $\vec{E}$, the electric energy density is $\left|-\vec{P}\cdot\vec{E}\right| = 10^3\ \mathrm{J/m^3}$, which is much higher than the electric energy density of the dielectric interaction of non-ferroelectric liquid crystals with electric field. This is one of the reasons for fast switching speed of ferroelectric liquid crystal devices.

Now we consider the dynamics of the switching of SSFLC. We only consider the rotation around the cone (Goldstone mode), as shown in Figure 4.10. The electric torque is

$$\vec{\Gamma}_e = \vec{P}_s \times \vec{E} = P_s E \sin\phi\hat{x}. \tag{4.47}$$

The viscosity torque is

$$\vec{\Gamma}_v = -\gamma\vec{n}_\perp \times \frac{\Delta\vec{n}}{\Delta t} = -\gamma\sin\theta\frac{\sin\theta\Delta\phi}{\Delta t}\hat{x} = -\gamma\sin^2\theta\frac{\partial\phi}{\partial t}\hat{x}, \tag{4.48}$$

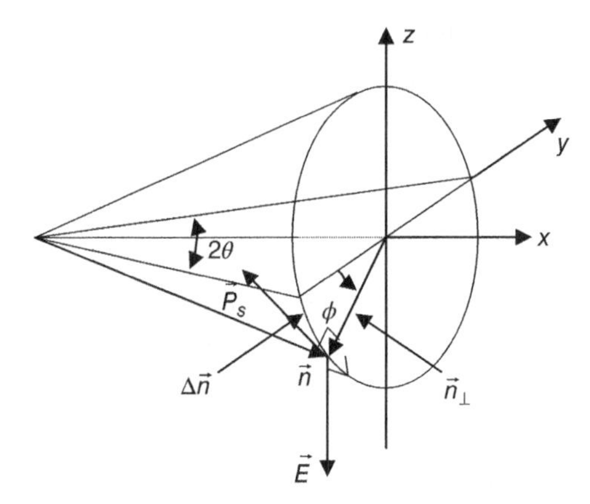

Figure 4.10 Schematic diagram showing the switching process in SSFLC.

where γ is the rotational viscosity coefficient. These two torques balance each other and the dynamic equation is

$$P_s E \sin\phi - \gamma \sin^2\theta \frac{\partial \phi}{\partial t} = 0. \tag{4.49}$$

The solution is

$$\phi(t) = 2\arctan\left[\tan\left(\frac{\phi_o}{2}\right) e^{t/(\gamma \sin^2\theta / P_s E}\right]. \tag{4.50}$$

The response time is

$$\tau = \frac{\gamma \sin^2\theta}{P_s E}. \tag{4.51}$$

The rotation around the (small) cone is another reason for the fast switching speed. For $P_s = 10^2$ nC/cm^2, $E = 1$ V/μm, $\gamma = 0.1$ poise and $\theta = 22.5°$, $\tau \sim 10$ μs

The fast switching speed is a merit of ferroelectric liquid crystal devices. Regarding the bistability, on the one hand, it is good because it enables multiplexed displays of the ferroelectric liquid crystal on passive matrices; on the other hand, the bistability is a problem because it makes it difficult to produce gray scales. Another issue with SSFLC is that it is more challenging to achieve uniform orientation in SSFLC than in nematic liquid crystals.

4.3.4 Electroclinic effect in chiral smectic liquid crystal

As discussed in Section 3.1, there is no ferroelectricity in chiral smectic-A crystals (smectic-A consisting of chiral molecules, denoted as smectic-A*). In a cell geometry of smectic-A* liquid crystal, as shown in Figure 4.11(b), at zero applied field, the liquid crystal director is perpendicular to the smectic layers. The transverse dipole moment has equal probability

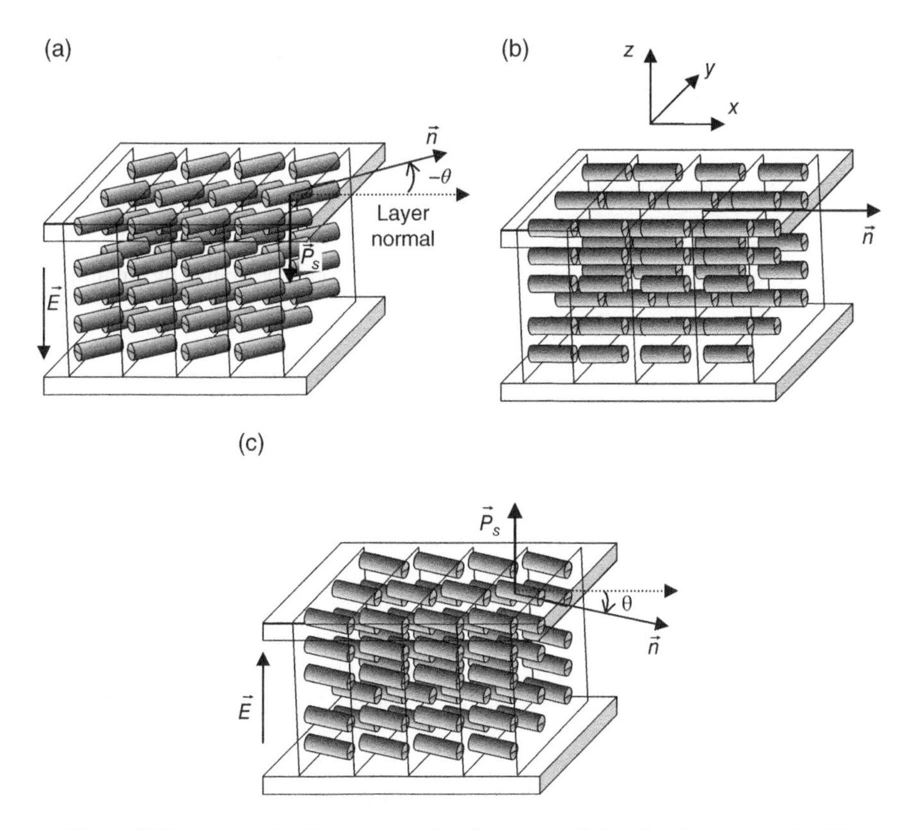

Figure 4.11 Schematic diagram showing the electroclinic effect in the smectic-A*.

of pointing any direction in the smectic layer plane because of the unbiased rotation of the molecule along its long molecular axis. When temperature is lowered toward the smectic-A*–smectic-C* transition, short length-scale and time-scale domains with smectic-C* order form, because the tilt of the director away from the layer normal direction does not cost much energy. This is known as the pretransition phenomenon, which was experimentally demonstrated and theoretically explained by Garoff and Meyer [17]. Within each domain, spontaneous polarization occurs. The macroscopic polarization, however, is still zero because the polarizations of the domains are random through the cell and fluctuate with time. When an electric field is applied across the cell, the temporal domains are stabilized and reorient, such that their polarizations become parallel to the applied field. Thus uniform macroscopic smectic-C* structure is established, and the tilt angle becomes non-zero. This effect of field-induced tilt of liquid crystal molecules in smectic-A* is known as the *electroclinic effect*. The same as in smectic-C*, the spontaneous polarization is perpendicular to the plane formed by the liquid crystal director $\vec{n}$ and the smectic layer normal $\vec{a}$. Therefore, in the induced smectic-C* structure, the $\vec{n}\vec{a}$ plane is perpendicular to the applied field. When a DC electric field pointing down is applied, smectic-C* structure, say, with positive tilt angle is induced as shown in Figure 4.11(a). When a DC electric field pointing up is applied, smectic-C structure with negative tilt angle is induced as shown in Figure 4.11(c).

As shown in Section 4.3.2, when the tilt angle is θ, the spontaneous polarization $\vec{P}_s$ is given by Equation (4.45), and the electric energy density is $-\vec{P}_s \cdot \vec{E} = -cE\theta$. The free energy density is [15]

$$f = f_o + \frac{1}{2}a(T - T_c)\theta^2 + \frac{1}{4}b\theta^4 - cE\theta. \tag{4.52}$$

The tilt angle as a function of the applied electric field E can be found by minimizing the free energy:

$$\frac{\partial f}{\partial \theta} = a(T - T_c)\theta + b\theta^3 - cE \equiv 0 \tag{4.53}$$

When the applied field is low, the tilt angle is small, and the cubic term in Equation (4.53) can be neglected. The tilt angle is

$$\theta = \frac{cE}{a(T - T^*)}. \tag{4.54}$$

The induced tilt angle is linearly proportional to the applied field. The tilt angle can be larger than 10° at the electric field of 10 V/μm for some liquid crystals.

In the tilting of the liquid crystal molecule, the torque produced by the electric field and the elastic force is $-\partial f/\partial \theta = -a(T - T_c)\theta + cE$, which is balanced by the viscosity torque $-\gamma \partial\theta/\partial t$:

$$-a(T - T_c)\theta + cE - \gamma \partial\theta/\partial t = 0 \tag{4.55}$$

The solution is

$$\theta(t) = \frac{cE}{a(T - T^*)}\left(1 - 2e^{-t/\tau}\right), \tag{4.56}$$

where $\tau = \gamma / a(T - T^*)$ is the response time. It is found by experiments that the response time is fast and on the order of a few tens microseconds. The advantages of the electroclinic effect of smectic-A* are that the tilt angle, and thus the electro-optical effect, is a linear function of the applied field, and the response time is fast. The disadvantage is that the electro-optical effect is temperature dependent.

Homework Problems

4.1 The electric field-induced orientational order in a nematic liquid crystal is given by Equation (4.7). The liquid crystal has the parameter $\beta = 0.3$. Numerically calculate the order parameter S as a function of the normalized temperature t in the region from -0.1 to 0.1 under various normalized electric fields $e = 0.0$, $0.5(\beta^3/27)$, $0.8(\beta^3/27)$, $(\beta^3/27)$, $1.2(\beta^3/27)$.

4.2 Consider the flexoelectric effect in the splay geometry as shown in Figure 4.5. The cell thickness h is 5 microns. The splay elastic constant K_{11} of the liquid crystal is 10^{-11} N. The flexoelectric coefficient e_s is 2×10^{-15} V. Calculate the tilt angle θ at the cell surface when the applied field is 1 V/μm.

4.3 *Flexoelectric effect in hybrid cell.* On the top of the cell ($z = h$) the liquid crystal is aligned homogeneously along the x direction while on the bottom ($z = 0$) of the cell the liquid crystal is aligned homeotropically. There is an induced polarization $\vec{P}$ because of the director deformation. When a DC electric field $\vec{E}$ is applied along the y direction, the liquid crystal is twisted to the y direction due to the interaction between $\vec{E}$ and $\vec{P}$. Under the one elastic constant approximation, show that the maximum twist angle (at $z = 0$) is given by $\phi(0) = -(e_s - e_b)Eh/\pi K$.

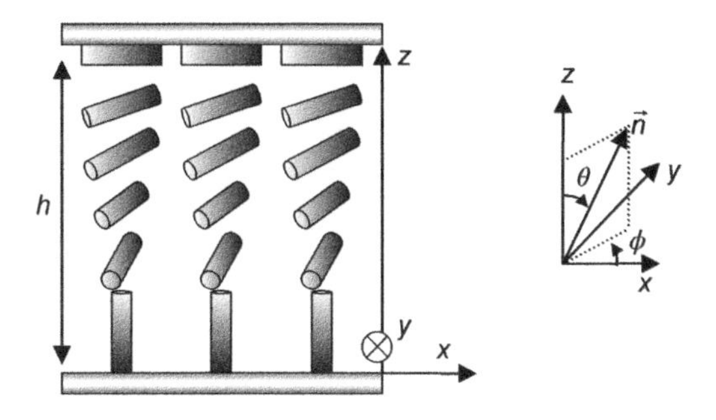

Figure 4.12 Illustration for Problem 4.3.

References

1. P. G. de Gennes and J. Prost, *The physics of liquid crystals* (Oxford University Press, New York, 1993).
2. L. M. Blinov and V. G. Chigrinov, *Electrooptical effects in liquid crystal materials* (Springer-Verlag, New York, 1994).
3. A. J. Leadbetter, Structure classification of liquid crystals, in *Thermotropic liquid crystals*, ed. G. W. Gray (John Wiley & Son, Chichester, 1987).
4. R. M. Hornreich, Landau theory of the isotropic–nematic critical point, *Phys. Lett.*, **109A**, 232 (1985).
5. I. Lelidis and G. Durand, Electric-field-induced isotropic–nematic phase transition, *Phys. Rev. E*, **48**, 3822 (1993).
6. R. B. Meyer, *Phys. Rev. Lett.*, **22**, 918 (1969).
7. A. I. Derzhanski, A. G. Petrov, and M. D. Mitov, *J. Phys., (Paris)*, **39**, 273 (1978).
8. D. Schmidt, M. Schadt and W. Z. Helfrich, *Naturforsch*, **A27**, 277 (1972).
9. E. L. Wood, G. P. Bryan-Brown, P. Brett, A. Graham, J. C. Jones and J. R. Hughes, Zenithal bistable device (ZBD) suitable for portable applications, *SID Intl. Symp. Digest Tech. Papers*, **31**, 124 (2000).
10. J. S. Patel and R. Meyer, Flexoelectric electro-optics of a cholesteric liquid crystals, *Phys. Rev. Lett.*, **58**, 1538 (1987).
11. G. Chilaya, Cholesteric liquid crystals: optics, electro-optics, and photo-optics, *Chirality in liquid crystals*, ed. H.-S. Kitzerow and C. Bahar (Springer, New York, 2001).
12. L. Komitov, S. T. Lagerwall, B. Stenler, and A. Strigazzi, Sign reversal of the linear electro-optical effect in the chiral nematic phase, *J. Appl. Phys.*, **76**, 3762 (1994).

13. J. S. Patel and S.-D. Lee, Fast linear electro-optic effect based on cholesteric liquid crystals, *J. Appl. Phys.*, **66**, 1879 (1987).
14. R. B. Meyer, L. Liebert, L. Strezelecki, P. Keller, *J. Phys. (Paris) Lett.*, **36**, L69 (1975).
15. J. W. Goodby, R. Blinc, N. A. Clark, S. T. Lagerwall, M. A. Osipov, S. A. Pikin, T. Sakurai, K. Yoshino, B. Žekš, Ferroelectric liquid crystals: Principle, properties and applications, *Ferroelectricity and related phenomena,* Vol. **7** (Gordon and Breach Publishers, Amsterdam, 1991)
16. N. A. Clark, S. T. Lagerwall, Submicrosecond bistable electro-optic switching in liquid crystals, *Appl. Phys. Lett.,* **36,** 899 (1980).
17. S. Garoff and R. B. Meyer, Electroclinic effect at the A-C phase change of a chiral smectic liquid crystal, *Phys. Rev. Lett.,* **38**, 848 (1977); *Phys. Rev. A.*, **19**, 388 (1979).

5

Fréedericksz Transition

Liquid crystals reorient in externally applied electric fields because of their dielectric anisotropies. The electric energy (a part of the free energy) of a liquid crystal depends on the orientation of the liquid crystal director in the applied electric field. Under a given electric field, the liquid crystal will be in the equilibrium state, where the total free energy is minimized.

5.1 Calculus of Variation

In a liquid crystal cell, under a given boundary condition and an externally applied field, the liquid crystal is in director field configuration $\vec{n}\left(\vec{r}\right)$ that minimizes the total free energy of the system. The free energy density has two parts: (1) the elastic energy, which depends on the spatial derivatives of $\vec{n}$, and (2) the dielectric electric energy which depends on $\vec{n}$. The total free energy is given by

$$F = \int f\left[\vec{n}(\vec{r}), \vec{n}\prime(\vec{r}), \vec{r}\right] d^3 r. \tag{5.1}$$

Mathematically F is referred to as the functional. In order to search for the director configuration $\vec{n}\left(\vec{r}\right)$ that minimizes the total free energy, we need the calculus of variation [1].

Fundamentals of Liquid Crystal Devices, Second Edition. Deng-Ke Yang and Shin-Tson Wu.

5.1.1 One dimension and one variable

5.1.1.1 Fixed boundary condition

We start with the simplest case, in which the liquid crystal is sandwiched between two parallel plates located at $z=0$ and $z=h$, respectively, as shown in Figure 5.1(a). The director $\vec{n}$ is described by an angle θ which is only a function of z, as shown in Figure 5.1 (b). The anchoring of the liquid crystal at the substrate surface is infinitely strong so that the orientation of the liquid crystal at the top and bottom surface is fixed. The boundary conditions are

$$\theta=\begin{cases}\theta_1 & at\ z=0,\\ \theta_2 & at\ z=h.\end{cases} \tag{5.2}$$

In this case, the total free energy (free energy per unit area) is given by

$$F=\int_0^h f(\theta,\theta',z)dz. \tag{5.3}$$

If the configuration given by the function $\theta=\theta(z)$ has the minimum or maximum free energy F_m, namely, F is stationary under $\theta=\theta(z)$, then for a small variation $\alpha\eta(z)$, where α is a constant and infinitely small and $\eta(z)$ is a function, the change of the free energy is zero to the first order of α [2]. Because of the fixed boundary condition, it is required that $\eta(z=0)=0$ and $\eta(z=h)=0$:

$$\begin{aligned}F(\alpha)&=\int_0^h f[\theta+\alpha\eta(z),\theta'+\alpha\eta'(z),z]dz\\&=\int_0^h\left\{f(\theta,\theta',z)+\frac{\partial f}{\partial\theta}[\alpha\eta(z)]+\frac{\partial f}{\partial\theta'}[\alpha\eta'(z)]+0(\alpha^2)\right\}dz\\&=\int_0^h f(\theta,\theta',z)dz+\alpha\int_0^h\left[\frac{\partial f}{\partial\theta}\eta(z)+\frac{\partial f}{\partial\theta'}\eta'(z)\right]dz+0(\alpha^2)\\&=F(\alpha=0)+\alpha\int_0^h\left[\frac{\partial f}{\partial\theta}\eta(z)+\frac{\partial f}{\partial\theta'}\eta'(z)\right]dz+0(\alpha^2)\end{aligned} \tag{5.4}$$

Therefore it is required that

$$\delta F=F(\alpha)-F(\alpha=0)=\alpha\int_0^h\left[\frac{\partial f}{\partial\theta}\eta(z)+\frac{\partial f}{\partial\theta'}\eta'(z)\right]dz=0.$$

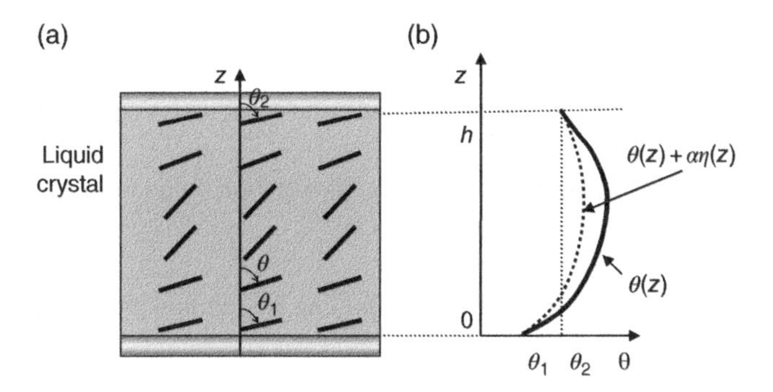

Figure 5.1 Schematic diagram of the 1-D liquid crystal director configuration.

Using partial integration, we have

$$\delta F=\alpha\int_0^h\left[\frac{\partial f}{\partial\theta}\eta(z)+\frac{\partial f}{\partial\theta'}\eta'(z)\right]dz=\alpha\int_0^h\left[\frac{\partial f}{\partial\theta}-\frac{d}{dz}\left(\frac{\partial f}{\partial\theta'}\right)\right]\eta(z)dz+\alpha\left[\eta(z)\frac{\partial f}{\partial\theta'}\right]\Bigg|_0^h=0. \tag{5.5}$$

The last term is 0 because of the boundary conditions for $\eta(z)$. Therefore it is required that

$$\int_0^h\left[\frac{\partial f}{\partial\theta}-\frac{d}{dz}\left(\frac{\partial f}{\partial\theta'}\right)\right]\eta(z)dz=0.$$

This should hold for any function $\eta(z)$ which satisfies the boundary condition. Hence it is required that

$$\frac{\delta f}{\delta\theta}\equiv\frac{\partial f}{\partial\theta}-\frac{d}{dz}\left(\frac{\partial f}{\partial\theta'}\right)=0. \tag{5.6}$$

This is the *Euler–Lagrange equation*. The solution $\theta(z)$ to this equation minimizes or maximizes the free total energy F.

Next we consider how to minimize the total free energy F under a constraint $G=\int_0^h g(\theta,\theta',z)dz=\sigma$, where σ is a constant. Now besides

$$\delta F=\int_0^h\frac{\delta f}{\delta\theta}\delta\theta dz=0, \tag{5.7}$$

it also is required that

$$\delta G=\int_0^h\frac{\delta g}{\delta\theta}\delta\theta dz=0, \tag{5.8}$$

because G is a constant. Therefore $\left(\frac{\delta f}{\delta \theta}\right)\Big/\left(\frac{\delta g}{\delta \theta}\right)$ must be a constant λ independent of z. λ is called the Lagrange multiplier. To minimize F under the constraint is equivalent to minimizing $\int_0^h [f(\theta,\theta',z)+\lambda g(\theta,\theta',z)]dz$ without the constraint. The solution found from $\delta(f+\lambda g)/\delta\theta=0$ will depend on λ. The value of λ can be found by substituting the solution into $G=\int_0^h g(\theta,\theta',z)dz=\sigma$.

5.1.1.2 Unfixed boundary condition

If the anchoring of the liquid crystal at the substrate surface is not infinitely strong but weak, the value of θ at the boundary is not fixed. The surface energy varies with the director configuration, and therefore must be included into the total free energy of the system:

$$F=\int_0^h f(\theta,\theta',z)dz+f_s[\theta(z=0)]+f_s[\theta(z=h)] \tag{5.9}$$

Under a small variation $\alpha\eta(z)$, the variation of the total free energy is

$$\begin{aligned}\delta F&=\left\{\int_0^h f[\theta+\alpha\eta(z),\theta'+\alpha\eta'(z),z]dz+f_s[\theta(z=0)+\alpha\eta(z=0)]+f_s[\theta(z=h)+\alpha\eta(z=h)]\right\}\\&\quad-\left\{\int_0^h f[\theta,\theta',z]dz+f_s[\theta(z=0)]+f_s[\theta(z=h)]\right\}\\&=\alpha\int_0^h\left[\frac{\partial f}{\partial\theta}-\frac{d}{dz}\left(\frac{\partial f}{\partial\theta'}\right)\right]\eta(z)dz+\alpha\left[\eta(z)\frac{\partial f}{\partial\theta'}\right]\Bigg|_0^h+\alpha\left(\eta\frac{\partial f_s}{\partial\theta}\right)\Bigg|_{z=0}+\alpha\left(\eta\frac{\partial f_s}{\partial\theta}\right)\Bigg|_{z=h}=0\\&=\alpha\int_0^h\left[\frac{\partial f}{\partial\theta}-\frac{d}{dz}\left(\frac{\partial f}{\partial\theta'}\right)\right]\eta(z)dz+\alpha\eta\left(\frac{\partial f_s}{\partial\theta}-\frac{\partial f}{\partial\theta'}\right)\Bigg|_{z=0}+\alpha\eta\left(\frac{\partial f_s}{\partial\theta}+\frac{\partial f}{\partial\theta'}\right)\Bigg|_{z=h}=0.\end{aligned} \tag{5.10}$$

If F is stationary under the director configuration $\theta(z)$, then besides Equation (5.6), it is also required that

$$\left(-\frac{\partial f}{\partial\theta'}+\frac{\partial f_s}{\partial\theta}\right)\Bigg|_{z=0}=0, \tag{5.11.a}$$

$$\left(\frac{\partial f}{\partial \theta'} + \frac{\partial f_s}{\partial \theta}\right)\bigg|_{z=h} = 0. \tag{5.11.b}$$

5.1.2 *One dimension and multiple variables*

If $\vec{n}$ is described by two angles θ and ϕ, which are functions of z, the total free energy of the system is given by

$$F = \int_0^h f(\theta, \theta', \phi, \phi', z)dz. \tag{5.12}$$

When the free energy is minimized or maximized, it is required that

$$\frac{\delta f}{\delta \theta} = \frac{\partial f}{\partial \theta} - \frac{d}{dz}\left(\frac{\partial f}{\partial \theta}\right) = 0 \tag{5.13.a}$$

and

$$\frac{\delta f}{\delta \phi} = \frac{\partial f}{\partial \phi} - \frac{d}{dz}\left(\frac{\partial f}{\partial \phi'}\right) = 0. \tag{5.13.b}$$

5.1.3 *Three dimensions*

If $\vec{n}$ is described by the angle θ, which is a function of more than one coordinate, say, x, y, and z, the total free energy is given by

$$F = \int f\left(\theta, \theta'_x, \theta'_y, \theta'_z, x, y, z\right). \tag{5.14}$$

When the free energy is minimized or maximized, it is required that

$$\frac{\delta f}{\delta \theta} = \frac{\partial f}{\partial \theta} - \frac{d}{dx}\left(\frac{\partial f}{\partial \theta'_x}\right) - \frac{d}{dy}\left(\frac{\partial f}{\partial \theta'_y}\right) - \frac{d}{dz}\left(\frac{\partial f}{\partial \theta'_z}\right) = 0. \tag{5.15}$$

When $\vec{n}$ is described by multiple variables which are a function of more than one coordinate, each variable must satisfy the Euler–Lagrange equation as Equation (5.15).

5.2 Fréedericksz Transition: Statics

When a nematic liquid crystal is confined, such as when sandwiched between two parallel substrates with alignment layers, in the absence of external fields, the orientation of the liquid crystal director is determined by the anchoring condition. When an external electric field is applied to the liquid crystal, it will reorient because of the dielectric interaction between the liquid crystal and the applied field. If the dielectric anisotropy is positive ($\Delta\varepsilon > 0$), the liquid crystal tends to align parallel to the applied field. If $\Delta\varepsilon < 0$, it tends to align perpendicular to the field. This field-induced reorientation of the liquid crystal is referred to as the *Fréedericksz transition* [3–5].

5.2.1 Splay geometry

The cell structure of the bend geometry is shown in Figure 5.2, which is popularly used for electrically controlled birefringence (ECB) devices. The liquid crystal is sandwiched between two parallel plates with the cell thickness h. The easy axis of the anchoring of the top and bottom alignment layers is parallel to the plates (the x axis). In the absence of fields, the liquid crystal director is uniformly aligned along the x axis, as shown in Figure 5.2(a). When a sufficiently high electric field is applied across the cell (in the z direction), the liquid crystal director will be tilted toward the cell normal direction, as shown in Figure 5.2(b). Because of the anchoring at the surface of the plates, the liquid crystal director in the distorted state is not uniform. This costs elastic energy and is against the transition. The liquid crystal director is always in the xz plane, provided $\Delta\varepsilon > 0$, and is given by

$$\vec{n} = \cos\theta(z)\hat{x} + \sin\theta(z)\hat{z}. \tag{5.16}$$

The divergence of $\vec{n}$ is $\nabla\cdot\vec{n} = \cos\theta\theta'$ and the curl is $\nabla\times\vec{n} = -\sin\theta\theta'\hat{y}$, where $\theta' = \partial\theta/\partial z$. The elastic energy is positive and given by

$$f_{elastic} = \frac{1}{2}K_{11}\cos^2\theta\theta'^2 + \frac{1}{2}K_{33}\sin^2\theta\theta'^2. \tag{5.17}$$

When the tilt angle is small, the splay elastic energy dominates and the cell geometry is called splay geometry. The electric energy is negative and is approximately given by

$$f_{electric} = -\frac{1}{2}\varepsilon_o\Delta\varepsilon\left(\vec{E}\cdot\vec{n}\right)^2 = -\frac{1}{2}\varepsilon_o\Delta\varepsilon E^2\sin^2\theta. \tag{5.18}$$

This is a good approximation when θ is small. The free energy density is

$$f = \frac{1}{2}\left(K_{11}\cos^2\theta + K_{33}\sin^2\theta\right)\theta'^2 - \frac{1}{2}\varepsilon_o\Delta\varepsilon E^2\sin^2\theta. \tag{5.19}$$

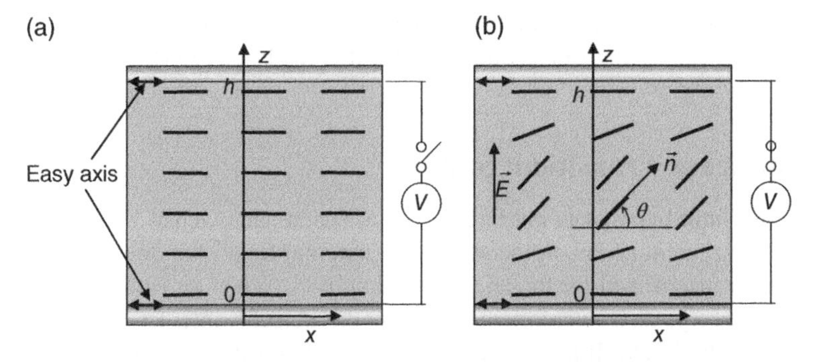

Figure 5.2 Schematic diagram of Fréedericksz transition in the splay geometry.

The total free energy (per unit area) of the system is

$$F=\int_0^h\left[\frac{1}{2}(K_{11}\cos^2\theta+K_{33}\sin^2\theta)\theta'^2-\frac{1}{2}\varepsilon_o\Delta\varepsilon E^2\sin^2\theta\right]dz. \tag{5.20}$$

Using the Euler–Lagrange method to minimize the free energy, we obtain

$$\begin{aligned}\frac{\delta f}{\delta\theta}&=\frac{\partial f}{\partial\theta}-\frac{d}{dz}\left(\frac{\partial f}{\partial\theta'}\right)\\&=-\varepsilon_o\Delta\varepsilon E^2\sin\theta\cos\theta-(K_{33}-K_{11})\sin\theta\cos\theta\theta'^2-(K_{11}\cos^2\theta+K_{33}\sin^2\theta)\theta''=0.\end{aligned} \tag{5.21}$$

When θ is small, we use the approximations: $\sin\theta=\theta$ and $\cos\theta=1$. Neglecting the second-order terms, Equation (5.21) becomes

$$-\varepsilon_o\Delta\varepsilon E^2\theta-K_{11}\theta''=0. \tag{5.22}$$

The general solution is

$$\theta=A\sin\left(\sqrt{\frac{\varepsilon_o\Delta\varepsilon E^2}{K_{11}}}z\right)+B\cos\left(\sqrt{\frac{\varepsilon_o\Delta\varepsilon E^2}{K_{11}}}z\right). \tag{5.23}$$

Now let us look at the boundary condition. Under infinitely strong anchoring, the boundary conditions are $\theta(z=0)=\theta(z=h)=0$. Therefore $B=0$ and

$$\sqrt{\frac{\varepsilon_o\Delta\varepsilon E^2}{K_{11}}}h=m\pi,\qquad m=1,2,3\ldots \tag{5.24}$$

When the applied field is low, it can only produce distortion with the longest wavelength π/h, and thus the threshold field E_c can be found using $\sqrt{\varepsilon_o\Delta\varepsilon E_c^2/K_{11}}h=\pi$. Therefore

$$E_c=\frac{\pi}{h}\sqrt{\frac{K_{11}}{\varepsilon_o\Delta\varepsilon}}, \tag{5.25}$$

which is inversely proportional to the cell thickness, because the elastic energy is higher with thinner cell thickness. The threshold voltage is

$$V_c=hE_c=\pi\sqrt{\frac{K_{11}}{\varepsilon_o\Delta\varepsilon}}, \tag{5.26}$$

which is independent of the cell thickness. In order to see the physical meaning of the existence of the threshold, let us consider the free energy when the tilt angle is small. Because of the boundary condition that θ is 0 at $z=0$ and $z=h$, $\theta = A\sin(\pi z/h)$. For small amplitude A, the average free energy density is approximately given by

$$f = \frac{1}{4}\left(\frac{K_{11}\pi^2}{h^2} - \varepsilon_o \Delta\varepsilon E^2\right)A^2. \tag{5.27}$$

When $E < E_c = \frac{\pi}{h}\sqrt{\frac{K_{11}}{\varepsilon_o \Delta\varepsilon}}$, in the reorientation of the liquid crystal, the decrease of the electric energy cannot compensate for the increase of the elastic energy. The free energy of the distorted state is positive and higher than the free energy of the undistorted state, which is zero, and therefore the transition cannot occur. When the applied field is increased, the electric energy decreases (becomes more negative). When $E > E_c$, the decrease of the electric energy can compensate for the increase of the elastic energy. The free energy of the system decreases as the transition takes place.

In the approximation discussed above, when $E < E_c$, the amplitude of the distortion is $A = 0$. When $E > E_c$, A suddenly explodes because larger A gives lower free energy. This will not happen in reality because it is not consistent with the assumption that θ is small. Now we calculate the precise solution. From Equation (5.21) we have

$$(K_{33} - K_{11})\sin\theta\cos\theta\theta'^2 + \left(K_{11}\cos^2\theta + K_{33}\sin^2\theta\right)\theta'' = -\varepsilon_o \Delta\varepsilon E^2 \sin\theta\cos\theta.$$

Multiplying both sides by $d\theta/dz$, we can get

$$\frac{d}{dz}\left[\frac{1}{2}(K_{11}\cos^2\theta + K_{33}\sin^2\theta)\left(\frac{d\theta}{dz}\right)^2\right] = -\frac{d}{dz}\left(\frac{1}{2}\varepsilon_o \Delta\varepsilon E^2 \sin^2\theta\right).$$

Integrating, we get

$$\left(K_{11}\cos^2\theta + K_{33}\sin^2\theta\right)\left(\frac{d\theta}{dz}\right)^2 = C - \varepsilon_o \Delta\varepsilon E^2 \sin^2\theta,$$

where C is the integration constant, which can be found by considering the fact that the distortion must be symmetric about the middle plane, $\frac{d\theta}{dz}(z = h/2) = 0$. The tilt angle at the middle plane is also the maximum angle θ_m [4]. Therefore

$$\left(\frac{d\theta}{dz}\right)^2 = \varepsilon_o \Delta\varepsilon E^2 \frac{\sin^2\theta_m - \sin^2\theta}{(K_{11}\cos^2\theta + K_{33}\sin^2\theta)}, \tag{5.28}$$

$$\sqrt{\varepsilon_o \Delta\varepsilon E^2}\, dz = \left(\frac{K_{11}\cos^2\theta + K_{33}\sin^2\theta}{\sin^2\theta_m - \sin^2\theta}\right)^{1/2} d\theta.$$

Using Equation (5.25) and integrating the above equation from 0 to z, over which θ changes from 0 to $\theta(z)$, we have

$$\left(\frac{E}{E_c}\right)\left(\frac{z}{h}\right)=\frac{1}{\pi}\int_0^{\theta(z)}\left(\frac{\cos^2\alpha+(K_{33}/K_{11})\sin^2\alpha}{\sin^2\theta_m-\sin^2\alpha}\right)^{1/2}d\alpha. \tag{5.29}$$

The maximum angle θ_m is given by

$$\left(\frac{E}{E_c}\right)\left(\frac{1}{2}\right)=\frac{1}{\pi}\int_0^{\theta_m}\left(\frac{\cos^2\alpha+(K_{33}/K_{11})\sin^2\alpha}{\sin^2\theta_m-\sin^2\alpha}\right)^{1/2}d\alpha. \tag{5.30}$$

For a given field E ($>E_c$), θ_m can be found by numerically solving this integration equation. In the calculation of θ_m as a function E, instead of calculating θ_m for each given E, it is much easier to calculate E for each given θ_m. Once θ_m is known, θ as a function of z can be calculated by using Equation (5.29). Instead of calculating θ for a given z, it much easier to calculate z for a given θ ($<\theta_m$). The solution is symmetrical about the middle plane, $\theta(z)=\theta(h-z)$ for $h/2\le z\le h$.

We can obtain some information on how the tilt angle depends on the applied field even without numerical calculation. For the purpose of simplicity, we assume that $K_{11}=K_{33}$, and then Equation (5.30) becomes

$$\left(\frac{E}{E_c}\right)=\frac{2}{\pi}\int_0^{\theta_m}\left(\frac{1}{\sin^2\theta_m-\sin^2\alpha}\right)^{1/2}d\alpha.$$

Using a new variable ψ defined by $\sin\alpha=\sin\theta_m\sin\psi$, we have

$$\left(\frac{E}{E_c}\right)=\frac{2}{\pi}\int_0^{\pi/2}\frac{1}{\sqrt{1-\sin^2\theta_m\sin^2\psi}}d\psi. \tag{5.31}$$

When the applied field is not much higher than the threshold, the tilt angle θ_m at the middle plane is small, and approximately we have

$$\left(\frac{E}{E_c}\right)\approx\frac{2}{\pi}\int_0^{\pi/2}\left(1+\frac{1}{2}\sin^2\theta_m\sin^2\psi\right)d\psi=1+\frac{1}{4}\sin^2\theta_m.$$

Therefore

$$\sin\theta_m=2\sqrt{(E-E_c)/E_c}, \tag{5.32}$$

which indicates that when the applied field is increased above the threshold, the increase of the tilt angle with the field is fast in the beginning, and then slows down. The numerically calculated tilt angle at the middle plane of the cell vs. the normalized field is shown in Figure 5.3. The numerically calculated tilt angle as a function of position at various applied fields is plotted in Figure 5.4.

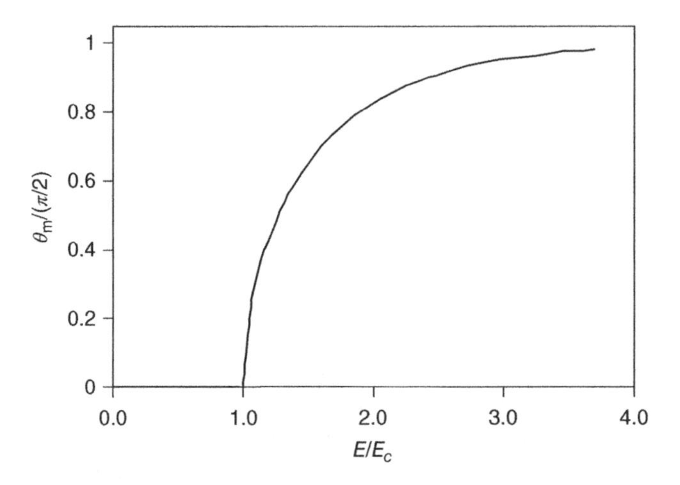

Figure 5.3 The tilt angle at the middle plane vs. the applied field in splay geometry. $K_{11} = 6.4 \times 10^{-12}$N and $K_{33} = 10 \times 10^{-12}$N are used.

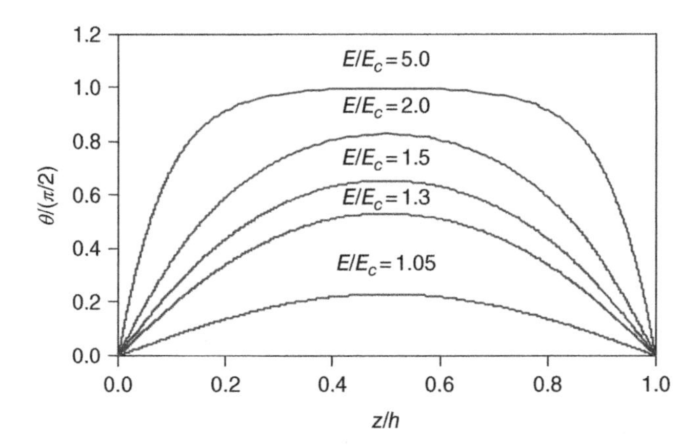

Figure 5.4 The tilt angle as a function of position at various fields in splay geometry. $K_{11} = 6.4 \times 10^{-12}$N and $K_{33} = 10 \times 10^{-12}$N are used.

5.2.2 *Bend geometry*

The cell structure for bend geometry is shown in Figure 5.5, where the liquid crystal is sandwiched between two parallel substrates with homeotropic alignment layer. In the field-off state, the liquid crystal is uniformly aligned perpendicular to the cell substrate because of the homeotropic anchoring condition of the alignment layer, as shown in Figure 5.5(a). When a sufficiently high electric field is applied parallel to the cell (in the x direction), the liquid crystal (with $\Delta\varepsilon > 0$) will be tilted toward the x direction, as shown in Figure 5.5(b). The liquid crystal director is always in the xz plane and is given by

$$\vec{n} = \sin\theta(z)\hat{x} + \cos\theta(z)\hat{z}. \tag{5.34}$$

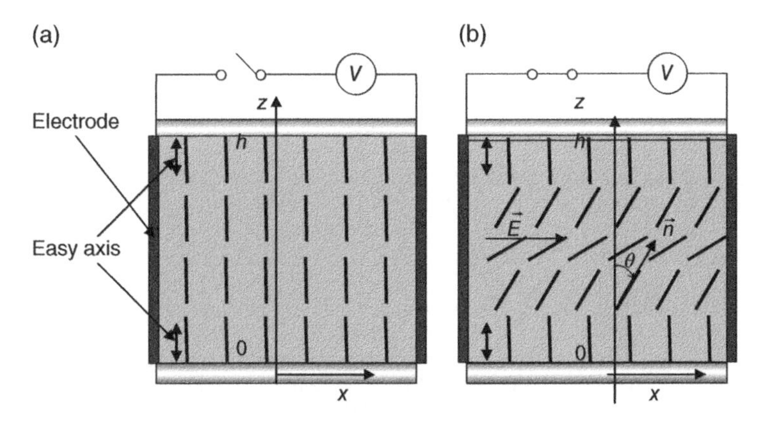

Figure 5.5 Schematic diagram of Fréedericksz transition in the bend geometry.

The elastic energy is given by

$$f_{elastic} = \frac{1}{2}K_{11}\sin^2\theta\theta'^2 + \frac{1}{2}K_{33}\cos^2\theta\theta'^2. \tag{5.35}$$

When the tilt angle is small, the bend elastic energy dominates, and the cell geometry is called bend geometry. The electric energy is negative and is approximately given by

$$f_{electric} = -\frac{1}{2}\varepsilon_o\Delta\varepsilon\left(\vec{E}\cdot\vec{n}\right)^2 = -\frac{1}{2}\varepsilon_o\Delta\varepsilon E^2\sin^2\theta. \tag{5.36}$$

The free energy density is

$$f = \frac{1}{2}\left(K_{11}\sin^2\theta + K_{33}\cos^2\theta\right)\theta'^2 - \frac{1}{2}\varepsilon_o\Delta\varepsilon E^2\sin^2\theta. \tag{5.37}$$

In a similar way as in the splay geometry, the threshold can be found to be

$$E_c = \frac{\pi}{h}\sqrt{\frac{K_{33}}{\varepsilon_o\Delta\varepsilon}}. \tag{5.38}$$

The tilt angle θ as a function of z under an applied field E ($> E_c$) can be calculated in a similar way as in the bend geometry.

The same phenomenon occurs if the liquid crystal has a negative dielectric anisotropy and the applied field is along the z direction (known as the vertical alignment mode), where the electric energy is

$$f_{electric} = -\frac{1}{2}\varepsilon_o(-|\Delta,\varepsilon|)\left(\vec{E}\cdot\vec{n}\right)^2 = \frac{1}{2}\varepsilon_o|\Delta,\varepsilon|E^2\cos^2\theta = \text{constant} - \frac{1}{2}\varepsilon_o|\Delta,\varepsilon|E^2\sin^2\theta.$$

The calculated transmittance vs. applied voltage of an AV mode liquid crystal display is shown in Figure 5.6. Light is incident normally on the liquid crystal cell. The parameters of the liquid

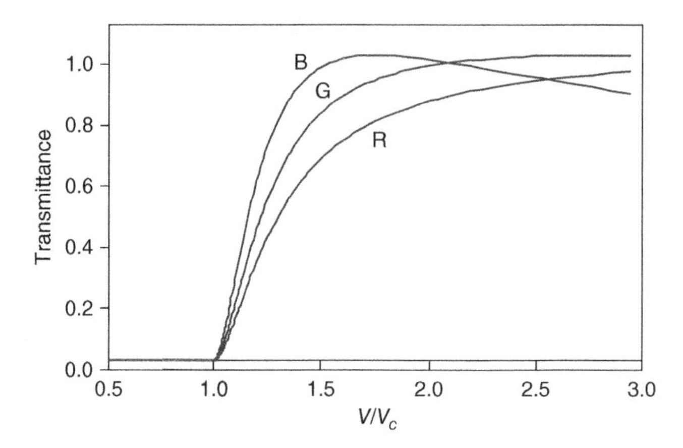

Figure 5.6 Transmittance of the VA mode liquid crystal display. R: 650 nm, G: 550 nm, B: 450 nm.

crystal are $K_{11}=6.4\times10^{-12}$N, $K_{33}=10\times10^{-12}$N, $\Delta\varepsilon=-3$, $n_e=1.57$, and $n_o=1.50$. The thickness of the cell is 5.0 μm. The liquid crystal is sandwiched between two crossed polarizer. In the voltage-activated state, the plane that the liquid crystal director lies on makes 45° with respect to the polarizers. The retardation is

$$\Gamma=\frac{2\pi}{\lambda}\int_0^h\left(\frac{n_e n_o}{\sqrt{n_e^2\cos^2\theta(z)+n_o^2\sin^2\theta(z)}}-1\right)dz.$$

The transmittance is $T=\sin^2(\Gamma/2)$. In the voltage-off state, the retardation is zero and the transmittance is zero for any wavelength. The retardation of the voltage activated states is wavelength-dependent and therefore the voltages for R, G, and B light to reach the maximum transmittance are different.

5.2.3 *Twist geometry*

The twist geometry is shown in Figure 5.7, where the liquid crystal is sandwiched between two parallel plates with a homogeneous alignment layer. In the absence of an external field, the liquid crystal director uniformly orients parallel to the cell surface in the y direction because of the homogeneous anchoring condition of the alignment layer, as shown in Figure 5.7(a). When a sufficiently high electric field is applied along the x direction, the liquid crystal (with $\Delta\varepsilon>0$) is tilted toward the field direction, as shown in Figure 5.7(b). The liquid crystal director is on the xy plane and is given by

$$\vec{n}=\sin\theta(z)\hat{x}+\cos\theta(z)\hat{y}. \tag{5.39}$$

In this geometry, only twist elastic energy is involved, and the elastic energy is given by

$$f_{elastic}=\frac{1}{2}K_{22}\theta'^2. \tag{5.40}$$

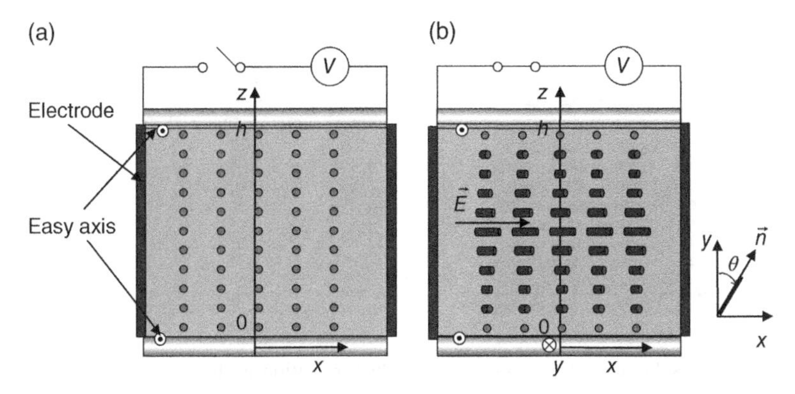

Figure 5.7 Schematic diagram of Fréedericksz transition in the bend geometry.

The electric energy is negative and is approximately given by

$$f_{electric} = -\frac{1}{2}\varepsilon_o \Delta\varepsilon\left(\vec{E}\cdot\vec{n}\right)^2 = -\frac{1}{2}\varepsilon_o \Delta\varepsilon E^2 \sin^2\theta. \tag{5.41}$$

The free energy density is

$$f = \frac{1}{2}K_{22}\theta'^2 - \frac{1}{2}\varepsilon_o \Delta\varepsilon E^2 \sin^2\theta. \tag{5.42}$$

In a similar way as in the splay geometry, the threshold can be found to be

$$E_c = \frac{\pi}{h}\sqrt{\frac{K_{22}}{\varepsilon_o \Delta\varepsilon}} \tag{5.43}$$

5.2.4 Twisted nematic cell

One of the most important liquid crystal displays is the twisted nematic (TN) shown in Figure 5.8. The liquid crystal is anchored parallel to the cell surface by the alignment layers. The angle between the two alignment directions is Φ, referred to as the total twist angle, which can be any value in general. In the particular case shown in Figure 5.8, $\Phi = 90^o$ and the twisting is counterclockwise when looking down from the top. In the absence of external electric fields, the liquid crystal is in the planar twisted state, where the liquid crystal director twists at a constant rate from the bottom to the top of the cell, as shown in Figure 5.8(a). The boundary conditions can also be satisfied if the director twists the complementary angle to Φ in the opposite twisting direction. If the liquid crystal is nematic, domains with the director twisting in both directions coexist in the cell. In order to achieve a single domain, usually a chiral dopant is added to the nematic host to select one twisting direction. The chirality of the mixture is $q_o = 2\pi(HTP)x_c$, where (HTP) and x_x are the helical twisting power and concentration of the chiral dopant, respectively. When a sufficiently high electric field is applied across the cell, the liquid crystal ($\Delta\varepsilon > 0$) is tilted toward the field direction as shown in Figure 5.8(b).

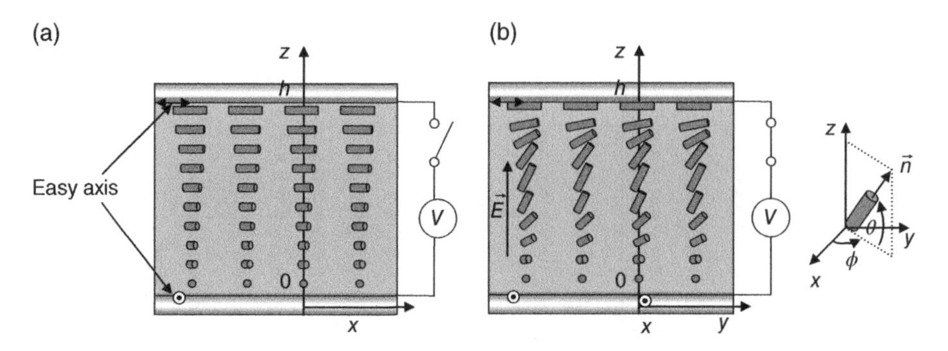

Figure 5.8 Schematic diagram of Fréedericksz transition in the twisted nematic cell.

The liquid crystal director is described by the polar angle θ and the azimuthal angle ϕ. Both angles are a function of z. The components of the director $\vec{n}$ are given by

$$n_x = \cos\theta(z)\cos\phi(z),\ n_y = \cos\theta(z)\sin\phi(z),\ n_z = \sin\theta(z). \tag{5.44}$$

We consider the case where the anchoring is finitely strong and the boundary conditions are

$$\phi(z=0) = \phi_1,\ \phi(z=h) = \phi_2, \tag{5.45}$$

where $\phi_2 - \phi_1 = \Phi$, and

$$\theta(z=0) = 0,\ \theta(z=h) = 0. \tag{5.46}$$

The divergence of $\vec{n}$ is

$$\nabla\cdot\vec{n} = \cos\theta\theta', \tag{5.47}$$

where $\theta' = \partial\theta/\partial z$. The curl is

$$\nabla\times\vec{n} = (\sin\theta\sin\phi\theta' - \cos\theta\cos\phi\phi')\hat{x} + (-\sin\theta\cos\phi\theta' - \cos\theta\sin\phi\phi')\hat{y}. \tag{5.48}$$

The free energy density is

$$f = \frac{1}{2}K_{11}\cos^2\theta\theta'^2 + \frac{1}{2}K_{22}\left(q_o - \cos^2\theta\phi'\right)^2 + \frac{1}{2}K_{33}\sin^2\theta\left(\theta'^2 + \cos^2\theta\phi\prime^2\right) - \frac{1}{2}\Delta\varepsilon\varepsilon_o E^2\sin^2\theta. \tag{5.49}$$

Using Euler–Lagrange method to minimize the total free energy,

$$-\frac{\delta f}{\delta\theta} = \left(K_{11}\cos^2\theta + K_{33}\sin^2\theta\right)\theta''$$
$$+ [(K_{33} - K_{11})\theta'^2 + (2K_{22}\cos^2\theta - K_{33}\cos 2\theta)\varphi\prime^2 + \Delta\varepsilon\varepsilon_o E^2 - 2K_{22}q_o\varphi']\sin\theta\cos\theta = 0, \tag{5.50}$$

$$-\frac{\delta f}{\delta\phi}=\left(K_{22}\cos^2\theta+K_{33}\sin^2\theta\right)\cos^2\theta\phi''$$

$$+2\left[\left(-2K_{22}\cos^2\theta+K_{33}\cos 2\theta\right)\phi'+K_{22}q_o\right]\theta'\sin\theta\cos\theta=0. \quad (5.51)$$

When the applied field is slightly above the threshold, θ is very small, and we have the approximations $\sin\theta\approx\theta$ and $\cos\theta\approx 1$. Keeping only first-order terms, Equation (5.51) becomes $-\delta f/\delta\phi=K_{22}\phi''=0$, whose solution is

$$\phi=\frac{z\Phi}{h}. \quad (5.52)$$

Equation (5.50) becomes

$$-\frac{\delta f}{\delta\theta}=K_{11}\theta''+\left[(2K_{22}-K_{33})\left(\frac{\Phi}{h}\right)^2+\Delta\varepsilon\varepsilon_o E^2-2K_{22}q_o\left(\frac{\Phi}{h}\right)\right]\theta=0. \quad (5.53)$$

Because of the boundary condition given in Equation (5.46), the solution is

$$\theta=A\sin\left(\frac{\pi}{h}z\right). \quad (5.54)$$

The dynamics of the transition is governed by

$$\gamma\frac{\partial\theta}{\partial t}=-\frac{\delta f}{\delta\theta}. \quad (5.55)$$

The physical meaning of the above equation is that viscosity torque, which is the product of the rotational viscosity coefficient γ and the angular speed $\partial\theta/\partial t$, is balanced by $-\delta f/\delta\theta$, which is the sum of the elastic and electric torques. Using Equations (5.53), (5.54) and (5.55), it can be obtained that

$$\gamma\frac{\partial A}{\partial t}=\left[(2K_{22}-K_{33})\left(\frac{\Phi}{h}\right)^2-K_{11}\left(\frac{\pi}{h}\right)^2+\Delta\varepsilon\varepsilon_o E^2-2K_{22}q_o\left(\frac{\Phi}{h}\right)\right]A=0. \quad (5.55)$$

Initially $\theta=0$, and therefore $A=0$. If the applied field is low, the coefficient on the right side of Equation (5.55) is negative, meaning that A cannot grow and remains at 0. When the applied field is sufficiently high, the coefficient becomes positive, meaning that A grows and therefore the transition takes place. Hence the field threshold can be found by setting

$$(2K_{22}-K_{33})\left(\frac{\Phi}{h}\right)^2-K_{11}\left(\frac{\pi}{h}\right)^2+\Delta\varepsilon\varepsilon_o E_c^2-2K_{22}q\left(\frac{\Phi}{h}\right)=0,$$

which gives [6,7]

$$E_c=\frac{\pi}{h}\sqrt{\frac{K_{11}}{\Delta\varepsilon\varepsilon_o}}\left[1+\frac{(K_{33}-2K_{22})}{K_{11}}\left(\frac{\Phi}{\pi}\right)^2+2\frac{K_{22}}{K_{11}}\left(\frac{hq_o}{\pi}\right)\left(\frac{\Phi}{\pi}\right)\right]^{1/2}. \quad (5.56)$$

The polar and azimuthal angles as a function of z under an applied field higher than the threshold can only be numerically calculated and will be discussed in Chapter 7.

5.2.5 *Splay geometry with weak anchoring*

If the anchoring of the liquid crystal at the boundary is not infinitely strong, the tilt angle θ at the boundary is no longer fixed but changes with the applied field. Now we must consider the surface energy in determining the equilibrium director configuration. The total free energy of the system is given by [5,8]

$$F=\int_0^h f\,dz+f_s|_{z=0}+f_s|_{z=h}$$

$$=\int_0^h\left[\frac{1}{2}(K_{11}\cos^2\theta+K_{33}\sin^2\theta)\theta'^2-\frac{1}{2}\varepsilon_o\Delta\varepsilon E^2\sin^2\theta\right]dz+\frac{1}{2}W\sin^2\theta_1+\frac{1}{2}W\sin^2\theta_2, \quad (5.57)$$

where W is the anchoring strength, and θ_1 and θ_2 are tilt angles at the bottom and top surfaces, respectively. The boundary conditions at the bottom and top surfaces are respectively

$$\left(-\frac{\partial f}{\partial\theta'}+\frac{\partial f_s}{\partial\theta}\right)\bigg|_{z=0}=-(K_{11}\cos^2\theta_1+K_{33}\sin^2\theta_1)\theta'+W\sin\theta_1\cos\theta_1=0, \quad (5.58)$$

$$\left(\frac{\partial f}{\partial\theta'}+\frac{\partial f_s}{\partial\theta}\right)\bigg|_{z=h}=(K_{11}\cos^2\theta_2+K_{33}\sin^2\theta_2)\theta'+W\sin\theta_2\cos\theta_2=0. \quad (5.59)$$

When the applied field is low, the liquid crystal director is homogeneously aligned along the x axis and the tilt angle is 0. When the applied field is increased above a threshold, the liquid crystal director begins to tilt. When the applied field is only slightly above the threshold, the tilt angle is small. Approximately, we have $\cos\theta\approx 1$ and $\sin\theta\approx\theta$. Equations (5.58) and (5.59) become

$$-K_{11}\theta'|_{z=0}+W\theta_1=0, \quad (5.60)$$

$$K_{11}\theta'|_{z=h}+W\theta_2=0. \quad (5.61)$$

The Euler–Lagrange equation of the minimization of the total bulk free energy is (Equation (5.22))

$$-\varepsilon_o\Delta\varepsilon E^2\theta-K_{11}\theta''=0. \quad (5.62)$$

Define the surface extrapolation length

$$L=K_{11}/W. \quad (5.63)$$

Also define the field coherence length

$$\xi=\left(\frac{K_{11}}{\varepsilon_o\Delta\varepsilon E^2}\right)^{1/2}, \quad (5.64)$$

whose physical meaning is that the applied field can produce a significant reorientation of the liquid crystal director over the distance ξ. When $K_{11} = 10^{-11}$N, $\Delta\varepsilon = 10$ and $E = 1$ V/μm, the field coherence length is $\xi = 0.3$ μm. Because the solution to Equation (5.62) must be symmetric about $z = h/2$, the solution is

$$\theta = B\cos\left(\frac{z - h/2}{\xi}\right) \tag{5.65}$$

Note that this is the solution for an applied field slightly above the threshold. The boundary condition equation (5.60) becomes $\frac{B}{\xi}\sin\left(\frac{h}{2\xi}\right) = \frac{B}{L}\cos\left(\frac{h}{2\xi}\right)$, namely,

$$\tan\left(\frac{h}{2\xi}\right) = \frac{\xi}{L}. \tag{5.66}$$

When the anchoring is infinitely strong, $W = \infty$ and thus $L = 0$; the solution to Equation (5.66) is $h/2\xi = \pi/2$, which gives the field threshold $E_c = (\pi/h)\sqrt{K_{11}/\varepsilon_o\Delta\varepsilon}$. When the anchoring is weak, for example, $W = 10^{-5}$j/m^2, the surface extrapolation length is $L = 10^{-11}$N/(10^{-5}j/m^2) = 10^{-1}μm. When the applied field is low, ξ is large and $\xi/L \gg 1$. At the field threshold, $h/2\xi$ is close to $\pi/2$, and thus we have the approximation

$$\tan\left(\frac{h}{2\xi}\right) = \tan\left[\frac{\pi}{2} - \left(\frac{\pi}{2} - \frac{h}{2\xi}\right)\right] \approx 1/\left(\frac{\pi}{2} - \frac{h}{2\xi}\right) = \frac{\xi}{L},$$

which gives $1/\xi = \pi/(h + 2L)$. Therefore the field threshold is

$$E_c = \frac{\pi}{(h + 2L)}\sqrt{\frac{K_{11}}{\varepsilon_o\Delta\varepsilon}}. \tag{5.67}$$

This result is the same as that when the anchoring is infinitely strong and the cell gap is increased from h to $h + 2L$.

5.2.6 Splay geometry with pretilt angle

In ECB liquid crystal devices, alignment layers with non-zero pretilt angle are usually used in order to avoid poly-domain structures resulting from opposite tilting of the liquid crystal director under externally applied fields. The rubbing directions of the bottom and top alignment layers are anti-parallel. Now we consider how the pretilt angle affects the reorientation of the liquid crystal under externally applied fields. The Euler–Lagrange Equation of the minimization of the total free energy is (Equation (5.21))

$$\frac{\delta f}{\delta\theta} = -\varepsilon_o\Delta\varepsilon E^2\sin\theta\cos\theta - (K_{33} - K_{11})\sin\theta\cos\theta\theta'^2 - \left(K_{11}\cos^2\theta + K_{33}\sin^2\theta\right)\theta'' = 0. \tag{5.68}$$

When the pretilt angle is θ_o, we define a new variable $\beta = \theta - \theta_o$. When the applied field is low, β is very small, and approximately we have $\sin(\beta + \theta_o) \approx \beta\cos\theta_o + \sin\theta_o$ and $\cos(\beta + \theta_o) \approx \cos\theta_o - \beta\sin\theta_o$. $\theta'' = \beta''$. Neglecting higher-order terms, Eq (5.68) becomes

$$-\varepsilon_o\Delta\varepsilon E^2\left(\frac{1}{2}\sin 2\theta_o+\beta\cos 2\theta_o\right)-\left(K_{11}\cos^2\theta_o+K_{33}\sin^2\theta_o\right)\beta''=0. \tag{5.69}$$

The general solution is

$$\beta=B\cos\left[\sqrt{\frac{\varepsilon_o\Delta\varepsilon E^2\cos(2\theta_o)}{(K_{11}\cos^2\theta_o+K_{33}\sin^2\theta_o)}}\left(z-\frac{h}{2}\right)\right]-\frac{1}{2}\tan(2\theta_o), \tag{5.70}$$

where B is a constant, which can be found from the boundary condition. Under infinitely strong anchoring, the boundary conditions are $\beta(z=0)=\beta(z=h)=0$. Therefore

$$B\cos\left[\sqrt{\frac{\varepsilon_o\Delta\varepsilon E^2\cos(2\theta_o)}{(K_{11}\cos^2\theta_o+K_{33}\sin^2\theta_o)}}\frac{h}{2}\right]=\frac{1}{2}\tan(2\theta_o).$$

Equation (5.70) becomes

$$\beta=\frac{1}{2}\tan(2\theta_o)\left\{\frac{\cos\left[\sqrt{\frac{\varepsilon_o\Delta\varepsilon E^2\cos(2\theta_o)}{(K_{11}\cos^2\theta_o+K_{33}\sin^2\theta_o)}}\left(z-\frac{h}{2}\right)\right]}{\cos\left[\sqrt{\frac{\varepsilon_o\Delta\varepsilon E^2\cos(2\theta_o)}{(K_{11}\cos^2\theta_o+K_{33}\sin^2\theta_o)}}\frac{h}{2}\right]}-1\right\}. \tag{5.71}$$

β is not zero for any non-zero applied field E. The pretilt angle breaks the reflection symmetry of the anchoring and eliminates the threshold. When E is very small, we have

$$\beta=\frac{\varepsilon_o\Delta\varepsilon E^2\sin(2\theta_o)}{4(K_{11}\cos^2\theta_o+K_{33}\sin^2\theta_o)}z(h-z). \tag{5.72}$$

The maximum tilt angle (at the middle plane) (when $K_{11}=K_{33}$)

$$\beta_m=\frac{\varepsilon_o\Delta\varepsilon E^2\sin(2\theta_o)h^2}{16K_{11}}. \tag{5.73}$$

When $E=E_c=\frac{\pi}{h}\sqrt{\frac{K_{11}}{\varepsilon_o\Delta\varepsilon}}$ (the threshold field when the pretilt angle is zero), the maximum tilt angle is $\beta_m=\beta_{mc}=\pi^2\sin(2\theta_o)/16$, which is small for small pretilt angle θ_o. For an applied field below E_c, $\beta_m=\beta_{mc}(E/E_c)^2$. For an applied field slightly above E_c, $\beta_m=\beta_{mc}+\sin^{-1}\left[2\sqrt{(E-E_c)/E_c}\right]$.

5.3 Measurement of Anchoring Strength

As discussed in the previous sections, anchoring strength of alignment layers in liquid crystal devices has profound effects on the electro-optical properties of the devices. Now we consider how to measure anchoring strengths.

5.3.1 Polar anchoring strength

An electrically controlled birefringence (ECB) display consists of two parallel substrates with a nematic liquid crystal ($\Delta\varepsilon > 0$) sandwiched between them, as shown in Figure 5.2. On the inner surface of the substrates there are anti-parallel rubbed homogeneous alignment layers. In the absence of applied voltage, the liquid crystal is in the uniform homogenous state, as shown in Figure 5.2(a). The polar angle of the liquid crystal director is zero everywhere. When a low voltage is applied across the cell, only the liquid crystal in the bulk will be rotated toward the cell normal direction. When a high voltage is applied, if the anchoring strength is finite, the liquid crystal in the bulk, as well as on the surfaces, will be rotated. The total free energy per unit area is given by Equation (5.57). When a very high field is applied, the liquid crystal in the bulk, except in a thin layer near the surface, is aligned along the cell normal direction. The field correlation length ξ, defined in Equation (5.64), is much shorter than the cell thickness and is comparable to the surface extrapolation length L, defined in Equation (5.63). Consider the case where the two alignment layers are the same. Using the approximation $K_{11} = K_{33} = K$, the total free energy becomes

$$F = 2\int_0^{h/2}\left[\frac{1}{2}K\theta'^2 - \frac{1}{2}\varepsilon_o\Delta\varepsilon E^2\sin^2\theta\right]dz + W\sin^2\theta_1 = (K/\xi)\int_0^{h/2\xi}\left[\left(\frac{\partial\theta}{\partial\eta}\right)^2 - \sin^2\theta\right]d\eta + (K/L)\sin^2\theta_1, \tag{5.74}$$

where $\eta = z/\xi$ and θ_1 is the tilt angle at the surface of the cell. Using the Euler–Lagrange method to minimize the free energy, we obtain

$$\frac{\delta f}{\delta\theta} = \frac{\partial f}{\partial\theta} - \frac{d}{d\eta}\left(\frac{\partial f}{\partial(\partial\theta/\partial\eta)}\right) = -\sin\theta\cos\theta - \frac{\partial^2\theta}{\partial\eta^2} = 0. \tag{5.75}$$

The boundary conditions at $z = h/2$ are $\theta = \pi/2$ and $\partial\theta/\partial\eta = 0$. The solution is

$$\frac{\partial\theta}{\partial\eta} = \cos\theta. \tag{5.76}$$

Integrating the above equation we have

$$\int_0^{\eta} d\eta = \int_{\theta_1}^{\theta}\frac{1}{\cos\theta}d\theta.$$

The solution is

$$\eta = \ln[\tan(\theta/2 + \pi/4)] - \ln[\tan(\theta_1/2 + \pi/4)]. \tag{5.77}$$

From Equation 5.11 we have the boundary condition at $z = 0$, which is

$$\frac{\partial f}{\partial(\partial\theta/\partial\eta)} = (K/\xi)\frac{\partial\theta}{\partial\eta} = (K/\xi)\cos\theta_1 = (K/L)\sin\theta_1\cos\theta_1. \tag{5.78}$$

Therefore

$$\sin\theta_1 = L/\xi, \tag{5.79}$$

where $L/\xi = (K/W)\Big/\left(\sqrt{K/\Delta\varepsilon\varepsilon_o E^2}\right) = \left(\sqrt{\Delta\varepsilon\varepsilon_o K}\right)(E/W)$. This is correct only when $\xi \geq L$. When $\xi < L$, $\theta = \pi/2$ everywhere. Using Equations (5.93) and (5.79) we get

$$\theta = -\frac{\pi}{2} + 2\arctan\left[\tan(\theta_1/2 + \pi/4)e^{z/\xi}\right]. \tag{5.80}$$

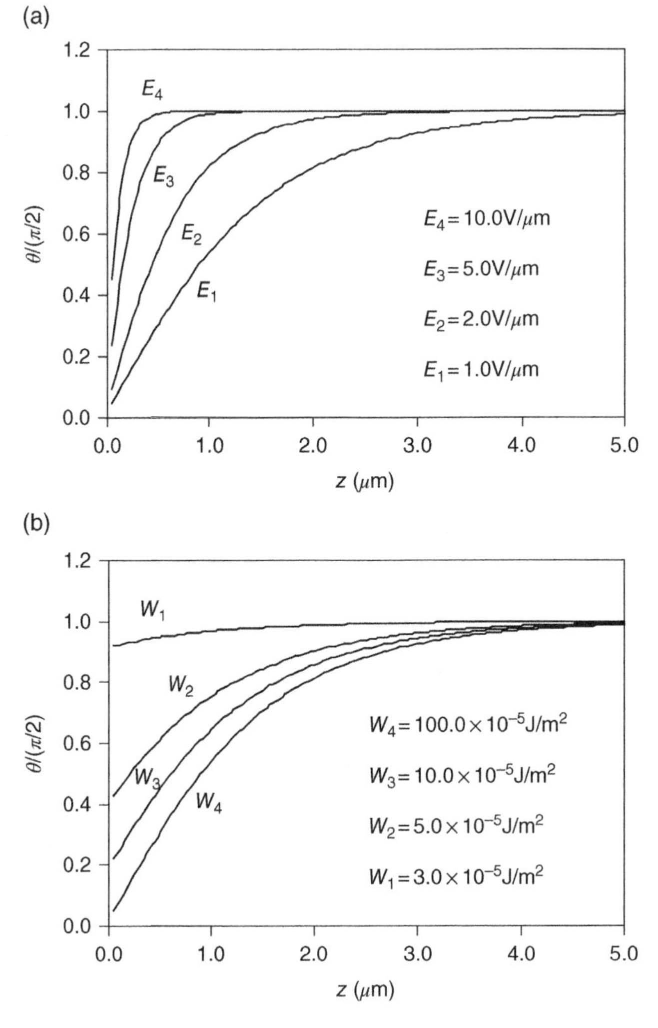

Figure 5.9 Tilt angle of liquid crystal director as a function of position under various applied electric fields and anchoring strengths. (a) anchoring strength $W = 10^{-3}$J/m^2. (b) applied field $E = 1$ V/μm.

The tilt angle of the liquid crystal director as a function of the z coordinate (which is perpendicular to the cell substrate) is shown in Figure 5.9. The cell thickness is 10 μm. The elastic constant is 10^{-11}N. The anchoring strength of rubbed polymers is usually in the region from 10^{-5}J/m^2 to 10^{-3}J/m^2. When the applied electric field is very high, the tilt angle at the surface of the alignment layer becomes large, and a quartic term may be needed in order to get the right anchoring energy. For a given electric field, the optical retardation of the cell can be theoretically calculated from $\theta(z)$ and thus the transmission of the cell, when sandwiched between two crossed polarizers, can be calculated. The transmission can also be experimentally measured. The anchoring strength can be obtained by using it as a fitting parameter in the comparison of the calculated and measured transmissions. This method of measurement of polar anchoring strength is called the high voltage method [9–14].

5.3.2 *Azimuthal anchoring strength*

When there is a torque to rotate the liquid crystal in the azimuthal direction, the liquid crystal director at the surface of the alignment layer may deviate from the easy direction. The anchoring strength can be measured from the resistance produced by the alignment layer in the azimuthal direction. Adding a chiral dopant into the liquid crystal can induce a change of the azimuthal angle of the liquid crystal. The chiral dopant causes the liquid crystal to twist through the cell while the alignment layer tries to keep the liquid crystal in the easy direction, against the twisting. The result of the competition between these two factors is that the liquid crystal in the bulk twists less than the intrinsic twist angle produced by the chiral dopant and on the surface of the cell the liquid crystal deviates from the easy direction, as shown in Figure 5.10(a). The free energy per unit area of the liquid crystal is given by

$$F=\int_0^h \frac{1}{2}K_{22}\left(\frac{\Phi}{h}-\frac{2\pi}{P}\right)^2 dz+\frac{1}{2}W_a\sin^2\phi_{o1}+\frac{1}{2}W_a\sin^2\phi_{o2}, \tag{5.81}$$

where Φ is the total twist angle from the bottom to the top of the cell and P is the intrinsic pitch, which can be calculated from the concentration and the helical twisting power of the doped chiral dopant. Because the top and bottom alignment layers are the same, the twist deviation angles at the top and bottom surface are the same: $\phi_{o1}=\phi_{o2}=\phi_o$. The pitch is chosen in such a way that $P>4h$ in order to avoid over 90° twist. From Figure 5.10(a) it can be seen that $\Phi=\phi_{o1}+\phi_{o2}=2\phi_o$. Equation (5.81) becomes

$$F=\frac{1}{2}K_{22}h\left(\frac{2\phi_o}{h}-\frac{2\pi}{P}\right)^2+W_a\sin^2\phi_o. \tag{5.81}$$

ϕ_o can be found by minimizing the free energy with respect to ϕ_o. When ϕ_o is small, $\sin\phi_o\approx\phi_o$, and minimizing the free energy leads to

$$\frac{\partial F}{\partial\phi_o}=K_{22}h\left(\frac{2\phi_o}{h}-\frac{2\pi}{P}\right)\frac{2}{h}+2W_a\phi_o\overset{\text{let}}{=}0. \tag{5.82}$$

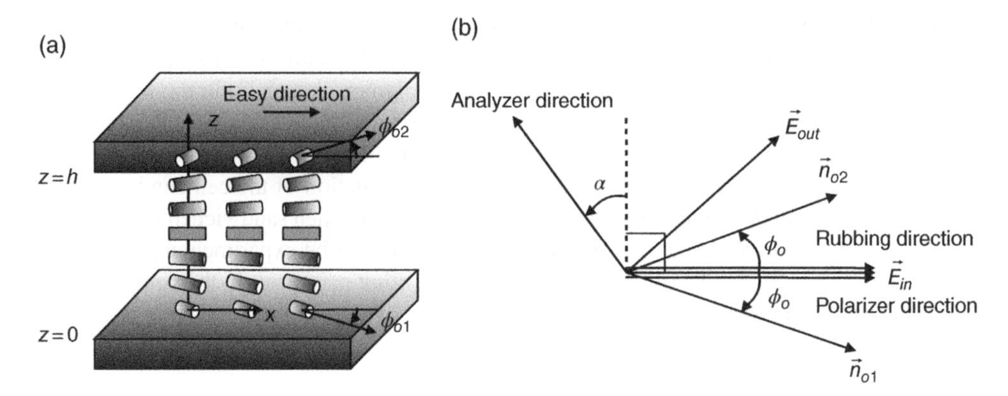

Figure 5.10 (a) Schematic diagram of the cell used for azimuthal anchoring measurement. (b) Diagram of the measurement of the twist deviation angle.

From the above equation, we get the anchoring strength

$$W_a = 2\left(\frac{\pi}{\phi_o} - \frac{P}{h}\right)\frac{K_{22}}{P}. \tag{5.83}$$

By measuring the twist angle, the azimuthal anchoring strength can be obtained. This method is called the twist angle method (TAM) [15,16].

In the measurement of the twist angle, the polarizer is fixed in the direction parallel to the rubbing direction, as shown in Figure 5.10(b). The analyzer makes the angle α with respect to the orthogonal direction of the polarizer. The electric field of the incident light can be decomposed into two components: e-mode $E_{in/e}$, parallel to the liquid crystal director $\vec{n}_{o1}$ at the entrance plane, and o-mode $E_{in/o}$, perpendicular to the liquid crystal director. $E_{in/e} = E_o \cos\phi_o$ and $E_{in/o} = E_o \sin\phi_o$. Because $(n_e - n_o)h/(2\pi/2\phi_o) \gg \lambda$ (Mauguin condition), when the light propagates through the liquid crystal cell, the polarizations of the e- and o-modes follow the twist of the liquid crystal director. Thus they remain parallel to and perpendicular to the liquid crystal director, respectively. Their phase changes are, however, different. The e component becomes $E_{out/e} = E_o \cos\varphi_o e^{-i2\pi n_e h/\lambda}$ and the o component becomes $E_{out/o} = E_o \sin\varphi_o e^{-i2\pi n_o h/\lambda}$. The sum of their projections along the analyzer is

$$\begin{aligned} E_a &= E_o \cos\varphi_o e^{-i2\pi n_e h/\lambda}\cos[\pi/2+\alpha-\varphi_o] + E_o \sin\varphi_o e^{-i2\pi n_o h/\lambda}\cos[\pi/2+\alpha-(\pi/2+\varphi_o)] \\ &= -E_o \cos\varphi_o e^{-i2\pi n_e h/\lambda}\sin(\alpha-\varphi_o) + E_o \sin\varphi_o e^{-i2\pi n_o h/\lambda}\cos(\alpha-\varphi_o). \end{aligned} \tag{5.84}$$

The transmittance is given by

$$T = \frac{I_{out}}{I_{in}} = \frac{E_a \cdot E_a^*}{E_o^2} = \sin^2\alpha - \sin(2\phi_o)\sin(2\alpha-2\phi_o)\cos^2(\pi\Delta n h/\lambda). \tag{5.85}$$

The angle α_m that gives the minimum transmittance is found by

$$\left.\frac{\partial T}{\partial \alpha}\right|_{\alpha_m}=2\sin\alpha_m\cos\alpha_m-2\sin(2\phi_o)\cos(2\alpha_m-2\phi_o)\cos^2(\pi\Delta nh/\lambda)=0,$$

which gives

$$\tan(2\alpha_m)=\frac{\sin(4\phi_o)\cos^2(\pi\Delta nh/\lambda)}{1-2\sin^2(2\phi_o)\cos^2(\pi\Delta nh/\lambda)}. \tag{5.86}$$

In the experiment, the analyzer is rotated to find the minimum transmittance and thus to determine α_m. ϕ_o is calculated from Equation (5.86).

5.4 Measurement of Pretilt Angle

In liquid crystal devices it is highly desirable that the alignment layer has a pretilt angle such that the reorientation of the liquid crystal is well controlled and uniform. For example, in a homogeneously aligned display cell, if the alignment layer has no pretilt angle (i.e. the pretilt angle is zero), when an electric field is applied in the cell normal direction, the liquid crystal (with $\Delta\varepsilon > 0$) can rotate either clockwise or counterclockwise. Consequentially the device has a slow response, and multi-domain structures form. As another example, in a homeotropically aligned display cell, if the alignment layer has no pretilt angle, when an electric field is applied in the cell normal direction, the liquid crystal (with $\Delta\varepsilon < 0$) can rotate in any azimuthal plane. Similarly the device has a long response time, and multi-domain structures form. Furthermore, alignment layers with proper pretilt angles can stabilize desired liquid crystal states such as optically compensated bend (OCB) mode.

Large pretilt angles (with alignment direction close to the normal direction of the cell) can be measured by using the magnetic 'null' method [17,18]. The liquid crystal cell is placed in a magnetic field and its transmittance is measured. If the magnetic field is not parallel to the liquid crystal orientation in the ground state (magnetic field-off state), a change of the magnetic field will result in a change of the liquid crystal orientation and thus produce a change of the transmittance. If the magnetic field is parallel to the liquid crystal orientation in the ground state, any change of the magnetic field does not change the liquid crystal orientation and thus the transmittance remains a constant value.

Small pretilt angles (with alignment direction close to the horizontal direction of the cell) can be measured by using the crystal rotation method, as shown in Figure 5.11 [19]. The liquid crystal cell is sandwiched between two crossed polarizers. The cell is rotated around an axis perpendicular to the liquid crystal director plane. The coordinate is defined in such a way that the rotation axis is parallel the y axis, and the light propagation direction is parallel to the z axis. The transmission axis of the polarizer is set to 45° with respect to the rotation axis (the y axis). The electric field of the incident light is decomposed into two components along the x and y axes, respectively:

$$E_{\text{in}/x}=\left(E_o/\sqrt{2}\right),\ E_{\text{in}/y}=\left(E_o/\sqrt{2}\right) \tag{5.87}$$

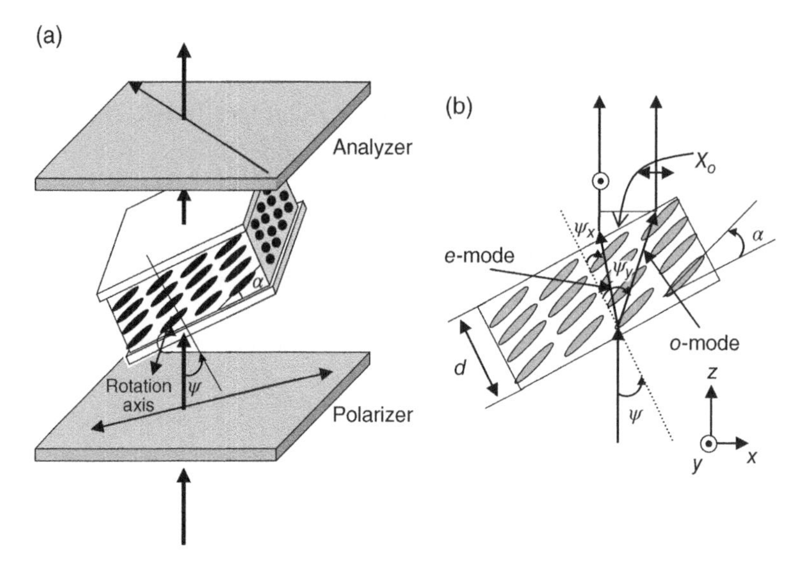

Figure 5.11 Schematic diagram of the crystal rotation method of pretilt angle measurement.

Inside the liquid crystal, the refractive indices for the x and y components are n_x and n_y, respectively. There is a double refraction at the air–liquid crystal interface: the refraction angles for the components are ψ_x and ψ_y, respectively, which are calculated by

$$\sin\psi = n_x \sin\psi_x, \quad \sin\psi = n_y \sin\psi_y. \tag{5.89}$$

From the above equations, we have

$$\cos\psi_x = \sqrt{1-\sin^2\psi_x} = \sqrt{1-\sin^2\psi/n_x^2}, \cos\psi_y = \sqrt{1-\sin^2\psi_y} = \sqrt{1-\sin^2\psi/n_y^2}. \tag{5.90}$$

The refractive indices are given by

$$n_x = \frac{n_e n_o}{\sqrt{n_e^2\sin^2(\alpha+\psi_x)+n_o^2\cos^2(\alpha+\psi_x)}}, n_y = n_o, \tag{5.91}$$

where n_o and n_e are ordinary and extraordinary refractive indices of the liquid crystal, respectively. Thus the x component of the light is called the e-mode and the y component of the light is called the o-mode. These two modes propagate through the liquid crystal with different speeds. The optical path length for the o-mode is

$$L_x = n_y d/\cos\psi_y = n_y d/\left(1-\sin^2\psi/n_y^2\right)^{1/2} = \frac{n_y^2 d}{\left(n_y^2-\sin^2\psi\right)^{1/2}}. \tag{5.92}$$

On the exit plane the e-beam and o-beam are separated by the distance given by

$$X_o = d\tan\psi_o - d\tan\psi_e = \frac{d\sin\psi_o}{\left(1-\sin^2\psi/n_y^2\right)^{1/2}} - \frac{d\sin\psi_e}{\left(1-\sin^2\psi/n_x^2\right)^{1/2}} = \frac{d\sin\psi}{\left(n_y^2-\sin^2\psi\right)} - \frac{d\sin\psi}{\left(n_x^2-\sin^2\psi\right)}. \tag{5.92}$$

The optical path length for the e-mode is

$$L_y = n_y d/\cos\psi_x + X_o\sin\psi = \frac{n_x^2 d}{\left(n_x^2-\sin^2\psi\right)^{1/2}} + \left(\frac{d\sin^2\psi}{\left(n_y^2-\sin^2\psi\right)^{1/2}} - \frac{d\sin^2\psi}{\left(n_x^2-\sin^2\psi\right)^{1/2}}\right). \tag{5.93}$$

After propagating through the liquid crystal cell, the phase difference of the e-beam and o-beam is

$$\Delta\phi = \frac{2\pi(L_x-L_y)}{\lambda} = \frac{2\pi}{\lambda}\left[\frac{n_x^2 d}{\left(n_x^2-\sin^2\psi\right)^{1/2}} - \frac{n_y^2 d}{\left(n_y^2-\sin^2\psi\right)^{1/2}} + \left(\frac{d\sin^2\psi}{\left(n_y^2-\sin^2\psi\right)^{1/2}} - \frac{d\sin^2\psi}{\left(n_x^2-\sin^2\psi\right)^{1/2}}\right)\right]$$

$$\Delta\phi = \frac{2\pi d}{\lambda}\left[\left(n_x^2-\sin^2\psi\right)^{1/2} - \left(n_y^2-\sin^2\psi\right)^{1/2}\right] = \frac{2\pi d}{\lambda}\left[\left(n_x^2-\sin^2\psi\right)^{1/2} - \left(n_o^2-\sin^2\psi\right)^{1/2}\right]. \tag{5.94}$$

The first term on the right side of the above equation is

$$\left(n_x^2-\sin^2\psi\right)^{1/2} = \left(n_x^2-n_x^2\sin^2\psi_x\right)^{1/2} = n_x\cos\psi_x \overset{\text{let}}{=} u. \tag{5.95}$$

From Equations (5.91) we have

$$n_e^2 n_x^2\sin^2(\alpha+\psi_x) + n_o^2 n_x^2\cos^2(\alpha+\psi_x) = n_e^2 n_o^2. \tag{5.95}$$

Expanding the above equation and using $\sin\psi = n_x\sin\psi_x$ and $u = n_x\cos\psi_x$, we get

$$\begin{aligned} &n_e^2[u^2\sin^2\alpha + 2u\sin\alpha\cos\alpha\sin\psi + \sin^2\psi\cos^2\alpha] \\ &+ n_o^2[u^2\cos^2\alpha - 2u\sin\alpha\cos\alpha\sin\psi + \sin^2\psi\sin^2\alpha] = n_e^2 n_o^2. \end{aligned} \tag{5.96}$$

Solving the above equation, we get

$$u = \frac{\left(n_o^2-n_e^2\right)\sin\alpha\cos\alpha\sin\psi + n_o n_e\left(n_o^2\cos^2\alpha + n_e^2\sin^2\alpha - \sin^2\psi\right)^{1/2}}{\left(n_o^2\cos^2\alpha + n_e^2\sin^2\alpha\right)}. \tag{5.97}$$

From Equations (5.94) and (5.97) we have the phase difference between the e-beam and o-beam

$$\Delta\phi = \frac{2\pi d}{\lambda}\left[\frac{\left(n_o^2 - n_e^2\right)\sin\alpha\cos\alpha\sin\psi + n_o n_e\left(n_o^2\cos^2\alpha + n_e^2\sin^2\alpha - \sin^2\psi\right)^{1/2}}{\left(n_o^2\cos^2\alpha + n_e^2\sin^2\alpha\right)} - \left(n_o^2 - \sin^2\psi\right)^{1/2}\right]. \tag{5.98}$$

From the projections of the electric fields of the o-beam and e-beam on the transmission axis of the analyzer we calculate the transmittance.

$$T = \left(-\frac{1}{\sqrt{2}}E_{in/x}e^{i\Delta\varphi} + \frac{1}{\sqrt{2}}E_{in/y}\right)\cdot\left(-\frac{1}{\sqrt{2}}E_{in/x}e^{i\Delta\varphi} + \frac{1}{\sqrt{2}}E_{in/y}\right)^* \Big/ E_{in}^2 = \sin^2(\Delta\varphi/2). \tag{5.99}$$

The calculated transmittance as a function of the incident angle ψ is shown in Figure 5.12. The following parameters are used: cell thickness $d = 10\ \mu$m, refractive indices $n_o = 1.5$, and $n_e = 1.7$. When the pretilt angle is 0°, the transmittance curve is symmetric around the 0° incident angle. When the pretilt angle is 5°, the transmittance curve is shifted by the angle of −15.8°, namely, the curve is symmetric around the angle of −16°. It can be seen that the transmittance vs. incident angle curve is symmetric around an incident angle at which the transmittance has an extreme value (maximum or minimum). Therefore

$$\left.\frac{\partial T}{\partial\psi}\right|_{\psi_s} = \sin(\Delta\phi/2)\cos(\Delta\phi/2)\left.\frac{\partial\Delta\phi}{\partial\psi}\right|_{\psi_s} = 0. \tag{5.100}$$

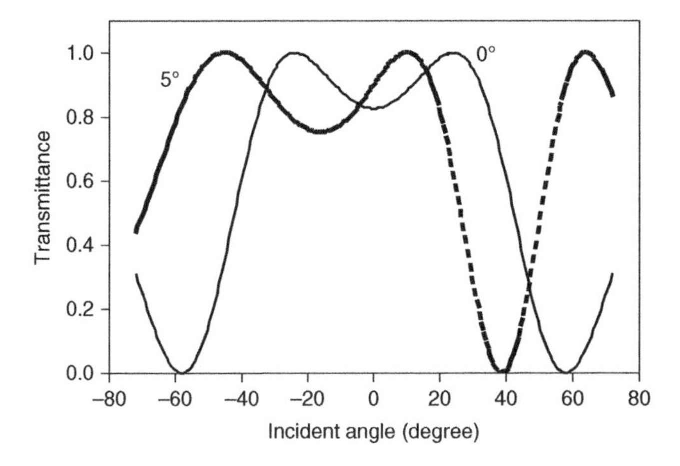

Figure 5.12 Transmittance vs. incident angle in the crystal rotation method of pretilt angle measurement.

From Equations (5.98) and (5.100) we have

$$\sin\psi_s\left[\frac{n_o n_e\left(n_o^2\cos^2\alpha+n_e^2\sin^2\alpha-\sin^2\psi_s\right)^{-1/2}}{\left(n_o^2\cos^2\alpha+n_e^2\sin^2\alpha\right)}-\left(n_o^2-\sin^2\psi_s\right)^{-1/2}\right]=\frac{\left(n_o^2-n_e^2\right)\sin\alpha\cos\alpha}{\left(n_o^2\cos^2\alpha+n_e^2\sin^2\alpha\right)}. \tag{5.101}$$

When the pretilt angle is 5°, the symmetry angle is calculated to be −16° from Equation (5.101). It can be seen that transmittance vs. incident angle curve is very sensitive to the pretilt angle. The curve is shifted by the angle which is three times larger than the pretilt angle.

The pretilt angle can be obtained in two ways. It can be obtained by comparing the theoretically calculated and experimentally measured transmittance vs. incident angle curve with the pretilt angle as the fitting parameter. It can also be obtained by experimentally determining the symmetry angle ψ_s and then calculating α from Equation (5.101).

5.5 Fréedericksz Transition: Dynamics

The dynamics of the rotation of liquid crystal molecules is very complicated in general because of the coupling between the rotational motion and the translational motion. A rotation of the liquid crystal molecules may generate a translational motion of the molecules. A gradient of the velocity of the translational motion produces a torque which in turn affects the rotation of the molecules. In this section we will proceed from some simple cases to complicated cases.

5.5.1 Dynamics of Fréedericksz transition in twist geometry

In the twist geometry, the rotation of the liquid crystal director is not coupled to the translational motion of the molecules. The rotation of the liquid crystal director is governed by the over-damped dynamics: the elastic and electric torques are balanced by the rotational viscosity torque and the inertial term can be neglected [4]. Mathematically we have

$$\gamma_r\frac{\partial\theta}{\partial t}=-\frac{\delta f}{\delta\theta}, \tag{5.102}$$

where θ is the twist angle as discussed in Section 2.3, γ_r is the rotational viscosity coefficient whose dimension is $\mathrm{N\cdot s/m^2}$. Using the free energy given by Equation (5.42), we can obtain

$$\gamma_r\frac{\partial\theta}{\partial t}=K_{22}\frac{\partial^2\theta}{\partial z^2}+\varepsilon_o\Delta\varepsilon E^2\sin\theta\cos\theta. \tag{5.103}$$

We first consider the turn-off time when the applied field is turned off from the distorted state. The anchoring is infinitely strong and the twist angle at the boundary is 0. The initial twist angle can be expanded in terms of all the possible modes:

$$\theta_o(z) = \sum_{m=1}^{\infty} A_n \sin\left(\frac{m\pi z}{h}\right) \tag{5.104}$$

The solution to the dynamic equation is

$$\theta(t) = \sum_{m=1}^{\infty} A_m \sin\left(\frac{m\pi z}{h}\right) e^{-t/\tau_m}, \tag{5.105}$$

where τ_m is the relaxation time of the mth mode and is given by

$$\tau_m = \frac{\gamma_r}{K_{22}}\left(\frac{h}{\pi}\right)^2 \frac{1}{m^2}. \tag{5.106}$$

The turn-off time τ_{off} is approximately equal to the relaxation time τ_1 of the slowest mode ($m = 1$),

$$\tau_{off} = \frac{\gamma_r}{K_{22}}\left(\frac{h}{\pi}\right)^2. \tag{5.107}$$

Now we consider the turn-on time. When the applied field is not much higher than the threshold, the twist angle is small and $m = 1$ is the only mode excited because it costs the least energy. The solution to the dynamic equation is

$$\theta(t) = A_1 e^{t/\tau} \sin\left(\frac{\pi z}{h}\right). \tag{5.108}$$

Using the approximation that $\sin\theta \approx \theta$ and $\cos\theta = 1$, and substituting Equation (5.108) into Equation (5.103), we have

$$\frac{\gamma_r}{\tau} = -K_{22}\left(\frac{\pi}{h}\right)^2 + \varepsilon_o \Delta\varepsilon E^2 = K_{22}\left(\frac{\pi}{h}\right)^2 \left[\left(\frac{E}{E_c}\right)^2 - 1\right]. \tag{5.109}$$

The turn-on time is

$$\tau_{on} = \frac{\gamma_r}{K_{22}}\left(\frac{h}{\pi}\right)^2 \frac{1}{(E/E_c)^2 - 1}. \tag{5.110}$$

Although the dynamics of Fréedericksz transition in splay geometry, bend geometry, and twisted geometry is more complicated, the response time is still of the same order and has the same cell thickness dependence. The rotational viscosity coefficient is of the order $0.1\,\mathrm{N \cdot s/m^2}$. When the elastic constant is 10^{-11}N and the cell thickness is 10μm, the response time is of the order 100 ms. Faster response times can be achieved by using thinner cell gaps.

5.5.2 *Hydrodynamics*

Now we consider the hydrodynamics of nematic liquid crystals [4,20,21]. For most phenomena in liquid crystals, it is a good assumption that liquid crystals are uncompressible

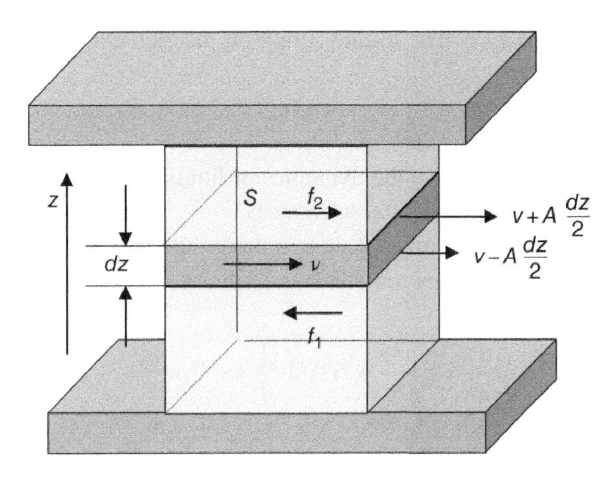

Figure 5.13 The viscosity is the force acting on the surface of the fluid element due to the velocity gradient.

fluids. We first consider viscosity in isotropic fluids. The viscosity of a fluid is an internal friction that hinders neighboring layers of the fluid from sliding with respect to each other. Consider a steady shear whose velocity is only a function of z, as shown in Figure 5.13. Look at a fluid element with the area of S and thickness dz. There are particle exchanges between the fluid element and the fluid above and below it. If the velocity is uniform, the net momentum transfer between the fluid element and the fluid above and below it is zero, and thus there is no force acting on it. If the velocity is not uniform but has a gradient, say $A(z)$ $(=\partial v/\partial z)$, the particles moving into the fluid above have a smaller momentum than the particles moving from the fluid above into the element, provided $A > 0$. The fluid element gains momentum from the fluid above [22]. This momentum gain can be described as a force acting on the surface of the element toward the right side, which is proportional to the area of the element and the velocity gradient (known as a Newtonian fluid):

$$f_2 = \eta A_2 S, \tag{5.111}$$

where η is the translational viscosity coefficient, which has the dimension $\{\mathrm{N}/[(\mathrm{m}\cdot\mathrm{s}^{-1}/\mathrm{m})(\mathrm{m}^2)]\} = [\mathrm{N}\cdot\mathrm{s}/\mathrm{m}^2]$. The commonly used unit is *poise,* which is equal to $0.1\mathrm{N}\cdot\mathrm{s}/\mathrm{m}^2$. The viscosity coefficient of water at room temperature is about $10^{-3}\mathrm{N}\cdot\mathrm{s}/\mathrm{m}^2$. In the same way, the particles moving into the fluid below have a larger momentum than the particles moving from the fluid below into the element. The fluid element loses momentum to the fluid below. This momentum loss can also be described as a force acting on the surface of the element toward the left side. $f_1 = \eta A_1 S$. The net force acting on the fluid element is

$$f = f_2 - f_1 = (A_2 - A_1)\eta S = [A(z + dz/2) - A(z - dz/2)]\eta S = \frac{dA}{dz}\eta S\cdot dz. \tag{5.112}$$

If $dA/dz = 0$, the net force is zero. The torque acting on the element, however, is not zero and is given by

$$\tau = \eta AS \cdot dz \tag{5.113}$$

This torque plays an important role in the dynamics of liquid crystals. A non-uniform translational motion will cause the liquid crystal to rotate.

Now we consider the hydrodynamics of an incompressible isotropic fluid. The velocity of the fluid is $\vec{v} = \vec{v}(\vec{r}, t)$. Applying Newton's law to a unitary volume of the fluid, we have [23]

$$\rho \frac{d\vec{v}}{dt} = \rho \left[\frac{\partial \vec{v}}{\partial t} + (\vec{v} \cdot \nabla) \vec{v} \right] = -\nabla p + \vec{f}_v, \tag{5.114}$$

where ρ is the mass density, p is the pressure, and $\vec{f}_v$ is the viscosity force. Equation (5.114) is known as the *Navier–Stokes* equation and can be rewritten as

$$\rho \left[\frac{\partial \vec{v}}{\partial t} + (\vec{v} \cdot \nabla) \vec{v} \right] = \nabla \cdot \overleftrightarrow{\sigma}, \tag{5.115}$$

where $\overleftrightarrow{\sigma}$ is the stress tensor defined by

$$\overleftrightarrow{\sigma} = -p \overleftrightarrow{I} + \overleftrightarrow{\sigma}'. \tag{5.116}$$

$\overleftrightarrow{\sigma}'$ is the viscous stress tensor, which arises from the velocity gradient. In order to see the relation between the viscous stress tensor and the velocity gradient, we consider a special case where the fluid rotates as a whole. When the angular velocity of the rotation is $\overleftarrow{\Omega}$, the velocity is $\vec{v} = \vec{\Omega} \times \vec{r}$. Introduce the following two new tensors: the symmetric part of the velocity gradient tensor $\overleftrightarrow{A}$ whose components are defined by [20]

$$A_{ij} = \frac{1}{2} \left(\frac{\partial v_j}{\partial x_i} + \frac{\partial v_i}{\partial x_j} \right), \tag{5.117}$$

where $i, j = 1, 2, 3$, $x_1 = x$, $x_2 = y$, and $x_3 = z$, and the anti-symmetric part of the velocity gradient tensor $\vec{W}$ whose components are defined by

$$W_{ij} = \frac{1}{2} \left(\frac{\partial v_j}{\partial x_i} - \frac{\partial v_i}{\partial x_j} \right). \tag{5.118}$$

In this special case, where the fluid rotates as a whole, $A_{ij} = 0$, $W_{xy} = \Omega_z$, $W_{yz} = \Omega_x$, and $W_{zx} = \Omega_y$. There should be no viscous stress. Therefore the viscous stress tensor must be proportional to $\overleftrightarrow{A}$:

$$\sigma'_{ij} = 2\eta A_{ij} = \eta \left(\frac{\partial v_j}{\partial x_i} + \frac{\partial v_i}{\partial x_j} \right), \tag{5.119}$$

where η is the translational viscosity coefficient. The vorticity of the flow (the angular velocity of the rotation of the fluid as a whole) is related to $\overleftrightarrow{W}$ by

$$\vec{\omega} = \frac{1}{2}\nabla \times \vec{v} = (W_{yz},\ W_{zx},\ W_{xy}). \tag{5.120}$$

Now we consider the viscous stress tensor of a nematic liquid crystal. The translational viscous stress depends on the orientation of the liquid crystal director, the flow direction, and the gradient direction. First we assume that the liquid crystal director is fixed. Consider a special case of shear flow as shown in Figure 5.14(a). The velocity is along the z axis and the gradient of the velocity is along the x direction, namely, $\vec{v} = u(x)\hat{z}$. The two non-zero components of the viscous stress are

$$\sigma'_{xz} = \sigma'_{zx} = \eta(\theta,\phi)\frac{\partial u}{\partial x}, \tag{5.121}$$

where $\eta(\theta, \phi)$ is the viscosity coefficient which depends on the orientation of the liquid crystal director and is given by

$$\eta(\theta,\phi) = (\eta_1 + \eta_{12}\cos^2\theta)\sin^2\theta\cos^2\phi + \eta_2\cos^2\theta + \eta_3\sin^2\theta\sin^2\phi. \tag{5.122}$$

$\eta_2 = \eta(\theta = 0)$ is the smallest viscosity coefficient, corresponding to the geometry shown in Figure 5.14(b). $\eta_1 = \eta(\theta = 90°, \varphi = 0°)$ is the largest viscosity coefficient, corresponding to

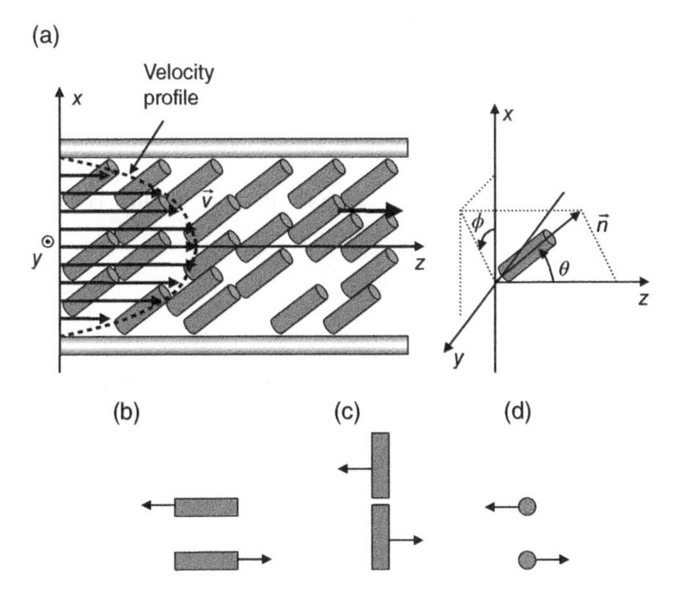

Figure 5.14 Diagram defining the orientation of the liquid crystal director with respect to the flow of the liquid crystal and special geometries of shear flows.

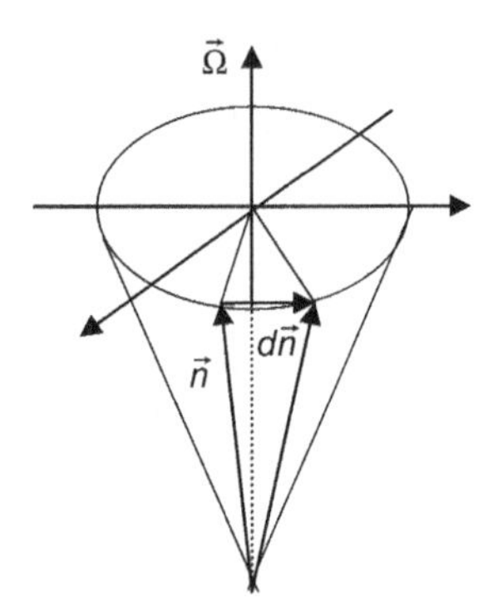

Figure 5.15 Schematic diagram showing the rotation of the liquid crystal director.

the geometry shown in Figure 5.14(c). $\eta_3 = \eta(90°, 90°)$ is the intermediate one, corresponding to the geometry is shown in Figure 5.14(d). η_{12} contributes most when $\theta = 45°$ and $\varphi = 0°$. At room temperature η_i $(i = 1, 2, 3)$ is about $10^{-2}\text{N} \cdot \text{s/m}^2$. η_{12} may be very small and can be neglected.

Now we consider the rotational motion of the liquid crystal director. The variation of the director $\vec{n}$ associated with a fluid element with respect to time is given by

$$\frac{d\vec{n}}{dt} = \frac{\partial \vec{n}}{\partial t} + (\vec{v} \cdot \nabla)\,\vec{n} \tag{5.123}$$

The second term of the above equation is due to the flow of the liquid crystal. As shown in Figure 5.15, the local angular velocity, $\vec{\Omega}$, of the director is related to $d\vec{n}/dt$ by $\vec{\Omega}\,dt \times \vec{n} = d\vec{n}$. Therefore

$$\vec{n} \times \left(\vec{\Omega} \times \vec{n}\right) = \vec{\Omega} = \vec{n} \times \frac{d\vec{n}}{dt}. \tag{5.124}$$

The dynamic equation of the rotation of the director per unit volume is

$$I\frac{d\vec{\Omega}}{dt} = \vec{\Gamma}, \tag{5.125}$$

where I is the moment of inertia per unit volume and $\vec{\Gamma}$ is the torque. The torque has three parts:

$$\vec{\Gamma} = \vec{\Gamma}_{mol} + \vec{\Gamma}_{vis} + \vec{\Gamma}_{flow} \tag{5.126}$$

$\vec{\Gamma}_{mol}$ is the molecular torque which is given by

$$\vec{\Gamma}_{mol} = \vec{n} \times \vec{h} = \vec{n} \times \left(-\frac{\delta f}{\delta \vec{n}}\right), \tag{5.127}$$

where $\vec{h} = -\delta f/\delta \vec{n}$ is the molecular field due to elastic distortion and the applied field. $\vec{\Gamma}_{vis}$ is the rotational viscosity torque which is given by

$$\vec{\Gamma}_{vis} = -\gamma_1 \vec{n} \times \vec{N}, \tag{5.128}$$

where γ_1 is the rotational viscosity coefficient and $\vec{N}$ is the net rotational velocity of the director, which equals the local angular velocity minus the angular velocity, $\vec{\omega}$, of the liquid crystal rotating as a whole:

$$\vec{N} = \left(\vec{\Omega} - \vec{\omega}\right) \times \vec{n} = \frac{d\vec{n}}{dt} - \vec{\omega} \times \vec{n} \tag{5.129}$$

$\vec{\Gamma}_{flow}$ is the torque produced by the viscosity of the translational motion as discussed before. Only the component of the velocity gradient parallel to the liquid crystal director can produce the torque to cause the director to rotate, therefore

$$\vec{\Gamma}_{flow} = -\gamma_2 \vec{n} \times \left(\overleftrightarrow{A} \cdot \vec{n}\right), \tag{5.130}$$

where γ_2 is referred to as the second rotational viscosity coefficient. γ_2 has a negative value with magnitude comparable to that of γ_1. In many liquid crystal phenomena, the dynamics is over-damped and the inertia term is negligible. Then the dynamic equation of the rotation of the director becomes

$$\vec{n} \times \left[\vec{h} - \gamma_1 \vec{N} - \gamma_2\left(\overleftrightarrow{A} \cdot \vec{n}\right)\right] = 0. \tag{5.131}$$

This equation generally means that $\left[\vec{h} - \gamma_1 \vec{N} - \gamma_2\left(\overleftrightarrow{A} \cdot \vec{n}\right)\right]$, not necessarily 0, is parallel to $\vec{n}$. The translational motion may induce a rotational motion of the liquid crystal molecules.

Now we consider how a rotational motion of the liquid crystal molecules induces a translational motion. As an example, initially the liquid crystal molecules are aligned vertically as shown in Figure 5.16(a). The average distance between the neighboring molecular centers on the vertical direction is larger than that in the horizontal direction. When the molecules rotate to the horizontal direction, as shown in Figure 5.16(b), the average distance between the centers of neighboring molecular in the horizontal direction becomes larger than that in the vertical direction. This means that the molecules move translationally.

In the Ericksen–Leslie theory, the viscous tress tensor is given by

$$\overleftrightarrow{\sigma}' = \alpha_1 (\vec{n}\vec{n})\left(\vec{n} \cdot \overleftrightarrow{A} \cdot \vec{n}\right) + \alpha_2 \vec{n}\vec{N} + \alpha_3 \vec{N}\vec{n} + \alpha_4 \overleftrightarrow{A} + \alpha_5 \vec{n}\left(\vec{n} \cdot \overleftrightarrow{A}\right) + \alpha_6\left(\vec{n} \cdot \overleftrightarrow{A}\right)\vec{n}. \tag{5.132}$$

The components are

$$\sigma'_{ij} = \alpha_1 n_i n_j n_k n_l A_{kl} + \alpha_2 n_i N_j + \alpha_3 n_j N_i + \alpha_4 A_{ij} + \alpha_5 n_i n_k A_{kj} + \alpha_6 n_j n_k A_{ki}, \tag{5.133}$$

where α_i $(i = 1, 2, 3, 4, 5, 6)$ are Leslie viscosity coefficients. The second and third terms describe the effect of director rotation on the translational motion. The relation between the Leslie coefficients and the translational and rotational viscosity coefficients are [20]:

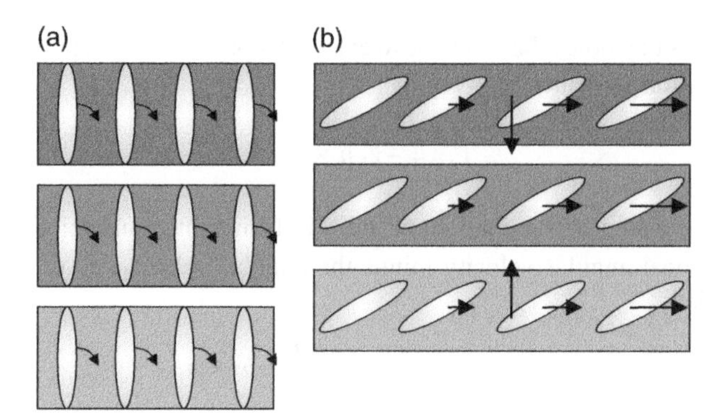

Figure 5.16 Schematic diagram showing how rotation of the liquid crystal molecules induces translational motion.

$$\eta_1 = \frac{1}{2}(-\alpha_2 + \alpha_4 + \alpha_5) \quad (5.134)$$

$$\eta_2 = \frac{1}{2}(\alpha_3 + \alpha_4 + \alpha_6) \quad (5.135)$$

$$\eta_3 = \frac{1}{2}\alpha_4 \quad (5.136)$$

$$\eta_{12} = \alpha_1 \quad (5.137)$$

$$\gamma_1 = \alpha_3 - \alpha_2 \quad (5.138)$$

$$\gamma_2 = \alpha_6 - \alpha_5 \quad (5.139)$$

$$\alpha_6 = \alpha_2 + \alpha_3 + \alpha_5 \quad (5.140)$$

5.5.3 Backflow

We consider the dynamics of the Fréedericksz transition in the splay geometry upon the removal of the applied field [24–27]. Initially the liquid crystal director is aligned vertically by the applied field, as shown in Figure 5.17(a). When the applied field is removed, the liquid crystal relaxes back to the homogeneous state. The rotation of the molecules induces a macroscopic translational motion known as the *backflow* effect. The velocity of the flow is

$$\vec{v} = [u(z), 0, 0]. \quad (5.141)$$

The y component of the velocity is 0 because of the symmetry of the cell. The z component is 0 because of the mass conservation and uncompressibility of the liquid crystal. The boundary condition of the velocity of the translational motion is

$$u(z = -h/2) = u(z = h/2) = 0. \quad (5.142)$$

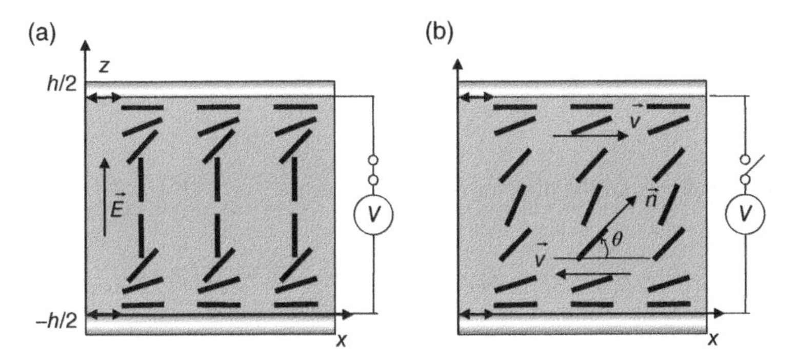

Figure 5.17 Schematic diagram showing the relaxation of the liquid crystal in the splay geometry.

The liquid crystal director is in the xz plane and is given by

$$\vec{n} = n_x\hat{x} + n_z\hat{z} = \cos\theta(z,t)\hat{x} + \sin\theta(z,t)\hat{z}. \tag{5.143}$$

The elastic energy density is

$$f = \frac{1}{2}K_{11}\left(\frac{\partial n_z}{\partial z}\right)^2 + \frac{1}{2}K_{33}\left(\frac{\partial n_x}{\partial z}\right)^2. \tag{5.144}$$

The molecular field is

$$\begin{aligned}\vec{h} &= \left[-\frac{\partial f}{\partial n_x} + \frac{\partial}{\partial z}\left(\frac{\partial f}{\partial(\partial n_x/\partial z)}\right)\right]\hat{x} + \left[-\frac{\partial f}{\partial n_z} + \frac{\partial}{\partial z}\left(\frac{\partial f}{\partial(\partial n_z/\partial z)}\right)\right]\hat{z} \\ &= K_{33}\frac{\partial^2 n_x}{\partial z^2}\hat{x} + K_{11}\frac{\partial^2 n_z}{\partial z^2}\hat{z} \\ &= -K_{33}\left[\sin\theta\frac{\partial^2\theta}{\partial z^2} + \cos\theta\left(\frac{\partial\theta}{\partial z}\right)^2\right]\hat{x} + K_{11}\left[\cos\theta\frac{\partial^2\theta}{\partial z^2} - \sin\left(\frac{\partial\theta}{\partial z}\right)^2\right]\hat{z}.\end{aligned} \tag{5.145}$$

The temporal change rate of the director is

$$\frac{d\vec{n}}{dt} = \dot{\theta}(-\sin\theta\hat{x} + \cos\theta\hat{z}), \tag{5.146}$$

where $\theta^{\cdot} = \partial\theta/\partial t$. The velocity gradient tensor is

$$\overleftrightarrow{A} = \begin{pmatrix} 0 & 0 & b \\ 0 & 0 & 0 \\ b & 0 & 0 \end{pmatrix}, \tag{5.147}$$

where $b = (1/2)\partial u/\partial z$. Thus

$$\overleftrightarrow{A}\cdot\vec{n} = \begin{pmatrix} 0 & 0 & b \\ 0 & 0 & 0 \\ b & 0 & 0 \end{pmatrix}\begin{pmatrix} \cos\theta \\ 0 \\ \sin\theta \end{pmatrix} = \begin{pmatrix} b\sin\theta \\ 0 \\ b\cos\theta \end{pmatrix}. \tag{5.148}$$

The angular velocity of the liquid crystal rotating as a whole is

$$\vec{\omega} = \frac{1}{2}\nabla \times \vec{v} = (0, b, 0). \tag{5.149}$$

The net rotation velocity of the director is

$$\vec{N} = \frac{d\vec{n}}{dt} - \vec{\omega} \times \vec{n} = \dot{\theta}(-\sin\theta\hat{x} + \cos\theta\hat{z}) - b\hat{y} \times (\cos\theta\hat{x} + \sin\theta\hat{z}) = \left(\dot{\theta} + b\right)[-\sin\theta\hat{x} + \cos\theta\hat{z}]. \tag{5.150}$$

The dynamic equation of the rotation of the liquid crystal director is

$$\vec{n} \times \left[\vec{h} - \gamma_1 \vec{N} - \gamma_2\left(\overleftrightarrow{A} \cdot \vec{n}\right)\right] = 0. \tag{5.151}$$

Therefore

$$\vec{h} - \gamma_1 \vec{N} - \gamma_2\left(\overleftrightarrow{A} \cdot \vec{n}\right) = c\vec{n}, \tag{5.152}$$

where c is a constant. In components, we have

$$-K_{33}\left[\sin\theta\frac{\partial^2\theta}{\partial z^2} + \cos\theta\left(\frac{\partial\theta}{\partial z}\right)^2\right] + \gamma_1\left(\dot{\theta} + b\right)\sin\theta - \gamma_2 b\sin\theta = c\cos\theta, \tag{5.153}$$

$$K_{11}\left[\cos\theta\frac{\partial^2\theta}{\partial z^2} - \sin\left(\frac{\partial\theta}{\partial z}\right)^2\right] - \gamma_1\left(\dot{\theta} + b\right)\cos\theta - \gamma_2 b\cos\theta = c\sin\theta. \tag{5.154}$$

Under the isotropic elastic constant assumption ($K_{11} = K_{33} = K$), Equation (5.154) $\times \cos\theta$ – Equation (5.152) $\times \sin\theta$, we can get

$$\gamma_1\frac{\partial\theta}{\partial t} = K\frac{\partial^2\theta}{\partial z^2} - \left[\gamma_1 + \gamma_2\left(\cos^2\theta - \sin^2\theta\right)\right]b. \tag{5.155}$$

Using Equations (5.138), (5.139) and (5.140), we then have

$$\gamma_1\frac{\partial\theta}{\partial t} = K\frac{\partial^2\theta}{\partial z^2} - 2\left(\alpha_3\cos^2\theta - \alpha_2\sin^2\theta\right)b. \tag{5.156}$$

Comparing Equation (5.155) with Equation (5.103), we can see the extra term which is from the translational motion. For translational motion, there is only motion in the x direction. Using the dynamic equation (Equation (5.115)), we have

$$\rho\frac{\partial u}{\partial t} = \frac{\partial}{\partial z}\sigma_{zx}. \tag{5.157}$$

When the gradient of pressure is small and can be neglected, we have

$$\sigma_{zx}=\alpha_1 n_z n_x(n_x n_z A_{xz}+n_z n_x A_{zx})+\alpha_2 n_z N_x+\alpha_3 n_x N_z+\alpha_4 A_{zx}+\alpha_5 n_z n_z A_{zx}+\alpha_6 n_x n_x A_{xz}$$
$$=[(2\alpha_1\cos^2\theta-\alpha_2+\alpha_5)\sin^2\theta+\alpha_4+(\alpha_6+\alpha_3)\cos^2\theta]b+(-\alpha_2\sin^2\theta+\alpha_3\cos^2\theta)\frac{\partial\theta}{\partial t}. \tag{5.158}$$

It is difficult to calculate the involved hydrodynamics analytically. Here we present a qualitative discussion. If the applied field E is sufficiently high, the liquid crystal is aligned in the cell normal direction, except very near the cell surfaces. In the equilibrium state under the field, the tilt angle θ is $\pi/2$ in most parts of the cell, as shown in Figure 5.18(a). We also have $K_{22}\frac{\partial^2\theta}{\partial z^2}=-\varepsilon_o\Delta\varepsilon E^2\sin\theta\cos\theta$, which is shown in Figure 5.18(b). Right after the applied high voltage is turned off ($t=0$), there is no flow. The variation rate of θ with respect to time is given by

$$\gamma_1\frac{\partial\theta}{\partial t}=K\frac{\partial^2\theta}{\partial z^2}<0. \tag{5.159}$$

The angle decreases with time. The non-zero components of the stress tensor are

$$\sigma_{zx}=\sigma_{xz}=(-\alpha_2\sin^2\theta+\alpha_3\cos^2\theta)\frac{\partial\theta}{\partial t}=\frac{K}{\gamma_1}(-\alpha_2\sin^2\theta+\alpha_3\cos^2\theta)\frac{\partial^2\theta}{\partial z^2}. \tag{5.160}$$

From Equation (5.157) we have

$$\rho\frac{\partial u}{\partial t}=\frac{\partial\sigma_{zx}}{\partial z}=\frac{K}{\gamma_1}\frac{\partial}{\partial z}\left[(-\alpha_2\sin^2\theta+\alpha_3\cos^2\theta)\frac{\partial^2\theta}{\partial z^2}\right], \tag{5.161}$$

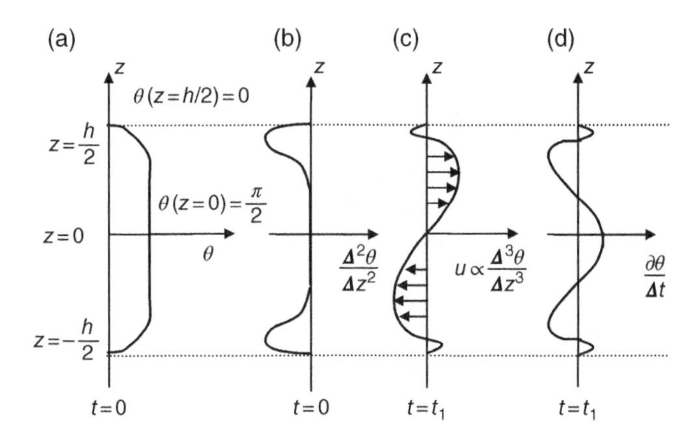

Figure 5.18 The profiles of the liquid crystal director and velocity in the relaxation of the bistable TN.

$$u = \frac{Kt}{\rho\gamma_1}\frac{\partial}{\partial z}\left[\left(-\alpha_2\sin^2\theta + \alpha_3\cos^2\theta\right)\frac{\partial^2\theta}{\partial z^2}\right] \propto \frac{\partial^3\theta}{\partial z^3}. \tag{5.162}$$

Consider the case of $\alpha_3 > 0$ (the final result is the same for negative α_3). At a time slightly later $(t = t_1)$, the profile of the velocity of the developed translational motion is shown in Figure 5.18(c). In the top half of the cell, $v_x = u > 0$, i.e., the flow is in the $+x$ direction. In the bottom half of the cell, $v_x = u < 0$, i.e., the flow is in the $-x$ direction. The translational motion will affect the rotation of the liquid crystal. From Equation (5.158) we have

$$\begin{aligned}\gamma_1\frac{\partial\theta}{\partial t} &= K\frac{\partial^2\theta}{\partial z^2} - \left(\alpha_3\cos^2\theta - \alpha_2\sin^2\theta\right)\left\{\frac{Kt}{\rho\gamma_1}\frac{\partial^2}{\partial z^2}\left[\left(-\alpha_2\sin^2\theta + \alpha_3\cos^2\theta\right)\frac{\partial^2\theta}{\partial z^2}\right]\right\} \\ &\approx K\frac{\partial^2\theta}{\partial z^2} - \left(\alpha_3\cos^2\theta - \alpha_2\sin^2\theta\right)^2\frac{Kt}{\rho\gamma_1}\frac{\partial^4\theta}{\partial z^4}.\end{aligned} \tag{5.163}$$

The second term on the right side of the above equation will make the angle increase in the middle of the cell. The translational motion makes the liquid crystal rotate in the opposite direction, which is known as the backflow effect. The strength of the backflow depends on the initial director configuration, which in turn depends on the initially applied field. If the applied field is high, the effect of the backflow is stronger.

In TN and ECB displays, the backflow slows the relaxation of the liquid crystal director from the distorted state under applied field to the undistorted state at zero field, and may even make the relaxation non-monotonic. The transmittance of a normal white TN as a function of time is shown in Figure 5.19 after the applied voltage is turned off [27]. When the applied voltage is 5 V, the liquid crystal is not well aligned homeotropically. When the applied field is removed, the backflow was not strong, and the transmittance increased monotonically as shown in Figure 5.19(a). When the applied voltage is 8 V, the liquid crystal is well aligned homeotropically. When the applied field is removed, the backflow is strong and the transmittance did not increase monotonically, as shown in Figure 5.19(b). The dip at 6 ms after the removal of the applied voltage is due the reverse rotation of the liquid crystal caused by the backflow.

Backflow can also be made use of in liquid crystal devices. Bistable TN is one such example [24–26], where the display cell has homogeneous alignment layers. Chiral dopant is added to the liquid crystal in such a way that the intrinsic pitch P is twice the cell thickness h. Initially a high voltage is applied across the cell and the liquid crystal is switched to the homeotropic state where there is no twist. If the applied voltage is removed slowly, the tilt angle of the liquid crystal director decreases slowly with time, and the backflow is small. The liquid crystal relaxes to the state where the liquid crystal director is aligned homogeneously, known as the 0° twist state. The free energy of this state is a local minimum and the liquid crystal remains in this state for quite a long period. If the applied voltage is removed quickly, the tilt angle of the liquid crystal director decreases quickly with time, and the backflow is large. The liquid crystal in the middle of the cell rotates in the direction opposite to the rotation direction of the director near the cell surfaces. The liquid crystal transforms into a state where the director twists 360° from the bottom to the top of the cell. The free energy of this state is also a local minimum, and the liquid crystal remains in this state for a long period.

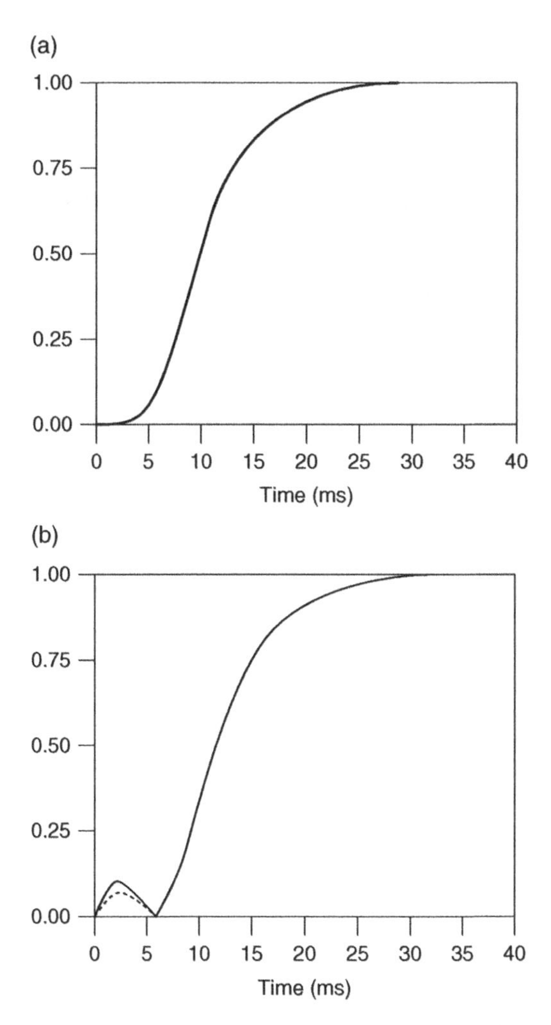

Figure 5.19 Transmittance vs. time of the normal white TN after the applied voltage is removed. (a) 5 V, (b) 8 V, [16]. Reproduced with permission from the American Institute of Physics.

Homework Problems

5.1 *Fréedericksz in the twist geometry shown in Figure 5.7.* The cell thickness is h. The free energy is given by Equation (5.42). Use the Euler–Lagrange equation of minimization of the total free energy to derive the twist angle as a function of the normalized position z/h when the applied field E is above the threshold E_c. The anchoring is infinitely strong and the twist angle at the bottom and top surfaces is 0.

5.2 Explain why the voltage threshold in the Fréedericksz transition is cell thickness-independent.

5.3 *Fréedericksz in splay geometry.* Use the parameters in Figure 5.3 and Equation (5.30) to calculate and plot the tilt angle at the middle plane as a function of the normalized field E/E_c.

5.4 *Fréedericksz in splay geometry*. The cell thickness is h. Use the parameters in Figure 5.4 and calculate and plot the tilt angle as a function of the normalized position z/h at various fields: $E/E_c = 1.2$, $E/E_c = 3.0$, and $E/E_c = 10$.
5.5 Calculate the transmittance of the VA mode liquid crystal display for R, G, and B light. The parameters of the liquid crystal are $K_{11} = 6.4 \times 10^{-12}$N, $K_{33} = 10 \times 10^{-12}$N, $\Delta\varepsilon = -3$, $n_e = 1.57$, and $n_o = 1.50$. The thickness of the cell is 5.0 μm. The wavelengths of the R, G, and B light are 650 nm, 550 nm, and 450 nm, respectively.
5.6 *Flow alignment angle*. Consider a shear as shown in Figure 5.20. The shear rate dv/dx = constant and the liquid crystal director is uniformly oriented in the xz plane. Show that when $|\gamma_1/\gamma_2| < 1$, the tilt angle is given by $\cos 2\theta = -\gamma_1/\gamma_2$ in the steady state.

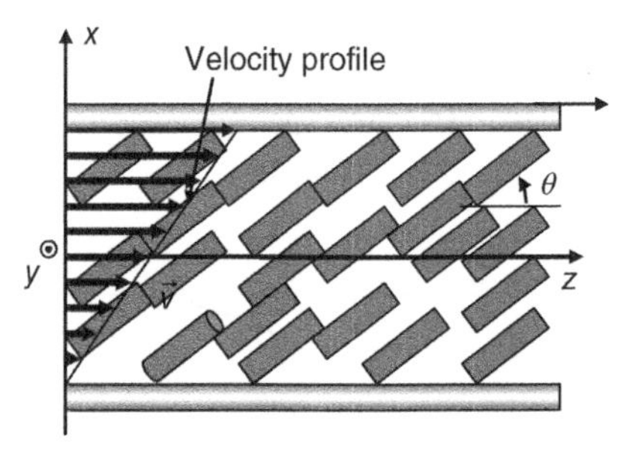

Figure 5.20 Diagram for Problem 5.6.

5.7 Show that in the twist geometry, it is possible to have director motion without any flow.

References

1. J. Mathews and R. L. Walker, *Mathematical methods of physics*, 2nd edn (W. A. Benjamin, Inc., Menlo Park, California, 1970).
2. R. Barberi and G. Barbero, Variational calculus and simple applications of continuum theory, Chapt. IX, in *Liquid crystal materials*, ed. I. C. Khoo (Gordon and Breach, Amsterdam, 1991).
3. P. G. de Gennes and J. Prost, *The physics of liquid crystals* (Oxford University Press, New York, 1993).
4. S. Chandrasekhar *Liquid crystals*, 2nd edn (Cambridge University Press, New York, 1997).
5. A. Strigazzi, Fréedericksz transition, Chapt. X, in *Liquid crystal materials*, ed. I. C. Khoo (Gordon and Breach, Amsterdam, 1991).
6. T. Scheffer and J. Nehring, Twisted nematic and super-twisted nematic mode LCDs, in *Liquid crystals – Applications and uses*, Vol. 1, ed. B. Bahadur (World Scientific, New Jersy, 1990).
7. E. P. Raynes, The theory of super-twist transitions, *Mol. Cryst. Liq. Cryst. Lett.*, **4**, 1 (1986).
8. S. Faetti, Anchoring effects in nematic liquid crystals, Chapt. XII, in *Liquid crystal materials*, ed. I. C. Khoo (Gordon and Breach, Amsterdam, 1991).
9. H. Yokoyama, in *Handbook of Liquid Crystal Research*, ed. P. J. Collings and J. S. Patel (Oxford University,Oxford, 1997), Chap. 6.
10. A. Yu., Nastishin, R. D. Polak, et al., *J. Appl. Phys.* **86**, 15 (1999).
11. S. Faetti and V. Palleschi, *J. Phys. (France) Lett.*, **45**, L-313 (1984).
12. H. Yokoyama and H. A. van Sprang, *J. Appl. Phys.*, **57**, 4520 (1985).
13. H. Yokoyama, S. Kobayashi, and H. Kamei, *J. Appl. Phys.*, **61**, 4501 (1987).
14. Yu. A. Nastishin, R. D. Polak, S. V. Shiyanovskii, and O. D. Lavrentovich, *Appl. Phys. Lett.*, **75**, 202 (1999).
15. R. Sun, X. Huang, K. Ma, Z. Wang, and M. Jiang, *Phys. Rev. E*, **50**, 1253 (1994).

16. Y. Cui, R. S. Zola, Y.-C. Yang, and D.-K. Yang, Alignment layers with variable anchoring strengths from Polyvinyl Alcohol, *Journal of Applied Physics,* **111**, 063520 (2012).
17. T. J. Scheffer and T. Nehring, *J. Appl. Phys.* **48**, 1783 (1977).
18. D. Subacius, V. M. Pergamenshchik, and O. D. Lavrentovicha, Measurement of polar anchoring coefficient for nematic cell with high pretilt angle, *Appl. Phys. Lett.* **67** (2), 214 (1995)
19. M.-P. Cuminal and M. Brunet, A technique for measurement of pretilt angles arising from alignment layers, *Liq. Cryst.*, **22**, 185 (1997).
20. W. H. de Jeu, Physical properties of liquid crystal materials, *Liquid crystal monographs*, vol. **1**, ed. G. W. Gray (Gordon Breach, London, 1980).
21. P. G. de Gennes and J. Prost, *The physics of liquid crystals* (Oxford University Press, New York, 1993).
22. D. Tabor, *Gases, liquids and solids and other states of matter*, 3rd edn (Cambridge University Press, 1991).
23. P. K. Kundu, and I. M. Cohen, *Fliud mechanics*, 2nd edn (Academic Press, San Diego, 2002).
24. D. W. Berreman and W. R. Heffner, New bistable cholesteric liquid-crystal display, *Appl. Phys. Lett.*, **37**, 109 (1980).
25. D. W. Berreman, Liquid-crystal twist cell dynamics with backflow, *J. Appl. Phys.*, **46**, 3746 (1975).
26. C. Z. van Doorn, Dynamic behaviour of twisted nematic, *J. Appl. Phys.*, **46**, 3738 (1975).
27. J. Kelly, S. Jamal, and M. Cui, Simulation of the dynamics of twisted nematic devices including flow, *J. Appl. Phys.*, **86**, 4091 (1999).

6

Liquid Crystal Materials

6.1 Introduction

Liquid crystal (LC) material, although only occupying a small portion of a display or photonic device, makes decisive contributions to the device performance. For instance, the device contrast ratio, response time, viewing angle, and operating voltage are all related to the LC materials and how they are aligned. The refractive indices and cell gap determine the phase retardation or phase change of the LC device employed for either amplitude or phase modulation. The dielectric constants and elastic constants jointly determine the threshold voltage. The viscosity, cell gap, and temperature determine the response time.

Absorption is another important factor affecting the physical properties of an LC material. Most of the conjugated LC compounds have strong absorption in the ultraviolet (UV) region due to allowed electronic transitions. These UV absorption bands and their corresponding oscillator strengths play important roles in affecting the LC refractive indices and photostability. In the visible region, the electronic absorption tail decays quickly so that the absorption effect is small and can be ignored. However, molecular vibrations appear in the mid and long infrared (IR) regions. The overtones of these vibration bands extend to the near IR (~1 μm). The material absorption affects the optical transmittance and especially the power handling capability of an LC-based optical phased array for steering a high-power infrared laser beam.

In this chapter, we will first describe the origins of the LC refractive indices and their wavelength and temperature dependencies, and then extend the discussions to dielectric constants, elastic constants, and viscosity.

Fundamentals of Liquid Crystal Devices, Second Edition. Deng-Ke Yang and Shin-Tson Wu.

6.2 Refractive Indices

The classic Clausius–Mossotti equation [1] correlates the permittivity (ε) of an *isotropic* media with molecular polarizability (α) as

$$\frac{\varepsilon-1}{\varepsilon+2}=\frac{4\pi}{3}N\alpha. \tag{6.1}$$

In Equation (6.1), N is the molecular packing density, or the number of molecules per unit volume. In the optical frequency regime, we substitute $\varepsilon = n^2$ and obtain the Lorentz–Lorenz equation [2]:

$$\frac{n^2-1}{n^2+2}=\frac{4\pi}{3}N\alpha \tag{6.2}$$

For an anisotropic LC media, there are two principal refractive indices, n_e and n_o, for the extraordinary ray and ordinary ray, respectively. In principle, each refractive index is supposedly related to the corresponding molecular polarizabilities, α_e and α_o. An early approach replaces both n^2 in Equation (6.2) by $n_{e,o}^2$ and α by $\alpha_{e,o}$. However, this model does not fit experimental results well. In 1964, Vuks made an assumption that the internal field in a crystal is the same in all directions [3]:

$$E_i=\frac{<n^2>+2}{3}E, \tag{6.3}$$

where E_i is the internal field, the average field that acts on a molecule, and E is the macroscopic electric field. With this assumption, Vuks derived the following equation for *anisotropic* media:

$$\frac{n_{e,o}^2-1}{<n^2>+2}=\frac{4\pi}{3}N\alpha_{e,o}, \tag{6.4}$$

where

$$\langle n^2\rangle=\left(n_e^2+2n_o^2\right)/3. \tag{6.5}$$

Equation (6.4) is different from Equation (6.2) in two aspects: (1) the n^2 term in the denominator of Equation (6.2) is replaced by $\langle n^2\rangle$, while the n^2 term in the numerator is replaced by $n_{e,o}^2$, and (2) the α is replaced by $\alpha_{e,o}$. The Vuks equation (6.4) has been validated experimentally using the refractive index data of several liquid crystal compounds and mixtures [4].

6.2.1 Extended Cauchy equations

In Equation (6.4), n_e and n_o are coupled together through the $\langle n^2\rangle$ term. By substituting Equation (6.5) to (6.4) and through a series expansion of the Vuks equation, n_e, n_o, and birefringence Δn (= $n_e - n_o$) can be expressed as [5]

$$n_e(\lambda,T)\approx n_i(\lambda)+GS\frac{\lambda^2\lambda^{*2}}{\lambda^2-\lambda^{*2}}, \tag{6.6}$$

$$n_o(\lambda,T) \approx n_i(\lambda) - \frac{GS}{2} \frac{\lambda^2\lambda^{*2}}{\lambda^2 - \lambda^{*2}}, \tag{6.7}$$

$$\Delta n(\lambda,T) \approx \frac{3GS}{2} \frac{\lambda^2\lambda^{*2}}{\lambda^2 - \lambda^{*2}}, \tag{6.8}$$

where $n_i(\lambda)$ is the LC refractive index in the isotropic phase, $<\alpha>$ is the average molecular polarizability, $G = \frac{2\sqrt{2}}{3}\pi gNZ\left(f_e^* - f_o^*\right)/\left(1 - \frac{4}{3}\pi N\langle\alpha\rangle\right)$ is a proportionality constant, and S is the order parameter. In the isotropic state, $n_i(\lambda)$ can be expressed by the traditional Cauchy equation:

$$n_i(\lambda) = A_i + \frac{B_i}{\lambda^2} + \frac{C_i}{\lambda^4}, \tag{6.9}$$

where A_i, B_i, and C_i are the Cauchy coefficients for the isotropic state. The temperature effect of N (molecular packing density) and $f_e^* - f_o^*$ (differential oscillator strength) is much smaller than that of S. Thus, we can assume that G is insensitive to the temperature. Equation (6.8) is identical to the single-band birefringence dispersion model [6],

In the off-resonance region, the right terms in Equations (6.6) and (6.7) can be expanded by a power series to the λ^{-4} term and then combined with Equation (6.9) to form the extended Cauchy equations for describing the wavelength-dependent refractive indices of *anisotropic* LCs as

$$n_{e,o} \cong A_{e,o} + \frac{B_{e,o}}{\lambda^2} + \frac{C_{e,o}}{\lambda^4}. \tag{6.10}$$

In Equation (6.10), $A_{e,o}$, $B_{e,o}$, and $C_{e,o}$ are three Cauchy coefficients. By measuring the refractive indices of an LC material at three wavelengths, these Cauchy coefficients can be determined. Afterwards, the refractive indices at any particular wavelength can be extrapolated.

Although the extended Cauchy equation fits experimental data well [7], its physical origin is not clear. A better physical meaning can be obtained by the three-band model which takes three major electronic transition bands into consideration.

6.2.2 *Three-band model*

The major absorption of an LC compound occurs in two spectral regions: ultraviolet (UV) and infrared (IR). The $\sigma \rightarrow \sigma^*$ electronic transitions take place in the vacuum UV (100–180 nm) region whereas $\pi \rightarrow \pi^*$ the electronic transitions occur in the UV (180–400 nm) region.

Figure 6.1 shows the measured polarized UV absorption spectra of 5CB [8]. To avoid saturation, only 1 wt% of 5CB was dissolved in a UV transparent nematic LC mixture, MLC-6815. A quartz cell without indium-tin-oxide (ITO) conductive coating was used. To produce homogeneous alignment, a thin SiO_2 layer was sputtered onto the quartz substrate. The cell gap was controlled at 6 μm. The λ_1 band which is centered at ~200 nm consists of two closely overlapped bands. The λ_2 band shifts to ~282 nm. The λ_0 band should occur in the vacuum UV region ($\lambda_0 \sim 120$ nm) which is not shown in Figure 6.1.

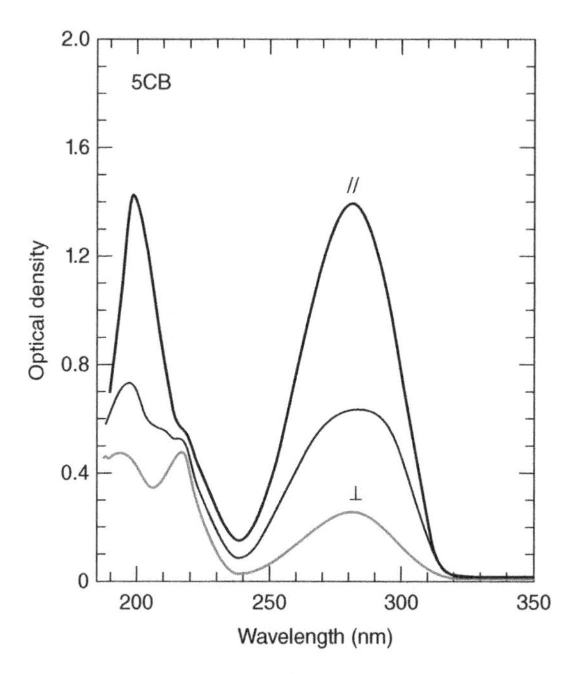

Figure 6.1 Measured polarized UV absorption spectra of 5CB using a homogeneous quartz cell without ITO. ∥ and ⊥ represent e-ray and o-ray, and the middle curve is for the unpolarized light. Sample: 1 wt% 5CB dissolved in ZLI-2359 (a UV transparent nematic mixture). Cell gap is 6 μm. $T = 23°C$. $\lambda_1 = 200$ nm and $\lambda_2 = 282$ nm.

If an LC compound has a longer conjugation, its electronic transition wavelength would extend to a longer UV wavelength. In the near IR region, some overtone molecular vibration bands appear [9]. The fundamental molecular vibration bands, such as CH, CN, and C = C, occur in the mid and long IR regions. Typically, the oscillator strength of these vibration bands is about two orders of magnitude weaker than that of the electronic transitions. Thus, the resonant enhancement of these bands to the LC birefringence is localized.

The three-band model takes one $\sigma \rightarrow \sigma^*$ transition (the λ_0 band) and two $\pi \rightarrow \pi^*$ transitions (the λ_1 and λ_2 bands) into consideration. In the three-band model, the refractive indices (n_e and n_o) are expressed as follows: [10,11]

$$n_{e,o} \cong 1 + g_{0e,o}\frac{\lambda^2\lambda_0{}^2}{\lambda^2-\lambda_0{}^2} + g_{1e,o}\frac{\lambda^2\lambda_1{}^2}{\lambda^2-\lambda_1{}^2} + g_{2e,o}\frac{\lambda^2\lambda_2{}^2}{\lambda^2-\lambda_2{}^2}, \tag{6.11}$$

In the visible region, $\lambda \gg \lambda_o$ (~120 nm) and the λ_0 band's contribution in Equation (6.11) can be approximated by a constant, $n_{0e,o}$, so that Equation (6.11) is simplified as

$$n_{e,o} \cong 1 + n_{0e,o} + g_{1e,o}\frac{\lambda^2\lambda_1{}^2}{\lambda^2-\lambda_1{}^2} + g_{2e,o}\frac{\lambda^2\lambda_2{}^2}{\lambda^2-\lambda_2{}^2}. \tag{6.12}$$

The three-band model clearly describes the origins of refractive indices of LC compounds. However, a commercial mixture usually consists of several compounds with different molecular structures in order to obtain a wide nematic range. Each individual λ_i is therefore different. Under such circumstances, Equation (6.12) would have too many unknowns to describe the refractive indices of an LC mixture.

To model the refractive indices of an LC mixture, we could expand Equation (6.12) into power series because in the visible and IR spectral regions, $\lambda \gg \lambda_2$. By keeping up to the λ^{-4} terms, the above extended Cauchy equation (6.10) is again derived.

Although Equation (6.10) is derived based on an LC compound, it can be extended easily to include eutectic mixtures by taking the superposition of each compound. From Equation (6.10), if we measure the refractive indices at three wavelengths, the three Cauchy coefficients ($A_{e,o}$, $B_{e,o}$, and $C_{e,o}$) can be obtained by fitting the experimental results. Once these coefficients are determined, the refractive indices at any wavelength can be calculated. From Equations (6.8) and (6.10), both refractive indices and birefringence decrease as the wavelength increases. In the long wavelength (IR and millimeter wave) region, n_e and n_o are reduced to A_e and A_o, respectively. The coefficients A_e and A_o are constants; they are independent of wavelength, but dependent on the temperature. That means that in the IR region the refractive indices are insensitive to wavelength, except for the resonance enhancement effect near the local molecular vibration bands. This prediction is consistent with much experimental evidence [12].

Equation (6.10) applies equally well to both high and low birefringence LC materials in the off-resonance region. For low birefringence ($\Delta n < 0.12$) LC mixtures, the λ^{-4} terms are insignificant and can be omitted. Thus, n_e and n_o each have only two fitting parameters. The two-coefficient Cauchy model has the following simple forms [13]:

$$n_{e,o} \cong A_{e,o} + \frac{B_{e,o}}{\lambda^2}, \tag{6.13}$$

By measuring the refractive indices at two wavelengths, we can determine $A_{e,o}$ and $B_{e,o}$. Once these two parameters are found, n_e and n_o can be calculated at any wavelength of interest.

For most LC displays [14], the cell gap is controlled at around 4 μm so that the required birefringence is smaller than 0.12. Thus Equation (6.13) can be used to describe the wavelength-dependent refractive indices. For infrared applications, high birefringence LC mixtures are required [15]. Under such circumstances, the three-coefficient extended Cauchy model (Equation (6.10)) should be used.

Figure 6.2 shows the fittings of experimental data of 5CB using the three-band model (solid lines) and the extended Cauchy equations (dashed lines). The fitting parameters are listed in Table 6.1. In the visible and near infrared regions, both models give excellent fits to the experimental data. In the UV region, the deviation between these two models becomes more apparent. The three-band model considers the resonance effect, but the extended Cauchy model does not. Thus, in the near resonance region the results from the three-band model are more accurate.

6.2.3 *Temperature effect*

The temperature effect is particularly important for projection displays [16] Due to the thermal effect of the lamp, the temperature of the display panel could reach 50°C. It is important to know the LC properties at the anticipated operating temperature beforehand. The thermal non-linearity of LC refractive indices is also very important for some new photonic applications, such as LC photonic bandgap fibers [17,18] and thermal solitons [19,20]

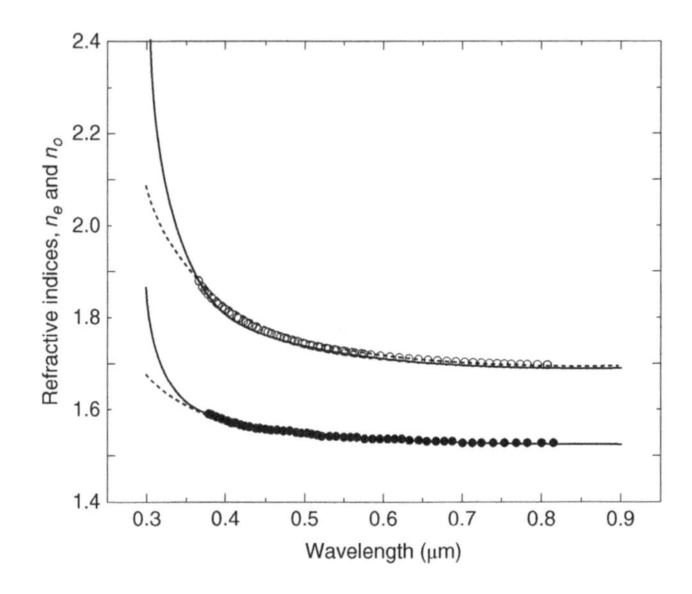

Figure 6.2 Wavelength-dependent refractive indices of 5CB at $T = 25.1°C$. Open and closed circles are experimental data for n_e and n_o, respectively. Solid line represents the three-band model and dashed lines are for the extended Cauchy model. The fitting parameters are listed in Table 6.1. Li and Wu 2004. Reproduced with permission from the American Institute of Physics.

Table 6.1 Fitting parameters for the three-band model and the extended Cauchy equations. LC: 5CB at $T = 25.1°C$. The units of Cauchy's B and C coefficients are μm^2 and μm^4, respectively.

Model		n_e			n_o	
three-band	n_{0e}	g_{1e}	g_{2e}	n_{0o}	g_{1o}	g_{2o}
model	0.4618	2.1042	1.4413	0.4202	1.2286	0.4934
Cauchy	A_e	B_e	C_e	A_o	B_o	C_o
model	1.6795	0.0048	0.0027	1.5187	0.0016	0.0011

Birefringence Δn is defined as the difference between the extraordinary and ordinary refractive indices, $\Delta n = n_e - n_o$ and the average refractive indices $<n>$ is defined as $<n> = (n_e + 2n_o)/3$. Based on these two definitions, n_e and n_o can be rewritten as

$$n_e = <n> + \frac{2}{3}\Delta n, \tag{6.14}$$

$$n_o = <n> - \frac{1}{3}\Delta n. \tag{6.15}$$

To describe the temperature-dependent birefringence, the Haller approximation has been commonly employed when the temperature is not too close to the clearing point [21]:

$$\Delta n(T) = (\Delta n)_o (1 - T/T_c)^{\beta} \tag{6.16}$$

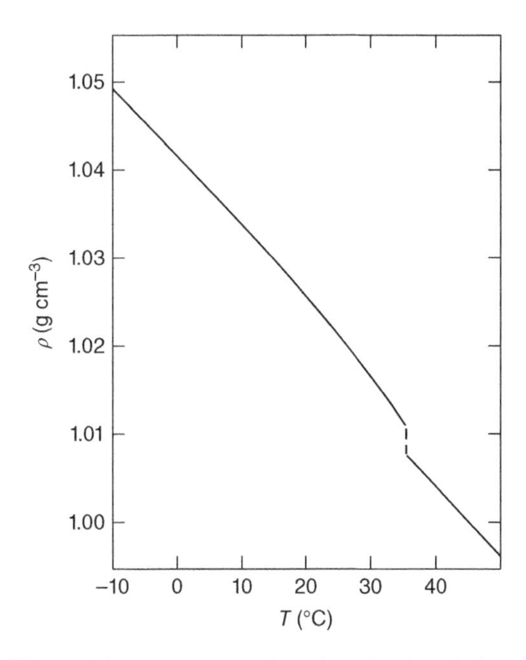

Figure 6.3 Temperature dependent density of 5CB.

In Equation (6.10), $(\Delta n)_o$ is the LC birefringence in the crystalline state (or $T = 0$ K), the exponent β is a material constant, and T_c is the clearing temperature of the LC material under investigation. On the other hand, the average refractive index decreases linearly with increasing temperature as [22]. Zeller 1982. Reproduced with permission from the American physical Society.

$$\langle n \rangle = A - BT, \tag{6.17}$$

because the LC density decreases with increasing temperature.

Figure 6.3 plots the temperature-dependent density of 5CB [23]. At room temperature, the density of 5CB is around 1.02 g/cm^3, slightly heavier than that of water because of its higher molecular weight. As the temperature increases, the density decreases almost linearly. Due to the second-order phase transition, a disrupt density change occurs at $T \sim 35.3$°C. In the isotropic state, the 5CB density continues to decrease linearly as the temperature increases.

By substituting Equations (6.16) and (6.17) back into Equations (6.14) and (6.15), the four-parameter model for describing the temperature dependence of the LC refractive indices is derived, as [24]

$$n_e(T) \approx A - BT + \frac{2(\Delta n)_o}{3}\left(1 - \frac{T}{T_c}\right)^{\beta}, \tag{6.18}$$

$$n_o(T) \approx A - BT - \frac{(\Delta n)_o}{3}\left(1 - \frac{T}{T_c}\right)^{\beta}. \tag{6.19}$$

The parameters [A, B] and [$(\Delta n)_o$, B] can be obtained respectively by two-stage fittings. To obtain [A, B], we fit the average refractive index $\langle n \rangle = (n_e + 2n_o)/3$ as a function of temperature using

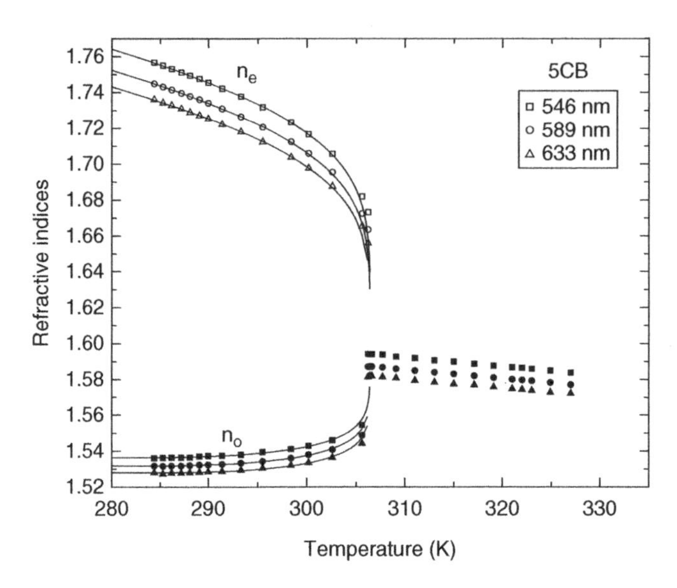

Figure 6.4 Temperature-dependent refractive indices of 5CB at $\lambda = 546$, 589, and 633 nm. Squares, circles and triangles are experimental data for refractive indices measured at $\lambda = 546$, 589 and 633 nm, respectively.

Equation (6.15). To find $[(\Delta n)_o, B]$, we fit the birefringence data as a function of temperature using Equation (6.14). Therefore, these two sets of parameters can be obtained separately from the same set of refractive indices but in different forms.

Figure 6.4 plots the temperature dependent refractive indices of 5CB at $\lambda = 546$, 589, and 633 nm. As the temperature increases, n_e decreases, but n_o gradually increases. In the isotropic state, $n_e = n_o$ and the refractive index decreases linearly with increasing temperature.

6.2.4 *Temperature gradient*

Based on Equations (6.18) and (6.19), we can derive the temperature gradient for n_e and n_o:

$$\frac{dn_e}{dT} = -B - \frac{2\beta(\Delta n)_o}{3T_c\left(1-\frac{T}{T_c}\right)^{1-\beta}} \tag{6.20}$$

$$\frac{dn_o}{dT} = -B + \frac{\beta(\Delta n)_o}{3T_c\left(1-\frac{T}{T_c}\right)^{1-\beta}} \tag{6.21}$$

In Equation (6.20), both terms on the right-hand side are negative, independent of temperature. This implies that n_e decreases as the temperature increases throughout the entire nematic range. However, Equation (6.21) consists of a negative term ($-B$) and a positive term which depends on the temperature. In the low temperature regime ($T \ll T_c$), the positive term could be smaller than the negative term, resulting in a negative dn_o/dT. As the temperature increases, the positive

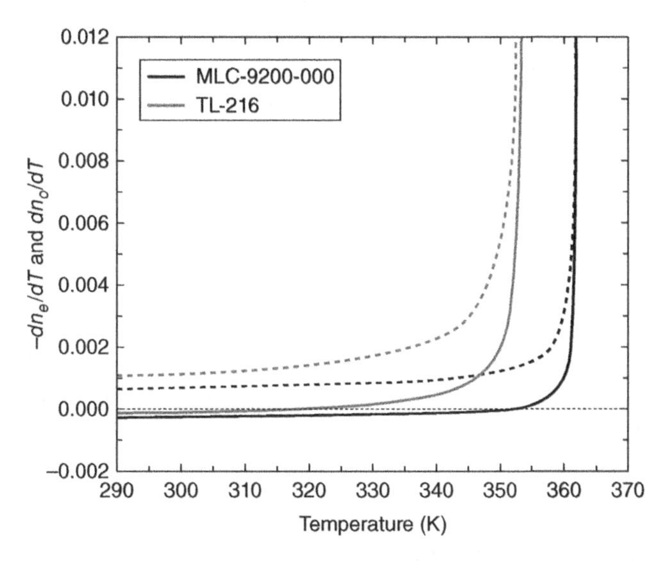

Figure 6.5 Temperature gradient for n_e and n_o of MLC-9200-000 and TL-216 at $\lambda = 546$ nm. Black and grey solid lines represent the calculated dn_o/dT curves for MLC-9200-000 and TL-216, respectively, while the dashed lines represent the calculated $-dn_e/dT$ curves. The crossover temperature for MLC-9200-000 and TL-216 are around 80.1 and 52.7°C, respectively. Li et al 2004. Reproduced with permission from the Optical Society of America.

term also increases. As T approaches T_c, dn_o/dT jumps to a large positive number. In the intermediate, there exists a transition temperature where $dn_o/dT = 0$. Let us define this temperature as the crossover temperature T_o for n_o. To find T_o, we simply solve $dn_o/dT = 0$ from Equation (6.21).

Figure 6.5 depicts the temperature-dependent values of $-dn_e/dT$ and dn_o/dT for two Merck LC mixtures, MLC-9200-000 and TL-216. The black and grey solid lines represent the calculated dn_o/dT curves for MLC-9200-000 and TL-216, respectively, while the dashed lines represent the calculated $-dn_e/dT$ curves. The crossover temperatures of MLC-9200-000 and TL-216 are around 80.1 and 52.7°C, respectively. In Figure 6.5, the $-dn_e/dT$ for both LC mixtures remain positive throughout their nematic range. That means that n_e, the extraordinary refractive index, decreases monotonously as the temperature increases in the entire nematic range. However, dn_o/dT changes sign at the crossover temperature T_o. The dn_o/dT is negative when the temperature is below T_o, whereas it becomes positive when the temperature is above T_o. This implies that n_o, the ordinary refractive index, decreases as temperature increases when the temperature is below T_o but increases with temperature when the temperature is above T_o.

6.2.5 *Molecular polarizabilities*

Since the Vuks equation correlates the macroscopic refractive index to the microscopic molecular polarizability, if we know refractive index, then we can calculate the molecular polarizability, or vice versa. For instance, if we know the n_e and n_o data of an LC, then we can calculate its α_e and α_o values at different temperatures and wavelengths.

In Equation (6.4), there is still an unknown parameter N; the number of molecules per unit volume. However, N is equal to $\rho N_A/M$, where ρ is the LC density, M is the molecular weight, and N_A is the Avogadro's number. Rearranging Equation (6.4), we find

$$\alpha_e = \frac{3M}{4\pi\rho N_A} \cdot \frac{n_e^2 - 1}{<n^2> + 2}, \tag{6.22}$$

$$\alpha_o = \frac{3M}{4\pi\rho N_A} \cdot \frac{n_o^2 - 1}{<n^2> + 2}. \tag{6.23}$$

Let us use 5CB (cyano-biphenyl) as an example to calculate the molecular polarizabilities. For 5CB, the molecular weight is $M = 249.3$ gm/mol and the density $\rho(T)$ is taken from Figure 6.3. Using the measured refractive indices at $\lambda = 589$ nm, we can calculate the α_e and α_o of 5CB from Equations (6.22) and (6.23).

Figure 6.6 plots the temperature-dependent α_e, α_o, and $\langle\alpha\rangle$ of 5CB at $\lambda = 589$ nm. The open and filled circles represent the calculated values for α_e and α_o, respectively. In the isotropic state, α_e and α_o are equal. The open triangles represent the calculated average polarizability $<\alpha>$ in the nematic phase. From Figure 6.6, α_e decreases while α_o increases as the temperature increases. However, the average polarizability $\langle\alpha\rangle$ is quite insensitive to the temperature. The average polarizability for 5CB at $\lambda = 589$ nm is found to be $\langle\alpha\rangle \sim 3.3 \times 10^{-23}\text{cm}^{-3}$, which agrees very well with the calculated value ($\langle\alpha\rangle \sim 3.25 \times 10^{-23}\text{cm}^{-3}$) published by Sarkar et al. [25].

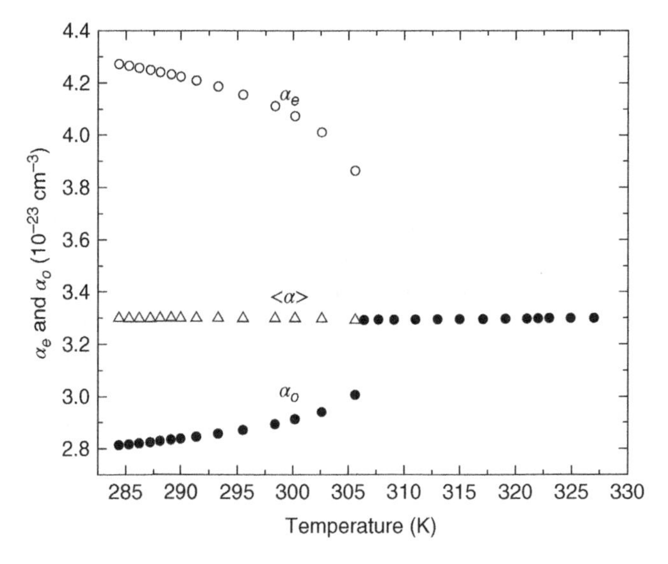

Figure 6.6 Temperature-dependent molecular polarizabilities, α_e and α_o, of 5CB at $\lambda = 589$ nm. Open and filled circles are the results for α_e and α_o, respectively. Triangles represent the average polarizability $<\alpha>$. In the isotropic phase, $\alpha_e = \alpha_o$.

6.3 Dielectric Constants

The dielectric constants of a liquid crystal affect the operation voltage, resistivity, and response time. For example, in a vertical alignment (VA) cell the threshold voltage (V_{th}) is related to dielectric anisotropy ($\Delta\varepsilon = \varepsilon_{//} - \varepsilon_{\perp}$) and bend elastic constant (K_{33}) as [26]:

$$V_{th} = \pi\sqrt{\varepsilon_o K_{33}/\Delta\varepsilon} \tag{6.24}$$

Thus, low threshold voltage can be obtained by either enhancing the dielectric anisotropy, reducing the elastic constant, or a combination of both. However, a smaller elastic constant slows down the response time because of the weaker restoring torque.

Dielectric constants of a liquid crystal are mainly determined by the dipole moment (μ), its orientation angle (θ) with respect to the principal molecular axis, and order parameter (S), as described by the Maier and Meier mean field theory: [27]

$$\varepsilon_{//} = NhF\{<\alpha_{//}> + (F\mu^2/3kT)[1-(1-3\cos^2\theta)S]\} \tag{6.25}$$

$$\varepsilon_{\perp} = NhF\{<\alpha_{\perp}> + (F\mu^2/3kT)[1+(1-3\cos^2\theta)S/2]\} \tag{6.26}$$

$$\Delta\varepsilon = NhF\{(<\alpha_{//}> - <\alpha_{\perp}>) - (F\mu^2/2kT)(1-3\cos^2\theta)S\} \tag{6.27}$$

Here, N stands for the molecular packing density, $h = 3\varepsilon/(2\varepsilon + 1)$ is the cavity field factor, $\varepsilon = (\varepsilon_{//} + 2\varepsilon_{\perp})/3$ is the averaged dielectric constant, F is the Onsager reaction field, $<\alpha_{//}>$ and $<\alpha_{\perp}>$ are the principal elements of the molecular polarizability tensor.

From Equation (6.27), for a non-polar compound, $\mu \sim 0$ and its dielectric anisotropy is very small ($\Delta\varepsilon < 0.5$). In this case, $\Delta\varepsilon$ is determined mainly by the differential molecular polarizability, i.e. the first term in Equation (6.27). For a polar compound, the dielectric anisotropy depends on the dipole moment, angle θ, temperature (T) and applied frequency. If an LC has more than one dipole, then the resultant dipole moment is their vector summation. In a phenyl ring, the position of the dipole is defined as

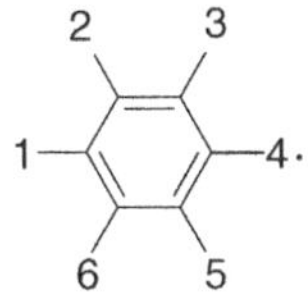

From Equation (6.27), if a polar compound has an effective dipole at $\theta < 55°$, then its $\Delta\varepsilon$ will be positive. On the other hand, $\Delta\varepsilon$ becomes negative if $\theta > 55°$.

Fluoro (F) [28] cyano (CN) [29], and isothiocyanato (NCS) [30] are the three commonly employed polar groups. Among them, the fluoro group possesses a modest dipole moment ($\mu \sim 1.5$ D), high resistivity, and low viscosity. However, its strong negativity compresses the electron clouds and subsequently lowers the compound's birefringence. As a result, the fluorinated compounds are more suitable for visible display applications where the required birefringence is around 0.1.

For infrared applications, a higher birefringence ($\Delta n > 0.3$) compound is needed in order to compensate for the longer wavelength. To obtain a higher birefringence, two approaches can be taken to enhance the electron conjugation length: (1) by elongating the core structure, such as tolane and terphenyl, and (2) by attaching an electron acceptor polar group, such as CN and NCS. The CN group has a larger dipole moment ($\mu \sim 3.9$ D) than NCS ($\mu \sim 3.7$ D) because of its linear structure. However, due to the very strong polarization of the carbon–nitrogen triple bond the Huckel charges of carbon and nitrogen are high and well localized. Accordingly, dimers are formed by the strong intermolecular interactions between the nitrile group and phenyl ring. Thus, a relatively high viscosity is observed in the cyano-based LC mixtures. On the other hand, the Huckel charges of nitrogen, carbon, and sulfur are smaller in the NCS group. The predicted intermolecular interactions by the NCS group in the isothiocyanato-benzene systems are smaller than those in the nitrile-based systems. The dimers are not formed and, therefore, the viscosity of such molecular systems is lower than that of nitrile-based ones. Due to the longer π-electron conjugation, the NCS-based LC compounds exhibit a higher birefringence than the corresponding CN compounds.

6.3.1 *Positive* $\Delta\varepsilon$ *liquid crystals for AMLCD*

Positive $\Delta\varepsilon$ LCs have been used in twisted nematic (TN) [31] and in-plane switching (IPS) [32,33] displays, although IPS can also use negative $\Delta\varepsilon$ LCs. For thin-film-transistor (TFT) based displays, the employed LC material must possess a high resistivity [34]. Fluorinated compounds exhibit a high resistivity and are the natural candidates for TFT LCD applications [35,36]. A typical fluorinated LC structure is shown below:

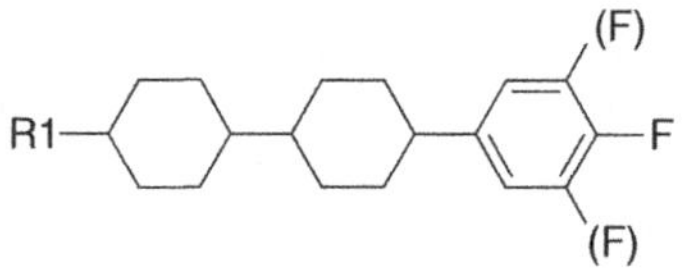

(F)
R1 — F
(F)
(I)

From Equation (6.27), to obtain the largest $\Delta\varepsilon$ for a given dipole, the best position for the fluoro substitution is along the principal molecular axis, i.e., in the 4 position. The single fluoro compound should have $\Delta\varepsilon \sim 5$. To further increase $\Delta\varepsilon$, more fluoro groups can be added. For example, compound (I) has two more fluoro groups in the 3 and 5 positions [37] Its $\Delta\varepsilon$ should increase to ~10, but its birefringence would slightly decrease (because of the lower molecular packing density) and viscosity increases substantially (because of the higher moment of inertia). Besides fluoro, the OCF_3 group is found to exhibit a fairly low viscosity. Low-viscosity LC is helpful for improving response time [38].

The birefringence of compound (I) is around 0.7. If a higher birefringence is needed, the middle cyclohexane ring can be replaced by a phenyl ring. The elongated electron cloud will increase the birefringence to about 0.12. The phase transition temperatures of an LC compound are difficult to predict beforehand. In general, the lateral fluoro substitution would lower the melting temperature of the parent compound because the increased molecular separation leads to a weaker intermolecular association. Thus, a smaller thermal energy is able to separate the molecules. That means, the melting point is decreased.

6.3.2 *Negative* $\Delta\varepsilon$ *liquid crystals*

For the vertical alignment [39], the LC employed should have a negative dielectric anisotropy. From Equation (6.27), in order to obtain a negative dielectric anisotropy, the dipoles should be in the lateral (2,3) positions. Similarly, in the interests of obtaining high resistivity, the lateral difluoro group is a favorable choice.

A typical negative $\Delta\varepsilon$ LC compound is shown below: [40]

F F
C_3H_7 — — OC_2H_5 (II)

Compound (II) has two lateral fluoro groups. Their components in the horizontal axis are perfectly cancelled. On the other hand, the vertical components add up. As a result, the $\Delta\varepsilon$ is negative. The neighboring alkoxy group also contributes to enhance the negative $\Delta\varepsilon$. However, its viscosity is somewhat larger than that of an alkyl group. The estimated $\Delta\varepsilon$ of compound (II) is about −4. To further increase $\Delta\varepsilon$, more fluoro groups need to be substituted. That would increase the viscosity unfavorably. This is a common problem of negative LCs. It is not easy to increase $\Delta\varepsilon$ value without trade-offs.

6.3.3 Dual-frequency liquid crystals

Dual-frequency liquid crystal (DFLC) [41,42] exhibits a unique feature that its $\Delta\varepsilon$ changes from positive at low frequencies to negative as the frequency passes the crossover frequency (f_c), as Figure 6.7 shows. The frequency that $\Delta\varepsilon = 0$ is called the crossover frequency. The major attraction of the DFLC device is fast response time. During the turn-on and turn-off processes, AC voltage bursts, with low and high frequencies, are applied. As a result, fast rise and decay times can be achieved [43].

In practice, a DFLC mixture is composed of some positive (with ester group) and negative $\Delta\varepsilon$ LC compounds, and its crossover frequency is around a few kilohertz, depending on the molecular structures and compositions [44]. The $\Delta\varepsilon$ of the ester compounds is frequency dependent, as shown in the top gray line in Figure 6.7. As the frequency increases, the $\Delta\varepsilon$ decreases gradually.

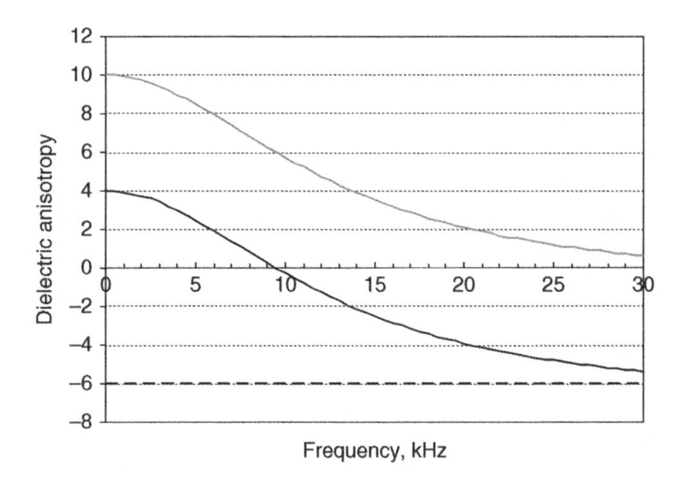

Figure 6.7 Frequency-dependent dielectric anisotropy of a positive LC mixture (A) whose $\Delta\varepsilon$ is frequency dependent (top), a negative $\Delta\varepsilon$ LC mixture (B, bottom curve), and a DFLC mixture (middle curve).

However, the $\Delta\varepsilon$ of the negative components of the DFLC mixture remains fairly flat, as depicted by the bottom dashed lines. The resultant $\Delta\varepsilon$ is frequency dependent, as plotted by the solid line. In this example, the crossover frequency occurs at about 9.2 kHz.

6.4 Rotational Viscosity

Viscosity, especially rotational viscosity (γ_1), plays a crucial role in the LCD response time. The response time of a nematic LC device is linearly proportional to γ_1 [45]. The rotational viscosity of an aligned liquid crystal depends on the detailed molecular constituents, structure, intermolecular association, and temperature. As the temperature increases, viscosity decreases rapidly. Several theories, rigorous or semi-empirical, have been developed in an attempt to account for the origin of the LC viscosity [46,47]. However, owing to the complicated anisotropic attractive and steric repulsive interactions among LC molecules, these theoretical results are not completely satisfactory [48,49].

A general temperature-dependent rotational viscosity can be expressed as

$$\gamma_1 = bS\exp(E/kT), \tag{6.28}$$

where b is a proportionality constant, which takes into account the molecular shape, dimension, and moment of inertia, S is the order parameter, E is the activation energy of molecular rotation, k is the Boltzmann constant, and T is the operating temperature. When the temperature is not too close to the clearing point (T_c), the order parameter can be approximated as

$$S = (1 - T/T_c)^{\beta}. \tag{6.29}$$

In Equation (6.29), β is a material parameter. Generally, rotational viscosity is a complicated function of molecular shape, moment of inertia, activation energy, and temperature. Among these factors, activation energy and temperature are the most crucial [50]. The activation energy depends on the detailed intermolecular interactions. An empirical rule is that for every 10 degrees of temperature rise, the rotational viscosity drops by about two times.

From the molecular structure standpoint, a linear LC molecule is more likely to have a low viscosity [51]. However, all other properties also need to be taken into account. For instance, a linear structure may lack flexibility and lead to a higher melting point. Within the same homologues, a longer alkyl chain will in general (except for the even–odd effect) have a lower melting temperature. However, its moment of inertia is increased. As a result, the homologue with a longer chain length is likely to exhibit higher viscosity.

6.5 Elastic Constants

There are three basic elastic constants (splay K_{11}, twist K_{22}, and bend K_{33}) involved in the electro-optics of an LC cell, depending on the molecular alignment [52]. Elastic constants affect an LC device through threshold voltage and response time. For example, the threshold voltage of a VA cell is expressed in Equation (5.38). A smaller elastic constant will result in a lower threshold voltage; however, the response time which is proportional to the visco-elastic coefficient – the ratio of γ_1/K_{ii} – is increased. Therefore, proper balance between threshold voltage and response time should be taken into consideration.

Several molecular theories have been developed for correlating the Frank elastic constants with molecular constituents. The commonly employed one is mean-field theory [53,54]. In the mean-field theory, the three elastic constants are expressed as

$$K_{ii} = aS^2, \tag{6.30}$$

where a is a proportionality constant.

For many LC compounds and mixtures studied, the magnitude of elastic constants has the following order: $K_{33} > K_{11} > K_{22}$. Therefore, LC alignment also plays an essential role for achieving fast response time. For example, a VA cell (K_{33} effect) should have faster response time than the IPS cell (K_{22} effect) owing to the elastic constant effect, provided that all the other parameters such as cell gap and viscosity remain the same. Usually, the lateral difluoro substitutions increase viscosity to a certain extent because of the increased molecular moment of inertia.

6.6 Figure-of-Merit (FoM)

To compare the performance among different LC materials, a figure-of-merit (FoM) has been defined as [55]

$$FoM = K(\Delta n)^2/\gamma_1. \tag{6.31}$$

In Equation (6.31), K is the elastic constant for a given molecular alignment. For example, $K = K_{33}$ for a vertical align cell, and $K = K_{11}$ for a planar (homogeneous) cell. Both K, Δn, and γ_1 are temperature dependent. Using Equations (6.16), (6.28), (6.29), and (6.30) for the corresponding temperature dependency, the figure-of-merit is derived as

$$FoM = (a/b)(\Delta n_o)^2(1 - T/T_c)^{3\beta}\exp(-E/kT). \tag{6.32}$$

Equation (6.32) has a maximum at an optimal operating temperature T_{op}:

$$T_{op} = \frac{E}{6\beta k_o}\left[\sqrt{1 + 12\beta k_o T_c/E} - 1\right] \tag{6.33}$$

The quantity $12\beta k_o T_c/E$ in Equation (6.33) is small, and the square root term can be expanded into a power series. Keeping the lowest order terms, we find:

$$T_{op} \sim T_c(1 - 3\beta k_o T_c/E + \ldots] \tag{6.34}$$

Figure 6.8 shows the temperature-dependent FoM of a Merck MLC-6608, a negative $\Delta\varepsilon$ LC mixture. The results were measured using a He–Ne laser with $\lambda = 633$ nm. The clearing temperature of MLC-6608 is $T_c = 92.1°C$. From fitting to Equation (6.32), $\beta = 0.272$ and $E = 325$ meV are obtained. At room temperature, the FoM is about 0.8. As the temperature increases, the FoM increases gradually and reaches a peak (~1.6) at $T_{op} \sim 70°C$ and then

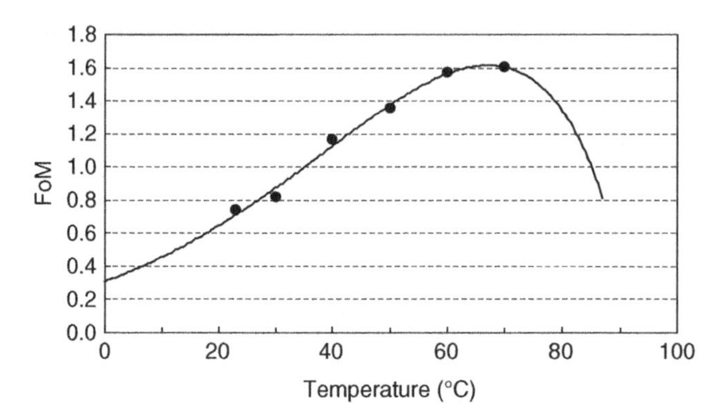

Figure 6.8 The temperature-dependent FoM of MLC-6608. $\lambda = 633$ nm. Dots are experimental data and the solid line is fitting using Equation (6.32) with $\beta = 0.272$ and $E = 325$ meV.

drops sharply. The optimal operating temperature is about 20 degrees below T_c. For an LCD TV application, about 50% of the backlight is absorbed by the polarizer, which is laminated onto the glass substrates. The absorbed light will be converted to heat. As a result, the LC temperature could reach about 35–40°C. From Figure 6.5, the FoM at $T \sim 40$°C is about 60% higher than that at room temperature.

6.7 Index Matching between Liquid Crystals and Polymers

Polymer-dispersed liquid crystal (PDLC) [56] and polymer-stabilized liquid crystal have been used for displays [57] and photonic devices [58]. In a PDLC, the refractive index difference between the LC droplets and polymer matrix plays an important role in determining the voltage-off and voltage-on state transmittance. In a normal-mode PDLC, the droplet size is controlled at ~1 μm, which is comparable with the visible light wavelength. In the voltage-off state, the droplets are randomly oriented. The index mismatch between the LC (whose average refractive index is given by: $\langle n \rangle = (n_e + 2n_o)/3$) and polymer matrix ($n_p$) affects the light scattering capability. For a given droplet size, the larger the index mismatch, the higher the light scattering. Conversely, in the voltage-on state, the LC directors inside the droplets are reoriented along the electric field direction so that the refractive index becomes n_o; the ordinary refractive index. If $n_o \sim n_p$, then the PDLC becomes isotropic and will have an excellent transmittance. Therefore, the preferred LC material for PDLC is not only high birefringence ($\Delta n = n_e - n_o$) but also good index match between n_o and n_p. In a polymer-stabilized LC system, polymer networks help to improve response time. A good index match would reduce light scattering.

6.7.1 Refractive index of polymers

NOA65 (Norland Optical Adhesive 65) is a commonly used photocurable polymer because its refractive index ($n_p \sim 1.52$) is close to the n_o of many commercial LC mixtures. Before UV curing, NOA65 is a clear and colorless liquid. The measurement of the monomer is fairly simple. However, in a practical device, such as a PDLC, all the monomers are cured to form a polymer

Table 6.2 The measured refractive index of cured NOA65 film at λ = 450, 486, 546, 589, 633, and 656 nm at different temperatures.

n / T°(C)	λ (mn)					
	450	486	546	589	633	656
20	1.5396	1.5352	1.5301	1.5275	1.5255	1.5243
25	1.5391	1.5347	1.5296	1.5270	1.5250	1.5239
30	1.5386	1.5343	1.5292	1.5266	1.5246	1.5235
35	1.5377	1.5335	1.5282	1.5254	1.5233	1.5225
40	1.5363	1.5324	1.5272	1.5245	1.5222	1.5214
45	1.5352	1.5311	1.5261	1.5235	1.5211	1.5204
50	1.5340	1.5305	1.5248	1.5223	1.5202	1.5192
55	1.5330	1.5298	1.5243	1.5217	1.5194	1.5187

matrix. Therefore, it is more meaningful to measure the refractive index of the cured polymers than the monomers.

To prepare a polymer film, the monomer is infiltrated into an empty cell with a 1 mm gap using capillary flow [59]. During the experiment, the cells were placed on a hot plate with a constant temperature ($T \sim 50°C$) and then illuminated with a uniform UV light ($I = 14\ mW/cm^2$, $\lambda \sim 365$ nm) for 40 minutes because of the large cell gap. Afterwards, the glass substrates were peeled off and the thick polymer film was removed at a high temperature ($T \sim 120°C$). The film was kept in an oven at a constant temperature of 50°C for 12 hours to age completely. The cured polymer film of NOA65 was quite flexible. In order to get an accurate measurement, the films were cut into rectangular parallelepipeds 15 mm long, 9 mm wide, and 1 mm thick. The bottom surface of the samples was polished so that it contacted the main prism surface of the Abbe refractometer completely. First, a small drop of contact liquid (monobromonaphthalene) was placed on the main prism before the sample. It is essential to spread the contact liquid evenly between the sample and the main prism and to get rid of any dust or bubbles between the solid sample and the main prism. A lighting glass was used to compensate for the weak light because the samples are thin. Similarly, a small amount of the contact liquid was spread on the top surface of the sample, and the lighting glass was placed on top of the contact liquid. The thin contact liquid should be spread evenly between the sample and the lighting glass. At this stage, the contact liquid is sandwiched as films between the main prism and the sample, and between the sample and the lighting glass. The incident light entered the sample slightly aslant from the upside.

Table 6.2 lists the measured refractive indices of NOA65 at various wavelengths and temperatures.

The refractive index of NOA65 was also measured in the monomer state. After UV curing, the refractive index of the cured polymers increases 1.7% for NOA65. This slight refractive index increase originates from the increased density of the polymer after cross-linking.

For a normal-mode PDLC, the light scattering in the voltage-off state depends on the LC birefringence; the higher the birefringence, the higher the scattering efficiency. In the voltage-on state, the transmittance depends on the refractive index match between the LC (n_o) and the polymer matrix (n_p). If $n_o \sim n_p$, then the on-state will be highly transparent. After having measured the n_p of NOA65, let us select two commercial high birefringence LC series with their n_o close to n_p. The two LC series are BL-series (BL038, BL006, and BL003) and E-series (E48, E44, and E7). To measure the refractive indices, the LCs are aligned perpendicular to the main and secondary prism surfaces of the Abbe refractometer by coating these two surfaces with a surfactant comprising of 0.294 wt% hexadecyletri-methyl-ammonium bromide (HMAB) in methanol solution.

6.7.2 *Matching refractive index*

For a linearly conjugated LC, a high n_o often leads to a high Δn. Most of the commercially available high-birefringence ($\Delta n \sim 0.20$–0.28) liquid crystals have $n_o \sim 1.50$–1.52. These are the mixtures of cyano-biphenyls and cyano-terphenyls, such as Merck E-series and BL-series. Only few high birefringence ($\Delta n \geq 0.4$) LCs have $n_o > 1.55$. These are mainly isothiocyanato-tolane mixtures. Thus, let us focus on the index matching phenomena of some Merck E-series (E7, E48, and E44) and BL-series (BL003, BL006, and BL038) liquid crystals with NOA65.

Figure 6.9 shows the measured refractive index of the UV-cured NOA65 and the ordinary refractive index of E48, E44, and E7, as a function of wavelength at $T = 20°\text{C}$. The filled circles, open squares, upward triangles, and downward triangles are the measured ordinary refractive index of NOA65, E48, E44, and E7, respectively. The respective solid lines represent the fittings of each material using the extended Cauchy model (Equation (6.10)). The fitting parameters A, B, and C are listed in Table 6.3. From Figure 6.9, E48, E44, and E7 all have a reasonably good index matching with NOA65. The mismatch is within 0.005 at $\lambda = 550$ nm. More specifically, E44 has the best match in the red spectral region while E7 and E48 have

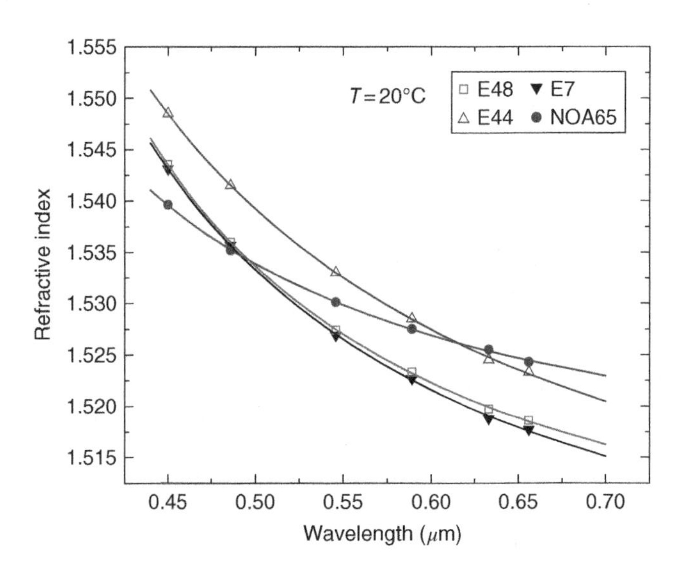

Figure 6.9 Wavelength-dependent refractive index of NOA65 and the ordinary refractive index of E48, E44, and E7 at $T = 20°\text{C}$. The open squares, upward-triangles, filled circles, and downward-triangles are the measured refractive index of E48, E44, NOA65 and E7, respectively. The solid lines represent the fittings using the extended Cauchy model (Equation (6.10)). The fitting parameters are listed in Table 6.3.

Table 6.3 The fitting parameters for the refractive index (Equation (6.10)) of NOA65 and the ordinary refractive index of E48, E44, E7, BL038, BL006, and BL003 at T = 20°C.

Cauchy coefficients	NOA65	E48	E44	E7	BL038	BL006	BL003
A	1.5130	1.5027	1.5018	1.4995	1.5042	1.5034	1.5056
B (μm^2)	0.0045	0.0055	0.0089	0.0068	0.0065	0.0085	0.0057
C (μm^2)	1.8×10^{-4}	5.6×10^{-4}	1.0×10^{-4}	4.1×10^{-4}	4.7×10^{-4}	1.9×10^{-4}	5.9×10^{-4}

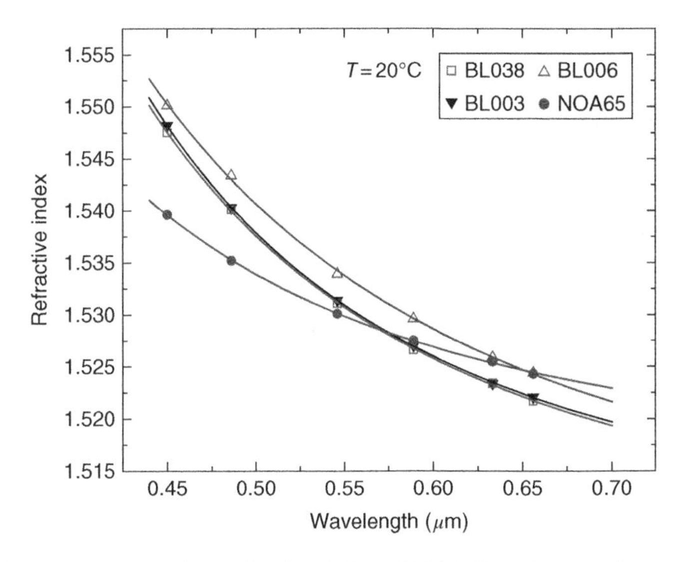

Figure 6.10 Wavelength-dependent refractive index of NOA65 and the ordinary refractive index of BL038, BL006, and BL003 at T = 20°C. The open squares, upward-triangles, filled circles, and downward-triangles are the measured refractive index of BL038, BL006, NOA65, and BL003, respectively. The solid lines represent the fittings using the extended Cauchy model (Equation (6.10)). The fitting parameters are listed in Table 6.3.

the best match in the blue region. In the green region (λ = 546 nm) where human eye is most sensitive, E44 has a slightly higher index, while E7 and E48 are slightly lower than NOA65, but the difference is in the third decimal.

Figure 6.10 shows the refractive index of the cured NOA65 and the ordinary refractive index of BL038, BL006, and BL003 as a function of wavelength at T = 20°C. The filled circles, open squares, upward-triangles, and downward-triangles are the measured ordinary refractive indices of NOA65, BL038, BL006, and BL003, respectively. The solid lines represent the fittings of each material using the extended Cauchy model (Equation (6.10)). The fitting parameters are also listed in Table 6.3. In Figure 6.10, BL038, BL006, and BL003 have a similar trend in $n_o(\lambda)$. The index matching with NOA65 is pretty good in the green and red spectral regions. A larger deviation is observed in the blue region, but the difference is still in the third decimal.

In a PDLC system, good index matching ($n_o \sim n_p$) between the employed polymer and liquid crystal helps to improve the transmittance in the voltage-on state. On the other hand, a larger index mismatch ($\langle n \rangle > n_p$) (i.e., a higher birefringence LC) enhances the light scattering efficiency in the voltage-off state. However, refractive index match or mismatch is not the only factor deciding the PDLC performance. The UV stability of liquid crystals and miscibility between polymers and liquid crystals also play an important role affecting the PDLC properties. From the above discussion, the E-series and BL-series liquid crystals all have a good index match with NOA65. All these six liquid crystals are good candidates for PDLC application when NOA65 is used. In the visible spectrum, E48 and E7 have very similar ordinary refractive index at $T = 20°C$, as do BL038 and BL003. However, E48 has a higher birefringence than E7, and BL038 has a higher birefringence than BL003. Therefore, E48 and BL038 are somewhat better choices than E7 and BL003 for the NOA65-based PDLC systems.

Homework problems

6.1 Given the following building blocks: two alkyl chains C_3H_7, C_5H_{11}, three polar groups: F, CN, and NCS, a cyclohexane ring, and a phenyl ring. Assuming that each alkyl chain and polar group can be used only once in each compound, except for the rings, then:
   a. construct a two-ring compound with the largest positive dielectric anisotropy.
   b. construct a two-ring compound with the largest negative dielectric anisotropy.
   c. construct a three-ring compound with $\Delta n \sim 0.05$ which is suitable for reflective TFT-LCDs.

6.2 Compare the physical properties (at the same reduced temperature) of the following compounds:

(a) C_3H_7–(cyclohexane)–(phenyl)–CN (b) C_3H_7–(cyclohexane)–(phenyl)–NCS (c) C_3H_7–(cyclohexane)–(phenyl)–F

   a. Dielectric anisotropy: ( ) > ( ) > ( )
   b. Birefringence: ( ) > ( ) > ( )
   c. Rotational viscosity: ( ) > ( ) > ( )

6.3 Prove that the extended Cauchy equations (Equation (6.10)) derived from LC compounds can be applied to liquid crystal mixtures. Hint: see Ref. 13.

6.4 A TFT LC mixture has $\Delta n = 0.090$ at $\lambda = 589$ nm, and 0.085 at $\lambda = 633$ nm. What is the extrapolated birefringence at $\lambda = 550$ nm?

6.5 Prove that n_e, n_o, and Δn all reach saturation values beyond near infrared region.

References

1. J. D. Jackson, *Classical Electrodynamics*, 2nd Ed. (New York: Wiley, 1962).
2. M. Born and E. Wolf, *Principle of Optics*, 6th edn. (New York: Pergamon Press, 1980).
3. M. F. Vuks, Determination of the optical anisotropy of aromatic molecules from the double refraction of crystals, *Opt. Spektrosk.* **20**, 644 (1966).

4. J. Li and S. T. Wu, Self-consistency of Vuks equations for liquid-crystal refractive indices, *J. Appl. Phys.* **96**, 6253 (2004).
5. J. Li and S. T. Wu, Extended Cauchy equations for the refractive indices of liquid crystals, *J. Appl. Phys.* **95**, 896 (2004).
6. S. T. Wu, Birefringence dispersions of liquid crystals, *Phys. Rev. A.* **33**, 1270 (1986).
7. H. Mada and S. Kobayashi, Wavelength and voltage dependences of refractive indices of nematic liquid crystals, *Mol. Cryst. Liq. Cryst.* **33**, 47 (1976).
8. S. T. Wu, E. Ramos and U. Finkenzeller, Polarized UV spectroscopy of conjugated liquid crystals *J. Appl. Phys.* **68**, 78–85 (1990).
9. S. T. Wu, Absorption measurements of liquid crystals in the ultraviolet, visible, and infrared, *J. Appl. Phys.* **84**, 4462 (1998).
10. S. T. Wu, A semiempirical model for liquid-crystal refractive index dispersions, *J. Appl. Phys.* **69**, 2080 (1991).
11. S. T. Wu, C. S. Wu, M. Warenghem, and M. Ismaili, Refractive index dispersions of liquid crystals, *Opt. Eng.* **32**, 1775 (1993).
12. S. T. Wu, U. Efron and L. D. Hess, Infrared birefringence of liquid crystals, *Appl. Phys. Lett.* **44**, 1033–35 (1984).
13. J. Li and S. T. Wu, Two-coefficient Cauchy model for low birefringence liquid crystals, *J. Appl. Phys.* **96**, 170 (2004).
14. S. T. Wu and D. K. Yang, Reflective Liquid Crystal Displays (Wiley, New York, 2001).
15. S. Gauza, H. Wang, C. H. Wen, et al., High birefringence isothiocyanato tolane liquid crystals, *Jpn. J. Appl. Phys.* **42**, 3463–6 (2003).
16. E. H. Stupp and M. S. Brennesholtz, *Projection Displays* (Wiley, New York, 1998).
17. T. T. Alkeskjold, A. Bjarklev, D. S. Hermann, and J. Broeng, Optical devices based on liquid crystal photonic bandgap fibers, *Opt. Express* **11**, 2589 (2003).
18. T. T. Alkeskjold, J. Lægsgaard, A. Bjarklev, et al., All-optical modulation in dye-doped nematic liquid crystal photonic bandgap fibers, *Opt. Express* **12**, 5857 (2004).
19. M. Warenghem, J. F. Henninot, and G. Abbate, Nonlinearly induced self waveguiding structure in dye doped nematic liquid crystals confined in capillaries, *Opt. Express* **2**, 483 (1998).
20. M. Warenghem, J. F. Henninot, F. Derrin, and G. Abbate, Thermal and orientational spatial optical solitons in dye-doped liquid crystals, *Mol. Cryst. Liq. Cryst.* **373**, 213 (2002).
21. I. Haller, Thermodynamic and static properties of liquid crystals, *Prog. Solid State Chem.* **10**, 103 (1975).
22. J. Li, S. Gauza, and S. T. Wu, High temperature-gradient refractive index liquid crystals, *Opt. Express* **12**, 2002 (2004).
23. H. R. Zeller, Dielectric relaxation in nematics and Doolittles law, *Phys. Rev.* A **26**, 1785 (1982).
24. J. Li and S. T. Wu, Temperature effect on liquid crystal refractive indices, *J. Appl. Phys.* **96**, 19 (2004).
25. P. Sarkar, P. Mandal, S. Paul, and R. Paul, *Liq. Cryst.* **30**, 507 (2003).
26. H. J. Deuling, *Solid State Phys. Suppl.* 14, Liquid Crystals, ed. L. Liebert (Academic, New York, 1978).
27. W. Maier and G. Meier, A simple theory of the dielectric characteristics of homogeneous oriented crystalline-liquid phases of the nematic type, *Z. Naturforsch,* A **16**, 262 (1961).
28. M. Schadt, Field-effect liquid-crystal displays and liquid-crystal materials – key technologies of the 1990s, *Displays* **13**, 11 (1992).
29. G. Gray, K. J. Harrison, and J. A. Nash, New family of nematic liquid crystals for displays, *Electron. Lett.* **9**, 130 (1973).
30. R. Dabrowski, Isothiocyanates and their mixtures with a broad range of nematic phase, *Mol. Cryst. Liq. Cryst.* **191**, 17 (1990).
31. M. Schadt and W. Helfrich, *Appl. Phys. Lett.* **18**, 127 (1971).
32. R.A. Soref, Transverse field effect in nematic liquid crystals, *Appl. Phys. Lett.* **22**, 165 (1973).
33. M. Oh-e and K. Kondo, Electro-optical characteristics and switching behavior of the in-plane switching mode, *Appl. Phys. Lett.,* **67**, 3895 (1995).
34. Y. Nakazono, H. Ichinose, A. Sawada, et al., *International Display Research Conference,* Toronto, Canada, 65 (1997).
35. R. Tarao, H. Saito, S. Sawada, and Y. Goto, Advances in liquid crystals for TFT displays, *SID Tech. Digest* **25**, 233 (1994).
36. T. Geelhaar, K. Tarumi, and H. Hirschmann, Trends in LC materials, *SID Tech. Digest* **27**, 167 (1996).

37. Y. Goto, T. Ogawa, S. Sawada and S. Sugimori, Fluorinated liquid crystals for active matrix displays, *Mol. Cryst. Liq. Cryst.* **209**, 1 (1991).
38. H. Saito, E. Nakagawa, T. Matsushita, et al., *IEICE Trans. Electron.*, E79-C, 1027 (1996).
39. M. F. Schiekel and K. Fahrenschon, Deformation of nematic liquid crystals with vertical orientation in electric fields, *Appl. Phys. Lett.* **19**, 391 (1971).
40. R. Eidenschink and L. Pohl, US patent 4,415,470 (1983).
41. H. K. Bücher, R. T. Klingbiel, and J. P. VanMeter., *Appl. Phys. Lett.* **25**, 186 (1974).
42. M. Schadt, Low frequency dielectric relaxation in nematic and dual frequency addressing of field effect, *Mol. Cryst. Liq. Cryst.* **89**, 77 (1982).
43. Y. Lu, X. Liang, Y. H. Wu, et al., Dual-frequency addressed hybrid-aligned nematic liquid crystal, *Appl. Phys. Lett.* **85**, 3354 (2004).
44. C. H. Wen and S. T. Wu, Dielectric heating effects of dual-frequency liquid crystals, *Appl. Phys. Lett.* **86**, 231104 (2005).
45. E. Jakeman and E. P. Raynes, Electro-optic response times of liquid crystals, *Phys. Lett.* **A39**, 69 (1972).
46. H. Imura and K. Okano, Temperature dependence of the viscosity coefficients of liquid crystals, *Jpn. J. Appl. Phys.* **11**, 1440 (1972).
47. A. C. Diogo and A. F. Martins, Thermal behavior of the twist viscosity in a series of homologous nematic liquid crystals, *Mol. Cryst. Liq. Cryst.* **66**, 133 (1981).
48. V. V. Belyaev, S. Ivanov, and M. F. Grebenkin, Temperature dependence of rotational viscosity of nematic liquid crystals, *Sov. Phys. Crystallogr.* **30**, 674 (1985).
49. S. T. Wu and C. S. Wu, Rotational viscosity of nematic liquid crystals, *Liq. Cryst.* **8**, 171 (1990).
50. M. A. Osipov and E. M. Terentjev, Rotational diffusion and rheological properties of liquid crystals, *Z. Naturforsch.* **A44**, 785 (1989).
51. S. T. Wu and C. S. Wu, Experimental confirmation of Osipov–Terentjev theory on the viscosity of liquid crystals, *Phys. Rev. A* **42**, 2219 (1990).
52. P. G. de Gennes, *The Physics of Liquid Crystals* (Clarendon, Oxford, 1974).
53. W. Maier and A. Saupe, A simple molecular statistical theory for nematic liquid crystal phase, Part II, *Z. Naturforsch. Teil* **A15**, 287 (1960).
54. H. Gruler, The elastic constants of a nematic liquid crystal, *Z. Naturforsch. Teil A* **30**, 230 (1975).
55. S. T. Wu, A. M. Lackner, and U. Efron, Optimal operation temperature of liquid crystal modulators, *Appl. Opt.* **26**, 3441 (1987).
56. P. S. Drzaic, *Liquid Crystal Dispersions* (World Scientific, New Jersey, 1995).
57. Y. H. Lin, H. W. Ren, and S. T. Wu, High contrast polymer-dispersed liquid crystal in a 90° twisted cell, *Appl. Phys. Lett.* **84**, 4083 (2004).
58. H. Ren, Y. H. Lin, Y. H. Fan, and S. T. Wu, Polarization-independent phase modulation using a polymer-dispersed liquid crystal, *Appl. Phys. Lett.* **86**, 141110 (2005).
59. J. Li, G. Baird, H. Ren, Y. H. Lin, and S. T. Wu, Refractive index matching between liquid crystals and photopolymers, *J. SID* **13**, 1017 (2005).

7

Modeling Liquid Crystal Director Configuration

Liquid crystal director configuration and optical modeling methods are well developed and reliable and widely used in liquid crystal device design [1–7]. In the modeling, the equilibrium director configuration is obtained by minimizing the total free energy of the system (elastic energy plus the electric energy). The popular numerical methods used in liquid crystal modeling are the *finite-difference method* (FDM) [1,8] and *finite-element method* (FEM) [7,9,10]. FDM is simple and easy to understand while FEM is versatile in modeling arbitrary liquid crystal device structures. We will only discuss FDM in this chapter.

7.1 Electric Energy of Liquid Crystals

In order to model the liquid crystal director configuration, we must first know how liquid crystals interact with externally applied electric fields. Many liquid crystal devices make use of uniaxial nematic liquid crystals which are dielectrics. We consider the electric energy of nematic liquid crystals in externally applied electric fields through dielectric interaction. A typical liquid crystal device cell is shown in Figure 7.1, where the liquid crystal is sandwiched between two parallel substrates with transparent electrodes. The electric energy of the liquid crystal is given by [11–13].

$$U_e = \int_V \frac{1}{2}\vec{E}\cdot\vec{D}\,d^3r, \tag{7.1}$$

where the volume integration is over the liquid crystal. The internal energy (including the electric energy) of the system is U. The change of internal energy dU in a process is equal to the sum

Fundamentals of Liquid Crystal Devices, Second Edition. Deng-Ke Yang and Shin-Tson Wu.

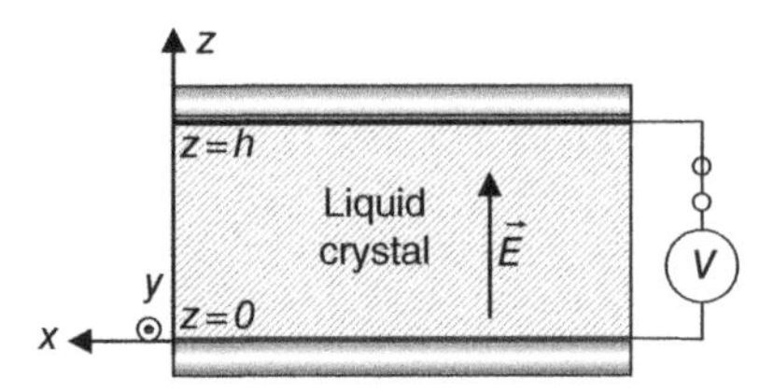

Figure 7.1 Schematic diagram of the liquid crystal cell connected to a voltage source.

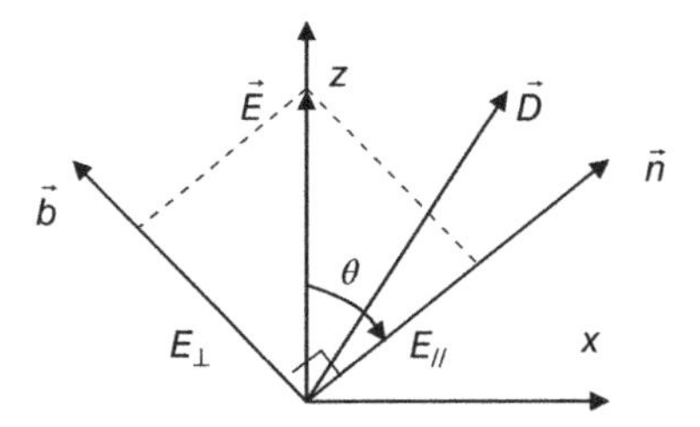

Figure 7.2 Schematic diagram showing the field decomposed into components parallel and perpendicular to the liquid crystal director.

of the heat absorbed dQ, the mechanical work dW_m done on the system, and the electric work dW_e done on the system by external sources:

$$dU = dQ + dW_m + dW_e \tag{7.2}$$

When the liquid crystal undergoes a change in its director configuration, the electric field may change and the electric work dW_e depends on whether the liquid crystal cell is connected to a voltage source or not. We will consider several cases in the following sections.

7.1.1 Constant charge

In the first case, the liquid crystal cell is disconnected from the voltage source. The free charge on the interface between the liquid crystal and the electrode is fixed, and this case is called fixed charge. The external voltage source does not do electrical work, that is $dW_e = 0$. At constant temperature and pressure, the Gibbs free energy,

$$G = \int \left(f_{elastic} + \frac{1}{2}\vec{E}\cdot\vec{D} \right) d^3r, \tag{7.3}$$

is minimized in the equilibrium state. As an example, we consider a one-dimensional case where the liquid crystal director $\vec{n}$ is confined in the xz plane and is only a function of the coordinate z that is parallel to the cell normal. Inside the electrode, $\vec{E} = 0$. The tangential boundary condition gives $E_x = E_y = 0$ everywhere. Therefore $\vec{E} = E(z)\hat{z}$. From Figure 7.2 the electric displacement is found to be

$$\vec{D}=\varepsilon_o\left\{\varepsilon_{//}\left(\vec{E}\cdot\vec{n}\right)\vec{n}+\varepsilon_\perp\left[\vec{E}-\left(\vec{E}\cdot\vec{n}\right)\vec{n}\right]\right\}=\varepsilon_o\left[\varepsilon_\perp\vec{E}+\Delta\varepsilon\left(\vec{E}\cdot\vec{n}\right)\vec{n}\right]=\varepsilon_o\left[\varepsilon_\perp\vec{E}+\Delta\varepsilon E\cos\theta\,\vec{n}\right]. \tag{7.4}$$

The surface free charge density $\sigma_1=D_z(z=0)=\sigma$ on the bottom surface and the surface free charge density $\sigma_2=-D_z(z=h)$ on the top surface are fixed. In the 1-D case here, $\vec{D}=\vec{D}(z)$ and $\nabla\cdot\vec{D}=\partial D_z/\partial z=0$; therefore D_z is a constant across the cell:

$$D_z(z)=D_z(z=h)=\varepsilon_o\varepsilon_\perp E+\varepsilon_o\Delta\varepsilon E\cos^2\theta=\sigma \tag{7.5}$$

The electric field is given by

$$E=\sigma/\varepsilon_o\left(\varepsilon_\perp+\Delta\varepsilon\cos^2\theta\right). \tag{7.6}$$

When the liquid crystal undergoes a configurational change, θ changes, and thus E changes. The electric energy density is

$$f_{electric}=\frac{1}{2}\vec{E}\cdot\vec{D}=\frac{1}{2}E\cdot D_z=\frac{\sigma^2}{2\varepsilon_o(\varepsilon_\perp+\Delta\varepsilon\cos^2\theta)}. \tag{7.7}$$

In the equilibrium state, the Gibbs free energy,

$$G=\int\left(f_{elastic}+\frac{1}{2}\vec{E}\cdot\vec{D}\right)dV=\int\left[f_{elastic}+\frac{\sigma^2}{2\varepsilon_o(\varepsilon_\perp+\Delta\varepsilon\cos^2\theta)}\right]d^3r, \tag{7.8}$$

is minimized. Please note that σ is a constant. The voltage across the cell is

$$V=\int_0^h E_z dz=\int_0^h\frac{\sigma}{\varepsilon_o(\varepsilon_\perp+\Delta\varepsilon\cos^2\theta)}dz, \tag{7.9}$$

which is not a constant and changes with the director configuration. If $\Delta\varepsilon>0$, when $\vec{n}\,//\,\vec{E}$, $\theta=0$, the electric energy is minimized; therefore the liquid crystal molecules tend to align parallel to the field. If $\Delta\varepsilon<0$ when $\vec{n}\perp\vec{E}$, $\theta=\pi/2$, the electric energy is minimized; therefore the liquid crystal molecules tend to align perpendicular to the field.

7.1.2 *Constant voltage*

Next we consider the second case, where the liquid crystal cell is connected to the voltage source such that the voltage applied V across the cell is fixed. The electric potential in the cell is $\phi(z)$. On top of the cell ($z=h$), the potential $\phi(z=h)=\phi_2$ is low. At the bottom of the cell ($z=0$), the potential $\phi(z=0)=\phi_1$ is high, and $\phi_1-\phi_2=V$. The electric field is in the $+z$ direction. The free surface charge density on the top surface of the liquid crystal cell is $\sigma_2 = -\sigma$, which is negative. The free surface charge density on the bottom surface of the liquid crystal cell is $\sigma_1 = \sigma$, which is positive. When the liquid crystal undergoes an orientational configuration change, σ may vary. During the reorientation of the liquid crystal, the total amount of charge $dQ = \int \delta\sigma dS$ is moved from the top plate to the bottom plate by the voltage source, and the electric work done by the voltage source to the liquid crystal is

$$dW_e = VdQ = (\phi_1 - \phi_2)dQ = (\phi_1 - \phi_2)\int \delta\sigma dS = \int_{S_2} \phi_2 \delta\sigma_2 dS + \int_{S_1} \phi_1 \delta\sigma_1 dS. \tag{7.10}$$

The electric energy of the liquid crystal is

$$\int_{Vol} \frac{1}{2}\vec{D}\cdot\vec{E}\,d^3r = \int_{Vol} \frac{1}{2}\vec{D}\cdot(-\nabla\phi)d^3r = -\int_{Vol} \frac{1}{2}\nabla\cdot\left(\phi\vec{D}\right)dV + \int_{Vol} \frac{1}{2}\phi\left(\nabla\cdot\vec{D}\right)d^3r. \tag{7.11}$$

There is no free charge inside the liquid crystal, $\nabla\cdot\vec{D} = 0$, and thus the last term on the right side of the above equation is zero. The surface normal vector points out of the liquid crystal. The first term on the right side of Equation (7.11) becomes

$$-\int_V \frac{1}{2}\nabla\cdot\left(\phi\vec{D}\right)dV = -\frac{1}{2}\oint_S \phi\vec{D}\cdot d\vec{S} = -\frac{1}{2}\int_{S_2} \phi_2 D_z(z=h)dS - \frac{1}{2}\int_{S_1} \phi_1 D_z(z=0)(-dS).$$

The free surface charge densities are given by $\sigma_2 = -D_z(z=h)$ and $\sigma_1 = D_z(z=0)$. Therefore

$$\int_{Vol} \frac{1}{2}\vec{D}\cdot\vec{E}\,d^3r = \frac{1}{2}\left(\int_{S_2} \phi_2\delta\sigma_2 dS + \int_{S_1} \phi_1\delta\sigma_1 dS\right). \tag{7.12}$$

Comparing Equations (7.10) and (7.12), we have

$$dW_e = \delta\int_{Vol} \vec{D}\cdot\vec{E}\,d^3r. \tag{7.13}$$

In a reorientation of the liquid crystal, as discussed in Chapter 1, the change of entropy of the system is given by

$$dS \geq dQ/T = (\delta U - dW_m - dW_e)/T = (\delta U + PdV - dW_e)/T.$$

At constant temperature and pressure $\delta(U - W_e + PV - TS) \leq 0$, that is

$$\delta\left[\int\left(f_{elastic} + f_{electric} - \vec{D}\cdot\vec{E}\right)d^3r\right] = \delta\left[\int\left(f_{elastic} - \frac{1}{2}\vec{D}\cdot\vec{E}\right)d^3r\right] \leq 0.$$

Therefore at the equilibrium state

$$R=\int\left(f_{elastic}-\frac{1}{2}\vec{D}\cdot\vec{E}\right)d^3r \tag{7.14}$$

is minimized. Using Equation (7.6), we have

$$R=\int\left[f_{elastic}-\frac{\sigma^2}{2\varepsilon_o(\varepsilon_\perp+\Delta\varepsilon\cos^2\theta)}\right]d^3r. \tag{7.15}$$

At first glance, it seems that, in order to minimize R, provided $\Delta\varepsilon>0$, $\vec{n}$ should be perpendicular to $\vec{E}$ ($\theta=\pi/2$), in contrast to the result of the constant charge case. The liquid crystal molecules do not know whether the charge is fixed or the voltage is fixed and the liquid crystal ($\Delta\varepsilon>0$) always tends to align parallel to the applied field. This paradox can be resolved by noting that, in Equation (7.15), σ is no longer a constant, because

$$V=\int_0^h E_z dz=\int_0^h\frac{\sigma}{\varepsilon_o(\varepsilon_\perp+\Delta\varepsilon\cos^2\theta)}dz=\frac{1}{A}\int\frac{\sigma}{\varepsilon_o(\varepsilon_\perp+\Delta\varepsilon\cos^2\theta)}d^3r=\text{a fixed constant}, \tag{7.16}$$

where A is the surface area of the cell. When $\theta(z)$ changes, σ must vary in order to keep V fixed. R is a functional of θ and σ. Equation (7.16) is the constraint under which R is minimized. The constraint can be removed if we use a Lagrange multiplier [14], and minimize

$$\Omega=\int\left(f_{elastic}-\frac{\sigma^2}{2\varepsilon_o(\varepsilon_\perp+\Delta\varepsilon\cos^2\theta)}+\lambda\cdot\frac{\sigma}{\varepsilon_o(\varepsilon_\perp+\Delta\varepsilon\cos^2\theta)}\right)d^3r, \tag{7.17}$$

where λ is the Langrange multiplier. Minimizing Ω with respect to σ_2, we have

$$\frac{\partial\Omega}{\partial\sigma}=-\frac{2\sigma}{2\varepsilon_o(\varepsilon_\perp+\Delta\varepsilon\cos^2\theta)}+\lambda\cdot\frac{1}{\varepsilon_o(\varepsilon_\perp+\Delta\varepsilon\cos^2\theta)}=0. \tag{7.18}$$

Hence $\lambda=\sigma$. We minimize

$$\Omega=G=\int\left(f_{elastic}+\frac{\sigma^2}{2\varepsilon_o(\varepsilon_\perp+\Delta\varepsilon\cos^2\theta)}\right)d^3r, \tag{7.19}$$

with σ treated as a constant and without the constraint given by Equation (7.16). This is the same as Equation (7.8) for the case of constant charge.

In reality, the voltage is usually fixed and known. We can take the following strategy to calculate the director configuration for a given voltage V. (1) Assume σ is fixed and has a trial value σ_t. (2) Use the Euler–Lagrange method to minimize the Gibbs free energy given by Equation (7.8) and find the solution $\theta=\theta(z)$. (3) Calculate the corresponding voltage

$V_{try} = \int_0^h \frac{\sigma_t}{\varepsilon_o(\varepsilon_\perp + \Delta\varepsilon\cos^2\theta)} dz$. If $V_{try} \neq V$, try a new surface charge density $\sigma_n = \sigma_t + \alpha(V - V_t)\varepsilon_o\varepsilon_\perp/h$, where α is a numerical constant which may be chosen to be 0.5. Repeat the above process until $|V_t - V|$ is sufficiently small.

7.1.3 Constant electric field

When the orientation of the liquid crystal is uniform in space, θ is a constant independent of z. For a fixed V, the electric field $E = E_z = V/h$ is a constant independent of the orientation of the liquid crystal. In the equilibrium state R, given by Equation (7.14), is minimized:

$$-(1/2)\vec{D}\cdot\vec{E} = -(1/2)\varepsilon_o\varepsilon_\perp E^2 - (1/2)\varepsilon_o\Delta\varepsilon\left(\vec{E}\cdot\vec{n}\right)^2$$

Because $-(1/2)\varepsilon_o\varepsilon_\perp E^2$ is a constant independent of the orientation of the liquid crystal, Equation (7.14) becomes

$$R = \int \left[f_{elastic} - (1/2)\varepsilon_o\Delta\varepsilon\left(\vec{n}\cdot\vec{E}\right)^2\right] d^3r. \tag{7.20}$$

In the equilibrium state R is minimized. If $\Delta\varepsilon > 0$, when $\vec{n} // \vec{E}$, $\theta = 0$, the electric energy is minimized; therefore the liquid crystal molecules tend to align parallel to the field. If $\Delta\varepsilon < 0$, when $\vec{n} \perp \vec{E}$, $\theta = \pi/2$, the electric energy is minimized; therefore the liquid crystal molecules tend to align perpendicular to the field.

7.2 Modeling Electric Field

Multiplexed displays have many pixels where liquid crystals are sandwiched between two substrates with conducting films. Electric fields are produced by applied electric voltage across the conducting coatings. The pixels are separated by gaps where the conducting coating is etched off. Fringe fields are produced at the edge of the pixels, which may cause serious problems in microdisplays where the gap between pixels is not much smaller than the pixel (linear) size. In this section we will consider how to numerically calculate electric fields in multiplexed displays.

Electric field $\vec{E}$ is related to electric potential φ by $\vec{E} = -\nabla\varphi$. Liquid crystals are dielectric media and there are usually no free charges inside them. From Maxwell's equation we have

$$\nabla\cdot\vec{D} = \nabla\cdot\left(\overleftrightarrow{\varepsilon}\cdot\vec{E}\right) = -\nabla\cdot\left(\overleftrightarrow{\varepsilon}\cdot\nabla\varphi\right) = 0. \tag{7.21}$$

We first consider a simple case where the medium is isotropic and uniform. Equation (7.21) becomes

$$\nabla^2\varphi = 0 \tag{7.22}$$

which is known as Laplace's equation. Several approaches have been developed to solve the Laplace's equation. The simple and dominant method is the *finite difference method* [1,15,16].

In the finite difference method, a regular mesh is used on the region in which a solution is to be found. As an example, the region to be considered is a rectangle with lengths L_x and L_y in x and y directions. Superimpose a mesh on the rectangle. The unit cell of the mesh is a square with the length Δ. At each lattice point of the mesh, the potential is $\varphi(i,j)$, and Laplace's equation is approximated by

$$\frac{\varphi(i+1,j)+\varphi(i-1,j)-2\varphi(i,j)}{\Delta^2}+\frac{\varphi(i,j+1)+\varphi(i,j-1)-2\varphi(i,j)}{\Delta^2}=0, \tag{7.23}$$

where $i=0, 1, 2, \ldots N_x$ $(N_x=L_x/\Delta)$ and $j=0, 1, 2, \ldots N_y$ $(N_y=L_y/\Delta)$. Rearranging Equation (7.23) we have

$$\varphi(i,j)=\frac{1}{4}[\varphi(i+1,j)+\varphi(i-1,j)+\varphi(i,j+1)+\varphi(i,j-1)]. \tag{7.24}$$

Usually the potential at the boundary is given. Assume an initial condition $\varphi_o(i,j)$ that is consistent with the boundary condition. The potential at any mesh point can be calculated from the assumed potentials at its nearest neighbor mesh points by using the above equation. In the calculation the latest available values of the potential are always used on the right side of the equation. The solution to the Laplace's equation can be iteratively calculated. This process has the shortcoming that it converges quite slowly to the solution. The rate of convergence can be improved by using the 'over-relaxation' method:

$$\varphi^{\tau+1}(i,j)=(1-\alpha)\varphi^{\tau}(i,j)+\frac{\alpha}{4}[\varphi^{\tau}(i+1,j)+\varphi^{\tau}(i-1,j)+\varphi^{\tau}(i,j+1)+\varphi^{\tau}(i,j-1)], \tag{7.25}$$

where τ is the order of the iteration and α is the relaxation constant, which should be a positive constant smaller than 1.5 in order to obtain a stable solution. Equation (7.25) can be rewritten as

$$\varphi^{\tau+1}(i,j)=\varphi^{\tau}(i,j)+\frac{\alpha}{4}[\varphi^{\tau}(i+1,j)+\varphi^{\tau}(i-1,j)-2\varphi^{\tau}(i,j)+\varphi^{\tau}(i,j+1)+\varphi^{\tau}(i,j-1)-2\varphi^{\tau}(i,j)],$$

that is

$$\varphi^{\tau+1}(i,j)=\varphi^{\tau}(i,j)+\frac{\alpha\Delta^2}{4}\left[\frac{\partial^2\varphi^{\tau}}{\partial x^2}(i,j)+\frac{\partial^2\varphi^{\tau}}{\partial y^2}(i,j)\right]. \tag{7.26}$$

As $\varphi^{(i,j)}(i,j)$ is approaching the actual solution to Laplace's equation, the change $\Delta\varphi^{\tau}(i,j)=\varphi^{\tau+1}(i,j)-\varphi^{\tau}(i,j)$ becomes smaller and smaller. When $\varphi^{\tau}(i,j)$ does not change any more, that is $\frac{\partial^2\varphi^{\tau}}{\partial x^2}(i,j)+\frac{\partial^2\varphi^{\tau}}{\partial y^2}(i,j)=0$, Laplace's equation is satisfied. In numerical calculation, the iteration is stopped when the maximum $|\Delta\varphi_m|$ of the absolute values of $\Delta\varphi^{\tau}(i,j)$ at the lattice points on the mesh is smaller than a specified value. For a 3-D non-uniform anisotropic medium, the potential can be numerically calculated by

$$
\begin{aligned}
\varphi^{\tau+1}(i,j,k) &= \varphi^{\tau}(i,j,k) + \frac{\alpha\Delta^2}{4}\nabla\cdot\left(\overleftrightarrow{\varepsilon}\cdot\nabla\varphi\right)^{\tau}(i,j,k) \\
&= \varphi^{\tau}(i,j,k) + \frac{\alpha\Delta^2}{4}\left[\frac{\partial}{\partial x}\left(\varepsilon_{11}\frac{\partial\varphi}{\partial x}\right)+\frac{\partial}{\partial x}\left(\varepsilon_{12}\frac{\partial\varphi}{\partial y}\right)+\frac{\partial}{\partial x}\left(\varepsilon_{13}\frac{\partial\varphi}{\partial z}\right)\right]^{\tau}(i,j,k) \\
&+ \frac{\alpha\Delta^2}{4}\left[\frac{\partial}{\partial y}\left(\varepsilon_{21}\frac{\partial\varphi}{\partial x}\right)+\frac{\partial}{\partial y}\left(\varepsilon_{22}\frac{\partial\varphi}{\partial y}\right)+\frac{\partial}{\partial y}\left(\varepsilon_{23}\frac{\partial\varphi}{\partial z}\right)\right]^{\tau}(i,j,k) \\
&+ \frac{\alpha\Delta^2}{4}\left[\frac{\partial}{\partial z}\left(\varepsilon_{31}\frac{\partial\varphi^{\tau}}{\partial x}\right)+\frac{\partial}{\partial z}\left(\varepsilon_{32}\frac{\partial\varphi^{\tau}}{\partial y}\right)+\frac{\partial}{\partial z}\left(\varepsilon_{33}\frac{\partial\varphi^{\tau}}{\partial z}\right)\right](i,j,k).
\end{aligned}
\tag{7.27}
$$

As an example, we calculate the electric field in a cell where the stripe electrode is along the y direction. The widths of the stripe electrode and the gap between electrodes are both 10 µm. The cell thickness is 5 µm, as shown in Figure 7.3. The dielectric constant of the glass substrate is $\varepsilon_G = 6.5$. The dielectric constant of the liquid crystal is assumed to be isotropic and equal to $\varepsilon_{LC} = 10.0$ in the calculation of the electric field. The voltage on the bottom electrode is 0 V and the voltage on the top electrode is 10 V. On the top substrate, in the gap regions between the electrodes, the electric potential is unknown, and the electric displacement is continuous because there is no free surface charge. An imagined boundary can be placed far away from the surface of the top substrate. The boundary condition at the imagined boundary can be either $\varphi = 0$ or $E_z = \partial\varphi/\partial z = 0$. Accurately speaking, the imagined boundary should be at $z = \infty$. Since the mesh cannot be infinite in the simulation, the imagined boundary is at $z = z_o = 50$ µm. If the boundary condition of $\varphi(z_o) = 0$ is used, the error will be on the order of $1/z_o$. If the boundary condition of $E_z(z_o) = \frac{\partial\varphi}{\partial z}(z_o) = 0$ is used, the error will be in the order of e^{-z_o}. Periodic boundary condition is used in the x direction.

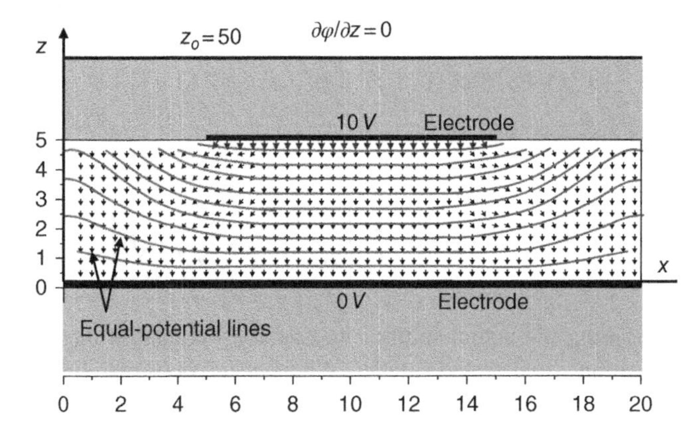

Figure 7.3 Electric field in the display cell with striped electrodes along the y direction. The unit of the length is micron.

7.3 Simulation of Liquid Crystal Director Configuration

In many liquid crystal devices the liquid crystal director configuration cannot be calculated analytically and must be numerically computed. Under a given external field and boundary condition, when a liquid crystal is in the equilibrium state, the total free energy is minimized. If the system is initially not in the equilibrium state, it will relax into a state with lower free energy. As the liquid crystal director configuration evolves, the free energy decreases. The change of the director configuration stops when the minimum free energy is reached. The dynamic equation for the change of the liquid crystal director can be used to numerically calculate the equilibrium director configuration, which is referred to as the relaxation method.

7.3.1 Angle representation

In some cases, it is simpler to describe the liquid crystal director $\vec{n}$ in terms of the polar angle θ and azimuthal angle ϕ. The angles may vary in one or two or three dimensions. We first consider a simple case: Fréedericksz transition in splay geometry. The liquid crystal director is represented by the tilt angle θ: $\vec{n} = \cos\theta(z)\hat{x} + \sin\theta(z)\hat{z}$, where the z axis is in the cell normal direction. The electric field is applied in the cell normal direction. From Equation (4.17) and (7.8) we have the free energy density

$$f = \frac{1}{2}\left(K_{11}\cos^2\theta + K_{33}\sin^2\theta\right)\theta'^2 + \frac{\sigma^2}{2\varepsilon_o\left(\varepsilon_\perp + \Delta\varepsilon\sin^2\theta\right)}, \tag{7.28}$$

where σ is the free surface charge density and $\theta' = \partial\theta/\partial z$. Note that here θ is the angle between the liquid crystal director and the x axis. In the equilibrium state, the total free energy $F = \int_0^h f dz$ is minimized, and the director configuration is $\theta_{eq}(z)$ which satisfies the Euler–Lagrange equation.

$$\begin{aligned} -\left.\frac{\delta f}{\delta\theta}\right|_{eq} &= -\left.\left[\frac{\partial f}{\partial\theta} - \frac{d}{dz}\left(\frac{\partial f}{\partial\theta'}\right)\right]\right|_{eq} \\ &= \left(K_{11}\cos^2\theta_{eq} + K_{33}\sin^2\theta_{eq}\right)\theta''_{eq} + \left(K_{33} - K_{11}\right)\sin\theta_{eq}\cos\theta_{eq}\theta'^2_{eq} + \frac{\sigma^2\Delta\varepsilon\sin\theta_{eq}\cos\theta_{eq}}{\varepsilon_o\left(\varepsilon_\perp + \Delta\varepsilon\sin^2\theta_{eq}\right)^2} \\ &\equiv u\left(\theta_{eq}\right)\theta''_{eq} + w\left(\theta_{eq}, \theta'_{eq}\right) = 0. \end{aligned} \tag{7.29}$$

If initially the system is not in the equilibrium state, $\theta(z) = \theta_{in}(z)$. It will relax toward the equilibrium state. The dynamic equation governing the relaxation of the system is given by Equation (4.74):

$$\gamma_r \frac{\partial \theta}{\partial t} = -\frac{\delta f}{\delta \theta} = -\frac{\partial f}{\partial \theta} + \frac{d}{dz}\left(\frac{\partial f}{\partial \theta'}\right) = u(\theta)\theta'' + w(\theta, \theta') \tag{7.30}$$

Using this equation, the angle at time $t + \Delta t$ can be calculated from the angle at time t as

$$\theta^{t+\Delta t} = \theta^t + \frac{\Delta t}{\gamma_r}\left[u(\theta^t)\theta''^t + w\left(\theta^t, \theta'^t\right)\right]. \tag{7.31}$$

When the system reaches the equilibrium state, the director configuration does not change any more, and $\partial\theta/\partial t = -\delta f/\delta\theta = 0$. Equation (7.30) may not describe the actual dynamic process because the hydrodynamic effect is not considered, but the final director configuration obtained is the actual one for the equilibrium state.

In the numerical calculation, the liquid crystal cell is discretized into a one-dimensional mesh with N lattice sites. The length of the lattice unit is $\Delta z = h/N$. At step τ, the tilt angle at the lattice site i is $\theta(i)$ $(i = 0, 1, 2, 3, \ldots\ldots, N)$. The derivatives are calculated by

$$\theta'^{\tau}(i) = \frac{\theta^{\tau}(i+1) - \theta^{\tau}(i-1)}{2\Delta z}, \tag{7.32}$$

$$\theta''^{\tau}(i) = \frac{\theta^{\tau}(i+1) + \theta^{\tau}(i-1) - 2\theta^{\tau}(i)}{(\Delta z)^2}. \tag{7.33}$$

The angle at step $\tau + 1$ can be calculated by

$$\theta^{\tau+1}(i) = \theta^{\tau}(i) + \Delta\theta^{\tau}(i), \tag{7.34}$$

$$\Delta\theta^{\tau}(i) = \alpha(\Delta z)^2\left\{u[\theta^{\tau}(i)]\theta''^{\tau}(i) + w[\theta^{\tau}(i), \theta'^{\tau}(i)]\right\}, \tag{7.35}$$

where α is a relaxation constant, which must be sufficiently small in order to avoid unstable solutions. In the numerical calculation, if first the change $\Delta\theta^{\tau}(i)$ at all the lattice sites is calculated using the angles at step τ, and then the angles at all the lattice sites are updated, α must be smaller than $0.5/u[\theta^{\tau}(i)]$ in order to avoid unstable solutions. If the change $\Delta\theta^{\tau}(i)$ at each lattice site is calculated and then the angle at that the lattice site is immediately updated, known as the over-relaxation method, an α larger than $1/u[\theta^{\tau}(i)]$ can be used. In the numerical calculation, the total change of the angle $T\theta = \sum_{i=0}^{N} |\Delta\theta^{\tau}(i)|$ in each step must be monitored. $T\theta$ decreases as the relaxation proceeds and becomes 0 when the equilibrium configuration is reached. In reality in the numerical calculation, $T\theta$ decreases but will never become exactly 0. When $T\theta$ becomes sufficiently small, the calculation can be stopped.

If the anchoring of the liquid crystal at the boundaries is infinitely strong, the angle at the boundary is fixed. If the pretilt angles at the boundaries $z = 0$ and $z = h$ are θ_1 and θ_2, respectively, the boundary conditions are $\theta(0) = \theta_1$ and $\theta(N) = \theta_2$. If the anchoring is weak with the anchoring energy W, the boundary conditions are $\theta'(0)=(W/K_{11})\theta(0)$ and $\theta'(N)=-(W/K_{11})\theta(N)$ (from Equations (5.60) and (5.61)), and this gives $\theta(0)=\theta(1)/(1+\Delta zW/K_{11})$ and $\theta(N)=\theta(N-1)/(1-\Delta zW/K_{11})$.

The angle representation is a valid method when the change of the angle within the cell is less than 90°. Otherwise it must be handled carefully in the case where the liquid crystal director at two neighboring lattice sites are anti-parallel. The numerical calculation may produce a large elastic energy while the actual elastic energy is 0, because $\vec{n}$ and $-\vec{n}$ are equivalent.

If two angles, say $\theta(z)$ and $\phi(z)$, are needed to describe the orientation of the liquid crystal director, the total free energy is given by

$$F=\int_0^h f[\theta,\phi,\theta',\phi',z)dz. \tag{7.36}$$

In the numerical relaxation method, the angles at the lattices sites can be calculated by

$$\theta^{\tau+1}(i)=\theta^{\tau}(i)+\alpha(\Delta z)^2\left[-\frac{\delta f^{\tau}}{\delta\theta}(i)\right], \tag{7.37}$$

$$\phi^{\tau+1}(i)=\phi^{\tau}(i)+\beta(\Delta z)^2\left[-\frac{\delta f^{\tau}}{\delta\phi}(i)\right], \tag{7.38}$$

where α and β are the relaxation constants.

As an example, we numerically calculate the polar angle θ and azimuthal angle ϕ in the Fréedericksz transition in the twisted nematic geometry. The parameters of the liquid crystal are $K_{11}=6.4\times10^{-12}$N, $K_{22}=3.0\times10^{-12}$N, $K_{33}=10.0\times10^{-12}$N and $\Delta\varepsilon=10$. The thickness h of the cell is 10 microns. The intrinsic pitch of the liquid crystal is $P=(2\pi/\Phi)h$, where Φ is the total twist angle. The polar angle is the angle between the liquid crystal director and the xy plane. When the twist angle Φ is 90°, the polar and azimuthal angles as a function of z at various applied voltages are shown in Figure 7.4. The voltage threshold calculated from Equation 5.56 is $V_{th}=0.996$ V.

The change of the polar angle as a function of the applied voltage depends on the total twist angle Φ, as shown in Figure 7.5. The threshold increases with increasing Φ for the two reasons. First, in the field-activated states, there is twist elastic energy that increases with Φ. Secondly, the bend elastic energy increases with Φ when the polar angle is small. The saturation voltage does not increase much with increasing Φ, because there is no bend deformation in the saturated state. Therefore the transition region (the region between the threshold voltage and the saturation voltage) decreases with increasing Φ. When Φ is increased above 270°, the polar angle has two different values for a given voltage, that is, there is a hysteresis in the Fréedericksz transition. TNs with twist angle larger than 90° are known as *super-twisted nematics* (STN*s*). Because of their steep transition, they are used to make multiplexed displays on passive matrices.

If the angle θ of the liquid crystal director varies in three dimensions, that is, $\theta=\theta(x,y,z)$, the total free energy is given by

$$F=\iiint f(\theta,\theta',x,y,z)dxdydz. \tag{7.39}$$

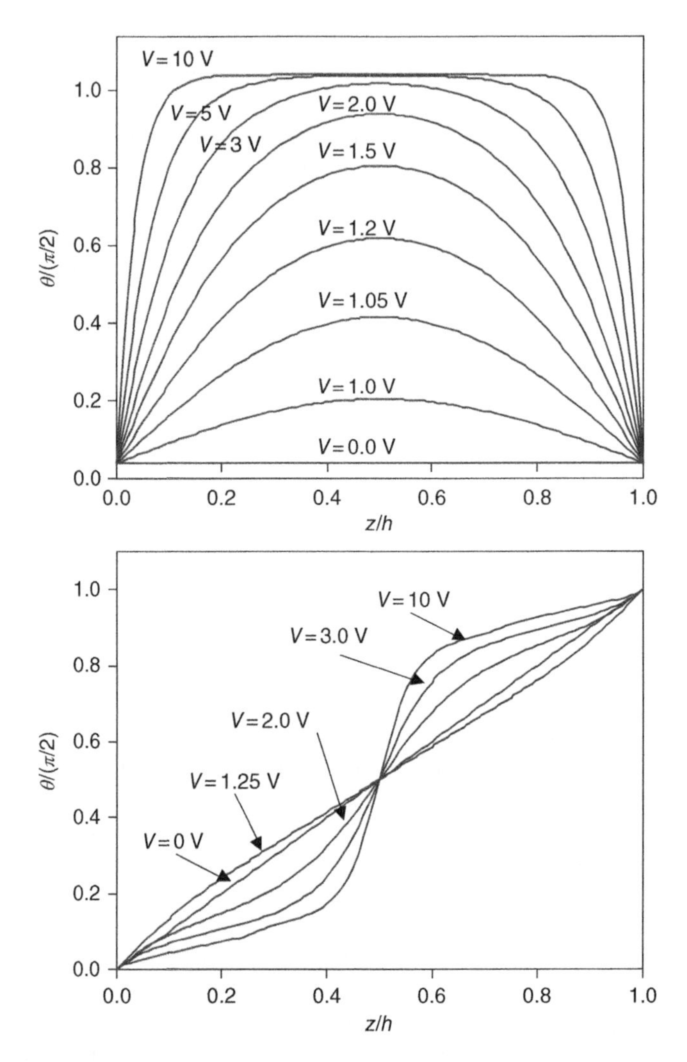

Figure 7.4 The polar and azimuthal angles of the liquid crystal director as functions of z in the 90° twisted nematic cell under various applied voltages.

In the numerical calculation, a 3-D mesh with unit cell size $(\Delta x)^3$ is used. The angle at the lattice sites (i, j, k) is calculated by

$$\theta^{\tau+1}(i,j,k) = \theta^{\tau}(i,j,k) + \alpha(\Delta x)^2 \left[-\frac{\delta f^{\tau}}{\delta \theta}(i,j,k) \right]. \tag{7.40}$$

The derivatives at step τ are calculated by

$$\theta_x'^{\tau}(i,j,k) = \frac{\theta^{\tau}(i+1,j,k) - \theta^{\tau}(i-1,j,k)}{2\Delta x}, \tag{7.41}$$

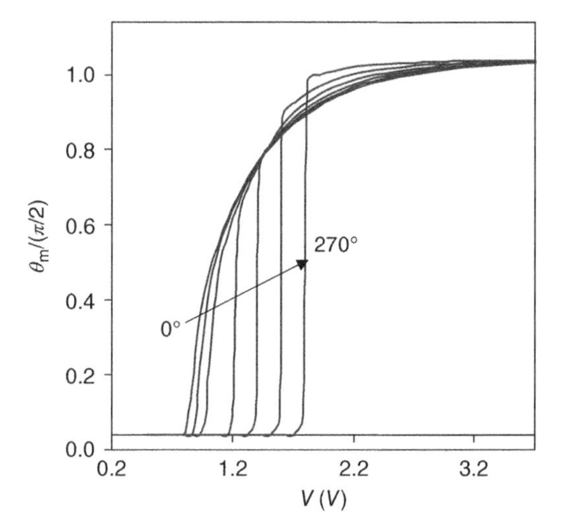

Figure 7.5 The polar angle at the middle plane vs. the applied voltage in twisted nematic cells with the twist angles: 0°, 45°, 90°, 135°, 180°, 225°, and 270°.

$$\theta_x''^\tau(i) = \frac{\theta^\tau(i+1,j,k) + \theta^\tau(i-1,j,k) - 2\theta^\tau(i,j,k)}{(\Delta x)^2}, \tag{7.42}$$

$$\theta_{xy}''^\tau(i) = \frac{\theta^\tau(i+1,j+1,k) + \theta^\tau(i-1,j-1,k) - \theta^\tau(i-1,j+1,k) - \theta^\tau(i+1,j-1,k)}{4\Delta x \Delta y}, \tag{7.43}$$

and so on.

7.3.2 Vector representation

The liquid crystal director $\vec{n}$ can also be specified by its three components (n_x, n_y, n_z). The free energy density (with constant voltage) is then expressed as a function of the components and their spatial derivatives:

$$f = \frac{1}{2}K_{11}(\nabla\cdot\vec{n})^2 + \frac{1}{2}K_{22}(\vec{n}\cdot\nabla\times\vec{n})^2 + \frac{1}{2}K_{33}(\vec{n}\times\nabla\times\vec{n})^2 + q_o K_{22}\,\vec{n}\cdot\nabla\times\vec{n} - \frac{1}{2}\vec{D}\cdot\vec{E} \tag{7.44}$$

In component form, we have

$$\nabla\cdot\vec{n} = \frac{\partial n_i}{\partial x_i}, \tag{7.45}$$

where $i = x, y, z$, and the convention of summing over repeating indices is used:

$$(\nabla\cdot\vec{n})^2 = \frac{\partial n_i}{\partial x_i}\cdot\frac{\partial n_j}{\partial x_j} \tag{7.46}$$

$$\nabla \times \vec{n} = e_{ijk} \frac{\partial n_k}{\partial x_j} \hat{x}_i, \tag{7.47}$$

where e_{ijk} is the Levi–Civita symbol ($e_{xyz} = e_{yzx} = e_{zxy} = -e_{xzy} = -e_{zyx} = -e_{yxz} = 1$ and all other $e_{ijk} = 0$).

$$\left(\nabla \times \vec{n}\right)^2 = \frac{\partial n_l}{\partial x_k}\frac{\partial n_l}{\partial x_k} - \frac{\partial n_k}{\partial x_l}\frac{\partial n_l}{\partial x_k}, \tag{7.48}$$

$$\vec{n} \cdot \nabla \times \vec{n} = e_{ijk} n_i \frac{\partial n_k}{\partial x_j}, \tag{7.49}$$

$$\vec{n} \times \nabla \times \vec{n} = e_{lmi} e_{ijk} n_m \frac{\partial n_k}{\partial x_j} \hat{x}_l = \left(n_k \frac{\partial n_k}{\partial x_l} - n_k \frac{\partial n_l}{\partial x_k}\right) \hat{x}_l = -n_k \frac{\partial n_l}{\partial x_k} \hat{x}_l. \tag{7.50}$$

Note that $n_k \frac{\partial n_k}{\partial x_l} = \frac{1}{2}\frac{\partial (n_k n_k)}{\partial x_l} = \frac{1}{2}\frac{\partial\left(n_x^2 + n_y^2 + n_z^2\right)}{\partial x_l} = \frac{1}{2}\frac{\partial (1)}{\partial x_l} = 0,$

$$\left(\vec{n} \times \nabla \times \vec{n}\right)^2 = \left(-n_k \frac{\partial n_l}{\partial x_k}\right)\left(-n_i \frac{\partial n_l}{\partial x_i}\right) = n_k n_i \frac{\partial n_l}{\partial x_k}\frac{\partial n_l}{\partial x_i}, \tag{7.51}$$

$$\left(\vec{n} \cdot \nabla \times \vec{n}\right)^2 = \left(\nabla \times \vec{n}\right)^2 - \left(\vec{n} \times \nabla \times \vec{n}\right)^2 = \left(\frac{\partial n_l}{\partial x_k}\frac{\partial n_l}{\partial x_k} - \frac{\partial n_k}{\partial x_l}\frac{\partial n_l}{\partial x_k}\right) - n_k n_i \frac{\partial n_l}{\partial x_k}\frac{\partial n_l}{\partial x_i}. \tag{7.52}$$

The electric energy is

$$-\frac{1}{2}\vec{E} \cdot \vec{D} = -\frac{1}{2}\vec{E} \cdot \left(\vec{\varepsilon} \cdot \vec{E}\right) = -\frac{1}{2}\vec{E} \cdot \left[\varepsilon_o \varepsilon_\perp \vec{E} + \varepsilon_o \Delta\varepsilon \left(\vec{E} \cdot \vec{n}\right) \vec{n}\right] = -\frac{1}{2}\varepsilon_o \varepsilon_\perp E^2 - \frac{1}{2}\varepsilon_o \Delta\varepsilon E_i E_j n_i n_j. \tag{7.53}$$

The first term on the right side of Equation (7.53) is a constant independent of n_i and thus it does not affect the orientation of the liquid crystal and can be omitted. The bulk free energy density becomes

$$f = \frac{1}{2}K_{11}\frac{\partial n_i}{\partial x_i} \cdot \frac{\partial n_j}{\partial x_j} + \frac{1}{2}K_{22}\left(\frac{\partial n_j}{\partial x_i}\frac{\partial n_j}{\partial x_i} - \frac{\partial n_i}{\partial x_j}\frac{\partial n_j}{\partial x_i}\right) + \frac{1}{2}(K_{33} - K_{22}) n_i n_j \frac{\partial n_k}{\partial x_i}\frac{\partial n_k}{\partial x_j} + q_o K_{22} e_{ijk} n_i \frac{\partial n_k}{\partial x_j} - \frac{1}{2}\varepsilon_o \Delta\varepsilon E_i E_j n_i n_j. \tag{7.54}$$

In the equilibrium state, the total free energy is minimized. The director components n_i ($i = x, y, z$) in space in the equilibrium state can also be numerically calculated by the relaxation method. At the lattice site (l_x, l_y, l_z) of the mesh the changes of the director components from step τ to step $(\tau + 1)$ are calculated from the values of the director components at step τ,

$$\Delta n_i^{\tau+1}\left(l_x, l_y, l_z\right) = \alpha (\Delta x)^2 \left(-\frac{\delta f}{\delta n_i}\right)^\tau \left(l_x, l_y, l_z\right). \tag{7.55}$$

The variation of the free energy with respect to n_i is

$$\frac{\delta f}{\delta n_i}=\frac{\partial f}{\partial n_i}-\frac{\partial}{\partial x_j}\left(\frac{\partial f}{\partial n'_{i,j}}\right), \tag{7.56}$$

where $n'_{i,j}=\partial n_i/\partial x_j$. Thus:

$$\frac{\partial f}{\partial n_i}=\frac{\partial}{\partial n_i}\left[\frac{1}{2}(K_{33}-K_{22})n_l n_j\frac{\partial n_k}{\partial x_l}\frac{\partial n_k}{\partial x_j}+q_o K_{22}e_{ljk}n_l\frac{\partial n_k}{\partial x_j}-\frac{1}{2}\varepsilon_o\Delta\varepsilon E_l E_j n_l n_j\right]$$

$$=\frac{1}{2}(K_{33}-K_{22})\left(\delta_{il}n_j\frac{\partial n_k}{\partial x_l}\frac{\partial n_k}{\partial x_j}+\delta_{ij}n_l\frac{\partial n_k}{\partial x_l}\frac{\partial n_k}{\partial x_j}\right)+q_o K_{22}\delta_{il}e_{ljk}\frac{\partial n_k}{\partial x_j}$$

$$-\frac{1}{2}\varepsilon_o\Delta\varepsilon\left(\delta_{il}E_l E_j n_j+\delta_{ij}E_l E_j n_l\right)$$

$$=\frac{1}{2}(K_{33}-K_{22})\left(n_j\frac{\partial n_k}{\partial x_i}\frac{\partial n_k}{\partial x_j}+n_l\frac{\partial n_k}{\partial x_l}\frac{\partial n_k}{\partial x_i}\right)+q_o K_{22}e_{ijk}\frac{\partial n_k}{\partial x_j}-\frac{1}{2}\varepsilon_o\Delta\varepsilon\left(E_i E_j n_j+E_l E_i n_l\right)$$

$$=(K_{33}-K_{22})n_j\frac{\partial n_k}{\partial x_i}\frac{\partial n_k}{\partial x_j}+q_o K_{22}e_{ijk}\frac{\partial n_k}{\partial x_j}-\varepsilon_o\Delta\varepsilon E_i E_j n_j \tag{7.57}$$

$$\frac{\partial f}{\partial n'_{i,j}}=\frac{\partial}{\partial n'_{i,j}}\left[\frac{1}{2}K_{11}\frac{\partial n_l}{\partial x_l}\cdot\frac{\partial n_m}{\partial x_m}+\frac{1}{2}K_{22}\left(\frac{\partial n_m}{\partial x_l}\frac{\partial n_m}{\partial x_l}-\frac{\partial n_l}{\partial x_m}\frac{\partial n_m}{\partial x_l}\right)\right.$$

$$\left.+\frac{1}{2}(K_{33}-K_{22})n_l n_m\frac{\partial n_k}{\partial x_l}\frac{\partial n_k}{\partial x_m}+q_o K_{22}e_{lmk}n_l\frac{\partial n_k}{\partial x_m}\right]$$

$$=K_{11}\delta_{ij}\frac{\partial n_m}{\partial x_m}+K_{22}\left(\frac{\partial n_i}{\partial x_j}-\frac{\partial n_j}{\partial x_i}\right)+(K_{33}-K_{22})n_j n_m\frac{\partial n_i}{\partial x_m}+q_o K_{22}e_{lji}n_l \tag{7.58}$$

$$\frac{\partial}{\partial x_j}\left(\frac{\partial f}{\partial n'_{i,j}}\right)=\frac{\partial}{\partial x_j}\left[K_{11}\delta_{ij}\frac{\partial n_m}{\partial x_m}+K_{22}\left(\frac{\partial n_i}{\partial x_j}-\frac{\partial n_j}{\partial x_i}\right)+(K_{33}-K_{22})n_j n_m\frac{\partial n_i}{\partial x_m}+q_o K_{22}e_{lji}n_l\right]$$

$$=K_{11}\frac{\partial^2 n_m}{\partial x_m\partial x_i}+K_{22}\left(\frac{\partial^2 n_i}{\partial x_j^2}-\frac{\partial^2 n_j}{\partial x_j\partial x_i}\right)+(K_{33}-K_{22})\left(n_j n_m\frac{\partial^2 n_i}{\partial x_m\partial x_j}+n_j\frac{\partial n_i}{\partial x_m}\frac{\partial n_m}{\partial x_j}+n_m\frac{\partial n_i}{\partial x_m}\frac{\partial n_j}{\partial x_j}\right)$$

$$+q_o K_{22}e_{lji}\frac{\partial n_l}{\partial x_j} \tag{7.59}$$

$$-\frac{\delta f}{\delta n_i}=(K_{11}-K_{22})\frac{\partial^2 n_j}{\partial x_j \partial x_i}+K_{22}\frac{\partial^2 n_i}{\partial x_j^2}+(K_{33}-K_{22})n_j n_k\frac{\partial^2 n_i}{\partial x_k \partial x_j}$$
$$+(K_{33}-K_{22})\left(n_j\frac{\partial n_i}{\partial x_k}\frac{\partial n_k}{\partial x_j}+n_k\frac{\partial n_i}{\partial x_k}\frac{\partial n_j}{\partial x_j}-n_j\frac{\partial n_k}{\partial x_i}\frac{\partial n_k}{\partial x_j}\right)-2q_o K_{22}e_{ijk}\frac{\partial n_k}{\partial x_j}+\varepsilon_o\Delta\varepsilon E_i E_j n_j \tag{7.60}$$

In this representation, we must ensure that $\vec{n}$ is a unit vector, namely, $n_i n_i = 1$. This issue can be taken care of by two methods: (1) Lagrange multiplier method, and (2) renormalization method. In the first method, the Lagrange multiplier $\lambda(n_i n_i - 1)$ should be added to the free energy density. In the second method, the values of the director components at step $\tau + 1$ are calculated by

$$n_i^{\tau+1}(l_x,l_y,l_z)=\frac{n_i^{\tau}(l_x,l_y,l_z)+\Delta n_i^{\tau+1}(l_x,l_y,l_z)}{\left\{\left[n_j^{\tau}(l_x,l_y,l_z)+\Delta n_j^{\tau+1}(l_x,l_y,l_z)\right]\cdot\left[n_j^{\tau}(l_x,l_y,l_z)+\Delta n_j^{\tau+1}(l_x,l_y,l_z)\right]\right\}^{1/2}}. \tag{7.61}$$

If the anchoring of the liquid crystal at the cell surface is infinitely strong, then $\vec{n}$ is fixed at the boundary. If the anchoring is weak, the surface energy must be considered in the minimization of the total free energy. Expressed in terms of the liquid crystal director, the surface energy is given by [17]

$$f_s=\frac{1}{2}W_{ij}n_i n_j, \tag{7.62}$$

where W_{ij} is the anchoring tensor, which is symmetric. In the principal frame of the anchoring, the anchoring tensor is diagonalized. As an example, in a cell with homogeneous anchoring along the x axis and the cell normal direction along the z axis, the anchoring matrix is given by

$$\overleftrightarrow{W}=\begin{pmatrix}0 & 0 & 0\\ 0 & W_a & 0\\ 0 & 0 & W_p\end{pmatrix}, \tag{7.63}$$

where W_p and W_a are the polar and azimuthal anchoring strengths, respectively. In this representation, the same problem occurs as in the angle representation: an incorrect free energy may be numerically calculated when the liquid crystal directors at two neighboring lattice sites are anti-parallel. In that case, the numerical calculation will generate a large elastic energy while the actual elastic energy is zero because $\vec{n}$ and $-\vec{n}$ are equivalent.

7.3.3 Tensor representation

In order to avoid the problem of incorrect calculation of the free energy when the liquid crystal directors at two neighboring lattice sites are anti-parallel, the tensor representation was introduced [5,18,19], where the orientation of the liquid crystal director is represented by the tensor defined by

$$\overleftrightarrow{Q} = \vec{n}\vec{n} - \frac{1}{3}\overleftrightarrow{I}, \tag{7.64}$$

where $\overleftrightarrow{I}$ is the identity tensor. Its components are given by

$$Q_{ij} = n_i n_j - \frac{1}{3}\delta_{ij}, \tag{7.65}$$

where δ_{ij} is the Kronecker delta. The elastic energy is calculated from $\overleftrightarrow{Q}$. When the liquid crystal directors at two neighboring lattice sites are anti-parallel, the Q tensor is the same. The actual zero elastic energy is calculated. The elastic energy (Equation (7.44)) has four terms, and therefore four terms of the derivatives of $\overleftrightarrow{Q}$ are needed:

$$G_1 = \frac{\partial Q_{jk}}{\partial x_l}\frac{\partial Q_{jk}}{\partial x_l} = \frac{\partial(n_j n_k)}{\partial x_l}\frac{\partial(n_j n_k)}{\partial x_l} = 2\frac{\partial n_j}{\partial x_l}\frac{\partial n_j}{\partial x_l} \tag{7.66}$$

Note that $\partial(n_j n_j)/\partial x_l = 0$. From Equations (7.45) and (7.50) we have

$$\begin{aligned} \nabla\cdot(\vec{n}\nabla\cdot\vec{n} + \vec{n}\times\nabla\times\vec{n}) &= \frac{\partial}{\partial x_l}\left(n_l\frac{\partial n_k}{\partial x_k} - n_k\frac{\partial n_l}{\partial x_k}\right) \\ &= \frac{\partial n_l}{\partial x_l}\frac{\partial n_k}{\partial x_k} - \frac{\partial n_k}{\partial x_l}\frac{\partial n_l}{\partial x_k} + n_l\frac{\partial^2 n_k}{\partial x_l \partial x_k} - n_k\frac{\partial^2 n_l}{\partial x_l \partial x_k} = \frac{\partial n_l}{\partial x_l}\frac{\partial n_k}{\partial x_k} - \frac{\partial n_k}{\partial x_l}\frac{\partial n_l}{\partial x_k}. \end{aligned} \tag{7.67}$$

Substituting Equations (7.46), (7.48), and (7.67) into Equation (7.66), we have

$$\begin{aligned} G_1 &= 2\left[(\nabla\cdot\vec{n})^2 + (\nabla\times\vec{n})^2 - \nabla\cdot(\vec{n}\nabla\cdot\vec{n} + \vec{n}\times\nabla\times\vec{n})\right] = \\ &= 2\left[(\nabla\cdot\vec{n})^2 + (\vec{n}\cdot\nabla\times\vec{n})^2 + (\vec{n}\times\nabla\times\vec{n})^2 - \nabla\cdot(\vec{n}\nabla\cdot\vec{n} + \vec{n}\times\nabla\times\vec{n})\right]. \end{aligned} \tag{7.68}$$

The last term on the right side of the above equation becomes a surface term when integrated over the volume and can usually be neglected.

$$G_2 = \frac{\partial Q_{jk}}{\partial x_k}\frac{\partial Q_{jl}}{\partial x_l} = \frac{\partial(n_j n_k)}{\partial x_k}\frac{\partial(n_j n_l)}{\partial x_l} = \frac{\partial n_k}{\partial x_k}\frac{\partial n_l}{\partial x_l} + n_k n_l\frac{\partial n_j}{\partial x_k}\frac{\partial n_j}{\partial x_l} = (\nabla\cdot\vec{n})^2 + (\vec{n}\times\nabla\times\vec{n})^2 \tag{7.69}$$

In obtaining this equation, we use Equations (7.46) and (7.51):

$$G_4 = e_{jkl}Q_{jm}\frac{\partial Q_{km}}{\partial x_l} = e_{jkl}\left(n_j n_m - \frac{1}{3}\delta_{jm}\right)\frac{\partial(n_k n_m)}{\partial x_l} = e_{jkl}n_j\frac{\partial n_k}{\partial x_l} = -\vec{n}\cdot\nabla\times\vec{n} \tag{7.70}$$

$$G_6 = Q_{jk}\frac{\partial Q_{lm}}{\partial x_j}\frac{\partial Q_{lm}}{\partial x_k} = \left(n_j n_k - \frac{1}{3}\delta_{jk}\right)\frac{\partial(n_l n_m)}{\partial x_j}\frac{\partial(n_l n_m)}{\partial x_k} = n_j n_k \frac{\partial(n_l n_m)}{\partial x_j}\frac{\partial(n_l n_m)}{\partial x_k} - \frac{1}{3}\frac{\partial(n_l n_m)}{\partial x_j}\frac{\partial(n_l n_m)}{\partial x_j}$$

$$= 2n_j n_k \frac{\partial n_m}{\partial x_j}\frac{\partial n_m}{\partial x_k} - \frac{1}{3}\frac{\partial(n_l n_m)}{\partial x_j}\frac{\partial(n_l n_m)}{\partial x_j} = 2\left(\vec{n}\times\nabla\times\vec{n}\right)^2 - \frac{1}{3}G_1 \tag{7.71}$$

From Equations (7.68), (7.69), (7.70), (7.71,) and (7.72) we get the free energy density:

$$f = \frac{1}{12}(-K_{11}+3K_{22}+K_{33})G_1 + \frac{1}{2}(K_{11}-K_{22})G_2 + \frac{1}{4}(-K_{11}+K_{33})G_6 - q_o K_{22} G_4$$
$$-\frac{1}{2}\varepsilon_o \Delta\varepsilon E_i E_j n_i n_j \tag{7.72}$$

The relaxation method is used in the numerical calculation. The change of the director component n_i on lattice site (l_x, l_y, l_z) of the mesh at step $\tau + 1$ is given by

$$\Delta n_i^{\tau+1}\left(l_x,l_y,l_z\right) = \alpha(\Delta x)^2\left[-\frac{\delta f^{\tau}}{\delta n_i}\left(l_x,l_y,l_z\right)\right]. \tag{7.73}$$

The variation of the free energy with respect to n_i can be expressed in terms of the variation of the free energy with respect to Q_{jk}:

$$\frac{\delta f}{\delta n_i} = \frac{\delta f}{\delta Q_{jk}}\frac{\partial Q_{jk}}{\partial n_i} = \frac{\delta f}{\delta Q_{jk}}\frac{\partial\left(n_j n_k\right)}{\partial n_i} = \frac{\delta f}{\delta Q_{jk}}\left(n_j\delta_{ik} + n_k\delta_{ij}\right) = 2n_j\frac{\delta f}{\delta Q_{ji}} \tag{7.74}$$

$$H_1 = \frac{\delta G_1}{\delta n_i} = 2n_j\frac{\delta G_1}{\delta Q_{ji}} = 2n_j\left[\frac{\partial G_1}{\partial Q_{ji}} - \frac{\partial}{\partial x_l}\left(\frac{\partial G_1}{\partial Q_{ji,l}}\right)\right]$$

$$= -2n_j\frac{\partial}{\partial x_l}\left[\frac{\partial(Q_{uv,w}Q_{uv,w})}{\partial Q_{ji,l}}\right] = -2n_j\frac{\partial\left(2Q_{uv,w}\delta_{ju}\delta_{iv}\delta_{lw}\right)}{\partial x_l} = -4n_j\frac{\partial^2 Q_{ji}}{\partial x_l\partial x_l} \tag{7.75}$$

$$H_2 = \frac{\delta G_2}{\delta n_i} = 2n_j\frac{\delta G_2}{\delta Q_{ji}} = -2n_j\left(\frac{\partial^2 Q_{jl}}{\partial x_i\partial x_l} + \frac{\partial^2 Q_{il}}{\partial x_j\partial x_l}\right) \tag{7.76}$$

$$H_4 = \frac{\delta G_4}{\delta n_i} = 2n_j\frac{\delta G_4}{\delta Q_{ji}} = -2n_j\left(e_{jkl}\frac{\partial Q_{li}}{\partial x_k} + e_{ikl}\frac{\partial Q_{lj}}{\partial x_k}\right) \tag{7.77}$$

$$H_6 = \frac{\delta G_6}{\delta n_i} = 2n_j\frac{\delta G_6}{\delta Q_{ji}} = -2n_j\left(2\frac{\partial Q_{kl}}{\partial x_k}\frac{\partial Q_{ji}}{\partial x_l} + 2Q_{kl}\frac{\partial^2 Q_{ji}}{\partial x_l\partial x_k} - \frac{\partial Q_{kl}}{\partial x_i}\frac{\partial Q_{kl}}{\partial x_j}\right), \tag{7.78}$$

where $Q_{jk,l} = \partial Q_{jk}/\partial x_l$. Expressed in terms of $H_1, H_2, H_4,$ and H_6, at each lattice site the change of the director component n_i is

$$\Delta n_i^{\tau+1} = \alpha(\Delta x)^2 \left[-\frac{1}{12}(-K_{11}+3K_{22}+K_{33})H_1^{\tau} - \frac{1}{2}(K_{11}-K_{22})H_2^{\tau} \right. \\ \left. -\frac{1}{4}(-K_{11}+K_{33})H_6^{\tau} + K_{22}q_o H_4^{\tau} + \Delta\varepsilon\varepsilon_o \left(E_i E_j n_j\right) \right]. \tag{7.79}$$

Because $\vec{n}$ is a unit vector, the director components must be renormalized by using Equation (7.61). In this representation, the problem of incorrect calculation of the free energy when the liquid crystal directors at two neighboring lattice sites are anti-parallel is avoided. A different problem, however, may exist that a real pi-wall defect is artificially removed [20]. Therefore the computer simulated results should be carefully compared with the experimental results in order to prevent mistakes.

As an example, we consider a vertical alignment (VA) mode microdisplay [21,22]. The pixel size is 15 μm and the inter-pixel gap is 0.9 μm. The cell thickness is 2.3 μm. Homeotropic alignment layers are coated on the inner surface of the cell. The parameters of the liquid crystal are: $K_{11} = 16.7 \times 10^{-12}$N, $K_{22} = 7 \times 10^{-12}$N, $K_{33} = 18.1 \times 10^{-12}$N, $\varepsilon_{//} = 3.6$, and $\varepsilon_{\perp} = 7.8$. The voltage applied across the on-pixel is $V_{on} = 5$ V and the voltage applied across the off-pixel is $V_{off} = 0.7$ V. The simulated director configuration is shown in Figure 7.6. In the field-off state, the liquid crystal is aligned homeotropically. When a sufficiently high field is applied across the cell, the liquid crystal director is tilted toward the x direction because of the negative dielectric anisotropy. Because of the fringing effect, the electric field is not exactly in the z direction near the fringes of the pixel. The liquid crystal director is tilted in opposite directions at the two edges of the pixel, and thus a defect wall is formed at the position x_b. If the liquid crystal director is confined in the xz plane, there would be a large splay and bend distortion. The figure shows that the liquid crystal director is escaped into the y direction because the small twist elastic constant.

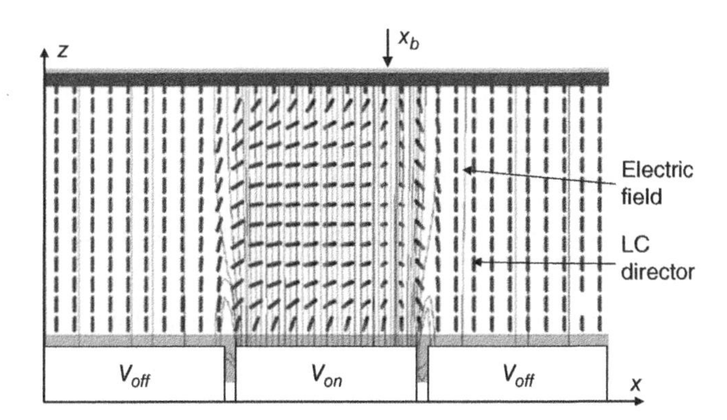

Figure 7.6 Simulated liquid crystal director configuration in the VA mode microdisplay.

Homework Problems

7.1 In the display cell shown in Figure 7.3, calculate and plot the electric field in the cell. Also calculate the equal-potential lines for the following voltages: 2 V, 4 V, 6 V, and 8 V.

7.2 Use the angle representation and numerically calculate the tilt angle in the splay geometry as a function of the coordinate z at the following applied fields: $E = 1.05E_c$, $E = 1.3E_c$, $E = 1.5E_c$, $E = 2.0E_c$, and $E = 5.0E_c$, where E_c is the threshold field of the Fréedericksz transition. The cell thickness is 5 μm. The elastic constants are $K_{11} = 6.4 \times 10^{-12}$N and $K_{33} = 10 \times 10^{-12}$N. Compare your results with Figure 5.4.

7.3 *90° twisted nematic display.* (1) Use the angle representation to numerically calculate the polar and azimuthal angle as a function of the coordinate z at the following applied voltages: 1.0 V, 1.20 V, 1.5 V, 2.0 V, and 5.0 V. The parameters of the cell and the liquid crystal are given in Figure 7.4. (2) Use the Jones matrix method to calculate the transmittance of the display as a function of applied voltage. The back polarizer is parallel to the liquid crystal director at the entrance plane and the front polarizer is parallel to the liquid crystal director at the exit plane. The refractive indices are $n_e = 1.6$ and $n_o = 1.5$.

7.4 Use the tensor representation to numerically calculate the liquid crystal director configuration of a cholesteric liquid crystal in a hybrid cell. The cell thickness and length are $L_z =$ 10 μm and $L_x = 20$ μm, respectively. The director is only a function of x (parallel to the cell surface) and z (perpendicular to the cell surface). On the top surface of the cell the liquid crystal is anchored homeotropically, while at the bottom of the cell the liquid crystal is anchored homogeneously. The pitch P of the liquid crystal is 5 μm. Initially the liquid crystal in the cell is in the isotropic state. The parameters of the liquid crystal are $K_{11} = 6 \times 10^{-12}$N, $K_{22} = 3 \times 10^{-12}$N, and $K_{33} = 10 \times 10^{-12}$N. Using the periodic boundary condition in the x direction, plot the director configuration in the xz plane.

7.5 Use the tensor representation to numerically calculate the liquid crystal director configuration of the cholesteric liquid crystal in the cell discussed in the above problem. A voltage of 5 V is applied across the cell. The dielectric constant of the liquid crystals are $\varepsilon_\perp = 5$ and $\varepsilon_{//} = 15$. Using the periodic boundary condition in the x direction, plot the director configuration in the xz plane.

References

1. J. E. Anderson, P. Watson and P. J. Bos, *LC3D: liquid crystal display 3-D director simulator, software and technology guide* (Reading, MA, Artech House, 1999).
2. DIMOS, Autronic-Melchers GmbH, http://www.autronic-melchers.com/index.html.
3. LCD Master, http://www.shinetech.jp.eng/index_e.html.
4. Techwitz LCD, http://www.sanayisystem.com/introduction.html.
5. LCD DESIGN, V. G. Chigrinov, H. S. Kwok, D. A. Yakpvlev, et al., Invited paper, LCD optimization and modeling, *SID Symp. Dig.* **28.1**, 982 (2004).
6. LCQuest, http://www.eng.ox.ac.uk/lcquest/.
7. Z. Ge, T. X. Wu, R. Lu, et al., Comprehensive three-dimension dynamic modeling of liquid crystal devices using finite element method, *J. Display Technology*, **1**, 194 (2005).
8. A. Taflove, *Computational electrodynamics: the finite-difference time domain method* (Reading, MA, Artech House, 1995).
9. J. Jin, *The finite element method in electromagnetics*, 2nd ed. (Piscataway, New Jersey, Wiley-IEEE Press, 2002).
10. Y. W. Kwon and H. Bang, *The finite element method using MATLAB* (CRC Press, BocaRoan, FlL, 2000).

11. R. Barberi and G. Barbero, Variational calculus and simple applications of continuum theory, Chap. IX, in *Liquid crystal materials*, ed. I. C. Khoo (Gordon and Breach, Amsterdam, 1991).
12. D. J. Griffiths, *Introduction to electrodynamics*, 2nd edn Prentice-Hall, Inc., New Jersey, 1989).
13. D. Jackson, *Classic electrodynamics*, 2nd edn (John Wiley & Sons, New York, 1975).
14. J. Mathews and R. L. Walker, *Mathematical methods of physics*, 2nd edn (W. A. Benjamin, Inc., Menlo Park, California, 1970).
15. J. R. Reitz, F. J. Milford, and R. W. Christy, *Foundations of electromagnetic theory* (Addison-Wesley Publishing Co., USA, 1993).
16. H. Mori, E. C. Gartland, Jr., J. R. Kelly, and P. J. Bos, Multidimensional director modeling using the Q tensor representation, *Jpn. J. Appl. Phys.*, **38**, 135 (1999)
17. S. V. Shiyanovskii, I. I. Smalyukh, and O. D. Lavrentovich, Computer simulations and fluorescence confocal polarizing microscopy of structures in cholesteric liquid crystals, p. 229, in *Defects in liquid crystals: computer simulations, theory and experiments* (Kluwer Academic Publishers, Netherland, 2001).
18. D. W. Berreman and S. Meiboom, Tensor representation of Oseen–Frank strain energy in uni-axial cholesterics, *Phys. Rev. A*, **30**, 1955 (1984).
19. S. Dickmann, J. Eschler, O. Cossalter, and D. A. Mlynski, Simulation of LCDs including elastic anisotropy and inhomogeneous fields, *SID Symposium Digest Tech. Paper*, **24**, 638 (1993).
20. J. E. Anderson, P. Watson, and P. J. Bos, Comparisons of the vector and tensor method for simulating liquid crystal devices, *Liq. Cryst*, **28**, 109 (2000).
21. K.-K F. Chiang, X. Zhu, S. T. Wu, and S. H. Chen, Eliminating fringing field effects of vertical aligned liquid-crystal-on-silicon by using circularly polarized light, *SID Symposium Digest Tech. Paper,* **36**, 1290 (2005).
22. K.-H. Fan-Chiang, S.-T. Wu, and S.-H. Chen, Fringing-field effects on high-resolution liquid crystal microdisplays, *J. Display Technology*, **1**, 309 (2005).

8

Transmissive Liquid Crystal Displays

8.1 Introduction

Three types of liquid crystal displays have been developed: (1) transmissive, (2) reflective, and (3) transflective. A transmissive LCD uses a backlight for illuminating the LCD panel, which results in high brightness and high contrast ratio. Some transmissive LCDs do not use phase-compensation films or a multi-domain approach so that their viewing angle is limited and they are more suitable for single viewer applications, such as notebook computers and games. With proper phase compensation, the direct-view transmissive LCDs exhibit a wide viewing angle and have been used extensively for multiple viewers, such as desktop computers and televisions. Transmissive LCDs can also be used for projection displays, such as data projectors. There, a high power arc lamp or a light-emitting diode (LED) array is used as a light source. To reduce the size of optics and save on the cost of the projection system, the LCD panel is usually made small (less than 25 mm in diagonal). Thus, poly-silicon thin-film transistors (TFTs) are commonly used.

Similarly, the reflective LCDs can be subdivided into direct-view and projection displays. In principle, a direct-view reflective LCD does not require a backlight so that it is light in weight and its power consumption is low. A major drawback is poor readability under weak ambient light. Thus, a reflective LCD is more suitable for projection TVs employing liquid-crystal-on-silicon (LCoS) microdisplay panels. In an LCoS, the reflector employed is an aluminum metallic mirror. The viewing angle is less critical in projection displays than in direct-view displays.

For outdoor applications, the displayed images of a transmissive LCD could be washed out by sunlight. A reflective LCD would be a better choice. However, such a reflective display is unreadable in dark ambient conditions. Transflective LCDs integrate the features

Fundamentals of Liquid Crystal Devices, Second Edition. Deng-Ke Yang and Shin-Tson Wu.

of a transmissive display and a reflective display. Thus, in dark ambient conditions the backlight is turned on and the display works in the transmissive mode. While at bright ambient conditions, the backlight is switched off and only the reflective mode is operational.

Two monographs have been dedicated to projection displays [1] and reflective displays [2]. Therefore, in this chapter we will focus on the mainstream TFT-addressed wide-view transmissive LCDs. We will start from introducing the twisted nematic (TN) mode, and then delve into in-plane switching (IPS) and multi-domain vertical alignment (MVA). Phase compensation methods for achieving wide viewing angle will be addressed.

8.2 Twisted Nematic (TN) Cells

The 90° twisted-nematic (TN) cell [3] has been used extensively for notebook computers where viewing angle is not too critical. Figure 8.1 shows the LC director configurations of the normally white TN cell in the voltage-off (left) and voltage-on (right) states.

In the voltage-off state, the top LC alignment is parallel to the optical axis of the top polarizer while the bottom LC directors are rotated 90° and parallel to the optical axis of the bottom analyzer. When the $d\Delta n$ of the LC layer satisfies the Gooch–Tarry's first minimum condition, [4] the incoming linearly polarized light will follow closely the molecular twist, and transmits the crossed analyzer. In the voltage-on state, the LC directors are reoriented to be perpendicular to the substrates, except the boundary layers. The incoming light experiences little phase change and is absorbed by the analyzer, resulting in a dark state. The beauty of the TN cell is that the boundary layers are orthogonal, so that their residual phases compensate for each other. As a result, the dark state occurs at a relatively low voltage.

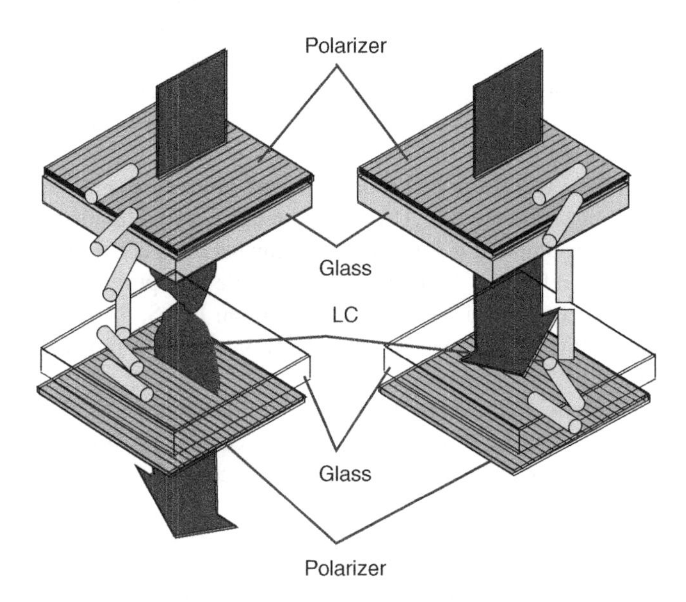

Figure 8.1 LC and polarizer configurations of a 90° TN cell. Left: $V=0$, and right $V \gg V_{th}$.

8.2.1 Voltage-dependent transmittance

To compare different operating modes, let us focus on the normalized transmittance by ignoring the optical losses from polarizers and indium–tin–oxide (ITO) layers, and the interface reflections from substrates. The normalized transmittance ($T_{\perp}$) of a TN cell can be described by the following Jones matrices as $T_{\perp} = |M|^2$ [5]:

$$M = \left| \cos\beta \quad \sin\beta \right| \begin{vmatrix} \cos\varphi & -\sin\varphi \\ \sin\varphi & \cos\varphi \end{vmatrix} \begin{vmatrix} \cos X - i\frac{\Gamma}{2}\frac{\sin X}{X} & \varphi\frac{\sin X}{X} \\ -\varphi\frac{\sin X}{X} & \cos X + i\frac{\Gamma}{2}\frac{\sin X}{X} \end{vmatrix} \begin{vmatrix} -\sin\beta \\ \cos\beta \end{vmatrix}. \tag{8.1}$$

Here β is the angle between the polarization axis and the front LC director, ϕ is the twist angle, $X = \sqrt{\phi^2 + (\Gamma/2)^2}$ and $\Gamma = 2\pi\, d\, \Delta n/\lambda$, where d is the cell gap. By simple algebraic calculations, the following analytical expression for $|M|$ [2] is derived:

$$|M|^2 = T_{\perp} = \left(\frac{\phi}{X}\cos\phi\sin X - \sin\phi\cos X\right)^2 + \left(\frac{\Gamma}{2}\frac{\sin X}{X}\right)^2 \sin^2(\phi - 2\beta) \tag{8.2}$$

Equation (8.2) is a general formula describing the light transmittance of a TN cell (without voltage) as a function of twist angle, beta angle, and $d\Delta n/\lambda$. For a 90° TN cell, $\phi = \pi/2$ and Equation (8.2) is simplified to

$$T_{\perp} = \cos^2 X + \left(\frac{\Gamma}{2X}\cos 2\beta\right)^2 \sin^2 X. \tag{8.3}$$

Equation (8.3) has a special solution, that is $\cos^2 X = 1$. When $\cos X = \pm 1$ (i.e. $X = m\pi$; $m =$ integer), then $\sin X = 0$ and the second term in Equation (8.3) vanishes. Therefore, $T_{\perp} = 1$, independent of β. By setting $X = m\pi$ and knowing that $\Gamma = 2\pi d\Delta n/\lambda$, the Gooch–Tarry's condition is found as

$$\frac{d\Delta n}{\lambda} = \sqrt{m^2 - \frac{1}{4}}. \tag{8.4}$$

For the lowest order $m = 1$, $d\Delta n/\lambda = \sqrt{3}/2$. This is the Gooch–Tarry's first minimum condition for the 90° TN cell. For the second order $m = 2$ and $d\Delta n = \sqrt{15}/2$. The second minimum condition is used only for low-end displays such as wrist watches, because a large cell gap is easier to fabricate, and the cyano-biphenyl LCs are less expensive. For notebook TFT-LCDs, the first minimum is preferred because fast response time is required.

Figure 8.2 depicts the normalized light transmittance ($T_{\perp}$) of the 90° TN cell at three primary wavelengths R = 650, G = 550, and B = 450 nm. Since the human eye is most sensitive at green, we normally optimize the cell design at $\lambda = 550$ nm. From Equation (8.4), the first $T_{\perp} = 1$ occurs at $d\Delta n \sim 480$ nm. The color dispersion (i.e. the wavelength dependency of the light transmittance) at $\beta = 0$ is not too sensitive to $d\Delta n/\lambda$ beyond the first minimum. Therefore, the TN cell can be treated as an achromatic half-wave plate. The response time depends on the cell gap and

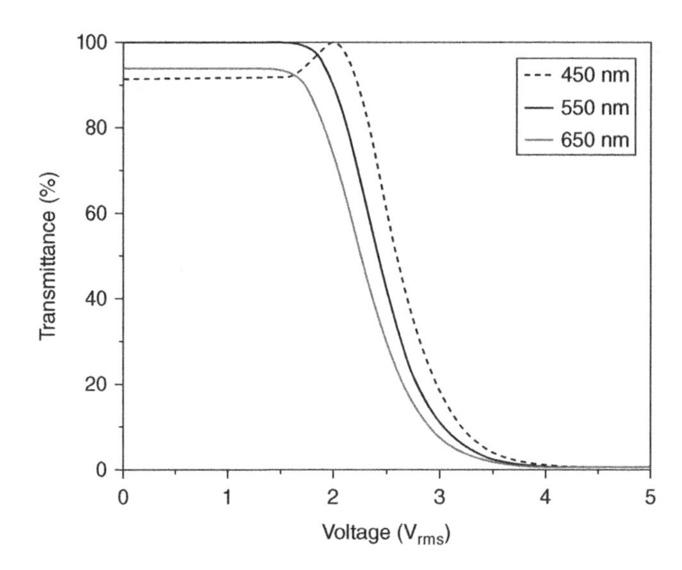

Figure 8.2 Voltage-dependent transmittance of a normally white 90° TN cell. $d\Delta n = 480$ nm.

the γ_1/K_{22} of the LC mixture employed. For a 4 μm cell gap, the optical response time is about 20–30 ms. At $V = 5\ V_{rms}$, the contrast ratio (CR) reaches about 400:1. These performances, although not perfect, are acceptable for notebook computer applications. A major drawback of the TN cell is its narrow viewing angle and gray scale inversion originated from the LC director's tilt. Because of this tilt, the viewing angle in the vertical direction is narrow and asymmetric [6].

8.2.2 Film-compensated TN cells

Figure 8.3 (left) shows the simulated iso-contrast contour of a TN LCD [7]. In the normal viewing direction, the TN cell exhibits a good contrast ratio, but the contrast rapidly decreases in the upper direction and in the lower diagonal directions. In the lower direction, the contrast remains high, but gray-scale inversion is observed (not shown in Figure 8.3). The narrow viewing angle of a TN LCD is caused by several factors, such as the optical anisotropy of liquid crystals, the off-axis light leakage from crossed polarizers, the light scattering on the surface of the polarizer or at the color filters, the collimation of the backlight, or the light diffraction from the cell structure.

In the on-state of a TN cell, the LC directors in the upper half are reoriented along the rubbing direction with almost no twist and the lower half has a similar structure with the director plane orthogonal to that of the upper half. Thus, a uniform phase compensation film, such as a uniaxial *a* plate, cannot compensate the upper and lower parts simultaneously. Instead, a pair of wide-view films need to be used separately on both sides of the TN LC cell in order to compensate each of the half layers. Fuji Photo has skillfully developed discotic LC films for widening the viewing angle of TN cells. The molecular structures of the wide-view (WV) discotic material are shown below in Figure 8.4.

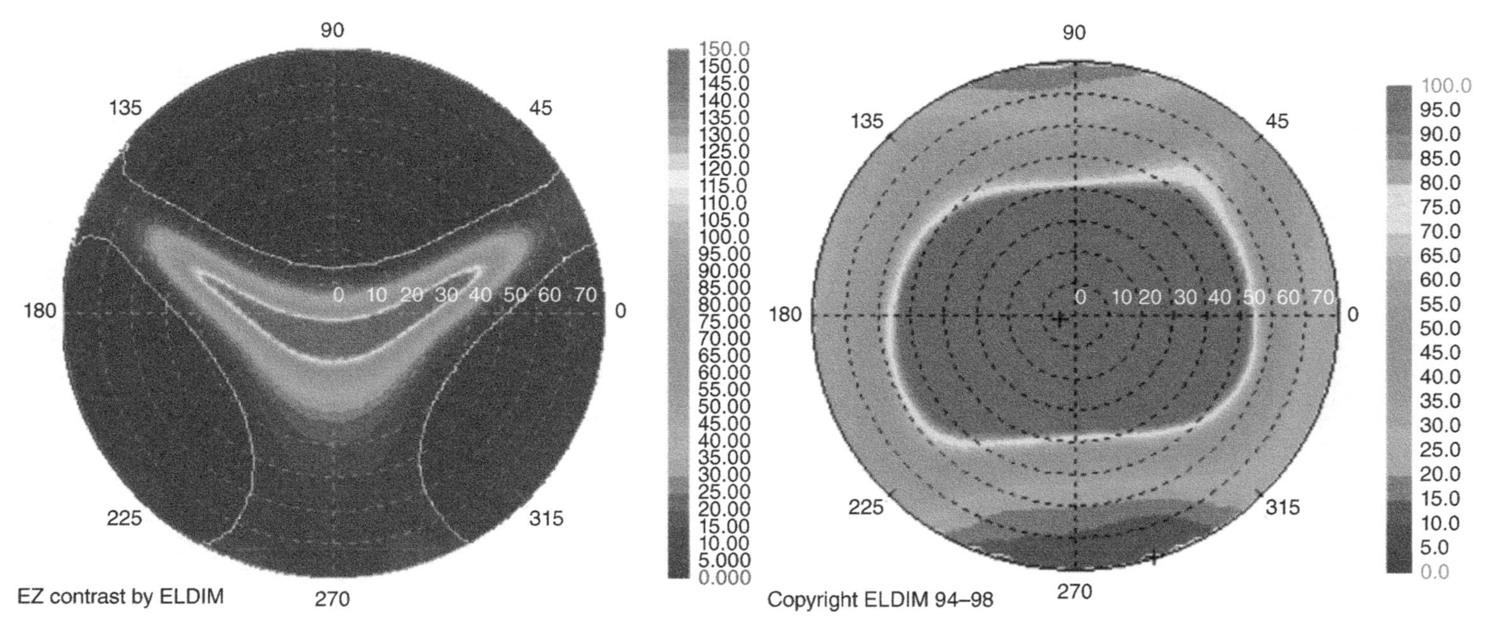

Figure 8.3 Measured iso-contrast plots for TN-LCDs without (left) and with (right) wide-view films. (Courtesy of Dr H. Mori, Fuji Photo Film).

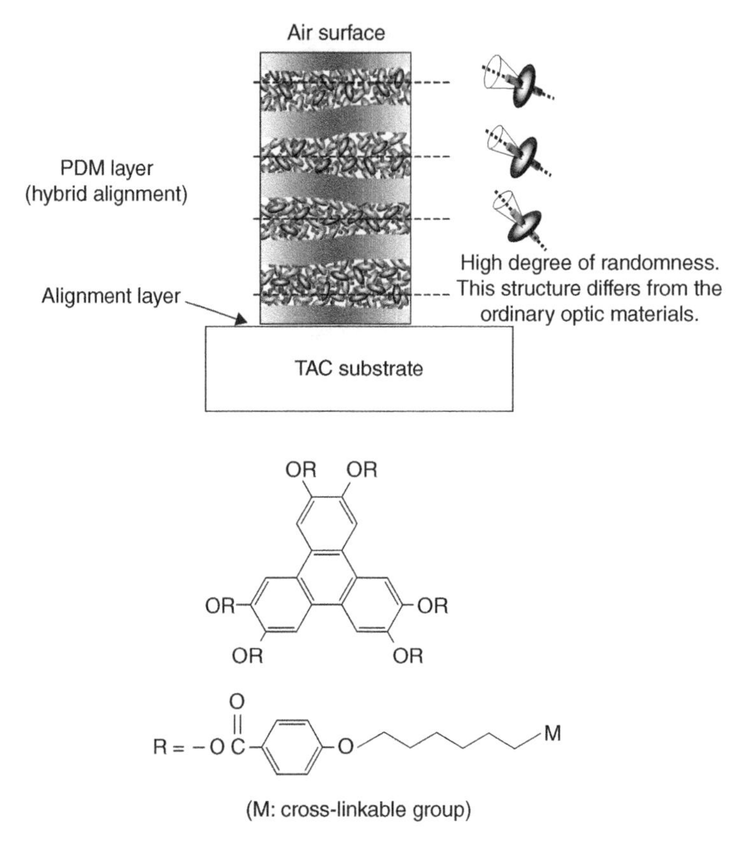

Figure 8.4 Structure of the WV film and the employed discotic compound (PDM = polymerized discotic material). Mori 2005. Reproduced with permission from IEEE.

Figure 8.4 shows the structure of the Fuji WV film. A discotic material (triphenylene derivatives) is coated on an alignment layer on a TAC (tri-acetyl cellulose) substrate. The discotic material has a hybrid alignment structure and three important features. (1) It has π electron spread in a disc-like shape, which gives to a high birefringence. (2) It takes on discotic nematic (N_D) phase at lower temperature than the temperature at which the TAC substrate starts to deform. This feature enables a uniform and monodomain film in a wide range of area without defects. (3) It has cross-linkable groups at all of six side chains to make the obtained film durable.

When heated, the discotic material takes on the N_D phase. The discotic material right next to the alignment layer has a high degree of randomness. And in the vicinity of the alignment layer, the discotic molecules tend to align with the molecular plane almost parallel to the alignment layer surface and have few degrees of pretilt angle in the rubbing direction of the alignment layer surface. On the other hand, in the vicinity of the air surface the discotic molecules tend to align with the molecular plane almost perpendicular to the air surface. With the pinned alignment on both sides, the discotic material exhibits a hybrid alignment structure in the N_D phase. When cured by UV light, the discotic material is polymerized and the hybrid alignment structure of the polymerized discotic material (PDM) layer is fixed even after it is cooled down to room temperature. Each film has a hybrid alignment structure in which the director continuously changes in the PDM layer thickness direction without twist while the direction of each discotic molecule fluctuates. This hybrid alignment structure consists of splay and bend deformations.

The azimuthal alignment direction of the PDM layer is parallel to the longitudinal direction of the film so that the WV film could be laminated on the polarizing film in a roll-to-roll process. Therefore, the WV film is used with the *o* mode in which the transmission axis of the polarizer is perpendicular to the adjacent rubbing direction of the TN cell. By contrast, the device configuration shown in Figure 8.1 is called *e* mode. The PDM layer exhibits a non-zero and asymmetric phase retardation value at all incident angles. This indicates that the PDM layer has an inhomogeneous alignment structure in the thickness direction. The TAC substrate also possesses a small birefringent, which plays an important role in optical compensation. The ideal TAC substrates should be isotropic.

8.2.3 Viewing angle

Figure 8.3 compares the viewing angle of a TN-LCD without (left side) and with (right side) a wide-view Fuji discotic compensation film. The viewing angle at a CR of 10:1 of the film-compensated TN LCD exceeds 80° in all azimuthal directions. The viewing angle is especially enlarged in the horizontal direction where both dark-state light leakage and yellowish color shift are reduced. The on-axis CR is also improved by 10%. The WV Fuji film remarkably improves the viewing angle of the TN LCDs without losing any light transmittance or deteriorating the image quality. No change in the panel process is required because the conventional polarizer is simply replaced with a new polarizer laminated with the compensation film. The discotic film is also a cost-effective approach for obtaining a wide viewing angle compared to in-plane switching (IPS) and multi-domain vertical alignment (MVA) modes. These features enable TN to penetrate into the larger-sized LCD market segment, say 20–25 inch diagonal. However, the reversed gray scale still exists in film-compensated TN LCDs, which ultimately limits their competitiveness with IPS and MVA-LCDs for large screen TVs.

8.3 In-Plane Switching Mode

In the TN cell shown in Figure 8.1, the applied electric field is in the longitudinal direction. The tilted LC directors in the bulk cause different phase retardation as viewed from the right or left direction. This causes narrow and asymmetric viewing angle in the vertical directions. To overcome the narrow viewing angle issue, an elegant driving scheme using transverse electric field was proposed in 1970s [8,9] and implemented in TFT-LCDs in 1990s [10,11]. The inter-digital electrodes are arranged in the same substrate, such that the generated fringing field is in the transverse plane. The LC directors are rotated in the plane. Thus, this driving scheme is often referred to as transverse field effect or in-plane switching (IPS).

In an IPS mode, the interdigitated electrodes are fabricated on the same substrate, and liquid crystal molecules are initially homogeneously aligned with a rubbing angle of ~10° with respect to the striped electrodes. The transmission axis of the polarizer can be set to be parallel (*e* mode) or perpendicular (*o* mode) to the LC directors, while the analyzer is crossed to the polarizer. The in-plane electric fields induced by the electrodes twist the LC directors, and so generate light transmission. However, due to the strong vertical electric field existing above the electrode surface, the LC directors in these regions mainly tilt rather than twist. As a result, the transmittance above the electrodes is greatly reduced. Overall, the conventional IPS mode has a light efficiency of about 76% of that of a twisted nematic (TN) LCD mode, when a positive dielectric anisotropy ($\Delta\varepsilon$) liquid crystal material is used. Although using a negative $\Delta\varepsilon$ liquid crystal in the IPS mode could enhance the light efficiency to above 85%, the required on-state driving voltage is increased. For TFT-LCDs, the preferred operating voltage is lower than 5.5 V_{rms}.

8.3.1 Voltage-dependent transmittance

Figure 8.5 depicts the basic device structure of the IPS mode, using a positive $\Delta\varepsilon$ LC [12,13]. The front polarizer is parallel to the LC directors, and the rear analyzer is crossed. In the voltage-off state, the incident light experiences no phase retardation so that the outgoing beam remains linearly polarized and is absorbed by the crossed analyzer. In a voltage-on state, the fringing field reorients the LC directors and causes phase retardation to the incoming light and modulates the transmittance through the analyzer.

As shown in Figure 8.5, above the electrodes (region I) the electric field is unable to twist the LC directors. As a result, the light transmittance is lower than in region II. The average transmittance is about 75% of the TN cell.

Based on the same operation principle, fringing field switching (FFS) [14] also utilizes the transverse electric field to switch the LC directors. The basic structure of FFS is similar to IPS except for the much smaller electrode gap ($\ell \sim 0$– 1 μm). In the IPS mode, the gap (ℓ) between

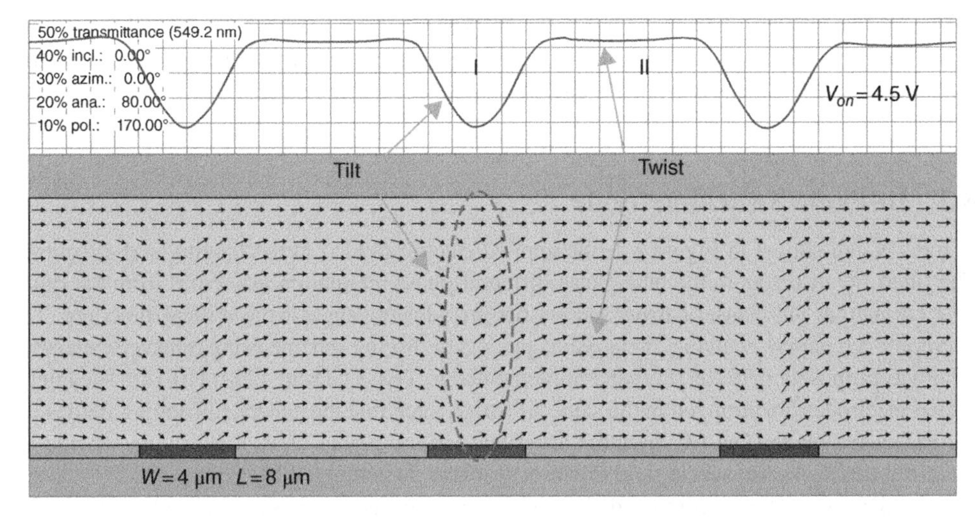

Figure 8.5 Device structure, simulated on-state LC director distribution, and corresponding light transmittance of an IPS cell. Electrode width $W = 4$ μm and electrode gap $L = 8$ μm.

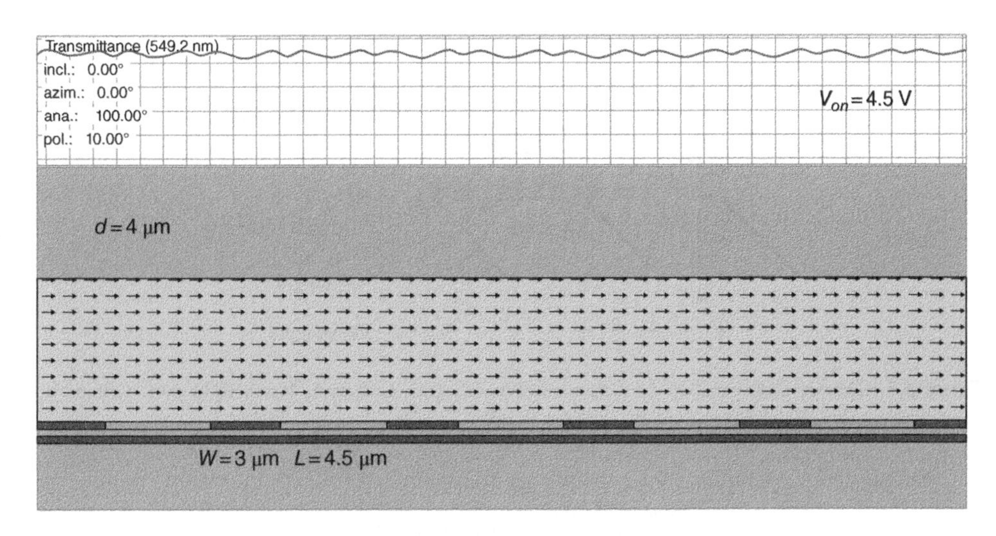

Figure 8.6 Device structure, simulated on-state LC director distribution, and corresponding light transmittance of a FFS cell. Electrode width $W = 3$ μm and electrode gap L = 4.5 μm.

electrodes is larger than the cell gap (d). The horizontal component of the electric field is dominant between the electrodes. However, in the FFS mode where $\ell < d$, the fringing field exists above the electrodes. The fringing fields are able to twist the LC directors above the electrodes. Therefore, high light transmittance is obtained. In an FFS mode, both positive and negative $\Delta\varepsilon$ liquid crystals can be used [15]. The FFS mode using a negative $\Delta\varepsilon$ material can achieve 98% transmittance of that of a TN cell. The idea of using a positive $\Delta\varepsilon$ LC material in the FFS mode for achieving high transmittance (~90% of TN mode) has also been attempted [16]. Positive LCs usually exhibit a larger $\Delta\varepsilon$ and lower viscosity than their corresponding negative $\Delta\varepsilon$ LCs because their polar group(s) are along the principal molecular axis. However, the FFS mode employing a positive $\Delta\varepsilon$ LC would require a high resolution photolithography to fabricate 1 μm electrode width and increase the on-state voltage to ~6.5 V_{rms} in order to generate sufficient twist to the LC directors.

In a FFS mode, the negative $\Delta\varepsilon$ LC tends to have a higher on-state transmittance than the positive LC because the directors of the positive $\Delta\varepsilon$ LC tend to align along the field so that it does not contribute to the phase retardation. Figure 8.6 shows an FFS structure with homogeneous alignment and positive $\Delta\varepsilon$ LC mixture. The fringing field covers both electrodes and gaps. Unlike in the IPS mode, there is no dead zone prohibiting light transmittance. Thus, the light transmittance is improved. Both IPS and FFS modes are normally black under the crossed-polarizer condition. The transmittance of the FFS mode reaches ~95% of the TN cell. The viewing characteristic of FFS is very similar to that of IPS; both are much wider than that of TN [17].

8.3.2 Response time

Figure 8.7 shows the electrode configuration of the IPS mode under study. The electrode gap is ℓ (~10 μm) and width ω (~5 μm). When backflow and inertial effects are ignored, the dynamics of liquid crystal director rotation is described by the following Erickson–Leslie equation [4,9]:

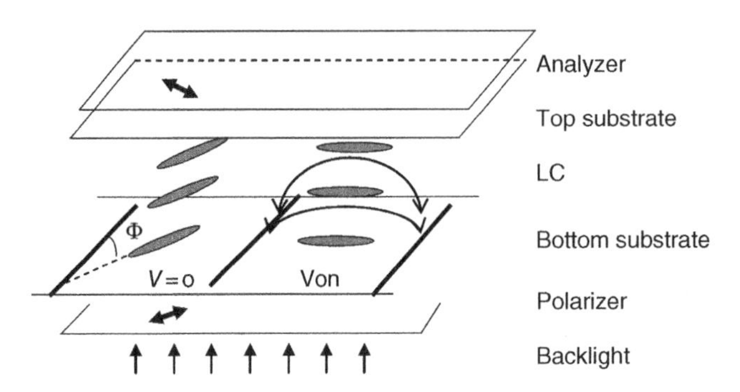

Figure 8.7 Device configuration of a transmissive IPS cell. Left part: $V=0$, right part: voltage-on. Φ = rubbing angle. Polarizer and analyzer are crossed.

$$\gamma_1 \frac{\partial \phi}{\partial t} = K_{22} \frac{\partial^2 \phi}{\partial z^2} + \varepsilon_o |\Delta\varepsilon| E^2 \sin\phi \cos\phi \tag{8.5}$$

In Equation (8.5), γ_1 is the rotational viscosity, K_{22} is the twist elastic constant, $\Delta\varepsilon$ is the dielectric anisotropy, E is the electric field strength, and ϕ is the LC rotation angle. The homogeneous LC layers having cell gap d are along the z axis.

For simplicity, let us assume that the surface anchoring strength is strong so that the bottom and top boundary layers are fixed at $\phi(0) = \phi(d) = \Phi$, where Φ is the LC alignment (or rubbing) angle with respect to the electrodes, as shown in Figure 8.7.

To solve the decay time, we set $E = 0$ in Equation (8.5). After some algebra, the decay time of the LC directors can be solved relatively easily. The decay time is independent of the initial rubbing angle Φ:

$$\tau_{off} = \gamma_1 d^2 / \pi^2 K_{22} \tag{8.6}$$

From Equation (8.6), the LC director's relaxation time is governed by the cell gap (d) and the LC visco-elastic coefficient (γ_1/K_{22}), and is independent of the rubbing angle. In a VA cell, the *optical* response time is about 50% of the LC *director's* response time.

From Equation (8.5), the rise (or turn-on) time is more difficult to solve because it depends on the applied voltage and the initial rubbing angle. When the rubbing angle $\Phi = 0$, the LC directors are perpendicular to the electric field and the Fréedericksz transition exists. Under these circumstances, the turn-on time can be solved [18]:

$$\tau_{on} = \frac{\gamma_1}{\varepsilon_o |\Delta\varepsilon| E^2 \frac{\sin(2\bar{x})}{2\bar{x}} - \frac{\pi^2}{d^2} K_{22}} \tag{8.7}$$

In Equation (8.7), $\bar{x} = \int_0^{\frac{d}{2}} x dz$ and $x = (\phi_m - \Phi) \sin\left(\frac{\pi z}{d}\right) \exp((t-\tau)/\tau)$, where $(\phi_m - \Phi)$ represents the twisted angle of the middle LC layer under the exerted electric filed. In principle, Equation (8.7) is not limited to the small signal regime. In the usual small angle approximation (i.e. the electric field is only slightly above threshold), $\bar{x} = 1$ and Equation (8.7) is reduced to the following commonly known equation:

$$\tau_{on} = \frac{\gamma_1}{\varepsilon_o |\Delta\varepsilon| E^2 - \frac{\pi^2}{d^2} K_{22}}. \tag{8.8}$$

When the rubbing angle $\Phi \neq 0$, τ_{on} has following complicated form:

$$\tau_{on} = \frac{\gamma_1}{\varepsilon_o |\Delta\varepsilon| E^2 \left(\cos(2\Phi) \frac{\sin(2\bar{x})}{2\bar{x}} + \sin(2\Phi) \frac{\cos(2\bar{x})}{2\bar{x}} \right) - \frac{\pi^2}{d^2} K_{22}} \tag{8.9}$$

At a given electric field, $\bar{x}$ can be obtained from ϕ_m which, in turn, is calculated from the following elliptical equation

$$\frac{Ed}{2} \sqrt{\frac{\varepsilon_0 |\Delta\varepsilon|}{K_2}} \sin\phi_m = \int_{\Phi}^{\phi_m} \frac{1}{\sqrt{1 - (\sin\phi / \sin\phi_m)^2}} d\phi. \tag{8.10}$$

Strictly speaking, when the rubbing angle Φ is not equal to zero the Fréedericksz transition threshold is smeared. However, in a normally black IPS mode, the transmittance is proportional to the phase retardation $\delta = 2\pi d \Delta n / \lambda$ of the LC cell as $T \sim \sin^2(\delta/2)$. In the small voltage regime, the phase retardation is small and transmittance exhibits a threshold-like transition.

This optical threshold voltage (V_{op}) can be derived by assuming that the rise time is approaching infinity at $V = V_{op}$. Thus, the denominator in Equation (8.9) should vanish:

$$\varepsilon_o |\Delta\varepsilon| E^2 \left(\cos(2\Phi) \frac{\sin(2\bar{x})}{2\bar{x}} + \sin(2\Phi) \frac{\cos(2\bar{x})}{2\bar{x}} \right) - \frac{\pi^2}{d^2} K_{22} \to 0 \tag{8.11}$$

From Equation (8.11), the optical threshold voltage is derived as

$$V_{op} = E \cdot \ell = \frac{\pi \ell}{d} \sqrt{\frac{K_{22}}{\varepsilon_o |\Delta\varepsilon| \left(\cos(2\Phi) \frac{\sin(2\bar{x})}{2\bar{x}} + \sin(2\Phi) \frac{\cos(2\bar{x})}{2\bar{x}} \right)}}. \tag{8.12}$$

From Equation (8.8), when $\Phi = 0$ and $\bar{x} \to 0$, the optical threshold is reduced to the Fréedericksz threshold. As the rubbing angle is increased, the optical threshold voltage is gradually decreased. At $\Phi = 45°$, the optical threshold voltage reaches a minimum; however, the on-state voltage is also increased.

To compromise for the response time and operating voltage, a typical rubbing angle is set at $\Phi \sim 10°$. As the rubbing angle is increased 30°, the rise time is reduced by two to three times, but the on-state voltage is slightly increased. An optimal rubbing angle is found to be around 20–30° [19].

8.3.3 Viewing angle

A common feature of IPS and FFS modes is that the LC cell is sandwiched between two crossed linear polarizers. At normal incidence, the LC layer in the voltage-off state does not modulate the polarization state of the incident linearly polarized light from the entrance polarizer. As a result, a good dark state is achieved since this linearly polarized light is completely absorbed by the crossed analyzer. However, at oblique angles the incident light leaks through the crossed polarizers, especially at the bisectors. This light leakage stems from two factors. (1) The absorption axes of the crossed polarizers are no longer orthogonal to each other under off-axis oblique view. As a result, the extinction ratio of these two crossed polarizers decreases and light leakage occurs. (2) In some cases, due to the effective birefringence effect of the LC layer the obliquely incident linearly polarized light is modulated and it becomes elliptically polarized after traversing through the LC layer. Consequently, the analyzer cannot completely absorb the elliptically polarized light leading to light leakage at off-axis. This light leakage in the dark state deteriorates the contrast ratio and thereby degrades the viewing angle performance.

To suppress the light leakage at oblique angles and further widen the viewing angle, several phase compensation schemes using uniaxial films [20–22] and biaxial films [23–25] have been proposed. Computer simulation and experimental results have been reported.

In this section, we focus on the analytical solutions for the uniaxial film-compensated wide view LCDs. With the analytical solutions, the interdependency between the LC cell and film parameters is clearly revealed. More importantly, analytical solutions provide a clear physical description of the compensation mechanisms.

8.3.4 Classification of compensation films

Table 8.1 lists some commercially available compensation films, classified by their refractive indices. Different LC modes need different types of compensation films in order to obtain satisfactory compensation effect. For example, the IPS mode may require a biaxial compensation film with $n_x > n_z > n_y$ [26], while the VA mode needs a compensation film with $n_x > n_y > n_z$ [27]. Theoretical analyses of biaxial film-compensated LCDs are rather difficult. Here, we focus on the uniaxial film-compensated wide view LCDs.

Uniaxial film is an anisotropic birefringent film with only one optical axis. For simplicity, let us limit our discussions to non-absorption uniaxial films. From the viewpoint of optical axis orientation, uniaxial films can be classified into *a* film and *c* film. An *a* film's optical axis is parallel to the film surface, while a *c* film's optical axis is perpendicular to the film surface.

Both *a* film and *c* film can be further divided into positive or negative films depending on the relative values of the extraordinary refractive index n_e and the ordinary refractive index n_o. Table 8.1 lists all the types of compensation films and their refractive index relationship. In our analyses, we focus on the uniaxial films. As a general rule, a positive uniaxial film means $n_e > n_o$, otherwise, $n_e < n_o$ for a negative uniaxial film.

8.3.5 Phase retardation of uniaxial media at oblique angles

Both uniaxial compensation film and nematic LC layer can be treated as uniaxial media. When a light propagates into a uniaxial film, generally two forward eigenwaves (one ordinary wave

Table 8.1 Different types of compensation films used for wide view LCDs.

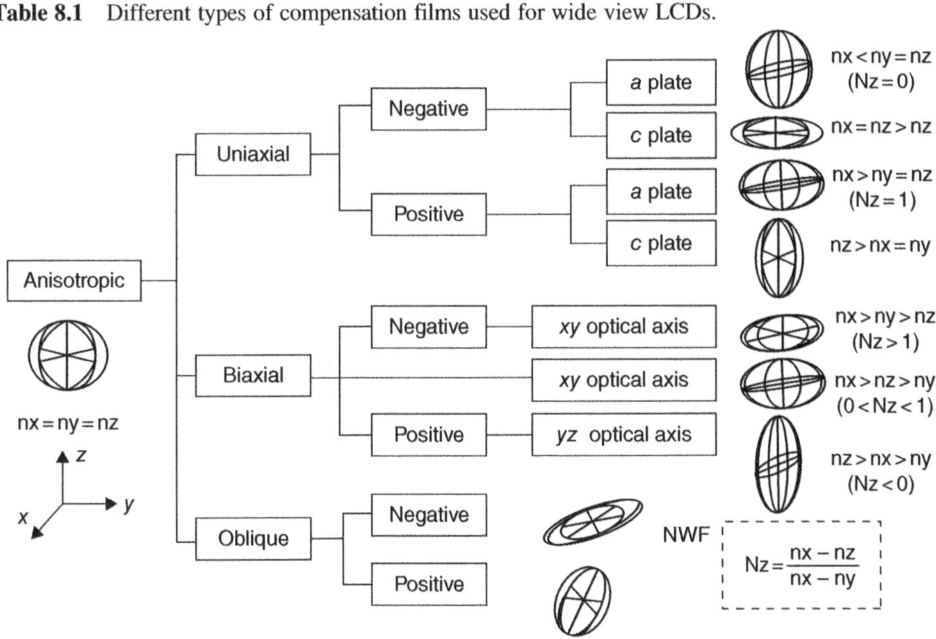

and one extraordinary wave) are evoked within the medium. After the light has passed through the uniaxial medium, phase retardation occurs between these two eigenwaves. Figure 8.8 shows an arbitrary oblique incident light with an incident angle θ_0 propagates in a uniaxial medium. Here, the *xy* plane is chosen to be parallel to the medium layer surface and *z* axis is along the surface normal. In such a coordinate system, the incident plane forms an angle ϕ_0 with respect to the *x* axis. The optical axis of the uniaxial medium is oriented at tilt angle θ_n and azimuthal angle ϕ_n, and the extraordinary and ordinary refractive indices of the uniaxial medium are n_e and n_o, respectively.

In general, the phase retardation of a uniaxial medium at oblique incidence can be expressed as [28]

$$\Gamma = (k_{e,z} - k_{o,z})d, \tag{8.13}$$

where *d* is the layer thickness of the uniaxial medium, $k_{e,z}$ and $k_{o,z}$ are the *z* axis components of wavevectors of extraordinary and ordinary waves, respectively. From Maxwell's equations, these two *z* axis components of wavevectors $k_{e,z}$ and $k_{o,z}$ can be solved and given by [29]

$$k_{e,z} = \frac{2\pi}{\lambda}\left[\frac{n_e n_o}{\varepsilon_{zz}}\sqrt{\varepsilon_{zz} - \left(1 - \frac{n_e^2 - n_o^2}{n_e^2}\cos^2\theta_n \sin^2(\phi_n - \phi_0)\right)\sin^2\theta_0} - \frac{\varepsilon_{xz}}{\varepsilon_{zz}}\sin\theta_0\right] \tag{8.14}$$

and

$$k_{o,z} = \frac{2\pi}{\lambda}\sqrt{n_o^2 - \sin^2\theta_0}, \tag{8.15}$$

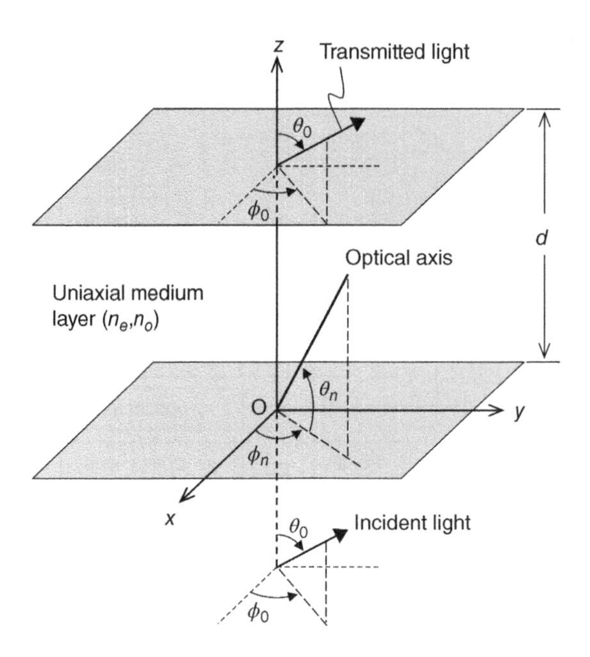

Figure 8.8 Schematic view of an arbitrary light impinging on a uniaxial medium. Zhu 2006. Reproduced with permission from IEEE.

with $\varepsilon_{xz} = (n_e^2 - n_o^2)\sin\theta_n \cos\theta_n \cos(\phi_n - \phi_0)$ and $\varepsilon_{zz} = n_o^2 + (n_e^2 - n_o^2)\sin^2\theta_n$. From Equations (8.13)–(8.15), one can easily obtain the phase retardation Γ of a general uniaxial medium at an arbitrary incident angle by using

$$\Gamma = \frac{2\pi}{\lambda} d \left[\frac{n_e n_o}{\varepsilon_{zz}} \sqrt{\varepsilon_{zz} - \left(1 - \frac{n_e^2 - n_o^2}{n_e^2} \cos^2\theta_n \sin^2(\phi_n - \phi_0)\right) \sin^2\theta_0} - \frac{\varepsilon_{xz}}{\varepsilon_{zz}} \sin\theta_0 - \sqrt{n_o^2 - \sin^2\theta_0} \right]. \tag{8.16}$$

From Equation (8.16), the phase retardation Γ is dependent on the optical axes orientation θ_n and ϕ_n as well as the beam incident direction θ_0 and ϕ_0.

In the uniaxial-film-compensated LCDs, both *a* and *c* films are commonly used. In these two special cases, Equation (8.16) can be further simplified.

1. *Phase retardation of a film:* For an *a* film, its optical axis lies in the plane parallel to the film surface, i.e. $\theta_n = 0°$. Consequently, the phase retardation of the *a* film at an arbitrary incident angle is given by:

$$\Gamma_a = \frac{2\pi}{\lambda} d \left[n_e \sqrt{1 - \frac{\sin^2\theta_0 \sin^2(\phi_n - \phi_0)}{n_e^2} - \frac{\sin^2\theta_n \cos^2(\phi_n - \phi_0)}{n_o^2}} - n_o \sqrt{1 - \frac{\sin^2\theta_0}{n_o^2}} \right]. \tag{8.17}$$

2. *Phase retardation of c film:* In a c film, its optical axis is perpendicular to the film surface, i.e. $\theta_n = 90°$. In this case, the phase retardation of the c film at any oblique incidence is

$$\Gamma_c = \frac{2\pi}{\lambda} n_o d \left(\sqrt{1 - \frac{\sin^2 \theta_0}{n_e^2}} - \sqrt{1 - \frac{\sin^2 \theta_0}{n_o^2}} \right). \tag{8.18}$$

From Equation (8.18), Γ_c is independent of the azimuthal angle (ϕ_0) of the incident light. This is because the c film's optical axis is perpendicular to its surface. Hence, the optical properties of a c film are axially symmetric around its optical axis.

8.3.6 Poincaré sphere representation

Poincaré sphere representation is an elegant geometrical means for solving problems involving the propagation of polarized light through birefringent and optically active media [30]. For an elliptically polarized light with long axis azimuthal angle α and ellipticity angle β, its polarization state can be represented by a point P on the Poincaré sphere with longitude 2α and latitude 2β, as shown in Figure 8.9. The radius of the sphere is one unit length. Here the long axis azimuthal angle α of the elliptically polarized light is with respect to the x axis. For a uniaxial film with its optical axis oriented at angle γ from the x axis, it can be represented by point A, which is located at longitude 2γ on the equator. If the abovementioned elliptically polarized light (point P) passes through the uniaxial film (point A), the overall effect on Poincaré sphere is equivalent to rotating the AO axis from point P to point Q by an angle Γ, which is determined by the phase retardation of the uniaxial film as expressed in Equation (8.16). From spherical triangle definition, the spherical angle PAQ is equal to the rotation angle Γ. It should be pointed out that if the uniaxial film has a positive birefringence ($\Delta n = n_e - n_o > 0$), then the abovementioned rotation from point P to point Q is clockwise; otherwise, the rotation is counterclockwise if the uniaxial layer has a negative birefringence ($\Delta n < 0$).

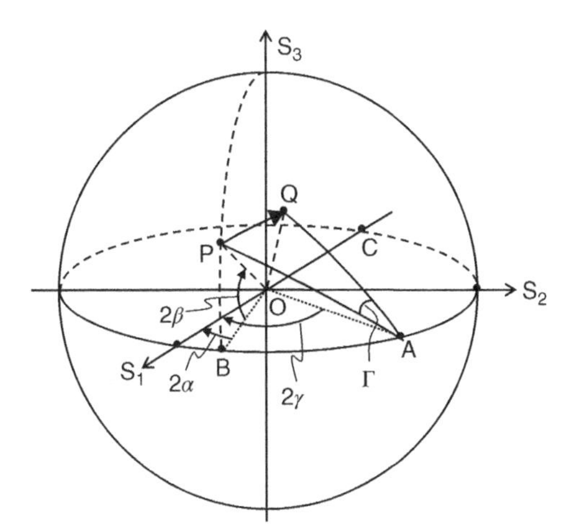

Figure 8.9 Schematic diagram of Poincaré sphere representation and the effect of uniaxial medium on the polarization state change of a polarized incident light.

For an *a* film, its optical axis lies in the plane parallel to the film surface. When an observer views the LCD panel from different azimuthal and polar angles, the effective optical axis on the wave plane will change with the viewing direction. As a result, its position on the equator of Poincaré sphere will also change accordingly. On the other hand, a *c* film's optical axis is perpendicular to the film surface. When an observer views the panel from different azimuthal and polar angles, the effective optical axis on the wave plane always forms 90° with respect to the horizontal reference. Therefore, its position on Poincaré sphere is always the intersection of equator and negative S_1 axis, which is denoted as point C in Figure 8.9.

8.3.7 Light leakage of crossed polarizers at oblique view

Considering a pair of crossed sheet polarizers with their absorption axes perpendicular to each other, the effective angle between their respective absorption axes varies with different viewing directions. Figure 8.10 shows the case when oblique light traverses through two sheet polarizers. The polarizer's absorption axis OM makes an angle ϕ_1 with respect to the x axis in the xy plane while the analyzer's absorption axis ON is oriented at angle ϕ_2. The shadow triangle OAB in Figure 8.10 denotes the plane of incidence. The light beam, denoted by the wave vector OK, propagates at azimuthal angle ϕ_k and polar angle θ_k inside the sheet polarizer.

1. *Effective polarizer angle on the wave plane:* Although these two linear polarizers form an angle $(\phi_2 - \phi_1)$ in the xy plane, their projections on the wave plane form another angle $\angle MKN$, as Figure 8.10 plots. Let us call this angle $\angle MKN$ the effective polarizer angle on the wave plane, which is expressed as φ hereafter. The extinction ratio of the crossed polarizers depends on this effective polarizer angle φ on the wave plane, rather than the absorption axes angle in the xy plane.

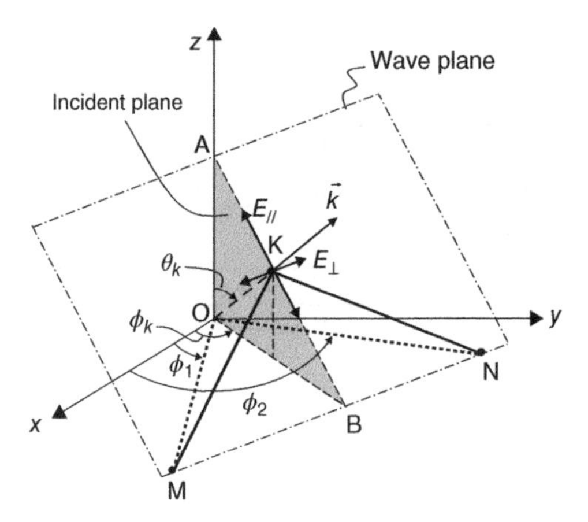

Figure 8.10 Schematic view of the effective polarizer angle φ of two sheet polarizers on the wave plane of an oblique incident light.

Based on the dot product of vectors, the effective polarizer angle ϕ can be expressed as [31]

$$\varphi = \cos^{-1}\left[\frac{\cos(\phi_2-\phi_1)-\sin^2\theta_k\cos(\phi_1-\phi_k)\cos(\phi_2-\phi_k)}{\sqrt{1-\sin^2\theta_k\cos^2(\phi_1-\phi_k)}\sqrt{1-\sin^2\theta_k\cos^2(\phi_2-\phi_k)}}\right], \tag{8.19}$$

where $\phi_k = \phi_0$ and $\theta_k = \sin^{-1}(\sin\theta_0/n_p)$. Here n_p (~1.5) is the average real refractive index of the sheet polarizer, and ϕ_0 and θ_0 are the azimuthal and incident angles of the incident light measured in air, respectively. In an LCD employing crossed polarizers, the absorption axes of the polarizer and the analyzer are perpendicular to each other. If we set $\phi_1 = 45°$ and $\phi_2 = -45°$, then the effective polarizer angle φ can be rewritten as

$$\varphi = \cos^{-1}\left[\frac{-\sin^2\theta_k\cos(\pi/4-\phi_k)\cos(\pi/4+\phi_k)}{\sqrt{1-\sin^2\theta_k\cos^2(\pi/4-\phi_k)}\sqrt{1-\sin^2\theta_k\cos^2(\pi/4+\phi_k)}}\right]. \tag{8.20}$$

As a quick verification, under normal view ($\theta_k = 0°$), the effective polarizer angle φ equals to 90°, which is identical to the absorption axes angle in the xy plane, i.e. $\phi_2 - \phi_1 = 90°$.

To find the tendency of φ when the azimuthal angle ϕ_0 ($= \phi_k$) changes, let us take the first-order derivative of φ with respect to ϕ_k and obtain

$$\frac{\partial\varphi}{\partial\phi_k} = -\frac{\sin^2\theta_k\cos\theta_k\sin 2\phi_k}{\cos^2\theta_k + {}^1/_4\sin^4\theta_k\cos^2 2\theta_k}. \tag{8.21}$$

Apparently, when $\phi_k = \phi_0 = 0°$, 90°, 180°, and 270° the effective polarizer angle φ reaches extrema. The second-order derivative

$$\frac{\partial^2\varphi}{\partial\phi_k^2} = -\frac{2\sin^2\theta_k\cos\theta_k\sin 2\phi_k}{\cos^2\theta_k + {}^1/_4\sin^4\theta_k\cos^2 2\phi_k} - \frac{\sin^6\theta_k\cos\theta_k\sin^2 2\phi_k\cos 2\varphi_k}{\left(\cos^2\phi_k + {}^1/_4\sin^4\theta_k\cos^2 2\phi_k\right)^2} \tag{8.22}$$

further reveals that φ reaches maxima at $\phi_k = \phi_0 = 0°$ and 180° and minima at $\phi_k = \phi_0 = 90°$ and 270°. By substituting $\phi_k = \phi_0 = 270°$ into Equation (8.20), we derive the effective polarizer angle φ at the lower bisector viewing position

$$\varphi = \cos^{-1}\left(\frac{\sin^2\theta_0/n_p^2}{2-\sin^2\theta_0/n_p^2}\right), \tag{8.23}$$

where θ_0 is the incident angle measured in air and n_p is the average real refractive index of the sheet polarizer.

Figure 8.11(a) plots the dependence of effective polarizer angle φ on viewing polar angle θ_0 and azimuthal angle ϕ_0 as calculated from Equation (8.20). During calculations, the average real refractive index of the sheet polarizer is taken to be $n_p = 1.5$. From Figure 8.11(a), at off-axis viewing directions, the effective polarizer angle φ deviates from 90°. Especially in all the bisector viewing directions, i.e. $\phi_0 = 0°$, 90°, 180°, and 270°, the effective polarizer angle φ deviates the farthest from 90° and reaches either maxima or minima. By contrast, in all

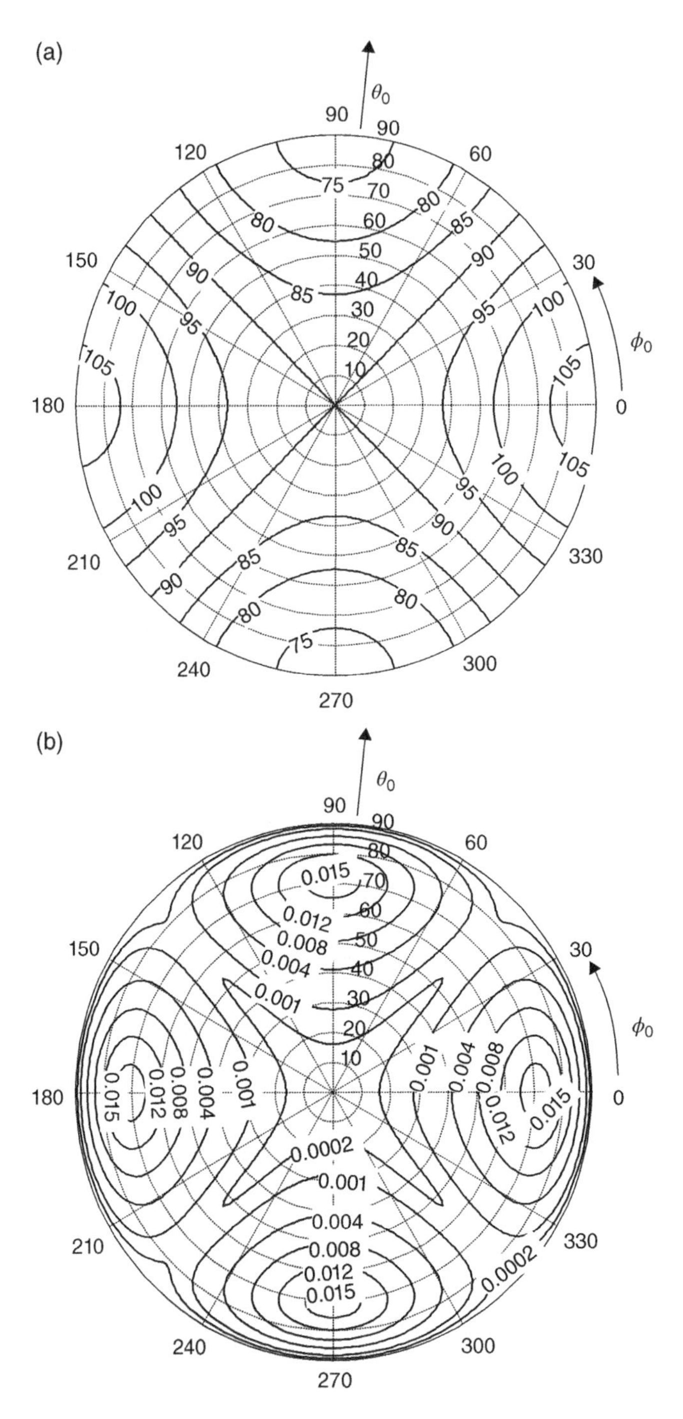

Figure 8.11 Dependence of (a) effective polarizer angle φ, and (b) dark-state light leakage of crossed polarizers, on the viewing azimuthal angle ϕ_0 and polar angle θ_0. The absorption axes of the crossed polarizers are set at 45° and −45°, respectively, and the incident light wavelength is $\lambda = 550$ nm. Zhu 2006. Reproduced with permission from IEEE.

on-axis viewing directions, i.e. $\phi_0 = 45°$, 135°, 225°, and 315°, the effective polarizer angle φ equals 90°, the same as the normal view.

The effective polarizer angle deviating from 90° at off-axis viewing directions causes dark-state light leakage which, in turn, degrades the device contrast ratio. As a typical example, Figure 8.11(b) shows the dark-state light leakage of crossed polarizers calculated by the extended Jones matrix method. In the calculation, both sheet polarizers are treated as anisotropic uniaxial media with complex refractive indices $n_e = 1.5 + j\,0.0022$ and $n_o = 1.5 + j0.000032$. As Figure 8.11(b) shows, the light leakage reaches maxima at the bisector viewing directions, i.e. $\phi_0 = 0°$, 90°, 180°, and 270°.

2) Crossed polarizers on Poincaré sphere: The dark-state light leakage of crossed polarizers can also be well explained on the Poincaré sphere, as shown in Figures 8.12(a) and (b). To facilitate the representation on the Poincaré sphere, we still set $\phi_1 = 45°$ and $\phi_2 = -45°$, keeping the absorption axes of these two sheet polarizers perpendicular to each other in the *xy* plane. Figure 8.12(a) represents the viewing from normal direction, while Figure 8.12(b) stands for an oblique view from the lower bisector, i.e. $\phi_0 = 270°$. In both figures, points P and A represent the effective absorption axis positions of polarizer and analyzer on the wave plane, respectively. The polarization state of the linearly polarized light after the polarizer, which is denoted by point T, is always orthogonal to the absorption axis of the polarizer on the wave plane. Therefore, on the Poincaré sphere, point T and point P are always located at the opposite sides along the diameter of the sphere.

As shown in Figure 8.12(a), under normal view the absorption axes of polarizer (point P) and analyzer (point A) are located at 90° and −90° on the equator of Poincaré sphere, respectively. Point T, the polarization state of the linearly polarized light after the polarizer, overlaps exactly with point A, the absorption axis of the analyzer, resulting in complete light absorption and no light leakage from normal viewing direction.

However, under oblique view from the lower bisector direction $\phi_0 = 270°$, the effective absorption axis positions of both polarizer and analyzer move toward the horizontal reference.

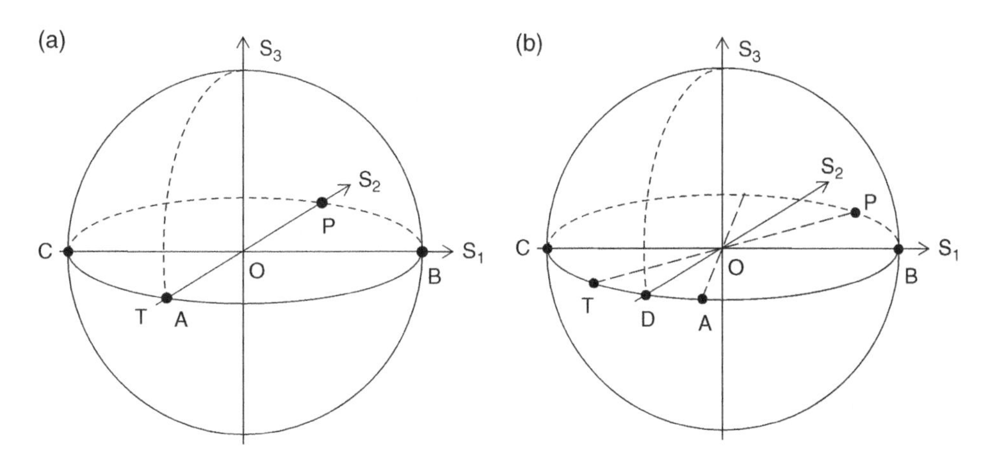

Figure 8.12 Demonstration of crossed polarizers on Poincaré sphere under (a) normal view and (b) oblique view at the lower bisector position $\phi_0 = 270°$. Here the absorption axes of polarizer and analyzer are set at $\phi_1 = 45°$ and $\phi_2 = -45°$, respectively. Zhu 2006. Reproduced with permission from IEEE.

Accordingly, from the lower bisector viewing direction $\phi_0 = 270°$, the effective polarizer angle φ becomes less than 90°, as illustrated in Figure 8.11(a). As a result, on the Poincaré sphere, points P and A are no longer located on the S_2 axis. Instead, point P is located between the S_1 and S_2 axes, while point A is located between the S_1 and negative S_2 axes, as Figure 8.12(b) shows. Moreover, point P is symmetric to point A about the S_1 axis, and angle $\angle POA$ is twice the effective polarizer angle φ, i.e. $\angle POA = 2\varphi$. Meanwhile, point T, representing the polarization state of light after the polarizer, is on the other end of the diameter passing through point P. Therefore, point T also deviates from the negative S_2 axis and is located symmetrically to point A with respect to the negative S_2 axis. Because point T no longer overlaps with point A, light leakage occurs from the bisector viewing direction of the crossed polarizers.

It is easy to figure out the relationship $\angle TOA = \pi - 2\varphi$ from Figure 8.12(b). The larger the angle $\angle TOA$, the more severe the light leakage. Since the effective polarizer angle ϕ deviates the farthest from 90° at all bisector viewing positions, the light leakage at bisectors is the severest, as depicted in Figure 8.11(b). If we can suppress the light leakage for all the bisector positions, the viewing angle of the LCD will be significantly enhanced. Thus, the goal of WVA LCDs, which incorporate compensation films into the panel design, is to move point T to point A for minimizing the light leakage from the analyzer. The introduced compensation film should improve the off-axis viewing performance but not affect the on-axis viewing performance.

In the following sections, let us analyze the compensation schemes of two uniaxial film-compensated WVA LCDs and derive the analytical solutions for each scheme. In the WVA LCDs with initially homogeneous alignment, such as IPS and FFS modes, let us assume that the stripe electrodes are in the bottom substrate and the electric fields are in the longitudinal direction. As the applied voltage exceeds the threshold voltage, i.e. $V > V_{th}$, the LC directors are gradually twisted from the anchored bottom boundary layer to the middle and then twisted back from the middle to the top (unaffected) boundary layer. Although the FFS mode can achieve a higher optical efficiency than IPS, their viewing angle performances are quite similar. For benchmarking, Figure 8.13 plots the calculated iso-contrast contours of an uncompensated IPS-LCD. In the calculation throughout this section, unless otherwise specified, we assume that at bright-state the middle layer LC directors are twisted 65° with respect to both boundary layers. Other parameters employed in simulations are listed in Table 8.2.

From Figure 8.13, without compensation films the IPS viewing angle at bisectors $\phi_0 = 0°$, 90°, 180°, and 270° are relatively poor. At bisectors, the 10:1 contrast ratio only extends to ~70° polar angle. This is due to the large dark-state light leakage at these bisector positions, as depicted in Figure 8.11(b).

In the following, we use IPS-LCD as an example to demonstrate two compensation schemes and provide each scheme with a comprehensive analytical solution. These compensation schemes are equally applicable to FFS LCDs.

8.3.8 IPS with a positive a *film and a positive* c *film*

Figure 8.14(a) shows the device configuration of an IPS-LCD using one positive *a* film and one positive *c* film for phase compensation. As shown in Figure 8.14(a), a positive *c* film and a positive *a* film are sandwiched between the analyzer and the homogeneous LC layer. More specifically, the positive *a* film, whose optic axis is oriented parallel to the absorption axis

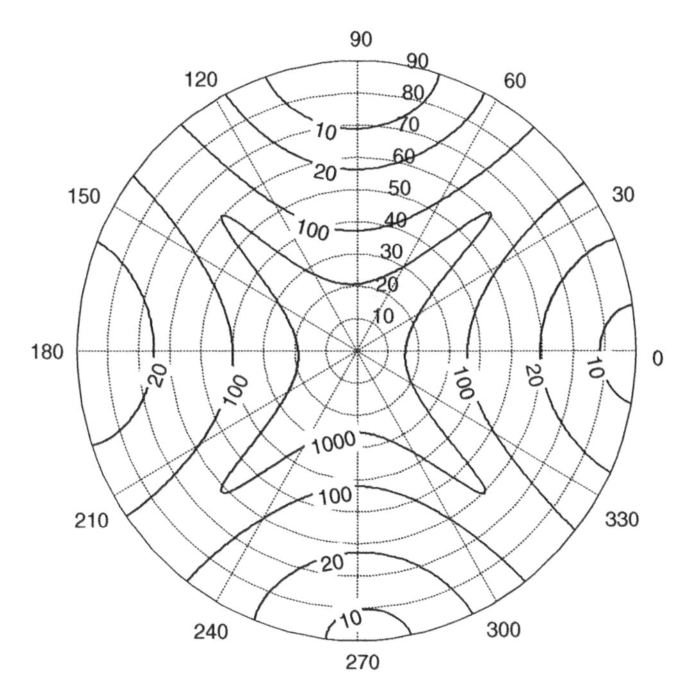

Figure 8.13 Iso-contrast contour of an uncompensated IPS-LCD at $\lambda = 550$ nm. Zhu 2006. Reproduced with permission from IEEE.

Table 8.2 Parameters used in simulating the IPS-LCD viewing angle performance.

Parameters	Description	Values
d_{LC}	Cell gap	4 μm
$\theta_{pretilt}$	Surface tilt angle	1°
$n_{LC,e}$	n_e of LC material	1.5649
$n_{LC,o}$	n_o of LC material	1.4793
$n_{p,e}$	n_e of sheet polarizer (complex)	$1.5 + j0.0022$
$n_{p,o}$	n_o of sheet polarizer (complex)	$1.5 + j0.000032$
$n_{c+,e}$	n_e of positive *c* film	1.5110
$n_{c+,o}$	n_o of positive *c* film	1.5095
$n_{c-,e}$	n_e of negative *c* film	1.5095
$n_{c-,o}$	n_o of negative *c* film	1.5110
$n_{a+,e}$	n_e of positive *a* film	1.5110
$n_{a+,o}$	n_o of positive *a* film	1.5095
$n_{a-,e}$	n_e of negative *a* film	1.5095
$n_{a-,o}$	n_o of negative *a* film	1.5110
ϕ_1	Absorption axis of polarizer	45°
ϕ_2	Absorption axis of analyzer	–45° (or 135°)
λ	Wavelength of incident light	550 nm

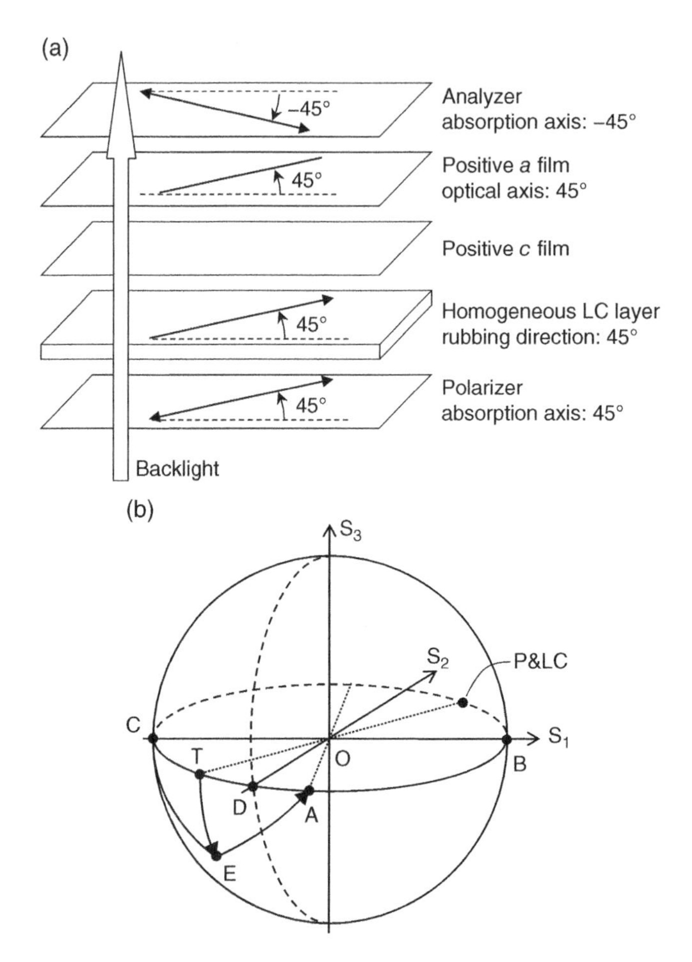

Figure 8.14 (a) Device structure and (b) compensation principle of an IPS-LCD with compensation of a positive *a* film and a positive *c* film. Zhu 2006. Reproduced with permission from IEEE.

of the polarizer, is adjacent to the analyzer. Figure 8.14(b) explains the compensation principle on a Poincaré sphere when the observer views the panel from an oblique angle at the lower bisector position $\phi_0 = 270°$.

The detailed compensation mechanism is explained as Figure 8.14(b) shows. When the unpolarized light from the backlight unit traverses the polarizer (point P), it becomes linearly polarized, and its polarization state is located at point T, which deviates from the absorption axis of the analyzer (point A). When such a linearly polarized light (point T) passes through the homogenous LC layer, whose position on the Poincaré sphere overlaps with point P, the linear polarization state still keeps the same (point T). Then, the linearly polarized light (point T) successively passes through the positive *c* film and the positive *a* film, whose effective optical axis positions on the Poincaré sphere are points C and P, respectively. When the linearly polarized light (point T) passes through the positive *c* film, its polarization state is rotated from point T to point E clockwise around the CO axis. This intermediate polarization state (point E), in general, is elliptical. By properly choosing the phase retardation values of the positive *a* and *c* films, we can always fulfill the following goal: when the elliptically polarized light (point E) passes through the positive *a* film, the polarization state can be rotated clockwise around the PO axis so that point E is moved to point A. As a result, at $V = 0$ the light is completely absorbed by the analyzer (point A) leading to a good dark state, even when viewed from an oblique angle at the bisectors.

To reach the abovementioned objective, we can easily determine from Figure 8.14(b) that the following two requirements must be satisfied: (1) the arc $\overline{\boldsymbol{TE}}$ should equal to the arc $\overline{\boldsymbol{TA}}$, and (2) the arc $\overline{\boldsymbol{TC}}$ should equal to the arc $\overline{\boldsymbol{EC}}$. Besides, from Figure 8.14(b) we can also obtain $\angle POB = \angle AOB = \varphi$, $\overline{\boldsymbol{TA}} = \pi - 2\varphi$, and $\overline{\boldsymbol{TC}} = \varphi$. Based on spherical trigonometry, we can find the following relationships from the spherical triangles *CTE* and *TEA*:

$$\angle TCE = 2\sin^{-1}(\text{ctg}\,\varphi), \tag{8.24a}$$

$$\angle CTE = 2\cos^{-1}(\text{ctg}^2\varphi), \tag{8.24b}$$

$$\angle ATE = \pi - \angle CTE, \tag{8.24c}$$

where φ, determined by Equation (8.23), is the effective polarizer angle on the wave plane from the lower bisector viewing position $\phi_0 = 270°$.

Since the required positive *c* film's phase retardation Γ_{c+} equals to the spherical angle $\angle \boldsymbol{TCE}$, then from (8.18) and (8.24a) the required positive *c* film's thickness d_{c+} can be derived as

$$d_{c+} = \lambda \frac{\sin^{-1}(\text{ctg}\varphi)/\pi}{n_{c+,o}\left(\sqrt{1-\dfrac{\sin^{-2}\theta_0}{n_{c+,e}^2}} - \sqrt{1-\dfrac{\sin^{-2}\theta_0}{n_{c+,o}^2}}\right)}. \tag{8.25}$$

On the other hand, the required positive *a* film's phase retardation Γ_{a+} equals to the spherical angle $\angle ATE$. Thus from (8.17), (8.24b), and (8.24c) we can derive the required positive *a* film's thickness d_{a+} as:

$$d_{a+} = \lambda \frac{1/2 - \cos^{-1}(\text{ctg}^2\varphi)/2\pi}{n_{a+,e}\sqrt{1-\dfrac{\sin^2\theta_0}{2n_{a+,e}^2} - \dfrac{\sin^2\theta_0}{2n_{a+,o}^2}} - n_{a+,o}\sqrt{1-\dfrac{\sin^2\theta_0}{n_{a+,o}^2}}}. \tag{8.26}$$

In the derivation of Equation (8.26), we substitute $\phi_n = 45°$ and $\phi_0 = 270°$ into Equation (8.17) because the positive *a* film's optical axis is oriented in the 45° direction, as Figure 8.14(a) shows, and the viewing direction is at $\phi_0 = 270°$ azimuthal angle.

As we can see from Equations (8.23), (8.25), and (8.26), the required film thickness depends on the incident angle θ_0, film's refractive indices $n_{c+,e}$, $n_{c+,o}$, $n_{a+,e}$, and $n_{a+,o}$, and the polarizer's average real refractive index n_p. Therefore, once we know both the refractive indices of the films and the polarizer, and the intended viewing angle (i.e. incident angle), we can determine the required film thickness from (8.23), (8.25), and (8.26). For instance, if we set $\theta_0 = 70°$ as the intended viewing angle where we would like to optimize our LCD designs, and use the parameters listed in Table 8.2, then we can calculate the required film thicknesses from

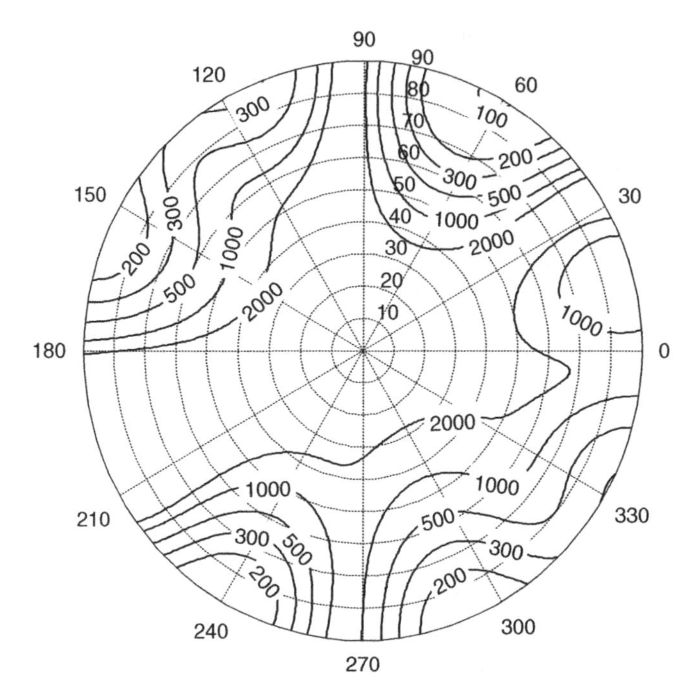

Figure 8.15 Iso-contrast contour of an IPS-LCD with a positive *a* film ($d_{a+} = 92.59$ μm) and a positive *c* film ($d_{c+} = 60.09$ μm) under $\lambda = 550$ nm. Zhu 2006. Reproduced with permission from IEEE.

Equations (8.25) and (8.26). The results are $d_{c+} = 60.09$ μm and $d_{a+} = 92.59$ μm. Based on these film thicknesses, Figure 8.15 depicts the calculated iso-contrast contour of an IPS-LCD with one positive *a* film and one positive *c* film for phase compensation. Comparing Figure 8.15 with Figure 8.13, we can clearly see that the viewing angle performance at off-axis viewing directions, especially the bisector positions $\phi_0 = 0°$, 90°, 180°, and 270°, is dramatically improved. Meanwhile, the contrast ratios at on-axis viewing directions $\phi_0 = 45°$, 135°, 225°, and 315° remain unchanged.

This compensation scheme can also be modified by exchanging the positions of the positive *c* film and *a* film, as shown in Figure 8.16(a). By contrast to Figure 8.14(a), now the positive *a* film is adjacent to the LC layer and its optical axis is parallel to the absorption axis of the analyzer. The compensation principle is demonstrated in Figure 8.16(b). When the linearly polarized light (point T) passes through the positive *a* film, the polarization state is rotated clockwise around the AO axis so that point T is moved to point E. Now the intermediate state (point E) is located on the upper hemisphere. The role of the positive *c* film is to rotate point E to point A clockwise around the CO axis. Although the process of polarization state change is different, the required film thicknesses are still identical to those of the previous case as determined by Equations (8.25) and (8.26). Its viewing angle is very similar to that shown in Figure 8.15.

As shown in Figure 8.15, the viewing angle is not very symmetrical although the contrast ratio of this compensation scheme exceeds 100:1 at any viewing direction. This is because the intermediate state (point E) is not located on the great circle passing through the S_2 and S_3 axes, as shown in Figures 8.14(b) and 8.16(b). To get a more symmetric viewing angle, it is essential

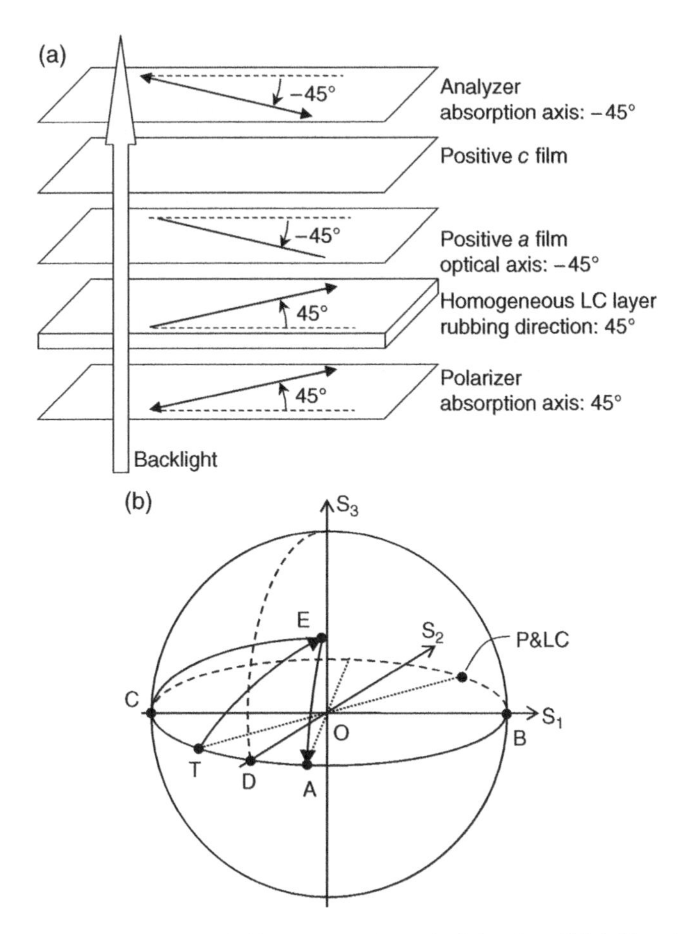

Figure 8.16 (a) Device structure and (b) compensation principle of an IPS-LCD using one positive *a* film and one positive *c* film. Zhu 2006. Reproduced with permission from IEEE.

to make the intermediate state located on the great circle that passes through S_2 and S_3 axes and bisects the arc $\overline{TA}$, as will be discussed in the following example.

8.3.9 *IPS with positive and negative* a *films*

Figure 8.17(a) shows the device configuration of an IPS-LCD with one positive *a* film and one negative *a* film [32,33]. As shown in the figure, a positive *a* film and a negative *a* film are sandwiched between the LC layer and the analyzer, with the positive *a* film adjacent to the LC layer. More specifically, the optic axis of the positive *a* film is parallel to the absorption axis of the analyzer, while the optical axis of the negative *a* film is parallel to the absorption axis of the polarizer. Figure 8.17(b) explains the compensation principle on the Poincaré sphere when the observer views the panel from an oblique angle at the lower bisector position, i.e. $\phi_0 = 270°$.

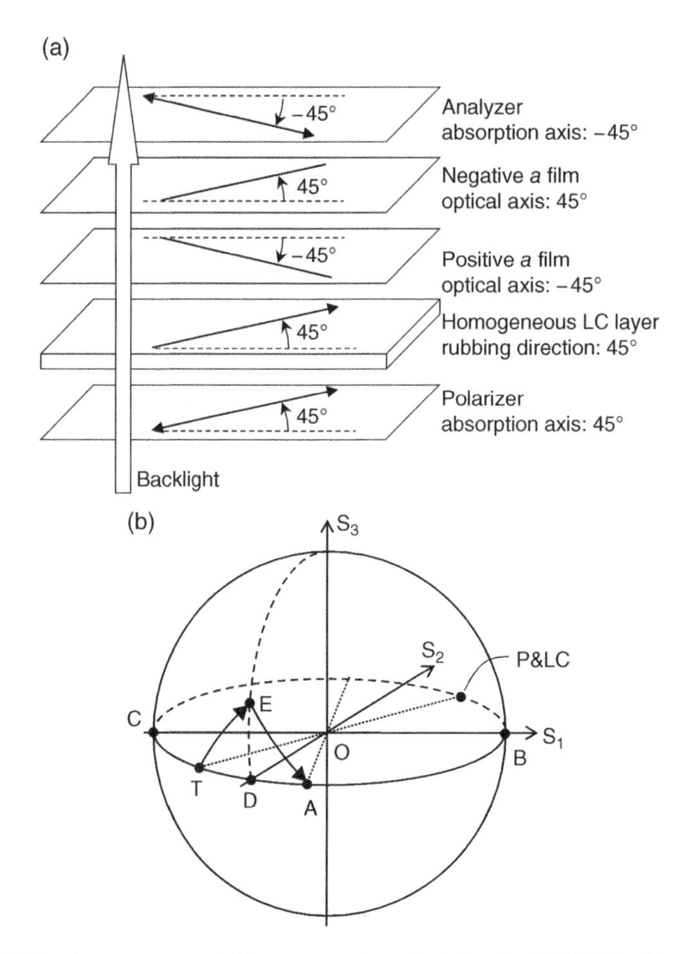

Figure 8.17 (a) Device structure and (b) compensation principle of an IPS-LCD with compensation of one positive *a* film and one negative *a* film.

The detailed compensation mechanism is explained as Figure 8.17(b) demonstrates. When the unpolarized light from backlight penetrates the polarizer (point P), it becomes linearly polarized and its polarization state is located at point T, which deviates from the absorption axis of the analyzer (point A). When such a linearly polarized light (point T) passes through the homogenous LC layer, whose position on the Poincaré sphere overlaps with point P, the linear polarization state remains the same (point T). Then, the linearly polarized light (point T) successively passes through the positive *a* film and the negative *a* film, whose positions on the Poincaré sphere are points A and P, respectively. When the linearly polarized light (point T) passes through the positive *a* film, its polarization state is rotated clockwise from point T to point E around the AO axis. This intermediate polarization state (point E), in general, is an elliptical polarization state. By properly choosing the phase retardation values of both positive *a* film and negative *a* film, we should be able to rotate point E to point A counterclockwise around the PO axis. As a result, in the voltage-off state the light is completely absorbed by the analyzer (point A) and a very good dark state is achieved even viewed from an oblique angle at the bisector positions.

To reach this objective, we can easily determine from Figure 8.17(b) that the following two requirements must be satisfied: (1) the arc $\overline{\boldsymbol{EA}}$ should be equal to the arc $\overline{\boldsymbol{TA}}$, and (2) the arc $\overline{\boldsymbol{TA}}$ should be equal to the arc $\overline{\boldsymbol{TE}}$. In other words, the spherical triangle ***ETA*** should be an equilateral spherical triangle. In addition, from the Poincaré sphere we can also obtain $\angle POB = \angle AOB = \varphi$ and $\overline{TA} = \pi - 2\varphi$. Based on the spherical trigonometry, the following relationship from the equilateral spherical triangle ***ETA*** can be derived:

$$\angle ETA = \angle EAT = \cos^{-1}\left(-\mathrm{ctg}\,\varphi \cdot \mathrm{ctg}^2\,\varphi\right), \tag{8.27}$$

where φ, determined from Equation (8.23), is the effective polarizer angle on the wave plane from the lower bisector viewing position, i.e. $\phi_0 = 270°$.

The required positive *a* film's phase retardation Γ_{a+} is equal to the spherical angle $\angle EAT$, thus from Equations (8.17) and (8.27) the required positive *a* film's thickness d_{a+} can be expressed as:

$$d_{a+} = \lambda \frac{\cos^{-1}(-\mathrm{ctg}\,\varphi \cdot \mathrm{ctg}2\varphi)/2\pi}{n_{a+,e}\sqrt{1 - \dfrac{\sin^2\theta_0}{2n_{a+,e}^2} - \dfrac{\sin^2\theta_0}{2n_{a+,o}^2}} - n_{a+,o}\sqrt{1 - \dfrac{\sin^2\theta_0}{n_{a+,o}^2}}}. \tag{8.28}$$

In the derivation of Equation (8.28), we substitute $\phi_n = -45°$ and $\phi_0 = 270°$ into Equation (8.17) because the positive *a* film's optical axis is oriented at −45° direction, as Figure 8.17(a) shows, and the viewing direction is from $\phi_0 = 270°$ azimuthal angle.

Similarly, the negative *a* film's phase retardation Γ_{a-} is equal to the negative spherical angle $\angle ETA$, i.e. $\Gamma_{a-} = -\angle ETA$. Here the minus sign denotes that the phase retardation of the negative *a* film is negative and the rotation around PO axis from point E to point A is counterclockwise. Thus, from Equations (8.17) and (8.27) we can obtain the negative *a* film's thickness d_{a-} as

$$d_{a-} = -\lambda \frac{\cos^{-1}(-\mathrm{ctg}\,\varphi \cdot \mathrm{ctg}2\varphi)/2\pi}{n_{a-,e}\sqrt{1 - \dfrac{\sin^2\theta_0}{2n_{a-,e}^2} - \dfrac{\sin^2\theta_0}{2n_{a-,o}^2}} - n_{a-,o}\sqrt{1 - \dfrac{\sin^2\theta_0}{n_{a-,o}^2}}}. \tag{8.29}$$

In the derivation of Equation (8.29), we substitute $\phi_n = 45°$ and $\phi_0 = 270°$ into Equation (8.17), because the negative *a* film's optical axis is oriented in the 45° direction, as Figure 8.17(a) shows, and the viewing direction is from $\phi_0 = 270°$ azimuthal angle.

From Equations (8.23), (8.28), and (8.29), the required film thicknesses depend on the incident angle θ_0, the film's refractive indices $n_{a+,e}$, $n_{a+,o}$, $n_{a-,e}$, and $n_{a-,o}$, and the polarizer's average real refractive index n_p. Therefore, once we know the refractive indices of both the films and the polarizer, as well as the intended viewing angle for LCD optimization, we can determine the required film thickness from Equations (8.23), (8.28), and (8.29). By using the parameters listed in Table 8.2 and choosing $\theta_0 = 70°$, the required film thicknesses as

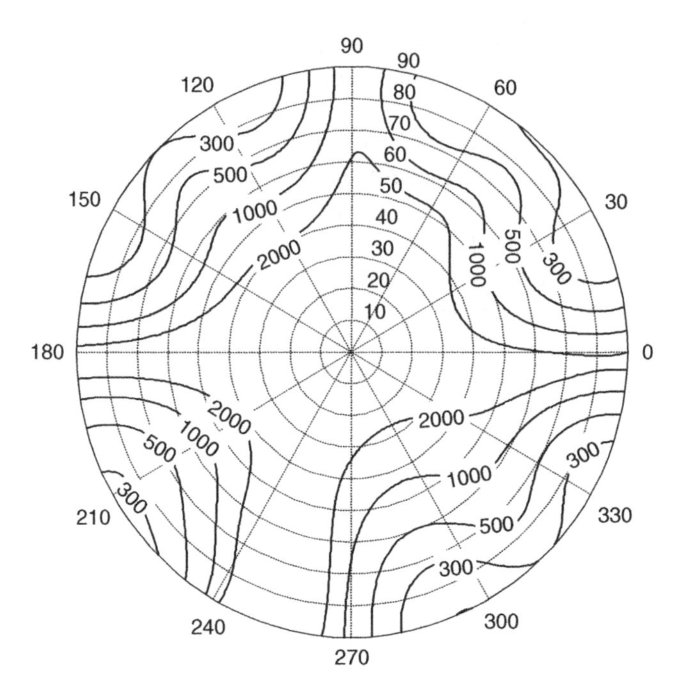

Figure 8.18 Iso-contrast contour of the IPS-LCD with a positive *a* film ($d_{a+} = 61.38$ μm) and a negative *a* film ($d_{a-} = 61.37$ μm) under $\lambda = 550$ nm.

calculated from Equations (8.28) and (8.29) are $d_{a+} = 61.38$ μm and $d_{a-} = 61.37$ μm. Based on these film thicknesses, Figure 8.18 plots the calculated iso-contrast contour of the IPS-LCD with one positive *a* film and one negative *a* film. Comparing Figure 8.18 with Figure 8.13, we see clearly that the viewing characteristic at off-axis directions, especially the bisector positions $\phi_0 = 0°$, 90°, 180°, and 270°, is dramatically improved. In the meantime, the contrast ratios along the horizontal and vertical axes ($\phi_0 = 45°$, 135°, 225°, and 315°) remain unchanged.

The positions of the positive *a* film and the negative *a* film shown in Figure 8.17(a) are exchangeable. Simulation results indicate that the required film thicknesses remain the same. The Poincaré representation is still similar, except that the intermediate polarization state (point E) is on the lower hemisphere. The required film thicknesses d_{a+} and d_{a-} are still the same as obtained in Equations (8.28) and (8.29). And finally, the viewing angle performance is almost identical to those shown in Figure 8.18.

As shown in Figure 8.18, the contrast ratio exceeds 200:1 from all viewing directions. This viewing angle is more symmetric than that shown in Figure 8.15 using one positive *a* film and one positive *c* film for compensation. This is due to the fact that in this compensation scheme the intermediate state (point E) is located on the great circle that passes through S_2 and S_3 axes and bisects the arc $\overline{TA}$. Another advantage of this compensation scheme is that it requires only uniaxial *a* films and does not require any *c* film or biaxial film. Since *a* film has a lower cost than *c* film and biaxial film, this compensation scheme has potentially lower cost while keeping excellent viewing characteristics.

8.3.10 *Color shift*

Color shift is another important issue for liquid crystal displays. In an IPS mode, yellowish color shift occurs at the $\phi = 45°$ azimuthal angle and bluish color shift occurs at $\phi = -45°$ due to the phase retardation difference. To suppress color shift, a chevron-shaped electrode similar to a two-domain structure has been proposed [34,35] Each pixel is divided into two domains, where the LC directors face in opposite directions and the color shift is compensated effectively.

8.4 Vertical Alignment Mode

Vertical alignment (VA), also called homeotropic alignment [36], is another common LC mode for direct-view transmissive and reflective projection displays. The vertical alignment exhibits the highest contrast ratio among all the LC modes developed; moreover, its contrast ratio is insensitive to the incident light wavelength, LC layer thickness, and operating temperature. Both projection [37,38] and direct view displays using homeotropic LC cells [39,40] have been demonstrated. Besides contrast ratio, homeotropic cell also exhibits a faster response time than its corresponding homogeneous or twisted nematic cell. Two factors contributing to the faster response time are elastic constant and cell gap. To achieve 1π phase retardation for a transmissive display, the required $d\Delta n$ for homogeneous and homeotropic cells is the same, that is $d\Delta n = \lambda/2$. However, for a 90° TN cell, the required $d\Delta n = 0.866\lambda$. On the other hand, the governing elastic constant for homogeneous, twisted, and homeotropic cells are splay (K_{11}), twist (K_{22}), and bend (K_{33}), respectively. From an elastic constant viewpoint, the following order $K_{33} > K_{11} > K_{22}$ holds for most LC mixtures. The response time of an LC layer is proportional to $\gamma_1 d^2/K\pi^2$ where γ_1 is the rotational viscosity and K is the corresponding elastic constant. Therefore, the homeotropic cell has the best response time and contrast ratio of the three compared. This has been proven by some wide angle direct-view displays employing a homeotropic cell [41,42]

One requirement of a VA cell is in the need for high resistivity LC mixtures to have negative dielectric anisotropy. High resistivity is required for active matrix LCD in order to avoid image flickering. Negative $\Delta\varepsilon$ is required for obtaining useful electro-optic effect. To obtain negative $\Delta\varepsilon$ LCs, the dipoles, in particular the fluoro groups, need to be in the lateral positions. Significant progress in material development has been obtained in the past decade. Nevertheless, the selection of negative $\Delta\varepsilon$ LC compounds is still far less than the positive ones. Besides, the lateral dipole groups often exhibit a higher viscosity than the axial compounds owing to the larger moment of inertia.

8.4.1 *Voltage-dependent transmittance*

Figure 8.19 shows the voltage-dependent optical transmittance of a VA cell with $d\Delta n = 350$ nm between crossed polarizers. For computer calculations, a single-domain VA cell employing Merck high resistivity MLC-6608 LC mixture is considered. Some physical properties of MLC-6608 are summarized as follows: $n_e = 1.558$, $n_o = 1.476$ (at $\lambda = 589$ nm and $T = 20°C$); clearing temperature $T_c = 90\,°C$; dielectric anisotropy $\Delta\varepsilon = -4.2$, and rotational viscosity $\gamma_1 = 186$ mPas at 20 °C. In principle, to obtain 100% transmittance for a transmissive VA

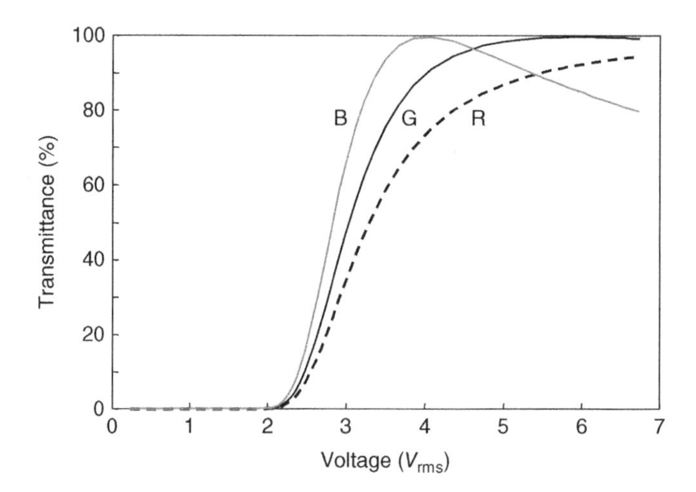

Figure 8.19 Voltage-dependent transmittance of a VA cell. LC: MLC-6608. $d\Delta n = 350$ nm, R = 650 nm, G = 550 nm, and B = 450 nm.

cell only requires $d\Delta n \sim \lambda/2$. Since the human eye is most sensitive in green ($\lambda = 550$ nm), the required $d\Delta n$ is around 275 nm. However, this is the minimum $d\Delta n$ value required because under such a condition the 100% transmittance would occur at $V \gg V_{th}$. Due to the finite voltage swing from TFT (usually below 6 V_{rms}), the required $d\Delta n$ should be increased to ~0.6λ, i.e. $d\Delta n$ ~330 nm.

From Figure 8.19, an excellent dark state is observed at normal incidence. As the applied voltage exceeds the Fréedericksz threshold voltage ($V_{th} \sim 2.1$ V_{rms}), LC directors are reoriented by the applied electric field, resulting in light transmission from the crossed analyzer. At ~6 V_{rms}, the normalized transmittance reaches 100% for the green light ($\lambda = 550$ nm).

8.4.2 *Optical response time*

When the backflow and inertial effects are ignored, the dynamics of the LC director reorientation is described by the following Erickson–Leslie equation [9,10]:

$$\left(K_{11}\cos^2\phi + K_{33}\sin^2\phi\right)\frac{\partial^2\phi}{\partial z^2} + (K_{33} - K_{11})\sin\phi\cos\phi\left(\frac{\partial\phi}{\partial z}\right)^2 + \varepsilon_o\Delta\varepsilon E^2\sin\phi\cos\phi = \gamma_1\frac{\partial\phi}{\partial t} \tag{8.30}$$

where γ_1 is the rotational viscosity, K_{11} and K_{33} represent the splay and bend elastic constants, respectively, $\varepsilon_o\Delta\varepsilon E^2$ is the electric field energy density, $\Delta\varepsilon$ is the LC dielectric anisotropy, and ϕ is the tilt angle of the LC directors. In general, Equation (8.30) can only be solved numerically. However, when the tilt angle is small ($\sin\phi \sim \phi$) and $K_{33} \sim K_{11}$ (the so-called small angle approximation), the Erickson–Leslie equation is reduced to

$$K_{33}\frac{\partial^2\phi}{\partial z^2}+\varepsilon_o\Delta\varepsilon E^2\phi=\gamma_1\frac{\partial\phi}{\partial t}. \tag{8.31}$$

Under such circumstances, both rise time and decay time have simple analytical solutions [43]:

$$\tau_{rise}=\frac{1}{2}\frac{\tau_o}{\left|\left(\frac{V}{V_{th}}\right)^2-1\right|}\ln\left(\frac{\frac{\delta_o/2}{\sin^{-1}\left(\sqrt{0.1}\sin\left(\frac{\delta_o}{2}\right)\right)}-1}{\frac{\delta_o/2}{\sin^{-1}\left(\sqrt{0.9}\sin\left(\frac{\delta_o}{2}\right)\right)}-1}\right), \tag{8.32a}$$

$$\tau_{decay}=\frac{\tau_o}{2}\ln\left(\frac{\sin^{-1}\left(\sqrt{0.9}\sin\left(\frac{\delta_o}{2}\right)\right)}{\sin^{-1}\left(\sqrt{0.1}\sin\left(\frac{\delta_o}{2}\right)\right)}\right). \tag{8.32b}$$

Here, both rise time and decay time are defined while transmittance (under crossed polarizers) changes from 10% to 90%. In the above equations, τ_o is the LC director reorientation time ($1 \rightarrow 1/e$) and δ_o is the net phase change from a bias voltage $V=V_b$ to $V=0$:

$$\tau_o=\frac{\gamma_1 d^2}{K_{33}\pi^2} \tag{8.33}$$

Equation (8.33) correlates the optical rise time and decay time to the LC director reorientation time (τ_o). Basically, it is a linear relationship except for the additional logarithm term of the phase dependence.

8.4.3 *Overdrive and undershoot voltage method*

From Equation (8.32), the rise time depends on the applied voltage (V), especially near the threshold region. Let us use a normally black VA cell as an example. Typically, the cell is biased at a voltage (V_b) which is slightly below V_{th}, in order to reduce the delay time incurred during the rising period and to keep a high contrast ratio. For some intermediate gray levels, the applied voltage is only slightly above V_{th}. Under such circumstances, the rise time would be very slow. To overcome the slow rise time, we could apply a high voltage for a short period and then hold the transmittance at the desired gray level, as shown in Figure 8.20. This is the so-called overdrive voltage method [44]. Meanwhile, during the decay period, the voltage is turned off for a short period and then a small holding voltage is applied to keep the LC at the desired gray level. This is the undershoot effect [45]. With voltage overdrive and undershoot the LC response time can be reduced by two to three times.

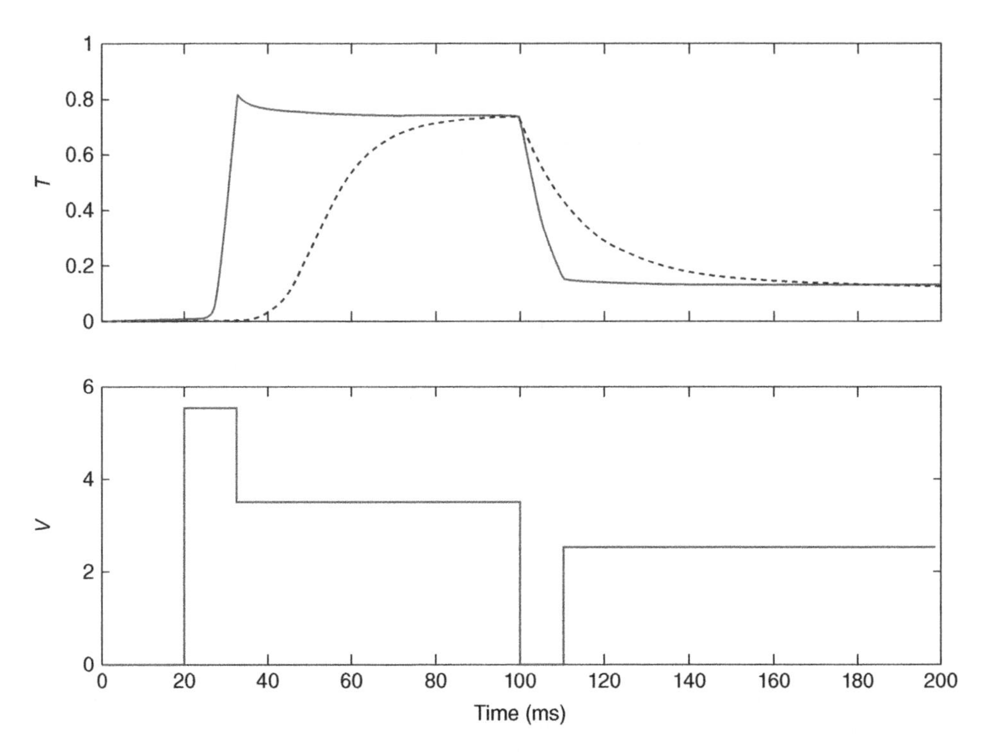

Figure 8.20 The overdrive and undershoot voltage method for speeding up LC rise and decay times. The top trace is the optical response and the bottom trace is the corresponding voltage waveforms. The dashed lines represent a normal driving and the solid lines are with overdrive and undershoot voltages.

8.5 Multi-Domain Vertical Alignment Cells

Single-domain VA has been used extensively in liquid-crystal-on-silicon (LCoS) [46,47] for projection displays because of its excellent contrast ratio. However, for direct-view display the single-domain VA has a relatively narrow viewing angle. To widen the viewing angle, multi-domain VA (MVA) has been developed. Fujitsu has developed protrusion-type MVA [48,49] and Samsung has developed patterned vertical alignment (PVA) [50,51] using slits to generate fringing fields. The operating mechanisms are alike, but PVA does not require any physical protrusions so that its contrast ratio is higher.

For simplicity but without losing generality, let us assume that in each pixel the LC directors form a four-domain orientation profile, as Figure 8.21(a) shows. Figure 8.21(b) depicts the calculated voltage-dependent transmittance curve of a typical MVA-LCD using Merck MLC-6608 LC material whose parameters are listed in Table 8.3. Here, the absorption loss of polarizers has been taken into consideration. In the film-compensated MVA cells, the refractive indices of the uniaxial films and polarizers are still the same as those listed in Table 8.2.

Figure 8.22 shows the calculated iso-contrast contour of the four-domain MVA-LCD without film compensation. In the contrast ratio calculation, we first use continuum theory [52] to calculate the LC director distribution at $V_{on} = 5\ V_{rms}$ and $V_{off} = 0$, respectively, and then use the

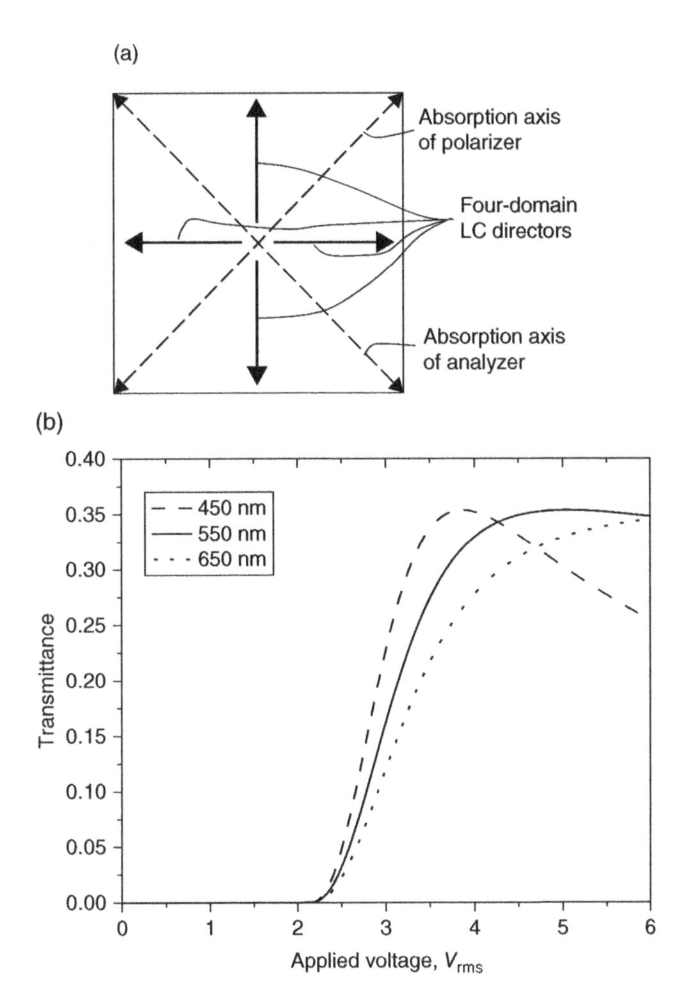

Figure 8.21 (a) Schematic top-view of the four-domain LC director distribution in the voltage-on state, and (b) the voltage dependent transmittance curve of an MVA-LCD. Zhu 2006. Reproduced with permission from IEEE.

Table 8.3 Parameters used in simulating the MVA-LCD viewing angle performance.

Parameters	Description	Values
d_{LC}	Cell gap	4.6 μm
$\theta_{pretilt}$	Surface tilt angle	89°
$n_{LC,e}$	n_e of LC material MLC – 6608	1.5606
$n_{LC,o}$	n_o of LC material MLC – 6608	1.4770
ϕ_1	Absorption axis of polarizer	45°
ϕ_2	Absorption axis of analyzer	–45° (or 135°)
λ	Wavelength of incident light	550 nm

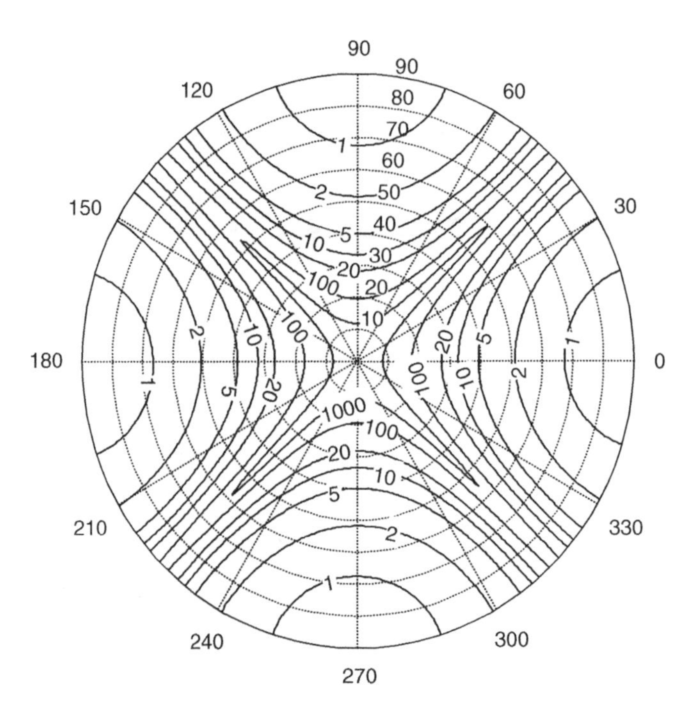

Figure 8.22 Simulated iso-contrast contour of a typical four-domain MVA-LCD under $\lambda = 550$ nm. Zhu 2006. Reproduced with permission from IEEE.

extended Jones matrix to calculate the optical transmittance for each domain and finally average all four domains up.

From Figure 8.22, without compensation films the viewing angle of the four-domain MVA cell at the bisector positions $\phi_0 = 0°$, 90°, 180°, and 270° are very poor. At these bisectors, the 10:1 contrast ratio only extends to ~30° polar angle. Two factors contribute to the narrow viewing angle: (1) the absorption axes of the crossed polarizers are no longer perpendicular to each other at off-axis oblique viewing directions, and (2) the vertically aligned LC layer behaves as a *c* film, which imposes a phase retardation on the obliquely incoming linearly polarized light and modulates its polarization state. The phase retardation Γ_{LC}, which is induced by the vertically aligned LC layer at an oblique incident light, can be easily obtained from Equation (8.16) as

$$\Gamma_{LC} = \frac{2\pi}{\lambda} n_{LC,o} d_{LC} \left(\sqrt{1 - \frac{\sin^2\theta_0}{n_{LC,e}^2}} - \sqrt{1 - \frac{\sin^2\theta_0}{n_{LC,o}^2}} \right). \tag{8.34}$$

From Equation (8.34), Γ_{LC} depends on the incident angle θ_0, the LC's refractive indices $n_{LC,e}$ and $n_{LC,o}$, and the LC layer's thickness d_{LC}. As an example, at incident angle $\theta_0 = 70°$ the corresponding Γ_{LC}, calculated from the parameters listed in Table 8.3, is 0.664π radians. Equation (8.34) will be frequently referred to in this section.

In the following, we use the four-domain MVA-LCD as an example to demonstrate some uniaxial-film compensation schemes and provide each scheme with a comprehensive analytical solution. All of these compensation schemes are equally applicable to PVA mode LCDs.

8.5.1 *MVA with a positive* a *film and a negative* c *film*

Figure 8.23(a) shows the schematic device configuration of an MVA-LCD with one positive *a* film and one negative *c* film. As shown in the figure, the positive *a* film and the negative *c* film are sandwiched between the polarizer and the MVA-LC layer. More specifically, the optical axis of the positive *a* film is parallel to the absorption axis of the analyzer. Figure 8.23(b) explains the compensation principle on the Poincaré sphere when the observer views the panel from an oblique angle at the lower bisector position $\phi_0 = 270°$.

The detail compensation mechanism is explained using the Poincaré sphere shown in Figure 8.23(b). When the unpolarized light from backlight source passes the polarizer (point P), it becomes linearly polarized and its polarization state is located at point T, which deviates from the absorption axis of the analyzer (point A). Then, such a linearly polarized light (point T)

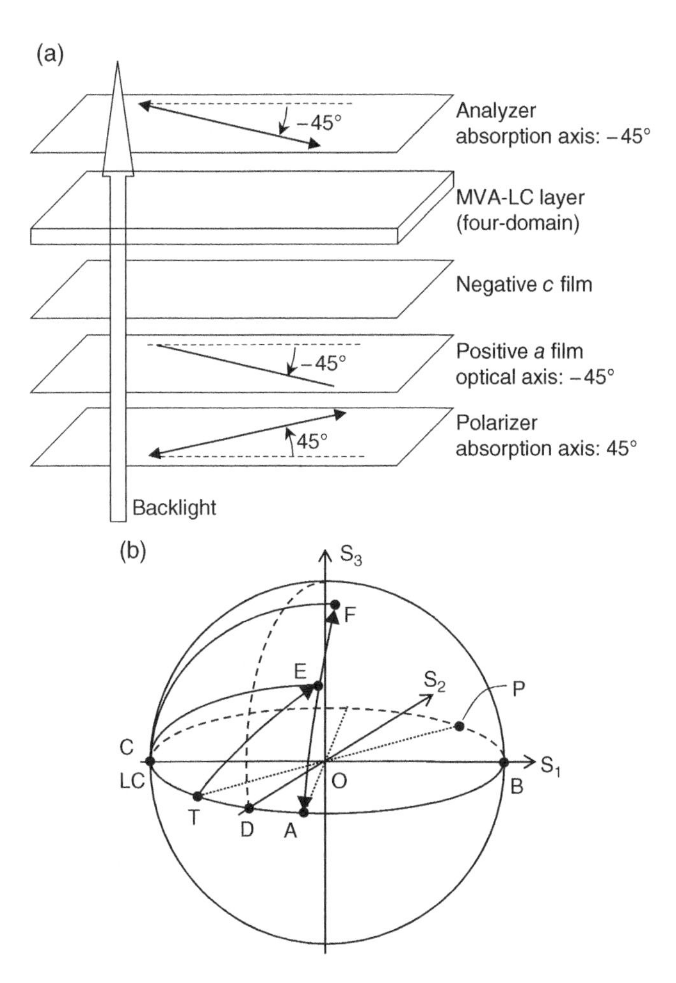

Figure 8.23 (a) Device structure and (b) compensation principle of an MVA-LCD using a positive *a* film and a negative *c* film. Zhu 2006. Reproduced with permission from IEEE.

successively passes through the positive a film and the negative c film, whose positions on the Poincaré sphere are points A and C, respectively. When the linearly polarized light (point T) passes through the positive a film, its polarization state is clockwise rotated from point T to point E around the AO axis. Point E is the first intermediate polarization state which, in general, is elliptical. When this elliptically polarized light traverses the negative c film, its polarization state is rotated counterclockwise from point E to point F around the CO axis. Point F is the second intermediate polarization state, which is also elliptical. Then this second intermediate elliptically polarized light passes through the unactivated MVA-LC layer, whose position on the Poincaré sphere overlaps with point C. Let us assume that we can find the proper phase retardations of the positive a film and negative c film such that when the second intermediate elliptically polarized light passes through the unactivated MVA-LC layer its polarization state is rotated clockwise from point F to point A around the CO axis. Consequently, at $V=0$ the light is completely absorbed by the analyzer (point A) and a good dark state is achieved even viewed from the oblique angle at the bisector positions.

To reach this objective, we can readily determine from Figure 8.23(b) that the following three requirements must be satisfied: (1) the arc $\overline{\boldsymbol{EA}}$ should be equal to the arc $\overline{\boldsymbol{TA}}$, (2) the arcs $\overline{\boldsymbol{AC}}$, $\overline{\boldsymbol{EC}}$, and $\overline{\boldsymbol{FC}}$ should all be equal to each other, and (3) the spherical angle $\angle ACF$ is the sum of the spherical angles $\angle ACE$ and $\angle ECF$. Besides, from Figure 8.23(b), we also find that $\angle POB=\angle AOB=\varphi$, $\overline{\boldsymbol{TA}}=\pi-2\varphi$, and $\overline{\boldsymbol{AC}}=\pi-\varphi$. Based on the spherical trigonometry, we can derive the following relationships from the spherical triangles CAE and CAF:

$$\angle EAC=\cos^{-1}\left(-\text{ctg}^2\varphi\right), \tag{8.35}$$

$$\angle ACE=2\sin^{-1}(\text{ctg}\,\varphi), \tag{8.36}$$

where ϕ, determined by Equation (8.23), is the effective polarizer angle on the wave plane from the lower bisector viewing position $\phi_0=270°$.

Since the required positive a film's phase retardation Γ_{a+} equals to the spherical angle $\angle EAT\ (=\angle EAC)$, then from Equations (8.16) and (8.35) the positive a film's thickness d_{a+} is found to be

$$d_{a+}=\lambda\frac{\cos^{-1}(-\text{ctg}^2\varphi)/2\pi}{n_{a+,e}\sqrt{1-\dfrac{\sin^2\theta_o}{2n_{a+,e}^2}-\dfrac{\sin^2\theta_o}{2n_{a+,o}^2}}-n_{a+,o}\sqrt{1-\dfrac{\sin^2\theta_o}{n_{a+,o}^2}}}. \tag{8.37}$$

In the derivation of Equation (8.37), we substitute $\phi_n=-45°$ and $\phi_0=270°$ into Equation (8.16) because the positive a film's optical axis is oriented at the $-45°$ direction, as Figure 8.23(a) shows, and the viewing direction is from $\phi_0=270°$ azimuthal angle.

On the other hand, the required negative c film's phase retardation Γ_{c-} equals the negative spherical angle $\angle ECF$, i.e. $\Gamma_{c-}=-\angle ECF=\angle ACE-\angle ACF$. The spherical angle $\angle ACF$ is equal to the unactivated MVA-LC layer's phase retardation Γ_{LC} as in Equation (8.34). This is because the function of the unactivated MVA-LC layer is to clockwise rotate around the CO axis from point F to point A. Thus, from Equations (8.18) and (8.36) we can obtain the required negative c film's thickness d_{c-} as

$$d_{c-} = \lambda \frac{[2\sin^{-1}(\text{ctg}\varphi) - \Gamma_{LC}]/2\pi}{n_{c-,o}\left(\sqrt{1 - \frac{\sin^2\theta_o}{n_{c-,e}^2}} - \sqrt{1 - \frac{\sin^2\theta_o}{n_{c-,o}^2}}\right)}, \tag{8.38}$$

where Γ_{LC} is given by Equation (8.34).

As we can see from Equations (8.23), (8.34), (8.37), and (8.38), for a given MVA-LC cell, the required film thicknesses depend on the incident angle θ_0, the film's refractive indices $n_{c-,e}$, $n_{c-,o}$, $n_{a+,e}$, and $n_{a+,o}$, and the polarizer's average real refractive index n_p. Therefore, once we know the refractive indices of both the films and the polarizer, as well as the intended viewing angle or incident angle for LCD optimization, we can determine the thickness of the compensation films from Equations (8.23), (8.34), (8.37), and (8.38). For example, if we want to optimize the LCD viewing angle at $\theta_0 = 70°$, then we can plug the parameters listed in Tables 8.2 and 8.3 into Equations (8.37) and (8.38) and find $d_{a+} = 92.59\ \mu\text{m}$ and $d_{c-} = 186.08\ \mu\text{m}$. Based on the obtained film thicknesses, Figure 8.24 depicts the calculated iso-contrast contour of an MVA-LCD compensated by one positive *a* film and one negative *c* film. Comparing Figure 8.24 with Figure 8.3, we can clearly see that the viewing angle performance at off-axis viewing directions, especially the bisector positions $\phi_0 = 0°$, 90°, 180°, and 270°, is dramatically improved. In the meantime, the contrast ratios at on-axis viewing directions ($\phi_0 = 45°$, 135°, 225°, and 315°) remain the same.

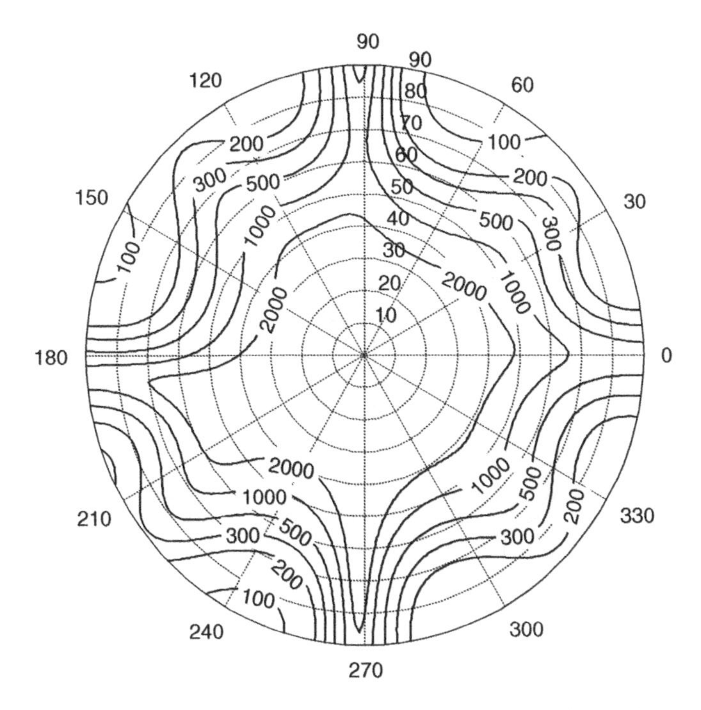

Figure 8.24 Simulated iso-contrast contour of an MVA-LCD with a positive *a* film ($d_{a+} = 92.59\ \mu\text{m}$) and a negative *c* film ($d_{c-} = 186.08\ \mu\text{m}$) under $\lambda = 550\ \text{nm}$.

This compensation scheme can be modified by exchanging the positions of the negative *c* film and the MVA-LC layer. In other words, the incident light passes through the MVA-LC layer first before it enters the negative *c* film. In this case, the required film thickness can still be determined by Equations (8.37) and (8.38), and the resultant viewing angle performance is nearly identical to Figure 8.24.

In addition, this compensation scheme can be further modified by placing the positive *a* film and negative *c* film between the MVA-LC layer and the analyzer, as Figure 8.25(a) shows. In contrast to Figure 8.23(a), now the positive *a* film is adjacent to the analyzer and its optical axis is perpendicular to the absorption axis of the analyzer. Figure 8.25(b) shows the compensation principle of the modified device. Following the same procedures, we can readily determine from Figure 8.25(b) that the required film thicknesses are identical to Equations (8.37) and (8.38). The resultant viewing angle is almost the same as that plotted in Figure 8.24.

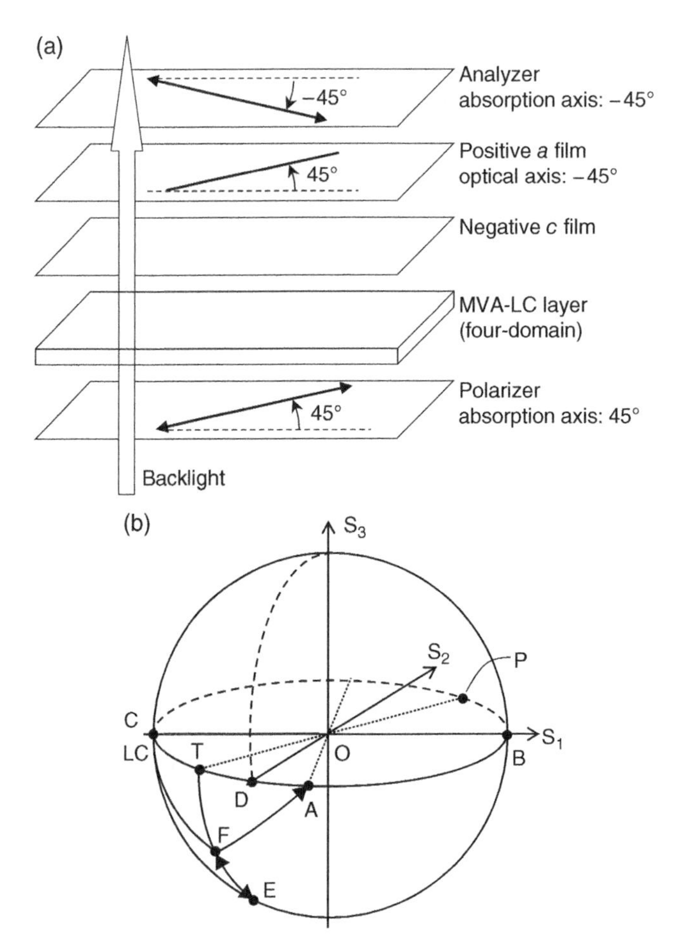

Figure 8.25 (a) An alternative device structure and (b) compensation principle of the MVA-LCD using a negative *c* film and a positive *a* film.

As shown in Figure 8.24, with film compensation the 100:1 contrast ratio barely exceeds ~75° polar angle. This viewing angle performance is not as good as that of the IPS-LCDs described in Section 8.5. This is due to the fact that the intermediate states, points E and F, are not located on the great circle that passes through S_2 and S_3 axes.

In the next example, we will describe a compensation scheme in which the intermediate states are located on the great circle passing through the S_2 and S_3 axes and bisecting the arc $\overline{TA}$.

8.5.2 *MVA with a positive* a*, a negative* a*, and a negative* c *film*

Figure 8.26(a) shows the schematic device configuration of an MVA-LCD with one positive *a* film, one negative *a* film, and one negative *c* film. As shown in the figure, the positive *a* film is located between the polarizer and the MVA-LC layer, while the negative *a* film and negative

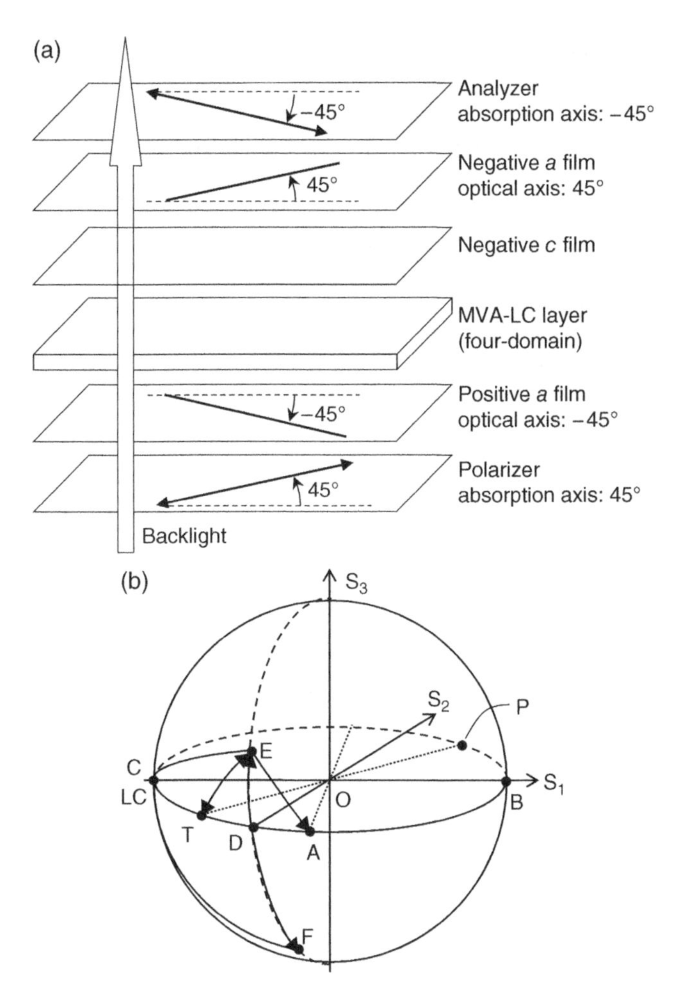

Figure 8.26 (a) Device structure and (b) compensation principle of an MVA-LCD with one positive *a* film, one negative *a* film, and one negative *c* film.

c film are sandwiched between the MVA-LC layer and the analyzer. More specifically, the optical axis of the positive a film is parallel to the absorption axis of the analyzer, and the optical axis of the negative a film is parallel to the absorption axes of the polarizer. Figure 8.26(b) explains the compensation principle on the Poincaré sphere when the observer views the LCD panel from an oblique angle at the lower bisector position $\phi_0 = 270°$.

The detailed compensation mechanism is explained using Figure 8.26(b). When the unpolarized light from backlight passes through the polarizer (point P), it becomes linearly polarized and its polarization state is located at point T, which deviates from the absorption axis of the analyzer (point A). Afterwards, the linearly polarized light (point T) traverses the positive a film, whose position on the Poincaré sphere overlaps with point A, and its polarization state is rotated clockwise from point T to point E around the AO axis. Point E is the first intermediate elliptical polarization state.

To obtain a symmetric viewing angle, we intentionally let the point E locate on the great circle passing through S_2 and S_3 axes and bisecting the arc $\overline{TA}$. Then this elliptically polarized light successively enters the unactivated MVA-LC layer and the negative c film, whose positions on the Poincaré sphere are both at point C. When the elliptically polarized light passes through the unactivated MVA-LC layer, its polarization state is clockwise rotated around the CO axis from point E to point F. Point F is the second intermediate elliptical polarization state, which is also located on the same great circle passing through S_2 and S_3 axes. Then this second intermediate elliptically polarized light hits the negative c film. If the phase retardation of the negative c film is so designed that when the second intermediate elliptically polarized light passes through the negative c film, its polarization state will be rotated counterclockwise around the CO axis from point F back to point E. Now point E represents the third intermediate elliptical polarization state. After that, this third elliptically polarized light passes through the negative a film, whose position on the Poincaré sphere overlaps with point P. If the phase retardation of the negative a film is properly chosen such that when the third intermediate elliptically polarized light (point E) passes through the negative a film, its polarization state can be rotated counterclockwise around the PO axis from point E to point A. Consequently, the light is completely absorbed by the analyzer (point A) and a good dark state is achieved even so it is viewed from the bisector directions.

To determine each film's thickness, from Figure 8.26(b) we find that the following two requirements must be satisfied: (1) the arcs of $\overline{EA}$, $\overline{ET}$, and $\overline{TA}$ should be all equal, and (2) the arc $\overline{EF}$ is located on the great circle passing through the S_2 and S_3 axes and bisecting the arc $\overline{TA}$. This implies that the spherical triangle ETA is an equilateral spherical triangle. Besides, from Figure 8.24(b), we also obtain $\angle POB = \angle AOB = \phi$ and $\overline{TA} = \pi - 2\phi$. Based on the spherical trigonometry, we derive the following relationships from the equilateral spherical triangle ETA:

$$\angle \mathrm{EAT} = \angle \mathrm{ETA} = \cos^{-1}(-\mathrm{ctg}\,\varphi \cdot \mathrm{ctg}\,2\varphi), \tag{8.39}$$

where φ, determined by Equation (8.23), is the effective polarizer angle on the wave plane as viewing from the lower bisector direction ($\phi_0 = 270°$).

The required positive a film's phase retardation Γ_{a+} is equal to the spherical angle $\angle EAT$, thus from Equations (8.17) and (8.39) the required positive a film's thickness d_{a+} can be expressed as

$$d_{a+} = \lambda \frac{\cos^{-1}(-\mathrm{ctg}\varphi \cdot \mathrm{ctg}2\varphi)/2\pi}{n_{a+,e}\sqrt{1-\frac{\sin^2\theta_o}{2n_{a+,e}^2}-\frac{\sin^2\theta_o}{2n_{a+,o}^2}} - n_{a+,o}\sqrt{1-\frac{\sin^2\theta_o}{n_{a+,o}^2}}}. \tag{8.40}$$

In the process of deriving Equation (8.40), we substitute $\phi_n = -45°$ and $\phi_0 = 270°$ into Equation (8.16) because the positive *a* film's optical axis is oriented in the –45° direction, as Figure 8.26(a) shows, and the viewing direction is from the $\phi_0 = 270°$ azimuthal angle.

Similarly, the required negative *a* film's phase retardation Γ_{a-} is equal to the negative spherical angle $\angle ETA$, i.e. $\Gamma_{a-} = -\angle ETA$. Thus, from Equations (8.17) and (8.39) we derive the negative *a* film's thickness d_{a-} as

$$d_{a-} = -\lambda \frac{\cos^{-1}(-\mathrm{ctg}\varphi \cdot \mathrm{ctg}2\varphi)/2\pi}{n_{a-,e}\sqrt{1-\frac{\sin^2\theta_o}{2n_{a-,e}^2}-\frac{\sin^2\theta_o}{2n_{a-,o}^2}} - n_{a-,o}\sqrt{1-\frac{\sin^2\theta_o}{n_{a-,o}^2}}}. \tag{8.41}$$

Here we substitute $\phi_n = 45°$ and $\phi_0 = 270°$ into Equation (8.17) because the negative *a* film's optical axis is oriented in the 45° direction, as Figure 8.26(a) shows, and the viewing direction is from the $\phi_0 = 270°$ azimuthal angle.

To obtain the negative *c* film's thickness, we need to find its phase retardation Γ_{c-} first. From the compensation mechanism, it is easy to find that $\Gamma_{c-} = -\Gamma_{LC}$ since the unactivated MVA-LC layer's role is to rotate clockwise around the CO axis from point E to point F. On the other hand, the negative *c* film's function is to rotate counterclockwise around the CO axis from point F back to point E. Therefore, from Equation (8.18) we derive the required negative *c* film's thickness d_{c-} as

$$d_{c-} = -\lambda \frac{\Gamma_{LC}/2\pi}{n_{c-,o}\left(\sqrt{1-\frac{\sin^2\theta_o}{n_{c-,e}^2}} - \sqrt{1-\frac{\sin^2\theta_o}{n_{c-,o}^2}}\right)}, \tag{8.42}$$

where Γ_{LC} is determined by Equation (8.34).

From Equations (8.23), (8.29), and (8.40), we find that for a given MVA-LC cell the required film thicknesses depend on the incident angle θ_0, the film's refractive indices $n_{a+,e}$, $n_{a+,o}$, $n_{a-,e}$, $n_{a-,o}$, $n_{c-,e}$, and $n_{c-,o}$, and the polarizer's average real refractive index n_p. Therefore, once we know the refractive indices of the films and the polarizer, and the intended viewing angle (θ_0) for optimizing the LCD panel, we can determine the thickness of the compensation films from Equations (8.23), (8.29), and (8.40). For instance, if we choose $\theta_0 = 70°$ and use the parameters listed in Tables 8.2 and 8.3, then we can calculate the required film thicknesses from Equations (8.40). The results are $d_{a+} = 61.38$ µm, $d_{a-} = 61.37$ µm, and $d_{c-} = 246.11$ µm. Based on these film thicknesses, Figure 8.27 plots the simulated iso-contrast contour for an MVA-LCD with one positive *a* film, one negative *a* film, and one negative *c* film. Comparing Figure 8.27 with Figure 8.24, we can clearly see that the viewing angle at off-axis viewing directions, especially the bisector positions ($\phi_0 = 0°, 90°, 180°$, and $270°$), is dramatically improved. In the meantime, the contrast ratios at on-axis, i.e. $\phi_0 = 45°, 135°, 225°$, and $315°$ remain unchanged.

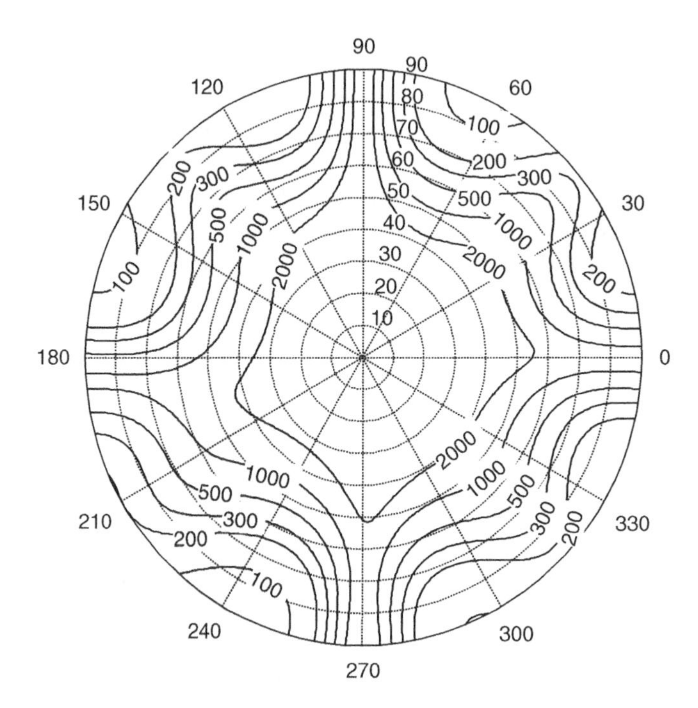

Figure 8.27 Iso-contrast contour of an MVA-LCD with one positive a film ($d_{a+} = 61.38\ \mu m$), one negative a film ($d_{a-} = 61.37\ \mu m$), and one negative c film ($d_{c-} = 246.11\ \mu m$) under $\lambda = 550\,nm$.

This compensation scheme can be modified by exchanging the positions of the negative c film and the MVA-LC layer while keeping both positive and negative a films unchanged. In other words, the incident light passes through the negative c film first before it enters the MVA-LC layer. Under these circumstances, the required film thickness can still be found from Equations (8.40)–(8.42), and the resultant iso-contrast contour is nearly identical to Figure 8.27.

Another alternative is to exchange the positions of both positive a film and negative a film, as shown in Figure 8.28(a). In contrast to the device configuration sketched in Figure 8.26(a), now the positive a film is adjacent to the analyzer, while the negative a film is adjacent to the polarizer. The corresponding compensation mechanism is illustrated in Figure 8.28(b). Following the same analysis as above, we can readily determine from Figure 8.28(b) that the required film thicknesses are identical to Equations (8.40)–(8.42). The resultant iso-contrast contour is almost the same as that shown in Figure 8.27.

From Figure 8.27, the 100:1 iso-contrast contours barely exceed ~75° polar angle. Although the two intermediate polarization states, points E and F, are located on the great circle which passes through the S_2 and S_3 axes, these two intermediate states are not symmetrically located with respect to the equator plane.

Examples using more sophisticated compensation schemes can be found in [31]. When more compensation films are used, there are more degrees of freedom to be used for optimization. However, the associated cost will be increased.

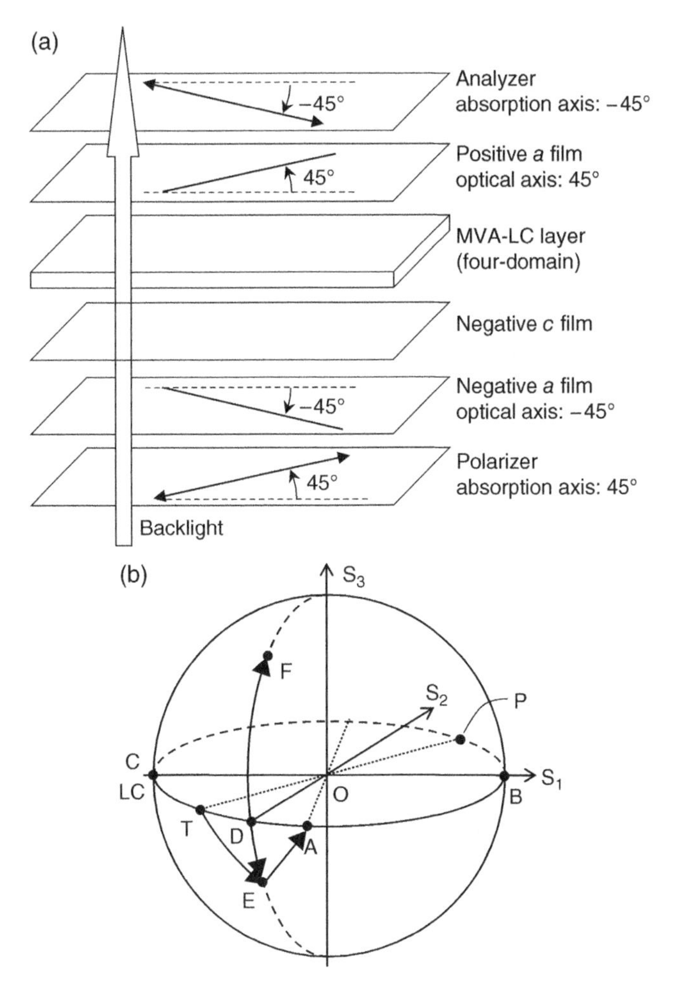

Figure 8.28 (a) An alternative device structure, and (b) compensation principle of an MVA-LCD with one positive *a* film, one negative *a* film, and one negative *c* film.

8.6 Optically Compensated Bend Cell

The optically compensated bend (OCB) mode utilizes a voltage-biased π cell compensated with phase retardation films. Its major advantages are two folds: (1) fast response time, and (2) symmetric and wide view angle. In a π-cell, [53] the pretilt angle in the alignment surfaces is in the opposite direction, as shown in Figure 8.29. The opposite pretilt angle exhibits two special features: (1) its viewing angle is symmetric, and (2) its bend director profile eliminates the backflow effect and therefore results in a fast response time.

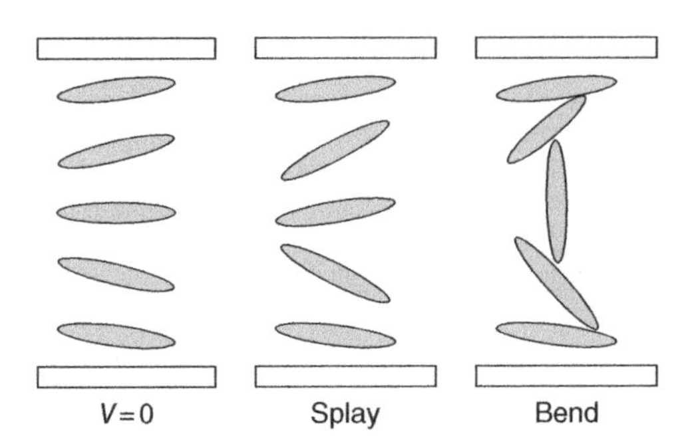

Figure 8.29 LC director configuration in a π cell.

8.6.1 Voltage-dependent transmittance

Figure 8.30 plots the voltage-dependent transmittance curves of a uniaxial film- compensated bend cell. To make the splay-to-bend transition, a critical voltage ($Vc \sim 1.0$–1.5 V) is biased to the π cell. Typically, the cell gap is around 6 μm and pretilt angle is 7–10° [54]. By adjusting the $d\Delta n$ value of the compensation film, both normally white and normally black modes can be achieved [55]. Figure 8.30 shows the *VT* curves of a normally white OCB at three primary wavelengths (R = 650, G = 550, B = 450 nm). The following parameters are used for simulations: LC $d\Delta n = 436$ nm, $\Delta\varepsilon = 10$, uniaxial *a* film $d\Delta n = 53.3$ nm, and pretilt angle = 7°. In reality, the uniaxial *a* film should be replaced by a biaxial film in order to widen the viewing angle. From Figure 8.30, a common dark state for RGB wavelengths appears at ~4.5 V_{rms}. Wavelength dispersion is a serious concern for any birefringence mode. To solve this problem, multiple cell gaps have to be used, that is, the $d\Delta n/\lambda$ value for all three primary wavelengths should be equal. For example, if $d = 6$ μm is used for the green pixels, then the gaps for red and blue pixels should be 7.1 and 4.9 μm, respectively. Here, the wavelength dispersion of the LC material is neglected [56]. Once these conditions are satisfied, the *VT* curves for R and B will overlap with that of G (dark line).

The fast response time of the OCB cell originates from three factors: bias voltage effect (also known as surface mode) [57], flow effect, and half-cell switching. The switching time between gray levels is less than 3 ms. Fast response time is particularly important for LCD TV applications, especially at cold ambient.

For other LCD modes, such as TN, multi-domain VA and IPS, flow in the LC layer slows the rotational relaxation process of the director when the applied voltage is changed. For the π cell, on the other hand, there is no conflict of torque exerted by flow and relaxation process of the director. The intrinsic wide viewing angle is due to the self-compensating structure. The retardation value stays almost the same, even when the incident angle is changed in the director plane. However, retardation is not self-compensated at incidence out of the director plane. In addition, the on-axis CR of the π cell is low because of residual retardation even at a high applied voltage. To obtain a high on-axis CR and a wide viewing angle, an optical compensation film is required.

To obtain the bend alignment structure of the π cell, a voltage above the splay-to-bend transition voltage must be applied. The transition from splay to bend takes time, typically, in the order of tens of seconds. The transition should be made faster than, say, one second.

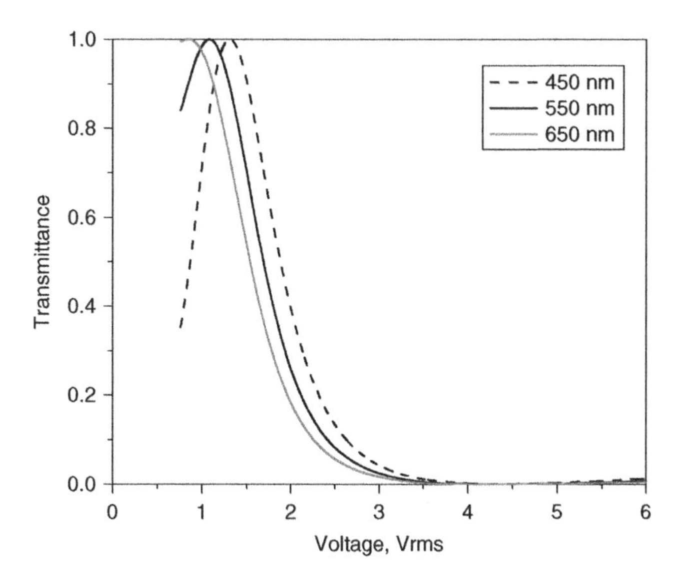

Figure 8.30 Voltage-dependent transmittance curves of a π cell. $d\Delta n = 436$ nm, uniaxial film $d\Delta n = 53.3$ nm and its optic axis is perpendicular to that of LC cell.

8.6.2 Compensation films for OCB

To obtain a comparable viewing angle with VA and IPS, OCB requires more sophisticated optical compensation based on a discotic material [58]. Figure 8.31 shows the compensation schemes for a normally white OCB mode. The fundamental idea is similar to that for TN. The retardation matching between the cell and the optical compensation film is especially important for the OCB mode, partially because the black state of the normally white OCB cell has a finite residual retardation value that must be compensated by an optical film. For example, any retardation fluctuation of the cell or the film is easily noticeable. The OCB system requires a high level of uniformity. And the cell parameters, as well as the film parameters, should be optimized in order to maximize the optical performance.

The polymer discotic material (PDM) developed by Fuji Photo Film has a hybrid alignment, which mimics half of the bend alignment structure of the OCB cell. In contrast to the discotic film developed for TN LCDs, the azimuthal alignment direction of the PDM layer is oriented at 45° to the transmission axis of polarizer, and the in-plane retardation of the PDM layer compensates for the in-plane retardation of the on-state OCB cell. The total in-plane retardation of the PDM layer should be the same as that of the on-state OCB cell so that the voltage-on state becomes black at a voltage lower than 5 V_{rms}.

Figure 8.32 shows the simulated iso-contrast contour plot of the OCB panel. It is seen that OCB has a comparable viewing angle performance to VA and IPS. In addition to faster response time, OCB has another advantage of less color shift at gray levels. Especially human skin looks good even at oblique incidence.

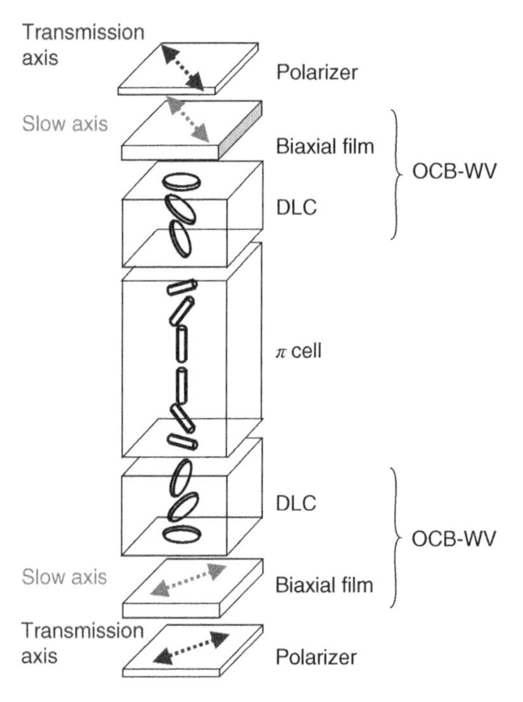

Figure 8.31 Idealized and simplified model of optical compensation for the π cell combined with the Fuji OCB films. Mori 2005. Reproduced with permission from IEEE.

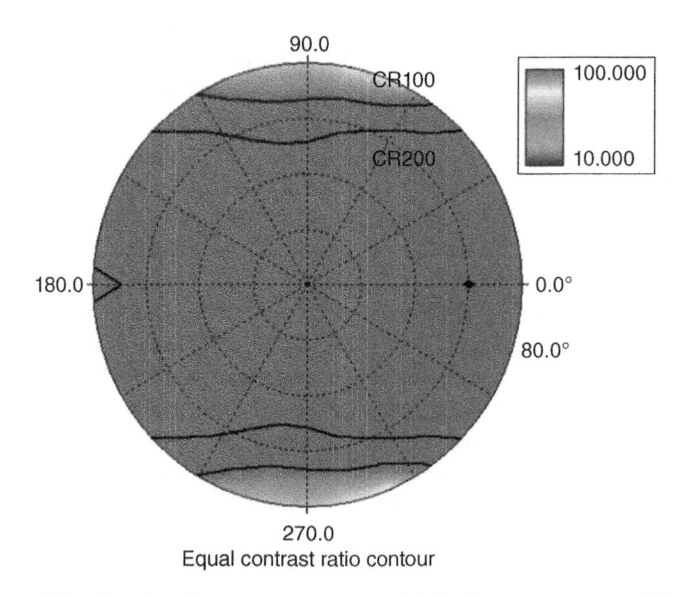

Figure 8.32 Simulated iso-contrast contour of Fuji-film-compensated OCB cell.

Homework Problems

8.1 *Twisted nematic cell*

a. A student has prepared two identical TN cells except that one has 90° twist angle and the other is 80°. How can he distinguish which is which?

b. A 90° TN cell is constructed using the following LC parameters: $\Delta\varepsilon = 8$, $\Delta n = 0.1$ at $\lambda =$ 550 nm, $K_{11} = 14$ pN, $K_{22} = 7$ pN, $K_{33} = 18$ pN, and $\gamma_1 = 0.2$ Pa s. What is the required cell gap to satisfy Gooch–Tarry's first minimum condition? Estimate the *optical* decay time (100–10%) of the TN LC cell.

8.2 *In-plane-switching cell*

The Figure 8.33 shows the device configuration of a transmissive IPS-LCD. The homogeneous alignment liquid crystal mixture is sandwiched between two substrates. The LC rubbing angle is 12° with respect to the IPS electrodes.

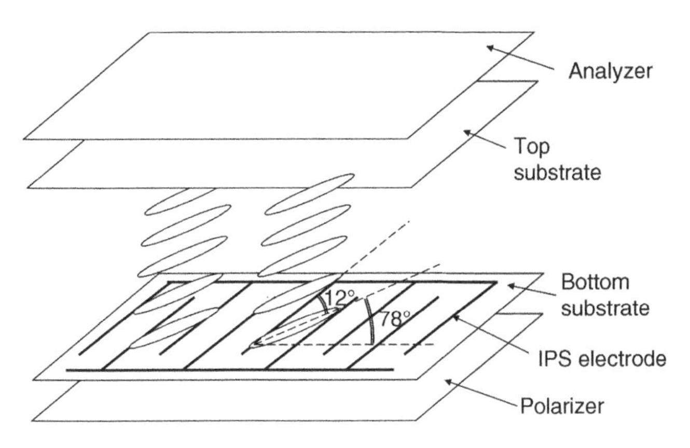

Figure 8.33 Diagram for Problem 8.2.

a. Draw the transmission axis of the polarizer and analyzer in the figure to obtain a normally black mode.

b. If the liquid crystal mixture has birefringence $\Delta n = 0.1$, what is the required minimum cell gap for obtaining high transmittance?

c. Does the above IPS cell work well under normally white condition? Explain.

d. What are the pros and cons if we increase the rubbing angle to 30°?

8.3 *Homeotropic cell*

a. Can a homeotropic cell be used for normally white LCD? Explain.

b. The voltage-dependent transmittance of a homeotropic cell, sandwiched between two crossed polarizers, is shown in Figure 8.34. Cell gap $d = 5$ μm, $\lambda = 633$ nm, and $K_{33} = 15$ pN. Estimate the birefringence (Δn) and dielectric anisotropy ($\Delta\varepsilon$) of the LC mixture.

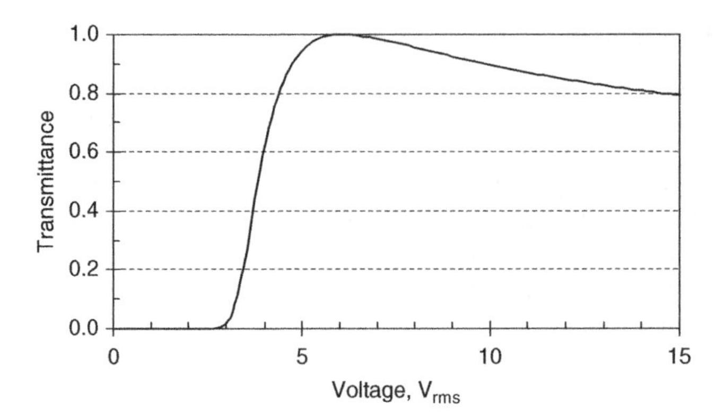

Figure 8.34 Diagram for Problem 8.3.

8.4 *Homogeneous cell*

A homogeneous cell is useful as a tunable phase retardation plate. Figure 8.35 plots the voltage-dependent transmittance curve of a homogeneous LC cell at $\lambda = 633$ nm. The polarizers are crossed and the angle between the front polarizer and the LC rubbing direction is 45°.

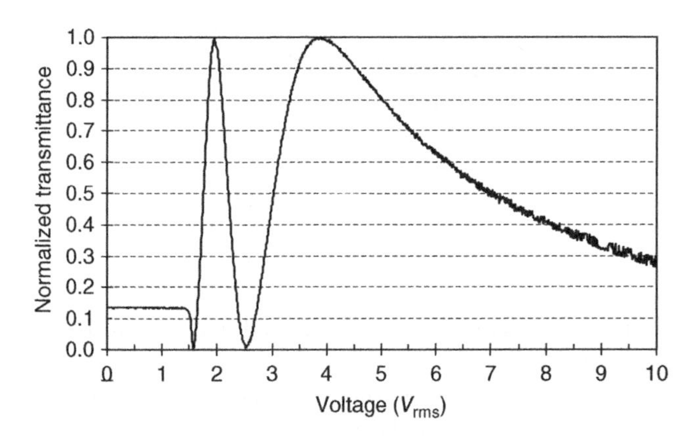

Figure 8.35 Diagram for Problem 8.4.

a. If the cell gap is $d = 5\ \mu m$, what is the birefringence of the LC?
b. At what voltages will the output beam (before the analyzer) be circularly polarized?
c. At what voltages will the output beam (before the analyzer) be linearly polarized?
d. If you want to switch from circular to linear polarization, which voltages do you use in order to obtain the fastest response time?

References

1. E. H. Stupp and M. Brennesholtz, *Projection Displays* (Wiley, New York, 1998).
2. S. T. Wu and D. K. Yang, *Reflective Liquid Crystal Displays* (Wiley, New York, 2001).
3. M. Schadt and W. Helfrich, Voltage-dependent optical activity of a twisted nematic liquid crystal, *Appl. Phys. Lett.* **18**, 127 (1971).
4. C. H. Gooch and H. A. Tarry, The optical properties of twisted nematic liquid crystal structures with twisted angles ≤90°, *J. Phys.* **D8**, 1575 (1975).
5. S. T. Wu and C. S. Wu, Mixed-mode twisted-nematic cell for transmissive liquid crystal display, *Displays* **20**, 231 (1999).
6. H. Mori, Y. Itoh, Y. Nishiura, et al., *Jpn. J. Appl. Phys.* **36**, 143 (1997).
7. H. Mori, The wide view film for enhancing the field of view of LCDs, *J. Display Technology* **1**, 179 (2005).
8. R. A. Soref, Transverse field effect in nematic liquid crystals, *Appl. Phys. Lett.* **22**, 165 (1973).
9. R. A. Soref, Field effects in nematic liquid crystals obtained with interdigital electrodes, *J. Appl. Phys.* **45**, 5466 (1974).
10. R. Kiefer, B. Weber, F. Windscheid, and G. Baur, In-plane switching of nematic liquid crystals, *Japan Displays '92*, p. 547 (1992).
11. M. Oh-e, M. Ohta, S. Arantani, and K. Kondo, Principles and characteristics of electro-optical behavior with in-plane switching mode, *Asia Display '95*, p. 577 (1995).
12. M. Oh-e, M. Yoneya, and K. Kondo, Switching of a negative and positive dielectric anisotropic liquid crystals by in-plane electric field, *J. Appl. Phys.* **82**, 528 (1997).
13. M. Ohta, M. Oh-e, and K. Kondo, Development of super-TFT-LCDs with in-plane switching display mode, *Asia Display '95*, p. 707 (1995).
14. S. H. Lee, S. L. Lee, and H. Y. Kim, Electro-optic characteristics and switching principle of a nematic liquid crystal cell controlled by fringe-field switching, *Appl. Phys. Lett.*, **73**, 2881 (1998).
15. Y. M. Jeon, I. S. Song, S. H. Lee, et al., Optimized electrode design to improve transmittance in the fringe-field switching liquid crystal cell, *SID Tech. Digest*, **36**, 328 (2005).
16. Y. M. Jeon, I. S. Song, S. H. Lee, et al., Optimized electrode design to improve transmittance in the fringe-field switching (FFS) liquid crystal cell, *SID Symp. Dig.*, **36**, 328, (2005)
17. K. H. Lee, S. H. Song, S. M. Yang, et al., CRT-like characteristics of 32" WXGA TFT-LCD by true vision advanced FFS pixel concept, *SID Tech. Digest* **36**, 1742 (2005).
18. Y. Sun, Z. Zhang, H. Ma, et al., Optimal rubbing angle for reflective in-plane-switching liquid crystal display, *Appl. Phys. Lett.*, **81**, 4907 (2002).
19. S. T. Wu, T. X. Wu, Q. Hong, et al., Fast-response in-plane-switching pi-cell liquid crystal displays, US patent 6,987,549 B2 (17 Jan. 2006).
20. J. Chen, K. H. Kim, J. J. Jyu, et al., Optimum film compensation modes for TN and VA LCDs, *SID Tech. Digest*, **29**, 315 (1998).
21. J. E. Anderson and P. J. Bos, Methods and concerns of compensating in-plane switching liquid crystal displays, *Jpn. J. Appl. Phys.*, Part 1, **39**, 6388 (2000).
22. Q. Hong, T. X. Wu, X. Zhu, et al., Extraordinarily high-contrast and wide-view liquid-crystal displays, *Appl. Phys. Lett.*, **86**, 121107, (2005).
23. Y. Saitoh, S. Kimura, K. Kusafuka, and H. Shimizu, Optimum film compensation of viewing angle of contrast in in-plane-switching-mode liquid crystal display, *Jpn. J. Appl. Phys.*, Part 1, **37**, 4822, (1998).
24. T. Ishinabe, T. Miyashita, T. Uchida, and Y. Fujimura, A wide viewing angle polarizer and a quarter-wave plate with a wide wavelength range for extremely high quality LCDs, *Proc. 21st Int'l Display Research Conference (Asia Display/IDW'01)*, 485 (2001).
25. T. Ishinabe, T. Miyashita, and T. Uchida, Wide-viewing-angle polarizer with a large wavelength range, *Jpn. J. Appl. Phys., Part 1*, **41**, 4553 (2002).
26. F. Di Pasqual, H. Deng, F. A. Fernandez, et al., Theoretical and experimental study of nematic liquid crystal display cells using the in-plane-switching mode, *IEEE Trans. Electron Devices*, **46**, 661 (1999).
27. K. Ohmuro, S. Kataoka, T. Sasaki, and Y. Koite, *SID Tech. Digest*, **26**, 845 (1997).
28. P. Yeh and C. Gu, *Optics of Liquid Crystal Displays* (Wiley, New York, 1999).
29. A. Lien, A detail derivation of extended Jones matrix representation for twisted nematic liquid crystal displays, *Liq. Cryst.* **22**, 171 (1997).

30. J. E. Bigelow and R. A. Kashnow, Poincaré sphere analysis of liquid crystal optics, *Appl. Opt.* **16**, 2090 (1977).
31. X. Zhu, Z. Ge, and S. T. Wu, Analytical solutions for uniaxial film-compensated wide-view liquid crystal displays, *J. Display Technology*, **2**, 3 (March, 2006).
32. X. Zhu and S. T. Wu, Super wide view in-plane switching LCD with positive and negative uniaxial a-films compensation, *SID Digest Tech. Papers,* **34**, 1164 (2005).
33. R. Lu, X. Zhu, S. T. Wu, et al., Ultrawide-view liquid crystal displays, *J. Display Technology*, **1**, 3 (2005).
34. W. S. Asada, N. Kato, Y. Yamamoto, et al., An advanced in-plane-switching mode TFT-LCD, *SID Tech. Digest* **28**, 929 (1997).
35. Y. Mishima, T. Nakayama, N. Suzuki, et al., Development of a 19" diagonal UXGA super TFT-LCM applied with super-IPS technology, *SID Tech. Digest* **31**, 260 (2000).
36. M. F. Schiekel and K. Fahrenschon, Deformation of nematic liquid crystals with vertical orientation in electric fields, *Appl. Phys. Lett.* **19**, 391 (1971).
37. J. Grinberg, W. P. Bleha, A. D. Jacobson, et al., Photoactivated birefringence liquid crystal light valve for color symbology display, *IEEE Trans. Electron Devices* **ED-22**, 775 (1975).
38. R. D. Sterling and W. P. Bleha, D-ILA technology for electronic cinema, *Soc. Infor. Display, Tech. Digest*, **31**, 310 (2000).
39. A. Takeda, S. Kataoka, T. Sasaki, et al., A super-high-image-quality multi-domain vertical alignment LCD by new rubbing-less technology, *SID Tech. Digest* **29**, 1077 (1997).
40. M. Oh-e, M. Yoneya, and K. Kondo, Switching of negative and positive dielectric anisotropic liquid crystals by in-plane electric fields, *J. Appl. Phys.* **82**, 528 (1997).
41. C. K. Wei, Y. H. Lu, C. L. Kuo, et al., A wide-viewing angle polymer-stabilized homeotropically aligned LCD, *SID Tech. Digest* **29**, 1081 (1998).
42. Y. Kume, N. Yamada, S. Kozaki, et al., Advanced ASM mode: Improvement of display performance by using a negative-dielectric liquid crystal, *SID Tech. Digest* **29**, 1089 (1998).
43. H. Wang, T. X. Wu, X. Zhu, and S. T. Wu, Correlations between liquid crystal director reorientation and optical response time of a homeotropic cell, *J. Appl. Phys.* **95**, 5502 (2004).
44. S. T. Wu and C. S. Wu, Small angle relaxation of highly deformed nematic liquid crystals, *Appl. Phys. Lett.* **53**, 1794 (1988).
45. S. T. Wu, A nematic liquid crystal modulator with response time less than 100 μs at room temperature, *Appl. Phys. Lett.* **57**, 986 (1990).
46. P. M. Alt, Single crystal silicon for high resolution displays, *Conference record of the Intl. Display Research Conf.*, M19-28, (1997).
47. H. Kurogane, K. Doi, T. Nishihata, et al., Reflective AMLCD for projection displays, *SID Tech. Digest* **29**, 33 (1998).
48. K. Ohmuro, S. Kataoka, T. Sasaki, and Y. Koike, Development of super-high-image-quality vertical alignment-mode LCD, *SID Tech. Digest* **28**, 845 (1997).
49. A. Takeda, S. Kataoka, T. Sasaki, et al., A super high image quality multi-domain vertical alignment LCD by new rubbing-less technology, *SID Tech. Digest* **29**, 1077 (1998).
50. J. O. Kwag, K. C. Shin, J. S. Kim, et al, Implementation of new wide viewing angle mode for TFT-LCDs, *SID Tech. Digest* **31**, 256 (2000).
51. S. S. Kim, The world's largest (82-in) TFT LCD, *SID Tech. Digest* **36**, 1842 (2005).
52. P. G. de Gennes, and J. Prost, *The Physics of Liquid Crystals*, 2nd edn, (New York: Oxford, 1995).
53. P. J. Bos, K. R. Koehler/Beran, The π-cell: a fast liquid crystal optical switching device, *Mol. Cryst. Liq. Cryst.* **113**, 329 (1984).
54. T. Uchida, Field sequential full color LCD without color filter by using fast response LC cell, *The 5th Int'l Display Workshops*, p. 151 (1998).
55. Y. Yamaguchi, T. Miyashita, and T. Uchida, Wide-viewing-angle display mode for the active-matrix LCD using bend-alignment liquid crystal cell, *SID Tech. Digest* **24**, 277 (1993).
56. S. T. Wu, Birefringence dispersion of liquid crystals, *Phys. Rev.* A **33**, 1270 (1986).
57. J. L. Fergason, Liquid crystal display with improved angle of view and response time, US patent **4**,385,806 (1983).
58. Y. Ito, R. Matsubara, R. Nakamura, et al., OCB-WV film for fast-response-time and wide viewing angle LCD-TVs, *SID Tech. Digest* **36**, 986 (2005).

9

Reflective and Transflective Liquid Crystal Displays

9.1 Introduction

As described in Chapter 8, transmissive liquid crystal displays (LCDs) have been widely used in laptop computers, desktop monitors, and high-definition televisions (HDTVs). The most commonly used transmissive 90° twisted-nematic (TN) LCD [1] exhibits a high contrast ratio due to the self phase compensation effect of the orthogonal boundary layers in the voltage-on state. However, its viewing angle is relatively narrow since the liquid crystal (LC) molecules are switched out of the plane and the oblique incident light experiences different phase retardations at different angles. For TV applications, wide viewing angle is highly desirable. Currently, in-plane switching (IPS) [2] and multi-domain vertical alignment (MVA) [3] are the mainstream approaches for wide-view LCDs. A major drawback of the transmissive LCD is that its backlight source needs to be kept on all the time, as long as the display is in use; therefore, the power consumption is relatively high. Moreover, the image of a transmissive LCD could be washed out by the strong ambient light, such as direct sunlight, because the panel's surface reflection from the direct sunlight is much brighter than the displayed images.

On the other hand, a reflective LCD has no built-in backlight unit; instead, it utilizes the ambient light for displaying images [4]. In comparison to transmissive LCDs, reflective LCDs have advantages in lower power consumption, lighter weight, and better outdoor readability. However, a reflective LCD relies on the ambient light and thus is inapplicable under low or dark ambient conditions.

In an attempt to overcome the above drawbacks and take advantage of both reflective and transmissive LCDs, transflective LCDs have been developed to use the ambient light when available and the backlight only when necessary [5]. A transflective LCD can display images in both transmissive mode (T-mode) and reflective mode (R-mode) simultaneously or independently. Since LC material itself does not emit light, the transflective LCD must rely on

Fundamentals of Liquid Crystal Devices, Second Edition. Deng-Ke Yang and Shin-Tson Wu.

either ambient light or backlight to display images. Under bright ambient circumstances, the backlight can be turned off to save power and therefore the transflective LCD operates in the R-mode only. Under dark ambient conditions, the backlight is turned on for illumination, and the transflective LCD works in the T-mode. In the low to medium ambient surroundings, the backlight is still necessary. In this case, the transflective LCD runs in both T- and R-mode simultaneously. Therefore, the transflective LCD can accommodate a large dynamic range. Currently, the applications of transflective LCD are mainly targeted to mobile display devices, such as cell phones, digital cameras, camcorders, personal digital assistants, pocket personal computers, and global position systems.

The major scientific and technological challenges for a transflective LCD are: transflector design, inequality in optical efficiency, color, and response time between the T-mode and R-mode. In this chapter, we first introduce the basic operation principles of reflective LCDs and then transflectors and their underlying operating principles. Afterwards, we analyze the factors affecting the image qualities. Finally, we describe the major problems of the current transflective LCD technologies and discuss potential solutions.

9.2 Reflective Liquid Crystal Displays

Two types of reflective liquid crystal displays (R-LCDs) have been developed: direct-view and projection. Direct-view R-LCDs use ambient light for reading the displayed images, but projection R-LCDs use an arc lamp or bright LEDs for projecting images onto a large screen. Direct-view R-LCDs are commonly used in games, signage, and some cell phones, while projection R-LCDs are used in liquid-crystal-on-silicon (LCoS) rear projection TVs. Although the involved panel resolution and optical systems for direct-view and projection displays are different, their underlying LC operation modes are quite similar. Two monographs have been devoted to projection [6] and direct-view reflective LCDs [7]. In this section, we will only cover the background material for the introduction of transflective LCDs.

Figure 9.1 shows a typical R-LCD device structure. The linear polarizer and a broadband quarter-wave film forms an equivalent 'crossed' polarizer. This is because the LC modes work better under crossed-polarizer conditions. The bumpy reflector not only reflects but also diffuses the ambient light to the observer. This is the most critical part in an R-LCD. The TFT is hidden beneath the bumpy reflector. Thus, the R-LCD can have a large aperture ratio (~90%). The light blocking layer (LBL) is used to absorb the scattered light from neighboring pixels.

Three popular LCD modes have been widely used for R-LCDs and transflective LCDs. They are: (1) vertical alignment (VA) cell, (2) film-compensated homogeneous cell, and (3) mixed-mode twisted nematic (MTN) cell. The VA and homogeneous cells utilize the phase retardation effect while the MTN cells use a combination of polarization rotation and birefringence effects. The VA cell has been described in detail in Section 8.4 for wide view LCDs. For reflective LCDs, the cell gap is reduced to one half of that of a transmissive LCD to account for the double pass of the incoming light. The voltage-dependent reflectance curves are basically the same as those shown in Figure 8.19 and will not be repeated in this chapter. The film-compensated VA cell is a favored choice for transflective LCDs because of its high contrast ratio and wide viewing angle. In the following two sections, we will briefly describe the film-compensated homogeneous cell and MTN cells.

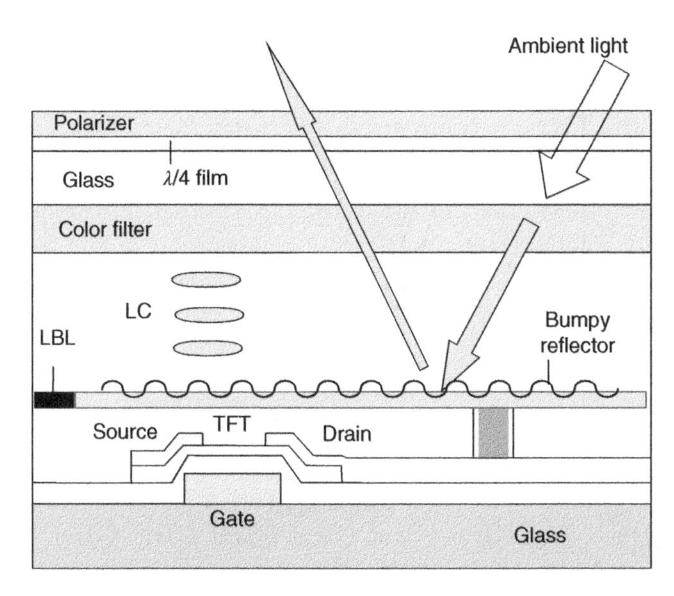

Figure 9.1 Device structure of a direct-view reflective LCD.

9.2.1 *Film-compensated homogeneous cell*

A homogeneous cell is not suitable for transmissive display because of its narrow viewing angle and lack of common dark state for RGB colors [8]. However, in a reflective display, the viewing angle is equivalent to a two-domain cell due to the mirror image effect [9]. For some handheld small-screen displays, the requirement for wide viewing angle is not so demanding, so the homogeneous cell can be still useful. But for some high-end transflective LCDs intended for playing videos and movies, wide view is a necessity.

For a homogeneous cell, to obtain a common dark state for full-color display a phase compensation film has to be used in order to cancel the residual phase retardation of the cell resulting from boundary layers [10]. To design a homogeneous cell for reflective display, the required minimal $d\Delta n$ value is $\lambda/4$, that is, the LC cell functions like a quarter-wave plate. Using $\lambda = 550$ nm, we find $d\Delta n = 137.5$ nm. The cell gap and birefringence can be chosen independently, depending on the need. One could choose a thinner cell gap for obtaining faster response time, or a lower birefringence LC mixture to maintain a reasonable cell gap for high yield manufacturing. Another design with $d\Delta n = \lambda/2$ has been found to have a weak color dispersion [11].

For the purpose of illustrating the design procedures, let us use a Merck LC mixture MLC-6297-000 as an example. The LC and cell parameters are listed as follows: the angle between the front polarizer and cell rubbing direction $\beta = 45°$, pretilt angle $\alpha = 2°$, elastic constants $K_{11} = 13.4$, $K_{22} = 6.0$, and $K_{33} = 19.0$ pN, dielectric constants $\varepsilon_{//} = 10.5$ and $\Delta\varepsilon = 6.9$, $\Delta n =$ 0.125, 0.127, and 0.129 for R = 650, G = 550, and B = 450 nm, respectively.

Figure 9.2(a) depicts the voltage-dependent light reflectance of a homogeneous cell with $d\Delta n = 137.5$ nm under crossed-polarizer condition. For the purpose of comparing intrinsic LC performance, we only consider the normalized reflectance; the optical losses from polarizer, substrate surfaces, indium-tin-oxide (ITO), and any other compensation films are neglected. From Figure 9.2(a), the bright state intensity variation among RGB colors is within 10%.

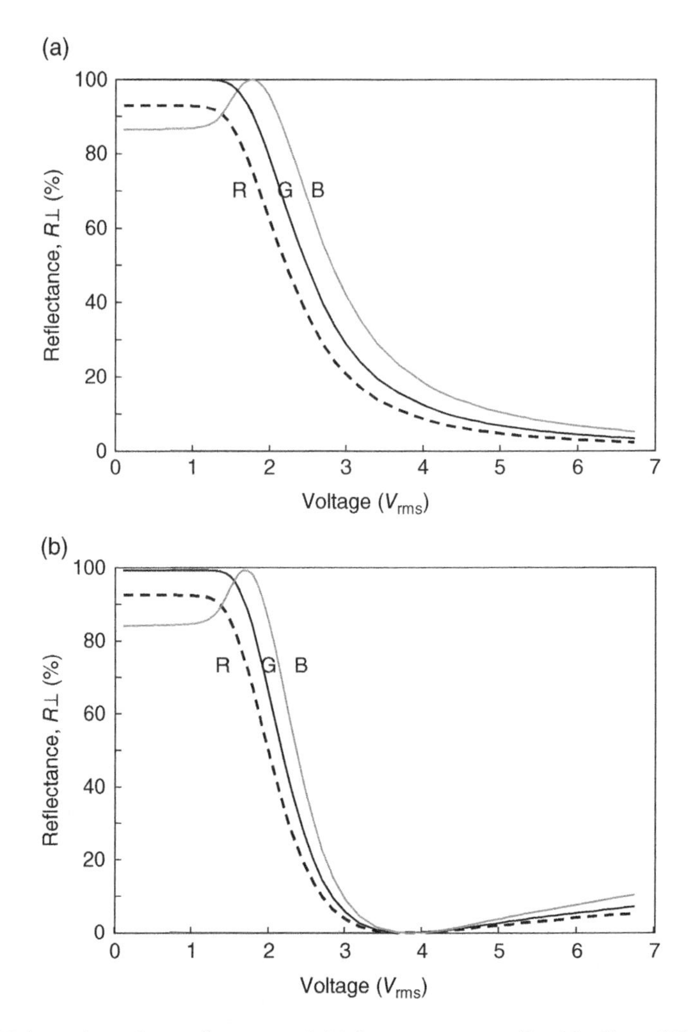

Figure 9.2 Voltage-dependent reflectance of (a) homogeneous cell with $d\Delta n = 137.5$ nm, and (b) film-compensated homogeneous cell with LC's $d\Delta n = 184$ nm and *a* film's $d\Delta n = -48$ nm.

In the high-voltage regime, the reflectance is monotonously decreasing. However, it is difficult to obtain a common dark state for the RGB colors. A uniaxial phase compensation film (also called *a* film) is needed. Figure 9.2(b) plots the voltage-dependent reflectance of a homogeneous cell ($d\Delta n = 184$ nm) compensated by an *a* film having $(d\Delta n)_{film} = -48$ nm. A positive *a* film can be used as well, as long as its optic axis is perpendicular to the LC's rubbing direction. This normally white mode has a relatively weak color dispersion and low dark state voltage. In the high-voltage regime, the residual LC phase is diminishing but the phase of the compensation film remains. As a result, some light leakage is observed. For display applications, the dark state voltage can be controlled by the driving circuit. An important consideration is on the width of dark state so that when temperature fluctuates, the display contrast would not be significantly affected. From simulations, a smaller *a* film's $d\Delta n$ value would lead to a broader dark state at a higher voltage. That means that the required voltage swing is larger.

9.2.2 *Mixed-mode twisted nematic (MTN) cells*

In a reflective cell shown in Figure 9.1, the incident light traverses the linear polarizer, $\lambda/4$ film, LC layer, and is reflected back by the embedded mirror in the inner side of the rear substrate. In the voltage-off state, the normalized reflectance can be obtained by the Jones matrix method [12]:

$$R_{\perp} = \left(\Gamma\frac{\sin X}{X}\right)^2 \left(\sin 2\beta \cos X - \frac{\phi}{X}\cos 2\beta \sin X\right)^2 \tag{9.1}$$

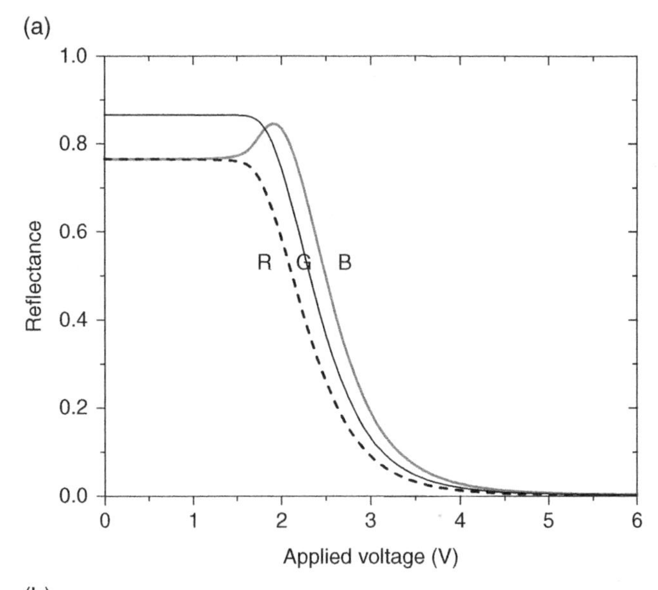

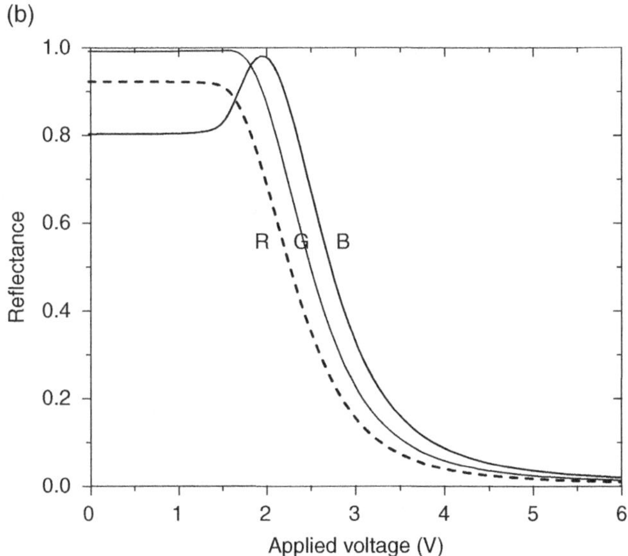

Figure 9.3 Voltage dependent reflectance of (a) 90° and (b) 75° MTN cells. The cell and LC material parameters are discussed in the text.

Here, $\Gamma = 2\pi\, d\, \Delta n/\lambda$, d is the cell gap, $X = \sqrt{\phi^2 + (\Gamma/2)^2}$, ϕ is the twist angle, and β is the angle between the polarization axis and the front LC director.

Several MTN modes with twist angle ϕ varying from 45° to 90° have been used for direct-view and projection displays, depending on the desired contrast ratio and optical efficiency. In transflective LCDs, the 75° and 90° MTN cells are frequently used. Therefore, we only discuss these two modes here.

Figures 9.3(a) and (b) show the voltage-dependent reflectance (VR) of the 90° and 75° MTN cells, respectively. For simulations, a Merck MLC-6694-000 ($\Delta n = 0.0857$ at $\lambda = 540$ nm) LC mixture and 2° pretilt angle are used. The 90° MTN cell has $d\Delta n = 240$ nm and $\beta = 20°$ and the 75° MTN cell has $d\Delta n = 250$ nm and $\beta = 15°$. Both MTN cells are broadband devices; that means their VR curves are insensitive to the wavelength. During simulations, the following bandwidths are considered: R = 620–680 nm, G = 520–560 nm, and B = 420–480 nm. For each mode, the reflectance is calculated at every 10 nm and then averaged over the entire band.

From Figure 9.3, the 90° MTN cell exhibits a good dark state, similar to a transmissive TN cell because of the self-phase compensation effect of the orthogonal boundary layers. However, its maximum reflectance is only ~88%. On the other hand, the 75° MTN cell has nearly 100% reflectance, but its dark state has a slight light leakage. The contrast ratio at 5 V_{rms} is around 100:1. This is because the boundary layers are not perfectly compensating each other.

For direct-view reflective displays, the outmost surface reflection (usually it is a plastic protective film without anti-reflection coating) limits the device contrast ratio. Thus, the ~100:1 contrast ratio of the 75° MTN is still adequate. However, in projection displays the contrast ratio needs to exceed 1000:1. The 90° MTN and the VA cells are the better choices.

9.3 Transflector

Since a transflective LCD should possess dual functions (transmission and reflection) simultaneously, a transflector is usually required between the LC layer and the backlight source. The main role of the transflector is to partially reflect the incident ambient light back, and to partially transmit the backlight to the viewer. From the device structure viewpoint, the transflector can be classified into four major categories: (1) openings-on-metal transflector, (2) half-mirror metal transflector, (3) multilayer dielectric film transflector, and (4) orthogonal polarization transflector, the first three of which are shown in Figure 9.4 and the last in Figure 9.5.

9.3.1 Openings-on-metal transflector

Figure 9.4(a) shows the schematic structure of the openings-on-metal transflector proposed by Ketchpel and Barbara [13]. The typical manufacturing steps include first forming wavy bumps on the substrate, then coating a metal layer, such as silver or aluminum, on the bumps, and finally etching the metal layer according to the predetermined patterns. After etching, those etched areas become transparent so that the incident light can transmit through, while those unaffected areas are still covered by the metal layer and serve as reflectors. The wavy bumps function as diffusive reflectors to steer the incident ambient light away from surface specular reflection. Thus, the image contrast ratio is enhanced and the viewing angle widened

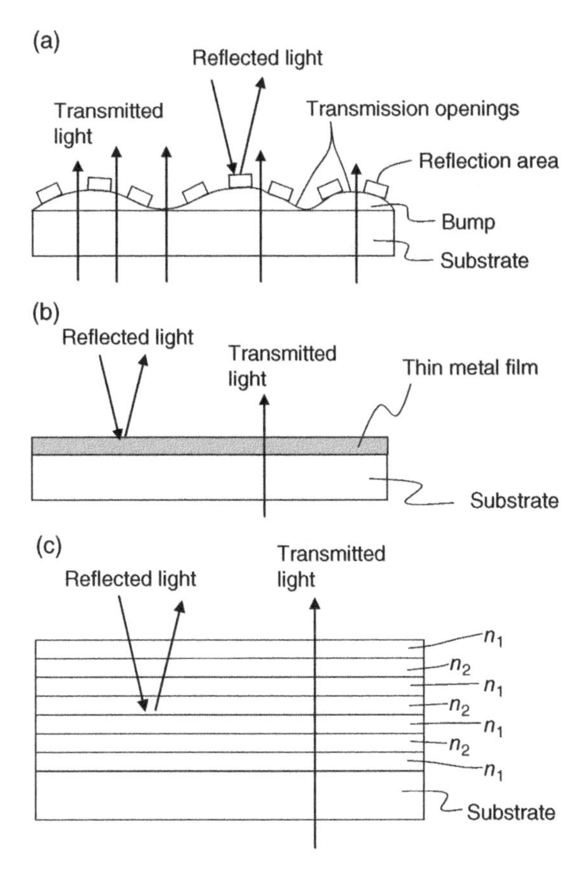

Figure 9.4 Schematic illustration of the first three major types of transflectors: (a) openings-on-metal transflector, (b) half-mirror metal transflector, and (c) multilayer dielectric film transflector. Zhu 2006. Reproduced with permission from IEEE.

in the R-mode. Due to the simple manufacturing process, low cost, and stable performance, this type of transflector is by far the most popularly implemented in the transflective LCD devices.

9.3.2 *Half-mirror metal transflector*

The half-mirror has been widely used in optical systems as a beam splitter. It was implemented into transflective LCDs by Borden [14] and Bigelow [15], with basic structure shown in Figure 9.4(b). When depositing a very thin metallic film on a transparent substrate, one can control the reflectance and transmittance by adjusting the metal film thickness. The film thickness could vary, depending on the metallic material employed. Typically, the film thickness is around a few hundred angstroms. Since the transmittance/reflectance ratio of such a half-mirror transflector is very sensitive to the metal film thickness, the manufacturing tolerance is very narrow and volume production is difficult. Consequently, this kind of transflector is not too popular for commercial products.

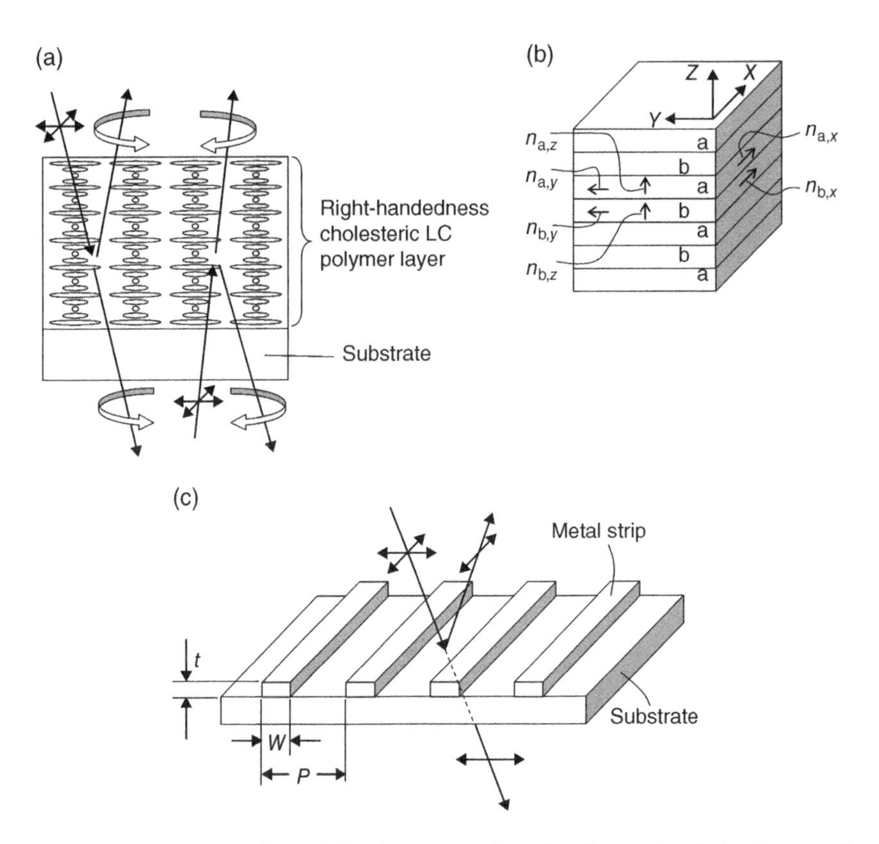

Figure 9.5 Schematic illustration of the three examples of orthogonal polarization transflectors: (a) cholesteric reflector, (2) birefringent interference polarizer, and (3) wire grid polarizer. Zhu 2006. Reproduced with permission from IEEE.

9.3.3 *Multilayer dielectric film transflector*

Multilayer dielectric film is a well-developed technique in thin-film optics, but only very recently was it incorporated in transflective LCDs [16]. As illustrated in Figure 9.4(c), two dielectric inorganic materials with refractive indices n_1 and n_2 are periodically deposited as thin films on the substrate. By controlling the refractive index and thickness of each thin layer, as well as the total number of layers, one can obtain the desired reflectivity and transmissivity. Similar to the half-mirror transflector, the transmittance/reflectance ratio of the multilayer dielectric film is sensitive to each layer's thickness. In addition, to produce several layers successively increases the manufacturing cost. Therefore, the multilayer dielectric film transflector is rarely used in commercial transflective LCDs.

9.3.4 *Orthogonal polarization transflectors*

The orthogonal polarization transflector has a special characteristic that the reflected and the transmitted polarized lights from the transflector have mutually orthogonal polarization states. For instance, if a transflector reflects a horizontal linearly (or right-handed circularly) polarized light, then it would transmit the complementary linearly (or left-handed circularly) polarized light. Figure 9.5 shows three such examples: cholesteric reflector [17], birefringent interference polarizer [18], and wire grid polarizer (WGP) [19].

Cholesteric LC molecules manifest as a planar texture with their helix perpendicular to the cell substrates when the boundary conditions on both substrates are tangential. If the incident wavelength is comparable to the product of the average refractive index and the cholesteric pitch, then the cholesteric LC layer exhibits a strong Bragg reflection [20]. Figure 9.5(a) shows the schematic configuration of a right-handedness cholesteric reflector, where the cholesteric LC polymer layer is formed on a substrate. For incident unpolarized light, the right-handed circularly polarized light which has the same sense as the cholesteric helix is reflected, but the left-handed circularly polarized light is transmitted.

The birefringent interference polarizer transflector consists of a multilayer birefringence stack with alternating low and high refractive indices, as shown in Figure 9.5(b). One way to produce such a transflector is to stretch a multilayer stack in one or two dimensions. The multilayer stack consists of birefringent materials with low/high index pairs [21]. The resultant transflective polarizer exhibits a high reflectance for the light polarized along the stretching direction and, meanwhile, a high transmittance for the light polarized perpendicular to the stretching direction. By controlling refractive indices of the three layers, n_x, n_y, and n_z, the desired polarizer behaviors can be obtained. For practical applications, an ideal reflective polarizer should have ~100% reflectance along one axis (the so-called extinction axis) and 0% reflectance along the other axis (the so-called transmission axis), at all the incident angles.

The wire grid polarizer has been widely used for infrared spatial light modulators [22,23]. It is constructed by depositing a series of parallel and elongated metal strips on a dielectric substrate, as Figure 9.5(c) shows. To operate in the visible spectral region, the pitch of metal strip P should be in the range of around 200 nm, which is approximately half of a blue wavelength [24]. In general, a WGP reflects light with its electric field vector parallel to the wires of the grid and transmits when it is perpendicular. In practice, the wire thickness t, wire width W, and grid pitch P play important roles in determining the extinction ratio and acceptance angle of the polarizer [25].

Unlike the first three transflectors discussed above, the entire area of the orthogonal polarization transflector can be utilized for reflection and transmission simultaneously. Nevertheless, the transmitted light and reflected light possess an orthogonal polarization state so that the reflective and transmissive images exhibit a reversed contrast. Although an inversion driving scheme may correct such a reversed contrast problem [19], the displayed images are still unreadable under moderate brightness surroundings when both ambient light and backlight are in use. Besides, the birefringent interference polarizer is difficult to implement inside the LC cell. As a result, the undesirable parallax problem occurs. Thus, the orthogonal polarization transflectors have not yet been widely adopted in the high-end transflective LCDs.

9.4 Classification of Transflective LCDs

Based on the light modulation mechanisms, transflective LCDs can be classified into four categories: (1) absorption, (2) scattering, (3) reflection, and (4) phase-retardation. The first three categories do not modulate the phase of the incident light; rather, they absorb, scatter, or reflect. In these cases, having one or zero polarizer is preferred from the viewpoint of achieving high brightness. As for the phase-retardation type, two polarizers are usually indispensable in order to make both transmissive and reflective modes work simultaneously.

9.4.1 *Absorption-type transflective LCDs*

To realize the absorption effect, a few percent (2–5 wt%) of dichroic dye is doped into a liquid crystal host. As the LC directors are reoriented by the electric field, the dye molecules follow. The dichroic dyes absorb light strongly (or weakly) when the incident light's polarization is parallel (or perpendicular) to the principal molecular axis. The ratio of these two absorption coefficients ($\alpha_{\|}/\alpha_{\perp}$) is called dichroic ratio, which significantly influences the contrast ratio of the display. Because of the dye's dichroism, the absorption of the LC cell is modulated by the electric field. This mechanism was first introduced in the nematic phase by Heilmeier and Zanoni [26] and later in the cholesteric phase by White and Taylor [27]. In the twisted or helical LC structure, the guest–host display does not require a polarizer. A major technical challenge of guest–host displays is the trade-off between reflectance/transmittance and contrast ratio. A typical contrast ratio for the guest–host LCD is ~5:1 with ~40–50% reflectance. The fairly low contrast ratio is limited by the dichroic ratio of the dye. Long-term stability of the dichroic dyes is another concern because of their strong absorption in the visible spectral region. So far, only few guest–host LCDs have been commercialized.

9.4.1.1 Nematic absorption transflective display

Figure 9.6(a) shows a transflective LCD structure using a half-mirror metallic transflector, two quarter-wave films, and nematic phase LC/dye mixtures [15]. In the figure, the upper and lower halves stand for the voltage-off and voltage-on states, respectively.

When there is no voltage applied, the LC/dye mixtures are homogeneously aligned by the anchoring energy of the cell. In the R-mode, the unpolarized incident ambient light becomes linearly polarized after passing through the LC/dye layer. Then, its polarization state turns into right-handed circularly polarized after traversing through the inner quarter-wave film. Upon reflection from the transflector, its polarization state becomes left-handed circularly polarized due to a π phase change. When the left-handed circularly polarized light passes through the inner quarter-wave film again, it becomes linearly polarized, whose polarization direction is parallel to the LC alignment direction. As a result, the light is absorbed by the doped dye molecules, and a dark state is achieved.

In the T-mode, the unpolarized light from the backlight unit becomes linearly polarized after the polarizer. Then it changes to left-handed circularly polarized light after emerging from the outer quarter-wave film. After penetrating the transflector, it still keeps the same left-handed circular polarization state. Thereafter, its travel path is identical to that of R-mode. Finally, the light is absorbed by the dye molecules, resulting in a dark state.

In the voltage-on state, the LC and dye molecules are reoriented nearly perpendicular to the substrates, as illustrated in the lower half of Figure 9.6(a). Therefore, the light passing through it experiences little absorption so that no change in the polarization state occurs. In the R-mode, the unpolarized ambient light passes through the LC/dye layer and the inner quarter-wave film successively without polarization state change. Upon reflection from the transflector, it is still

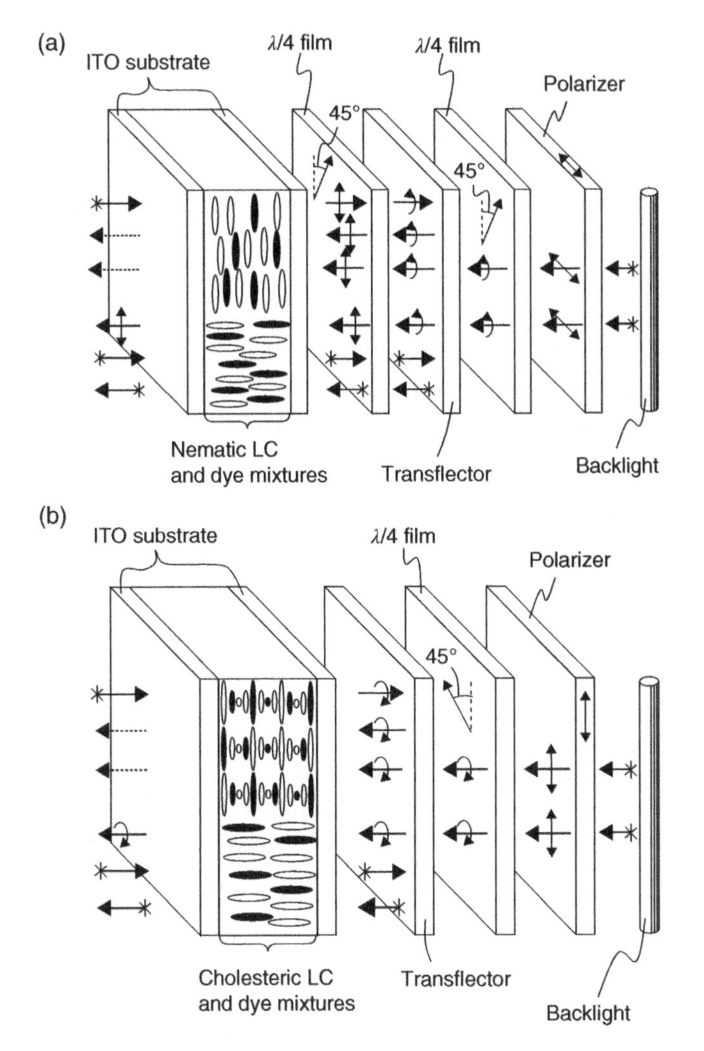

Figure 9.6 Schematic configurations and operating principles of two absorption-type transflective LCDs with (a) nematic phase LC (host) and dye (guest) mixtures, and (b) cholesteric phase LC (host) and dye (guest) mixtures. Zhu 2006. Reproduced with permission from IEEE.

unpolarized light and then goes all the way out of the transflective LCD. Consequently, a bright state with little attenuation is achieved. In the T-mode, the unpolarized backlight becomes linearly polarized after passing through the linear polarizer, the outer quarter-wave film, the transflector, and the inner quarter-wave film, successively. Since the dye molecules are reoriented perpendicularly, the absorption loss is small. As a result, the linearly polarized light emerges from the transflective LCD, which leads to a bright state.

In the abovementioned transflective guest–host LCD, the inner quarter-wave film is put between the transflector and the guest–host layer. There are two optional positions for the transflector. If the transflector is located inside the LC cell, then the quarter-wave film should also be sandwiched inside the cell. Nevertheless, it is difficult to fabricate such a quarter-wave film and assemble it inside the cell. The final process of the cell is polyimide coating, baking, and rubbing. The post-baking temperature of polyimide is typically at ~250 °C for 1 h. The polymeric quarter-wave film may not be able to sustain such a high-temperature processing. Therefore, the external transflector is preferred. If the transflector is located outside the cell, then both quarter-wave film and transflector can be laminated on the outer surface of the LC cell. In this case, however, a serious parallax problem would occur.

9.4.1.2 Cholesteric absorption transflective display

To eliminate the quarter-wave film between the transflector and the LC layer, a transflective LCD design using a half-mirror metallic transflector and cholesteric LC/dye mixture is proposed [28]. The device structure is illustrated in Figure 9.6(b). From Figure 9.6(b), only one quarter-wave film is employed, which is located between the transflector and the linear polarizer. Consequently, the quarter-wave film can be put outside of the cell, while the transflector can be sandwiched inside the cell. As a result, no parallax occurs. The upper and lower portions of this figure demonstrate the voltage-off and voltage-on states, respectively. In the voltage-off state, the LC/dye molecules render a right-handed planar texture with its helix perpendicular to the substrates. In the R-mode, the unpolarized light is largely attenuated by the LC/dye layer and only weak light passes through it. Upon reflection from the transflector, it is further absorbed by the guest dye molecules, resulting in a dark state. In the T-mode, the unpolarized backlight first becomes linearly polarized and then right-handed circularly polarized after passing through the polarizer and, in turn, the quarter-wave film. The circularly polarized light is further attenuated after it penetrates the transflector. Such weak right-handed circularly polarized light is absorbed by the same handedness cholesteric LC/dye mixture, resulting in a dark state.

In the voltage-on state, both LC and dye molecules are reoriented perpendicular to the substrates. As a result, little absorption occurs to the incident light. In the R-mode, the unpolarized light is unaffected throughout the whole path, resulting in a very high reflectance. In the T-mode, the unpolarized backlight becomes right-handed circularly polarized after passing through the polarizer, the quarter-wave film, and the transflector. It finally penetrates the LC/dye layer with little attenuation. Again, a bright state is obtained.

In the abovementioned two absorption-type transflective LCDs, only one polarizer is employed instead of two. Therefore, the overall image in both T- and R-modes is relatively bright. However, due to the limited dichroic ratio of the employed dye molecules (DR ~ 15:1), a typical contrast ratio of the guest–host LCD is around 5:1, which is inadequate for high-end full color LCD applications [29]. Thus, the absorption-type transflective LCDs only occupy a small portion of the handheld LCD market.

9.4.2 Scattering-type transflective LCDs

Polymer-dispersed LC (PDLC) [30], polymer-stabilized cholesteric texture (PSCT) [31], and LC gels [32] all exhibit optical scattering characteristics and have wide applications in displays and optical devices. The LC gel-based reflective LCD can also be extended to transflective

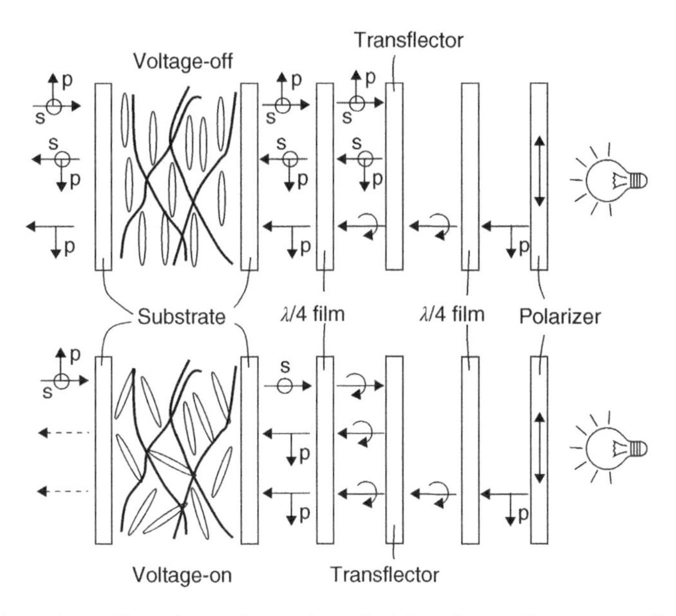

Figure 9.7 Schematic configuration and operating principles of scattering type transflective LCD with homogeneously aligned LC gel. Zhu 2006. Reproduced with permission from IEEE.

LCDs [33]. Figure 9.7 shows the schematic structure and operating principles of the LC gel-based transflective LCD. The device is composed of an LC gel cell, two quarter-wave films, a transflector, a polarizer, and a backlight. The cell was filled with homogeneously aligned nematic LC and monomer mixture. After UV-induced polymerization, polymer networks are formed and the LC molecules are confined within the polymer networks.

When there is no voltage applied, the LC molecules exhibit a homogeneous alignment. Consequently, the LC gels are highly transparent for the light traveling through, as illustrated in the upper portion of Figure 9.7. In the R-mode, the unpolarized ambient light remains unpolarized all the way from entering to exiting the LC cell. As a result, a fairly bright state is obtained. In the T-mode, the unpolarized backlight turns into a linearly polarized p-wave after the polarizer. After passing the first quarter-wave film, penetrating the transflector, and the second quarter-wave film whose optic axis is orthogonal to that of the first one, the p-wave remains linearly polarized. Since the LC gel is highly transparent in the voltage-off state, the linearly polarized p-wave finally comes out of the display panel, resulting in a bright output.

However, when the external applied voltage is high enough, the LC molecules deviate from the original homogeneous alignment by the exerted torque of electric field. Therefore, micro-domains are formed along the polymer chains such that the extraordinary ray (i.e. the linear polarization along the cell rubbing direction) is scattered, provided that the domain size is comparable to the incident light wavelength. In the mean time, the ordinary ray would pass through the LC gels without being scattered. In the R-mode, the unpolarized ambient light becomes a linearly polarized s-wave after passing the activated LC cell, since the p-wave is scattered. After a round trip of passing the quarter-wave film, being reflected by the transflector, and passing the quarter-wave film again, the s-wave is converted into a p-wave. Due to the scattering of LC gels, this p-wave is scattered again. Consequently, a scattering translucent state is achieved.

In the T-mode, the unpolarized backlight turns into a linearly polarized p-wave after passing the polarizer, the second quarter-wave film, the transflector, and the first quarter-wave film, successively. Thereafter, similar to the case of R-mode, the p-wave is scattered by the activated LC gels, resulting in a scattering translucent output.

This scattering type transflective LCD only needs one polarizer, so it can achieve a very bright image. However, there are three major drawbacks in the above LC gel-based transflective LCD. First, light scattering mechanism usually leads to a translucent state rather than a black state. Therefore, the image contrast ratio is low and highly dependent on the viewing distance to the display panel. Although doping the LC gels with a small concentration of black dye can help to achieve a better dark state, the contrast ratio is still quite limited due to the limited dichroic ratio of dye dopant. Second, the insertion of the first quarter-wave film will cause a similar parallax problem to the absorption-type transflective LCD using cholesteric LCs. Third, the required driving voltage is usually over 20 V, due to the polymer network constraint, which is beyond the capability of current thin-film transistors developed for LCD applications. Therefore, these drawbacks hinder the scattering type transflective LCD from widespread applications.

9.4.3 Scattering and absorption type transflective LCDs

A white paper scatters and diffuses light in the bright state, resulting in a wide viewing angle. When printed, the ink absorbs light and the printed areas become dark. To mimic the display shown in a white paper, we should combine light scattering and absorption mechanisms together in a dye-doped LC gel system [34].

Both dual-frequency liquid crystal and negative $\Delta\varepsilon$ LC gels have been demonstrated. Here, we only describe the dye-doped negative LC gel system because the TFT-grade negative $\Delta\varepsilon$ LC has been commonly employed in vertical alignment. For example, Merck ZLI-4788 has $\Delta\varepsilon = -5.7$ at $f = 1$ kHz and $\Delta n = 0.1647$ at $\lambda = 589$ nm. To form a gel, we can mix ZLI-4788, a diacrylate monomer (bisphenol-A-dimethacrylate) and a dichroic dye S428 (Mitsui, Japan) at 90:5:5 wt% ratios. To make the device independent of polarization, the ITO glass substrates should have polyimide (PI) coating, but *without* rubbing treatment. The PI layer provides vertical alignment for the LC molecules. The cell gap is controlled at about 5 μm. To form a gel, the filled LC cell is cured by a UV light ($\lambda \sim 365$ nm, $I \sim 15$ mW/cm [2]) at 13 °C for 2 h. After photo-polymerization, the formed chain-like polymer networks are along the cell gap (z) direction because the LC directors are aligned perpendicular to the glass substrates during the UV curing process, as Figure 9.8(a) shows.

Figures 9.8(a) and (b) illustrate the light modulation mechanisms of the dye-doped negative LC gel. At $V = 0$, the cell does not scatter light, and the absorption is rather weak because the LC and dye molecules are aligned perpendicular to the substrate surfaces. At this stage, the display has the highest reflectance. When a high voltage at $f = 1$ kHz is applied to the LC gel, the liquid crystals and dye molecules are reoriented in the xy plane, as Figure 9.8(b) shows. The polymer network scatters light strongly. Since the alignment layer has no rubbing treatment, the absorption has no preferred direction. Therefore, the display is polarization insensitive. Due to multiple scattering and absorption, a descent dark state can be obtained.

Figure 9.9 depicts the voltage-dependent reflectance of the dye-doped LC gel using a linearly polarized green diode laser ($\lambda = 532$ nm) instead of a white light source for characterizing the

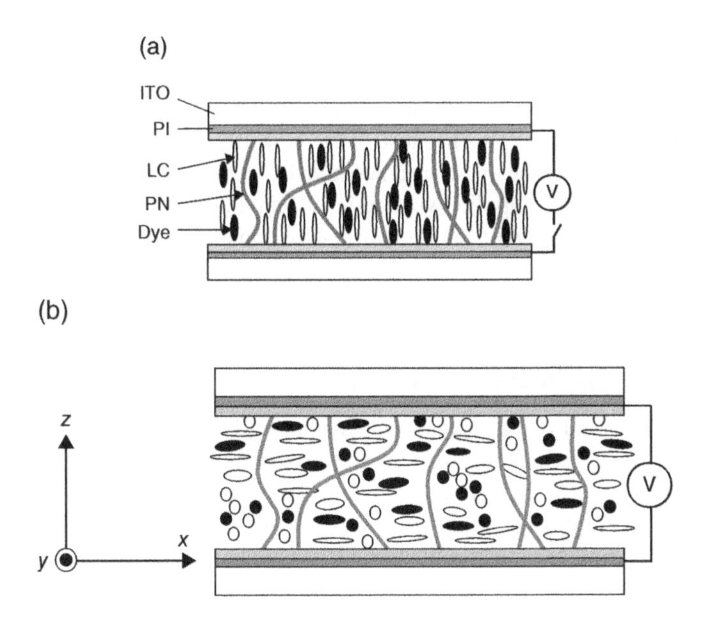

Figure 9.8 Operating principle of the dye-doped negative LC gel. (a) Voltage-off state, and (b) voltage-on state. The PI has no rubbing treatment. PN: polymer network.

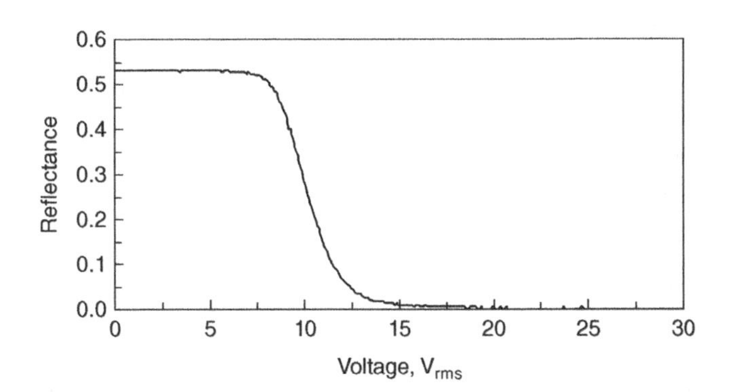

Figure 9.9 Voltage-dependent reflectance of a dye-doped negative LC gel at $f = 1$ kHz. Cell gap $d = 5$ μm and $\lambda = 532$ nm.

device performances because the guest–host system appears dark red rather than black. A dielectric mirror was placed behind the cell so that the laser beam passed through the cell twice. A large area photodiode detector was placed at ~25 cm (the normal distance for viewing a mobile display) behind the sample which corresponds to ~2° collection angle. The curve shown in Figure 9.9 is independent of the laser polarization. The maximum reflectance reaches ~52% in the low voltage regime and decreases gradually as $V > V_{th}$ because the employed LC has a negative $\Delta\varepsilon$ and LC directors are in homeotropic structure at $V = 0$. Because of the formed polymer networks, the threshold voltage is increased to ~7.5 V_{rms}. At $V = 20$ V_{rms}, the measured contrast ratio of the dye-doped negative LC gel exceeds ~200:1. By contrast, a typical guest–host LCD has a contrast ratio of about 5:1.

Response time is another important issue for guest–host displays. A typical response time of a guest–host display is around 50 ms because of the bulky dye molecules. Due to the polymer network, the response time of the dye-doped negative LC gel is fast. The rise time is 1.0 ms and decay time is 4.5 ms when the applied voltage is switched between 0 and 20 V_{rms}. This dye-doped LC gel can also be configured to a polarizer-free transflective display using the dual cell gap approach.

9.4.4 Reflection-type transflective LCDs

A cholesteric liquid crystal (CLC) layer exhibits a strong Bragg reflection with its reflection band centered at $\lambda_o = nP_o$, where n and P_o are the average refractive index and the cholesteric helix pitch, respectively. The reflection bandwidth $\Delta\lambda_o = \Delta nP_o$ is proportional to the birefringence Δn of the CLC employed. Apparently, to cover the whole visible spectral range, a high birefringence ($\Delta n > 0.6$) CLC material is needed, assuming the pitch length is uniform [35]. Because the transmitted and reflected circular polarization states are orthogonal to each other, the CLC layer must rely on some additional elements to display a normal image without the reversed contrast ratio. By adopting an image-enhanced reflector (IER) on the top substrate as well as a patterned ITO and a patterned absorption layer on the bottom substrate, the transflective cholesteric LCD can display an image without reversed contrast ratio [36,37], as shown in Figures 9.10(a) and (b). The opening areas of the patterned absorption layer on the bottom substrate match the IER on the top substrate. In addition, right above the patterned absorption layer and below the IER is the opening area of the patterned ITO layer. Therefore, the CLC molecules below the IER are not reoriented by the external electric filed.

In operation, when there is no voltage applied, the CLC layer exhibits a right-handed planar helix texture throughout the cell, as Figure 9.10(a) shows. In the R-mode, when an unpolarized ambient light enters the CLC cell, the left-handed circularly polarized light passes through the right-handed CLC layer and is absorbed by the patterned absorption layer. At the same time, the right-handed circularly polarized light is reflected by the same sense CLC layer and the bright state results. In the T-mode, when the unpolarized backlight enters the CLC layer, similarly, the right-handed circularly polarized light is reflected and it is either absorbed by the patterned absorption layer or recycled by the backlight unit. In the meantime, the left-handed circularly polarized passes through the CLC layer and impinges on the IER. Due to a π phase change upon reflection, it is converted to a right-handed circularly polarized light, which is further reflected by the CLC layer to the reviewer. Consequently, a bright state occurs.

In the voltage-on state, the planar helix texture above the bottom-patterned ITO layer becomes a focal conic texture, while those LC molecules between the IER and the opening area of the bottom-patterned ITO layer are still unaffected, as shown in Figure 9.10(b). The focal conic texture, if the domain size is well controlled, exhibits a forward scattering for the incident light. In the R-mode, the unpolarized incident ambient light is forward scattered by the focal conic textures. It is then absorbed by the patterned absorption layer, resulting in a dark state. In the T-mode, the unpolarized light still experiences a right-handed planar helix texture before it reaches the IER on the top substrate. Thus, the right-handed polarized light is reflected back and it is either absorbed by the patterned absorption layer or recycled by

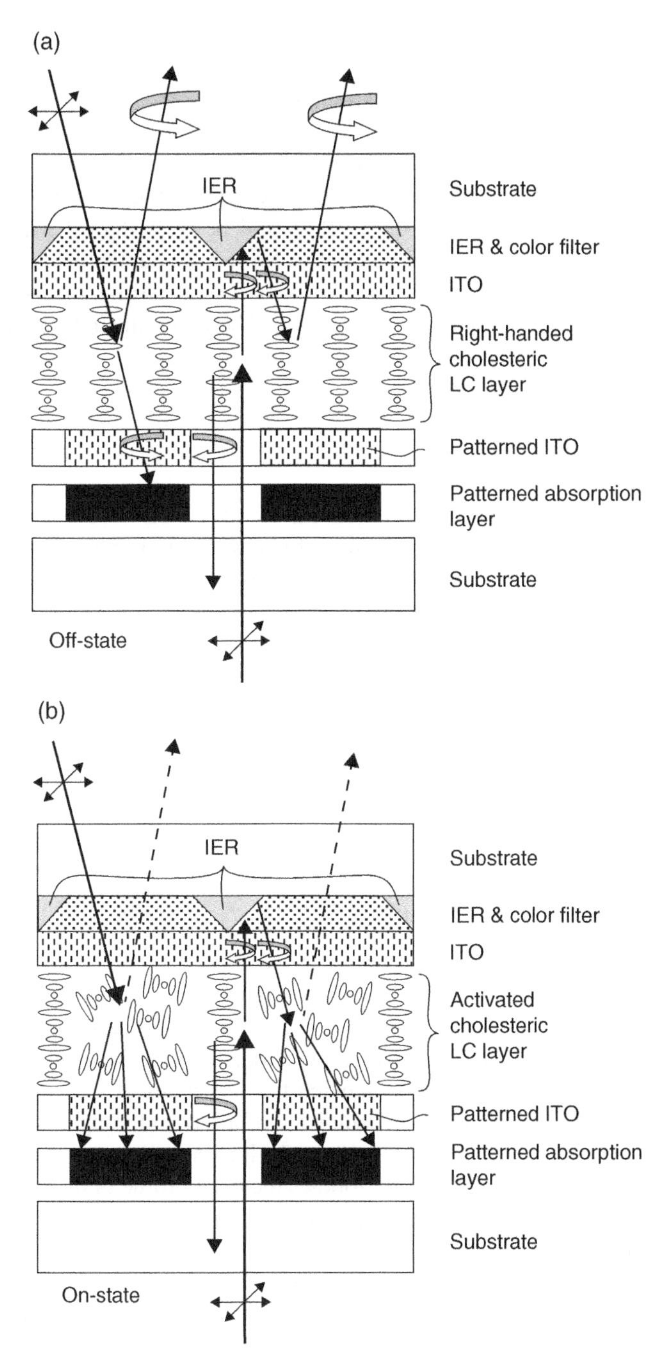

Figure 9.10 Schematic configuration of reflection type transflective cholesteric LCD and its operating principles at (a) voltage-off state, and (b) voltage-on state. Zhu 2006. Reproduced with permission from IEEE.

the backlight unit. At the same time, the left-handed circularly polarized light passes through the planar texture and impinges on the IER. Upon reflection, it turns into a right-handed circularly polarized light. Then it is forward scattered by the focal conic texture and finally absorbed by the patterned absorption layer. As a result, the dark state is obtained.

In the abovementioned transflective cholesteric LCD, no polarizer is employed. Therefore, its light efficiency is high. However, to produce the IER array on the top substrate increases the manufacturing complexity. In addition, the IER should be well aligned with the patterned absorption layer; otherwise, light leakage will occur. More importantly, the forward scattering of the focal conic texture is incomplete. Some backward scattered light causes a translucent dark state, which deteriorates the image contrast ratio. Therefore, the transflective cholesteric LCD is not yet popular for high-end LCD applications.

9.4.5 Phase retardation type

The operation principle of the phase-retardation transflective LCDs, including homogeneous, VA and TN cells, is based on the voltage-induced LC reorientation. Strictly speaking, the TN cells should belong to the polarization rotation effect. Since a transflective LCD consists of both T- and R-modes, two polarizers are usually required. Compared to the absorption, scattering, and reflection types, the phase-retardation type transflective LCDs have the advantages of higher contrast ratio, lower driving voltage, and better compatibilities with the current volume manufacturing techniques. Therefore, the phase-retardation type transflective LCDs dominate the current commercial products, such as cellular phones and digital cameras.

In the following sections, we will describe the major transflective LCD approaches based on the phase-retardation mechanism.

9.4.5.1 Transflective TN and STN LCDs

The 90° TN cell can be used not only in transmissive and reflective LCDs [38], but also in transflective LCDs [39]. Figure 9.11(a) shows the device configuration of a transflective TN LCD. A 90° TN LC cell, which satisfies the Gooch–Tarry minima conditions [40], is sandwiched between two crossed polarizers. In addition, a transflector is laminated at the outer side of the bottom polarizer and a backlight is intended for dark ambient.

In the null voltage state, the LC directors exhibit a uniform twist throughout the cell from the lower substrate to the upper substrate. In the T-mode, the incoming linearly polarized light which is generated by the bottom polarizer, closely follows the twist profile of the LC molecules and continuously rotates 90° with respect to its original polarization state. This is known as the polarization rotation effect of the TN cell. Thus the linearly polarized light can pass through the top polarizer, resulting in a bright output known as a normally white (NW) mode. While in the R-mode, the incoming linearly polarized light is rotated by 90° as it passes through the TN LC layer. It then penetrates the bottom polarizer and reaches the transflector. A portion of the linearly polarized light is reflected back by the transflector and passes the bottom polarizer again. This linearly polarized light then follows the twisted LC molecules, and its polarization axis is rotated by 90°, that is, parallel to the transmission axis of the top polarizer. Accordingly, a bright state is achieved.

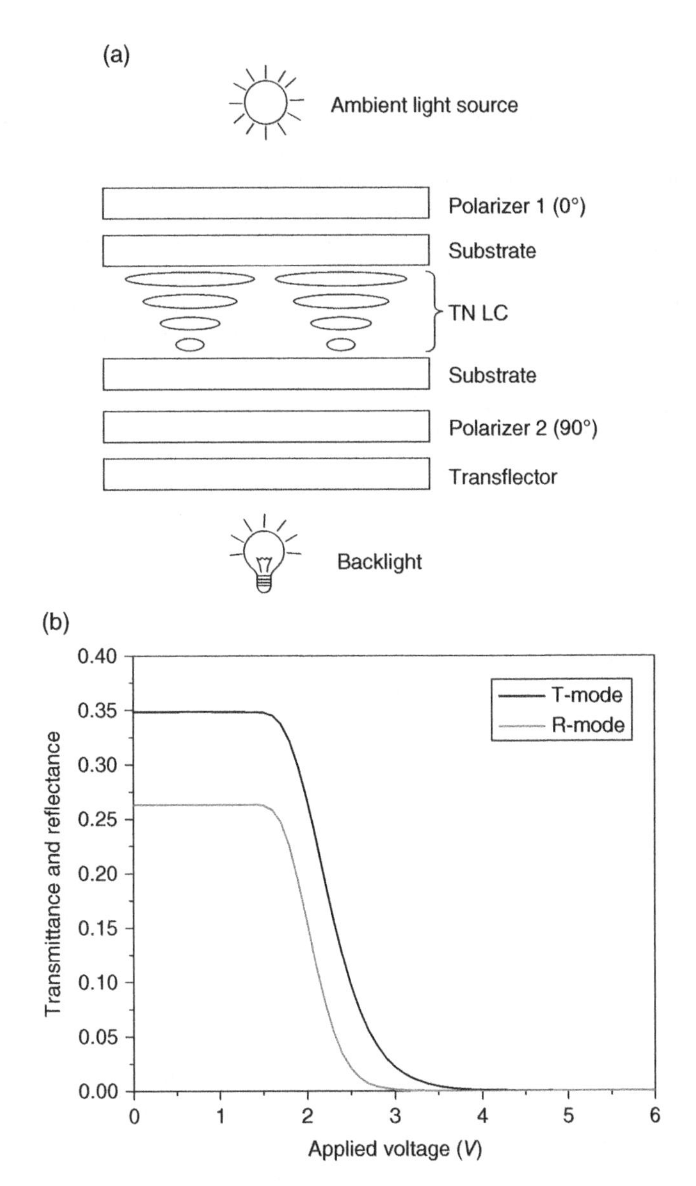

Figure 9.11 Transflective TN LCDs: (a) schematic device configuration, and (b) voltage-dependent transmittance and reflectance curves. Zhu 2006. Reproduced with permission from IEEE.

In the voltage-on state, the bulk LC directors are reoriented substantially perpendicular to the substrate. The two boundary layers are orthogonal. The perpendicularly aligned bulk LC molecules do not modulate the polarization state of the incoming light. Meanwhile, those two orthogonal boundary layers compensate for each other. Consequently, the incoming linearly polarized light still keeps the same polarization state after it passes through the activated TN LC layer. In the T-mode, the linearly polarized light which is generated by the bottom polarizer propagates all the way to the top polarizer without changing its polarization state. Therefore, it is blocked by the top polarizer, resulting in a dark state. In the R-mode, the linearly polarized light produced by the top polarizer passes through the activated LC layer without changing its polarization state. Consequently, it is absorbed by the bottom polarizer and no light returns to the viewer's side. This is the dark state of the display.

To have a better understanding of the underlying operation principle and electro-optical (EO) performance, we carried out numerical simulations based on extended Jones matrix method [41]. Hereafter, unless otherwise specified, we assume that (1) the LC material is MLC-6694-000 (from Merck), (2) the polarizer is a dichroic linear polarizer with complex refractive indices $n_e = 1.5 + i \times 0.0022$ and $n_o = 1.5 + i \times 0.000032$, (3) the transflector does not depolarize the polarization state of the impinging light upon reflection and transmission, (4) the transflector does not cause any light loss upon reflection and transmission, (5) the ambient and backlight enters and exits from the panel in the normal direction, and (6) the light wavelength is $\lambda = 550$ nm.

Figure 9.11(b) plots the voltage dependent transmittance and reflectance curves of a typical transflective TN LCD. Here, twist angle $\phi = 90°$ and the first Gooch–Tarry minimum condition $d\Delta n = 476$ nm are employed, where d is the cell gap and Δn is the LC birefringence. The gray-scales of both T and R modes overlap well with each other. This is because the reflection beam in the R-mode experiences the bottom polarizer, LC layer, and top polarizer successively in turn, as the transmission beam does in the T-mode.

Compared to the conventional transmissive TN LCD, the above transflective TN LCD only requires one additional transflector between the bottom polarizer and the TN LC layer. Naturally, this transflective LCD device configuration can also be extended to an STN-based transflective LCD [42]. In contrast to the so-called polarization rotation effect in TN LCD, the STN LCD uses the birefringence effect of the super-twisted nematic LC layer [43]. Therefore, a larger twist angle (180 ~ 270°), a thicker LC cell gap, and a different polarizer/analyzer configuration are required.

The abovementioned TN and STN type transflective LCDs have advantages in simple device structure and matched gray-scales; however, their major drawbacks are in parallax and low reflectance.

Parallax is a deteriorated shadow image phenomenon in the oblique view of a reflective LCD [44]. Similarly, it also occurs in some transflective LCDs, such as the above-described transflective TN and STN LCDs. Figure 9.12 demonstrates the cause of parallax in the R-mode of a transflective TN LCD when the polarizer and transflector are laminated at the outer side of the bottom substrate. The switched-on pixel does not change the polarization state of the incident light because the LC molecules are reoriented perpendicular to the substrate. From the observer side, when a pixel is switched on, it appears dark, as designated by a′b′ in the figure. The dark image a′b′, generated by the top polarizer, actually comes from the incident beam ab. Meanwhile, another incident beam cd passes through the activated pixel and does not change its linear polarization state as well. Therefore, it is absorbed by the bottom polarizer, resulting in no light reflection. Accordingly, a shadow image c′d′ occurs from the observer viewpoint. Unlike the dark image a′b′, which is generated by the top polarizer, the shadow image c′d′ is actually caused by the bottom polarizer. This is why the shadow image c′d′ appears to be under the dark image a′b′. Because the bottom polarizer and transflector are laminated outside the bottom substrate, the ambient light must traverse the bottom substrate before it is reflected back. Due to the thick bottom substrate, the reflection image beams a′b′ and c′d′ are shifted away from the pixel

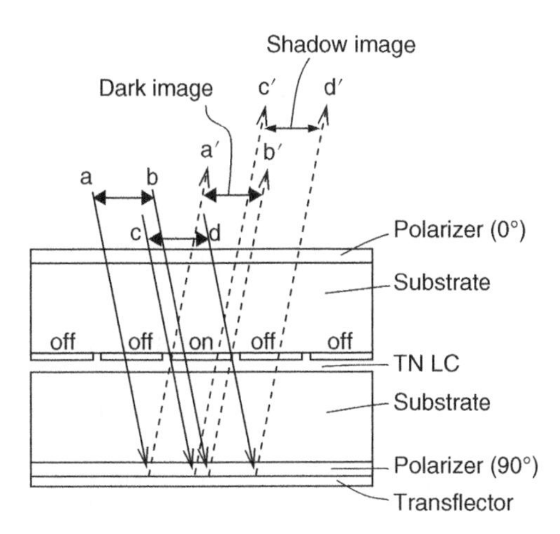

Figure 9.12 Schematic view of the cause of parallax phenomenon in the R-mode of a transflective LCD with polarizer and transflector laminated outside of the bottom substrate. Zhu 2006. Reproduced with permission from IEEE.

area that the incoming beams ab and cd propagate in, resulting in a shadow image phenomenon called parallax. Such a parallax problem becomes more serious with the decrease of pixel size. Therefore, the transflective TN and STN LCDs with abovementioned structures are not suitable for the high resolution full color transflective LCD devices.

To overcome the parallax problem in transflective TN and STN LCDs, the bottom polarizer and transflector must be located inside the LC cell. A burgeoning in-cell polarizer technology, based on thin crystal film growth from aqueous lyotropic LC of supramolecules, has attracted a certain amount of attention in the transflective LCD industry [45]. By depositing both transflector and polarizer inside the cell, the abovementioned annoying parallax problem can be significantly reduced.

Nevertheless, the transflective TN and STN LCDs still have another shortcoming, which is low reflectance in the R-mode. As shown in Figure 9.11(b), although the gray-scales of both modes overlap reasonably well with each other, the reflectance in the R-mode is much lower than the transmittance in the T-mode. This is because the light accumulatively passes through polarizers four times in the R-mode but only twice in the T-mode. Due to the absorption of polarizers, the light in the R-mode suffers much more loss than that in the T-mode. Accordingly, the reflectance of the R-mode is reduced substantially.

9.4.5.2 Transflective MTN LCDs

To overcome the parallax and low reflectance problems of the transflective TN and STN LCDs, the bottom polarizer for the R-mode should be removed to the outer surface of the bottom substrate. Thus, the transflector can be implemented on the inner side of the LC cell, acting as an internal transflector. With such a device configuration, the R-mode operates as a single-polarizer reflective LCD. More importantly, both ambient light and backlight pass through the polarizer twice; therefore, both T- and R-modes experience the same light absorption. Nevertheless, the conventional TN LC cell does not work well in the single-polarizer

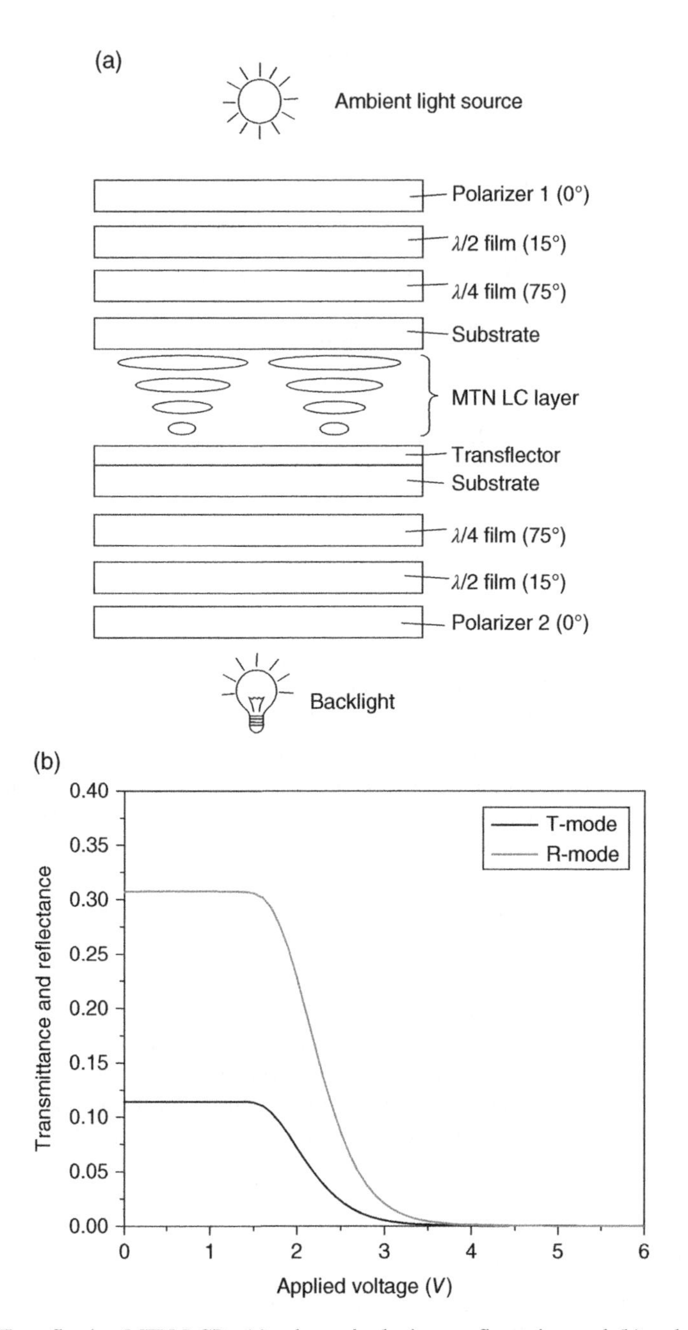

Figure 9.13 Transflective MTN LCD: (a) schematic device configuration and (b) voltage-dependent transmittance and reflectance curves. Zhu 2006. Reproduced with permission from IEEE.

reflective LCD. This is because, after the light travels a round-trip in the LC layer, the light polarization state in the voltage-on state is identical to that in the voltage-off state.

By reducing the $d\Delta n$ value of the TN LC layer to around one half of that required in a conventional transmissive TN LCD, the MTN mode overcomes the problem mentioned above [46]. Unlike the TN LCD, the twist angle of MTN mode can vary from 0° to 90° and its operating mechanism is based on the proper mixing between the polarization rotation and birefringence effects. Molsen and Tillin of Sharp Corp. incorporated the MTN mode into their transflective LCD design [47], as shown in Figure 9.13(a). Compared to the transflective TN LCD shown in Figure 9.11(a), this transflective MTN LCD exhibits two different features. First, the transflector is located inside the LC cell, so no parallax problem occurs. Second, a half-wave film and a quarter-wave film are inserted in each side of the MTN LC cell. These two films together with the adjacent linear polarizer function as a broadband circular polarizer over the whole visible spectral range [48]. Thereby, a good dark state can be guaranteed over the whole visible range for the R-mode.

In the voltage-off state, the MTN LC layer is equivalent to a quarter-wave film. In the R-mode, the incident unpolarized ambient light is converted into a linearly polarized light after passing through the top polarizer. After penetrating the top two films and the MTN LC layer, the linearly polarized light still keeps its linear polarization except that it has been rotated 90° from the original polarization direction. Upon reflection from the transflector, it experiences the MTN LC layer and the top two films once again. Hence its polarization state is restored to the original one, resulting in a bright output from the top polarizer. In the T-mode, the unpolarized backlight turns into linearly polarized after passing through the bottom polarizer. After it passes through the bottom two films, penetrates the transflector, and continues to traverse the MTN LC layer and the top two films, it becomes circularly polarized light. Finally, a partial transmittance is achieved from the top polarizer.

In the voltage-on state, the bulk LC directors are almost fully tilted up and those two unaffected boundary layers compensate each other in phase. Therefore, the LC layer does not affect the polarization state of the incident light. In the R-mode, the linearly polarized light generated by the top polarizer turns into an orthogonal linearly polarized light after a round-trip in the top two films and the activated LC layer. Accordingly, this orthogonal linearly polarized light is blocked by the top polarizer, leading to a dark state. In the T-mode, the linearly polarized light, caused by the bottom polarizer, passes through the bottom two films, penetrates the transflector, then continues to pass through the activated LC layer and the top two films. Before it reaches the top polarizer, its linear polarization state is rotated by 90°, which is perpendicular to the transmission axis of the top polarizer, and the dark state results.

As an example, Figure 9.13(b) depicts the voltage dependent transmittance and reflectance curves of a transflective MTN LCD with $\phi = 90°$ and $d\Delta n = 240$ nm. Here both T- and R-modes operate in an NW mode. Generally speaking, for the TN- or MTN-based LCDs, the NW display is preferred to the normally black (NB) because the dark state of the NW mode is controlled by the on-state voltage. Thus, the dark state of the NW mode is insensitive to the cell gap variation. A large cell gap tolerance is highly desirable for improving manufacturing yield.

By comparing Figure 9.13(b) with Figure 9.11(b), we see two distinctions between the transflective MTN LCD and the transflective TN LCD. First, without the absorption from the bottom polarizer the reflectance of the transflective MTN LCD is higher than that of the transflective TN LCD. Second, the transmittance of the transflective MTN LCD is much lower than that of the transflective TN LCD. This is because the maximum obtainable normalized transmittance is

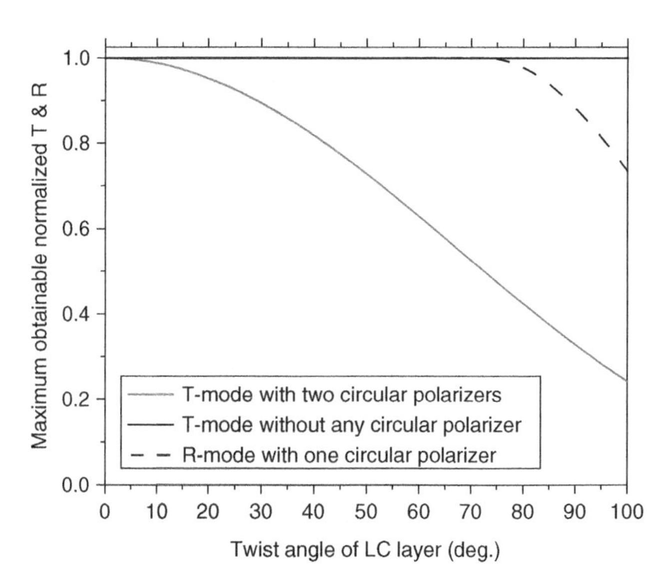

Figure 9.14 The maximum obtainable normalized reflectance and transmittance in the transflective MTN LCD and the transflective TN LCD as a function of twist angle. Zhu 2006. Reproduced with permission from IEEE.

always less than 100% for a transmissive TN cell sandwiched between two circular polarizers [49]. Figure 9.14 shows the maximum obtainable normalized reflectance and transmittance in optimized transflective MTN and TN LCDs as a function of twist angle. Here the normalized reflectance and transmittance represent only the polarization state modulation efficiency is taken into consideration; the light losses caused by the polarizers and reflector are all neglected. Due to the effect of circular polarizer on both sides of the MTN cell, as long as the twist angle is larger than 0°, the maximum obtainable normalized transmittance gradually decreases in spite of the $d\Delta n$ value of the MTN LC layer, as represented by the solid gray line in Figure 9.14. For instance, in the 90° MTN cell with a circular polarizer on both sides, the maximum obtainable normalized transmittance is ~33%. On the other hand, the dark dashed line shows that the maximum obtainable normalized reflectance steadily keeps 100% until the twist angle goes beyond 73°. In short, although the transflective MTN LCD overcomes the parallax problem, its maximum obtainable transmittance in the T-mode is too low. A low transmittance demands a brighter backlight which, in turn, would consume more battery power and reduce its lifetime.

9.4.5.3 Patterned-retarder transflective LCDs

If we can remove both circular polarizers in the T-mode, the maximum transmittance can be boosted to 100% for any twist angle from 0° to 100°, as designated by the solid dark line shown in Figure 9.14. Without the circular polarizers, the T-mode operates at the same way as a conventional transmissive TN LCD. The Philips research group proposed a dual-cell-gap transflective MTN/TN LCD using patterned phase retarders. Figure 9.15(a) shows the schematic device structure. Each pixel is divided into a transmission region and a reflective region by a derivative openings-on-metal type transflector. A patterned broadband phase retarder is deposited on the

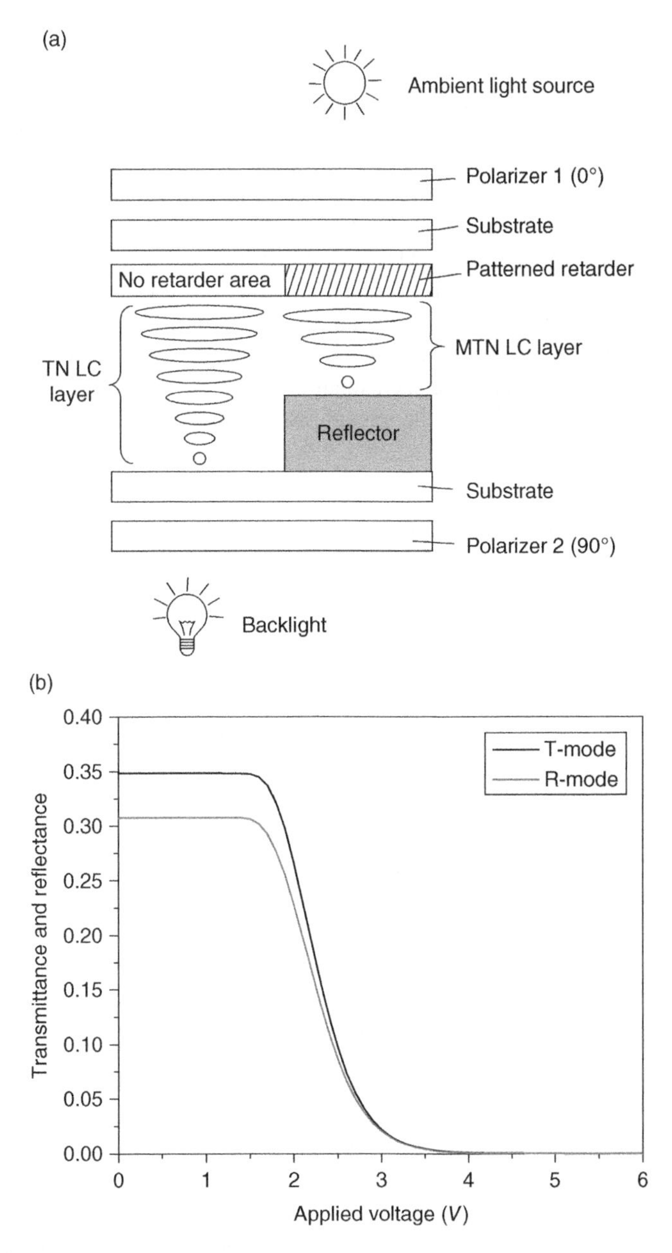

Figure 9.15 Patterned retarder transflective MTN/TN LCD: (a) schematic device configuration and (b) voltage dependent transmittance and reflectance curves. Zhu 2006. Reproduced with permission from IEEE.

inner side of the top substrate. More specifically, the patterned phase retarder is located right above the reflection region, while no phase retarder exists above the transmission region. In addition, the cell gap in the transmission region is around twice that of the reflection region, and the LC layer twists 90° in both regions. The patterned phase retarder actually comprises a half-wave film and a quarter-wave film fabricated by wet coating techniques [50]. In the transmissive region, the cell is identical to the traditional transmissive TN LCD, while in the reflective region, it is an MTN mode. Figure 9.15(b) shows the voltage dependent transmittance and reflectance curves with $d\Delta n = 476$ nm in the transmission region and $d\Delta n = 240$ nm in the reflective region. From the figure, both T- and R-modes have a very good gray-scale overlapping. Since the maximum normalized reflectance of the 90° MTN mode is around 88%, the reflectance is slightly lower than the transmittance.

The patterned retarder transflective MTN/TN LCD has advantages in the matched gray-scale, high contrast ratio, and low color dispersion. However, under oblique incident angles, the ambient light might not pass through the patterned retarder; likewise, the backlight might pass through the patterned retarder. Thus, a deteriorated image may arise when viewed from an oblique angle. In addition, to fabricate such small-scale patterned phase retarders on the glass substrate and aligned them well with the transmission/reflection pixels is a challenging task.

9.4.5.4 Transflective mixed-mode LCDs

To compensate the intrinsic optical path differences between the transmission and reflection regions, Sharp Corp. proposed an approach to generate different director configurations simultaneously in both regions [51]. The different director configurations can be realized, for example, by applying different alignment treatments, exerting different driving voltages, generating different electric fields, or producing different cell gaps in both regions. Thus, the transmission region may, in principle, operate in a different LC mode from the reflection region, which leads to the name of transflective mixed-mode LCDs.

If two circular polarizers are indispensable in both sides of the cell, one solution to maximizing the normalized transmittance is to decrease the LC twist angle to 0° in the transmission region while still maintaining a twist profile in the reflection region. Thus the transmission region can operate in electrically controlled birefringence (ECB) mode while the reflection region still runs in MTN mode. Figure 9.16(a) shows the device configuration of a transflective MTN/ECB LCD using the opening-on-metal transflector [52]. The top substrate is uniformly rubbed while the bottom substrate has two rubbing directions: in the reflective region the LC layer twists 75°, while in the transmission region the LC layer has zero twist, that is, homogenous alignment. Therefore, the reflective region works in the 75° MTN mode while the transmission region operates in the ECB mode. Coincidently, their $d\Delta n$ requirements are very close to each other, so a single cell gap device configuration is adopted in both regions. Figure 9.16(b) plots the voltage dependent transmittance and reflectance curves with $d\Delta n = 278$ nm in both regions. Both T- and R-modes in the transflective MTN/ECB LCD almost simultaneously reach their maximum light efficiency through such a dual-rubbing process. Even so, the T-mode might be slightly lower light efficiency than the R-mode. This is because the $d\Delta n$ requirement for the T- and R-modes is slightly different, and a compromise is taken to optimize the R-mode.

Besides the above-demonstrated dual-rubbing transflective MTN/ECB LCD, other similar dual-rubbing transflective mixed-mode LCDs are possible, such as dual-rubbing transflective VA/HAN LCD [53] and dual-rubbing transflective ECB/HAN LCD [54]. The common characteristic of these dual-rubbing transflective LCDs is that different rubbing directions or different alignment layers are required on at least one of the substrates. This leads to two obstacles for its widespread applications. First, the dual-rubbing requirement needs a complicated manufacturing process and hence an increased cost. But more seriously, the dual rubbing

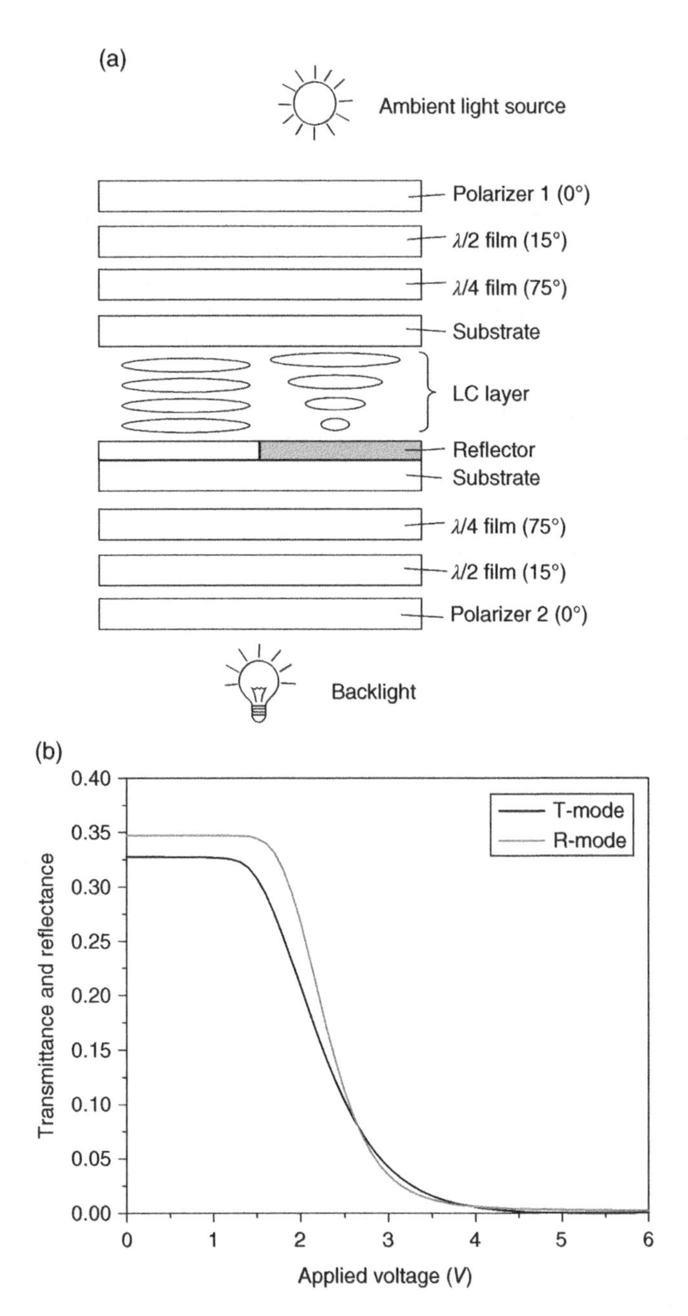

Figure 9.16 Dual-rubbing transflective MTN/ECB LCD: (a) schematic device configuration, and (b) voltage dependent transmittance and reflectance curves. Zhu 2006. Reproduced with permission from IEEE.

usually introduces a disclination line on the border between different rubbing regions, which lowers the image brightness and also deteriorates the contrast ratio.

To avoid the dual-rubbing process while still maintaining a single cell gap device configuration, an alternative way to achieve different director configurations in the two regions is to introduce different electric field intensities in each region. For example, the transflective VA LCD uses periodically patterned electrodes to generate different LC tilt angle profiles in the two regions [55]. Nevertheless, the metal reflector there is insulated from its surrounding ITO electrodes, which increases the manufacturing complexity. On the other hand, the patterned reflector is either connected with the common electrode or electrically floated, which results in either a dead zone in the reflection region or charge stability uncertainties. Another example is transflective IPS LCD, which uses the different twist angle profiles along the horizontal direction of interdigitated electrodes for the transmission and reflection regions [56]. In this design, an in-cell retarder is used between the transflector and LC layer. When there is no voltage applied, the LC layer is homogeneously aligned. The LC cell together with the in-cell retarder acts as a broadband quarter-wave film. Such a design has two shortcomings. First, unlike the conventional transmissive IPS LCD, here the dark state is very sensitive to the LC layer thickness. Second, the in-cell retarder is difficult to fabricate inside the cell.

9.5 Dual-Cell-Gap Transflective LCDs

Unless identical display modes are adopted in both T- and R-modes, otherwise there are always some discrepancies between their voltage dependent transmittance and reflectance curves. This is the reason that none of the abovementioned transflective mixed-mode LCDs has perfectly matched voltage dependent transmittance and reflectance curves. Different from the mixed display modes employed between transmission and reflection regions as described above, Sharp Corp. also introduced the dual-cell-gap concept for transflective LCDs [57].

Figure 9.17(a) shows the schematic device configuration of a dual-cell-gap transflective ECB LCD. Similar to the case of dual-rubbing transflective MTN/ECB LCD, this dual-cell-gap transflective ECB LCD also uses a circular polarizer on both sides of the cell. The role of the circular polarizer is to make the display operate in a NW mode so that its dark state is not too sensitive to the cell gap variation. Each pixel is divided into a transmission region with cell gap d_T and a reflection region with cell gap d_R. The LC directors are all homogeneously aligned within the cell; therefore, no dual rubbing process is necessary and both regions operate identically in the ECB mode. Since the homogeneously aligned LC layer only imposes pure phase retardation on the incident polarized light, d_R is set to be around half of d_T to compensate the optical path difference between ambient light and backlight. Figure 9.17(b) depicts the voltage dependent transmittance and reflectance curves with $d_R\Delta n = 168$ nm and $d_T\Delta n = 336$ nm. As expected, both curves perfectly match with each other and both modes reach the highest transmittance and reflectance simultaneously. Here $d_R\Delta n$ and $d_T\Delta n$ are designed to be slightly larger than $\lambda/4$ and $\lambda/2$, respectively, in order to reduce the on-state voltage.

The downside of the dual-cell-gap approach is threefold. First, due to the cell gap difference the LC alignment is distorted near the boundaries of the transmissive and reflective pixels. This area should be covered by black matrices in order to retain a good contrast ratio. Second, the thicker cell gap in the transmission region results in a slower response time than the reflective region. Fortunately, the dynamic response requirement in mobile applications is not as strict as those for video applications. This response time difference, although not perfect, is still tolerable. Third, the view angle of the T-mode is rather narrow because the LC directors are tilted up along one direction by the external electric field. By substituting the quarter-wave film with a biaxial film on each side of the cell, the viewing angle can be greatly improved [58]. Because the manufacturing

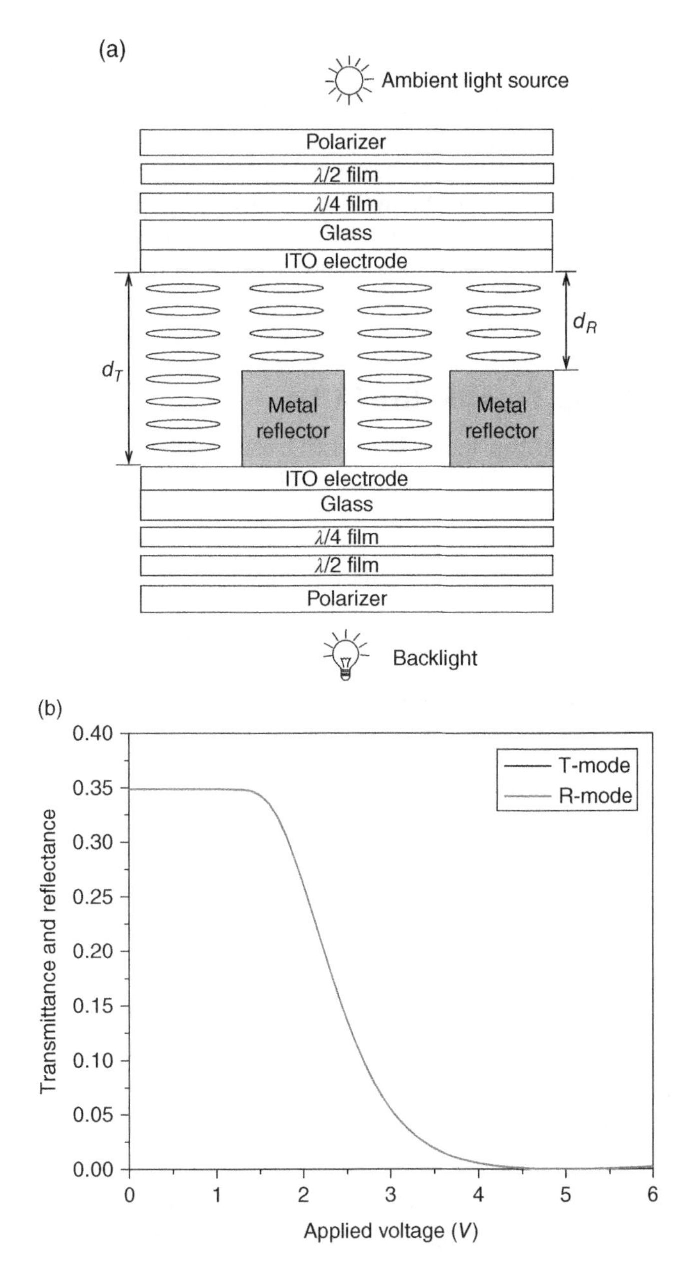

Figure 9.17 (a) Schematic device configuration and (b) voltage dependent transmittance and reflectance curves of the dual-cell-gap transflective ECB LCD. Zhu 2006. Reproduced with permission from IEE.

process is compatible with the state-of-the-art LCD fabrication lines, the dual-cell-gap transflective ECB LCD is so far the mainstream approach for the commercial transflective LCD products.

Beside the above dual-cell-gap transflective ECB LCD, others dual-cell-gap transflective LCDs are also proposed, such as dual-cell-gap transflective VA LCD [59], dual-cell-gap transflective HAN LCD [60], and dual-cell-gap transflective FFS (fringe-field switching) LCD [61]. [62]' Similar to the dual-cell-gap transflective ECB LCD, both dual-cell-gap transflective VA LCD and dual-cell-gap transflective HAN LCD also operate in ECB mode although the initial LC alignment is different. On the other hand, in the dual-cell-gap transflective FFS LCD, LC molecules are switched in the plane parallel to the supporting substrates. Its dark state is achieved by a half-wave film and the initially homogeneously aligned LC layer. Consequently, the dark state is very sensitive to the LC cell gap variation, which causes difficulties to maintain a good dark state in both transmission and reflection regions due to the dual-cell-gap device configuration.

9.6 Single-Cell-Gap Transflective LCDs

Different from the dual-cell-gap transflective LCD, the single-cell-gap transflective LCD renders a uniform cell gap profile throughout the cell. Therefore, the dynamic responses of the T- and R-modes are close to each other. For instance, a single-cell-gap transflective LCD using an IER is proposed [63], which is similar to the structure described in Figure 9.10. In this design, the backlight is reflected by the IER to the reflection area; as a result, the transmitted beam from the backlight traverses a similar optical path to that of the ambient beam, which leads to the same color saturation in both T- and R-modes. However, similar to the transflective cholesteric LCD, to produce an IER on the top substrate increases the manufacturing complexity. Besides, a mismatch between the IER and bottom pixel layout may cause light leakage.

In fact, several transflective LCDs described in the above sections also belong to this single-cell-gap category, such as transflective TN and STN LCDs, transflective MTN LCD, dual-rubbing transflective MTN/ECB LCD, dual-rubbing transflective VA/HAN LCD, dual-rubbing transflective ECB/HAN LCD, transflective VA LCD utilizing periodically patterned electrodes, and transflective IPS LCD. Because the ambient light travels twice while the backlight propagates only once in the LC layer, the light efficiency of T- and R-modes cannot reach maximum simultaneously unless mixed display modes are employed. This leads to the transflective mixed-mode LCDs as described in Figure 9.16. As discussed there, the transflective mixed-mode LCDs require either a dual-rubbing process or complicated electrode designs. Consequently single-cell-gap transflective mixed-mode LCDs have not been commercialized yet.

9.7 Performance of Transflective LCDs

We have just described the basic operating principles of some main transflective LCDs. The simulation results are based on some ideal assumptions. It is understandable that many other factors can affect the display image qualities, such as color balance, image brightness, and viewing angle.

9.7.1 Color balance

Because the reflection beam passes through the color filter (CF) twice, while the transmission beam only passes once, generally speaking, the transflective LCD experiences an unbalanced color between the T- and R-modes. To solve the color imbalance problem, different CF approaches have been developed. Sharp Corp. proposed a multi-thickness color filter

(MT-CF) design for the transflective LCDs [64]. In this design, the CF thickness in the reflection region is around one half of that in the transmission region. The ambient beam passes through the thinner CF twice, while the transmission beam passes through the thicker CF once, so these two beams experience almost the same spectral absorption. Therefore, such a CF thickness difference ensures almost identical color saturation between the transmission and reflection regions, resulting in a good color balance between T- and R-modes.

In addition to the MT-CF design, a pinhole type CF design was also proposed by Sharp. There, the thicknesses of the CF in the two regions are equal, but the CF in the reflection region is punched with some pinholes. Therefore, a portion of the ambient light does not 'see' the CF; instead, it passes through the pinholes directly. The problem of such a pinhole type CF is its narrow color reproduction area because the ambient light spectrum is mixed with the RGB primary colors, which causes the color impurity.

An alternative approach to obtaining the same color balance between the T- and R-modes is to fill some scattering materials in the CF in the reflection region [65]. The filled scattering materials serve two purposes: First, the equivalent CF thickness in the reflection region is decreased to around one half of that in the transmission region. Second, the scattering materials can steer the reflection beam from the specular reflection direction; therefore, a pure flat metal reflector can be used in the reflection region, which greatly simplifies the manufacturing process.

9.7.2 *Image brightness*

Image brightness is a very important feature for transflective LCDs, and there are many factors that decrease the overall image brightness. For instance, the red, green, and blue color filters have different light attenuation, which affects the overall brightness of the display panel. Besides, the reflection region of the openings-on-metal transflector, usually made from aluminum, has ~92% reflectivity over the visible spectral region [66], which leads to a slightly lower light efficiency in the R-mode.

In the openings-on-metal transflector, the transflector area is intended for either the reflection region or the transmission region. To increase the backlight utilization efficiency while still keeping the ambient light efficiency unchanged, a transflective LCD design using a microtube array below the transmission pixels region has been proposed [67]. The microtube structure, which is funnel-shaped, allows most of the backlight to enter from a larger lower aperture and to exit from a smaller upper aperture. Consequently, the backlight utilization efficiency can be greatly enhanced, provided that the transmission/reflection area ratio remains unchanged. After optimization, the average backlight utilization efficiency is improved by ~81%.

9.7.3 *Viewing angle*

Although the display panel size for most transflective LCDs is not too large, viewing angle is another important concern. The user of a cell phone would like to see clear images from every angle. The future cell phone will have expanded functions, such as videos and movies. The dual-cell-gap transflective ECB LCD has a relatively narrow viewing angle in the T-mode. But by substituting the quarter-wave film with a biaxial film on each side of the cell, the viewing angle of the T-mode can be greatly widened. In the R-mode, surface reflection is the main factor deteriorating the image contrast ratio and viewing angle. To solve this problem, a bumpy reflector in the reflection region is commonly employed. The bumpy reflectors serve two purposes: (1) to diffuse the reflected light which is critical for widening viewing angle, and (2) to steer the reflected light away from the specular reflection so that the images are not overlapped with the surface reflections. To design bumpy reflectors [68], we need to consider the fact that the incident beam and reflected beam might form different angles with respect to the panel normal. In the optical modeling for the R-mode, the asymmetric incident and exit angles feature should be taken into consideration [69].

Homework Problems

9.1 Let us design a normally black reflective LCD using a 45° twisted nematic cell for projection displays. We want the display to be independent of beta angle (β); the angle between the top LC rubbing direction; and the incoming polarization axis. What is the required $d\Delta n$ value at $\lambda = 550$ nm?

9.2 In a transflective LCD, the reflective part usually has a lower contrast ratio and narrower viewing angle than the transmissive part. Why?

9.3 In most of the transflective LCDs, the reflective part usually has a broadband quarter-wave film. Explain why.

9.4 Sketch the device configuration of a transflective LCD using a double-cell-gap VA cell. The LC parameters are: $\Delta n = 0.1$ ($\lambda = 550$ nm), $\Delta\varepsilon = -4$, $K_{11} = 10$ pN, $K_{22} = 6$ pN, $K_{33} = 20$ pN, and $\gamma_1 = 0.1$ Pa·s. (a) What are the cell gaps for the R and T regions? (b) Estimate the optical decay time, and (c) sketch the expected voltage-dependent transmittance and reflectance curves.

9.5 Sketch the device configuration of a transflective LCD using a double-cell-gap TN cell. The LC parameters are: $\Delta n = 0.1$ ($\lambda = 550$ nm), $\Delta\varepsilon = 10$, $K_{11} = 10$ pN, $K_{22} = 6$ pN, $K_{33} = 20$ pN, and $\gamma_1 = 0.1$ Pa·s. (a) What are the cell gaps for the R and T regions? (b) Estimate the optical decay time, and (c) sketch the expected voltage-dependent transmittance and reflectance curves.

9.6 Given two linear sheet polarizers, a vertical aligned cell filled with a negative $\Delta\varepsilon$ LC, a wire grid polarizer (WGP), and a backlight, construct a normally black transflective LCD. Assume that the WGP can be deposited at any portion of the pixel. (a) Sketch the display configuration and show how it works. (b) Sketch the voltage-on state LC configuration and find the required $d\Delta n$ value for achieving maximum reflectance.

9.7 Given two linear sheet polarizers, a 90° TN cell, a wire grid polarizer (WGP), and a backlight, construct a normally white transflective LCD. Assume that the WGP can be deposited at any portion of the pixel. (a) Sketch the display configuration and show how it works. (b) Sketch the voltage-off state LC configuration and find the required $d\Delta n$ value for achieving maximum reflectance and transmittance.

References

1. M. Schadt and W. Helfrich, Voltage-dependent optical activity of a twisted nematic liquid crystal, *Appl. Phys. Lett.*, **18**, 127 (1971).
2. M. Oh-e and K. Kondo, Electro-optical characteristics and switching behavior of the in-plane switching mode, *Appl. Phys. Lett.*, **67**, 3895 (1995).

3. K. Ohmuro, S. Kataoka, T. Sasaki, and Y. Koike, Development of super-high-image-quality vertical alignment mode LCD, *SID Tech. Digest* **28**, 845 (1997).
4. A. R. Kmetz, A single-polarizer twisted nematic display, *Proc. SID*, **21**, 63 (1980).
5. X. Zhu, Z. Ge, T. X. Wu, and S. T. Wu, Transflective liquid crystal displays, *J. Display Technology*, **1**, 15 (2005).
6. E. H. Stupp and M. S. Brennesholtz, *Projection Displays* (Wiley, New York, 1998).
7. S. T. Wu and D. K. Yang, *Reflective Liquid Crystal Displays* (Wiley, New York, 2001).
8. S. T. Wu and C. S. Wu, Optimization of film-compensated homogeneous cells for liquid crystal displays, *Liq. Cryst.* **24**, 811 (1998).
9. C. L. Kuo, C. K. Wei, S. T. Wu and C. S. Wu, Reflective display using mixed-mode twisted nematic liquid crystal cell, *Jpn. J. Appl. Phys.* **36**, 1077 (1997).
10. S. T. Wu and C. S. Wu, Comparative studies of single-polarizer reflective liquid crystal displays, *J. SID* **7**, 119 (1999).
11. T. H. Yoon, G. D. Lee, G. H. Kim, et al., Reflective liquid crystal display using 0o-twisted half-wave cell, *SID Tech. Digest* **31**, 750 (2000).
12. S. T. Wu and C. S. Wu, Mixed twisted-nematic mode for reflective liquid crystal displays, *Appl. Phys. Lett.* **68** (1996) 1455.
13. R. D. Ketchpel and S. Barbara, Transflector, US Patent **4**,040,727 (Aug. 9, 1977).
14. H. C. Borden Jr., Universal transmission reflectance mode liquid crystal display, US Patent **3**,748,018 (Jul. 24, 1973).
15. J. E. Bigelow, Transflective liquid crystal display, US Patent **4**,093,356 (Jun. 6, 1978).
16. H. Furuhashi, C. K. Wei, and C. W. Wu, Transflective liquid crystal display having dielectric multilayer in LCD cells, US Patent **6**,806,934 (Oct. 19, 2004).
17. D. R. Hall, Transflective LCD utilizing chiral liquid crystal filter/mirrors, US Patent **5**,841,494 (Nov. 24, 1998).
18. W. J. Schrenk, V. S. Chang, and J. A. Wheatley, Birefringent interference polarizer, US Patent 5,612,820 (18 Mar. 1997).
19. D. P. Hansen, J. E. Gunther, Dual mode reflective/transmissive liquid crystal display apparatus, US Patent **5**, 986,730 (Nov. 16, 1999).
20. P. G. de Gennes and J. Prost, The Physics of Liquid Crystals, 2nd edn (Oxford university press, New York, 1993).
21. J. Ouderkirk, S. Cobb Jr., B. D. Cull, M. F. Weber, D. L. Wortman, Transflective displays with reflective polarizing transflector, US Patent 6,124,971 (26 Sept. 2000).
22. M. Bass, E. W. Van Stryland, D. R. Williams, and W. L. Wolfe, *Handbook of Optics, vol. II, Devices, Measurements, & Properties*, 2nd edn, pp. 3.32–3.35 (McGraw-Hill, New York, 1995).
23. S. T. Wu, U. Efron, and L. D. Hess, Birefringence measurements of liquid crystals, *Appl. Opt.* **23**, 3911 (1984).
24. R. T. Perkins, D. P. Hansen, E. W. Gardner, et al., Broadband wire grid polarizer for the visible spectrum, US Patent 6,122,103 (19 Sept. 2000).
25. X. J. Yu and H. S. Kwok, Optical wire-grid polarizers at oblique angles of incidence, *J. Appl. Phys.* **93**, 4407 (2003).
26. G. H. Heilmeier and L. A. Zanoni. Guest-host interactions in nematic liquid crystals. A new electro-optic effect, *Appl. Phys. Lett.*, **13**, 91 (1968).
27. D. L. White and G. N. Taylor, New absorptive mode reflective liquid-crystal display device", *J. Appl. Phys.*, **45**, 4718 (1974).
28. H. S. Cole, Transflective liquid crystal display, US Patent 4,398,805 (16 Aug. 1983).
29. S. Morozumi, K. Oguchi, R. Araki, et al., Full-color TFT-LCD with phase-change guest–host mode, *SID Tech. Digest* **16**, 278 (1985).
30. J. W. Done, N. A. Vaz, B.-G. Wu, and S. Zumer, Field controlled light scattering from nematic microdroplets, *Appl. Phys. Lett.*, **48**, 269 (1986).
31. D. K. Yang, J. W. Doane, Z. Yaniv, and J. Glasser, Cholesteric reflective display: Drive scheme and contrast, *Appl. Phys. Lett.*, **64**, 1905 (1994).
32. R. A. M. Hikmet, Electrically induced light scattering from anisotropic gels, *J. Appl. Phys.*, **68**, 4406 (1990).
33. H. Ren and S.-T. Wu, Anisotropic liquid crystal gels for switchable polarizers and displays, *Appl. Phys. Lett.*, **81**, 1432 (2002).

34. Y. H. Lin, H. Ren, S. Gauza, et al., Reflective direct-view displays using dye-doped dual-frequency liquid crystal gel, *J. Display Technology*, **1**, 230 (2005).
35. Q. Hong, T. X. Wu, and S. T. Wu, Optical wave propagation in a cholesteric liquid crystal using the finite element method, *Liq. Cryst.* **30**, 367 (2003).
36. Y. P. Huang, X. Zhu, H. Ren, et al., Full-color transflective Ch-LCD with image-enhanced reflector, *SID Tech. Digest* **35**, 882 (2004).
37. Y. P. Huang, X. Zhu, H. Ren, et al., Full-color transflective cholesteric LCD with image-enhanced reflector, *J. SID* **12**, 417 (2004).
38. F. J. Kahn, Reflective mode, 40-character, alphanumeric twisted-nematic liquid crystal displays, *SID Tech. Digest*, **9**, 74 (1978).
39. W. H. McKnight, L. B. Stotts, and M. A. Monahan, Transmissive and reflective liquid crystal display, US Patent 4,315,258 (9 Feb. 1982).
40. C. H. Gooch and H. A. Tarry, The optical properties of twisted nematic liquid crystal structures with twist angles ≤90 degrees, *J. Phys. D: Appl. Phys.*, **8**, 1575 (1975).
41. A. Lien, Extended Jones matrix representation for the twisted nematic liquid-crystal display at oblique incidence, *Appl. Phys. Lett.*, **57**, 2767 (1990).
42. K. Kawasaki, K. Yamada, R. Watanabe, and K. Mizunoya, High-display performance black and white supertwisted nematic LCD, *SID Tech. Digest* **18**, 391 (1987).
43. T. J. Scheffer and J. Nehring, A new, highly multiplexable liquid crystal display, *Appl. Phys. Lett.*, **45**, 1021 (1984).
44. T. Maeda, T. Matsushima, E. Okamoto, et al., Reflective and transflective color LCDs with double polarizers, *J. SID*, **7**, 9 (1999).
45. T. Ohyama, Y. Ukai, L. Fennell, et al., TN mode TFT-LCD with in-cell polarizer, *SID Tech. Digest* **35**, 1106 (2004).
46. S. T. Wu and C. S. Wu, Mixed-mode twisted nematic liquid crystal cells for reflective displays, *Appl. Phys. Lett.*, **68**, 1455 (1996).
47. H. Molsen, and M. D. Tillin, Transflective liquid crystal displays, International patent application No. PCT/JP99/05210, International publication No. WO 00/17707 (30 Mar. 2000).
48. S. Pancharatnam, Achromatic combinations of birefringent plates: Part I. An achromatic circular polarizer, *Proc. of the Indian Academy of Science*, Section A, **41**, 130 (1955).
49. S. J. Roosendaal, B. M. I. van der Zande, A. C. Nieuwkerk, et al., Novel high performance transflective LCD with a patterned retarder, *SID Tech. Digest* **34**, 78 (2003).
50. B. M. I. van der Zande, A. C. Nieuwkerk, M. van Deurzen, et al., Technologies towards patterned optical foils, *SID Tech. Digest* **34**, 194 (2003).
51. M. Okamoto, H. Hiraki, and S. Mitsui, Liquid crystal display, US Patent 6,281,952 (28 Aug. 2001).
52. T. Uesaka, E. Yoda, T. Ogasawara, and T. Toyooka, Optical design for wide-viewing-angle transflective TFT-LCDs with hybrid aligned nematic compensator, *Proc. 9th Int'l. Display Workshops*, pp. 417-420 (2002).
53. S. H. Lee, K. H. Park, J. S. Gwag, et al., A multimode-type transflective liquid crystal display using the hybrid-aligned nematic and parallel-rubbed vertically aligned modes, *Jpn. J. Appl. Phys.*, part 1, **42**, 5127 (2003).
54. Y. J. Lim, J. H. Song, Y. B. Kim, and S. H. Lee, Single gap transflective liquid crystal display with dual orientation of liquid crystal, *Jpn. J. Appl. Phys.*, part 2, **43**, L972 (2004).
55. S. H. Lee, H. W. Do, G. D. Lee, et al., A novel transflective liquid crystal display with a periodically patterned electrode, *Jpn. J. Appl. Phys.*, part 2, **42**, L1455 (2003).
56. J. H. Song and S. H. Lee, A single gap transflective display using in-plane switching mode, *Jpn. J. Appl. Phys.*, part 2, **43**, L1130 (2004).
57. M. Shimizu, Y. Itoh, and M. Kubo, Liquid crystal display device, US Patent 6,341,002 (22 Jan. 2002).
58. M. Shibazaki, Y. Ukawa, S. Takahashi, et al., Transflective LCD with low driving voltage and wide viewing angle, *SID Tech. Digest*, **34**, 90 (2003).
59. H. D. Liu and S. C. Lin, A novel design wide view angle partially reflective super multi-domain homeotropically aligned LCD, *SID Tech. Digest* **23**, 558 (2002).
60. C. L. Yang, Electro-optics of a transflective liquid crystal display with hybrid-aligned liquid crystal texture, *Jpn. J. Appl. Phys.*, part 1, **43**, 4273 (2004).

61. T. B. Jung, J. C. Kim, and S. H. Lee, Wide-viewing-angle transflective display associated with a fringe-field driven homogeneously aligned nematic liquid crystal display, *Jpn. J. Appl. Phys.*, part 2, **42**, L464 (2003).
62. T. B. Jung, J. H. Song, D.S. Seo and S. H. Lee, Viewing angle characteristics of transflective display in a homogeneously aligned liquid crystal cell driven by fringe-field, *Jpn. J. Appl. Phys.*, part 2, **43**, L1211 (2004).
63. Y. P. Huang, M. J. Su, H. P. D. Shieh and S. T. Wu, A single cell-gap transflective color TFT-LCD by using image-enhanced reflector, *SID Tech. Digest* **34**, 86 (2003).
64. K. Fujimori, Y. Narutaki, Y. Itoh, et al., New color filter structures for transflective TFT-LCD, *SID Tech. Digest* **33**, 1382 (2002).
65. K. J. Kim, J. S. Lim, T. Y. Jung, et al., A new transflective TFT-LCD with dual color filter, *Proc. 9th Int'l. Display Workshops*, pp. 433-436 (2002).
66. M. Bass, E. W. Van Stryland, D. R. Williams, and W. L. Wolfe, *Handbook of Optics, vol. II, Devices, Measurements, & Properties*, 2nd edn (McGraw-Hill, New York, 1995), Ch. 35.
67. H. P. D. Shieh, Y. P. Huang, and K. W. Chien, Micro-optics for liquid crystal displays applications, *J. Display Technology*, **1**, 62 (2005).
68. K. Nakamura, H. Nakamura, and N. Kimura, Development of high reflective TFT, *Sharp Technical Journal* **69**, 33 (1997).
69. Z. Ge, T. X. Wu, X. Zhu, and S. T. Wu, Reflective liquid crystal displays with asymmetric incidence and exit angles, *J. Opt. Soc. Am. A*, **22**, 966 (2005).

10

Liquid Crystal Display Matrices, Drive Schemes and Bistable Displays

Liquid crystal displays are a dominant display technology. They are used in electronic watches, calculators, handheld devices such as cellular phones, head-mounted displays, laptop and desktop computers, direct view and projection TV, and electronic papers and books. They have the advantages of flat panel, low weight, energy-saving and low drive voltage. In display applications, the liquid crystals modulate light intensity because of their birefringence. Liquid crystals can also be reoriented by externally applied electric fields because of their dielectric anisotropies or ferroelectricity, which makes it possible to show spatial images when patterned electric fields are applied.

10.1 Segmented Displays

In order for a liquid crystal display to display images, multi-elements are needed. The simplest multi-element displays are the segmented displays where each element has its own electrode that is separated from the electrodes of other elements [1]. Voltages can be applied to each element independently. As an example, a segmented numerical liquid crystal display is shown in Figure 10.1. When an appropriate voltage is applied between the common electrode and a segmented electrode, an electric field is generated in the region between the electrodes, and the liquid crystal in that region is switched to the field-on state. For example, when the voltage is applied to electrode 1, 2, 4, 6, and 7, the numeral '5' is displayed. Segmented displays are also referred to as direct drive displays. This type of display is only good for low information content displays because one electrode is needed for each element.

Fundamentals of Liquid Crystal Devices, Second Edition. Deng-Ke Yang and Shin-Tson Wu.

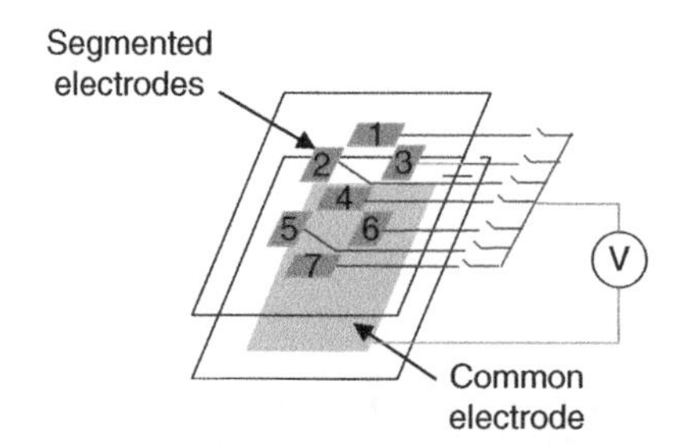

Figure 10.1 Schematic diagram of the segmented numerical liquid crystal display.

10.2 Passive Matrix Displays and Drive Scheme

In order to display high information content images, an *xy* matrix must be used [1,2]. There are striped electrodes on the substrates, as shown in Figure 10.2. The overlapped region between a front electrode and a rear electrode is a display element referred to as pixel. For a display consisting of N rows and M columns, there are $N \times M$ elements, but there are only $N + M$ electrodes. In the *xy* matrix, the structure of the electrode is greatly simplified. Driving the pixels, however, becomes complicated. It is impossible to apply a voltage to a pixel without affecting the other pixels in the matrix. For the purpose of simplicity, let us consider a 2×2 matrix display as shown in Figure 10.3(a). The equivalent circuit is shown in Figure 10.3(b). The liquid crystals in the pixels can be regarded as capacitors. If we want to switch Pixel 11, we apply a voltage V to Column 1 and we ground Row 1. The voltage across Pixel 11 is V. As can be seen from Figure 10.3(b), there is also a voltage $V/3$ applied to the other three pixels, which may partially switch the pixels. This undesired voltage that is applied to the other pixels and partially switches the pixels is referred to as *crosstalk*.

Now we consider how to address an $N \times M$ (N rows and M columns) *xy* matrix display. The display is addressed one row at a time. The row electrodes are called scanning electrodes and the column electrodes are called signal electrodes or data electrodes. The state of the pixels is controlled by the voltages applied to the column electrodes. There are three issues that must be considered. The first is that there are voltages applied to the pixels on the not-being-addressed rows because of the column voltages. The second is that the voltages across the pixels on the row after addressing are not retained. The third is the frame time T_f. If the time interval to address a row is Δt, the frame time is $T_f = N\Delta t$. The frame time must be not only shorter than the response time (~40 ms) of the human eye but also shorter than the relaxation time of the liquid crystal. When the frame time is shorter than the relaxation time of the liquid crystal, the state of the liquid crystal in a pixel is determined by the averaged (over the frame time) rms voltage applied across the pixel.

Let us consider a normal black liquid crystal display. At zero volts, the transmittance of the liquid crystal display is 0 and the display is black. At the voltage V_{on}, the display is switched to the bright state. A pixel of the display to be addressed into the bright state is called the *selected pixel,* and a pixel to be addressed into the dark state is called a *non-selected pixel.* We consider a simple drive scheme. The column voltage to select the bright state is $-V/b$ and the column voltage to select the black state is V/b, where b is a constant [2–4]. The row voltage to the being-addressed row is $(b-1)V/b$ and the row voltage to the not-being-addressed rows is 0. For the selected pixel, the applied voltage is $(b-1)V/b-(-V/b)=V$ when being addressed and $0-(\pm V/b)=\mp V$

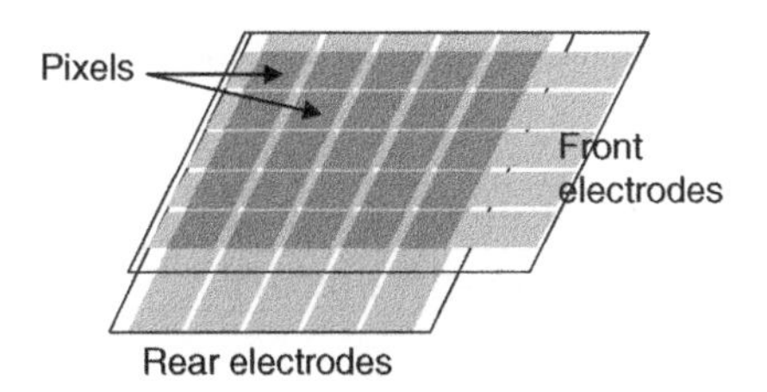

Figure 10.2 Schematic diagram of *xy* matrix.

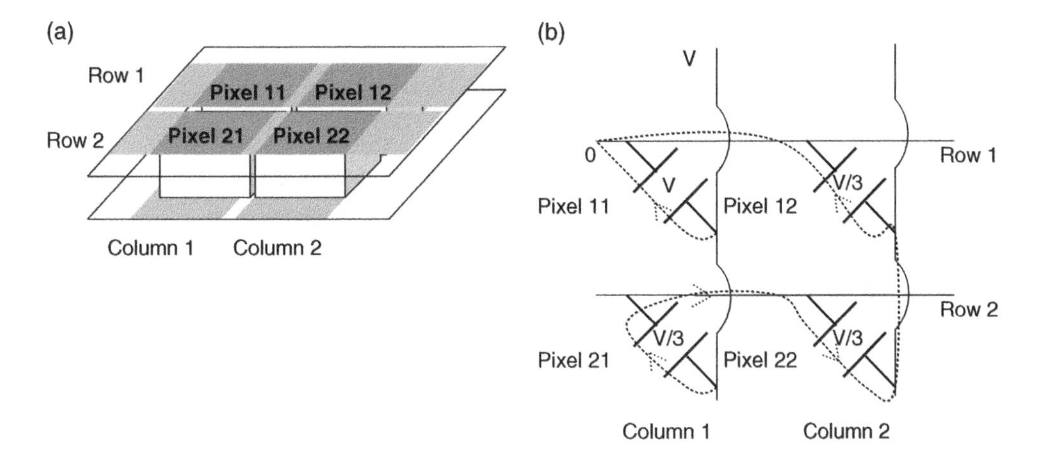

Figure 10.3 (a) Schematic diagram of the 2 × 2 matrix display. (b) The equivalent circuit.

afterward. For the non-selected pixel, the applied voltage is $(b-1)V/b-(V/b)=(b-2)V/b$ when being addressed and $0-(\pm V/b)=\mp V$ afterward. The rms voltage on the selected pixel is

$$\overline{V}_s=\left\{\frac{1}{N}\left[1\cdot V^2+(N-1)\left(\frac{V}{b}\right)^2\right]\right\}^{1/2}, \tag{10.1}$$

The rms voltage on the non-selected pixel is

$$\overline{V}_{ns}=\left\{\frac{1}{N}\left[1\cdot\left(\frac{b-2}{b}V\right)^2+(N-1)\left(\frac{V}{b}\right)^2\right]\right\}^{1/2} \tag{10.2}$$

The ratio between these two voltages is

$$R=\frac{\overline{V}_s}{\overline{V}_{ns}}=\left[\frac{b^2+(N-1)}{(b-2)^2+(N-1)}\right]^{1/2}$$

In order to optimize the performance of the display, the parameter b should be chosen to maximize R:

$$\frac{\partial R^2}{\partial b}=\frac{4[-b^2+2b+(N-1)]}{(b-2)^2+(N-1)}\equiv 0, \tag{10.3}$$

which gives

$$b=\sqrt{N}+1. \tag{10.4}$$

The maximized ratio is

$$R_m=\left(\frac{\sqrt{N}+1}{\sqrt{N}-1}\right)^{1/2} \tag{10.5}$$

R_m as a function of N is shown in Figure 10.4. The larger the number of rows, the smaller the difference between the selected voltage and the non-selected voltage becomes.

In passive matrix displays, the quality of the displayed images depends on the electro-optical response of the liquid crystal material. Consider a liquid crystal display whose voltage–transmittance curve is shown in Figure 10.5. In order to achieve high contrast, the non-selected voltage $\overline{V}_{ns}$ should be set at below V_{10}, such that the transmittance of the black pixel is less than $T_{min}+0.1\Delta T$. The maximum selected voltage is $\overline{V}_s=R_m\overline{V}_{ns}$. If $\overline{V}_s$ is lower than V_{90}, the transmittance of the bright pixel is lower than $T_{max}-0.1\Delta T$, which is clearly not good. Therefore the quality of the displayed images depends on the steepness of the voltage–transmittance curve of the liquid crystal and the number of rows of the display. The steepness of the voltage–transmittance curve can be characterized by the parameter γ defined as

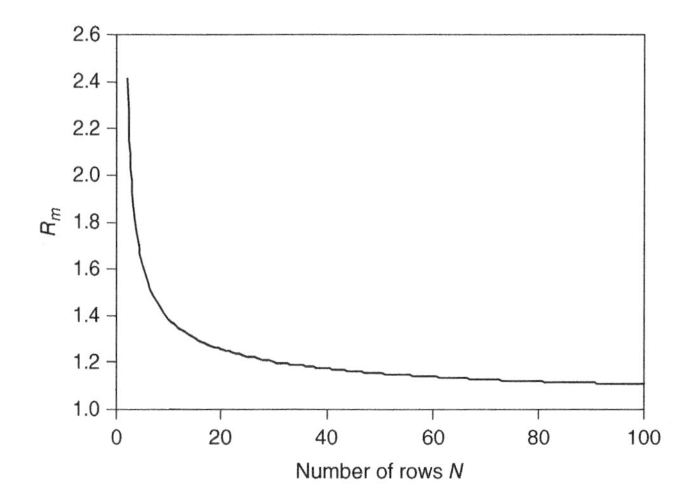

Figure 10.4 The ratio between the selected voltage and non-selected voltage as a function of the number of rows of the display.

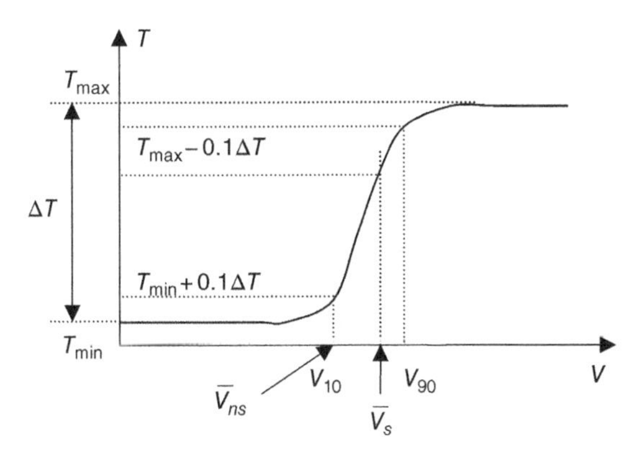

Figure 10.5 The voltage–transmittance curve of the TN liquid crystal display.

$$\gamma = \frac{V_{90}}{V_{10}}. \tag{10.6}$$

If $\gamma \leq R_m$, images with good contrast and high brightness can be displayed. In other words, for a given liquid crystal display, γ is fixed. The maximum number of rows that the passive matrix display can have is given by

$$N_{max} = \left(\frac{\gamma^2+1}{\gamma^2-1}\right)^2 \tag{10.7}$$

$1/N_{max}$is sometimes referred to as the duty ratio. For TN LCDs, $\gamma \sim 1.4$, $N_{max} \sim 9$. This number of rows is only suitable for displays on simple calculators.

In order to increase the number of rows of twisted nematic liquid crystal passive matrix displays, the steepness of the voltage–transmittance curve must be increased. The steepness is measured as the difference between the saturation voltage V_s and the threshold voltage V_{th}. Smaller $\Delta V = V_s - V_{th}$ generates steeper voltage–transmittance curves. The steepness can be achieved by the following methods [4]. (1) Increase the twist angle, because the threshold voltage increases with the twist angle, as shown in Figure 7.5, (2) Decrease pretilt angle, because the threshold voltage increases with decreasing pretilt angle as discussed in Chapter 5. (3) Decrease the ratio between the cell thickness h and the pitch P of the liquid crystal, because the saturation voltage decreases with increasing pitch. The twist elastic energy is smaller for larger pitch. (4) Increase K_{33}/K_{11}, because the threshold voltage increases with K_{33}. There is bend deformation in the field-activated states with small tilt angles but not in the saturated state. (5) Decrease K_{22}/K_{11}, because the saturation voltage decreases with K_{22}. (6) Decrease $\Delta\varepsilon/\varepsilon_{\perp}$.

Usually AC voltage waves are used in addressing liquid crystal displays in which the liquid crystal molecules interact with applied electric fields through dielectric interaction. AC voltages can prevent injection of ions into the liquid crystals, which degrade the displays through long-term effects. The waveforms of the addressing voltages and the corresponding transmittance of the display are schematically shown in Figure 10.6, where the labeled voltages are the voltage of the first half of the voltage pulses [2]. If the column voltage is varied between $-V/b$ and $+V/b$, gray-scale transmittances can be obtained.

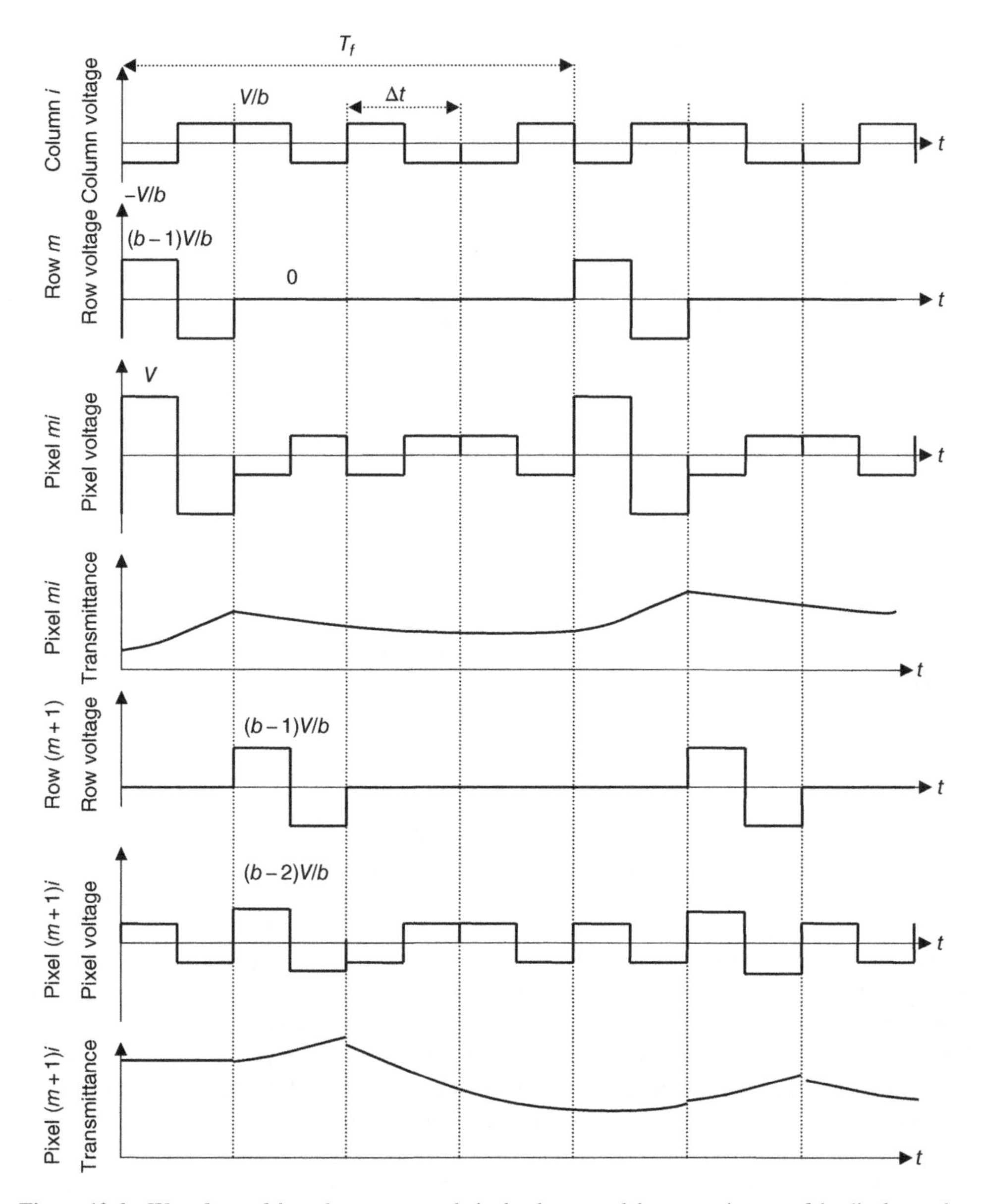

Figure 10.6 Waveform of the column, row, and pixel voltages and the transmittance of the display under the pixel voltages.

10.3 Active Matrix Displays

Active matrix displays using thin film transistors (TFTs) as electrical switches to control the transmission state of liquid crystal pixels offer excellent image quality and are commonly employed for direct-view displays [5,6]. Figure 10.7 shows the device structure of a

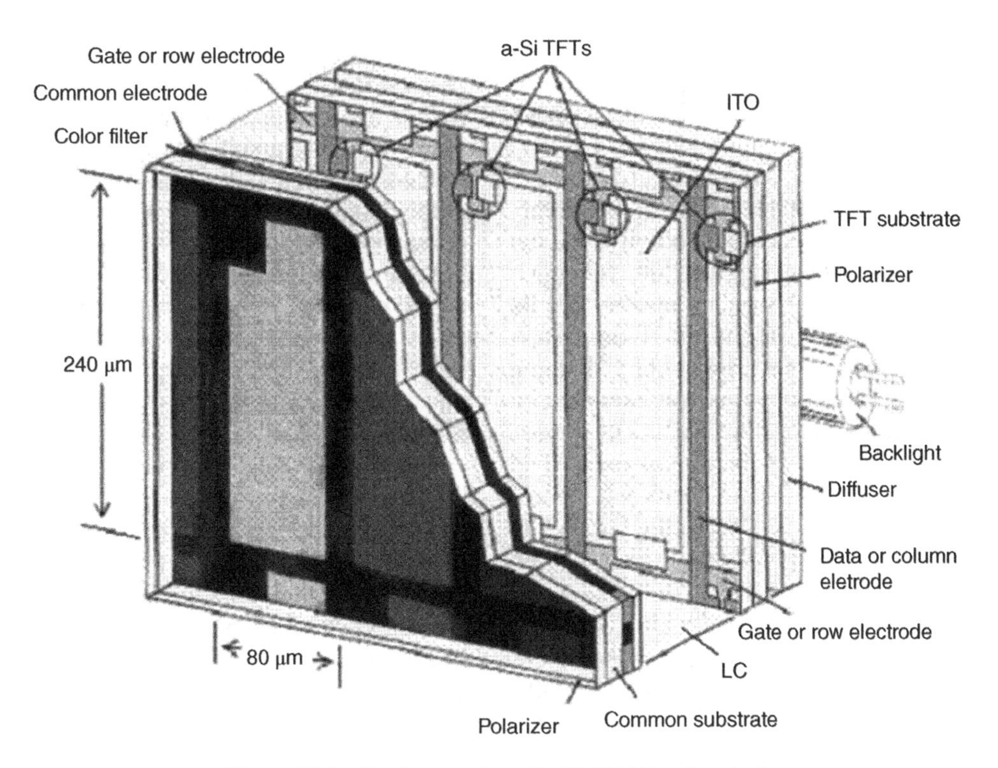

Figure 10.7 Device structure of a TFT LCD color pixel.

transmissive TFT-LCD using amorphous silicon (a-Si) transistors for large screen displays. Since liquid crystals do not emit light, a backlight is needed. A diffuser is used to homogenize the backlight. Since most LCDs require a linearly polarized light for achieving a high contrast ratio, two sheets of stretched dichroic polarizers are commonly used for large screen direct-view displays. The first glass substrate contains TFT arrays which serve as light switches. Each display pixel is independently controlled by a TFT. Since a-Si exhibits photoelectric characteristics, TFTs need to be protected from backlight (by gate metal lines) and ambient light (by black matrices). Because of the black matrices, the actual aperture ratio (the transparent indium-tin-oxide electrode area) drops to 80–0%, depending on the device resolution and panel size. The LC layer is sandwiched between two substrates. The cell gap is usually controlled at around 4 μm for transmissive LCDs. The performance of the display, such as light throughput, response time, and viewing angle, are all determined by the LC mode employed.

For direct-view displays, such as notebook computers and desktop monitors, compact size and light weight are critically important. Under such circumstances, color filters are usually imbedded on the inner side of the second substrate. Some development efforts are attempting to integrate color filters on the TFT substrate. Three subpixels (red, green, and blue) form a color pixel. Each subpixel transmits only one color; the rest is absorbed. Thus, the transmittance of each color filter alone is less than 33%. The color filters are made of pigment materials. Their transmittance at each color band is about 80–85%. Thus, the final transmittance of each color filter is ~27%. After having considered polarizers, color filters, and TFT aperture ratio, the overall system optical efficiency is only about 6–7% for a direct-view LCD panel. Low optical efficiency implies high power consumption. For portable displays, low power consumption is desirable because it lengthens the battery's operating time. For LCD TVs, although the power consumption issue is not as important at this stage, it will be addressed eventually.

For large-screen direct-view LCDs, amorphous silicon (a-Si) TFT is a preferred choice because of its simpler manufacturing process and lower cost than the poly-silicon (p-Si) TFT. However, the electron mobility of a-Si is about two orders of magnitude lower than that of p-Si, and the required pixel size is larger in order to maintain a good storage capacitance. A typical subpixel size for an a-Si TFT LCD is ~80 × 240 μm. It takes three subpixels (RGB) to form a color pixel. Therefore, the pixel size of each color pixel is about 240 × 240 μm. On the other hand, p-Si has higher electron mobility than a-Si, so that its pixel size can be made smaller and its device resolution is therefore higher. This advantage is particularly important for the small-screen LCDs, where the aperture is an important issue.

10.3.1 TFT structure

The most commonly used TFT is the inverse-staggered (called bottom-gate) type, as shown in Figure 10.8. The ohmic layer (n + a-Si) in the channel region can either be etched directly or etched by forming a protective film on the a-Si thin film. Each method has its own merits and demerits. The inverse-staggered structure offers a relatively simple fabrication process and its electron mobility is ~30% larger than that of the staggered type. These advantages make the bottom-gate TFT structure a favored choice for TFT-LCD applications.

Because a-Si is photosensitive, the a-Si TFT must be protected from incident backlight and ambient light, especially if the backlight is quite strong. Furthermore, the a-Si layer should be kept as thin as possible in order to minimize the photo-induced current, which would degrade the signal-to-noise ratio. In the bottom-gate TFTs, an opaque gate electrode is first formed at the TFT channel region, where it also serves as a light-shield layer for the backlight. On the color filter substrate, a black matrix shields the TFT from ambient light irradiation. In Figure 10.8,

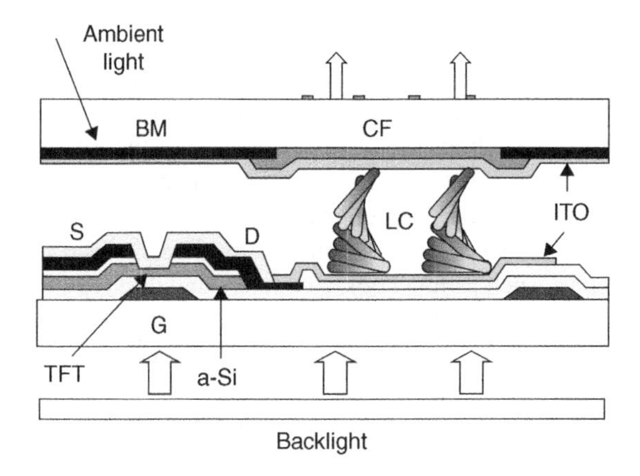

Figure 10.8 The bottom-gate TFT structure. S stands for source, G for gate, D for drain, BM for black matrix, and CF for color filter.

the drawing is not in scale. The TFT and black matrix parts should be much smaller than the transparent ITO part. A more realistic dimension is shown in Figure 10.7.

10.3.2 TFT operation principles

Figure 10.9 shows the equivalent circuit of a single pixel of a TFT LCD. Once the gate voltage exceeds a threshold, the TFT switch is open. The source (S) data voltage is transferred to the drain (D), which is connected to the bottom pixel electrode (ITO). The bottom pixel electrode and a gate line form a storage capacitor (Cs) which plays an important role in holding the voltage. If the voltage across the LC layer is higher than the threshold voltage of the employed LC material, the LC directors will be reoriented by the voltage resulting in light modulation of the backlight. The detailed transmission characteristics depend on whether the LC is in normally white mode (twisted nematic) or normally black mode (in-plane switching and multi-domain vertical alignment). In both situations, the polarizers are crossed.

DC voltage would induce undesirable electrochemical degradation of the organic LC molecules and should be avoided. Therefore the polarity of the voltage has to be alternated every other frame. In a normal operation, the gate voltage is set at 20 V for switch-on or at −5 V for switch-off state. Under these operating conditions, the a-Si TFT exhibits an on/off current ratio larger than 10^6. Figure 10.10 illustrates the TFT operating principle.

Let us assume that the common ITO electrode is biased at +5 V. For a given gray-scale, the data voltage is at +8 V. When the gate is open, the TFT is turned on and the current flows through the channel and charges up the storage capacitor. The drain terminal has the same voltage as the data terminal, i.e. Vd = +8 V. Since the drain is connected to the bottom pixel electrode, the effective voltage across the LC cell is +3 V, as shown in the top left quadrant (defined as the first quadrant) of Figure 10.10. If the gate voltage is removed or below threshold, the TFT is turned off for a frame time which is 16.7 ms (60 Hz frame rate). In this period, the storage capacitor holds the charges so that the pixel voltage remains at +3 V. To balance the DC

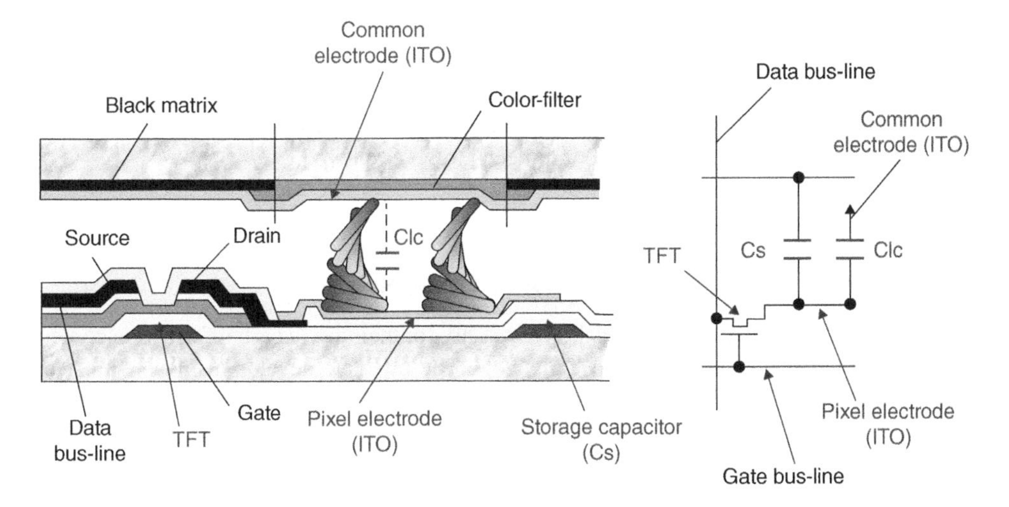

Figure 10.9 Equivalent circuit of a single pixel of TFT LCD.

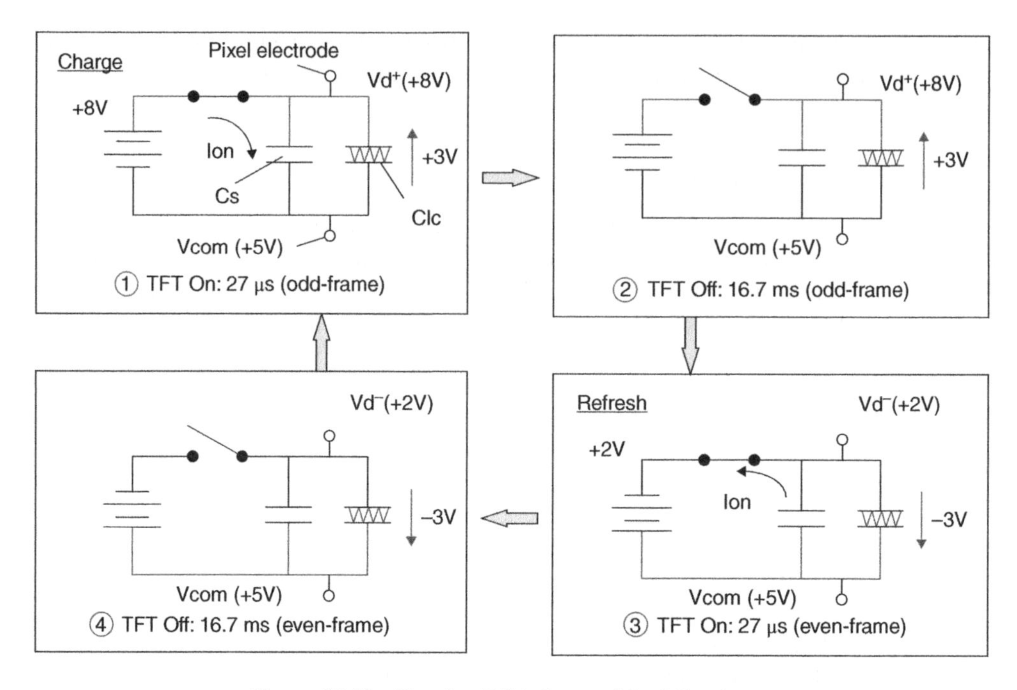

Figure 10.10 Keeping DC balance of the LC voltage.

voltage, in the next frame the data voltage is reduced to +2 V, as shown in the third quadrant (clockwise). When the TFT is turned on, the voltage across the LC cell is reversed to −3 V, which is opposite to the +3 V shown in the previous frame. When the TFT is turned off, as shown in the fourth quadrant, the storage capacitor holds the charge, and the LC voltage remains at −3 V.

The LC reorientation dynamics depends on the square of the electric field, that is it is independent of the polarity of the electric field. However, if the LC is biased at a DC voltage for too long, then the ions would be swept to the polyimide alignment layer interface and stay there to form a thin layer to shield the voltage. The gray-scale voltage will be misrepresented. Therefore, to reduce the undesirable DC voltage effect, the polarity of the DC pulses needs to be alternated and a high resistivity LC mixture needs to be employed.

10.4 Bistable Ferroelectric LCD and Drive Scheme

Multiplexibility of liquid crystal materials is necessary for their use in high information content displays. There are three ways to achieve this goal. The first is to develop displays with steep voltage–transmittance curve as discussed in the first section. The second way is using active matrices where the voltage on a pixel can be controlled independently. The third way is developing bistable liquid crystals, the subject of the rest of this chapter.

As discussed in Chapter 4, surface stabilized ferroelectric liquid crystals (SSFLCs) have two stable states at zero field. The two states have different planar orientational angles as shown in

Figure 4.9. In the SSFLC display, the liquid crystal is sandwiched between two crossed polarizers. The transmission axis of the entrance polarizer is parallel to the orientation direction of one of the stable states, say, state 1. The transmittance of state 1 is then 0. When the liquid crystal is in the other state, say, state 2, the liquid crystal director makes the angle 2θ with the entrance polarizer, and the transmittance is $T = \sin^2(4\theta)\sin^2(\pi\Delta nh/\lambda)$, where 2θ is the cone angle. When 2θ is near $\pi/4$ and $\pi\Delta nh/\lambda$ is close to $\pi/2$, the transmittance of state 2 is 1. The liquid crystal is switched between the two states by DC voltage pulses. When a voltage pulse with positive polarity is applied across the liquid crystal, say, the liquid crystal is switched into state 1. After the pulse, the liquid crystal remains in state 1. When a voltage pulse with negative polarity is applied, the liquid crystal is switched into state 2 and remains there afterward.

Experiments show that FLCs with high spontaneous polarizations respond accumulatively to voltage pulses [6–8]. The switching between the two stable states is determined by the 'voltage-time-area' A defined by $A = \int_{t_1}^{t_2} V(t)dt$, where the voltage is applied from time t_1 to time t_2. When the voltage is applied sufficiently long or its amplitude is sufficiently high so that A is larger than a threshold A_{th}, the liquid crystal starts to transform from one state to the other state, as shown in Figure 10.11. If the FLC is initially in state 1 with low transmittance T_1, when the voltage-time-area is below the threshold A_{th}, the liquid crystal remains in state 1. When the voltage-time-area is increased above A_{th}, the liquid crystal starts to transform into state 2 and the transmittance increases. When the voltage-time-area is increased above A_s, the liquid crystal is completely switched to state 2 and the transmittance reaches the maximum value T_2. When the applied voltage is removed, the liquid crystal remains in state 2. When a voltage with negative polarity is applied, the voltage-time-area is negative. When the voltage-time-area is decreased below $-A_{th}$, the liquid crystal starts to transform back to state 1 and the transmittance begins to decrease. When the voltage-time-area is decreased below $-A_s$, the liquid crystal is completely switched back to state 1 and the transmittance decreases to the minimum value T_1.

A drive scheme for the SSFLC display is shown in Figure 10.12. At the beginning of each frame, the liquid crystal is reset to the dark state by applying a positive/negative (P/N) voltage pulse to all the rows. The column voltage to select the bright state is P/N and the column voltage to retain the dark state is negative/positive (N/P). An N/P voltage pulse is applied to the being-addressed row. If the column voltage is P/N, the pixel voltage-time-area of the second half of the pulse is higher than A_s, and thus the bright state is selected. If the column voltage is N/P, the pixel voltage-time-area of the second half of the pulse is lower than A_{th}, and thus the dark state

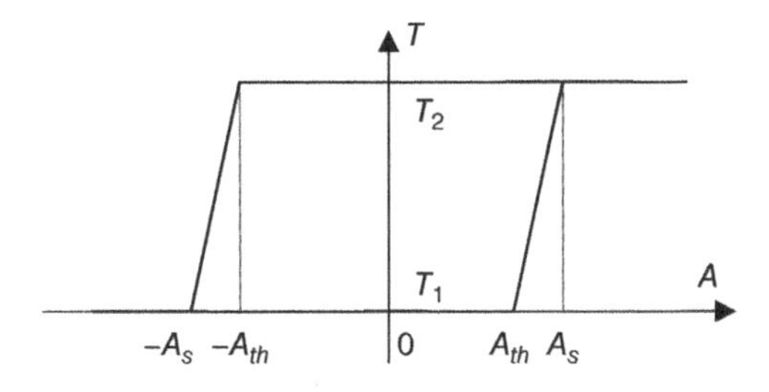

Figure 10.11 Schematic diagram showing the transmittance of the SSFLC display as a function of voltage–area.

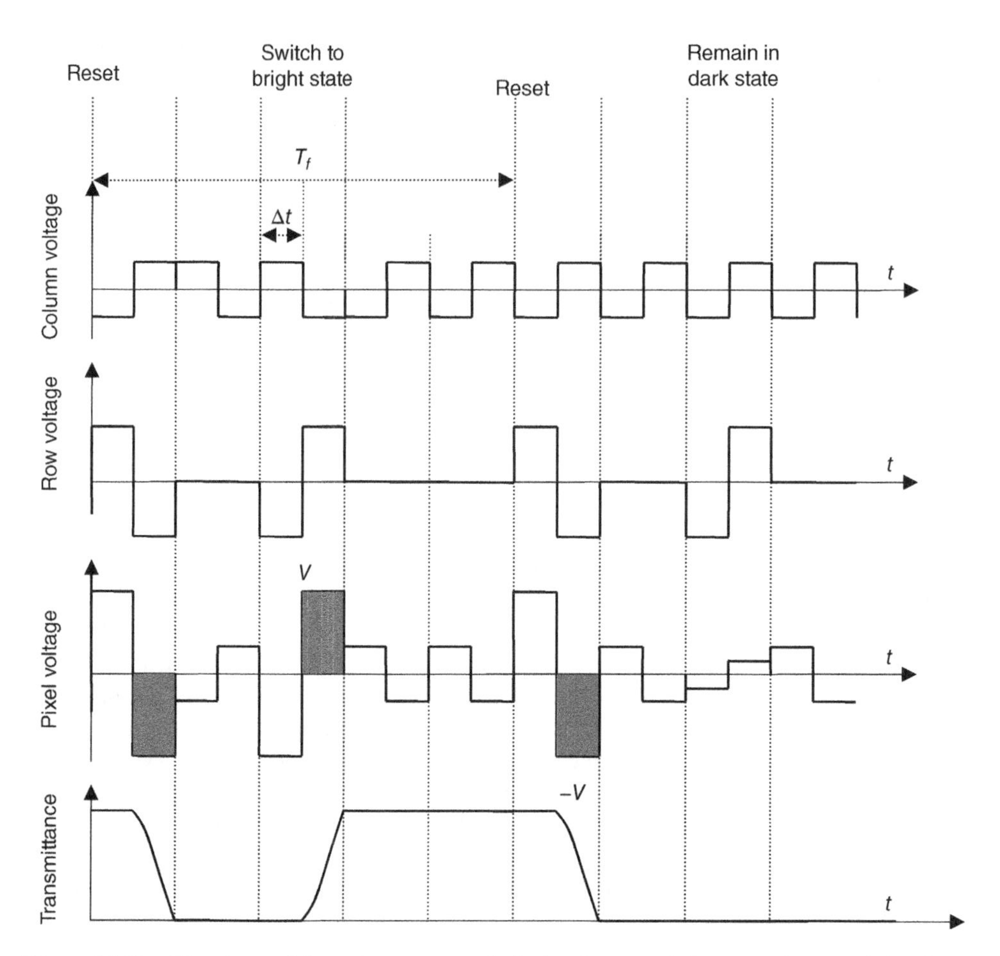

Figure 10.12 Waveform of the column, row and pixel voltages and the transmittance of the display under the pixel voltages.

is retained. The voltage applied to the not-being-addressed rows is 0. The voltage-time-areas of the voltages applied to the pixels on the not-being-addressed rows are higher than $-A_{th}$ but less than A_{th}, and therefore their states do not change. AC voltage pulses are used to reduce undesired ionic effects.

10.5 Bistable Nematic Displays

10.5.1 Introduction

There are several types of bistable nematic LCDs that have good performance characteristics. They are twisted-untwisted bistable nematic LCD [9–13], zenithal bistable nematic LCD [14,15], surface-induced bistable nematic LCD [16–19], mechanically bistable nematic LCD and bistable STN-LCD [20–24]. A bistable nematic material has two bistable states with different optical properties. Once the liquid crystal is driven into a bistable state, it remains there. It can be used to make highly multiplexed displays on passive matrices. There is no limitation on the information content. Bistable nematic LCDs exhibit high contrast ratios and large viewing angles. The drawback is that most bistable nematic LCDs do not have gray-scale capability.

10.5.2 Twisted-untwisted bistable nematic LCDs

In the bistable twisted-untwisted nematic (BTN) (also called 2π bistable) LCD, the two bistable states are selected by making use of the hydrodynamic motion of the liquid crystal, as discussed in Chapter 5 [9,11,25]. Under one hydrodynamic condition, the liquid crystal is switched to one twisted state; under another, the liquid crystal is switched to the other twisted state. An example is shown in Figure 10.13. One stable state is the 0° twist state shown by Figure 10.13(a) and the other stable state is the 360° state shown by Figure 10.13(c). Besides this particular design, there are other possible designs [26,27]. Generally speaking, the twist angles of the two bistable states are ϕ and $\phi + 2\pi$, respectively. ϕ is the angle between the alignment directions of the alignment layers on the bottom and top substrates of the cell. The angle ϕ is usually in the region between $-\pi/2$ and $\pi/2$. The twist angle difference between the two bistable states is 2π.

10.5.2.1 Bistability and switching mechanism

We first consider the 0°/360° BTN, where the alignment directions on the two cell surfaces are parallel, as shown in Figure 10.13. Chiral dopants are added to the nematic liquid crystal to

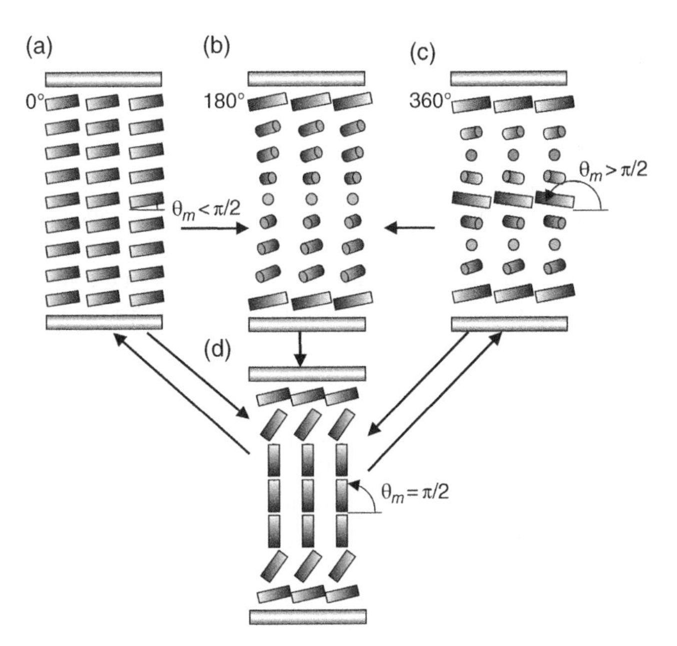

Figure 10.13 The liquid crystal director configurations of the states in the bistable TN.

obtain the intrinsic pitch P such that $h/P = 1/2$, where h is the cell thickness. The real stable state is the 180° twist state, as shown in Figure 10.13(b); it has a free energy lower than both the 0° and 360° twist states. The 0° and 360° states are actually metastable.

The hydrodynamics effect is used to switch the liquid crystal between the bistable states in the bistable TN. As discussed in Chapter 5, the rotational motion of the liquid crystal director and the translational motions of the liquid crystal are coupled [28,29]. On one hand, a rotation of the liquid crystal produces a viscous stress that results in a translational motion. On the other hand, a translational velocity gradient produces a viscous torque and affects the rotation of the director. In the BTN ($\Delta\varepsilon > 0$), when an electric field slightly higher than the threshold V_{th} of the Fréedericksz transition is applied, the liquid crystal is switched to the homeotropic state, as shown in Figure 10.13(d). In this state, the liquid crystal is aligned homeotropically only in the middle of the cell and has no twisting. The liquid crystal near the surface of the cell has some twisting. Once the field is turned off, the liquid crystal relaxes into the 0° twist state, because the 0° twist state and the homeotropic state are topologically the same while the 180° twist state is topologically different. If a very high field, higher than a saturation voltage V_{sa}, is applied to the liquid crystal, the liquid crystal in most regions except very near the cell surfaces is aligned homeotropically and has low elastic energy. The liquid crystal director changes orientation rapidly in space near the surface and has a very high elastic energy. When the applied field is removed suddenly, in the region near the cell surface the liquid crystal director rotates very quickly because of the high elastic torque, while the liquid crystal director in the middle rotates slowly because of the low elastic torque. Thus a translational motion is induced, which will affect the rotation of the liquid crystal in the middle in such a way that the tilt angle is increasing instead of decreasing. If the liquid crystal is a nematic without chiral agents, the opposite rotations of the director near the surface and the director in the middle produce a distortion of the director, which is not energetically favored. The angle of the director in the middle eventually decreases again, resulting in the backflow phenomenon [30–32]. If the liquid crystal has an intrinsic twist, the angle of the director in the middle can increase further, and the liquid crystal is switched into the 360° twist state. In order for the angle of the liquid crystal at the middle to reach a value close to π, the liquid crystal in the middle must gain sufficient momentum at the beginning. If the initially applied voltage is not sufficiently high, or a bias voltage is applied when the high voltage is turned off, or the applied voltage is removed slowly, the liquid crystal at the middle cannot obtain sufficiently high angular velocity to transform into the 360° twist state, and therefore the liquid crystal ends in the 0° twist state.

The parameters controlling the bistability are the angle ϕ between the aligning directions of the alignment layers and intrinsic pitch P of the liquid crystal. So far bistability has been observed for $-\pi/2 \le \phi \le \pi/2$. The intrinsic pitch of the liquid crystal should be chosen heuristically in such a way that the $(\phi + \pi)$ twist state has the minimum free energy, that is $(\phi + \pi)$ is the intrinsic twist. Hence $2\pi(h/P) = \phi + \pi$[26].

A 0°/180° bistable TNLD has also been reported [33,34]. In this bistable TN, the two stable states are the 0° twisted and 180° twisted states. The chiral dopant concentration is chosen such that the two states have the same energy. The switching between the two states also makes use of the hydrodynamic effect. When a sufficiently high voltage is applied, the liquid crystal is switched to the homeotropic state. If the applied voltage is turned off slowly, the liquid crystal relaxes into the 0° twisted state. If the applied voltage is turned off abruptly, the liquid crystal relaxes into the 180° twisted state. With the employment of one tilted strong anchoring alignment layer and one weak planar anchoring alignment layer, the time interval of the addressing pulse can be reduced to microseconds.

10.5.2.2 Optical properties

When a bistable TN display is optimized, the transmittance of one of the stable states should be 0 and the transmittance of the other stable state should be 1. The parameters of the display are the twist angles (ϕ, $2\pi+\phi$) of the stable states, the angle α_i of the entrance polarizer, the angle α_o of the exit polarizer, and the retardation Γ of the liquid crystal. As discussed in Chapter 3, the transmittance of a uniformly twisted nematic display in the geometry shown in Figure 3.3 is

$$T=\cos^2(\alpha_o-\alpha_i-\Phi)-\sin^2\Theta\sin\left[2(\alpha_o-\Phi)\right]\sin\left(2\alpha_i\right)$$
$$-\frac{\Phi^2}{\Theta^2}\sin^2\Theta\cos\left[2(\alpha_o-\Phi)\right]\cos\left(2\alpha_i\right)-\frac{\Phi}{2\Theta}\sin\left(2\Theta\right)\sin\left[2(\alpha_o-\alpha_i-\Phi)\right], \quad (10.8)$$

where Φ is the twist angle, h is cell thickness, $\Gamma=\frac{2\pi}{\lambda}(n_e-n_o)h$ is the total phase retardation angle, and $\Theta=[\Phi^2+(\Gamma/2)^2]^{1/2}$. As an example, we consider how to choose the parameters for the (0°, 360°) bistable TN. Put the entrance polarizer at 45° with respect to the liquid crystal at the entrance plane: $\alpha_i=\pi/4$. Put the exit polarizer at −45° with respect to the liquid crystal at the entrance plane: $\alpha_o=-\pi/4$. When the liquid crystal is in the state with the twist angle of 0°, namely $\Phi=0$, the transmittance is

$$T(0^o)=\sin^2\left(\frac{\Gamma}{2}\right). \quad (10.9)$$

When the liquid crystal is in the state with a twist angle of 360°, namely, $\Phi=2\pi$, the transmittance is

$$T(360^o)=\sin^2\left\{\left[(2\pi)^2+\left(\frac{\Gamma}{2}\right)^2\right]^{1/2}\right\}. \quad (10.10)$$

In order to find the retardation Γ, which maximizes the contrast, the difference of the transmittances is calculated and plotted in Figure 10.14. Good performance is achieved when the transmittance difference is maximized. For that case, the transmittances of the stable states and the corresponding retardation are listed in Table 10.1. The good choices are: (1) $\Gamma=0.972\pi$ which generates $T(0°)=0.998$ and $T(360°)=0.033$ [10], (2) $\Gamma=6.924\pi$ which generates $T(0°)=0.986$ and $T(360°)=0$.

The angles of the polarizers as well as the twist angle ϕ can also be varied to achieve good performance. For example, consider a (−90°, 270°) bistable TN where $\phi=-90°$. The angles of

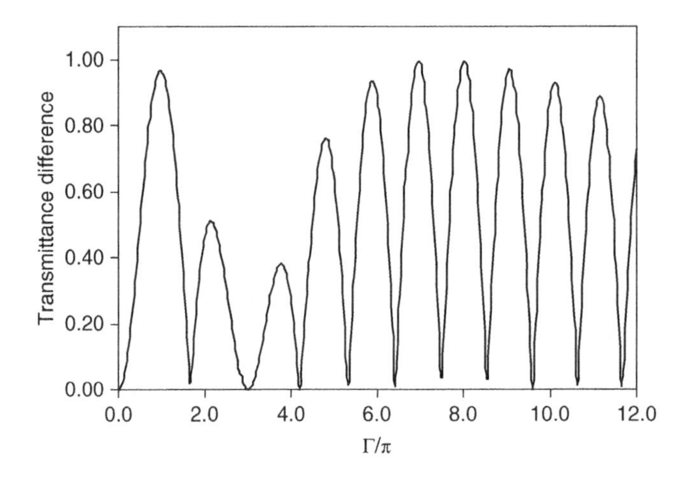

Figure 10.14 Transmittance difference of the two stable states of the bistable TN vs. the retardation of the liquid crystal.

Table 10.1 Some of the retardations with which the performance of the bistable TN is optimized.

Γ/π	$T(0°)$	$T(360°)$	0° twist state	360° twist state
0.972	0.998	0.033	Bright state	Dark state
2.000	0	0.491	Dark state	Bright state
4.000	0	0.267	Dark state	Bright state
6.000	0	0.894	Dark state	Bright state
6.924	0.986	0	Bright state	Dark state
8.000	0	0.993	Dark state	Bright state

the polarizers are $-\alpha_i = \alpha_o = \pi/4$. When the liquid crystal is in the $-90°$ twist state, $\Phi = -\pi/2$. The transmittance is

$$T(-90°) = 1 - \sin^2\left\{\left[\left(\frac{\pi}{2}\right)^2 + \left(\frac{\Gamma}{2}\right)^2\right]^{1/2}\right\} = \cos^2\left\{\left[\left(\frac{\pi}{2}\right)^2 + \left(\frac{\Gamma}{2}\right)^2\right]^{1/2}\right\}. \quad (10.11)$$

When the liquid crystal is in the 270° twist state, $\Phi = 3\pi/2$. The transmittance is

$$T(270°) = 1 - \sin^2\left\{\left[\left(\frac{3\pi}{2}\right)^2 + \left(\frac{\Gamma}{2}\right)^2\right]^{1/2}\right\} = \cos^2\left\{\left[\left(\frac{3\pi}{2}\right)^2 + \left(\frac{\Gamma}{2}\right)^2\right]^{1/2}\right\}. \quad (10.12)$$

When $\Gamma = 2\sqrt{2}\pi$, $T(-90°) = 0$ and $T(270°) = 0.965$ [26]. With the help of the simplified Mueller matrix method and the Poincaré sphere, a general condition for optimized performance can be derived. The optimization considered here is only for one wavelength. In reality, the transmission spectra of the two stable states are wavelength-dependent because the phase retardation angle Γ is wavelength-dependent. Therefore in designing bistable TN displays, the wavelength dispersion of the transmission spectra must be considered [35–39]. The bistable TN liquid crystals can also be used to make reflective displays [35].

10.5.2.3 Drive schemes

Bistability of a display material does not guarantee multiplexibility. Proper drive schemes must be designed in order to make multiplexed displays on a passive matrix. A good drive scheme should possess the properties of fast addressing speed, low drive voltage, no crosstalk, and a simple waveform. According to the number of phases in the addressing, there are three major types of drive scheme for the bistable TN: (1) one-phase drive scheme, (2) two-phase drive scheme, and (3) three-phase drive schemes.

a. **One-phase drive scheme** The one-phase drive scheme is shown schematically in Figure 10.15. The state of the liquid crystal is changed by one voltage pulse [10,26,40]. A low voltage V_L addressing pulse switches the material to the low twisted state while a high voltage V_H addressing pulse switches it into the high twisted state. The low voltage is slightly higher than the threshold V_{th} of the Fréedericksz transition. When the low voltage is applied, independent of the initial state of the liquid crystal, the liquid crystal is switched to a homeotropic state where the liquid crystal is aligned homeotropically only in a small region in the middle of the cell. When the low voltage is turned off, the liquid crystal relaxes into the low twisted state because it does not have a sufficiently high potential. The high voltage is higher than the saturation voltage V_{sa}, which is much higher than the threshold of the Fréedericksz transition. When the high voltage is applied, independent of the initial state, the liquid crystal is switched to a homeotropic state where the liquid crystal is aligned homeotropically in most regions of the cell except very close to the cell surfaces, and gains a high potential. When the high voltage is turned off, the liquid crystal relaxes into the high twisted state because of the hydrodynamic effect.

In addressing the display, the row voltage for the being-addressed-row is $V_{rs} = (V_L + V_H)/2$; the row voltage for the not-being-addressed row $V_{rns} = 0$ V. The column voltage is $V_{con} = -(V_H - V_L)/2$ to select the high twisting state and $V_{coff} = (V_H - V_L)/2$ to select the low twisting

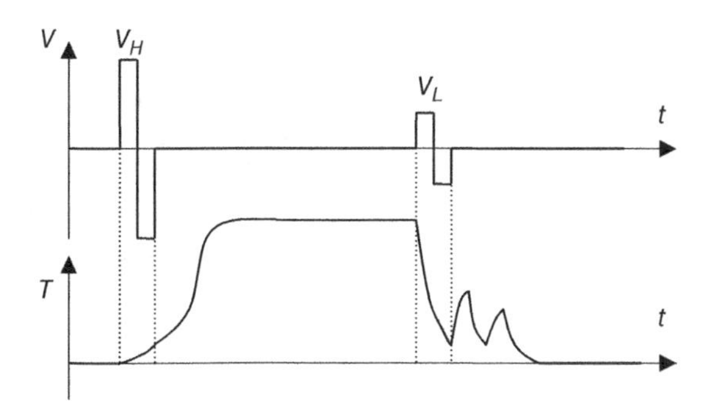

Figure 10.15 The schematic diagram of the one-phase drive scheme and the response of the bistable TN.

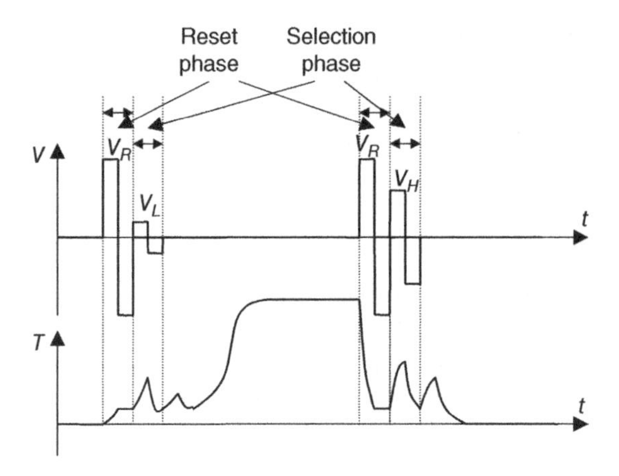

Figure 10.16 The schematic diagram of the two-phase drive scheme and the response of the bistable TN.

state. The threshold of the Fréedericksz transition of the liquid crystal must be higher than $(V_H - V_L)/2$ in order to prevent crosstalk. The problem of this drive scheme is that the time interval to address one line is on the order of 10 ms, and thus the addressing speed is slow.

b. **Two-phase drive scheme** The two-phase drive scheme is shown in Figure 10.16 [26]. In the reset phase, a high voltage V_R ($>V_{sa}$) is applied to switch the liquid crystal to the homeotropic state. When the reset voltage is turned off, the liquid crystal begins to relax. In the selection phase, if the selection voltage V_S is the low voltage V_L, there is no hindrance to the rotation of the liquid crystal molecules in the middle of the cell; the tilt angle at the middle plane increases, and the liquid crystal relaxes to the high twisted state. If the selection voltage V_S is the high voltage V_H, the applied voltage hinders the rotation of the liquid crystal molecules in the middle of the cell; the tilt angle at the middle plane decreases, and the liquid crystal relaxes to the low twisted state. The time interval of the reset phase is on the order of 10 ms while the time interval of the selection phase is on the order of 1 ms. $(V_H - V_L)/2$ must be lower than the threshold of the Fréedericksz transition of the liquid crystal in order to prevent crosstalk. Although the reset phase is long, multiple lines can be put into the reset phase such that the time is shared, which is known as the *pipeline algorithm*. Therefore the addressing speed of the two-phase drive scheme is faster than the one-phase drive scheme.

c. **Three-phase drive scheme** The three-phase drive scheme is shown in Figure 10.17. It consists of three phases: reset, delay, and selection [41]. The physics behind this drive scheme is that in the beginning of the relaxation after the reset phase, the liquid crystal is allowed to relax freely, and the hydrodynamic effect can be controlled by a voltage in the late stage of the relaxation. Therefore the time interval of the selection phase is reduced. In the reset phase, the high voltage V_R ($>V_{sa}$) switches the liquid crystal into the homeotropic texture. In the delay phase, the applied voltage V_D is 0, and the liquid crystal starts to relax. In the selection phase, if the applied voltage V_S is the high voltage V_H, the rotation of the liquid crystal molecules in the middle of the cell is hindered. After the selection phase, the liquid crystal is addressed to the low twisted state. If the applied voltage V_S is the low voltage V_L, the rotation of the liquid crystal molecules in the middle of the cell is not hindered and the tilt angle increases. After the selection phase, the liquid crystal is addressed to the high twisted

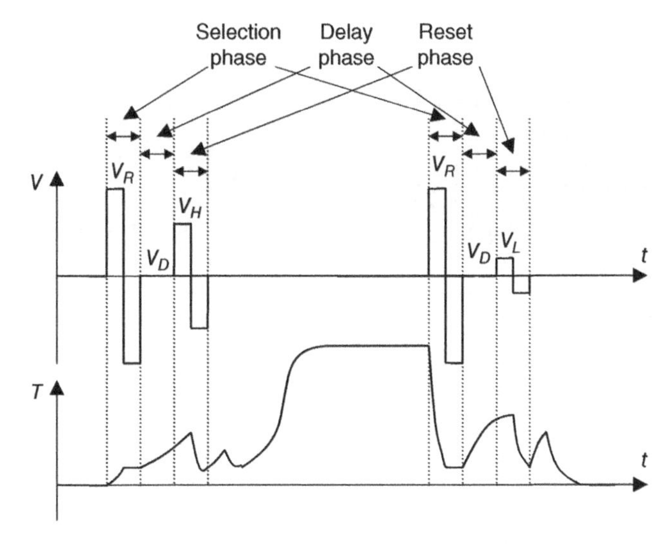

Figure 10.17 The schematic diagram of the three-phase drive scheme and the response of the bistable TN.

state. $(V_H - V_L)/2$ must be lower than the threshold of the Fréedericksz transition of the liquid crystal in order to prevent crosstalk. The time intervals of the reset and delay phases can be shared using the pipeline algorithm. The time interval of the selection phase can be as short as 100 μs. Therefore the addressing speed is increased dramatically, and video rate display becomes possible with this drive scheme.

10.5.3 Surface-stabilized nematic liquid crystals

In LCDs, liquid crystals are usually sandwiched between two substrates. A certain alignment of the liquid crystal at the surface of the substrates is usually necessary in order for a display to operate properly. Bistable nematic liquid crystals can be created by using surface alignment layers. They are divided into two categories: zenithal bistable TN and azimuthal TN.

10.5.3.1 Zenithal bistable TN

The zenithal bistable (Z-bistable) nematic liquid crystal was developed by G. P. Bryan-Brown, et al., using surface stabilization [15]. One substrate of the cell has an alignment layer with homeotropic anchoring and the other substrate is a one-dimensional grating as shown in Figure 10.18. The groove of the grating is along the *y* direction. The grooves are made from a photoresist. The non-symmetric profile of the grooves is obtained by using UV light incident obliquely at 60° in the photolithography. A surfactant is coated on top of the grooves to obtain homeotropic anchoring.

The liquid crystal in the Z-bistable cell has two stable states at zero field. One is the high-tilt state shown in Figure 10.18(a). The other is the low-tilt state shown in Figure 10.18(b). The flexoelectric effect plays an important role in switching the liquid crystal between the two stable

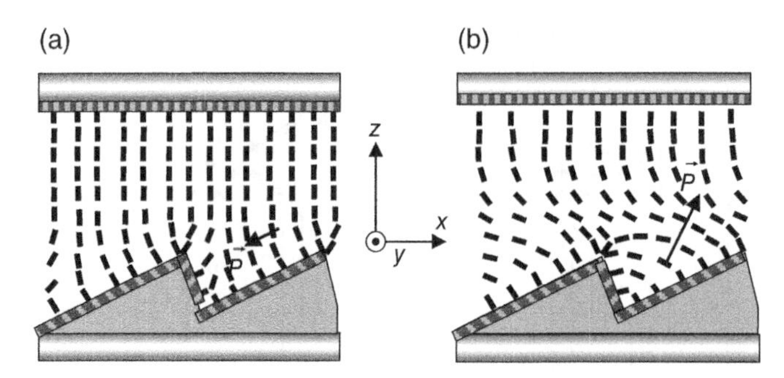

Figure 10.18 Schematic diagram of the liquid crystal director configurations of the two bistable states of the Z-bistable nematic liquid crystal.

states. As discussed in Chapter 4, the polarization produced by flexoelectric effect is given by $\vec{P} = e_1\,\vec{n}\,(\nabla\cdot\vec{n}) + e_2(\nabla\times\vec{n})\times\vec{n}$. In the cell geometry shown in Figure 10.18, the liquid crystal director deformation occurs mainly near the grating surface; the bend deformation is dominant. The bending directions in the two states are different and therefore the induced polarization is upward in one of the states and downward in the other state. The liquid crystal has a positive dielectric anisotropy. Besides flexoelectric interaction, there is dielectric interaction when a voltage is applied across the cell. When a sufficiently high voltage of one polarity is applied, the liquid crystal is switched to a homeotropic state with the liquid crystal near the grating substrate having a configuration similar to that in the high-tilt state; it relaxes into the high-tilt state after the applied voltage is removed. When a sufficiently high voltage with the opposite polarity is applied, the liquid crystal is switched to a homeotropic state with the liquid crystal near the grating substrate having a configuration similar to that in the low-tilt state; it relaxes into the low-tilt state after the applied voltage is removed.

In building a transmissive Z-bistable display, crossed polarizers are used. The polarizers make the angle of 45° with the grating groove direction. Hence the plane containing the liquid crystal director makes the angle of 45° with the polarizers. The cell thickness and birefringence of the liquid crystal is chosen in such a way that the retardation of the low-tilt state is π, and therefore the transmittance of the low-tilt state is high. The retardation of the high-tilt state is small and therefore its transmittance is low. The Z-bistable display can be addressed by DC voltage pulses. The width of the addressing voltage pulse is about 100 μs for a field about 10 V/μm, with which video rate is possible. The relaxation time from the field-on state to the low-tilt state is about 20 ms and the relaxation time from the other field-on state to the low-tilt state is about 1 ms. The material can also be used to make reflective displays with the retardation adjusted properly.

The Z-bistable nematic liquid crystal can also be used to make displays by using a different geometry: the cell is made of the grating substrate and another substrate with a homogeneous anchoring [14,42]. The aligning direction of the homogeneous anchoring makes the angle of 90° with respect to the liquid crystal director near the grating surface in the low-tilt state. Thus a hybrid TN is formed. In making a transmissive display, two crossed polarizers are used. The groove of the grating is arranged parallel to one of the polarizers. When the liquid crystal is in the low-tilt state, the material acts as a polarization guide and the transmittance of the display is high. When the liquid crystal is in the high-tilt state, the polarization of the incident light is rotated only slightly and therefore the transmittance is low. The selection of the states is made by using DC voltage pulses, as already described. In this design, higher contrast is achieved. Furthermore, written images are retained at zero field even if the display is squeezed.

10.5.3.2 Azimuthal bistable nematic liquid crystal

The alignment of a liquid crystal at the cell surface is due to the intermolecular interaction between the molecules of the alignment layer and the liquid crystal molecules as well as the geometrical shape of the surface of the alignment layer through the elastic energy of the liquid crystal. For an alignment layer having unidirectional grooves (grating) on the alignment layer, the liquid crystal is aligned along the groove direction. For an alignment layer having grooves in two perpendicular directions (bi-grating), two alignment directions can be created with properly controlled groove amplitude and pitch. The liquid crystal can be anchored along either direction. Thus two bistable orientation states can be achieved [43]. In order to be able to select the two states by applying voltage in the cell normal direction, the pretilt angles of the two anchoring directions must be different.

Alignment layers with two anchoring directions and different pretilt angles can be produced by obliquely evaporating SiO on glass substrates twice. The blaze direction of the first evaporation is in the xz plane (with the azimuthal angle $\varphi = 0°$). The blaze direction of the second evaporation is in the yz direction (with the azimuthal angle $\varphi = 90°$). The resulting alignment layer has two alignment directions: one has the azimuthal angle $\varphi = 45°$ and a non-zero pretilt angle and the other has the azimuthal angle $\varphi = -45°$ and zero pretilt angle. In a cell with two such alignment layers on the two substrates, there are two stable states as shown in Figure 10.19. Because of the non-zero pretilt angle, there are splay deformations in the two bistable states, which induce flexoelectric polarizations. In the state shown in Figure 10.19(a), the flexoelectric polarization is upward while in the state shown in Figure 10.19(b), the flexoelectric polarization is downward. Therefore these states can be selected by using DC voltages applied across the cell.

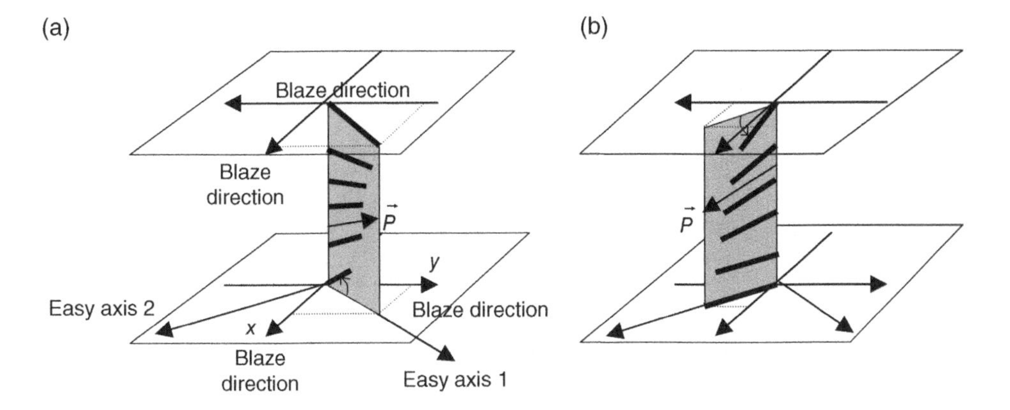

Figure 10.19 Schematic diagram of the azimuhtal bistable nematic liquid crystal.

10.6 Bistable Cholesteric Reflective Display

10.6.1 Introduction

Cholesteric (Ch) liquid crystals have a helical structure in which the liquid crystal director twists around a perpendicular axis named the helical axis [29]. The distance along the helical axis for the director to twist 2π is called the pitch and is denoted by P_o. In this section we only discuss cholesteric liquid crystal with short pitches (in visible and IR light regions). The optical properties of a cholesteric liquid crystal depend on the orientation of the helical axis with respect to the cell surface. There are four states as shown in Figure 10.20 [44]. When a cholesteric liquid crystal is in the planar state (also called planar texture) where the helical axis is perpendicular to the cell surface, as shown in Figure 10.20 (a), the material reflects light. A microphotograph of the planar state is shown in Figure 10.21(a). The dark lines are the disclination lines called oily streaks [45–47], where

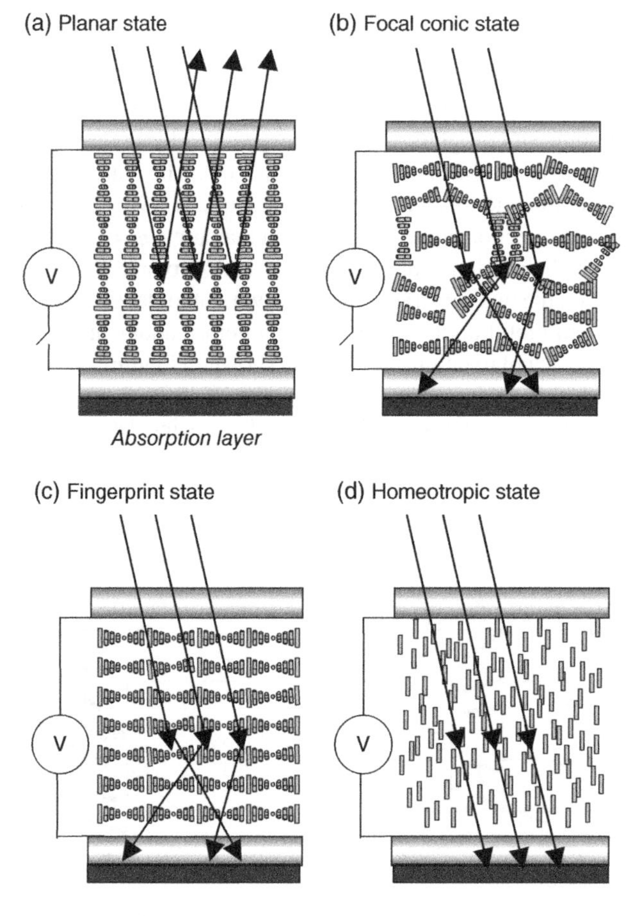

Figure 10.20 Schematic diagram of the cholesteric states.

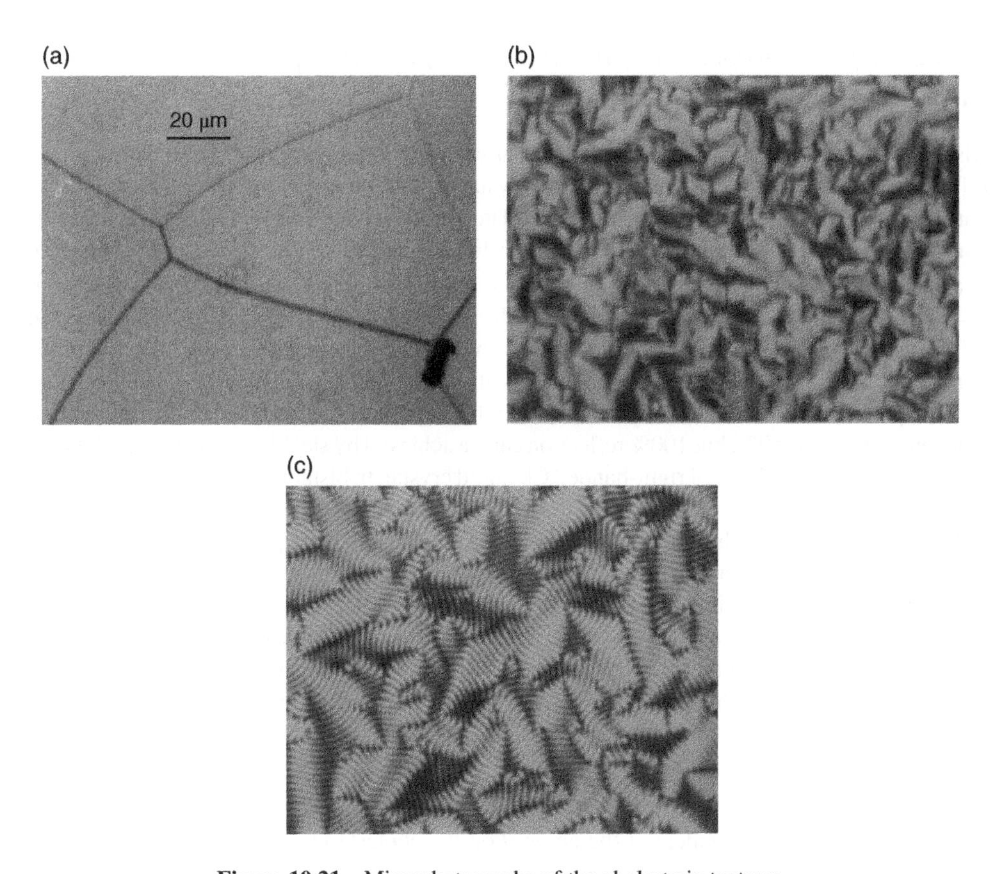

Figure 10.21 Microphotographs of the cholesteric textures.

the cholesteric layers are bent. When the liquid crystal is in the focal conic state (texture), the helical axis is more or less random throughout the cell as shown in Figure 20(b). It is a multi-domain structure and the material is scattering. A microphotograph of the focal conic state is shown in Figure 10.21(b), which is similar to the focal conic texture of smectic-A because the cholesteric liquid crystal can be regarded as a layered structure [46]. When an intermediate electric field is applied across the cell, the liquid crystal is switched to the fingerprint state (texture), the helical axis is parallel to the cell surface, as shown in Figure 10.20(c). A microphotograph of the fingerprint state is shown in Figure 10.21(c). When a sufficiently high field is applied across the cell, the liquid crystal ($\Delta\varepsilon > 0$) is switched to the homeotropic state where the helical structure is unwound with the liquid crystal director perpendicular to the cell surface, as shown in Figure 10.20(d) [48]. The material is transparent in this state. When homeotropic alignment layers or weak homogeneous alignment layers are used (known as surface stabilization) or a small amount of polymer is dispersed in the liquid crystal (known as polymer stabilization), both the planar state and the focal conic state can be stable at zero field [49–54].

10.6.2 Optical properties of bistable Ch reflective displays

10.6.2.1 Reflection

Bistable Ch reflective displays are operated between the reflecting planar state and the non-reflecting focal conic state. When a Ch liquid crystal is in the planar texture, the refractive index varies periodically in the cell normal direction. The refractive index oscillates between the ordinary refractive index n_o and the extraordinary refractive index n_e. The period is $P_o/2$ because $\vec{n}$ and $-\vec{n}$ are equivalent. The liquid crystal exhibits Bragg reflection at the wavelength $\lambda_o = 2\bar{n}(P_o/2) = \bar{n}P_o$ for normally incident light [28], where $\bar{n} = (n_e + n_o)/2$ is the average refractive index. The reflection bandwidth is given by ΔnP_o, where $\Delta n = n_e - n_o$ is the birefringence. The reflected light is circularly polarized with the same handedness as the helical structure of the liquid crystal. If the (normally) incident light is unpolarized, then the maximum reflection from one Ch layer is 50%, but 100% reflection can be achieved by stacking a layer of left-handed Ch liquid crystal and a layer of right-handed Ch liquid crystal. In bistable Ch reflective displays, a color absorption layer is coated on the bottom substrate. When the liquid crystal is in the planar state, the reflection of the display is the sum of the reflection from the liquid crystal and the reflection from the absorption layer. When the liquid crystal is in the focal conic state, the reflection of the display is only contributed by the reflection of the absorption layer. If the absorption layer is black and the liquid crystal reflects green light, the planar state appears green while the focal conic state is black [55]. If the absorption layer is blue and the liquid crystal reflects yellow light, the planar state appears white and the focal conic state appears blue [56].

10.6.2.2 Viewing angle

When light is obliquely incident at the angle θ on the cholesteric liquid crystal in the planar state, the central wavelength of the reflection band is shifted to $\lambda = \bar{n}P_o \cos\theta$. This shift of the reflection band is undesirable in display applications if the liquid crystal is in the perfect planar state, because the color of the reflected light changes with viewing angle and the reflected light is only observed at the corresponding specular angle. This problem can be partially solved by dispersing a small amount of polymer in the liquid crystal or by using an alignment layer which gives weak homogeneous anchoring or homeotropic anchoring. The dispersed polymer and the alignment layer produce defects and create a poly-domain structure as shown in Figure 10.22(a). In this imperfect planar state, the helical axis of the domains is no longer exactly parallel to the cell normal but distributed around the normal. For an incident light at one angle, light reflected from different domains is in different directions, as shown in Figure 10.22(b). Under room light conditions where light is incident at all angles, at one viewing angle, light reflected from different domains has different colors. Because the observed light is a mixture of different colors, the colors observed at different viewing angle are not much different. The poly-domain structure of the imperfect planar state and the isotropic incidence of room light is responsible for the large viewing angle of the cholesteric display [57].

10.6.2.3 Polymer stabilize black-white Ch display

The deviation of the helical axis from the cell normal direction in the poly-domain planar texture depends on the amount of the dispersed polymer. In the regular polymer stabilized cholesteric

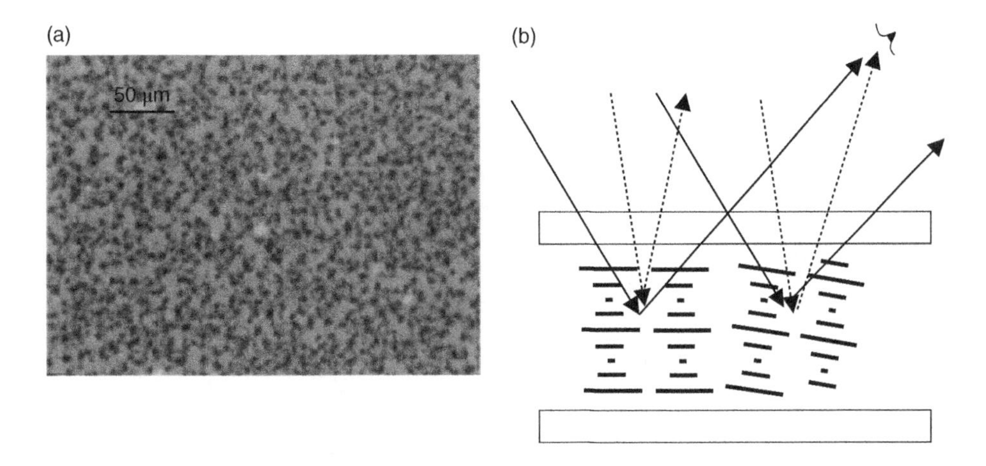

Figure 10.22 (a) Microphotograph of the imperfect planar state. (b) The reflection from the imperfect planar state under room light condition.

display, the polymer concentration is low and the deviation is small. The reflection spectrum of the planar texture is not very wide, as shown in Figure 10.23(a). The color of the reflected light is pure. The reflection spectra of surface-stabilized cholesteric displays are similar. The reflection of the focal conic texture is low, as shown in Figure 10.23(a). If the polymer concentration is high, the deviation becomes large. When $\bar{n}P_o$ equals the wavelength of red light, the reflection spectrum of the planar texture becomes very broad, as shown in Figure 10.23(b). The planar texture has a white appearance. The display is called polymer stabilized black-white cholesteric display [58,59]. In this display, the scattering of the focal conic texture is stronger than that of the focal conic texture of the regular polymer stabilized cholesteric displays.

10.6.2.4 Gray scale

The bistable Ch displays exhibit gray-scale memory states because of their multi-domain structure [55,60,61]. For each domain, it is bistable, that it is either in the planar state or in the focal conic state. For different domains, the voltages to switch them from the planar state to the focal conic state are different. Once a domain is switched to the focal conic state, it remains there even after the applied voltage is turned off. If initially the liquid crystal is in the planar state, under a low applied voltage, few domains are switched to the focal conic state and the resulting reflectance is high; under a high applied voltage, many domains are switched to the focal conic state and the resulting reflectance is low. Therefore gray-scale reflectances are possible in bistable Ch reflective displays. The domain has a size around 10 μm and cannot be observed by the naked eye.

10.6.2.5 Multiple color Ch displays

In a cholesteric display with a single layer of cholesteric liquid crystal, only a single color can be displayed. In order to make multiple color displays, Ch LCs with a variety of pitches must be

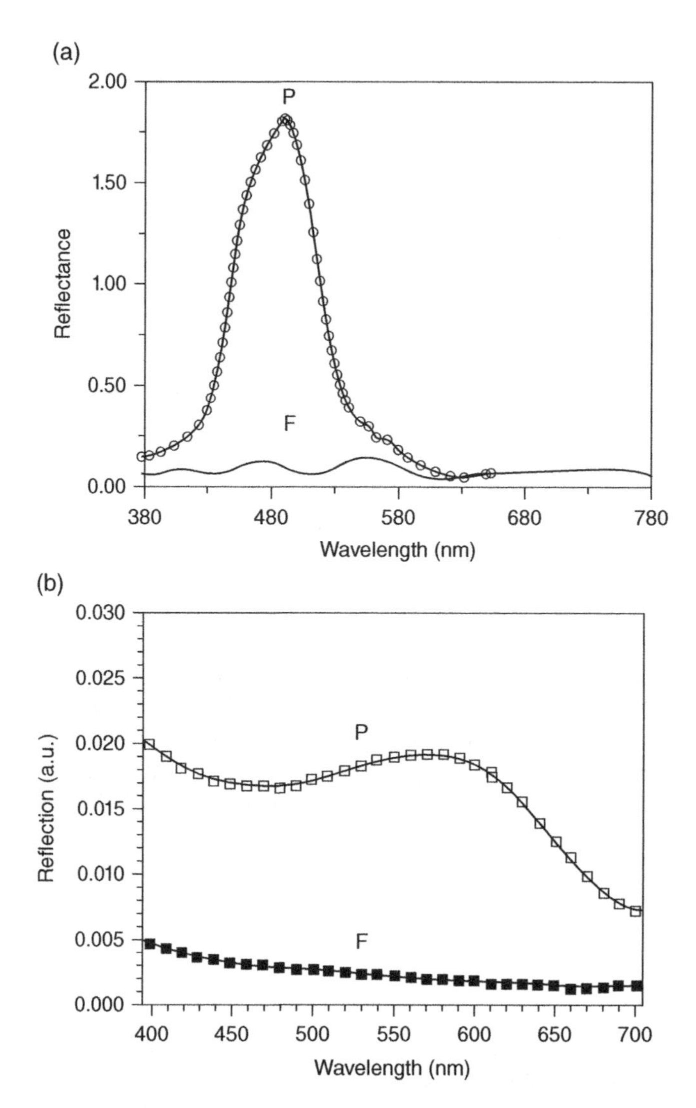

Figure 10.23 The reflection spectra of the cholesteric displays. P: planar texture, F: focal conic texture. (a) The regular polymer stabilized cholesteric display, and (b) The polymer stabilized black-white display.

used. This can be done either by stacking multiple layers of Ch LCs with different pitches or by using one layer of Ch LCs with different pitches partitioned in plane.

Multiple color displays from one layer can be made from pixelation of colors. The displays have three alternating types of stripes of Ch liquid crystals with three different pitches reflecting blue, green and red light. Partition or some other means of preventing inter-stripe diffusion must be used. Polymer walls, especially field-induced polymer walls, are good candidates. The different pitches can be achieved by two methods. In the first method, three cholesteric liquid crystals with different pitches are filled into empty cells with partitions. The second method is photo color tuning [62,63]. A photosensitive chiral dopant is added to the liquid crystal. The dopant undergoes chemical reaction under UV irradiation and thus its chirality changes, and the pitch of the liquid crystal changes. After the mixture is filled into display cells, the cells are irradiated by UV light with photo masks. By varying irradiation time, different pitches are achieved. In this method, partitions are fabricated either before or after the photo color tuning. A polymer dispersing technique with large liquid crystal droplets can also be used with this method [64]. The major drawback of one-layer multiple color displays is that the reflection is low.

Multiple color displays from multiple layers are made by stacking three layers of cholesteric liquid crystals with pitches reflecting blue, green, and red light [65–68]. Single-layer displays with the three colors are fabricated first. Then they are laminated together. In order to decrease parallax, thin substrates, preferably substrates with conducting coating on both sides, should be used to decrease the distance between the liquid crystal layers. Because of the scattering of the cholesteric liquid crystals, experiments show that the best stacking order from bottom to top is red, green, and blue.

10.6.3 Encapsulated cholesteric liquid crystal displays

When cholesteric liquid crystals are encapsulated in droplet form, the bistability can be preserved when droplet size is much larger than the pitch [64]. There are two methods which are used to encapsulate Ch liquid crystals: phase separation and emulsification. In phase separation [69], the Ch liquid crystal is mixed with monomers or oligomers to make a homogeneous mixture. The mixture is coated on plastic substrates and then another substrate is laminated on. The monomers or oligomers are then polymerized to induce phase separation. The liquid crystal phase separates from the polymer to form droplets. In the emulsification method [70–73], the Ch liquid crystal, water, and a water dissolvable polymer are placed in a container. Water dissolves the polymer to form a viscous solution, which does not dissolve the liquid crystal. When this system is stirred by a propeller blade at a sufficiently high speed, micron-size liquid crystal droplets are formed. The emulsion is then coated on a substrate and the water is allowed to evaporate. After the water evaporates, a second substrate is laminated to form the Ch display.

The encapsulated cholesteric liquid crystals are suitable for flexible displays with plastic substrates. They have much higher viscosities than pure cholesteric liquid crystals and can be coated on substrates in roll-to-roll process [71,72]. The polymers used for the encapsulation have good adhesion to the substrates and can make the materials self-adhesive to sustain the cell thickness. Furthermore, the encapsulated Ch liquid crystals can no longer flow when squeezed, which solves the image-erasing problem in displays from pure cholesteric liquid crystals where squeezing causes the liquid crystal to flow and to be switched to the planar state.

10.6.4 Transition between cholesteric states

The state of a cholesteric liquid crystal is mainly determined by surface anchoring, cell thickness, and applied fields. The liquid crystal can be switched from one state to another by applying electric fields. There are many possible transitions among the states, as shown in Figure 10.24 [50,54]. In order to design drive schemes for the bistable Ch reflective display,

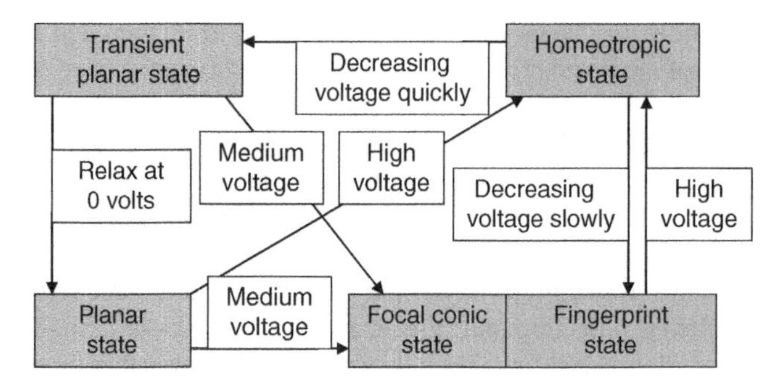

Figure 10.24 Schematic diagram showing the transitions among the cholesteric states.

it is essential to understand the transitions. The cholesteric liquid crystals considered here have positive dielectric anisotropies unless otherwise specified.

10.6.4.1 Transition between planar state and focal conic state

Under a given electric field and boundary condition, a liquid crystal system is in the state with the minimum free energy. In considering the state of the Ch liquid crystal in the bistable Ch display, the energies involved are the elastic energy of the deformation of the liquid crystal director, electric energy and surface energy. In both the planar state and the focal conic state, the helical structure is preserved. In the planar state, the elastic energy is zero because there is no director deformation, while in the focal conic state, the elastic energy is positive because of the bend of the Ch layers. The electric energy is given by $-(1/2)\Delta\varepsilon\varepsilon_o\left(\vec{E}\cdot\vec{n}\right)^2$, which depends on the orientation of the liquid crystal director. In the planar state, the electric energy is zero because the liquid crystal director $\vec{n}$ is perpendicular to the field everywhere, while in the focal conic state, the electric energy is negative because the liquid crystal is parallel to the applied field in some regions. The elastic energy is against the planar–focal conic (P–F) transition while the electric energy favors the transition. When the applied field is sufficiently high, the planar state becomes unstable and the liquid crystal transforms from the planar state to the focal conic state. There are two possible mechanisms for the transition from the planar state to the focal conic state. One mechanism is the oily streaks, shown in Figure 10.25 [45,47], which are bent cholesteric layers whose structure is shown in Figure 10.25(b). When the applied field is higher than a threshold E_{oily}, the oily streaks nucleate from nucleation seeds such as impurities, guest particles, and surface irregularities; they grow with time until the whole system is switched into the focal conic state. The other mechanism is Helfrich deformation, as shown in Figure 10.26, which is a two-dimensional undulation in the plane parallel to the cell surface [29,45,74,75]. The structure of the liquid crystal in a vertical plane is shown in Figure 10.26(b). The wavelength of the undulation is $\lambda = (2K_{33}/K_{22})^{1/4}(hP_o)^{1/2}$. When the applied field is above a threshold $E_{Helfrich}$, the cholesteric layers start to undulate. Helfrich deformation is a homogeneous process and can take place simultaneously everywhere, and therefore it is much faster than the process of the oily streak. Once the applied field is above the threshold $E_{Helfrich}$, the amplitude of the undulation increases with increasing voltage, and eventually the amplitude diverges and the liquid crystal transforms into the focal conic state.

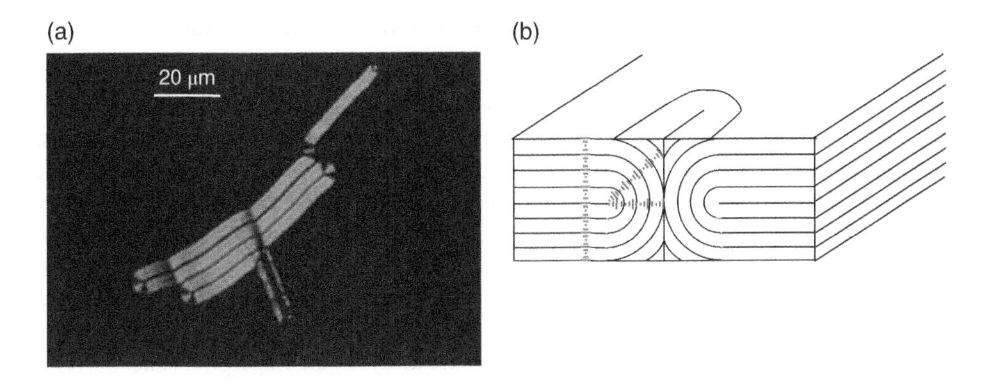

Figure 10.25 (a) Microphotograph of the oily streak in the cholesteric liquid crystal. The bright finger is the oily streak. The dark background is the planar texture. (b) Schematic diagram showing the structure of the oily streak on a cross section.

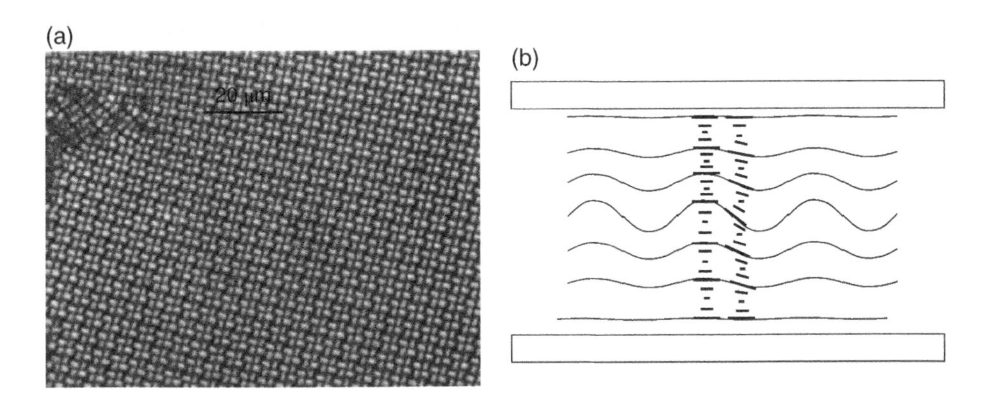

Figure 10.26 (a) Microphotograph of Helfrich deformation, (b) Schematic diagram showing the structure of Helfrich deformation in a plane perpendicular to the cell surface.

In bistable Ch reflective display applications, it is desirable that the threshold of the transition from the planar state to the focal conic state be high, so that the cholesteric liquid crystal can remain in the planar state and the display does not exhibit flicker under column voltage in addressing.

Once the cholesteric liquid crystal is in the focal conic state, it may remain there, depending on the surface anchoring condition. In bistable Ch reflective displays where either weak tangential or homeotropic alignment layers are used, or polymers are dispersed in the liquid crystal, the liquid crystal remains in the focal conic state when the applied voltage is turned off. In order to switch the liquid crystal from the focal conic state back to the planar state, a high voltage must be applied to switch it to the homeotropic texture, then it relaxes back to the planar state after the high voltage is removed. This will be discussed in more detail later. If the cell has strong homogeneous alignment layers, the focal conic texture is not stable and the liquid crystal relaxes slowly back to the planar texture.

10.6.4.2 Transition between the fingerprint state and homeotropic state

When the liquid crystal is in the focal conic state and the externally applied electric field is increased, more liquid crystal molecules are aligned parallel to the field. The liquid crystal is gradually switched to the fingerprint state. There is no sharp boundary between the focal conic state and the fingerprint state. When the applied field is increased further, the pitch of the liquid crystal becomes longer, as shown in Figure 10.27. When the applied field is above a threshold E_C, the helical structure is unwound [29,48], the pitch becomes infinitely long, and the liquid crystal is switched to the homeotropic state.

We first consider the unwinding of the helical structure. As the applied field is increased, the pi-walls (the narrow regions in which the liquid crystal director rotates by π) are propelled apart horizontally and annihilated at the boundaries far away. In the fingerprint state, the liquid crystal director is given by $n_x = \sin\theta(y)$, $n_y = 0$, $n_z = \cos\theta(y)$. The free energy is given by

$$f = \frac{1}{2}K_{22}\left[\vec{n}\cdot(\nabla\times\vec{n}) + q_o\right]^2 - \frac{1}{2}\Delta\varepsilon\varepsilon_o\left(\vec{n}\cdot\vec{E}\right)^2 = \frac{1}{2}K_{22}(\theta' - q_o)^2 + \frac{1}{2}\Delta\varepsilon\varepsilon_o E^2\sin^2\theta + \text{constant}, \tag{10.13}$$

where $\theta' = \partial\theta/\partial y$. The constant in the above equation does not affect the director configuration and can be omitted. Using the dimensionless variables: $\psi = f/K_{22}q_o^2$, $\xi = q_o y$, $e = E/E_o$, where $E_o = \frac{\pi}{2}q_o\sqrt{\frac{K_{22}}{\varepsilon_o\Delta\varepsilon}} = \frac{\pi^2}{P_o}\sqrt{\frac{K_{22}}{\varepsilon_o\Delta\varepsilon}}$, we have the dimensionless free energy density

$$\psi = \frac{1}{2}\left(\frac{d\theta}{d\xi} - 1\right)^2 + \frac{1}{2}\left(\frac{\pi e}{2}\right)^2\sin^2\theta. \tag{10.14}$$

Using the Euler–Lagrange equation to minimize the free energy, we obtain

$$\frac{d\theta}{d\xi} = \left[\left(\frac{\pi e}{2}\sin\theta\right)^2 + A\right]^{1/2}, \tag{10.15}$$

where A is the integration constant which is field-dependent. When $e = 0$, $\frac{d\theta}{d\xi} = 1$, then $A = 1$. When $e \geq e_c = E_C/E_o$, $\frac{d\theta}{d\xi} = 0$ and $\theta = 0$, hence$A = 0$. Hence as the applied field is increased from

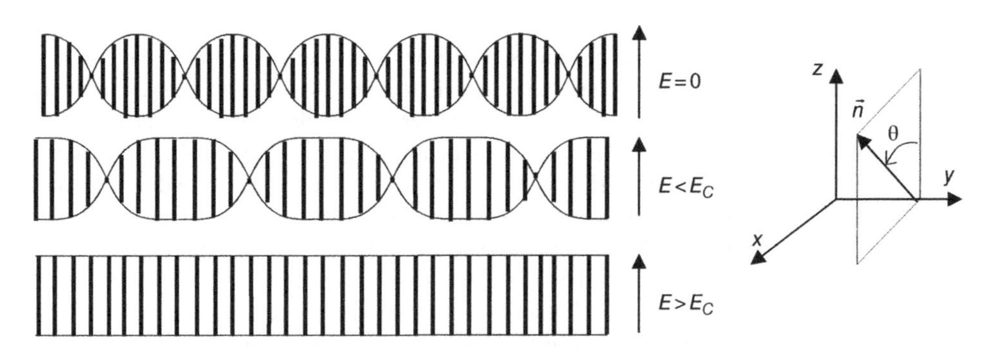

Figure 10.27 Schematic diagram showing the process of unwinding the helical structure in the fingerprint–homeotropic transition.

0 to e_C, A changes from 1 to 0. The normalized periodicity of the fingerprint state is $(P/2)q_o$ and is given by

$$(P/2)q_o = \int_0^{\pi} \left[A + \left(\frac{\pi e}{2}\sin\theta\right)^2\right]^{-1/2} d\theta. \tag{10.16}$$

By substituting Equation (10.15) into Equation (10.14), we have the free energy density

$$\psi = \frac{1}{2}(1+A) - \frac{d\theta}{d\xi} + \left(\frac{\pi e}{2}\right)^2 \sin^2\theta. \tag{10.17}$$

The free energy density is a periodic function of ξ with the period of $(P/2)q_o$. The averaged free energy density is given by

$$\begin{aligned}
\bar{\psi} &= \frac{\int_0^{P/2} \left[\frac{1}{2}(1+A) - \frac{d\theta}{d\xi} + \left(\frac{\pi e}{2}\sin\theta\right)^2\right] d\xi}{(P/2)q_o} \\
&= \frac{\int_0^{\pi} \left[\frac{1}{2}(1+A) - \frac{d\theta}{d\varsigma} + \left(\frac{\pi e}{2}\sin\theta\right)^2\right] \frac{d\xi}{d\theta} d\theta}{\int_0^{\pi} \left[A + \left(\frac{\pi e}{2}\sin\theta\right)^2\right]^{-1/2} d\theta} \\
&= \frac{-\pi + \int_0^{\pi} \left\{\left[A + \left(\frac{\pi e}{2}\sin\theta\right)^2\right]^{1/2} + \frac{1}{2}(1-A)\left[A + \left(\frac{\pi e}{2}\sin\theta\right)^2\right]^{-1/2}\right\} d\theta}{\int_0^{\pi} \left[A + \left(\frac{\pi e}{2}\sin\theta\right)^2\right]^{-1/2} d\theta}.
\end{aligned} \tag{10.18}$$

Minimizing $\bar{\psi}$ with respect to A, $\partial\bar{\psi}/\partial A = 0$, which gives

$$\int_0^{\pi} \left[A + \left(\frac{\pi}{2}e\sin\theta\right)^2\right]^{1/2} d\theta = 2\int_0^{\pi/2} \left[A + \left(\frac{\pi}{2}e\sin\theta\right)^2\right]^{1/2} d\theta = \pi. \tag{10.19}$$

At any applied field e, the value of A can be found by solving Equation (10.19). Once the value of A is known, the helical pitch P can be calculated from Equation (10.16). At the threshold e_C, $A = 0$. From Equation (10.19), it can be obtained that $e_C = 1$. At this field, the pitch is

$$(P/2)q_o = \int_0^{\pi} \left[A + \left(\frac{\pi e_c}{2}\sin\theta\right)^2\right]^{-1/2} d\theta = \int_0^{\pi} \left(\frac{\pi}{2}\sin\theta\right)^{-1} d\theta = \infty.$$

Therefore the critical field unwinding the helical structure is

$$E_C = E_o = \frac{\pi}{2} q_o \sqrt{\frac{K_{22}}{\varepsilon_o \Delta\varepsilon}} = \frac{\pi^2}{P_o} \sqrt{\frac{K_{22}}{\varepsilon_o \Delta\varepsilon}}. \tag{10.20}$$

In reality, the pi-walls in the fingerprint–homeotropic transition are not parallel to each other. Instead, they form circles. The pi-wall circles shrink with increasing field. They annihilate at a threshold, which depends on the cell thickness and anchoring condition, slightly higher than E_c.

The fingerprint–homeotropic (F–H) transition is reversible. The liquid crystal can transform directly from the homeotropic state back to the fingerprint. The transition is, however, a nucleation process and therefore is slow (on the order of 100 ms). There is also a hysteresis that the transition occurs only when the applied field is decreased below a threshold which is lower than the threshold to unwind the helical structure. If there are no nucleation seeds, the formation of helical structure in the middle of the homeotropic state always causes the free energy to increase, namely, there is an energy barrier against the homeotropic–fingerprint transition. Experiments have shown that the threshold E_{HF} of the homeotropic–fingerprint transition is about $0.9E_C$. The hysteresis plays an important role in the dynamic drive scheme, which will be discussed shortly.

10.6.4.3 Transition between the homeotropic state and the planar state

For the liquid crystal in the homeotropic state, when the applied field is turned down, there are two relaxation modes. One is the H–F mode in which the liquid crystal relaxes into the fingerprint state (and then to the focal conic state) as discussed in the previous section. The other is the H–P mode in which the liquid crystal relaxes into the planar state [76,77]. The rotation of the liquid crystal in the H–P mode is shown in Figure 10.28. The liquid crystal forms a conic helical

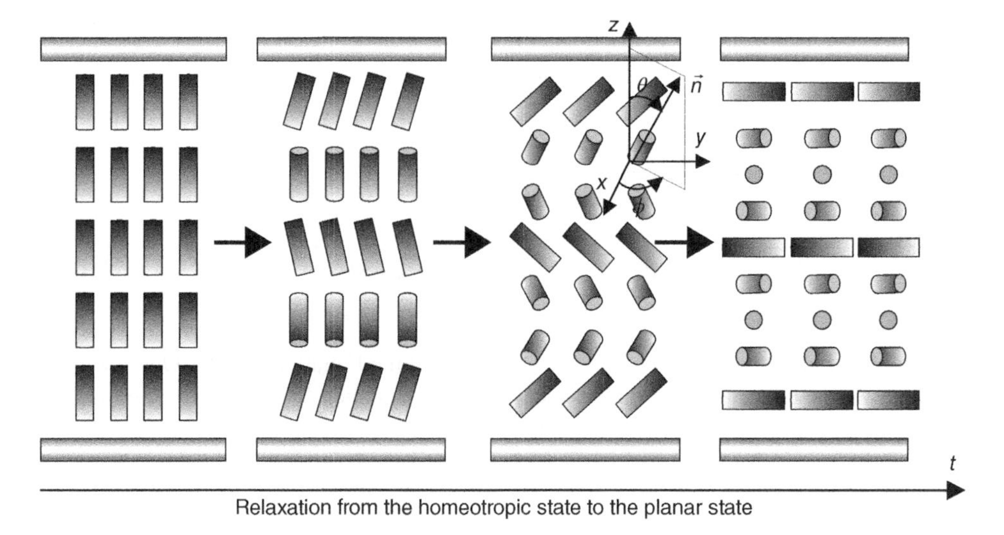

Figure 10.28 Schematic diagram showing the rotation of the liquid crystal in the H–P relaxation mode.

structure with the helical axis in the cell normal direction. As the relaxation takes place, the polar angle θ increases. When the polar angle θ is zero, the liquid crystal is in the homeotropic state. When the polar angle is $\pi/2$, the liquid crystal is in the planar state.

We now consider the static conic helical structure. It is assumed that the polar angle θ is a constant independent of z, and the azimuthal angle ϕ varies along z with a constant rate q, that is, the twisting is uniform. The components of the director $\vec{n}$ are given by $n_x = \sin\theta\cos(qz)$, $n_y = \sin\theta\sin(qz)$, and $n_z = \cos\theta$. The free energy is given by

$$f = \frac{1}{2}K_{22}\left(q_o - q\sin^2\theta\right)^2 + \frac{1}{2}K_{33}q^2\sin^2\theta\cos^2\theta + \frac{1}{2}\Delta\varepsilon\varepsilon_o E^2\sin^2\theta. \tag{10.21}$$

With the dimensionless variables:$K_3 = K_{33}/K_{22}$, $\lambda = q/q_o$, $\psi = f/K_{22}q_o^2$, and $e = E/E_c$, the free energy becomes

$$\psi = \frac{1}{2}\left(1-\lambda\sin^2\theta\right)^2 + \frac{1}{2}K_3\lambda^2\sin^2\theta\cos^2\theta + \frac{1}{2}\left(\frac{\pi}{2}e\right)^2\sin^2\theta. \tag{10.22}$$

By minimizing ψ with respect to λ, we obtain

$$\lambda = \frac{1}{\sin^2\theta + K_3\cos^2\theta}. \tag{10.23}$$

That is $q = \frac{q_o}{\sin^2\theta + K_3\cos^2\theta}$. When the polar angle θ is very small, $q = \frac{q_o}{K_3} = \frac{K_{22}q_o}{K_{33}}$, and the pitch is $P = \frac{K_{33}}{K_{22}}P_o$. For most liquid crystals, $K_{33}/K_{22} \approx 2$. Hence the pitch P of the conic helical structure with small polar angle is about twice the intrinsic pitch P_o. When the polar angle θ is $\pi/2$, $q = q_o$, that is,$P = P_o$. After minimization with respect to λ, the free energy is

$$\psi = \frac{1}{2} + \frac{1}{2}\sin^2\theta\left[\left(\frac{\pi}{2}e\right)^2 - \frac{1}{K_3 + (1-K_3)\sin^2\theta}\right]. \tag{10.24}$$

We then examine whether there is any stable conic helical structure. We minimize the free energy with respect to the polar angle θ,

$$\frac{\partial\psi}{\partial\left(\sin^2\theta\right)} = \frac{1}{2}\left(\frac{\pi}{2}e\right)^2 - \frac{K_3}{2\left[K_3 + (1-K_3)\sin^2\theta\right]^2}, \tag{10.25}$$

$$\frac{\partial^2\psi}{\partial\left(\sin^2\theta\right)^2} = \frac{K_3(1-K_3)}{\left[K_3 + (1-K_3)\sin^2\theta\right]^3}. \tag{10.26}$$

Because $K_3 > 1$, the second-order derivative is negative, and therefore there is no minimum free energy state in the region $0 < \theta < \pi/2$; therefore, there is no stable conic helical structure. The liquid crystal is either in the homeotropic state with $\theta = 0$ or in the planar state with $\theta = \pi/2$. In Figure 10.29 the free energy of the conic helical structure given by Equation (10.24) is plotted as a function of $\sin^2\theta$ at three different fields. $e_{eq} = 2/\pi$ is the field at which the planar state and

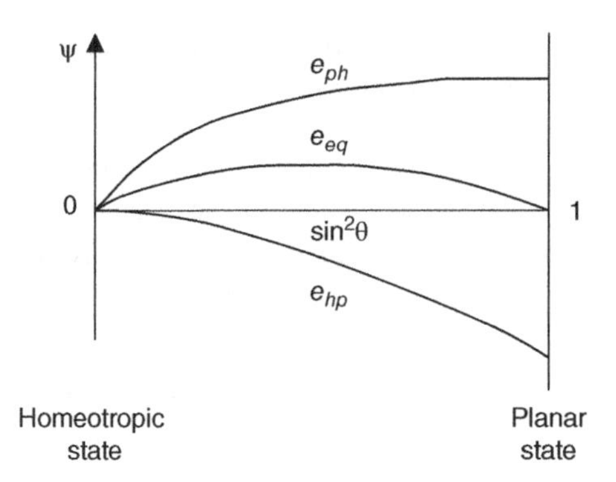

Figure 10.29 The free energy of the conic helical structure as a function of the polar angle θ at various applied fields.

the homeotropic state have the same free energy. At this field there is, however, an energy barrier between the two states. When the applied field is increased from e_{eq}, the free energy of the homeotropic state becomes lower than that of the planar state, but the energy barrier persists. The energy barrier becomes lower with increasing field. When the field is sufficiently high, the energy barrier decreases to zero, and the planar state will become absolutely unstable. The critical field $e_{ph} = \frac{2}{\pi}\sqrt{K_3}$ can be obtained from the equation $\partial\psi/\partial(\sin^2\theta)|_{\theta=\pi/2} = 0$. The un-normalized critical field for the planar–homeotropic transition is

$$E_{ph} = \frac{2}{\pi}\sqrt{\frac{K_{33}}{K_{22}}}E_c.$$

For a liquid crystal with $K_{33}/K_{22} = 2$, $E_{hp} = 0.9E_c$, which is slightly lower than the critical field E_c to unwind the helical structure in the fingerprint state. If the field is decreased from e_{eq}, the free energy of the planar state becomes lower than that of the homeotropic state, but the energy barrier persists. The energy barrier becomes lower with decreasing field. When the field is sufficiently low, the energy barrier decreases to zero, and the homeotropic state will become absolutely unstable. The critical field $e_{hp} = \frac{2}{\pi}\sqrt{1/K_3}$ can be obtained from the equation $\left.\frac{\partial\psi}{\partial\left(\sin^2\theta\right)}\right|_{\theta=0} = 0$. The un-normalized critical field under which the H–P relaxation can occur is $E_{hp} = \frac{2}{\pi}\sqrt{\frac{K_{22}}{K_{33}}}E_c$. For a liquid crystal with $K_{33}/K_{22} = 2$, $E_{hp} = 0.45E_c$, which is much lower than the threshold E_{hf} for the homeotropic–fingerprint relaxation mode. Detailed dynamic studies show that in the homeotropic–planar transition, the polar angle changes quickly but not the twisting rate. When the polar angle changes from 0 to $\pi/2$, the twisting rate is still around $(K_{22}/K_{33})q_o$, which corresponds to the pitch $(K_{33}/K_{22})P_o$. This planar state with the pitch $(K_{33}/K_{22})P_o$ is named the *transient planar state*. The transition from the homeotropic state to the transient planar state is a homogeneous transition with the transition time $T_{hp} \approx \gamma P_o^2/K_{22}$. For a liquid crystal with $\gamma = 5\times10^{-2}\frac{\text{N}\cdot\text{S}}{\text{m}^2}$, $K_{22} = 10^{-11}$N and $P_o = 0.5$ μm, $T_{hp} \sim 1$ ms. The transient planar state is unstable because its elastic energy is still high and the liquid crystal will relax through a nucleation process into the stable planar state with the intrinsic pitch P_o [78]. The transition time is of the order of 100 ms.

In summary, if the liquid crystal is in the homeotropic state and the applied field is reduced, there are two possible relaxation modes. If the applied field is reduced to the region $E_{hp} < E < E_{hf}$, the liquid crystal relaxes slowly into the fingerprint state and then to the focal conic state when the applied field is reduced further. If the applied field is reduced below E_{hp}, the liquid crystal relaxes quickly into the transient planar state and then to the stable planar state. In bistable Ch reflective displays, the way to switch the liquid crystal from the focal conic state to the planar state is by first applying a high field to switch it to the homeotropic state, and then turning off the field quickly to allow it to relax to the planar state.

10.6.5 *Drive schemes for bistable Ch displays*

As discussed in previous sections, cholesteric liquid crystals exhibit two bistable states at zero field: the reflecting planar state and the non-reflective focal conic state. They can be used to make multiplexed displays on passive matrices. In this section, we consider the drive schemes for the bistable Ch displays.

10.6.5.1 Response of bistable Ch material to voltage pulses

In order to design drive schemes, we first must know the electro-optical response of the bistable Ch liquid crystals to voltage pulses. A typical response of a bistable Ch liquid crystal to voltage pulse is shown in Figure 10.30 [51]. The horizontal axis is the amplitude of the voltage pulse. The vertical axis is the reflectance measured *not during the pulse but a few hundreds of*

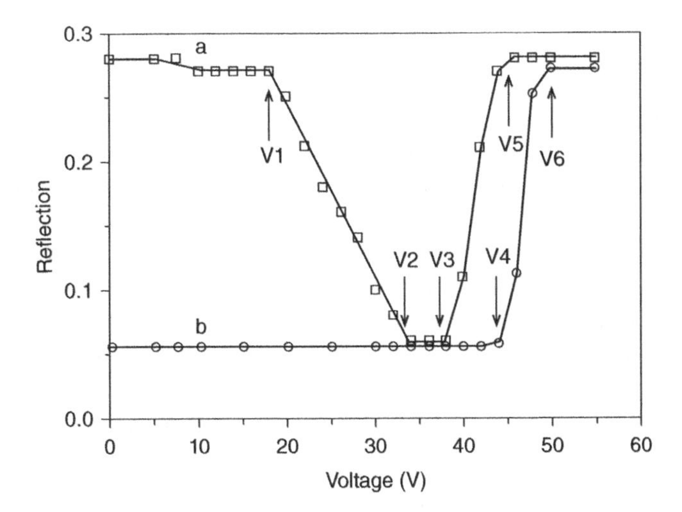

Figure 10.30 The response of the bistable Ch liquid crystal to 40 ms wide voltage pulses. a: initially in the planar state, b: initially in the focal conic state.

milliseconds after the removal of the voltage pulse, when the reflectance no longer changes. The response depends on the initial state of the Ch material. Curve (a) represents the response of the material initially in the planar state which is obtained by applying a voltage pulse higher than V_6. When the voltage of the pulse is below $V_1 = 18$ V, the stability threshold voltage, the Ch material remains in the planar state during and after the pulse. When the voltage of the pulse is increased above V_1, some domains are switched into the focal conic state during the pulse and stay in the focal conic state after the pulse, and thus the reflectance after the pulse decreases. The higher the voltage of the pulse, the more domains are switched to the focal conic state. When the voltage of the pulse reaches $V_2 = 34$ V, all the domains are switched to the focal conic state and the minimum reflectance is reached. The region from V_1 and V_2 is the best region to achieve gray-scale reflectance. When the voltage of the pulse is increased above $V_3 = 38$ V, some domains are switched to the homeotropic state and the remaining domains are switched to the focal conic state during the pulse. The domains switched to the homeotropic state relax to the planar state after the pulse, and therefore the reflectance increases again. When the voltage of the pulse is increased above $V_5 = 46$ V, all domains are switched to the homeotropic state during the pulse and relax to the planar state after the pulse, and the maximum reflectance is obtained. Curve (b) represents the response of the material initially in the focal conic state, which is obtained by applying an intermediate voltage pulse with a voltage, say, V_3. When the voltage of the pulse is below $V_4 = 44$ V, the Ch material remains in the focal conic state during and after the pulse. When the voltage of the pulse is increased above V_4, some domains are switched to the homeotropic state and the remaining domains stay in the focal conic state during the pulse. The domains switched to the homeotropic state relax to the planar state after the removal of the pulse, and therefore the reflectance increases. When the voltage of the pulse is increased above $V_6 = 52$ V, all the domains are switched to the homeotropic state during the pulse.

10.6.5.2 Conventional drive scheme for bistable Ch displays

Because of the bistability and high stability threshold of Ch liquid crystals, they can be used to make multiplexed displays on passive matrix. In the conventional drive scheme for the bistable Ch display, the display is addressed one line at a time [51,54]. A high-voltage pulse switches the liquid crystal into the reflecting planar state and a low-voltage pulse switches the liquid crystal into the non-reflecting focal conic state. For the Ch liquid crystal whose response to voltage pulse is shown in Figure 10.30, for the row being addressed, the applied voltage is $V_a = (V_6 + V_3)/2 = 45$ V, as shown in Figure 10.31. The column voltage to select the planar state is $-\frac{1}{2}\Delta V = -\frac{1}{2}(V_6 - V_3) = -7\,\text{V}$ (out of phase with respect to the row voltage). The voltage across the pixel to be addressed to the planar state is $V_a - \left(-\frac{1}{2}\Delta V\right) = V_6$. The column voltage to select the focal conic state is $\frac{1}{2}\Delta V = \frac{1}{2}(V_6 - V_3) = +7\,\text{V}$ (in phase with respect to the row voltage). The voltage across the pixel to be addressed to the focal conic state is $V_a - \frac{1}{2}\Delta V = V_3$. If the column voltage is varied between $-\frac{1}{2}\Delta V$ and $\frac{1}{2}\Delta V$, gray-scale reflectance can be obtained [79]. For the rows not being addressed, the applied voltage is $V_{na} = 0$. For the pixels on the row not being addressed, the absolute value of the voltage applied across them is $\left|\frac{1}{2}\Delta V\right| = 7\,\text{V}$, which is lower than the stability threshold voltage $V_1 = 18$ V, as shown in Figure 10.30. Therefore the state of the Ch liquid crystal in these pixels remains unchanged. In this drive scheme, the time

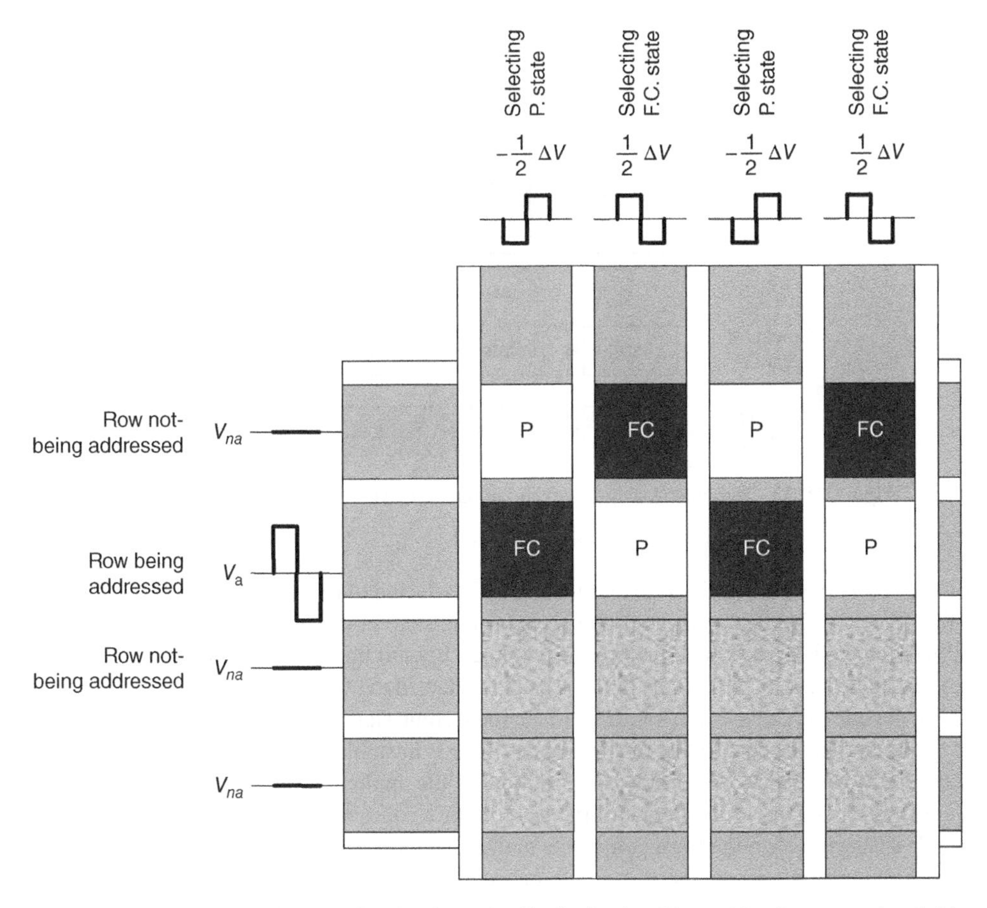

Figure 10.31 Schematic diagram showing how the Ch display is addressed by the conventional drive scheme.

interval to address one line is usually a few tens of milliseconds. Therefore this drive scheme is only suitable for low information content displays.

10.6.5.3 Dynamic drive scheme for bistable Ch displays

The dynamic drive scheme makes use of the dynamic process of the homeotropic-planar transition and the hysteresis in the focal conic-homeotropic transition, and is much faster. The dynamic drive scheme consists of three phases: preparation, selection, and evolution, as shown in Figure 10.32 [54,80]. In the preparation (with the time interval $\Delta t_P \sim 50$ ms), a high voltage pulse is applied to switch the Ch liquid crystal into the homeotropic state. In the selection phase (with the time interval $\Delta t_S \sim 1$ ms), if the applied voltage is V_H, which is higher than V_{hp}, the liquid crystal remains in the homeotropic state because of the high voltage and the short time interval. If the applied voltage is V_L, which is lower than V_{hp}, the liquid crystal relaxes into the transient planar state. In the evolution phase (with the time interval $\Delta t_S \sim 50$ ms), the applied

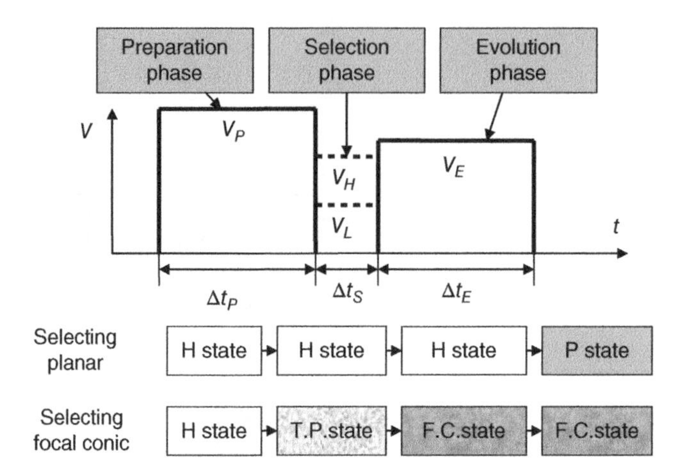

Figure 10.32 Schematic diagram of the dynamic drive scheme for the Ch display.

voltage is between V_{hf} and V_c. If the liquid crystal is selected to remain in the homeotropic state in the selection phase, it remains in the homeotropic state in the evolution phase because the applied voltage is higher than the threshold voltage V_{hf} and thus the material cannot relax into the focal conic state [50,81]. If the liquid crystal is selected to transform into the transient planar state in the selection phase, it is switched to the focal conic state but not the homeotropic state in the evolution phase because the transient planar state is unstable under the field and the applied voltage is lower than V_c. After the evolution phase, the applied voltage is reduced to 0. If the liquid crystal remains in the homeotropic state in the evolution phase, it relaxes to the planar state. If it is switched to the focal conic state in the evolution phase, it remains in the focal conic state. Although time intervals of the preparation and evolution phase are long, time can be shared by simultaneously putting multiple lines in the preparation and evolution phases. The time (frame time) needed to address a N line display is $\Delta t_P + N \times \Delta t_S + \Delta t_E$, which is much shorter than the frame time of the conventional drive scheme. The addressing speed of the dynamic drive scheme is fast enough for electronic book and paper applications, where an updating rate of one page per second is required, but not fast enough for video rate applications.

Homework Problems

10.1 *Multiplexed STN LCD on a passive matrix.* In order to have one hundred rows, at least how steep must the voltage–transmittance curve be (find the γ value defined by Equation (10.6))?

10.2 Consider a (0°, 180°) bistable TN where the entrance polarizer is at 45° with respect to the liquid crystal at the entrance plane and the exit polarizer is at –45° with respect to the liquid crystal at the entrance plane. Find the three values of the retardation with which the display has contrast ratios. Give the transmittances of the bright and dark states under those retardation values.

10.3 In the Helfrich deformation shown in Figure 10.26, the cell thickness of the cell is h. The cholesteric liquid crystal has the pitch P and dielectric anisotropy $\Delta\varepsilon$ (>0). For small undulation, calculate the field threshold $E_{Helfrich}$ and the wavelength λ of the undulation.

10.4 In the fingerprint–homeotropic transition shown in Figure 10.27, calculate the integration constant A and the normalized pitch P/P_o as a function of the normalized electric field E/E_c.

References

1. A. R. Kmetz, Matrix addressing of non-emissive displays, in *Nonemissive electrooptical displays*, ed. A. R. Kmetz and F. K. von Willisen (Plenum Press, New York, 1976).
2. E. Kaneko, Liquid crystal TV: principles and applications of liquid crystal displays, (KTK Scientific Publishers, Tokyo, 1987).
3. P. M. Alt and P. Pleshko, Scanning limitations of liquid crystal displays, *IEEE Trans. Electron Device*, **ED-21**, 146 (1974).
4. T. Scheffer and J. Nehring, Twisted nematic and supertwisted nematic mode LCDs, in *Liquid crystals–applications and uses*, Vol. 1, ed. B. Bahadur (World Scientific, New Jersy, 1990).
5. F. C. Luo, Active matrix LC Displays, Chapt. 15, *Liquid crystals–applications and uses,* Vol. 1, ed. B. Bahadur (World Scientific, New Jersey, 1990).
6. E. Lueder, *Liquid crystal displays: addressing schemes and electro-optical effects*, (John Wiley & Sons, Chichester, 2001).
7. J. W. Goodby, R. Blinc, N. A. Clark, et al., Ferroelectric liquid crystals: Principle, properties and applications, *Ferroelectricity and related phenomena,* Vol. 7 (Gordon and Breach Publishers, Amsterdam, 1991).
8. J. Dijon, Ferroelectric LCDs, Chapt. 13, *Liquid crystals–applications and uses*, Vol. 1, ed. B. Bahadur (World Scientific, New Jersey, 1990).
9. D. W. Berreman and W. R. Heffner, New bistable Ch liquid-crystal display, *Appl. Phys. Lett.*, **37**, 109 (1980).
10. T. Tanaka, Y. Sato, A. Inoue, et al., A bistable twisted nematic (BTN) LCD driven by a passive-matrix addressing, *Proc. Asia Display* **95**, 259 (1995).
11. T. Tanaka, Y. Sato, T. Obikawa, et al., Physical and electro-optical properties of bistable twisted nematic (BTN) LCD, *Proc. Intnl. Display Research Conf,* M-64 (1997).
12. T.-Z. Qian, Z.-L. Xie, H.-S. Kwok and P. Sheng, Dynamic flow and switching bistability in nematic liquid crystal cells, *Appl. Phys. Lett.*, **71**, 596 (1997).
13. Y. J. Kim, S. M. Park, I. Lee, et al., Numerical modeling and optical switching characteristics of a bistable TN-LCD, *Proc. EuroDisplay* **96**, 337 (1996).
14. G. P. Brown, Ultra low power bistable LCDs, *Proc. Intnl. Display Research Conf. 00*, 76 (2000).
15. G. P. Bryan-Brown, C. V. Brown, J. C. Jones, et al., Grating aligned bistable nematic device, *SID Intl. Symp. Digest Tech. Papers*, **28**, 37 (1997).
16. R. Barberi and G. Durand, Electrically controlled bistable surface switching in nematic liquid crystals, *Appl. Phys. Lett.*, **58**, 2907 (1991).
17. R. Barberi, M. Giocondo and G. Durand, Flexoelectrically controlled bistable surface switching in nematic liquid crystals, *Appl. Phys. Lett.*, **60**, 1085 (1992).
18. R. Barberi, M. Giocondo, J. Li and R. Bartolino, Fast bistable nematic display with gray scale, *Appl. Phys. Lett.*, **71**, 3495 (1997).
19. R. Barberi and G. Durand, Controlled textural bistability in nematic liquid crystals, in *Handbook of liquid crystal research*, ed. P. J. Collings and J. S. Patel (Oxford University Press, New York, 1997).
20. G. D. Boyd, J. Cheng, and P. D. T. Ngo, Liquid-crystal orientational bistability and nematic storage effects, *Appl. Phys. Lett.*, **36**, 556 (1980).
21. R. N. Thurston, J. Cheng, and G. D. Boyd, Mechanically bistable liquid crystal display structures, *IEEE Trans. Elec. Dev.*, **ED-27**, 2069 (1980).

22. J. Cheng and R. N. Thurston, The propagation of disclinations in bistable switching, *J. Appl. Phys.*, **52**, 2766 (1981).
23. P. A. Breddels and H. A. van Sprang, An analytical expression for the optical threshold in highly twisted nematic systems with nonzero tilt angles at the boundaries, *J. Appl. Phys.*, **58**, 2162 (1985).
24. H. A. van Sprang and P. Breddels, Numerical calculations of director patterns in highly twisted nematic configurations with nonzero pretilt angles, *J. Appl. Phys.*, **60**, 968 (1986).
25. J. C. Kim, G.-J. Choi, Y.-S. Kim, et al., Numerical modeling and optical switching characteristics of a bistable TN-LCD, *SID Intl. Symp. Digest Tech. Papers*, **28**, 33 (1997).
26. Z. L. Xie and H. S. Kwok, New bistable twisted nematic liquid crystal displays, *J. Appl. Phys. Lett*, **84**, 77 (1998).
27. Z. L. Xie, Y. M. Dong, S. Y. Xu, et al., $\pi/2$ and $5\pi/2$ twisted bistable nematic liquid crystal display, *J. Appl. Phys.* **87**, 2673 (2000).
28. W. H. de Jeu, Physical properties of liquid crystalline materials (Gordon and Breach, New York, 1980).
29. P. G. de Gennes and J. Prost, *The physics of liquid crystals* (Oxford University Press, New York, 1993).
30. D. W. Berreman, Liquid-crystal twist cell dynamics with backflow, *J. Appl. Phys.*, **46**, 3746 (1975).
31. C. Z. van Doorn, Dynamic behaviour of twisted nematic, *J. Appl. Phys.*, **46**, 3738 (1975).
32. J. Kelly, S. Jamal, and M. Cui, Simulation of the dynamics of twisted nematic devices including flow, *J. Appl. Phys.*, **86**, 4091 (1999).
33. I. Dozov, M. Nobili, and G. Durand, Fast bistable nematic display using monostable surface switching, *Appl. Phys. Lett.*, **70**, 1179 (1997).
34. P. Martinot-Lagrade, I. Dozov, E. Polossat, et al., Fast bistable nematic display using monostable surface anchoring switching, *SID Intl. Symp. Digest Tech. Papers*, **28**, 41 (1997).
35. S. T. Tang, H. W. Chiu, and H. S. Kwok, Optically optimized transmittive and reflective bistable twisted nematic liquid crystal display, *J. Appl. Phys.*, **87**, 632 (2000).
36. Z. L. Xie, H. J. Gao, S. Y. Xu, and S. H. Kwok, Optimization of reflective bistable nematic liquid crystal displays, *J. Appl. Phys.*, **86**, 2373 (1999).
37. H. Cheng and H. Gao, Optical properties of reflective bistable twisted nematic liquid crystal display, *J. Appl. Phys.* **87**, 7476 (2000).
38. Z. L. Xie, H. J. Gao, B. Z. Chang and S. Y. Xu, A new BTN LCD with high contrast ratio and large cell gap, *Proc. Asia Display* **98**, 303 (1998).
39. F. Zhou and D.-K. Yang, Analytical solution of film compensated bistable twisted nematic liquid crystal displays, *J. Display Tech.*, **1**, 217 (2005).
40. G.-D. Lee, K.-H. Park, K.-C. Chang, et al., Optimization of drive scheme for matrix addressing of a bistable twisted nematic LCD, *Proc. Asia Display* **98**, 299 (1998).
41. T. Tanaka, T. Obikawa, Y. Sato, et al., An advanced driving method for bistable twisted nematic (BTN) LCD, *Proc. Asia Display* **98**, 295 (1998).
42. E. L. Wood, G. P. Bryan-Brown, P. Brett, et al., Zenithal bistable device (ZBD) suitable for portable applications, *SID Intl. Symp. Digest Tech. Papers,* **31**, 124 (2000).
43. G. P. Bryan-Brown, M. J. Towler, M. S. Bancroft, and D. G. McDonnell, Bistable nematic alignment using bigratings, *Proc. Intnl. Display Research Conf.* **94**, 209 (1994).
44. L. M. Blinov and V. G. Chigrinov, Electrooptical effects in liquid crystal materials (Springer-Verlag, New York, 1994).
45. S. Chandrasekhar Liquid crystals, 2nd edn. (Cambridge University Press, New York, 1997).
46. M. Kleman and O. D. Lavrentovich, *Soft matter physics, Introduction*, (Springer-Verlag, New York, 2003).
47. O.D. Lavrentovich and D.-K. Yang, Cholesteric cellular patterns with electric-field -controlled line tension, *Phys. Rev. E, 57, Rapid Communications,* R6269 (1998).
48. R. B. Meyer, Distortion of a cholesteric structure by a magnetic field, *Appl. Phys. Lett.*, **14**, 208(1969).
49. W. Greubel, U. Wolf, and H. Kruger, Electric field induced texture changes in certain nematic/cholesteric liquid crystal mixtures, *Mol. Cryst. Liq. Cryst.* **24**, 103 (1973).
50. S.-T. Wu and D.-K. Yang, *Reflective liquid crystal displays*, John Wiley & Sons, Ltd., (2001).
51. D.-K. Yang and J.W. Doane, Cholesteric liquid crystal/polymer gel dispersions: reflective displays, *SID Intl. Symp. Digest Tech. Papers,* **23**, 759 (1992).
52. Z.-J. Lu, W.D. St. John, X.-Y. Huang, et al., Surface modified reflective cholesteric displays, *SID Intl. Symp. Digest Tech. Papers*, **26**, 172 (1995).
53. D.-K. Yang, J.L. West, L.C. Chien and J.W. Doane, Control of the reflectivity and bistability in displays based on cholesteric liquid crystals, *J. Appl. Phys.*, **76**, 1331 (1994).

54. D.-K. Yang, X.Y. Huang, and Y.-M. Zhu, Bistable cholesteric reflective displays: material and drive schemes, *Annual Review of Materials Science,* **27,** 117 (1996).
55. J. W. Doane, D.-K. Yang, and Z. Yaniv, Front-lit flat panel display from polymer stabilized cholesteric textures, *Proc. Japan Display* **92**, 73 (1992).
56. M.H. Lu, H. J. Yuan, and Z. Yaniv, Color reflective liquid crystal display, US Patent 5,493,430, 1996.
57. D.-K. Yang, J.W. Doane, Z. Yaniv, and J. Glasser, Cholesteric reflective display: drive scheme and contrast, *Appl. Phys. Lett.*, **65**, 1905 (1994).
58. R.Q. Ma and D.-K. Yang, Polymer stabilized bistable black-white cholesteric reflective display, *SID Intl. Symp. Digest Tech. Papers*, **28**, 101 (1997).
59. R.Q. Ma and D.-K. Yang, Optimization of polymer stabilized bistable black-white cholesteric reflective display, *J. SID.* **7**, 61 (1999).
60. X.-Y. Huang, N. Miller, A. Khan, et al., Gray scale of bistable reflective cholesteric displays, *SID Intl. Symp. Digest Tech. Papers,* **29**, 810 (1998).
61. M. Xu and D.-K. Yang, Optical properties of the gray-scale states of cholesteric reflective displays, *SID Intl Symp. Digest Tech. Papers*, **30**, 950 (1999).
62. L.-C. Chien, U. Muller, M.-F. Nabor, and J. W. Doane, Multicolor reflective cholesteric displays, *SID Intl. Symp. Digest Tech. Papers*, **26**, 169 (1995).
63. F. Vicentini and L.-C. Chien, Tunable chiral materials for multicolor reflective cholesteric displays, *Liq. Cryst.*, **24**, 483 (1998).
64. D.-K. Yang, Z.J. Lu, L.C. Chien, and J. W. Doane, Bistable polymer dispersed cholesteric reflective display, *SID Intl Symp. Digest Tech. Papers*, **34**, 959–961 (2003).
65. K. Hashimoto, M. Okada, K. Nishguchi, et al., Reflective color display using cholesteric liquid crystals, *SID Intl. Symp. Digest Tech. Papers*, **29**, 897 (1998).
66. D. Davis, A. Kahn, X.-Y. Huang, and J. W. Doane, Eight-color high-resolution reflective cholesteric LCDs, *SID Intl. Symp. Digest Tech. Papers*, **29**, 901 (1998).
67. J. L. West and V. Bodnar, Optimization of stacks of reflective cholesteric films for full color displays, *Proc. 5th Asian Symp. on Information Display*, **29** (1999).
68. D. Davis, K. Hoke, A. Khan, et al., Multiple color high resolution reflective cholesteric liquid crystal displays, *Proc. Intnl. Display Research Conf.*, 242 (1997).
69. T. Schneider, F. Nicholson, A. Kahn, and J. W. Doane, Flexible encapsulated cholesteric LCDs by polymerization induced phase separation, *SID Intl Symp. Digest Tech. Papers*, **36**, 1568–1571 (2005).
70. I. Shiyanovskaya, S. Green, G. Magyar, and J. W. Doane, Single substrate encapsulated cholesteric LCDs: coatable, drapable and foldable, *SID Intl Symp. Digest Tech. Papers*, **36**, 1556–1559 (2005).
71. S. W. Stephenson, D. M. Johnson, J. I. Kilburn, et al., Development of a flexible electronic display using photographic technology, *SID Intl Symp. Digest Tech. Papers*, **35**, 774–777 (2004).
72. G. T. McCollough, C. M. Johnson, and M. L. Weiner, Roll-to-roll manufacturing considerations for flexible, cholesteric liquid crystal (ChLC) display media, *SID Intl Symp. Digest Tech. Papers*, **36**, 64–47 (2005).
73. N. Hiji, T. Kakinuma, M. Araki, and Y. Hikichi, Cholesteric liquid crystal micro-capsules with perpendicular alignment shell for photo-addressable electronic paper, *SID Intl Symp. Digest Tech. Papers*, **36**, 1560–1563 (2005).
74. W. Helfrich, Deformation of cholesteric liquid crystals with low threshold voltage, *Appl. Phys. Lett.*, **17**, 531 (1970).
75. J. P. Hurault, Static distortions of a cholesteric planar structure induced by magnetic or ac electric fields, *J. Chem. Phys.*, **59**, 2068 (1973).
76. D.-K. Yang and Z.-J. Lu, Switching mechanism of bistable Ch reflective displays, *SID Intl. Symp. Digest Tech. Papers*, **26**, 351 (1995).
77. M. Kawachi, O. Kogure, S. Yosji, and Y. Kato, Field-induced nematic-cholesteric relaxation in a small angle wedge, *Jpn. J. Appl. Phys.*, **14**, 1063 (1975).
78. P. Watson, J. E. Anderson, V. Sergan, and P. J. Bos, The transition mechanism of the transient planar to planar director configuration change in cholesteric liquid crystal displays, *Liq. Cryst.*, **26**, 1307 (1999).
79. J. Gandhi, D.-K. Yang, X.-Y. Huang, and N. Miller, Gray scale drive schemes for bistable Ch reflective displays, *Proc. Asia Display* **98**, 127 (1998).
80. X.-Y. Huang, D.-K. Yang, P. Bos, and J. W. Doane, Dynamic drive for bistable reflective cholesteric displays: a rapid addressing scheme, *SID Intl. Symp.Digest Tech. Papers*, **26**, 347 (1995).
81. X-Y. Huang, D.-K. Yang, and J.W. Doane, Transient dielectric study of bistable reflective cholesteric displays and design of rapid drive scheme, *Appl. Phys. Lett.* **69**, 1211 (1995).

11

Liquid Crystal/Polymer Composites

11.1 Introduction

Liquid crystal/polymer composites (LCPCs) are a relatively new class of materials for use in displays, light shutters, optical fiber telecommunications, and switchable windows [1–5]. They consist of low molecular weight liquid crystals and high molecular weight polymers, which are phase separated. According to the morphology, LCPCs can be divided into two subgroups: polymer dispersed liquid crystals (PDLCs) and polymer-stabilized liquid crystals (PSLCs). In a PDLC, the liquid crystal exists in the form of micron and submicron size droplets which are dispersed in the polymer binder. The concentration of the polymer is comparable to that of the liquid crystal. The polymer forms a continuous medium. The liquid crystal droplets are isolated from one another. A scanning electron microscope (SEM) picture of a PDLC sample is shown in Figure 11.1(a). In a PSLC, the polymer forms a sponge-like structure. The concentration of the liquid crystal is much higher than that of the polymer. The liquid crystal forms a continuous medium. An SEM picture of a PSLC is shown in Figure 11.1(b). Liquid crystal/polymer composites can also be divided into two subgroups according to the application: scattering device and non-scattering device. In a scattering device, the polymer produces or helps to produce a poly-domain structure of the liquid crystal in one field condition. The domain size is comparable to the wavelength of the light to be scattered. The material is highly scattering because of the large birefringence of the liquid crystal. In another field condition the liquid crystal is aligned unidirectionally along the applied field and the material becomes transparent. Two scattering devices from LCPCs are shown in Figure 11.2. In a non-scattering device, the liquid crystal is used to stabilize states of the liquid crystal.

Fundamentals of Liquid Crystal Devices, Second Edition. Deng-Ke Yang and Shin-Tson Wu.

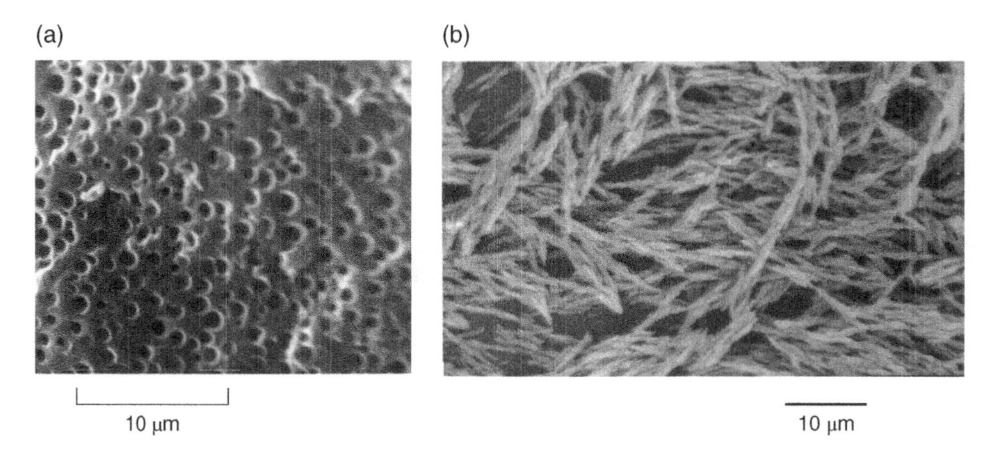

Figure 11.1 (a) SEM picture of a PDLC. It was taken after the PDLC sample was fractured and the liquid crystal was extracted. The dark circles correspond to the liquid crystal droplets. (b) SEM picture of a PSLC. The picture was taken after the cell was split and the liquid crystal was extracted.

Figure 11.2 (a) Photo of the light shutter from polymer stabilized cholesteric texture (PSCT) normal material, (b) Photo of the display from PSCT reverse-mode material.

11.2 Phase Separation

Liquid crystal/polymer composites are phase-separated systems. In order to understand the phase separation involved, we will first discuss the basics of phase separation. Composites (or mixtures) have two or more components. They can be divided into two classes: homogeneous mixtures and heterogeneous mixtures. In a homogeneous mixture, the constituents are mixed on an atomic (or molecular) scale to form a single phase. Conversely, a heterogeneous mixture contains two or more distinct phases. Whether a composite forms a homogeneous single phase, or the phase separates to form a heterogeneous mixture is determined by the free energy. If the homogeneous mixture has a lower free energy than the heterogeneous mixture, the composite is in the homogeneous phase. If the heterogeneous mixture has a lower free energy than the homogeneous mixture, the composite is in the heterogeneous phase. The mixing free energy F_m is defined as the free energy F_{homo} of the homogeneous mixture minus the free energy F_{hetero} of the completely phase-separated system, namely, $F_m = F_{\text{homo}} - F_{\text{hetero}}$. Phase separation depends on the details of the curve of the mixing free energy as a function of concentrations of the components.

11.2.1 Binary mixture

The simplest mixture is a binary mixture that has two components A and B. In order to understand phase separation in this system, we consider its mixing free energy, which is composed of two parts: mixing interaction energy U_m and mixing entropy S_m.

$$F_m = U_m - TS_m \tag{11.1}$$

where T is the temperature. The mixing interaction energy is the interaction energy of the homogeneous mixture minus the interaction energy of the completely phase-separated heterogeneous mixture. In the calculation of the mixing interaction energy, we assume that the molecules only interact with their nearest neighbor molecules. The interaction energies are u_{AA} for the interaction between A and A, u_{BB} for the interaction between B and B, and u_{AB} for the interaction between A and B. The total number of particles of the system is N, of which there are N_A particles of A and N_B particles of B. The molar fraction of component B is $x_B = x = N_B/N$ and the molar fraction of component A is $x_A = 1 - x = N_A/N$. The number of nearest neighbors is P. In the homogeneous phase, among the nearest neighbors, the average number of A particles is $(1 - x)P$ and the average number of B particles is xP. The total interaction energy of the homogeneous mixture is [6]

$$\begin{aligned} U_{\text{homo}} &= \frac{1}{2}\{N_A[(1-x)Pu_{AA} + xPu_{AB}] + N_B[(1-x)Pu_{AB} + xPu_{BB}]\} \\ &= \frac{N}{2}P\{(1-x)[(1-x)u_{AA} + xu_{AB}] + x[(1-x)u_{AB} + xu_{BB}]\}, \end{aligned} \tag{11.2}$$

where the factor 1/2 is used to take care of double counting the interaction energy. When they are completely phase separated, the total interaction energy is

$$U_{\text{hetero}} = \frac{1}{2}N_APu_{AA} + \frac{1}{2}N_BPu_{BB} = \frac{1}{2}NP[(1-x)u_{AA} + xu_{BB}]. \tag{11.3}$$

Therefore

$$U_m = U_{\text{homo}} - U_{\text{hetero}} = \frac{1}{2}PNx(1-x)(2u_{AB} - u_{AA} - u_{BB}). \tag{11.4}$$

The average mixing interaction energy per particle is

$$u_m = \frac{U_{\text{m}}}{N} = \frac{1}{2}Px(1-x)(2u_{AB} - u_{AA} - u_{BB}). \tag{11.5}$$

Now we consider the mixing entropy. Imagine putting the N_A particles of A and N_B particles of B into a lattice with $N = N_A + N_B$ lattice sites. The number of distinct states (arrangements of the particles) is [6]

$$G = \frac{N!}{N_A! \cdot N_B!} = \frac{N!}{(N-N_B)! \cdot N_B!}. \tag{11.6}$$

When the system is completely phase separated, the number of states is close to 1. Therefore the mixing entropy is

$$S_m = k_B \ln G - k_B \ln 1 = k_B[\ln N! - \ln N_B! - \ln(N-N_B)!]. \tag{11.7}$$

For a macroscopic system, $N \gg 1$, $N_A \gg 1$, and $N_B \gg 1$. Using the Sterling approximation that $\ln N! = N(\ln N - 1)$, Equation (7) becomes

$$S_m = -Nk_B[(1-x)\ln(1-x) + x\ln x]. \tag{11.8}$$

The mixing entropy per particle is

$$s_m = -k_B(1-x)\ln(1-x) - k_B x \ln x. \tag{11.9}$$

The mixing free energy per particle is

$$f_m = u_m - Ts_m = ax(1-x) + k_B T[(1-x)\ln(1-x) + x\ln x], \tag{11.10}$$

where $a = \frac{1}{2}P(2u_{AB} - u_{AA} - u_{BB})$. In future discussion, the subscript m is omitted and when we say the free energy, we mean the mixing free energy. The entropic part of the mixing free energy is always negative, and therefore always favors mixing. If the mixing interaction energy is negative, then the mixing free energy is negative and the system will be in the homogeneous phase. If the mixing interaction energy is positive, we have to look at the details of the f–x curve in order to see whether the system will phase separate or not. From Equation (11.10) we have

$$\frac{\partial^2 f}{\partial x^2} = -a + k_B T\left(\frac{1}{x} + \frac{1}{1-x}\right). \tag{11.11}$$

The second-order derivative has the minimum value at $x = 0.5$, which is $(\partial^2 f/\partial x^2)_{\min} = -a + 4k_B T$. When the temperature T is higher than $a/4k_B$, the second-order derivative is positive at any fraction x. The f–x curve is shown by Curve (1) in Figure 11.3. The system is in homogeneous phase.

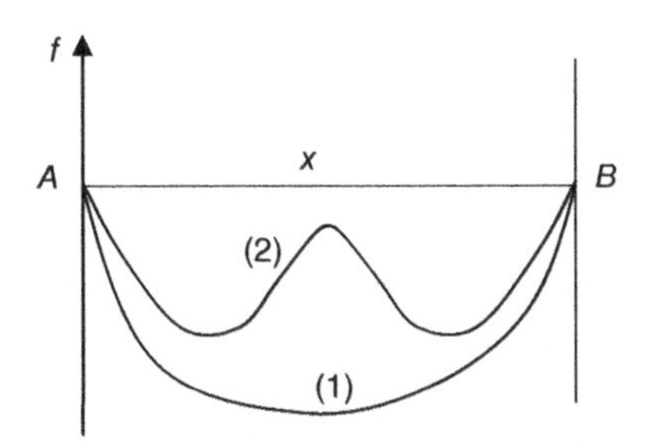

Figure 11.3 Schematic diagram showing the mixing free energy f as a function of the fraction x at two temperatures.

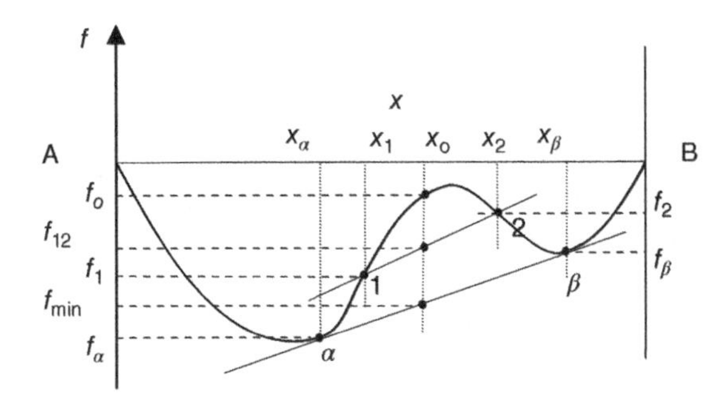

Figure 11.4 Schematic diagram showing how the mixing free energy f changes in the phase separation.

When the temperature T is lower than $a/4k_B$, $\partial^2 f/\partial x^2 < 0$ in some region. The f vs. x curve is shown by curve (2). We will show that in this case the system phase separates into an A-rich phase (more A particles in the phase) and a B-rich phase (more B particle in the phase). We say that the system is partially mixing.

Now we examine the partial mixing case in more detail. Consider a mixture with N particles and the fraction (of B particle) x_o. The number of A particles in the mixture is $N_A = (1-x)N$ and the number of B particles in the mixture is $N_B = xN$. Its free energy per particle is f_o, as shown in Figure 11.4. Assume that it phase separates into two new phases: phase 1 with the fraction x_1 and phase 2 with the fraction x_2. N_1 particles are in phase 1 and N_2 particles are in phase 2. Because of the conservation of particles, we have

$$N_A = (1-x_o)N = (1-x_1)N_1 + (1-x_2)N_2, \tag{11.12}$$

$$N_B = x_o N = x_1 N_1 + x_2 N_2. \tag{11.13}$$

Solving these two equations, we have

$$N_1 = \frac{(x_2 - x_o)}{(x_2 - x_1)} N, \tag{11.14}$$

$$N_1 = \frac{(x_o - x_1)}{(x_2 - x_1)} N. \tag{11.15}$$

The average free energy per particle after phase separation is

$$f_{12}=\frac{1}{N}[N_1f_1+N_2f_2]=f_1+\frac{(x_0-x_1)}{(x_2-x_1)}(f_2-f_1), \tag{11.16}$$

which is the free energy value of the intersection point of the straight line $\overline{12}$ and the vertical line at x_o. It can be seen from the figure that f_{12} is lower than f_o. Therefore the phase separation can take place. It can also be seen that the free energy can be lowered even further if the mixture phase separates into phase α and phase β. Points α and β are the tangential points of the straight line $\alpha\beta$ to the free energy curve [6]. Now the average free energy per particle after the mixture phase separating into phases α and β is

$$f_{\min}=f_\alpha+\frac{(x_\beta-x_o)}{(x_\beta-x_\alpha)}(f_\beta-f_\alpha). \tag{11.17}$$

Therefore if $x_\alpha<x<x_\beta$, the single homogeneous phase is not stable and the mixture phase separates into phase α and phase β. If $0<x<x_\alpha$ and $x_\beta<x<1$, the free energy increases if the mixture phase separates, and therefore the mixture does not phase separate and the single homogeneous phase with the initial fraction is stable.

The criterion for determining partial phase separation is the second-order derivative of the free energy with respect to the concentration. If $\partial^2 f/\partial x^2>0$ for any x, it is impossible to draw a straight line that is tangential to the free energy curve at two points, and there is no phase separation. If $\partial^2 f/\partial x^2<0$ in some region, the free energy curve must have a local maximum at a fraction within this region, and it is possible to draw a straight line tangential to the free energy curve at x_α and x_β. The single homogeneous phase is not stable in the region (x_α, x_β) and phase separation will take place.

We next discuss how the mixing free energy changes as phase separation takes place. Consider a composite with the initial fraction x_o; before phase separation, the free energy is $f_o=f(x_o)$. As shown in Figure 11.5, if x_o is in the region (x_α, x_β), the homogeneous phase is not stable and the system phase separates into two phases: phase 1 with the fraction $x_1=x_o-\Delta x_1$ and phase 2 with the fraction $x_2=x_o+\Delta x_2$. When the phase separation just begins to take place, Δx_1 and Δx_2 are very small. Because of particle conservation, the percentage of the material in phase 1 is $[\Delta x_2/(\Delta x_1+\Delta x_2)]$ and the percentage of the material in phase 2 is $[\Delta x_1/(\Delta x_1+\Delta x_2)]$. The change of the free energy is (keeping up to second-order terms):

$$\Delta f=\frac{\Delta x_2}{(\Delta x_1+\Delta x_2)}f(x_o-\Delta x_1)+\frac{\Delta x_1}{(\Delta x_1+\Delta x_2)}f(x_o+\Delta x_2)-f(x_o)$$

$$\Delta f=\frac{\Delta x_2}{(\Delta x_1+\Delta x_2)}\left[f(x_o)+\left.\frac{\partial f}{\partial x}\right|_{x_o}(-\Delta x_1)+\left.\frac{1}{2}\frac{\partial^2 f}{\partial x^2}\right|_{x_o}(-\Delta x_1)^2\right]$$

$$+\frac{\Delta x_1}{(\Delta x_1+\Delta x_2)}\left[f(x_o)+\left.\frac{\partial f}{\partial x}\right|_{x_o}(\Delta x_2)+\left.\frac{1}{2}\frac{\partial^2 f}{\partial x^2}\right|_{x_o}(\Delta x_2)^2\right]-f(x_o)$$

$$\Delta f=\frac{1}{2}\Delta x_1\Delta x_2\left.\frac{\partial^2 f}{\partial x^2}\right|_{x_o} \tag{11.18}$$

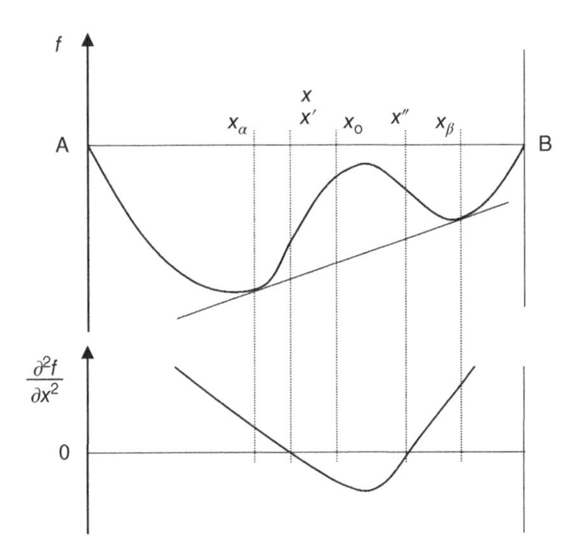

Figure 11.5 Schematic diagram showing how the mixing free energy f changes as phase separation takes place.

There are two types of phase separation in the region (x_α, x_β). When $x' < x_o < x''$, $\partial^2 f/\partial x^2 < 0$, the free energy decreases as phase separation takes place. The phase separation is known as spinodal phase separation. The phase separation occurs simultaneously throughout the system, referred to as homogeneous transition. When $x_\alpha < x_o < x'$ or $x'' < x_o < x_\beta$, $\partial^2 f/\partial x^2 > 0$, the free energy increases in the initial stage of the phase separation process, which means that there is an energy barrier against the transition. The phase separation is known as the binodal phase separation. When the deviation of the fractions of the phase-separated phases from the initial fraction is sufficiently large, the free energy will then decrease. In the phase separation, nucleation seeds are required to overcome the energy barrier, referred to as nucleation transition.

11.2.2 Phase diagram and thermal induced phase separation

As mentioned earlier, the free energy of a binary mixture depends on the temperature. Therefore the phase separation also depends on the temperature. The phase diagram of a binary composite is shown in Figure 11.6. At low temperatures, the entropy term of the free energy does not contribute much to the free energy: $\partial^2 f/\partial x^2 < 0$ for fraction x in some region. The free energy is lowered when the system phase separates into two phases. For example, when the temperature is T_1, the system phase separates into phase 1 with the fraction and phase 2 with the fraction x_2. The fractions x_1 and x_2 are the values of the intersection points of the horizontal line at T_1, referred to as the *tie line*, with the phase boundary of the two phase region. As the temperature is increased, the entropy term plays a more important role, which tends to make the system homogeneously mixing. The two-phase region becomes narrower. The summit point (x_c, T_c) of the phase boundary curve is the critical point. At this temperature, the minimum value of the second-order derivative becomes 0.

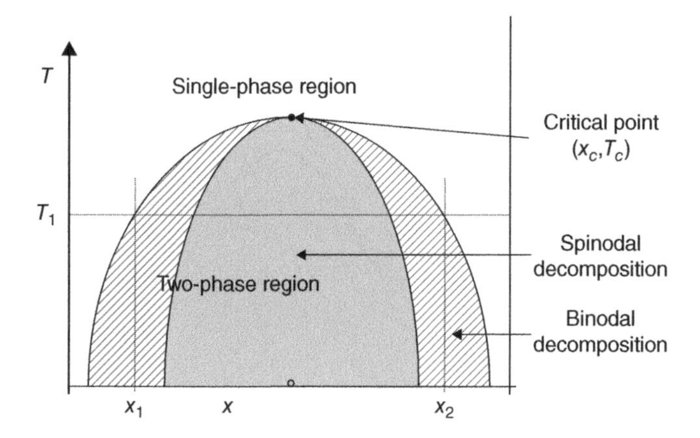

Figure 11.6 Phase diagram of the binary composite.

$$\frac{\partial^2 f}{\partial x^2}=0 \tag{11.19}$$

Because the second-order derivative has the minimum value at the critical point, we have

$$\frac{\partial}{\partial x}\left(\frac{\partial^2 f}{\partial x^2}\right)=\frac{\partial^3 f}{\partial x^3}=0. \tag{11.20}$$

Equation (11.20) gives the location where the second-order derivative has a minimum. Equation (11.19) means that the minimum of the second-order derivative is 0. When the temperature is above T_c, $\partial^2 f/\partial x^2>0$ for any fraction x. The system is in a single phase where the components are homogeneously mixed. When the temperature is decreased from the single-phase region into the two-phase region, the system transforms from a single phase into two phases. This method can be used to produce phase separation in LCPCs, which is referred to as *thermally induced phase separation (TIPS)* [3,7].

TIPS is used to make PDLCs. For example, 50% thermoplastic poly methyl methacrylate (PMMA) is mixed with 50% nematic liquid crystal E7 (from Merck). The glass transition temperature of PMMA is 105 °C and the isotropic–nematic phase transition (the clearing point) temperature is 60.5 °C. The materials can be initially uniformly mixed with the help of a common solvent, such as chloroform, in a bottle (more details will be given in the section on solvent-induced phase separation). Spacers can be added to help achieve uniform thickness of the PDLC film when the material is later sandwiched between two substrates. After mixing, the material is cast on a substrate and solvent is allowed to evaporate in an open space. After the chloroform evaporation, a cover glass plate is put on top of the PDLC. Pressure can be applied to the substrates to obtain uniform PDLC film thickness. At high temperature (>70 °C), the mixture is in a single homogeneous isotropic phase and the film appears clear. At room temperature, the liquid crystal phase separates from the thermoplastic to form droplets, and the film is opaque. The droplet size can be controlled by the cooling rate with smaller droplets formed at faster cooling rate. Thermally induced phase separation is rarely used in manufacturing large area PDLC films because it is difficult to achieve uniform thickness. It is, however, very useful in scientific investigations because PDLCs from thermal plastics can be thermally cycled many times. Different droplet sizes can be obtained in one sample using different cooling rates. It should be noted that phase separation in LCPCs is complicated because of the involvement of mesophases.

11.2.3 Polymerization induced phase separation

The mixing entropy given by Equation (11.9) is only valid when both components of the binary composite are small molecules. When one (or both) of the components is a polymer, the number of distinct states decreased because of the constraint that consecutive monomers in a polymer chain must be in the neighboring lattice sites. Therefore the mixing entropy decreases [8].

We consider a system consisting of a polymer and a solvent (low molecular weight molecule such as liquid crystal). There are n polymer molecules and m solvent molecules. The degree of polymerization (number of monomers) of the polymer is x. One solvent molecule and one monomer of the polymer occupy the same volume. We calculate the number of distinct configurations of the system, from which the mixing entropy can be calculated. Imagine putting the polymer and solvent molecules into a lattice with $N = nx + m$ lattice sites. Z is the number of the nearest neighbors in the lattice. The polymer molecules are put into the lattice one by one. When putting the $(i + 1)$th polymer molecule into the lattice, the first monomer of the polymer can be put into one of the $N - ix$ lattice sites, because ix lattice sites have been occupied by the first i polymer molecules. Therefore the number of ways of putting the monomer into the lattice is $N - ix$. The second monomer can only be put into one of the Z nearest neighboring lattice sites of the first monomer as shown in Figure 11.7. These lattice sites, however, may have been occupied by the first i polymer molecules. The number of lattice sites occupied by the first i polymers and the first monomer of the $(i + 1)$th polymer is $(ix + 1)$. The probability that a lattice site is not occupied is $(N - ix - 1)/N$. Therefore the average number of ways to put the second monomer into the lattice is $Z[(N - ix)/N]$. The third monomer can only be placed on the nearest neighboring lattice sites of the second monomer. One of these nearest neighboring sites is occupied by the first monomer. Now the probability that a lattice site is not occupied is $(N - ix - 2)/N$. Therefore the number of ways to put the third monomer into the lattice is

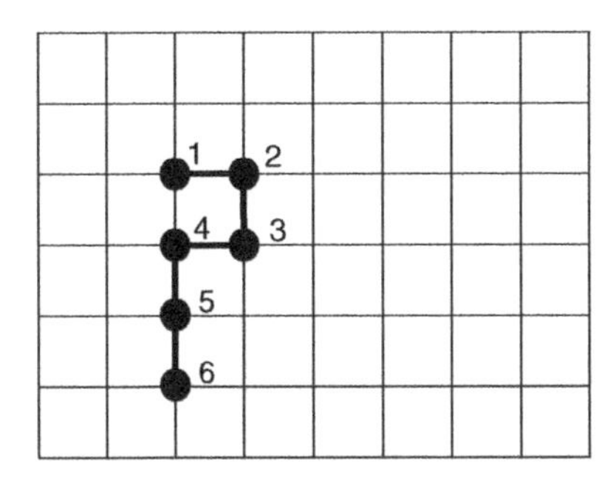

Figure 11.7 Schematic diagram showing how to put a polymer molecule into the lattice.

$(Z-1)[(N-ix-2)/N]$. In such a way the remaining monomers of the $(i+1)$th polymer can be put into the lattice. The number of ways to put the $(i+1)$th polymer into the lattice is

$$\Omega_{i+1}=\frac{1}{2}[(N-ix)]\cdot\left[Z\frac{(N-ix-1)}{N}\right]\cdot\left[(Z-1)\frac{(N-ix-2)}{N}\right]\cdot\ldots\ldots\left\{(Z-1)\frac{[N-(i+1)x+1]}{N}\right\}$$

$$=\frac{1}{2}\frac{Z}{N}\cdot\left(\frac{Z-1}{N}\right)^{x-2}\frac{(N-ix)!}{(N-ix-x)!},\qquad(11.21)$$

where the factor 1/2 takes care of the fact that either of the two end monomers can be chosen as the first monomer. The number of ways to put the n polymer molecules into the lattice is

$$G=\frac{1}{n!}\prod_{i=1}^{n}\Omega_i=\frac{1}{n!}\left[\frac{Z}{2N}\cdot\left(\frac{Z-1}{N}\right)^{x-2}\right]^n\frac{N!}{(N-x)!}\cdot\frac{(N-x)!}{(N-2x)!}\cdot\frac{(N-2x)!}{(N-3x)!}\cdot\ldots\ldots\cdot\frac{[N-(n-1)x]!}{[N-nx]!}$$

$$=\frac{1}{n!}\left[\frac{Z}{2N}\cdot\left(\frac{Z-1}{N}\right)^{x-2}\right]^n\frac{N!}{(N-nx)!}=\frac{1}{n!}\left[\frac{Z}{2N}\cdot\left(\frac{Z-1}{N}\right)^{x-2}\right]^n\frac{N!}{m!}$$

$$=\frac{1}{n!}\left[\frac{Z}{2N}\cdot\left(\frac{Z-1}{N}\right)^{x-2}\right]^n\frac{N!}{m!},\qquad(11.22)$$

where the factor $1/n!$ takes care of the fact that the polymer molecules are not distinguishable. Because the solvent molecules are also not distinguishable, there is only one way to put them into the lattice sites not occupied by the polymer molecules. Therefore the number of distinct configurations is G. When the polymer and solvent are completely phase separated, the number of distinct configurations of the n polymer molecules is (equivalently putting the polymer molecules into a lattice with nx lattice sites)

$$G_o=\frac{1}{n!}\left[\frac{Z}{2(nx)}\cdot\left(\frac{Z-1}{nx}\right)^{x-2}\right]^n\frac{(nx)!}{0!}.\qquad(11.23)$$

The total mixing entropy is

$$\Delta S_m=k_B\ln G-k_B\ln G_o$$

$$=k_B\ln\left[N^{-n(x-1)}\frac{N!}{m!}\right]-k_B\ln\left[(nx)^{-n(x-1)}(nx)!\right]$$

$$=k_B\left[-n\ln\left(\frac{nx}{N}\right)-m\ln\left(\frac{m}{N}\right)\right].\qquad(11.24)$$

The volume fractions of the polymer and solvent are $\phi_n = nx/N = nx/(m+nx) \equiv \phi$ and $\phi_m = m/N = m/(m+nx) = 1-\phi$, respectively. Therefore total mixing entropy is [8–10]

$$\Delta S_m = -k_B N\left[\frac{\phi}{x}\ln\phi + (1-\phi)\ln(1-\phi)\right]. \tag{11.25}$$

The mixing entropy per particle is

$$\Delta s_m = -k_B\left[\frac{\phi}{x}\ln\phi + (1-\phi)\ln(1-\phi)\right]. \tag{11.26}$$

The mixing interaction energy per particle is (see Equation (11.5))

$$\Delta u_m = \frac{1}{2}Z\phi(1-\phi)(2u_{nm} - u_{nn} - u_{mm}), \tag{11.27}$$

where u_{nn} is the interaction energy between two monomers, u_{mm} is the interaction energy between two solvent molecules, and u_{nm} is the interaction energy between a monomer and a solvent molecule. The mixing free energy is

$$f = \Delta u_m - T\Delta s_m = k_B T\left[\frac{Q}{T}\phi(1-\phi) + (1-\phi)\ln(1-\phi) + \frac{\phi}{x}\ln\phi\right], \tag{11.28}$$

where $Q = Z(2u_{nm} - u_{nn} - u_{mm})/2k_B$. This equation is known as the Flory–Huggins Equation. The phase diagram of a polymer/liquid crystal system is shown in Figure 11.8. At high temperatures, the entropy dominates in the free energy and the system is in a homogeneous phase. At low temperatures, the mixing entropy cannot compensate for the mixing interaction energy, and the system phase separates into two phases. The summit point of the phase boundary between the single-phase region and the two-phase region is the critical point. Thus the critical

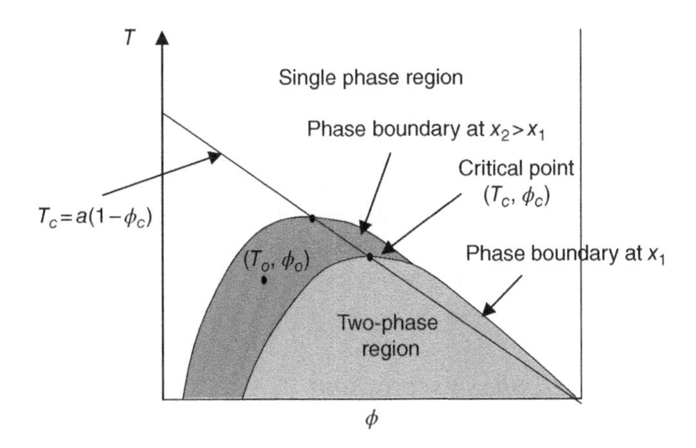

Figure 11.8 Phase diagram of the polymer/liquid crystal system.

point (T_c, ϕ_c) is an indication of where the phase boundary is. The critical point can be calculated by using

$$\left.\frac{\partial^2(f/k_BT)}{\partial\phi^2}\right|_{\phi_c,T_c} = \frac{1}{1-\phi_c} + \frac{1}{x\phi_c} - 2\frac{Q}{T_c} = 0, \tag{11.29}$$

$$\left.\frac{\partial^3(f/k_BT)}{\partial\phi^3}\right|_{\phi_c,T_c} = \frac{1}{(1-\phi_c)^2} - \frac{1}{x\phi_c^2} = 0. \tag{11.30}$$

From these two equations we have

$$\phi_c = \frac{1}{1+\sqrt{x}}, \tag{11.31}$$

$$T_C = 2Q \cdot \frac{x}{(1+\sqrt{x})^2}. \tag{11.32}$$

For large x ($\gg 1$), then $\phi_c = 1/\sqrt{x}$, $T_c = 2Q(1-2\sqrt{x}) = 2Q(1-2\phi_c)$. With increasing degree of polymerization x, the critical volume fraction of the polymer decreases and the critical temperature increases, which indicates that the phase boundary moves upward and the two-phase region becomes larger. For example, consider a system with polymer fraction ϕ_o at temperature T_o as shown in Figure 11.8. Initially the degree of polymerization is x_1, which is low. The mixing entropy is large. The system is in the homogeneous single phase. As the polymerization takes place, the degree of polymerization increases, and the entropy decreases. The two-phase region expands. When the degree of polymerization becomes x_2, the point (T_o, ϕ_o) is now in the two-phase region. The liquid crystal and the polymer phase separate. This method can be used to produce phase separation in LCPCs, which is referred to as *polymerization-induced phase separation* (PIPS) [7]. The size of the formed liquid crystal droplets in this method depends on the polymerization rate. Under faster polymerization rate, smaller droplets are formed.

This method, especially with photo-polymerization, is suitable for large volume manufacture. In photo-polymerization, monomers with acrylate or methacrylate end groups, which have a double bond, are used. Some photo-initiators are also added. Upon absorbing a photon, the photo-initiator becomes a free radical, which reacts with the acrylate group. The opened double bond reacts with another acrylate group. The chain reaction propagates until the opened double bond reacts with another free radical or open double bond, and then the polymerization stops. As an example, Norland 66 (which is a combination of acrylate monomers and photo-initiators) and E7 (EM Chemicals) are mixed with the ratio 1:1. The mixture is sandwiched between two substrates with electrodes and then cured under the irradiation of UV light of a few mW/cm^2. The polymerization rate is determined by the photo-initiator concentration and UV intensity. Smaller droplets are formed under higher UV irradiation or higher photo-initiator concentrations.

11.2.4 Solvent-induced phase separation

It is difficult to mix liquid crystals directly with polymers because polymers are usually in solid state at room temperature. A solvent can be added to mix liquid crystals with polymers. We now

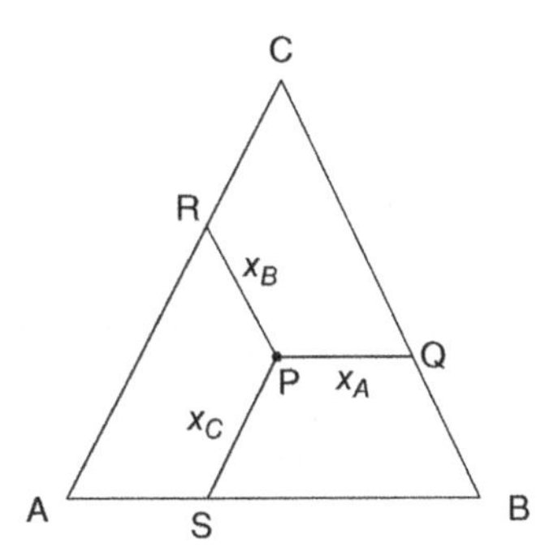

Figure 11.9 Equilateral triangle representation of ternary mixture.

consider phase separation in ternary mixtures with three components: liquid crystal (A), polymer (B), and solvent (C). We consider only the case where components A and B are immiscible in some region or fraction. Components A and C are miscible in any proportion and components B and C are also miscible in any proportion.

A ternary mixture is specified by the three fractions, x_A, x_B, and x_C, of the three components A, B, and C. Because $x_A + x_B + x_C = 1$, only the fractions of two of the components have to be specified. Therefore a ternary mixture is represented by a point on a 2-D surface. The common way to represent ternary mixtures is the *equilateral triangle diagram,* as shown in Figure 11.9. The length of each of the three sides, AB, BC, and AC, of the triangle is 1. The three corners, A, B, and C, correspond to the three components. For a composite represented by a point P in the triangle, the fraction of a component is equal to the length of the line drawn from P to the side opposite to the corner representing the component, which is parallel to one of the sides from the corner. The fraction of A is $PQ = x_A$, the fraction of B is $PR = x_B$, and the fraction of C is $PS = x_C$. It can be shown from geometry that $x_A + x_B + x_C = 1$. The equilateral triangle diagram has the following properties:

1. All the points on a line drawn from one corner of the triangle to the opposite side represent the composites in which the relative fractions of the two components represented by the other two corners remain unchanged.
2. When a mixture represented by point P phase separates into two phases represented by points U and V, the three points, P, U, and V, lie on a straight line, and point P is between points U and V. The percentages of the material in the phases U and V are equal to *PU*/*UV* and *PV*/*UV*, respectively.

The mixing free energy of a ternary mixture is given by

$$f = \frac{1}{2}p[x_A x_B(2u_{AB} - u_{AA} - u_{BB}) + x_A x_C(2u_{AC} - u_{AA} - u_{CC}) + x_B x_C(2u_{BC} - u_{BB} - u_{CC}]$$
$$+ k_B T[x_A \ln x_A + K_B x_C \ln x_B + K_B x_C \ln x_C], \quad (11.33)$$

where u_{AA} is the interaction energy between A and A, u_{BB} is the interaction energy between B and B, u_{CC} is the interaction energy between C and C, u_{AB} is the interaction energy between A and B,

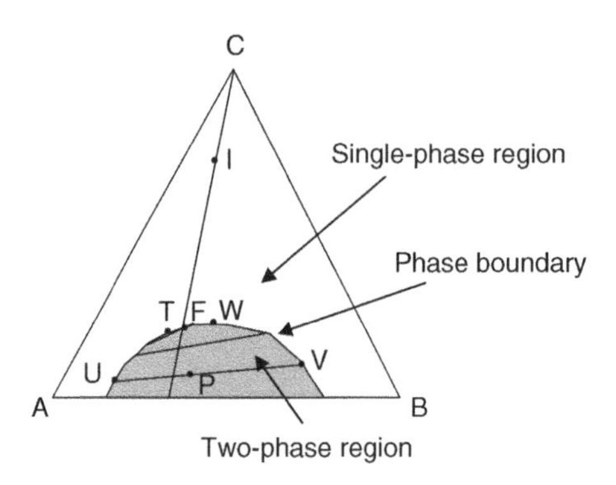

Figure 11.10 Phase diagram of the ternary mixture.

u_{AC} is the interaction energy between A and C, u_{BC} is the interaction energy between B and C, and p is the number of nearest neighbors. We consider the case where $(2u_{AB} - u_{AA} - u_{BB}) > 0$, $(2u_{AC} - u_{AA} - u_{CC}) < 0$, and $(2u_{BC} - u_{BB} - u_{CC}) < 0$. When x_C is large, $\Delta u_m < 0$, the three components are homogeneously mixed. When x_C becomes small, Δu_m becomes positive. At sufficiently low temperatures, the system phase separates into two phases. The phase diagram is shown in Figure 11.10. For a composite, represented by point P, in the two-phase region, it will phase separate into two phases represented by points U and V, which are the intersection points of a straight line (referred to as the tie line) with the phase boundary of the two-phase region. Point P lies on the straight line UV. UV is not necessarily parallel to AB, and in this case the critical point T does not coincide with the summit point W. If the initial composite has a high fraction of the solvent, it is in the homogeneous single phase; say, it is represented by point I. When the solvent is allowed to evaporate, the point representing the composite moves along the straight line CI toward the side AB, because the ratio between the fractions of the liquid crystal and polymer is fixed. When the fraction of the solvent becomes sufficiently low, the representing point moves across the two-phase region boundary at point F into the two-phase region; the system will phase separate into two phases. This method is also used to make PDLCs and is referred to as *solvent-induced phase separation (SIPS)*. The droplet size can be controlled by the solvent evaporation rate with smaller droplets formed at higher evaporation rates.

For example, 5% nematic liquid crystal E7 (Merck), 5% PMMA, and 90% chloroform are put into a closed bottle to mix. Then the homogeneous mixture, which is clear in appearance, is cast on a glass plate. The glass plate is put into a chamber with inject and vent holes. Air is blown into the chamber and then vented out at controlled rates. After a sufficient amount of chloroform has evaporated, the material changes to opaque when liquid crystal droplets begin to form. After all of the chloroform has evaporated, another glass plate is put on top of the first glass plate to sandwich the PDLC. In practice, the SIPS method is rarely used, because it is difficult to control the solvent evaporation rate. This method is, however, very useful in preparing the initial mixtures for thermally induced phase separation.

11.2.5 Encapsulation

PDLCs formed by encapsulation are also called emulsion-based PDLCs [1,11,12]. In this method, a nematic liquid crystal, water, and a water-dissolvable polymer such as polyvinyl alcohol, are put into a container. Water and the polymer dissolve each other to form a viscous solution. This aqueous solution does not dissolve the liquid crystal. When this system is stirred by a propeller blade at a sufficiently high speed, micron-size liquid crystal droplets are formed. Smaller liquid crystal droplets form at higher stirring speeds. Then the emulsion is coated on a substrate and the water is allowed to evaporate. After evaporation of the water, a second substrate is laminated to form PDLC devices.

11.3 Scattering Properties of LCPCs

The scattering of LCPCs is caused by the spatial variation of refractive index in the materials [13]. It is similar to the scattering of clouds in which water droplets are dispersed in air, or milk in which fat particles are dispersed in water. In PDLCs, the scattering is due to the refractive index mismatch between the liquid crystal and the polymer. In PSLCs, the scattering is due to the refractive index mismatch between the liquid crystal and the polymer network as well as that between liquid crystal domains. A precise calculation of the scattering of LCPCs is very difficult because of the birefringence of the liquid crystals, the dispersion in domain size, and the irregularity of domain shape. Here we will include only some qualitative discussion of the Rayleigh–Gans scattering theory.

When a light is propagating in a medium, the electric field of the light induces a dipole moment at each point, which oscillates with the frequency of the light [13]. Each oscillating dipole radiates light in all directions, and the net electric field at any point is the vector sum of the fields produced by all the dipole radiators, as schematically shown in Figure 11.11. The incident light is collimated and the electric field at the source point $\vec{r}$ is

$$\vec{E}_{in}(\vec{r},t) = \vec{E}_o e^{-i\vec{K}_o\cdot\vec{r}+i\omega t}. \tag{11.34}$$

The total induced dipole moment in the volume element d^3r is given by

$$d\vec{P}(\vec{r},t) = \overleftrightarrow{\alpha}\cdot\vec{E}_{in} = \overleftrightarrow{\alpha}\cdot\vec{E}_o e^{-i\vec{K}_o\cdot\vec{r}+i\omega t}, \tag{11.35}$$

where $\overleftrightarrow{\alpha}$ is the polarizability and $\vec{K}_o$ is the wavevector of the incident light. $\overleftrightarrow{\alpha}$ is related to the dielectric tensor by $\overleftrightarrow{\alpha}(\vec{r}) = \varepsilon_o\left[\overleftrightarrow{\varepsilon}(\vec{r}) - \overleftrightarrow{I}\right]$, where $\overleftrightarrow{I}$ is the unit matrix. $\vec{K}_o$ is related to the

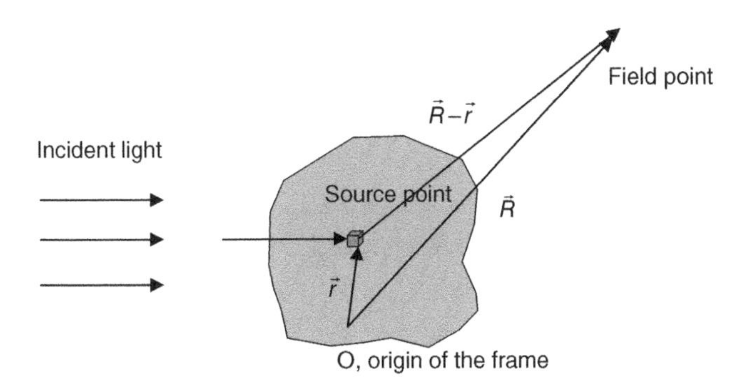

Figure 11.11 Schematic diagram showing the scattering of a medium.

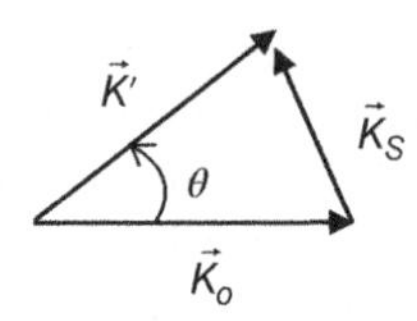

Figure 11.12 Schematic diagram showing the wavevectors of the incident and scattered light.

frequency ω by $\vec{K}_o = K_o\hat{k}_o = (\omega/c)\hat{k}_o = (2\pi/\lambda)\hat{k}_o$, where c and λ are the speed and wavelength of light in vacuum, respectively, and $\hat{k}_o$ is a unit vector along the incident direction. The wavevector of the scattered light is $\vec{K}' = K_o\hat{k}'$, as shown in Figure 11.12. Then

$$\vec{K}_s = \vec{K}' - \vec{K}_o, \quad K_s = 2K_o\sin(\theta/2), \tag{11.36}$$

where θ is the scattering angle.

The scattered field radiated by the dipole moment in the volume element $d^3\vec{r}$ is given by

$$\begin{aligned} d\vec{E}_s\left(\vec{R},t\right) &= \frac{\hat{k}' \times \left[\hat{k}' \times d\vec{P}(\vec{r},t)\right]}{4\pi\varepsilon_o c^2\left|\vec{R}-\vec{r}\right|} e^{-i\vec{K}'\cdot\left(\vec{R}-\vec{r}\right)} \\ &= \frac{\hat{k}' \times \left\{\hat{k}' \times \left[-\omega^2 \overleftrightarrow{\alpha}(\vec{r})\cdot\vec{E}_o(\vec{r},t)d^3r\right]\right\}}{4\pi\varepsilon_o c^2\left|\vec{R}-\vec{r}\right|} e^{i\omega t - i\vec{K}_o\cdot\vec{r}} e^{-i\vec{K}'\cdot\left(\vec{R}-\vec{r}\right)} \\ &= \frac{-\omega^2\hat{k}' \times \left\{\hat{k}' \times \left[\overleftrightarrow{\alpha}(\vec{r})\cdot\vec{E}_o\right]\right\}}{4\pi\varepsilon_o c^2\left|\vec{R}-\vec{r}\right|} e^{i\vec{K}_s\cdot\vec{r}} e^{i\omega t - i\vec{K}'\cdot\vec{R}} d^3r. \end{aligned} \tag{11.37}$$

For far field point, $R \gg r$, $\left|\vec{R}-\vec{r}\right| \approx R$. Equation (11.37) becomes

$$d\vec{E}_s\left(\vec{R},t\right) = \frac{-K_o^2\hat{k}' \times \left\{\hat{k}' \times \left[\overleftrightarrow{\alpha}(\vec{r})\cdot\vec{E}_o\right]\right\}}{4\pi\varepsilon_o R} e^{i\vec{K}_s\cdot\vec{r}} e^{i\omega t - i\vec{K}'\cdot\vec{R}} d^3r. \tag{11.38}$$

The total scattered field is given by

$$\begin{aligned} \vec{E}_s\left(\vec{R},t\right) &= -\frac{K_o^2}{4\pi\varepsilon_o R} e^{i\omega t - i\vec{K}'\cdot\vec{R}} \int \hat{k}' \times \left\{\hat{k}' \times \left[\overleftrightarrow{\alpha}(\vec{r})\cdot\vec{E}_o\right]\right\} e^{i\vec{K}_s\cdot\vec{r}} d^3r \\ &= -\frac{\pi}{\varepsilon_o R\lambda^2} V e^{i\omega t - i\vec{K}'\cdot\vec{R}} \hat{k}' \times \left\{\hat{k}' \times \left[\overleftrightarrow{\alpha}\left(\vec{K}_s\right)\cdot\vec{E}_o\right]\right., \end{aligned} \tag{11.39}$$

where $\overleftrightarrow{\alpha}\left(\vec{K}_s\right)=\frac{1}{V}\int\overleftrightarrow{\alpha}\left(\vec{r}\right)e^{i\vec{K}_s\cdot\vec{r}}d^3r$ is the Fourier component of the polarizability and V is the volume of the scattering medium. The intensity of the the scattered light

$$I_S=\left|\vec{E}_s\right|^2\propto\frac{1}{\lambda^4}\left|\frac{1}{4\pi\varepsilon_o}\overleftrightarrow{\alpha}\left(\vec{K}_s\right)\right|^2=\frac{1}{\lambda^4}\left|\frac{1}{4\pi}\overleftrightarrow{\varepsilon}\left(\vec{K}_s\right)\right|^2. \tag{11.40}$$

The factor $1/\lambda^4$ describes the wavelength-dependence of the Rayleigh light scattering, which is responsible for the blue background of sky. When sunlight propagates through the atmosphere, blue light is scattered more than red light by air density fluctuation due to the factor $1/\lambda_4$. The factor $\left|\overleftrightarrow{\varepsilon}\left(\vec{K}_s\right)\right|$ describes the wavelength-dependence of the light scattering due to the structure of the medium. If the medium consists of many domains with linear size around D, the Fourier component of the refractive index peaks at $2\pi/D$, which means that light with wavelength around D is scattered strongly.

We define the coordinate for the incident light in such a way that the z axis is parallel to the incident direction, and the x axis is in the plane defined by $\vec{K}_o$ and $\vec{K}'$ and the coordinate for the scattered light in such a way that the z' axis is parallel to the scattering direction and the x' axis is also in the plane defined by $\vec{K}_o$ and $\vec{K}'$, as shown in Figure 11.13. If the incident field $\vec{E}_o$ is linearly polarized in a direction making an angle α with the x axis, in matrix form, $\vec{E}_o=\begin{pmatrix}E_{lo}\\E_{ro}\end{pmatrix}=E_o\begin{pmatrix}\cos\alpha\\\sin\alpha\end{pmatrix}$ defined in frame xyz. The scattered field is $\vec{E}_s=\begin{pmatrix}E_{ls}\\E_{rs}\end{pmatrix}$ defined in frame $x'y'z'$. Rewrite Equation (11.39) in matrix form

$$\vec{E}_s=-\frac{K_o^2}{4\pi\varepsilon_oR}Ve^{-i\omega t+i\vec{K}'\cdot\vec{R}}\hat{k}'\times\left\{\hat{k}'\times\left[\overleftrightarrow{\alpha}\left(\vec{K}_s\right)\cdot\vec{E}_o\right]\right\}\equiv\frac{1}{iK_oR}e^{-i\omega t+i\vec{K}'\cdot\vec{R}}\overleftrightarrow{S}\cdot\vec{E}_o, \tag{11.41}$$

where $\overleftrightarrow{S}=\overleftrightarrow{S}(\theta,\alpha)=\begin{pmatrix}S_{ll}&S_{lr}\\S_{rl}&S_{rr}\end{pmatrix}$ is the scattering matrix. The two components of the differential scattering cross section are

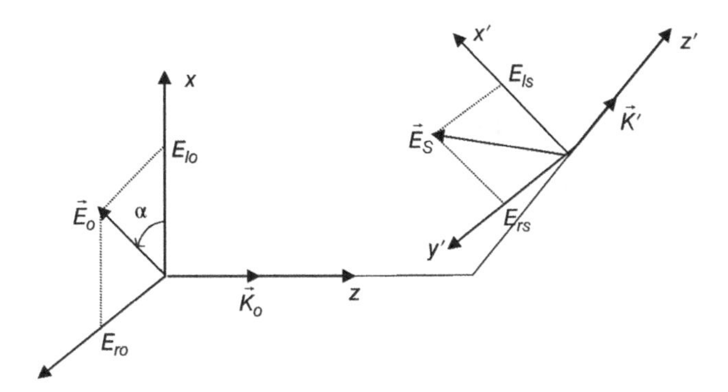

Figure 11.13 The coodinates for the incident and scattered light.

$$\left(\frac{d\sigma}{d\Omega}\right)_l = \frac{|E_{ls}|^2}{|E_o|^2}R^2 = \frac{1}{K_o^2}|S_{ll}\cos\alpha + S_{lr}\sin\alpha|^2, \tag{11.42}$$

$$\left(\frac{d\sigma}{d\Omega}\right)_r = \frac{|E_{rs}|^2}{|E_o|^2}R^2 = \frac{1}{K_o^2}|S_{lr}\cos\alpha + S_{rr}\sin\alpha|^2. \tag{11.43}$$

The total differential scattering section is $\frac{d\sigma}{d\Omega} = \left(\frac{d\sigma}{d\Omega}\right)_l + \left(\frac{d\sigma}{d\Omega}\right)_r$.

As an example, let us first consider the scattering of an isotropic spherical droplet with the refractive index n and radius a. $\overleftrightarrow{\alpha}\left(\vec{K}_S\right)\cdot\vec{E}_o = \alpha\left(\vec{K}_S\right)\vec{E}_o$:

$$\begin{aligned}
E_{ls} &= \hat{x}'\cdot\left\{-i\frac{1}{4\pi\varepsilon_o}K_o^3V\alpha\left(\vec{K}_s\right)\hat{k}'\times\left[\hat{k}'\times\vec{E}_o\right]\right\} \\
&= -i\frac{1}{4\pi\varepsilon_o}K_o^3V\alpha\left(\vec{K}_s\right)\hat{x}'\cdot\left[\hat{k}'\left(\hat{k}'\cdot\vec{E}_o\right)-\vec{E}_o\right] \\
&= i\frac{1}{4\pi\varepsilon_o}K_o^3V\alpha\left(\vec{K}_s\right)\hat{x}'\cdot\vec{E}_o \\
&= i\frac{1}{4\pi\varepsilon_o}K_o^3V\alpha\left(\vec{K}_s\right)(\cos\theta\hat{x} + \cos\theta\hat{z})\cdot(E_{lo}\hat{x} + E_{ro}\hat{y}) \\
&= i\frac{1}{4\pi\varepsilon_o}K_o^3V\alpha\left(\vec{K}_s\right)\cos\theta E_{lo}
\end{aligned} \tag{11.44}$$

$$\begin{aligned}
E_{ls} &= \hat{y}'\cdot\left(-i\frac{1}{4\pi\varepsilon_o}K_o^3V\alpha\left(\vec{K}_s\right)\hat{k}'\times\left[\hat{k}'\times\vec{E}_o\right]\right) \\
&= i\frac{1}{4\pi\varepsilon_o}K_o^3V\alpha\left(\vec{K}_s\right)\hat{y}\cdot(E_{lo}\hat{x} + E_{ro}\hat{y}) \\
&= i\frac{1}{4\pi\varepsilon_o}K_o^3V\alpha\left(\vec{K}_s\right)E_{ro}
\end{aligned} \tag{11.45}$$

Therefore the scattering matrix is

$$\overleftrightarrow{S} = i\frac{1}{4\pi\varepsilon_o}K_o^3V\alpha\left(\vec{K}_s\right)\begin{pmatrix}\cos\theta & 0 \\ 0 & 1\end{pmatrix}. \tag{11.46}$$

Now we calculate the Fourier component of the polarizability:

$$\begin{aligned}
\alpha\left(\vec{K}_s\right) &= \frac{1}{V}\int_{\text{whole space}}\alpha(\vec{r})e^{-i\vec{K}_s\cdot\vec{r}}d^3r \\
&= \frac{1}{V}\int_{\text{inside droplet}}\alpha(\vec{r})e^{-i\vec{K}_s\cdot\vec{r}}d^3r + \frac{1}{V}\int_{\text{outside droplet}}\alpha(\vec{r})e^{-i\vec{K}_s\cdot\vec{r}}d^3r
\end{aligned}$$

$$=\frac{1}{V}\int_{\text{whole space}}\alpha_{out}e^{-i\vec{K}_s\cdot\vec{r}}d^3r+\frac{1}{V}\int_{\text{inside droplet}}(\alpha_{in}-\alpha_{out})e^{-i\vec{K}_s\cdot\vec{r}}d^3r$$

The first integral is zero. $\alpha_{in}=\varepsilon_o(\varepsilon-1)=\varepsilon_o(n^2-1)$. The medium outside the droplet is also isotropic with the refractive index n_o. $\alpha_{out}=\varepsilon_o\left(n_o^2-1\right)$. Hence

$$\alpha\left(\vec{K}_s\right)=\frac{1}{V}\varepsilon_o(2\bar{n}\Delta n)\int_{\text{inside droplet}}e^{-i\vec{K}_s\cdot\vec{r}}d^3r, \tag{11.47}$$

where $\bar{n}=(n-n_o)$ and $\Delta n=(n-n_o)$. Define $Q(\theta)=\frac{1}{V}\int_{\text{inside droplet}}e^{-i\vec{K}_s\cdot\vec{r}}d^3r$. For the integration, we use a polar coordinate with $\vec{K}_s$ in the polar direction:

$$Q(\theta)=\frac{2\pi}{V}\int_0^{\pi}d\beta\int_0^{a}e^{-iK_s r\cos\beta}\sin\beta r^2dr=\frac{4\pi}{VK_s^3}[\sin(aK_s)-aK_s\cos(aK_s)] \tag{11.48}$$

The differential scattering cross section for unpolarized incident light is given by

$$\frac{d\sigma}{d\Omega}=\frac{K_o^6}{K_o^2}\frac{1}{16\pi^2}(2\bar{n}\Delta\mathrm{n})^2\left(\cos^2\theta<\cos^2\alpha>+<\sin^2\alpha>\right)\left\{\frac{4\pi}{K_s^3}[\sin(aK_s)-aK_s\cos(aK_s)]\right\}^2. \tag{11.49}$$

We know that $K_s=2K_o\sin(\theta/2)$. Let $A=aK_o$, then

$$\frac{d\sigma}{d\Omega}=\pi a^2(2\bar{n}\Delta\mathrm{n})^2\frac{(\cos^2\theta+1)}{128\pi A^2}\left\{\frac{\sin[2A\sin(\theta/2)]-2A\sin(\theta/2)\cos[2A\sin(\theta/2)]}{\sin^3(\theta/2)}\right\}^2. \tag{11.50}$$

The light scattered in forward direction is given by

$$\sigma_{\text{forward}}=2\pi\int_0^{\pi/2}\frac{d\sigma}{d\Omega}\sin\theta d\theta$$

$$=\pi a^2(2\bar{n}\Delta\mathrm{n})^2\int_0^{\pi/2}\frac{(\cos^2\theta+1)}{64A^2}\left\{\frac{\sin[2A\sin(\theta/2)]-2A\sin(\theta/2)\cos[2A\sin(\theta/2)]}{\sin^3(\theta/2)}\right\}^2\sin\theta d\theta. \tag{11.51}$$

The light scattered in backward direction is given by

$$\sigma_{\text{backward}} = 2\pi \int_{\pi/2}^{\pi} \frac{d\sigma}{d\Omega} \sin\theta d\theta$$

$$= \pi a^2 (2\bar{n}\Delta \mathrm{n})^2 \int_{\pi/2}^{\pi} \frac{(\cos^2\theta + 1)}{64A^2} \left\{ \frac{\sin[2A\sin(\theta/2)] - 2A\sin(\theta/2)\cos[2A\sin(\theta/2)]}{\sin^3(\theta/2)} \right\}^2 \sin\theta d\theta. \tag{11.52}$$

The scattering cross sections of the materials with $n_o = 1.5$ and $n = 1.7$ are plotted in Figure 11.14, where the unit of the vertical axis is πa^2. When the droplet size a is smaller than the wavelength λ ($A = 2\pi a/\lambda$), the forward and backward scattering cross sections are about the same. When the droplet size is larger than the wavelength, most of the scattered light is in forward directions.

The scattering of liquid crystal droplets can be calculated in the same way except that the dielectric tensor has to be used [14–16]. The calculation is more complicated and is not presented here. Readers interested in the detailed calculation of the scattering of PDLCs are referred to the papers published by Zumer, Kelly, et al. [14,17–19]. Most of the incident light is scattered in the forward direction. The formulation presented in this section is called Rayleigh–Gans scattering, which uses the following three assumptions: (1) $|n/n_o - 1| \ll 1$; therefore refraction at the droplet interface can be neglected, (2) $2K_o a|n - n_o| \ll 1$, and (3) $\sigma \ll 1$; therefore there is no multiple scattering inside the droplet and the incident light intensity at any point inside the medium is the same.

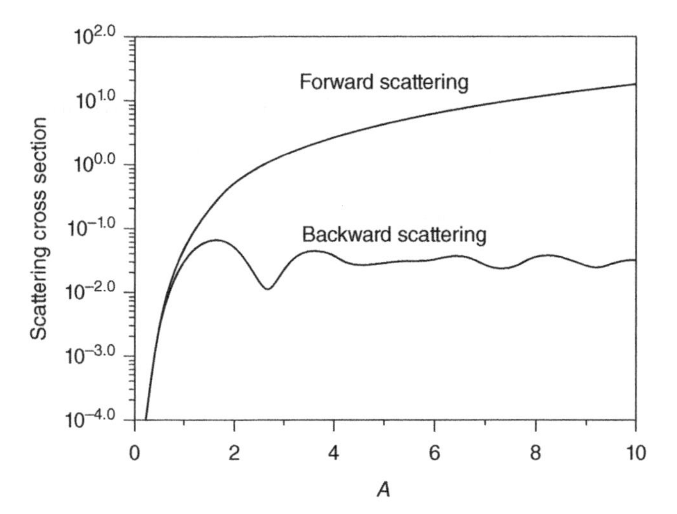

Figure 11.14 The scattering cross sections vs. A.

11.4 Polymer Dispersed Liquid Crystals

In polymer dispersed liquid crystals (PDLCs), the concentrations of polymer and liquid crystal are comparable. Liquid crystals form isolated droplets embedded in polymers. We will consider first the liquid crystal director configuration inside droplets, and then the electro-optical properties of PDLCs, as well as their applications.

11.4.1 Liquid crystal droplet configurations in PDLCs

The liquid crystal dispersed in the polymer of a PDLC can be in one of the many possible liquid crystal phases such as nematic, cholesteric, smectic-A, and smectic-C* [4]. The most common PDLC is polymer-dispersed nematic liquid crystal, which is one we will discuss here. Inside a nematic droplet, the director configuration is determined by the droplet shape and size, the anchoring condition on the droplet surface and externally applied field as well as the elastic constants of the liquid crystal. There are four main types of nematic droplets, as shown in Figure 11.15. When the anchoring condition is tangential, there are two types of droplets. One is the bipolar droplet, as schematically shown in Figure 11.15(a) [20,21] and the other is the toroidal droplet, Figure 11.15(b) [1,22]. When the anchoring condition is perpendicular, there are also two types of droplets: the radial droplet, Figure 11.15(c) [23], and the axial droplet, Figure 11.15(d) [23]. When a droplet is bigger than 5 μm in diameter, it is possible to identify the droplet configuration using an optical microscope.

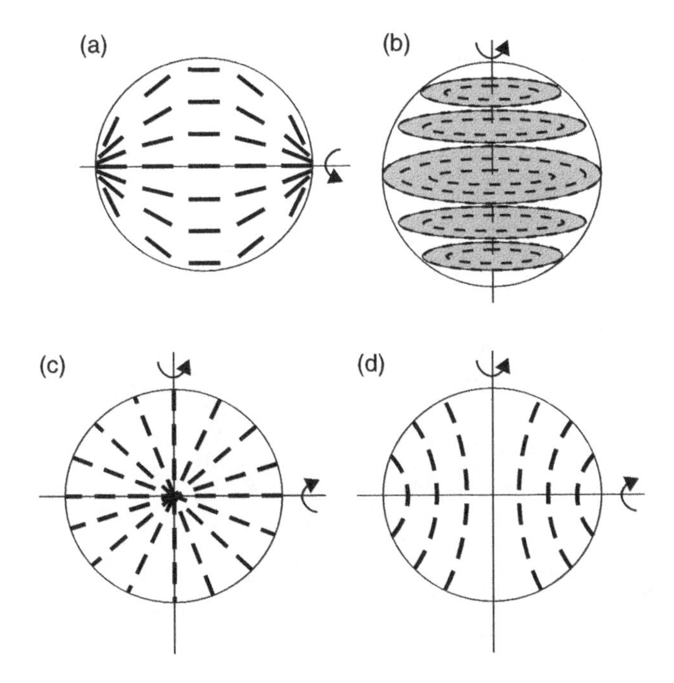

Figure 11.15 Liquid crystal director configurations confined in droplets in PDLCs: (a) bipolar droplet, (b) toriodal droplet, (c) radial droplet, and (d) axial droplet.

In the bipolar droplet, the rotation symmetry axis is referred to as the bipolar axis. The droplet director $\vec{N}$ is defined as a unit vector along the bipolar axis. In the bipolar droplet with strong anchoring, there are two point defects at the ends of the diameter along the bipolar axis. The director field on a plane containing the bipolar axis is shown in Figure 11.15(a). The liquid crystal director is tangential to the circle along the circumference, and parallel to the bipolar axis along the diameter. At other places inside the droplet, the director is oriented in such a way that the total free energy is minimized. There is a rotational symmetry of the director around the bipolar axis. A typical optical microphotograph of a sample with bipolar droplets under a microscope with crossed polarizers is shown in Figure 11.16(a). The dark splashes are the region where the liquid crystal director is parallel or perpendicular to the polarizers. The elastic deformations involved are splay and bend. When there is no externally applied field, the orientation of $\vec{N}$ is arbitrary for perfect spherical droplets. In practice, the droplets are usually deformed. The deviation of the droplet from spherical shape results in a particular orientation of $\vec{N}$ [24,25]. Preferred deformed bipolar droplets can be made by applying stresses or external fields during the formation of the droplets. When a sufficiently high external electric field is applied, the liquid crystal ($\Delta\varepsilon > 0$) is reoriented with the bipolar axis parallel to the field.

In the toroidal droplet, the liquid crystal director is aligned along concentric circles on planes perpendicular to a diameter, as shown in Figure 11.15(b). There is a line defect along the diameter of the droplet. There is a rotational symmetry around the defect line. The bend elastic deformation is the only one involved. Toroidal droplets exist when bend elastic constant is smaller than the splay elastic constant; otherwise the droplets take the bipolar configuration. Toroidal droplets rarely exist because for most liquid crystals the bend elastic constant is usually larger than the splay elastic constant. Nevertheless, toroidal droplets have been

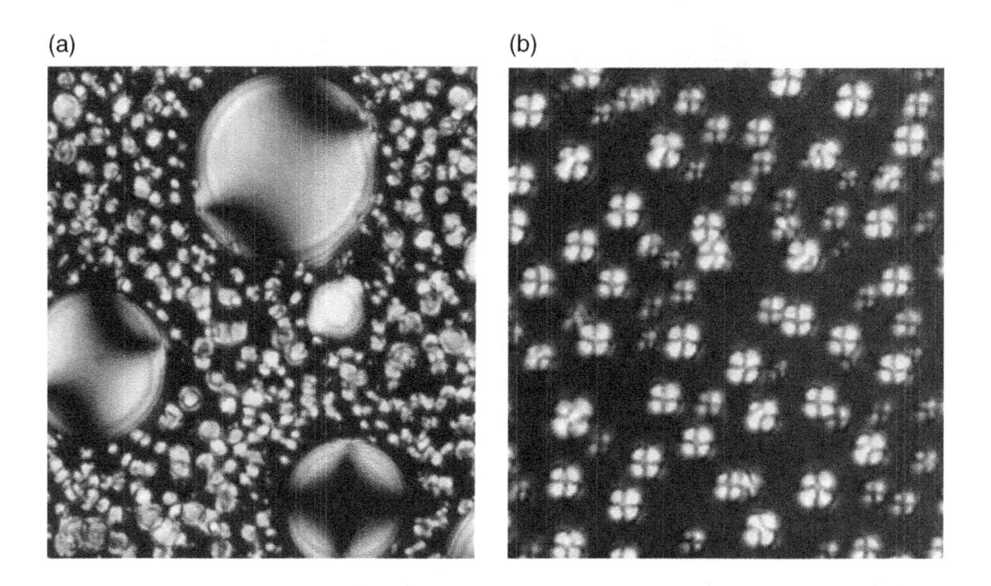

Figure 11.16 Microphotographs of PDLCs: (a) bipolar droplets, (b) radial droplets.

reported [22]. When the droplet is sufficiently large or an external field is applied, the director near the defect line will escape in the direction parallel to the symmetry axis so that the total free energy is reduced.

In the radial droplet, the director everywhere is along the radial direction [23], and there is a point defect in the center of the droplet. The director configuration on a plane cut through the droplet center is shown in Figure 11.15(c). There is rotational symmetry around any diameter of the droplet. Figure 11.16(b) shows an optical microphotograph of a sample with radial droplets with crossed polarizers. The dark cross is the region where the liquid crystal director is parallel or perpendicular to the polarizers. One striking feature of radial droplets is that they have only one texture, which does not change when the sample is rotated with the microscope stage. Note that the dark cross at the center is narrow, because only in a small region the director is parallel to the propagation direction of the light. Splay elastic deformation is the only one present in the radial droplet.

In the axial droplet, there is a line defect along an equator. The director on the plane perpendicular to the equator is shown in Figure 11.15(d) [21]. There is a rotational symmetry around the diameter perpendicular to the equator plane. The texture of the axial droplet with the symmetry axis parallel to the propagation direction of the light is similar to that of the radial droplet, except that the dark cross at the center is wider, indicating that the director is aligned closer to the symmetry axis in the center. Both splay and bend elastic deformations exist in the axial droplet. The splay elastic energy of the axial droplet is lower than that of a radial droplet if all material parameters and droplet size are identical. The axial droplet is more stable than the radial droplet if the anchoring is weak or the droplet is small or an external field is applied. If the liquid crystal is in the radial configuration at zero electric field, when a sufficiently high electric field is applied it can be switched to the axial configuration with the symmetry axis parallel to the field.

Besides the four droplet configurations discussed above, other droplet configurations could exist under appropriate conditions. For example, there is a twisted bipolar droplet where the splay and bend elastic energy is reduced by introducing twist deformation. The point defect in the radial droplet could escape away from the center in an effort to reduce the total free energy.

11.4.2 Switching PDLCs

Devices from PDLCs make use of two states of the liquid crystal droplets. One of the states is the field-off state in which either the droplet director orients randomly throughout the PDLCs or the liquid crystal director inside the droplet orients in all directions. The liquid crystal is in the directoral configuration in which the sum of the elastic and surface energies is minimized. The other state is the field-on state in which the droplet director is aligned uniformly along the applied field. If the droplets are bipolar, then in the field-off state the bipolar axes (droplet director) of the droplets orient randomly throughout the cell, while in the field-on state, the bipolar axes of the droplets are aligned along the applied field. If the droplets are axial droplets, in the field-off state the symmetry axes (droplet director) of the droplets orient randomly throughout the cell; in the field-on state the symmetry axes of the droplets are aligned parallel to the applied field. If the droplets are radial, in the field-off state there is no droplet director; in the field-on state, they are switched to the axial droplet with the symmetry axis parallel to the applied field.

In order to have the transition from the first state to the second in all those cases, the applied field must be sufficiently high so that the decrease of the electric energy can compensate for the increase of the elastic energy and surface energy. The threshold field, above which the applied field can produce the transition, depends on: (1) droplet size, (2) droplet shape, (3) anchoring condition, and (4) the material parameters, such as elastic constants and dielectric anisotropy, of the liquid crystal [3,24,25].

We discuss qualitatively the effects of droplet size on the drive voltage. Let us consider a PDLC with droplet size D. In the switching of a PDLC droplet, there are three energies involved: elastic energy, surface energy, and electric energy. The elastic energy density f_{elas} is proportional to $K(1/D)^2$, and the total elastic energy F_{elas} of the droplet is proportional to $D^3 \cdot K(1/D)^2 = KD$, where K is the elastic constant. The surface energy density is proportional to the anchoring strength W. The surface extrapolation length is defined by $d_e = K/W$. The total surface energy of the droplet is proportional to $WD^2 = (K/d_e)D^2 = (D/d_e)KD$. The total electric energy of the droplet is proportional to $-\Delta\varepsilon E^2 D^3$. Depending on the droplet size and anchoring strength, there are three possible cases.

11.4.2.1 $D \gg d_e$, corresponding to large droplet or strong anchoring

The liquid crystal on the surface of the droplet orients along the anchoring easy direction; the change of the surface energy in the switching is negligible. Because the surface of the droplet is curved, the director inside the droplet cannot be uniform, and therefore elastic energy KD dominates. In the field-off state, the liquid crystal inside the droplet is in the state where the elastic energy is minimized. In the field-on state, the elastic energy is higher and the electric energy is lower. At the threshold field, E_c, the decrease of the electric energy, $-\Delta F_{electric}$, can compensate for the increase of the elastic energy, $\Delta F_{elastic}$:

$$\Delta F_{elastic} = a_{elastic} KD \sim \Delta F_{electric} = a_{electric} \Delta\varepsilon E_c^2 D^3, \tag{11.53}$$

where $a_{elastic}$ and $a_{electric}$ are constants. The threshold field is given by

$$E_c = \left(\frac{a_{elastic}}{a_{electric}\Delta\varepsilon}\right)^{1/2} \frac{1}{D} \propto \frac{1}{D}. \tag{11.54}$$

The threshold field is sometimes referred to as the switching field, and the corresponding voltage is referred to as the switching voltage. When the droplet size is varied in a relatively small region, the switching field will change in such a way that $E_c D$ remains fixed.

11.4.2.2 $D \ll d_e$, corresponding to small droplet or weak anchoring

The liquid crystal inside the droplet is uniformly oriented along the direction $\vec{N}$ at the expense of the surface energy. The change of elastic energy in the switching is small and negligible, and the surface energy dominates. In the field-off state, $\vec{N}$ is in the direction such that the surface energy is minimized. In the field-on state, $\vec{N}$ reorients parallel to the applied field. The surface energy is higher, and the electric energy is lower. At the threshold field, E_c, the decrease of the electric energy, $-\Delta F_{electric}$, can compensate for the increase of the surface energy, $\Delta F_{elastic}$:

$$\Delta F_{surface} = a_{surface} W D^2 \sim \Delta F_{electric} = a_{electric} \Delta\varepsilon E_c^2 D^3, \tag{11.55}$$

where $a_{surface}$ is a constant. The threshold field is given by

$$E_c = \left(\frac{a_{surface} W}{a_{electric} \Delta\varepsilon}\right)^{1/2} \frac{1}{\sqrt{D}} \propto \frac{1}{\sqrt{D}}. \tag{11.56}$$

When the droplet size is varied in a relatively small region, the switching field will change in such a way that $E_c \sqrt{D}$ remains fixed.

11.4.2.3 $D \sim d_e$, corresponding to medium droplet size and medium anchoring strength

In this case, the liquid crystal in the bulk of the droplet is not aligned uniformly along one direction, nor is the liquid crystal on the surface of the droplet aligned along the anchoring easy direction. Both elastic energy and surface energy are involved. At the threshold field, the decrease of the electric energy must compensate for the increase of the elastic energy and surface energy:

$$\Delta F_{surface} + \Delta F_{surface} = a_{elastic} K D + a_{surface} W D^2 \sim \Delta F_{electric} = a_{electric} \Delta\varepsilon E_c^2 D^3 \tag{11.57}$$

$$E_c = \left(\frac{a_{elastic}}{a_{electric} \Delta\varepsilon} \frac{1}{D^2} + \frac{a_{surface} W}{a_{electric} \Delta\varepsilon} \frac{1}{D}\right)^{1/2} \tag{11.58}$$

The droplet size dependence of the switching field can be used to obtain information on the droplet size and anchoring strength.

In PDLCs with strong anchoring, the drive voltage is usually high. One example is the PDLC made from PVA and ZLI2061 (from Merck) using the NCAP method [26]. The drive voltage V_d is approximately linearly proportional to $1/D$, as shown in Figure 11.17 where the cell thickness is 13 μm.

In PDLCs with weak anchoring, the drive voltage is usually low. One example is the PDLC made from E7 (from Merck) and NOA65 (Norland Optical Adhesive) by photo-polymerization-induced phase separation [27]. The cell thickness is 12 μm. The square of the drive voltage V_d is approximately linearly proportional to $1/D$, as shown in Figure 11.18.

11.4.3 Scattering PDLC devices

We now consider the working principle of scattering PDLC devices. As an example, we will look at a PDLC with bipolar droplets. At zero field, the droplet director $\vec{N}$ is oriented randomly throughout the cell, as shown in Figure 11.19(a). For normal incident light with linear

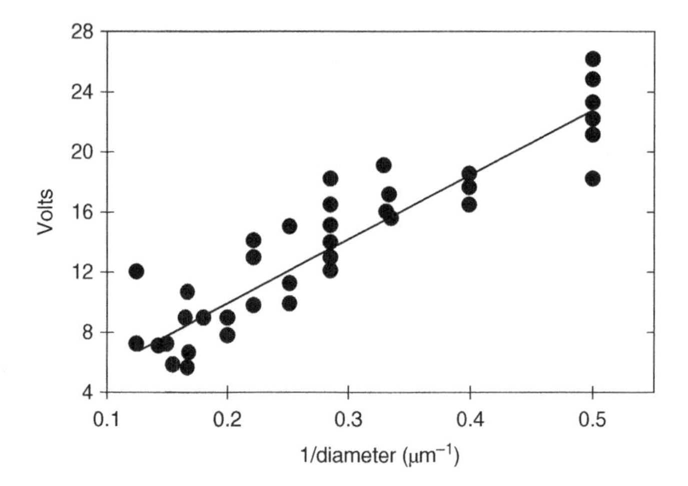

Figure 11.17 Drive voltage of the PDLC with strong anchoring as a function of the droplet sizes.

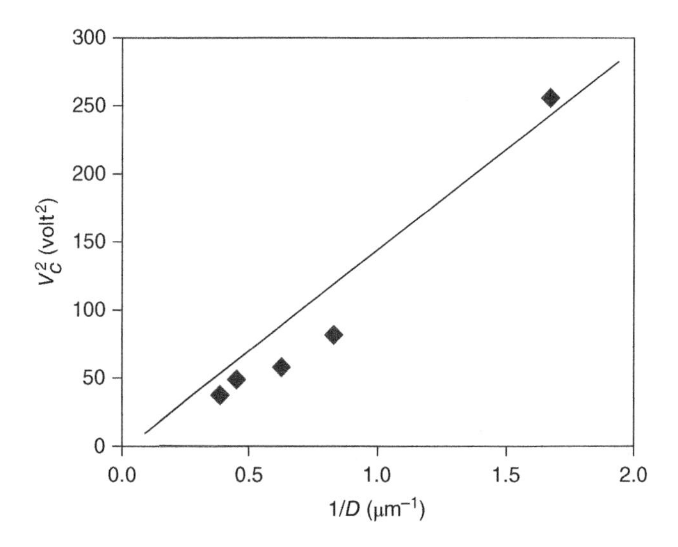

Figure 11.18 Square of the drive voltage of the PDLC with weak anchoring as a function of the droplet size.

polarization in the plane defined by the propagation direction of the light and the bipolar axis, when it propagates through a droplet whose droplet director makes an angle θ with the normal of the cell, it encounters a refractive index given approximately by

$$n(\theta) = \frac{n_{//}n_{\perp}}{\left(n_{//}^2\cos^2\theta + n_{\perp}^2\sin^2\theta\right)^{1/2}}, \tag{11.59}$$

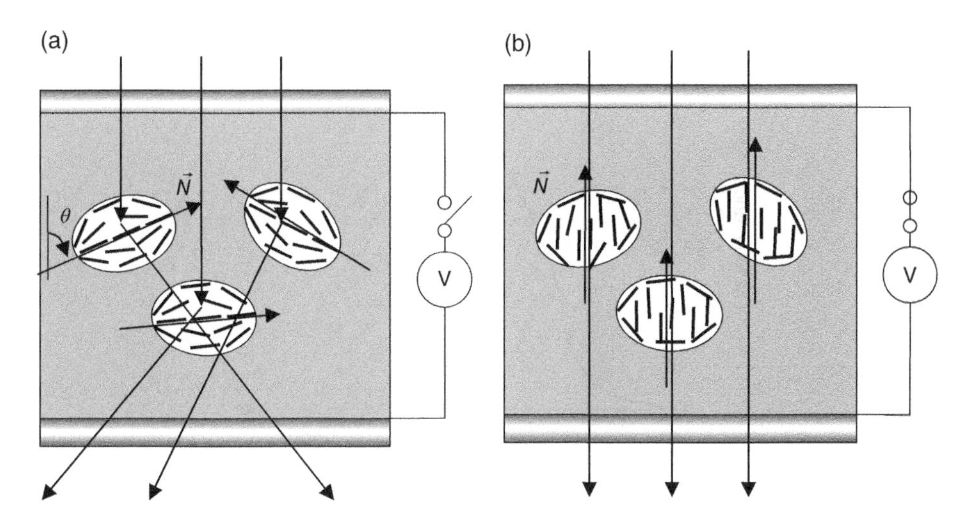

Figure 11.19 Orientation of liquid crystal inside the PDLC droplets in the field-off and field-on states. Drzaic 1996. Reproduced with permission from the American institute of Physics.

where $n_{//}$ and $n_{\perp}$ are the refractive indices for light polarized parallel and perpendicular to the liquid crystal director, respectively. The (isotropic) polymer is chosen such that its refractive index n_p is the same as $n_{\perp}$. The light encounters a different refractive index when it propagates through the polymer and the liquid crystal droplet. Therefore the PDLC is a non-uniform optical medium and the light is scattered when it goes through it. When a sufficiently high electric voltage is applied across the PDLC cell, the droplets are reoriented with their droplet director $\vec{N}$ parallel to the normal of the cell, as shown in Figure 11.19(b): $\theta = 0$ for all the droplets. Now, when the normal incident light propagates through the droplets, it encounters the refractive index $n_{\perp}$, which is the same as the encountered refractive index when it propagates through the polymer. The PDLC is a uniform optical medium for the light. Therefore the light goes through the PDLC without scattering. The PDLC discussed here is a normal-mode light shutter in the sense that it is opaque in the field-off state and transparent in the field-on state.

A typical voltage–transmittance curve of PDLCs is shown in Figure 11.20 [28]. At 0 V, the material is in the scattering state, and the transmittance is low. As the applied voltage is increased, the droplet director is aligned toward the cell normal direction and the transmittance increases. The drive voltage (at which the transmittance reaches 90% of the maximum value) is about 25 V. The maximum transmittance is about 90% (normalized to the transmittance of an empty cell).

The measured transmittance of PDLCs in the scattering state depends on the collection angle of the detection, as defined in Figure 11.21(b) [29]. A typical scattering profile $S(\theta)$ of PDLCs as a function of the polar angle θ defined with respect to the incident direction is shown in Figure 11.21(a) [30]. The scattering is independent of the azimuthal angle. The full width at half maximum (FWHM) of the scattering profile of the scattering state is about 30°. When the linear collection angle is 2δ, the measured light intensity is

$$I = \pi \int_{-\delta}^{+\delta} S(\theta) \sin\theta d\theta. \tag{11.60}$$

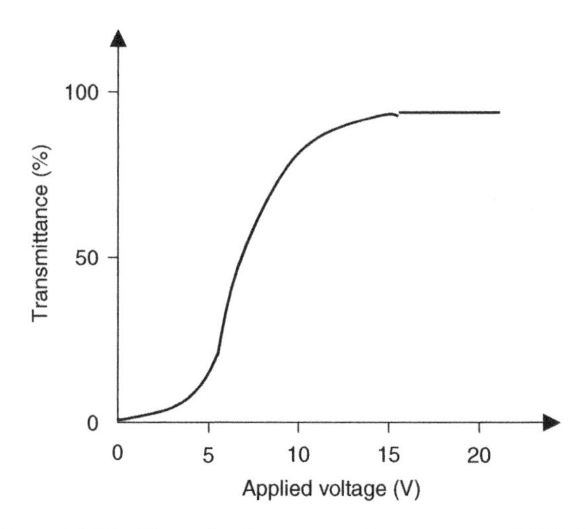

Figure 11.20 Typical voltage–transmittance curve of PDLCs.

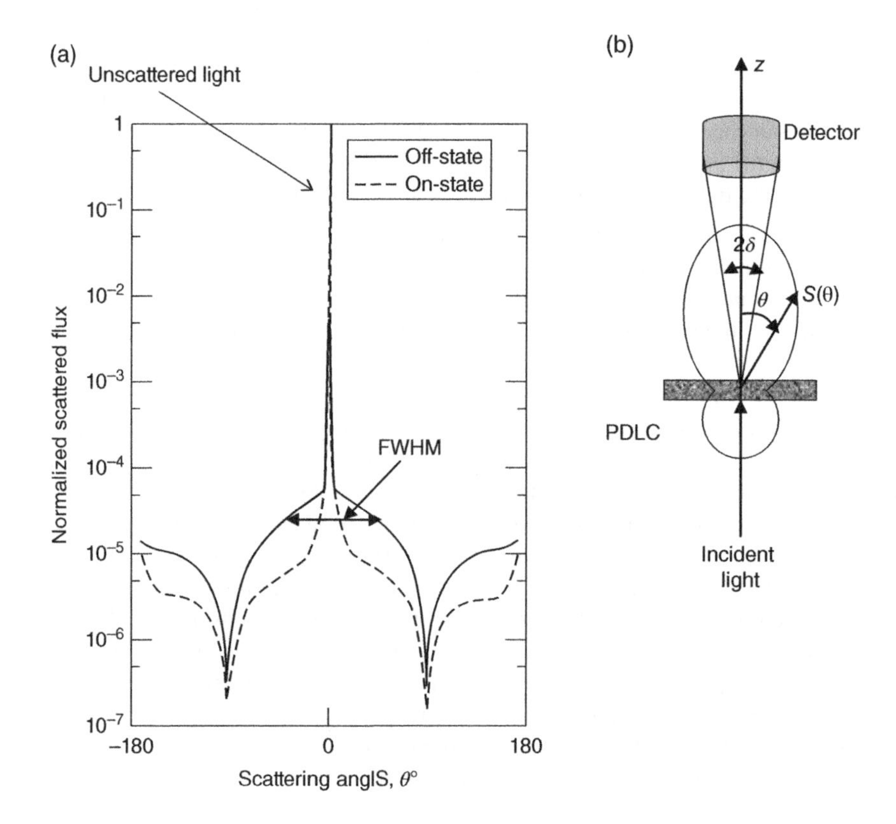

Figure 11.21 (a) The scattering profile of the PDLC as a function of the polar angle θ in the field-off and field-on states, AND (b) Measurement geometry. Vaz, Smith and Montgomery 1987. Reproduced with permission from Taylor and Francis.

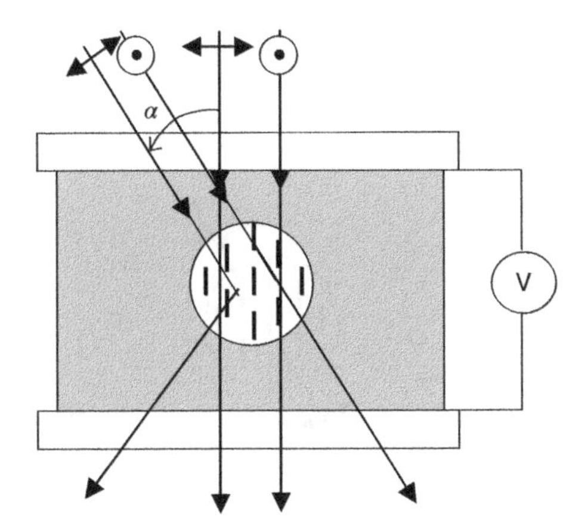

Figure 11.22 Schematic diagram showing the scattering of the PDLC in field-on state.

The contrast ratio is

$$C = \int_{-\pi}^{+\pi} S(\theta)\sin\theta d\theta \Big/ \int_{-\delta}^{+\delta} S(\theta)\sin\theta d\theta. \tag{11.61}$$

The larger the collection angle is, the more scattered light will be collected, and thus the lower the contrast ratio.

In the field-on state, the transmittance of the PDLC depends on the incident angle of light. For normally incident light, the encountered refractive index in the liquid crystal droplet is $n_\perp$, which is matched to that of the polymer, so there is no scattering. For obliquely incident light, as shown in Figure 11.22, if the polarization of the incident light is perpendicular to the incident plane, the light encounters the refractive index $n_\perp$, and therefore is not scattered. If the polarization is in the incident plane, when the incident angle is α, the encountered refractive index is $n(\alpha) = n_{//}n_\perp / \left(n_{//}^2\cos^2\alpha + n_\perp^2\sin^2\alpha\right)^{1/2}$, which is different from the refractive index of the polymer, and so it is scattered. The larger the incident angle is, the more the refractive index encountered in the liquid crystal droplet is mismatched to the refractive index of the polymer, and the stronger the scattering is. The transmittance of the light with this polarization as a function of the incident angle α is shown in Figure 11.23 [20]. The transmittance decreases to half when the incident angle is increased to 30°. If the incident light is unpolarized, the component with the parallel polarization is scattered at oblique angles, which make the PDLC milky. This limitation on the viewing angle can be eliminated when a linear polarizer is laminated on the PDLC with the trade-off that the on-state transmittance is decreased to half.

11.4.4 Dichroic dye-doped PDLC

Dichroic dyes can be incorporated into PDLCs [31,32]. The dye molecules are usually elongated and have low molecular weight as liquid crystals and have good solubility in the liquid

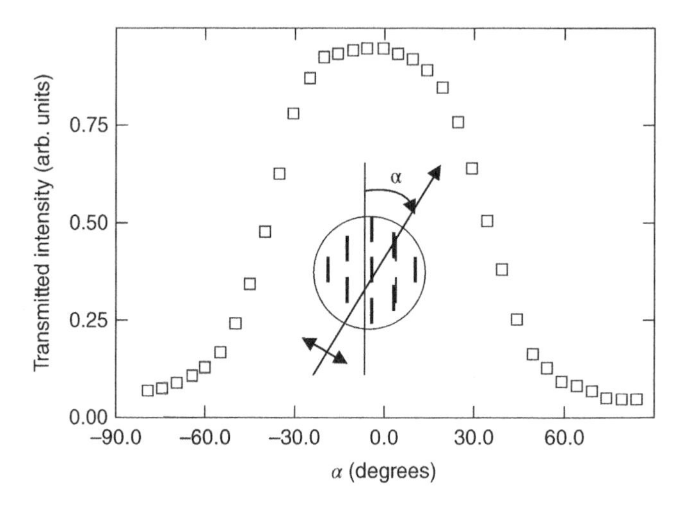

Figure 11.23 Angular dependence of the transmittance of the PDLC in the on-state for incident light polarized in the incident plane.

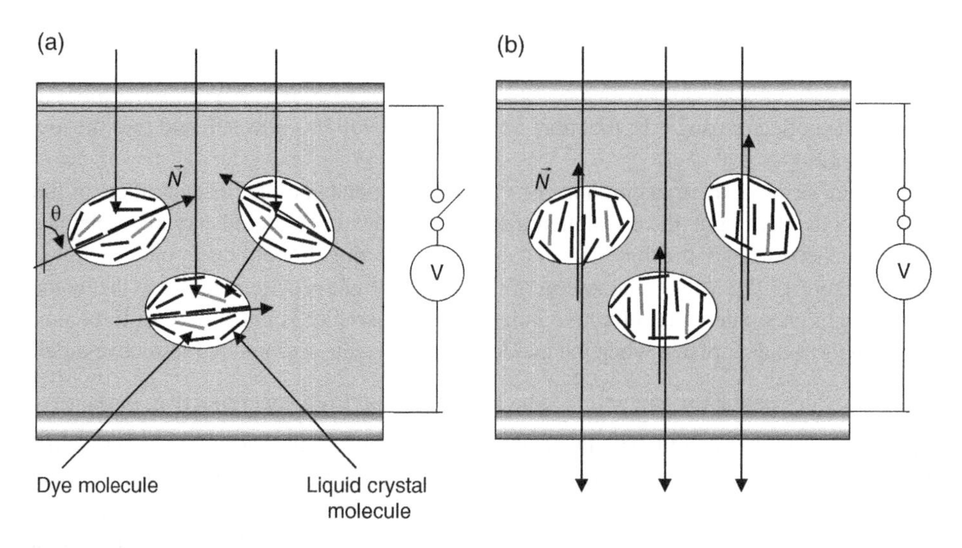

Figure 11.24 Schematic diagram showing how the dichroic dye doped PDLC works: (a) absorbing state, (b) transparent state.

crystal, but not in the polymer. The dye and the liquid crystal phase separates from the polymer binder. The dye molecules are inside the droplet and can be switched. The used dye must be a positive type in the sense that the absorption transition dipole is along the long molecular axis. When the polarization of the incident light is parallel to the long axis of the dye molecules, the light is absorbed. When the polarization of the incident light is perpendicular to the long axis of the dye molecules, the light is not absorbed. In the field-off state, the dye molecules are randomly oriented with the droplets, as shown in Figure 11.24(a). When the cell is sufficiently thick, there are droplets oriented in every direction. The unpolarized incident light is absorbed. In the field-on state, the dye molecules are aligned in the cell normal direction with the liquid crystal, as shown in Figure 11.24(b). They are always perpendicular to the polarization of normally incident light. Therefore the light ideally passes through the cell without absorption. In practice, there is some absorption even in the field-on state, because of the thermal fluctuation of the dye molecules and the anchoring of the curved surface of the droplet.

There are a few points worth noting. (1) The dye molecules dissolved in the polymer do not change orientation under the applied field, and so they tend to decrease the contrast of the PDLC. Therefore it is desirable that the solubility of the dye in the polymer is as low as possible. (2) An oblate droplet shape is desirable, because inside such a droplet, the dye molecules are oriented more in the plane parallel to the cell surface and so absorb the light more strongly. (3) In the field-off state, the scattering of the material increases the optical path length of the light inside the cell, and therefore enhances the absorption. (4) Dye-doped PDLCs do not need polarizers, because of the random orientation of the droplets in the field-off state, which is an advantage over nematic dichroic dye displays. (5) Dye-doped PDLCs have gray levels, because as the applied field is increased, the droplets are gradually aligned toward the cell normal direction, which is an advantage over cholesteric dichroic dye displays.

11.4.5 Holographic PDLCs

In PDLCs formed by polymerization-induced phase separation, spatial variations in structure can be achieved when non-uniform polymerization conditions are introduced. Such an example is holographically formed PDLC [33–40]. The mixture of a liquid crystal and a photo-polymerizable monomer is sandwiched between two glass substrates. A coherent laser light is used to initiate the polymerization. In the polymerization, the cell is irradiated by the laser light from both sides as shown in Figure 11.25. The two incident lights interfere with each other inside the cell and form the intensity pattern as shown on the right side of the figure. In the region where the light intensity is high, more free radicals are produced, which initiates polymerization. When monomers migrate into that region, they will likely be polymerized and will not come out. The net effect is that monomers are attracted into the high light intensity regions to form polymer, and the liquid crystal molecules are pushed out. Thus alternating polymer-rich and liquid-crystal-rich layers are formed. The period d (the thickness of one layer polymer plus the thickness of one layer of liquid crystal) is determined by the wavelength and incident angle of the laser light.

The liquid crystal and polymer are chosen such that the ordinary refractive index n_o of the liquid crystal is equal to the refractive index n_p of the polymer. At zero field, the liquid crystal has a random orientation structure, as shown in Figure 11.26(a), and the cell has a periodic refractive index. If the incident light satisfies the Bragg condition $\lambda = d\cos\theta$, it will be reflected. When a sufficiently high external electric field is applied across the cell, the liquid crystal ($\Delta\varepsilon > 0$) will be aligned perpendicular to the layers, as shown in Figure 11.26(b). The incident light encounters the same refractive index in the polymer-rich and liquid-crystal-rich layers, and passes through the material without reflection. Thus holographic PDLC can be used for switchable mirrors and reflective displays.

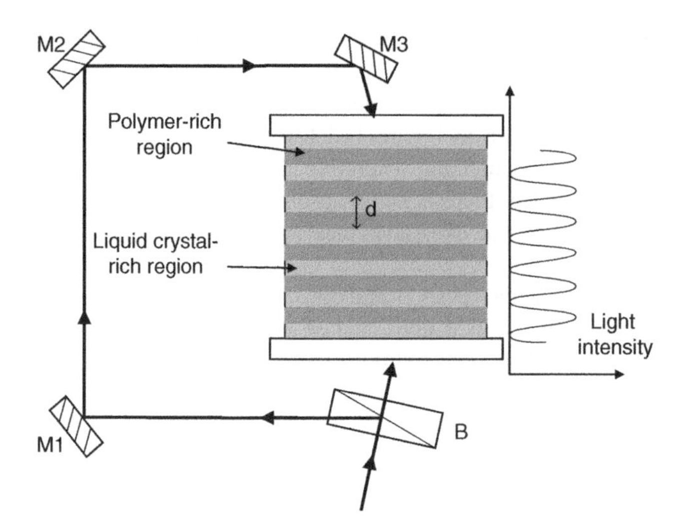

Figure 11.25 Schematic diagram showing how the holographic PDLC is formed. B: beam splitter, M: mirror.

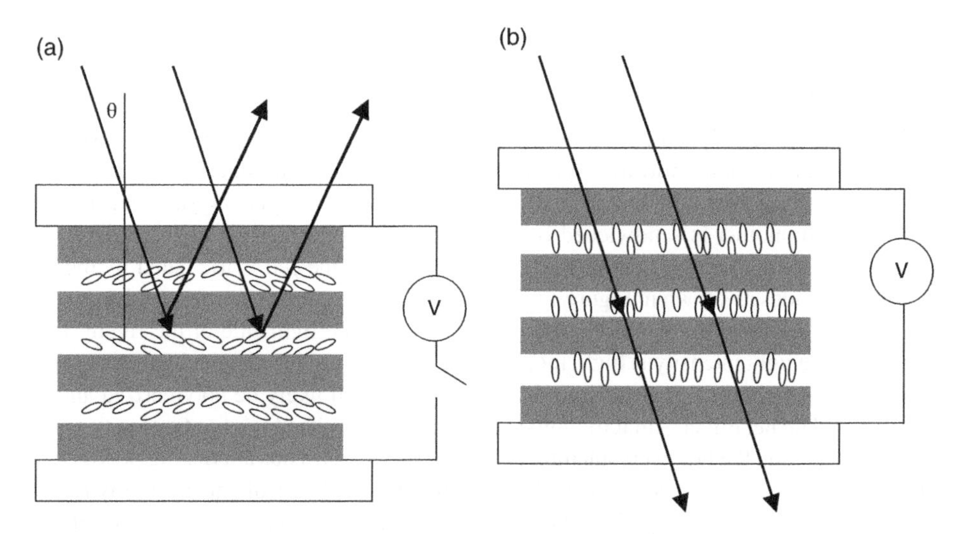

Figure 11.26 Schematic diagram showing how the holographic PDLC is used for reflective displays.

The spectral response of a holographic PDLC to applied electric fields is shown in Figure 11.27, where white incident light is used [37]. At 0 V, due to the periodic refractive index, the cell has a high narrow reflection peak. When the applied voltage is increased, the liquid crystal is aligned toward the layer normal direction. The amplitude of the oscillation of the refractive index decreases and the reflection of the cell also decreases. The drive voltage is approximately equal to the product of the field threshold of the Fréedericksz transition of the liquid crystal layer and the cell thickness.

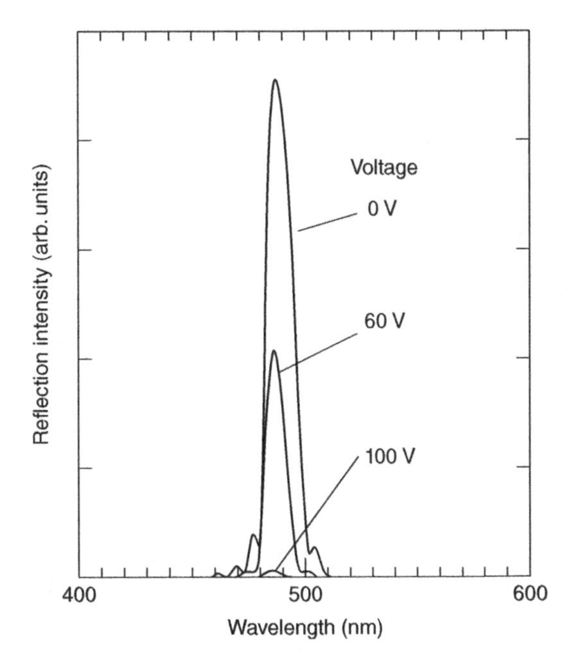

Figure 11.27 Reflection spectra of the holographic PDLC under various applied voltages. Reproduced with permission from Wiley.

11.5 PSLCs

In polymer-stabilized liquid crystals (PSLCs), the polymer concentration is usually less than 10%. The monomer used may be mesogenic with rigid cores similar to those of liquid crystal molecules [2,41–44]. Before polymerization, the mixture of the monomer and the liquid crystal is in a liquid crystal phase. The polymerization environment is anisotropic, due to the aligning effect of the liquid crystal on the monomer and the anisotropic diffusion of the monomer in the liquid crystal. Anisotropic fiber-like polymer networks are usually formed, which mimic the structure of the liquid crystal during polymerization. Because the liquid crystal and monomer are homogeneously mixed and in liquid crystal phase during polymerization, external fields and surface alignment techniques can be applied to create various polymer network structures. Therefore many fascinating structures can be achieved.

After polymerization, polymer networks tend to stabilize the state in which they are formed. In a PSLC, the liquid crystal near the polymer network is aligned along the polymer network. The strength of the interaction between the liquid crystal and the polymer network is proportional to the surface area of the polymer network. The surface area of the polymer network can be increased by using higher polymer concentrations or producing smaller lateral size polymer networks.

11.5.1 Preparation of PSLCs

PSLCs are usually made from mixtures of liquid crystals and monomers. The monomer can be directly dissolved in the liquid crystal. Although any type of polymerization method can be used, photo-initiated polymerization is fast and is usually used. The monomer is usually acrylate or methacrylate because of their fast reaction rate. In order to form stable polymer networks, the functionality of the monomer must be larger than 1. A small amount of photo-initiator is added to the mixture. The concentration of the photo-initiator is typically 1–5% of the monomer. When irradiated under UV light, the photo-initiator produces free radicals which react with the double bonds of the monomer and initiate the chain reaction of polymerization.

When the mixture of the liquid crystal, monomer, and photo-initiator is irradiated by UV light, the monomer is polymerized to form a polymer network. The UV intensity is usually a few mW/cm^2 and the irradiation time is on the order of minutes. The morphology of polymer networks in polymer-stabilized liquid crystals has been studied using SEM [45,46], neutron scattering, confocal microscopy [47], birefringence study and the Fréedericksz transition technique [48,49]. The results suggest a bundle structure for the polymer networks. The lateral size of the bundle, as shown in Figure 11.1(b), is at the submicron level. The bundle consists of polymer fibrils, with lateral size around a few nanometers, and liquid crystals. The morphology of the polymer network is affected by the following: structure of the monomer, UV intensity, photo-initiator type and concentration, and the temperature. The lateral size of polymer networks is determined by the polymerization rate, mobility, and concentration of monomers [48]. Polymer networks with smaller lateral sizes are obtained with higher polymerization rates which can be achieved with higher UV intensities or high photo-initiator concentrations in photo-polymerization. Polymer networks with smaller lateral sizes are also obtained with low mobility of monomers, which can be achieved with lower polymerization temperature. For example, 96.7% nematic liquid crystal E7, 3% monomer BAB6 {4,4′-bis[6-(acryloyloxy)-hexy]-1,1′-biphenylene} and 0.3% BME (benzoin methyl ether) are mixed. The mixture is in nematic phase at room temperature. The viscosity of the mixture is comparable to that of the nematic liquid crystal and can be easily placed into cells in a vacuum chamber. The cells are then irradiated under UV light for the monomer to form a polymer network.

Monomers to be used in PSLCs preferably have a rigid core and flexible tails. They form anisotropic fibril-like networks. If the monomer does not have flexible tails, it forms a bead-like structure which is not stable under perturbations such as externally applied fields. If the monomer does not have a rigid core and is flexible, it can still form anisotropic networks.

11.5.2 Working modes of scattering PSLCs

Polymer networks formed in liquid crystals are anisotropic and affect the orientation of liquid crystals. They tend to align the liquid crystal in the direction of the fibrils. They are used to stabilize desired liquid crystal configurations and to control the electro-optical properties of liquid crystal devices. Polymer networks have been used to improve the performance, such as drive voltage and response times, of conventional liquid crystal devices such as TN and IPS displays.

11.5.2.1 Polymer-stabilized nematic liquid crystals

a. *Polymer-stabilized homogeneously aligned nematic LC light shutter* – The polymer-stabilized homogeneously aligned nematic liquid crystal light shutter is made from a mixture

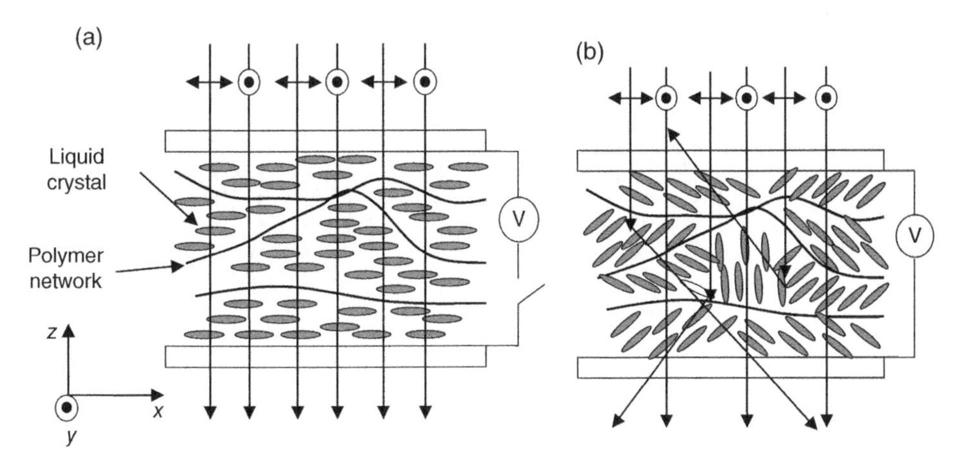

Figure 11.28 Schematic diagram showing how the polymer-stabilized homogeneously nematic liquid crystal light shutter works.

of a nematic liquid crystal and diacrylate liquid crystal monomer [41,50,51]. The mixture is placed into cells with anti-parallel homogeneous alignment layers and then photo-polymerized in the nematic phase. The polymer network formed is parallel to the cell surface, as is the liquid crystal. The nematic liquid crystal has a positive dielectric anisotropy. Figure 11.28 schematically shows how the shutter works. In Figure 11.28(a), when there is no applied voltage across the cell, the liquid crystal and the polymer network are homogeneously aligned in the x direction. When light goes through the material, it encounters the same refractive index in the liquid crystal and polymer regions, and therefore it passes through the material without scattering. In Figure 11.28(b), the voltage applied across the cell tends to align the liquid crystal in the z direction, while the polymer network tries to keep the liquid crystal in the x direction. As a result of the competition between the applied field and the polymer network, the liquid crystal is switched into a multi-domain structure. The liquid crystal molecules orient along the same direction within each domain but this varies from domain to domain. The directions of the domains are random in the xz plane. For light polarized in the x direction, when it goes through the cell, it encounters different refractive indices in different domains, and therefore is scattered. For light polarized in the y direction, when it goes through the cell, it always encounters the ordinary refractive index of the liquid crystal because the liquid crystal is oriented in the xz plane, and therefore it can pass through the cell without scattering. If the incident light is unpolarized, this shutter does not work well.

b. *Polymer-stabilized homeotropic nematic LC light shutter* – In order to overcome the problem that only one polarization component of unpolarized incident light is scattered in the polymer-stabilized homogeneously aligned nematic LC light shutter, polymer-stabilized homeotropically aligned nematic LC was introduced [52]. The LC has a negative dielectric anisotropy ($\Delta\varepsilon < 0$). It is mixed with a small amount of diacrylate monomer and placed into cells with homeotropic alignment layers. The cells are irradiated with UV light for photo-polymerization in the homeotropically aligned state. Thus the formed polymer network is perpendicular to the cell surface.

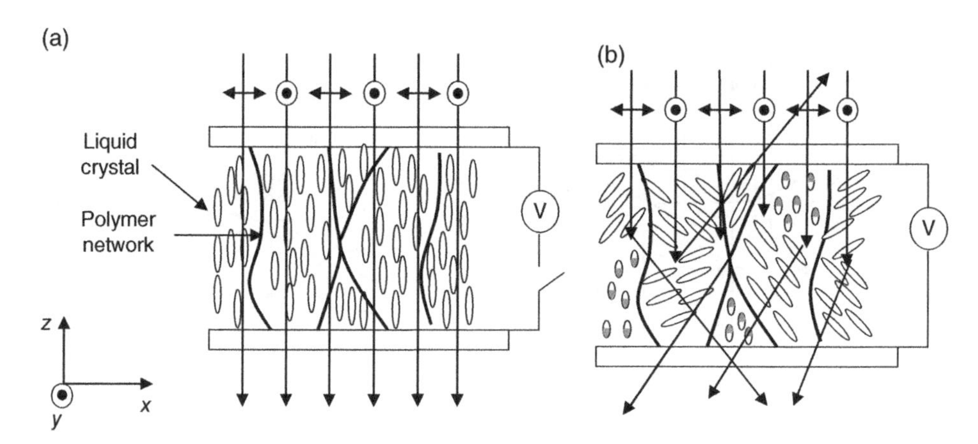

Figure 11.29 Schematic diagram showing how the polymer-stabilized homeotropical nematic liquid crystal light shutter works.

At zero field, the liquid crystal is in the uniform homeotropic state, as shown in Figure 11.29(a); the material is a homogeneous optical medium. The light propagates through the material without scattering. When an electric field is applied, the liquid crystal molecules are tilted away from the field direction, because of their negative dielectric anisotropy. The material is switched to a multi-domain structure, as shown in Figure 11.29(b). When light propagates through the cell, it encounters different refractive indices in different domains because of the different tilts of the liquid crystal. The material is optically non-uniform and therefore it is scattering. When the liquid crystal molecules tilt, they tilt toward the x direction in some domains but tilt toward the y direction in the other domains. Therefore, light polarized in both x and y directions is scattered. The polymer-stabilized homeotropically aligned nematic LC is a reverse-mode light shutter in the sense that it is transparent in the field-off state and scattering in the field-on state.

11.5.2.2 Polymer-stabilized cholesteric liquid crystals

Cholesteric liquid crystals (CLCs) have a helical structure where the liquid crystal director twists around a perpendicular axis – the helical axis. They exhibit three main textures (also referred to as states), depending on the boundary condition and the applied field. When a CLC is in the planar texture (also referred to as Grandjean texture), the helical axis is perpendicular to the cell surface, and the material reflects light around the wavelength $\bar{n}P$, where $\bar{n}$ is the average refractive index and P is the pitch of the liquid crystal. When the CLC has a focal conic texture, the helical axis is more or less random throughout the cell, and the material is usually optically scattering. When a sufficiently high field is applied across the cell (along the cell normal direction), the CLC ($\Delta\varepsilon > 0$) is switched to the homeotropic texture where the helical structure is unwound and the liquid crystal director is aligned in the cell normal direction. The material is transparent. Polymer networks can be used to stabilize the planar texture or the focal conic texture at zero field.

a. *PSCT normal-mode light shutter (PSCT)* – The polymer stabilized cholesteric texture (PSCT) normal-mode material is made from a mixture of CLC and a small amount of

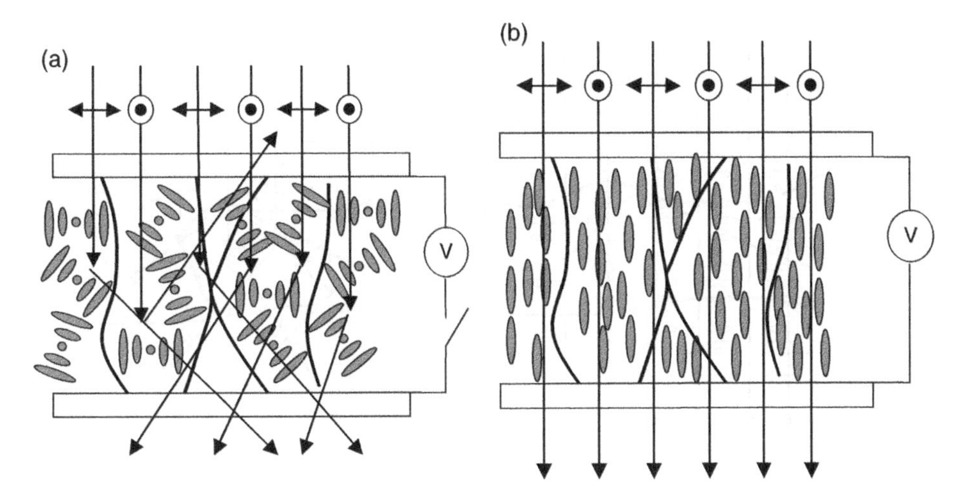

Figure 11.30 Schematic diagram showing how the polymer-stabilized cholesteric texture normal-mode light shutter works.

monomers [43,53]. The pitch of the liquid crystal is a few microns (~0.5–5 μm). No special cell surface treatment is needed. The mixture has a homeotropic texture in the presence of an external field when the monomers are polymerized. The polymer network formed is perpendicular to the cell surface, as shown in Figure 11.30.

When a PSCT normal-mode light shutter is in zero-field state, the liquid crystal tends to have a helical structure, while the polymer network tends to keep the liquid crystal director parallel to it. The competition between these two factors results in the focal conic texture shown in Figure 11.30(a). In this state, the material is optically scattering. When a sufficiently high electric field is applied across the cell, the liquid crystal ($\Delta\varepsilon > 0$) is switched to the homeotropic texture, as shown in Figure 11.30(b), and therefore it becomes transparent. Because the concentration of the polymer is low and both the liquid crystal and the polymer are aligned in the cell normal direction, the PSCT normal-mode light shutter is transparent at any viewing angle. A photograph of a PSCT normal-mode light shutter is shown in Figure 11.2(a).

In order to scatter visible light strongly, the focal conic domain size has to be around the wavelength of the light. The main factors affecting the domain size are the pitch, polymer concentration, and curing UV intensity. The drive voltage is mainly determined by the pitch and the dielectric anisotropy of the liquid crystal. Faster response can be achieved with shorter-pitch CLCs. There is a hysteresis in the transition between the focal conic texture and the homeotropic texture, which also exists in pure CLCs.

b. *PSCT reverse-mode light shutter* – This is also made from a mixture of cholesteric liquid crystal ($\Delta\varepsilon > 0$) and a small amount of monomer. The pitch of the liquid crystal is a few microns (~3–15 μm). The mixture is placed into a cell with homogeneous alignment layers. The mixture is in the planar texture at zero field because of the alignment layers. The monomers are polymerized in the planar texture. The polymer network formed is parallel to the cell surface [43,53,54].

At zero field, the material is in the planar texture as shown in Figure 11.31(a). Because the pitch is in the infrared region, the material is transparent for visible light. When an external

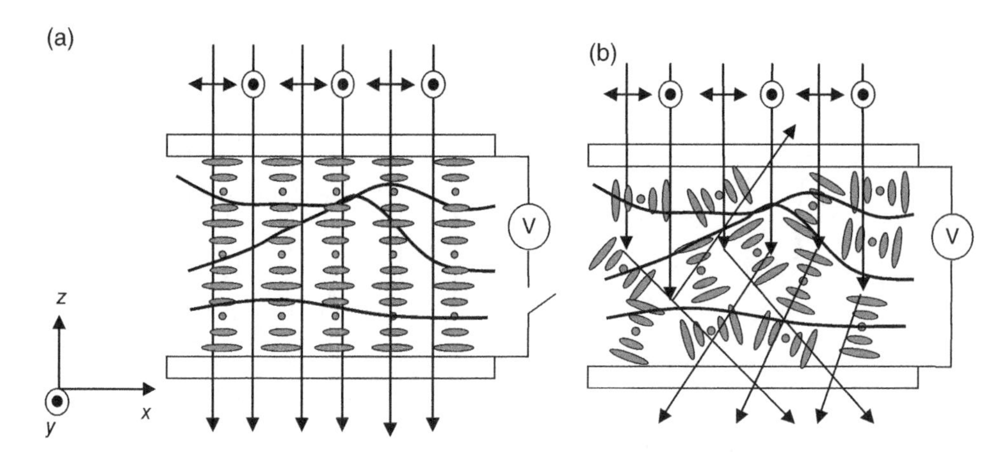

Figure 11.31 Schematic diagram showing how the polymer-stabilized cholesteric texture reverse-mode light shutter works.

field is applied across the cell, the field tends to align the liquid crystal in the cell normal direction, while the polymer network tends to keep the liquid crystal in the planar texture. As a result of the competition between these two factors, the liquid crystal is switched to the poly-domain focal conic texture as shown in Figure 11.31(b), and the material becomes scattering. The polymer concentration must be sufficiently high and the polymerization rate should be low enough so that the polymer network is strong and would not be damaged (reoriented) under the applied field. The photograph of a PSCT reverse-mode light shutter is shown in Figure 11.2(b).

11.6 Scattering-Based Displays from LCPCs

Scattering liquid crystal/polymer composites (LCPCs), besides being used for switchable privacy windows, can also be used to make displays. Without polarizers, they can only be used for reflective displays and projection displays because in both the transparent and scattering states, most of the incident light still comes out in forward directions, except that in the scattering state, light is deviated from its original propagation direction. With polymerizers, they can be used for transmissive direct-view displays. These displays may not have the best optical performance, but they are compatible with flexible plastic substrates because of their adhesion to plastic substrates. They can be manufactured in a roll-to-roll process.

11.6.1 Reflective displays

There are several designs of scattering LCPCs for reflective displays. The simplest design is shown in Figure 11.32(a). The display consists of a layer of LCPC and a black absorbing layer [55]. When the LCPC in a pixel is in the transparent state, the incident light reaches the black layer and is absorbed, and the pixel appears black. When the LCPC is in the scattering state, some of the incident light is scattered backward and is observed by the reader's eyes, and the

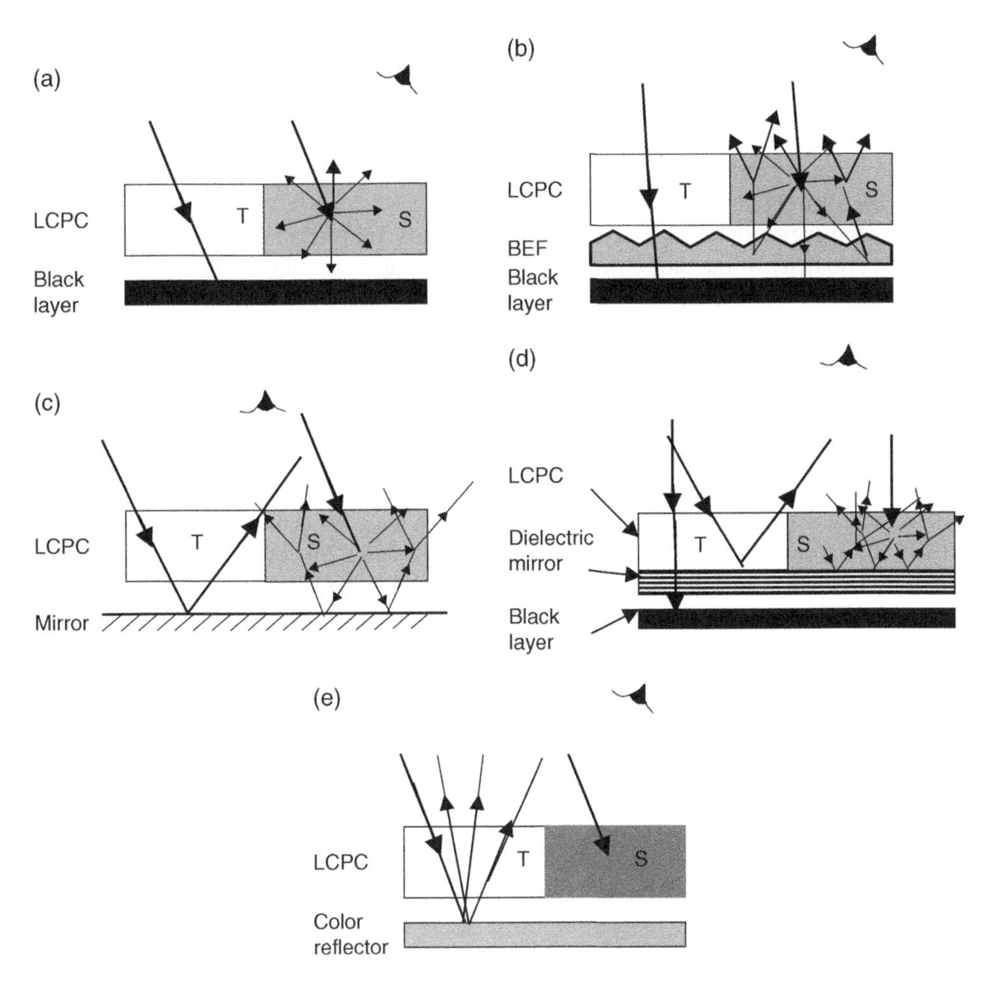

Figure 11.32 Schematic structures of the reflective displays from scattering LCPCs. T: transparent state, S: scattering state.

pixel appears gray–white. The problem with this design is that the reflectivity (the percentage of light scattered in backward direction) is usually less than 25% [55].

One way to improve the reflectivity is to insert a brightness-enhancing film (BEF) [56,57], as shown in Figure 11.32(b), which transmits light with small incident angles (with respect to the film normal) but reflects light with large incident angles. When the LCPC in a pixel is in the transparent state, incident light with small incident angles goes through the LCPC and BEF, and reaches the black layer and is absorbed. When the LCPC is in the scattering state, incident light with small incident angles is scattered by the LCPC, and reaches the BEF with large incident angles and therefore is reflected back by the BEF. Therefore the reflectivity of the display is greatly increased. The trade-off is that the contrast and viewing angle are decreased.

The third design is shown in Figure 11.32(c), where a mirror is used to replace the black layer. This display works well only when light is incident on it at one incident angle. When the LCPC in a pixel is in the transparent state, the light passes the LCPC and is reflected by the mirror. The outcoming light is at the specular angle. If the reader looks at the display not at the specular angle, the pixel has a black appearance. When the LCPC is at the scattering state, some of the incident light is scattered backward and the rest is scattered forward by the LCPC. The forward-scattered light reaches the mirror and is reflected in all directions. Therefore the pixel has a white appearance. The reflectivity of the white is very high. The problem with this design is the viewing angle is small.

The viewing angle can be improved by using a dielectric mirror, as shown in Figure 11.32(d) [58]. The periodicity of the dielectric mirror is in the long (near IR) wavelength region. For normally incident light, it only reflects long-wavelength light. For obliquely incident light, the reflection band is shifted to the short-wavelength region. The reader looks at the display in the normal direction. When the LCPC in a pixel is in the transparent state, light with small incident angles and short wavelength passes the LCPC and the dielectric mirror, and reaches the absorbing layer. The pixel has a black appearance. When the LCPC is in the scattering state, the incident light is scattered by the LCPC and reaches the dielectric mirror at large incident angle, and therefore is reflected. The pixel has a white appearance.

Black dichroic dye-doped LCPCs can also be used to make reflective displays [59]. Behind the LCPC film, there is a color reflector, as shown in Figure 11.32(e). When the LCPC in a pixel is in the transparent state, light passes the LCPC and reaches the reflector and is reflected. The pixel shows the color of the reflector. When the LCPC is in the scattering absorbing state, the incident light is absorbed by the LCPC, and the pixel has a black appearance.

11.6.2 Projection displays

Scattering LCPCs can be used to make projection displays [29,60]. They do not need polarizers, and therefore have high light efficiency. A simple projection display shown in Figure 11.33 is used to demonstrate the operating principle. The optical design is similar to that of a slide projector, except that an additional aperture is placed at the focal plane of the objective lens. The condenser lens generates a collimated light parallel to the principal axis. The objective lens produces an image on the screen of the display panel made from the LCPC. When the LCPC in a pixel (on-pixel) is in the transparent state, the collimated light passes the material without scattering, and then passes the aperture and reaches the screen. The corresponding area on the screen is bright. When the LCPC in a pixel is in the scattering state

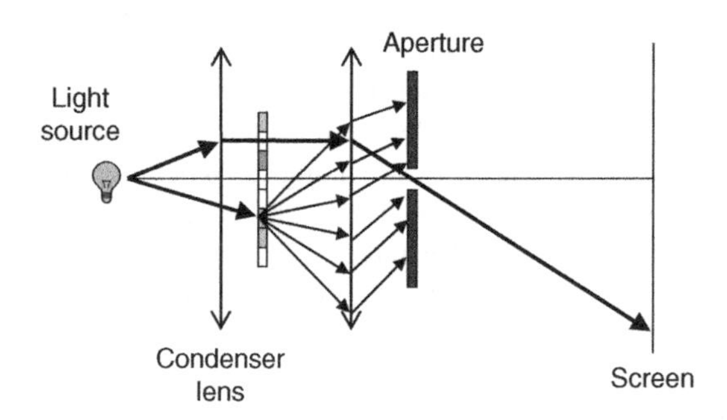

Figure 11.33 Schematic diagram of the projection display from the scattering liquid crystal/polymer composite.

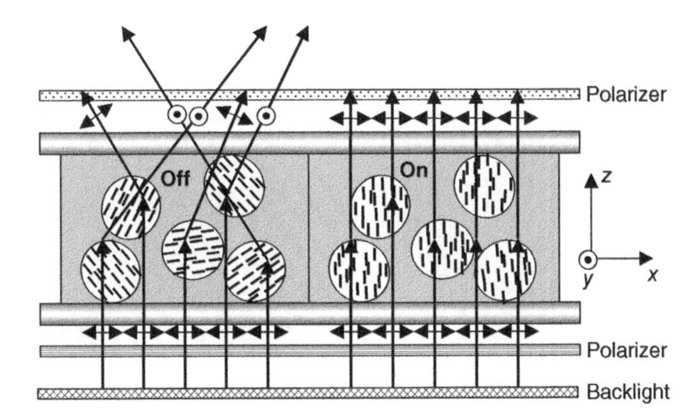

Figure 11.34 34 Schematic diagram showing how the transmissive direct view display from the scattering LCPC works.

(off-pixel), the collimated incident light is scattered in all directions. Only a small part of the incident light can pass the aperture and reach the screen. The corresponding part on the screen is dark. In order to achieve high contrast, the aperture should be small so that little of the scattered light can go through it. The trade-off of the small aperture is that light efficiency is low because incident light is never perfectly collimated if the light source is not a point light source but a filament of finite size. When the aperture is small and the incident light is not perfectly collimated, some light cannot pass the aperture and thus is lost even if a pixel is in the on-state.

11.6.3 Transmissive direct-view displays

Scattering LCPCs can also be used to make transmissive direct view displays if polarizers are used. An example is shown in Figure 11.34 where an LCPC is sandwiched between two crossed polarizers along the x and y directions, respectively [61]. The light from the backlight becomes linearly polarized along the x direction after passing through the bottom polarizer. When the LCPC in a pixel (on-pixel) is in the scattering state, the linearly polarized incident light is depolarized with 50% of the light polarized along the x direction and 50% of the light polarized along the y direction. The light polarized along the y direction passes the top polarizer. The pixel is bright. When the LCPC in a pixel (off-pixel) is in the transparent state, the polarization of the incident light does not change when propagating through the material, and the light is absorbed by the top polarizer. The pixel is dark. Besides its suitability for flexible displays, this display has a large viewing angle when compensated with a negative c plate. The drawback is that the maximum light efficiency is only 25%.

11.7 Polymer-Stabilized LCDs

As described in previous sections, polymer stabilized liquid crystals (PSLCs) are made from mixtures of liquid crystals and monomers. The monomers are usually mesogenic with rigid cores and flexible tails, similar to the structures of liquid crystal molecules, and bifunctional.

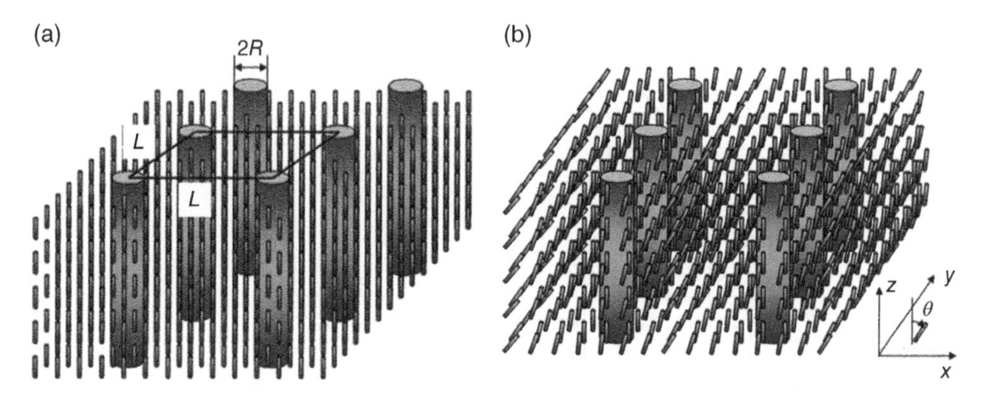

Figure 11.35 Schematic diagram of the polymer-stabilized liquid crystal.

In the polymerization of the monomers, the mixture is in a liquid crystal phase. The polymerization environment is anisotropic due to the aligning effect of the liquid crystal on the monomer and the anisotropic diffusion of the monomer in the liquid crystal. After the monomers are polymerized, they phase separate from the liquid crystals to form anisotropic fiber-like polymer networks which mimic the structure of the liquid crystal during the polymerization. The polymer networks consist of cylindrical shape fibers with submicron lateral diameter and hundreds of microns in longitudinal length. The inter-distance between the fibers is typically a few microns.

After polymerization, the polymer network in a polymer-stabilized liquid crystal tends to stabilize the liquid crystal in the state in which the polymer network is formed. Because of it being highly cross-linked, the polymer network is strong and can stand up to variations of temperature and externally applied fields. The liquid crystal molecules near the polymer network tend to be aligned parallel to the longitudinal direction of the polymer network. Polymer networks can be used to improve the performance of liquid crystal displays such as IPS and VA LCDs. They can also be used to stabilize desired liquid crystal states such as blue phases. They can dramatically increase the blue phase temperature region from milli-degrees to a few tens of degrees.

The aligning effect of the polymer network can be understood in terms of the boundary condition imposed by the polymer fibers. The monomers in the polymer fiber orient along its longitudinal direction. The anisotropic interaction between the monomers and the liquid crystal molecules on the surface of the networks tends to make liquid crystal molecules aligned in the same direction. This imposed boundary condition keeps the liquid crystal in bulk aligned in that direction. We try to estimate the aligning effect of polymer network in the polymer stabilized nematic liquid crystal, as shown in Figure 11.35 [62,63]. For the purpose of simplicity, we assume that the polymer fibers are periodically located in the liquid crystal. The period is L and the radius of the polymer fibers is R. The volume fraction C of the polymer network is given by

$$C = \pi R^2 / L^2. \tag{11.62}$$

For low concentrations of polymer network, $R \ll L$. In the absence of external field, the liquid crystal is aligned uniformly parallel to the polymer fibers, as shown by in Figure 11.35(a).

When an external electric field E higher than a threshold field E_c is applied in the x direction, the liquid crystal will reorient as shown in Figure 11.35(b). Because of the periodical structure, we only have to consider the liquid crystal director configuration in the region ($0 \le x \le L,\ 0 \le y \le L$). The liquid crystal director $\vec{n}$ varies in space and is described by

$$\vec{n} = \sin\theta(x,y)\hat{x} + \cos\theta(x,y)\hat{z}, \tag{11.63}$$

where θ is the angle between the liquid crystal director and the z axis. We use the assumption of isotropic elastic constant ($K_{11} = K_{22} = K_{33} = K$), the total free energy (elastic energy and electric energy) of one unit (in 2-D) is given by

$$F = \int_0^L\int_0^L \left\{\frac{1}{2}K\left[\left(\frac{\partial\theta}{\partial x}\right)^2 + \left(\frac{\partial\theta}{\partial y}\right)^2\right] - \frac{1}{2}\varepsilon_o\Delta\varepsilon E^2\sin^2\theta\right\}dxdy. \tag{11.64}$$

When the applied field is slightly above the threshold field E_c, the tilt angle θ is small. Equation (11.64) becomes

$$F = \int_0^L\int_0^L \left\{\frac{1}{2}K\left[\left(\frac{\partial\theta}{\partial x}\right)^2 + \left(\frac{\partial\theta}{\partial y}\right)^2\right] - \frac{1}{2}\varepsilon_o\Delta\varepsilon E^2\theta^2\right\}dxdy. \tag{11.65}$$

Using the Euler–Lagrange method to minimize the free energy, we get

$$\frac{\partial^2\theta}{\partial x^2} + \frac{\partial^2\theta}{\partial y^2} + \frac{1}{\xi^2}\theta = 0, \tag{11.66}$$

where $\xi = \sqrt{K/\varepsilon_o\Delta\varepsilon E^2}$ is the field correlation length. We consider the case when the anchoring strength of the polymer fibers is infinitely strong, and thus the boundary conditions are

$$\theta(x=0,y=0) = \theta(x=0,y=L) = \theta(x=L,y=0) = \theta(x=L,y=L) = 0. \tag{11.67}$$

From the symmetry of the system we can see that the liquid crystal orientation must be symmetric about the $x = L/2$ and $y = L/2$ lines. An approximate solution of the liquid crystal tilt angle is

$$\theta = \frac{\theta_m}{2}\left\{\cos\left[\frac{\pi}{L}\left(x - \frac{L}{2}\right)\right] + \cos\left[\frac{\pi}{L}\left(y - \frac{L}{2}\right)\right]\right\}. \tag{11.68}$$

Substituting Equation (11.68) into Equation (11.65) we get

$$F = \frac{K\theta_m^2}{8}\left(\frac{L}{\pi}\right)^2\left\{\pi^2\left(\frac{\pi}{L}\right)^2 - \frac{(\pi^2+8)}{\xi^2}\right\} = \frac{KL^2}{8}\left\{\left(\frac{\pi}{L}\right)^2 - \left(1 + \frac{8}{\pi^2}\right)\varepsilon_o\Delta\varepsilon E^2/K\right\}\theta_m^2. \tag{11.69}$$

When the applied field is low, the coefficient of θ_m^2 in Equation (11.69) is positive. The free energy increases when the tilt angle increases. Therefore the reorientation of the liquid crystal cannot occur. When the applied field is increased above a threshold value E_c, the coefficient of θ_m^2 becomes negative. The free energy decreases with the tilt angle, so the reorientation takes place. The threshold field can be obtained from

$$\left(\frac{\pi}{L}\right)^2 - \left(1+\frac{8}{\pi^2}\right)\varepsilon_o \Delta\varepsilon E_c^2/K = 0,$$

$$E_c = \left(\frac{\pi^2}{8+\pi^2}\right)^{1/2}\frac{\pi}{L}\sqrt{\frac{K}{\varepsilon_o \Delta\varepsilon}}. \tag{11.69}$$

Note that the threshold field here is smaller than the threshold field of the Fréedericksz transition in the regular liquid crystal cell consisting of two parallel substrates with the same cell gap, for the following reason. In the regular cell the liquid crystal is anchored by the two-dimensional surface of the substrate, while in the polymer-stabilized liquid crystal cell here, the liquid crystal is anchored by the one-dimensional polymer fiber.

We introduce an effective aligning field E_P, which equals E_c, to describe the aligning effect of the polymer network. From Equations (11. 62) and (11.69), we get

$$E_P = \left(\frac{\pi^2}{8+\pi^2}\right)^{1/2}\frac{\pi}{L}\sqrt{\frac{K}{\varepsilon_o \Delta\varepsilon}} = \left(\frac{\pi^3 K}{(8+\pi^2)\varepsilon_o \Delta\varepsilon}\right)^{1/2}\frac{\sqrt{c}}{R}. \tag{11.70}$$

Note that the effective aligning field E_P of the polymer network depends only on the density of the polymer fibers, but not on the dielectric anisotropy. Equation (11.70) shows that if the dielectric anisotropy of the liquid crystal is $\Delta\varepsilon$, an electric field higher than E_P must be applied in order to overcome the aligning effect of the polymer network such that the liquid crystal can reorient. The direction of $\vec{E}_P$ is parallel to the polymer fiber. As an example, for a liquid crystal with $K = 10^{-11} N$ and $\Delta\varepsilon = 10$, when the polymer concentration is $c = 1\%$ and the radius of the fiber is $R = 0.1$ μm, the effective aligning field of the polymer network is $E_P = 0.4$ V/μm. The energy density of the interaction between the polymer network and the liquid crystal can be effectively described as

$$f_p = -\frac{1}{2}\varepsilon_o \Delta\varepsilon\left(\vec{E}_P \cdot \vec{n}\right)^2. \tag{11.71}$$

Now we consider the Fréedericksz transition in a polymer-stabilized nematic liquid crystal in the splay geometry. The liquid crystal director is given by

$$\vec{n} = \cos\theta(z)\hat{x} + \sin\theta(z)\hat{z}, \tag{11.72}$$

where θ is the tilt angle defined with respect to the surface, and the z axis is perpendicular to the cell surface. The polymer network is parallel to the cell surface (the x axis) and thus $\vec{E}_P$ is parallel to the x axis. The applied electric field is along the z direction. The free energy density is

$$f=\frac{1}{2}K_{11}\left(\nabla\cdot\vec{n}\right)^2-\frac{1}{2}\varepsilon_o\Delta\varepsilon\left[\left(\vec{E}_P\cdot\vec{n}\right)^2+\left(\vec{E}\cdot\vec{n}\right)^2\right]$$

$$f=\frac{1}{2}K_{11}\left(\frac{\partial\theta}{\partial z}\right)^2-\frac{1}{2}\varepsilon_o\Delta\varepsilon\left(E^2-E_P^2\right)\sin^2\theta-\frac{1}{2}\varepsilon_o\Delta\varepsilon E_P^2. \tag{11.73}$$

The threshold field is then given by

$$\left(E_c^2-E_P^2\right)=\left(\frac{\pi}{d}\right)^2\frac{K_{11}}{\varepsilon_o\Delta\varepsilon},$$

$$E_c=\sqrt{\left(\frac{\pi}{d}\right)^2\frac{K_{11}}{\varepsilon_o\Delta\varepsilon}+E_P^2}. \tag{11.74}$$

Where d is the cell thickness. The turn-off time is given by

$$\tau=\frac{\gamma}{\left[\varepsilon_o\Delta\varepsilon E_P^2+K_{11}\left(\pi/h\right)^2\right]}. \tag{11.75}$$

The turn-off time can be reduced by using the polymer network. The trade-off is that the switching voltage is increased.

M. J. Escuti et al. used polymer network to improve the switching time of IPS LCD [64]. When the polymer concentration was 0%, the saturation field was 0.7 V/μm and the turn-off time was 70 ms. When 2% polymer was added, the saturation field was increased to 1.5 V/μm and the turn-off time reduced to 35 ms. J.-I. Baek et al. used a polymer network to improve VA LCD [65]. When the polymer concentration was 0%, the saturation voltage was 5 V and the turn-off time was 20 ms. When 5% polymer was added, the saturation voltage was increased to 10 V and the turn-off time reduced to 10 ms. Y.-Q Lu et al. used polymer stabilization to improve TN light shutter [66]. When the polymer concentration was 0%, the saturation voltage was 2 V and the turn-off time was 43 ms. When 3% polymer was added, the saturation voltage was increased to 12 V and the turn-off time reduced to 9 ms.

One must pay attention to light scattering in polymer-stabilized polarizer-based devices. In the dark state, if there was light scattering, depolarization of light would occur, which causes light leakage. Two factors cause light scattering. One is refractive index mismatch between the polymer network and the liquid crystal. The other one is that the polymer network may create poly liquid crystal domains. The light scattering can be reduced by using liquid crystals and polymers whose ordinary and extraordinary refractive indices are the same, respectively. The light scattering is a minimum when the liquid crystal and the polymer are aligned in the same direction in the dark state.

Homework Problems

11.1 Consider a ternary mixture consisting of A, B, and C, three different molecules. Their molar fractions are x_A, x_B, and x_C, respectively, where $x_A+x_B+x_C=1$. The intermolecular interaction energies are u_{AA} for the interaction between A and A, u_{BB} for the interaction between B and B, u_{CC} for the interaction between C and C, u_{AB} for the interaction between A and B, u_{AC} for the interaction between A and C, u_{BC} for the interaction between B and C. Calculate the mixing interaction energy.

11.2 Consider a binary mixture consisting of A and B. The molar fraction of B is x. The interaction energies between them are: $u_{AA} = -0.15$ eV, $u_{BB} = -0.17$ eV and $u_{AB} = -0.13$ eV. The number P of nearest neighbors is 6. (1) Determine the phase diagram numerically. (2) Find the critical point (T_c, x_c) of the phase separation. (3) If the fraction of B in the homogeneous mixture is $x_o = 0.3$, at what temperature T_s will the system separate into two phases? (4) At temperature $T = 800$ K, what is the fractions of B in the two-phase separated phases?

11.3 Consider a ternary mixture consisting of three types of molecules (A, B, C). The interaction energies between them are: $u_{AA} = -0.15$ eV, $u_{AB} = -0.05$ eV, $u_{BB} = -0.15$ eV, $u_{AC} = -0.20$ eV, $u_{BC} = -0.175$ eV, $u_{CC} = -0.15$ eV. We consider only the interaction between nearest neighbors. The number of nearest neighbors is 6. The temperature is $k_B T = 0.2$ eV. (1) Find the phase diagram represented as an equilateral triangle. (2) Determine the critical point. (3) For a mixture with the molar fractions $x_a = 0.4$, $x_B = 0.45$, and $x_C = 0.15$, will it phase separate? If it does, determine the fractions of the components in the two phases, and the percentages of molecules in the two phases.

11.4 *Phase diagram of ternary mixture*. Consider a ternary mixture consisting of three types of molecules A, B, and C. A and C dissolve each other; B and C dissolve each other; but A and B do not dissolve each other. The interaction energy between them are $u_{AA} = -0.15$ eV, $u_{AB} = -0.05$ eV, $u_{AC} = -0.20$ eV, $u_{BB} = -0.14$ eV, $u_{BC} = -0.175$ eV, and $u_{CC} = -0.15$ eV. We consider only the interaction between nearest neighbors. The number of nearest neighbors is 6. The temperature is $k_B T = 0.2$ eV. (1) Find the phase diagram represented as an equilateral triangle. (2) Determine the critical point. (3) If the initial mixture has the fractions: $x_a = 0.4$, $x_B = 0.45$, and $x_C = 0.15$, will it phase separate? If it does, determine the fractions of three components in the two phases, and the percentages of the molecules in the two phases.

The phase diagram of a ternary mixture at a given temperature is determined in the following way:

a. Choose an initial mixture with the fractions $x_{Ao}, x_{Bo}, x_{Co}(=1 - x_{Ao} - x_{Bo})$ which is a point on a straight line drawn from the corner corresponding to component C to the opposite side of the equilateral triangle. Calculate the average free energy per molecule $f_h = f(x_{Ao}, x_{Bo})$ of the homogeneous single phase with the initial fractions.
b. Assume the initial mixture phase separates into two phases: phase 1 with fractions (x_{A1}, x_{B1}, x_{C1}) and phase 2 with the fractions (x_{A2}, x_{B2}, x_{C2}). The percentage of molecules in phase 1 is w and the percentage of particles in phase 2 is $(1 - w)$. Because of particle conservation, we have $x_{Ao} = wx_{A1} + (1 - w)x_{A2}$ and $x_{Bo} = wx_{B1} + (1 - w)x_{B2}$. Therefore among these variables $x_{A1}, x_{B1}, x_{A2}, x_{B2}, w$, only three are independent. Choose x_{A1}, x_{B1}, w as the independent variables, and then we have $x_{C1} = 1 - x_{A1} - x_{B1}$, $x_{A2} = (x_{Ao} - wx_{A1})/(1 - w)$, $x_{B2} = (x_{Bo} - wx_{B1})/(1 - w)$, and $x_{C2} = (x_{Co} - wx_{C1})/(1 - w)$. Note that $0 \le w \le 1$, $0 \le x_{A1} \le 1$, $0 \le x_{B1} \le 1$, and $0 \le x_{A1} + x_{B1} \le 1$. Find the average free energy per molecule $f_t = wf(x_{A1}, x_{B1}) + (1 - w)f(x_{A2}, x_{B2})$ of the phase separated system.

c. Find the x_{A1}, x_{B1}, w, which gives the minimum free energy f_{tm}. If $f_{tm} < f_h$, the system will phase separate into two phases corresponding to the fractions (x_{A1}, x_{B1}, x_{C1}) and (x_{A2}, x_{B2}, x_{C2}). These two points are on the phase boundary curve. If $f_{tm} > f_h$, then there is no phase separation.

d. Repeat the above steps with a different initial mixture. After sufficiently large number of initial mixtures are tested, the points representing the phase-separated phases form the phase boundary.

11.5 Consider a brightness-enhancing film (BEF) shown in Figure 11.36. Calculate the incident angular region within which the incident light will be reflected from the bottom surface of the BEF by total internal reflection.

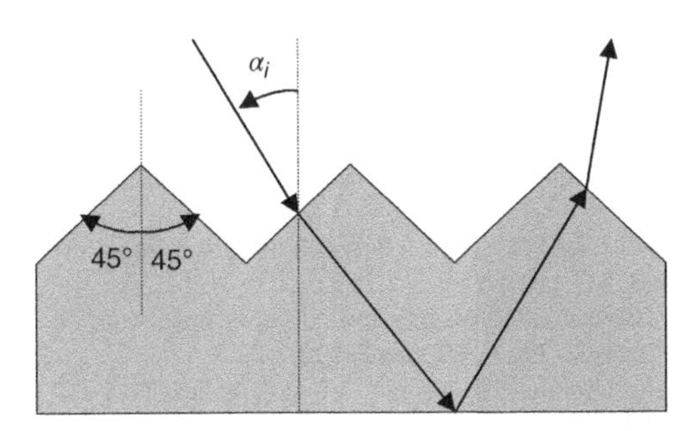

Figure 11.36 Structure of BEF for Problem 11.5.

11.6 Calculate the director configuration inside a bipolar droplet with the radius $R = 5$ μm R under the following externally applied electric fields: 0, 1 V/μm, 1 0 V/μm. The elastic constants are $K_{11} = K_{33} = 10^{-11}$ N and the dielectric anisotropy is $\Delta\varepsilon = 10$. The electric is applied along the bipolar axis.

11.7 The scattering profile of a PDLC in scattering state is described by $I(\theta) = (2/\pi)(1 + \cos\theta)$, where θ is the polar angle with the incident light direction. The incident light is collimated and normal to the cell surface. When the cell is in transparent state, the light goes through the cell without scattering. The detector has a circular detection surface with the linear collection of 10°. What is the contrast ratio of the PDLC?

References

1. P. S. Drzaic, *Liquid crystal dispersions*, (World Scientific, NJ, 1995).
2. G. P. Crawford and S. Zumer, *Liquid crystals in complex geometries*, (Taylor & Francis, London, 1996).
3. J. W. Doane, *Polymer dispersed liquid crystal displays, Liquid crystals, applications and uses*, Vol. 1, Chapter 14, ed. B. Bahadur (World Scientific, Singapore, 1990).
4. H.-S. Kitzerow, Polymer-dispersed liquid crystals, from the nematic curvilinear aligned phase to ferroelectric films, *Liq. Cryst.* **16**, 1 (1994).
5. G. P. Crawford, J. W. Doane, and S. Zumer, Chapter 9, Polymer dispersed liquid crystals: nematic droplets and related systems in *Handbook of liquid crystal research*, ed. P. J. Collings and J. S. Patel (Oxford University Press, New York, 1997).
6. C. Kittel and H. Kroemer, *Thermal physics*, 2nd edn (W. H. Freeman and Company, San Francisco, 1980).

7. J. L. West, Phase separation of liquid crystals in polymers, *Mol. Cryst. Liq. Cryst.* **157**, 427 (1988).
8. U. Eisele, *Introduction to polymer physics*, (Springer-Verlag, Berlin Heidelberg, 1990).
9. P. J. Flory, Thermodynamics of high polymer solutions, *J. Chem. Phys.*, **10**, 51 (1942).
10. M. L. Huggins, Thermodynamic properties of solutions of long-chain compounds, *Ann. New York Acad. Sci.*, **43**, 1 (1942).
11. J. L. Fergason, Polymer encapsulated nematic liquid crystals for scattering and light control applications, *SID Intnl. Symp. Digest Tech. Papers*, **16**, 68 (1985).
12. P. Becher, Emulsions, theory and practice, 3rd edn (Oxford University press, Oxford, 2001).
13. H. C. van de Hulst, *Light scattering by small particles*, (Dover Publications, New York, 1957).
14. S. Zumer and J. W. Doane, Light scattering from small nematic droplet, *Phys. Rev.* A, **34**, 3373 (1986).
15. G. P. Montgomery, J., Angle-dependent scattering of polarized light by polymer dispersed liquid-crystal films, *J. Opt. Am.* B, **5**, **774** (1988).
16. G. P. Montgomery, J. and N. Vaz, Light-scattering analysis of the temperature-dependent transmittance of a polymer-dispersed liquid-crystal film in its isotropic phase, *Phys. Rev.* A, **40**, 6580 (1989).
17. S. Zumer, Light scattering from nematic droplets: anomalous-diffraction approach, *Phys. Rev.* A, **37**, 4006 (1988).
18. J. R. Kelly, W. Wu, and P. Palffy-Muhoray, Wavelength dependence of scattering in PDLC film: droplet size effect, *Mol. Cryst. Liq. Crys.*, **223**, 251 (1992).
19. J. R. Kelly and W. Wu, Multiple-scattering effects in polymer-dispersed liquid-crystals, *Liq. Cryst.*, **14**, 1683 (1993).
20. J. W. Doane, A. Golemme, J. L. West, et al., Polymer dispersed liquid crystals for display application, *Mol. Cryst. Liq. Cryst.* **165**, 511 (1988).
21. R. Ondris-Crawford, E. P. Boyko, B. G. Wagner, et al., Microscope textures of nematic droplets in polymer dispersed liquid crystals, *J. Appl. Phys.*, **69**, 6380 (1991).
22. P. S. Drzaic, A new director alignment for droplets of nematic liquid crystal with low bend-to-splay ratio, *Mol. Cryst. Liq. Cryst.*, **154**, 289 (1988).
23. J. H. Erdmann, S. Zumer, and J. W. Doane, Configuration transition in a nematic liquid crystal confined to a small cavity, *Phys. Rev. Lett.*, **64**, 1907 (1990).
24. B. G. Wu, J. H. Erdmann, and J. W. Doane, Response times and voltages for PDLC light shutters, *Liq. Cryst.*, **5**, 1453 (1989).
25. H. Lin, H. Ding, and J. R. Kelly, The mechanism of switching a PDLC film, *Mol. Cryst. Liq. Cryst.*, **262**, 99 (1995).
26. P. S. Drzaic, Polymer dispersed nematic liquid crystal for large area displays and light valves, *J. Appl. Phys.* **60**, 2142 (1986).
27. W. Wu, Single and multiple light scattering studies of PDLC films in the presence of electric fields, Dissertation, Kent State University, 1999.
28. J. W. Doane, N. A. Vaz, B.-G. Wu, and S. Zumer, Field controlled light scattering from nematic microdroplets, *Appl. Phys. Lett.*, **48**, 269 (1996).
29. A. Tomita, P. Jones, Projection displays using nematic dispersions, *SID Intl. Symp. Digest Tech. Papers*, **23**, 579 (1992).
30. N. A. Vaz, G. W. Smith, and G. P. Montgomery, J., A light control film composed of liquid crystal droplets dispersed in a UV-curable polymer, *Mol. Cryst. Lid. Cryst.*, **146**, 1 (1987).
31. P. S. Drzaic, Nematic droplet/polymer films for high-contrast colored reflective displays, *Display*, 2–13 (1991).
32. J. L. West and R. Ondris-Crawford, Characterization of polymer dispersed liquid crystal shutters by ultraviolet/visible and infrared absorption spectroscopy, *J. Appl. Phys.*, **70**, 3785 (1991).
33. R. L. Sutherland, Bragg scattering in permanent nonlinear-particle composite gratings, *J. Opt. Soc. Am.* B, **8**, 1516 (1991).
34. R. L. Sutherland, V. P. Tondiglia, and L. V. Natarajan, Electrically switchable volume gratings in polymer-dispersed liquid crystal, *Appl. Phys. Lett.*, **64**, 1074 (1994).
35. T. J. Bunning, L. V. Natarajan, V. P. Tondiglia, et al., Holographic polymer-dispersed liquid crystals (H-PDLCs), *Annu. Rev. Mater. Sci.*, **30**, 83 (2000).
36. T. J. Bunning, L. V. Natarajan, V. P. Tondiglia, et al., Morphology of reflection holograms formed in situ using polymer-dispersed liquid crystals, *Polymer*, **14**, 3147 (1996).

37. K. Tanaka, K. Kato, S. Tsuru, and S. Sakai, Holographically formed liquid-crystal/polymer device for reflective color display, *J. SID*, **2**, 37 (1994).
38. G. P. Crawford, T. G. Fiske, and L. D. Silverstein, Reflective color LCDs based on H-PDLC and PSCT technologies, *SID Intl. Symp. Digest Tech Papers*, **27**, 99 (1996).
39. M. J. Escuti, P. Kosyrev, C. C. Bowley, et al., Diffuse H-PDLC reflective displays: an enhanced viewing-angle approach, *SID Intl. Symp. Digest Tech. Papers*, **31**, 766 (2000).
40. C. C. Bowley, A. K. Fontecchio, and G. P. Crawford, Electro-optical investigations of H-PDLCS: the effect of monomer functionality on display performance, *SID Intl. Symp. Digest Tech. Papers*, **30**, 958 (1999).
41. R. A. M. Hikmet, Anisotropic gels obtained by photopolymerization in the liquid crystal state, in *Liquid crystals in complex geometries*, ed. G. P. Crawford and S. Zumer (Taylor & Francis, London, 53–82, 1996).
42. D. J. Broer, Networks formed by photoinitiated chain cross-linking, in *Liquid crystals in complex geometries*, ed. G. P. Crawford and S. Zumer (Taylor & Francis, London, 239–255, 1996).
43. D.-K. Yang, L.-C. Chien, and Y. K. Fung, Polymer stabilized cholesteric textures: materials and applications, in *Liquid crystals in complex geometries*, ed. G.P. Crawford and S. Zumer (Taylor & Francis, London, 103–143,1996).
44. D. J. Broer, R. G. Gossink, and R. A. M. Hikmet, Oriented polymer networks obtained by photopolymerization of liquid crystal-crystalline monomers, *Die Angewandte Makromolekulare Chemie*, **183**, 45 (1990).
45. Y.K. Fung, D.-K. Yang, Y. Sun, et al., Polymer networks formed in liquid crystals, *Liq. Cryst.*, **19**, 797–901 (1995).
46. I. Dierking, L. L. Kosbar, A. C. Lowe, and G. A. Held, Two-stage switching behavior of polymer stabilized cholesteric textures, *J. Appl. Phys.*, **81**, 3007 (1997).
47. G. A. Held, L. L. Kosbar, I. Dierking, et al., Confocal microscopy study of texture transitions in a polymer stabilized cholesteric liquid crystal, *Phys. Rev. Lett.*, **79**, 3443 (1997).
48. R. Q. Ma and D.-K. Yang, Fréedericksz Transition in Polymer Stabilized Nematic Liquid Crystals, *Phys. Rev. E.* **61**, 1576 (2000).
49. Y. K. Fung, A. Borstnik, S. Zumer, et al., Pretransitional nematic ordering in liquid crystals with dispersed polymer networks, *Phys. Rev.* E, **55**, 1637 (1997).
50. R. A. M. Hikmet, Anisotropic gels and plasticised networks formed by liquid crystal molecules, *Liq. Cryst.*, **9**, 405 (1991).
51. R. A. M. Hikmet and H. M. J. Boots, Domain structure and switching behavior of anisotropic gels, *Phys. Rev.* E, **51**, 5824 (1995).
52. R. A. M. Hikmet, Electrically induced light scattering from anisotropic gels with negative dielectric anisotropy, *Mol. Cryst. Liq. Cryst.*, **213**, 117 (1992).
53. D.-K. Yang, L.C. Chien, and J.W. Doane, Cholesteric liquid crystal/polymer gel dispersion for haze-free light shutter, *Appl. Phys. Lett.*, **60**, 3102 (1992).
54. R.Q. Ma and D.-K. Yang, Polymer stabilized cholesteric texture reverse-mode light shutter: Cell design, *J. SID*, **6**, 125 (1998).
55. P. Nolan, M. Tillin, D. Coates, et al., Reflective mode PDLC displays – paper white display, *Proc. Euro-Display*, **93**, 397 (1993).
56. J. D. LeGrange, T. M. Miller, P. Wiltzius, et al., Brightness enhancement of reflective polymer-dispersed LCDs, *SID Intl. Symp. Digest Tech. Papers*, **26**, 275 (1995).
57. A. Kanemoto, Y. Matsuki, and Y. Takiguchi, Back scattering enhancement in polymer dispersed liquid crystal display with prism array sheet, *Proc. Intnl. Display Research Conf.*, **183** (1994).
58. H. J. Cornelissen, J. H. M. Neijzen, F. A. M. A. Paulissen, and J. M. Schlangen, Reflective direct-view LCDs using polymer dispersed liquid crystal (PDLC) and dielectric reflectors, *Proc. Intl. Display Research Conf.* **97**, 144 (1997).
59. P. S. Drzaic, Light budget and optimization strategies for display applications of dichroic nematic droplet/polymer films, *Proc. SPIE*, **1455**, 255 (1991).
60. Y. Ooi, M. Sekine, S. Niiyama, et al., LCPC project display system for HDTV, *Proc. Japan Display*, **92**, 113–116 (1992).
61. H. Yoshida, K. Nakamura, H. Tsuda, et al., Direct-view polymer-dispersed LCD with crossed Nicols and uniaxial film, *J. Soc. Inf. Display*, **2**, 135 (1994).
62. D.-K. Yang, Polymer stabilized liquid crystal displays, in *Progress in liquid crystal science and technology*, World Scientific, (2012).

63. D.-K. Yang, Y. Cui, H. Nemati, et al., Modeling aligning effect of polymer network in polymer stabilized nematic liquid crystals, *J of Appl. Phys.* **114**, 243515 (2013).
64. M. J. Escuti, C. C. Bowley, S. Zumer, and G. P. Crawford, Model of the fast-switching, polymer-stabilized IPS configuration, *SID Intl Symp. Journal of the SID*, **7**, 285–288 (1999).
65. J.-I. Baek, K.-H. Kim, J. C. Kim, et al., Fast switching of vertical alignment liquid crystal cells with liquid crystalline polymer networks, *Jpn. J. of Appl. Phys.* **48**, 056507 (2009).
66. Y.-Q. Lu, F. Du, Y.-H. Lin, and S.-T. Wu, Variable optical attenuator based on polymer stabilized twisted nematic liquid crystal, *Optics Express*, **12**, 1222 (2004).

12

Tunable Liquid Crystal Photonic Devices

12.1 Introduction

In addition to displays, liquid crystals have also been used extensively in tunable photonic devices, such as optical phased array for laser beam steering, variable optical attenuator (VOA) for telecommunications, tunable-focus lens for camera zoom lens, LC-infiltrated photonic crystal fibers [1,2], diode laser-pumped dye-doped LC laser, just to mention a few.

The performance criteria for displays and photonics are quite different. First, most displays such as computers and TVs are operating in the visible spectral region, but many photonic applications are aimed for the infrared. The laser beam steering for free-space communications is operated at $\lambda = 1.55\ \mu m$, and so are VOAs and photonic fibers. Second, most displays use amplitude modulation, but some photonics use phase modulation. For intensity modulation, the required phase retardation is 1π at $\lambda = 550$ nm, but for phase modulation, say beam steering, the minimum phase change is 2π at $\lambda = 1550$ nm. If we use the same LC material, then the required cell gap for a laser beam steerer is about six times thicker than that for a display device. A thicker cell gap also implies a slower response time because the response time of an LC device is proportional to the cell gap square. Third, for display applications the 'detector' is the human eye but for beam steering or fiber communication the detector is a solid state diode. The latter has a much faster response time than the former. Based on the abovementioned three key performance factors, the technical challenges for a near infrared phase modulator is at least one order of magnitude harder than for visible displays. Of course, most displays require a wide viewing angle, which is not so demanding in beam steering and VOA.

In this chapter, we only select four topics to illustrate the potential applications of liquid crystals in photonics and their technical challenges. The four representative subjects selected are: (1) laser beam steering, (2) variable optical attenuator, (3) tunable-focus lens, and (4) polarization-independent LC devices.

Fundamentals of Liquid Crystal Devices, Second Edition. Deng-Ke Yang and Shin-Tson Wu.

12.2 Laser Beam Steering

Laser beam steering is an important subject for free-space communications, military, optical interconnects, projection displays, and other general industrial applications. The goal is to deliver and control precisely the laser beams to a desired location. The most common technique is to reflect the light by mechanically controlled mirrors. Because of the nature of mechanical movement, the speed of the system is limited. It is always desirable to develop compact and lightweight non-mechanical beam-steering devices to replace the large and bulky mechanical systems. The other well-established beam steering device is the acousto-optic modulator, which has a severely limited angular range. Many new solid state/micro-component technologies such as optical micro-electro-mechanical system (MEMS), patterned liquid crystals, diffractive micro-optics, and photonic crystals have been investigated for building small, ultra-light, rapidly steered laser beam subsystems.

Two types of liquid crystal electro-optic beam steering devices have been developed: diffractive [3] and prismatic types [4,5]. A conventional simple grating structure produces several diffraction orders (first and higher orders). The theoretical diffraction efficiency of the first-order beam is about 34% [6]. The laser holographic blazing process shapes the grooves of the grating to concentrate the light into the first order. The result is a much brighter spectrum. A well-designed LC prism grating could reach 100% diffraction efficiency for the first-order beam. A simple method for fabricating an LC blazed grating is to use a glass substrate with a sawtooth surface structure, as Figure 12.1 shows [7]. The LC layer thickness in this structure is

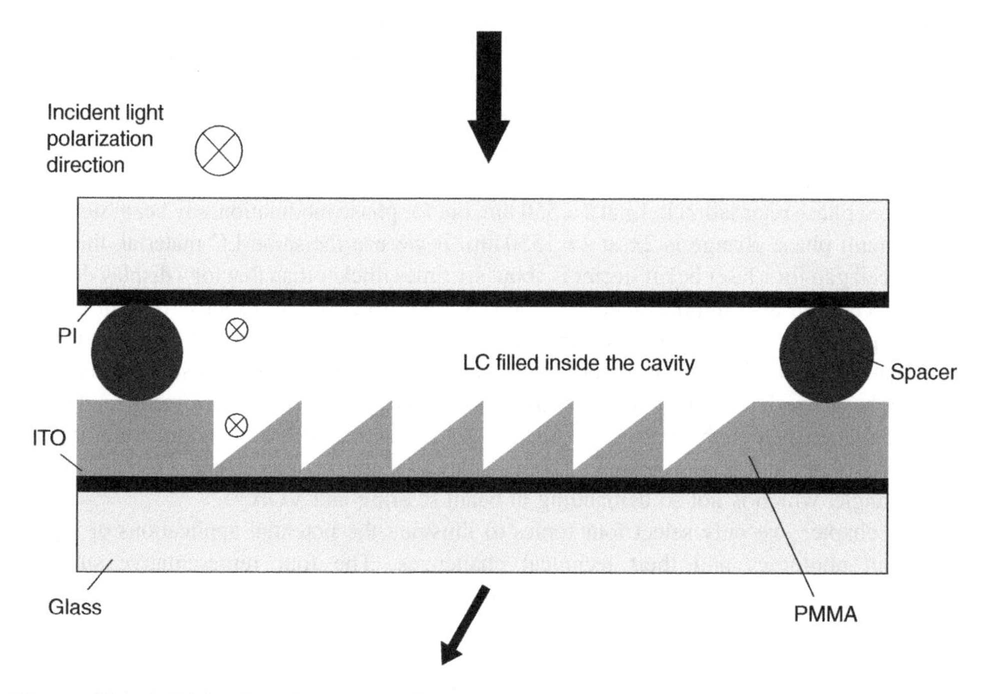

Figure 12.1 LC blazed-grating beam deflector by using a glass substrate with a sawtooth surface structure. PI: polyimide; ITO: indium-tin-oxide; PMMA: poly(methyl-methacrylate).

varying periodically and asymmetrically. The sawtooth substrate is a poly(methyl-methacrylate) (PMMA) blazed grating made by electron-beam lithography. The refractive index of PMMA is close to the ordinary index of the selected LC, but smaller than the extraordinary index. Therefore, in the voltage-off state, the in-plane distribution of the optical path length of the incident light has a sawtooth profile. The blazed grating would be switched off when the applied voltage is sufficiently high.

12.2.1 Optical phased array

The above surface relief grating causes distortion in the LC alignment and might degrade the diffraction efficiency. It is better to use a sawtooth electric field to generate an LC blazed grating on a uniform LC layer, as shown in Figure 12.2. Several approaches have been reported to achieve the sawtooth electric field distribution, such as LC spatial light modulators, beam steerers with a stripe electrode [8,9], and a combination of low- and high-resistive electrodes [10].

Figure 12.2 illustrates a transmission-mode optical phased array beam steerer composed of a one-dimensional (1-D) array of equal-spaced liquid crystal phase shifters [11]. The inner surface of the bottom transparent substrate is photo-lithographically patterned with transparent, conducting, striped electrodes having the desired spacing for the phase shifters in the array. The entire active aperture of the array is so patterned. The inner side of the top substrate is coated with a uniform transparent conducting ground electrode. On top of the electrodes, a thin SiO_2 layer is deposited to align the LC molecules. For phase-only modulation, homogeneous alignment with a small (~2–3°) pretilt angle is preferred. When the applied voltage between any given striped electrode and the underlying ground plane exceeds the Fréedericksz transition

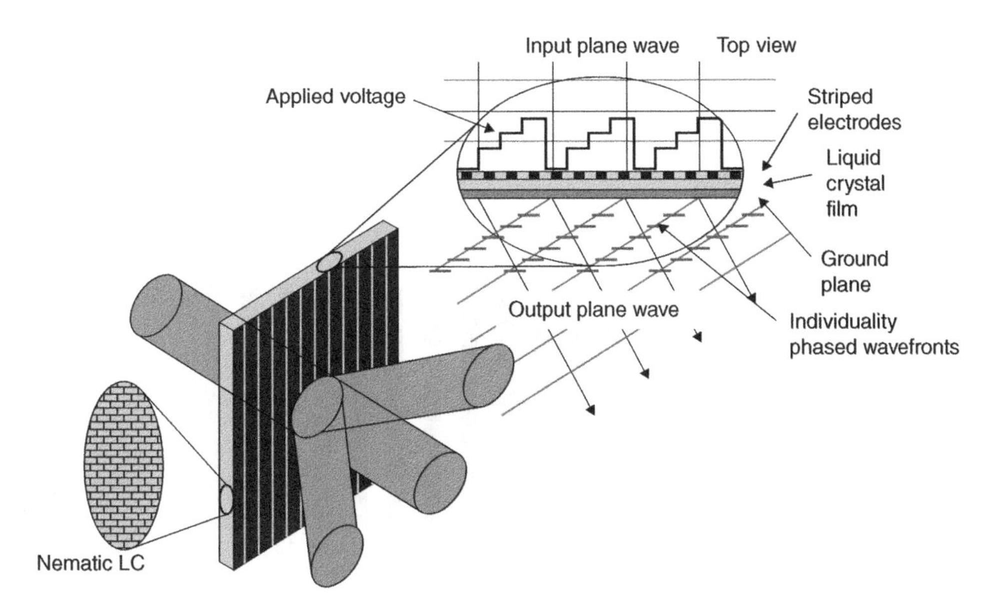

Figure 12.2 Schematic drawing of a 1-D OPA and a conceptual imposed phase shift.

threshold, a 1-D phase shifter in the liquid crystal volume underlying the patterned electrode is created.

The degree of nematic LC reorientation depends on the applied voltage [12]. Therefore, by controlling the applied voltage to each pixel, a stair-step blazed phase grating can be generated. Application of a periodic sequence of staircase voltage ramps of period Λ across the array aperture creates a corresponding periodic staircase profile of phase, as shown in the expanded portion of Figure 12.2. With properly weighted voltage steps, chosen to compensate for the non-linear phase-voltage profile of a typical liquid crystal, a linearly increasing phase profile can be produced. If the maximal phase shift on each staircase ramp of N voltage steps is $2\pi(N-1)/N$, the periodic (modulo 2π) phase profile is equivalent to a single staircase phase ramp across the aperture. An electronically adjustable prism is simulated, and the beam is steered to a new direction θ (relative to the phased array boresite) given by the general grating equation,

$$\sin\theta + \sin\theta_{inc} = \lambda_o/\Lambda, \tag{12.1}$$

where θ_{inc} is the incident angle of the beam, and Λ is the period of the programmed grating. For a normally incident laser beam, $\theta_{inc} = 0$ and Equation (12.1) is simplified. The steering direction depends on the periodicity (and sign) of the applied voltage ramp. The diffraction efficiency η of a grating with a stair-step blaze designed to maximize energy in the first order is related to the number of steps N as [13]

$$\eta = [\sin(\pi/N)/(\pi/N)]^2. \tag{12.2}$$

Figure 12.3 plots the diffraction efficiency as a function of N. The efficiency grows rapidly with N and reaches ~95% for an eight-stair-step OPA. The remaining 5% is diffracted to higher orders, called sidelobes. Although increasing the number of stair steps would enhance the diffraction efficiency slightly, its fabrication complicity also increases. In practice, eight stair steps are normally used.

The OPA can be operated in reflective mode, provided that the bottom transparent electrode is replaced by a reflector. To achieve the same phase change, the required cell gap can be

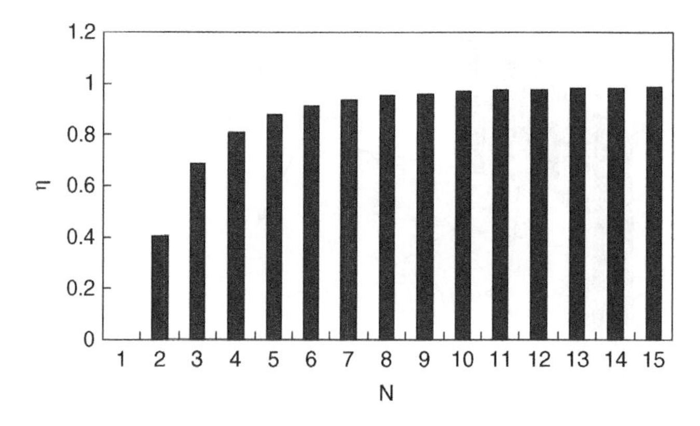

Figure 12.3 Diffraction efficiency of a blazed grating as a function of the number of phase steps.

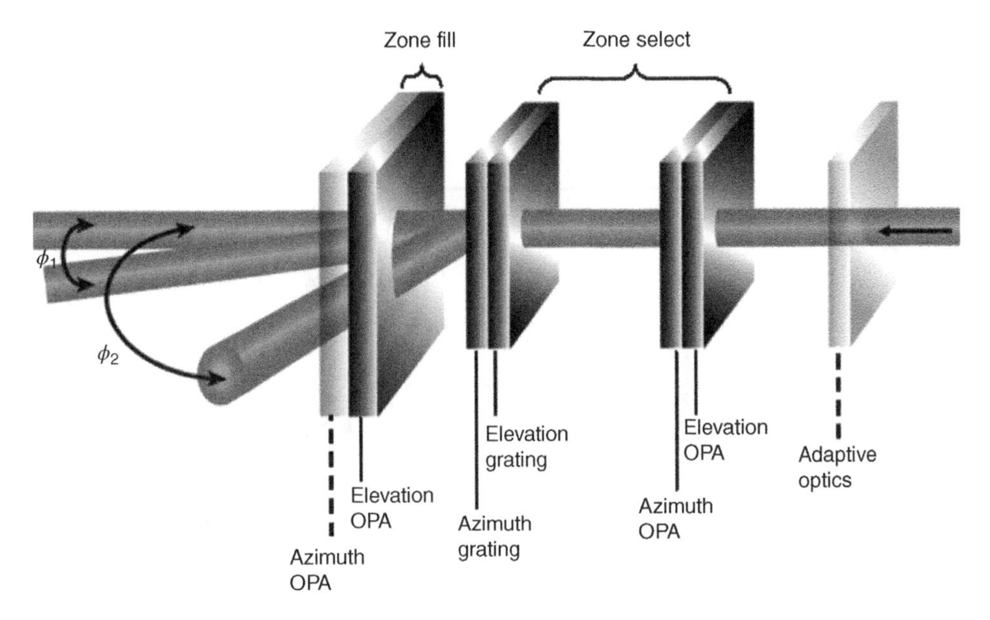

Figure 12.4 A 2-D beam steerer using zone select and zone fill OPAs in conjunction with two gratings.

reduced to half because the double pass of the incident beam. As a result, the response time is four times faster.

Figure 12.4 shows a two-dimensional (2-D) steering by cascading two 1-D steering arrays with crossed electrode patterns: one for azimuth (AZ), and one for elevation (EL). The cascading can use individual discrete steering devices. Relay lenses can be used to avoid beam walkoff between the devices, as is sometimes done with mechanical steerers [14]. However, the inherently thin (<0.5 mm) format of these LC cells has obviated the need for such relay lenses. The availability of both reflection- and transmission-mode variants facilitates the design of cascaded systems. An attractive prospect is the integration of AZ and EL steering units into a single thin cell, potentially conformal with an airframe. Current OPA designs are polarization dependent; they require the input beam to be linearly polarized along the direction of quiescent liquid crystal alignment for maximal efficiency. Later in this chapter, we will discuss two potential approaches that are polarization independent.

Since an OPA has a relatively small steering angle, to achieve large-angle steering an auxiliary diffraction grating is needed. In Figure 12.4, the first two OPAs in the incident beam side and the double grating made of photo-thermal refractive glass [15,16], form the large angle (ϕ_2) beam steerer for zone selection (also called a coarse beam steerer). Once the laser beam is steered to a designated zone, the last two OPAs are used to fill the zone (also called fine beam steerer).

12.2.2 Prism-based beam steering

Figure 12.5 shows a prism-based beam steering device using a birefringent prism with wedge angle α and a switchable 90° TN cell as a polarization rotator. At $V = 0$, the TN cell rotates the input linearly polarized light by 90°, which acts as an extraordinary light (refractive index is n_e)

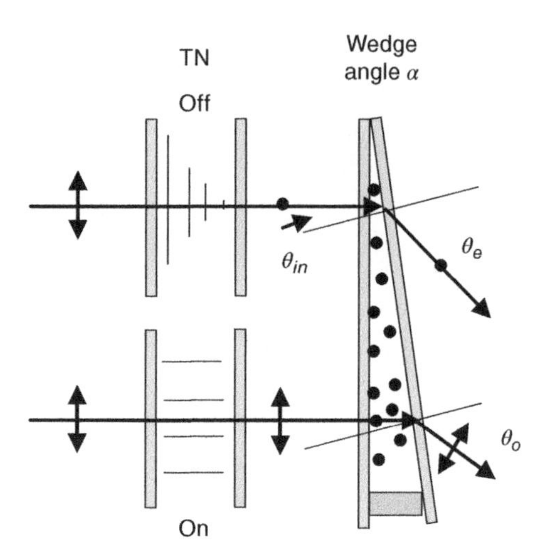

Figure 12.5 The operation principles of a prism-type beam steering device using a 90° TN cell as a polarization rotator.

to the prism. The beam is deflected to an angle θ_e. When the TN cell is activated, the LC directors are reoriented perpendicular to the substrate surfaces so that the polarization rotation effect vanishes. The polarization of the incident light is not affected by the TN cell. As a result, the outgoing light behaves like an ordinary beam (refractive index is n_o) to the prism and is deflected to an angle θ_o.

From Snell's law, we can calculate the exit angles θ_e and θ_o:

$$n_e \sin\theta_{in} = \sin\theta_e \tag{12.3a}$$

$$n_o \sin\theta_{in} = \sin\theta_o \tag{12.3b}$$

The difference between θ_e and θ_o is the steering angle. The prism with a larger birefringence would lead to a larger steering angle.

As shown in Figure 12.5, both beams walk off from the original beam path. To correct this walk-away phenomenon, a birefringent/isotropic bi-prism concept has been developed [17]. Figure 12.6 shows the device structure and operating mechanisms of the birefringent/isotropic bi-prism. The refractive index of the top isotropic prism is chosen to be equal to the n_o of the birefringent prism. Thus, the impinging ordinary ray is not deflected while the extraordinary ray is deflected. Similarly, a 90° TN cell is used to control the input polarization. Using Snell's law, it is fairly easy to correlate the steering angle with the prism's wedge angle and the refractive indices of the birefringent prism as follows:

$$\delta = \sin^{-1}\left(n_o \sin\left[\left(\sin^{-1}\left((n_e/n_o)\sin\alpha\right)\right) - \alpha\right]\right. \tag{12.4}$$

Let us assume that the refractive indices of the birefringent prism (it can be a crystal or liquid crystal) are $n_e = 1.7$ and $n_o = 1.5$. Then, in order to get 1° steering angle, the required prism angle should be 5°. By stacking N basic units whose α values are in binary sequence, then we can obtain 2^N beam steering positions [18].

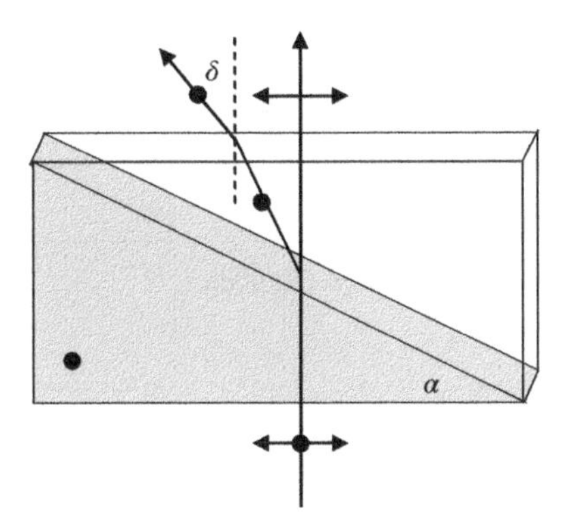

Figure 12.6 Device structure of a birefringent/isotropic bi-prism beam steerer.

12.3 Variable Optical Attenuators

LC-based variable optical attenuators (VOAs) have been developed for fiber optic communications at $\lambda = 1.55$ μm because of their low cost, low loss, and low power consumption [19–21]. For most telecommunications applications, a fast response time and a large dynamic range (>30 dB) are required. Two types of LC-VOAs have been developed, nematic and ferroelectric [22]. Ferroelectric liquid crystal (FLC) is attractive for its microsecond response time, but there are several challenging issues remain to be solved: (1) it is a bistable device, (2) its ultra-thin cell gap ($d < 2$ μm), (3) residual DC voltage, (4) mechanical robustness, and (5) long-term reliability. Because FLC is a bistable device, to obtain gray scales a pulse width modulation method has to be implemented. The thin-cell requirement lowers the manufacturing yield and the residual DC voltage causes gray-scale instability. Moreover, the molecular alignment in an FLC cell is sensitive to mechanical disturbance. By contrast, nematic VOA is easy to fabricate and it has natural gray scales. The major disadvantage is a slow response time. To achieve a fast response time, the following have been investigated: a small cell gap [23], the high-temperature effect [24,25], dual-frequency LC materials [26–28], the voltage effect [29], and a polymer-network LC [30]. A typical nematic LC-based VOA has a response time of approximately 5–15 ms, which is still slower than a mechanical shutter whose response time is approximately 1 ms. To outperform the mechanical shutter, the nematic VOA should have a submillisecond response time at room temperature while maintaining a wide dynamic range and low operating voltage (≤20 V_{rms}).

In Chapter 6, we introduced dual-frequency liquid crystal (DFLC) materials and the operating mechanisms for achieving fast response time. In this section, we will demonstrate a fast-response and wide-dynamic-range nematic VOA using a high-birefringence and low-viscosity DFLC together with the overdrive and undershoot voltage method described in Chapter 8. To achieve a submillisecond response time at room temperature (T ~ 21 °C), we use a low-frequency ($f = 1$ kHz) overdrive voltage to decrease rise time, and a high-frequency ($f = 30$ kHz) undershoot voltage to accelerate the decay process. The measured dynamic range exceeds 40 dB at $\lambda = 1.55$ μm.

Figure 12.7 shows a schematic diagram of the LC-based VOA, where two polarization beam displacers and an LC cell are sandwiched between two identical fiber collimators with an 80 mm working distance. The light from the input fiber is collimated by the first gradient index

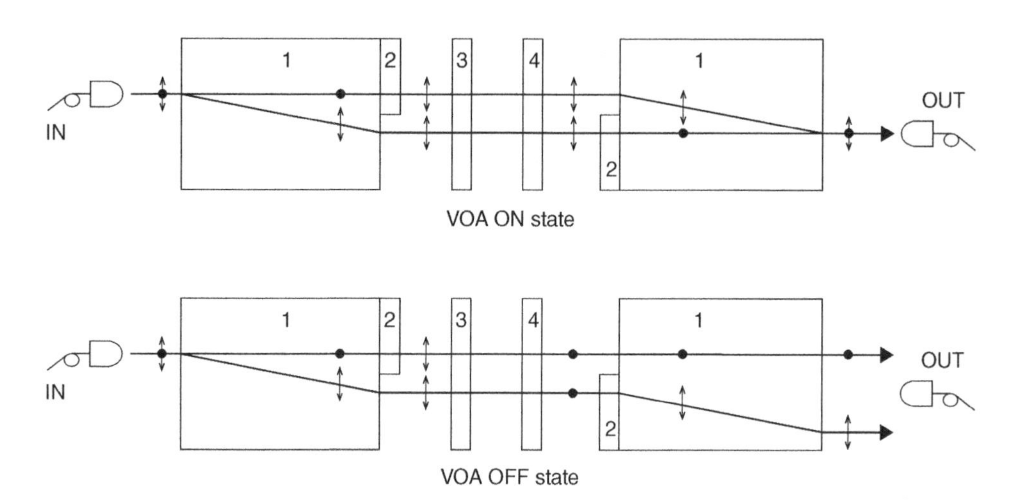

Figure 12.7 Schematic diagram of DFLC-based variable optical attenuator (1 = polarization beam displacer, 2 = half-wave plate, 3 = master LC cell, and 4 = compensation cell).

(GRIN) lens collimator. When the light is incident to the first polarization beam displacer (PBD) – a calcite crystal, 10 mm thick, with a 45° cut – it is separated into an ordinary beam and an extraordinary beam. A quartz half-wave plate (HWP) is laminated to the calcite beam displacer to rotate the polarization state of the top beam. Therefore, both beams have the same polarization before entering the LC cell which is a 3.7 μm homogeneous cell with its rubbing direction oriented at 45° to the input light polarization. Its phase retardation ($\delta = 2\pi d\Delta n/\lambda$) is approximately 1.2π at $\lambda = 1.55$ μm. To ensure a high transmittance at $V = 0$, an identical LC cell (i.e. with the same cell gap, LC material and alignment, but with no voltage applied) is placed behind the master LC cell to act as a phase compensation cell. The rubbing direction of the compensation cell is orthogonal to that of the master cell so that the net phase retardation at $V = 0$ is zero. This master compensation cell configuration has an excellent wavelength tolerance [31]. Under such circumstances, the top and bottom beams are recombined by the second PBD and HWP and then coupled into the collecting fiber collimator, as shown in the upper part of Figure 12.7. This is the high-transmittance state of the VOA. To make the device compact, the compensation LC cell can be replaced with a polymeric film, which is used in display devices to increase the viewing angle [32].

When a proper voltage is applied to the master LC cell to make a π phase change, the incident beams could not retain their original polarizations. As a result, they are separated by the second beam displacer. No light is coupled into the collecting fiber collimator, and the off-state results, as shown in Figure 12.7 (lower). By tuning the master cell voltage, different gray scales can be obtained. If the LC cell gaps are all uniform, then the VOA should have no polarization-dependent loss (PDL) and no polarization mode dispersion (PMD).

The VOA performance is mainly determined by the LC material employed. To achieve a fast response time, a DFLC mixture is chosen for the demonstration. The key feature of a DFLC is that it exhibits a crossover frequency (f_c). In the $f < f_c$ region, the dielectric anisotropy ($\Delta\varepsilon$) is positive, while in the $f > f_c$ region the $\Delta\varepsilon$ becomes negative. In the low-frequency region, the electric-field-induced torque reorients the LC molecules along the field direction. This leads to

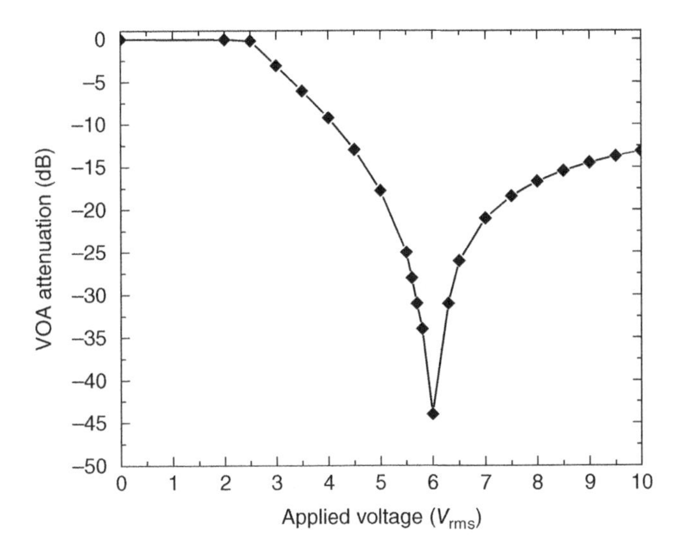

Figure 12.8 Measured VOA attenuation on a dB scale as a function of applied voltage. The VOA is addressed by a 1 kHz square-wave AC source.

the fast rise time. During the relaxation period, a high-frequency electric field is applied to the cell. Because the $\Delta\varepsilon$ is negative, the high-frequency electric field helps to accelerate the relaxation of the LC molecules to their original positions. As a result, a fast decay time is achieved.

Most of the commercially available DFLC mixtures have low birefringence, high viscosity, and small $|\Delta\varepsilon|$ values. Due to their low birefringence, a thick LC layer is required, particularly for the 1.55 μm infrared wavelength. The thick LC layer leads to a slow response time and a high operating voltage. To overcome these drawbacks, S.-T. Wu's group developed a high-birefringence and low-viscosity DFLC mixture using 30% biphenyl esters and 70% lateral difluoro tolanes [33]. The physical properties of the DFLC mixture at room temperature (T = 21 °C) are summarized as follows: crossover frequency $f_c \approx 4$ kHz; $\Delta n = n_e - n_o = 0.25$ at $\lambda = 1.55$ μm; $\Delta\varepsilon = 4.73$ at $f = 1$ kHz; and $\Delta\varepsilon = -3.93$ at $f = 30$ kHz.

For VOA demonstration, an Ando AQ4321-D tunable laser operated at $\lambda = 1.55$ μm is used as a light source. The output fiber is connected to an Ando AQ8201-21 power monitor for measuring transmittance. A computer-controlled LabVIEW system is used for data recording and processing. The insertion loss of the DFLC VOA at $V = 0$ is about −2.0 dB (without connector). The PDL remains less than 0.1 dB over the whole International Telecommunication Union (ITU) C-band (1.53–1.57 μm) as expected. Although the measured insertion loss is still not sufficiently low, the actual fiber-to-fiber coupling loss is only ~0.8 dB when taking into account the ~1.2 dB propagation loss which is mainly contributed by the uncoated LC cells.

Figure 12.8 plots the measured voltage-dependent VOA attenuation. The VOA is addressed by square waves at $f = 1$ kHz. Because of the positive $\Delta\varepsilon$ in the low-frequency region, the LC directors are reoriented along the electric field direction as the voltage exceeds 2.5 V_{rms} (threshold voltage). At $V = 6\ V_{rms}$, which corresponds to a π phase change, an off-state with −43 dB attenuation is achieved. As shown in Figure 12.8, this off-state is quite stable. Within ±0.3 V_{rms} voltage variation, the measured attenuation remains at over −30 dB, which is important if

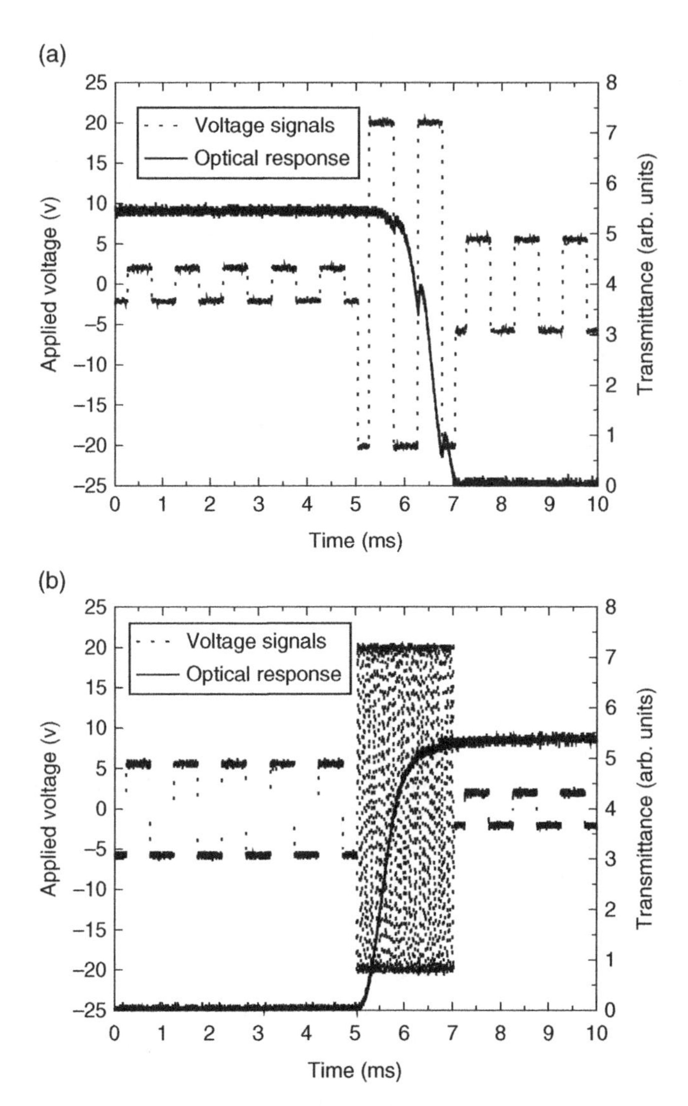

Figure 12.9 Rise (a) and decay (b) time of the DFLC VOA using overdrive and undershoot voltages and corresponding optical responses. $\lambda = 1.55\ \mu m$.

this VOA is to be used as a light switch or wavelength blocker. As the applied voltage exceeds 6 V_{rms}, the net phase change due to the orthogonal master and compensation cells is deviated from 1π so that the optimal off-state condition is no longer satisfied. As a result, the VOA attenuation gradually decreases.

Figure 12.9 shows the turn-on and turn-off times of the DFLC cell with the overdrive and undershoot voltages applied during the rise and decay periods. The commercial LC-VOA is normally driven by a 20 V AC voltage source. For a fair comparison, the overdriving voltage of the DFLC cell is also limited to 20 V_{rms}. Figure 12.9(a) shows the applied waveforms and the corresponding optical signals. The 20 V_{rms} low-frequency voltage burst is applied for 2 ms between the 2 V_{rms} bias and 6 V_{rms} holding voltages. The turn-on time (90–10%) is 0.73 ms for the 3.7 μm DFLC cell.

During the relaxation process, the constant bias voltage exerts a torque to resist LC molecules returning to their original positions. To overcome this bottleneck, a high-frequency (f = 30 kHz) voltage is imposed before applying the bias voltage, as depicted in Figure 12.9(b). This is known as the undershoot effect. The decay time (10–90%) is reduced to 0.78 ms, which is approximately an order of magnitude faster than that of a commercial LC-VOA.

The response times shown above are all between the VOA's on- and off-states. However, the dual frequency overdriving and undershooting can also apply to the fast gray-scale transition between two arbitrary attenuation states. A high-voltage (e.g. 20 V_{rms}) burst of suitable frequency and duration may be inserted between the initial and target states to accelerate the LC director's rise or decay.

12.4 Tunable-Focus Lens

A mechanical zoom lens typically consists of two groups of lenses. It would be ideal if these lenses were replaced by a single tunable-focus LC lens. To make an LC lens, the gradient refractive index profile has to be created. Generally speaking, there are three approaches to generating the desired gradient refractive index: (1) homogeneous LC layer with inhomogeneous electric field, (2) inhomogeneous LC layer with homogeneous electric field, and (3) inhomogeneous LC layer with inhomogeneous electric field. To achieve these goals, various efforts such as surface relief profile [34–36], line- or hole-patterned electrode [37–40], Fresnel zone type [41,42], modal control [43], as well as polymer network LC technique [44,45], have been tried. Of these approaches, the surface relief lens which combines a passive solid-state lens and an LC modulator possesses some attractive features, such as simple fabrication, single electrode, and its being easy to realize a spherical phase profile within the LC layer. However, the LC lens with curved inner surfaces would scatter light due to the poor molecular alignment. Moreover, its focus tunable range is rather limited.

12.4.1 Tunable-focus spherical lens

In this section, we use an example to illustrate the fabrication procedure and operation principle of a tunable-focus spherical lens. Unlike the surface relief LC lens, the present lens has planar substrates and a uniform LC layer. To create an inhomogeneous electric field, one of the flat substrates has an imbedded spherical electrode and the other has a planar electrode. The electric field from the spherical and planar electrodes induces a centro-symmetric gradient refractive index distribution within the LC layer which, in turn, causes the focusing effect. The electric field strength will affect the LC alignment and then change the refractive index profile. As a result, the focal length can be tuned by the applied voltage. Unlike the non-uniform LC layer approach, this lens exhibits a uniform optical response across the lens aperture due to the homogeneous cell gap. No light scattering or diffraction occurs because of the homogeneous LC alignment and continuous electrode. Both positive and negative lenses can be realized by simply reversing the shape of the spherical electrode.

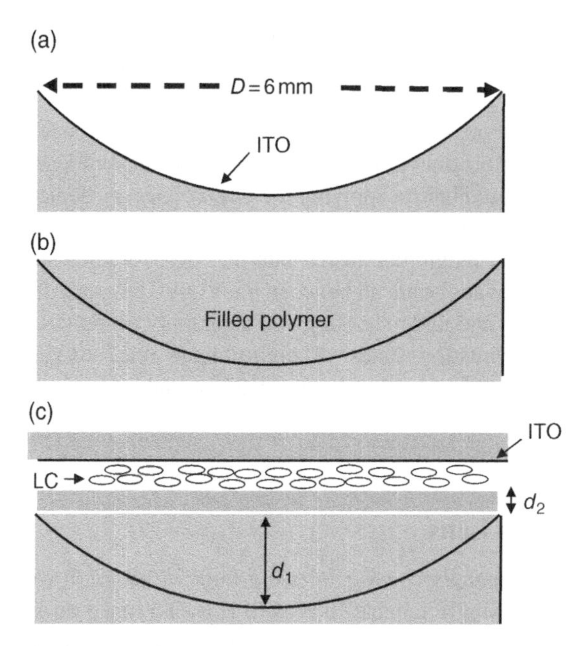

Figure 12.10 Fabrication process of a spherical LC lens: (a) Deposit ITO on a concave glass lens, (b) fill the sag area with polymer, and (c) assemble the LC lens cell with another flat glass substrate. Ren 2004. Reproduced with permission from the American Institute of Physics.

Figure 12.10 illustrates the fabrication procedures of such a positive spherical LC lens. The concave surface of the bottom glass substrate is coated with a transparent indium-tin-oxide (ITO) electrode – Figure 12.10(a). Next, the sag area could be matched by a convex glass lens with the same curvature or filled with a polymer having the same refractive index as the employed glass substrate to form a planar substrate – Figure 12.10(b). In this case, let us assume the sag area is filled with a UV curable prepolymer. The glass substrate which is in contact with the prepolymer has $d_2 = 0.55$ mm but no ITO electrode. When the prepolymer is cured by UV, the lens and the LC cell are attached together. The inner surfaces of the LC cell are coated with polyimide alignment layers and rubbed in an anti-parallel direction. The pretilt angle is ~3°. The empty LC mixture is filled with a high birefringence LC mixture and a homogeneous alignment is induced by the buffed polyimide layers – Figure 12.10(c).

In Figure 12.10, the concave glass lens has a radius $R = -9.30$ mm, aperture $D = 6$ mm, and sag $d_1 = 0.34$ mm (BK7 glass, $n_g = 1.517$). The refractive index of the filled polymer NOA65 ($n_p \sim 1.524$, Norland Optical Adhesive) will affect the initial focal length of the LC lens. If $n_p \approx n_g$, then the LC device would not focus light in the voltage-off state. But if n_p is much smaller than n_g, then the device will have an initial focus. To reduce the cell gap for keeping a reasonably fast response time, a high birefringence LC mixture designated as UCF-2 ($\Delta n = 0.4$ at $\lambda = 633$ nm) [46] is used. The cell gap is 40 μm.

Figure 12.11 shows the CCD images of the lens at three voltage states: $V = 0$, 23, and 35 V_{rms}. At V = 0, the observed He–Ne laser beam is not very uniform due to its Gaussian intensity distribution. The peak intensity is ~6×10^3 arbitrary units. As the voltage reaches 23 V_{rms}

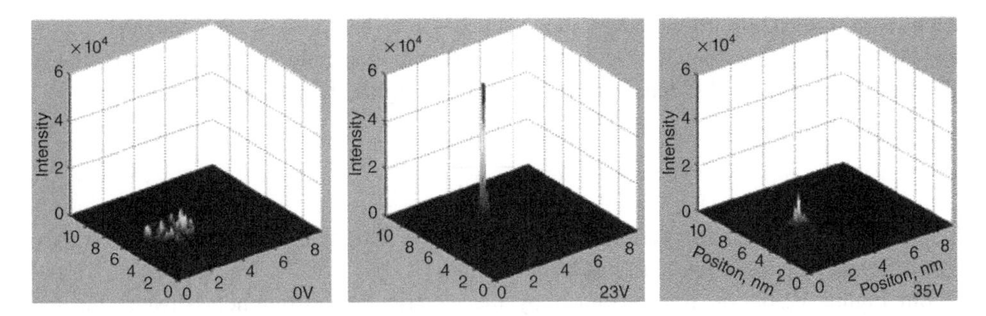

Figure 12.11 CCD images of the measured He–Ne laser beam intensity profile at V = 0, 23, and 35 V_{rms}. Ren 2004. Reproduced with permission from the American Institute of Physics.

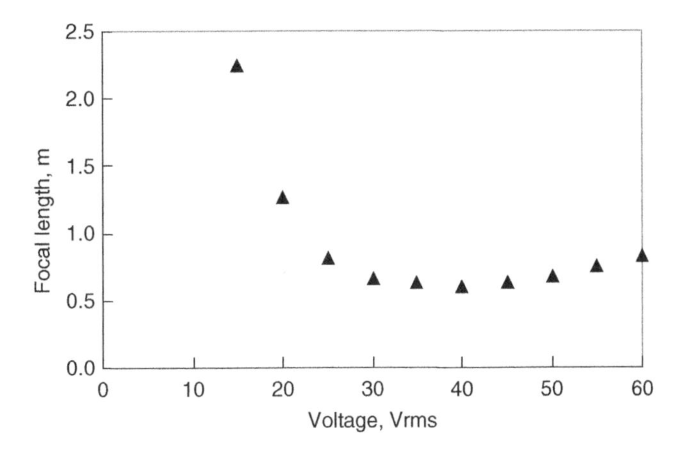

Figure 12.12 Voltage-dependent focal length of the flat LC spherical lens. Lens aperture D = 6 mm, LC: UCF-2, cell gap d = 40 μm and λ = 633 nm. Ren 2004. Reproduced with permission from the American Institute of Physics.

the focusing effect manifests. The measured intensity at the CCD focal plane exceeds 6.5×10^4 arbitrary units. As the voltage is further increased, the peak intensity of the outgoing beam tends to decrease. At V = 35 V_{rms}, the peak intensity drops to 1.7×10^4 arbitrary units. This is because the LC directors in the bulk have been reoriented by the electric field and the curvature of the refractive index profile is gradually flattened. As a result, the focal length of the lens increases and the measured light intensity at the CCD focal plane decreases.

Figure 12.12 plots the voltage-dependent focal length of the flat LC spherical lens. At $V = 0$, LC directors are aligned homogeneously due to the surface anchoring effect from the substrates. Thus, no focusing effect occurs, that is the focal point is at infinity. As the voltage increases, the focal length is reduced because of the established gradient refractive index. At V ~ 40 V_{rms}, the focal length reaches a minimum (f ~ 60 cm). Further increasing the voltage would cause the focal length to bounce back, but at a different rate. The response time of the lens is around 1–2 s at room temperature. This is somewhat too slow. For practical applications, the switching time should be less than 5 ms to avoid image blurring during focus change.

The focal length of an LC lens can be calculated using the Fresnel's approximation:

$$f = \frac{r^2}{2\delta n d_{LC}}, \tag{12.5}$$

where $r = D/2$ (D is the lens aperture), d_{LC} is the LC layer thickness, and δn is the refractive index difference between the lens center and border.

From Figure 12.10, δn is determined by the electric field difference between the lens center and the border. When a voltage V is applied to the lens cell, the electric field in the center (E_{center}) and at the border (E_{border}) is expressed as follows:

$$E_{center} = \frac{V/\varepsilon_{LC}}{\frac{d_{LC}}{\varepsilon_{LC}} + \frac{d_2}{\varepsilon_2} + \frac{d_1}{\varepsilon_1}}, \tag{12.6}$$

$$E_{border} = \frac{V/\varepsilon_{LC}}{\frac{d_{LC}}{\varepsilon_{LC}} + \frac{d_2}{\varepsilon_2}}, \tag{12.7}$$

where ε_{LC}, ε_2, and ε_1 represent the dielectric constant of the LC, medium 2, and medium 1, respectively. In an ideal case, the glass substrate which is closer to the spherical electrode should be eliminated, that is, $d_2 \sim 0$. In such a condition, the electric field shielding effect resulting from the glass substrate is minimized and the required operating voltage is lowered. From Equation (12.5), the shortest focal length occurs when $\delta n = \Delta n$, that is, the LC directors in the border are completely reoriented by the electric field while those in the center are not yet reoriented owing to the weaker electric field.

Can the abovementioned flat spherical lens be used for eyeglasses? For an eyeglass, the aperture (D) needs to be at least 3 cm and the focal length should be around 25 cm. From Equation (12.5), if we use an LC material with $\delta n \sim 0.4$, the required cell gap is ~1.1 mm. For such a thick LC cell, the response time would be very sluggish. Moreover, the LC inside the cell will probably not align well so that light scattering will occur. Thus, the lens design shown in Figure 12.10 may not be practical for large aperture eyeglasses. It is more realistic for millimeter-sized lens apertures.

12.4.2 Tunable-focus cylindrical lens

A cylindrical lens focuses light into one dimension. It can be used for stretching an image, focusing light into a slit, converging light for a line scan detector or correcting a low-order aberration. For a solid cylindrical lens, its focal length is fixed. To get a variable focal length, a group of lenses (e.g. a mechanical zoom lens) is often necessary. However, this makes the optical system bulky and costly. An alternative approach for obtaining a variable focal length is through the use of LC-based cylindrical lenses for which several methods have been considered and proposed [47–49], Among them, lenses with a slit electrode are particularly

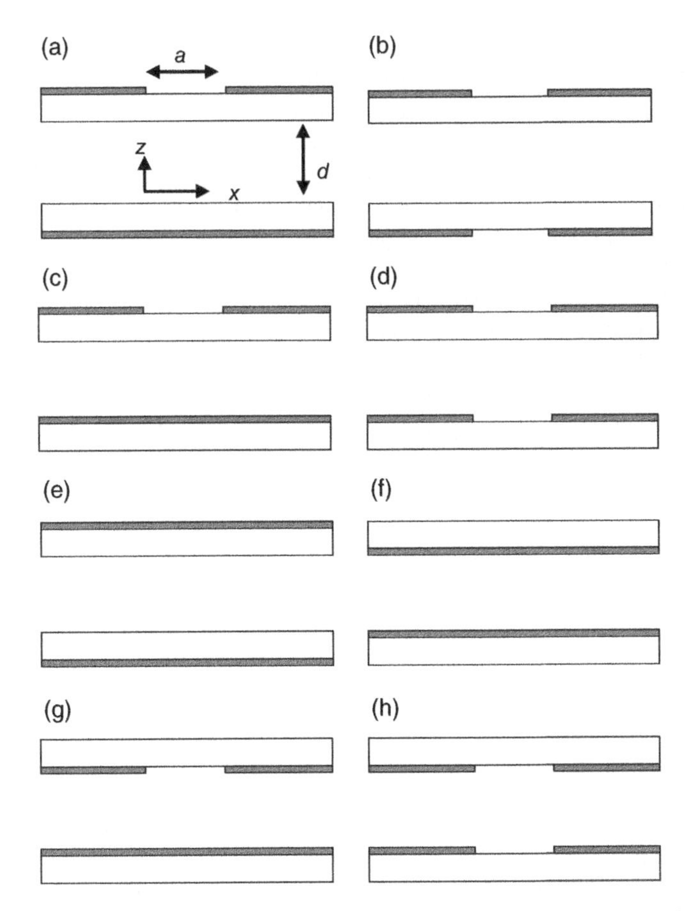

Figure 12.13 The eight possible electrode configurations considered for making a cylindrical LC lens. a = slit width and d = cell gap. The rubbing direction is along y axis and the polarization of the incident light is along the rubbing direction.

interesting, due to their simple fabrication, simple operation, and the possibility of widening the aperture size.

Figure 12.13 shows eight possible electrode configurations for generating electric fields for creating cylindrical lenses [50]. However, configurations E and F are not suitable for forming lenses due to the lack of inhomogeneous electric field. The structures in G and H are more suitable for making cylindrical microlens arrays than a single large-aperture lens due to their narrow electrode gaps. The aperture size of a micro-sized cylindrical LC lens is usually in the 100 μm range. Thus, its application is limited to microlenses or a microlens array.

In order to have a large aperture size, four possible configurations have been considered as shown in A to D. In configuration A, a slit electrode is coated on the outer surface of the top substrate, whereas a continuous electrode is coated on the outer surface of the bottom substrate. In configuration B, a slit electrode is coated on the outer surface of both top and bottom substrates. The two slits are parallel and symmetrical. In configuration C, a slit electrode is coated on the outer surface of the top substrate whereas a continuous electrode is coated on the inner surface of the bottom substrate. In configuration D, a slit electrode is coated on the outer surface of the top substrate whereas another slit electrode is coated on the inner surface of the bottom substrate. The two slits are also parallel and symmetrical. Each of the LC cells has the same cell gap $d = 40$ μm. The slit spacing, called the aperture width, of the slit electrode is $a = 2$ mm. The inner surfaces of both top and bottom substrates are coated with polyimide and buffed in anti-parallel directions along the slit direction. This rubbing direction is helpful, to avoid disclination lines during device operation. In the interest of using a thinner cell gap, a high birefringence and low viscosity LC mixture should be used.

To optimize the lens design, we need to know the LC director profiles of the above four device configurations. Several commercial software packages are available and can be used to calculate the LC directors distribution. The parameters used in the simulations are: $\varepsilon_{//} = 14.9$, $\varepsilon_{\perp} = 3.3$, $K_{11} = 20.3$ pN, $K_{33} = 33.8$ pN, $n_e = 1.9653$, and $n_o = 1.5253$. First, we need to calculate the effective extraordinary refractive index (n_{eff}). The LC director profile for every LC layer can be extracted using the following equation:

$$\frac{1}{n_{eff}(\theta(V))^2} = \frac{\sin^2\theta(V)}{n_o^2} + \frac{\cos^2\theta(V)}{n_e^2}, \tag{12.8}$$

where $\theta(V)$ is the tilt angle of the LC layer at a given applied voltage V. Second, we need to calculate the refractive index difference $dn(V) = n_{eff}(\theta(V)) - n_o$ in each of the LC layers and then average them. The effective focal length f of a cylindrical LC lens is related to the lens radius (r), wavelength (λ), and phase difference ($\Delta\delta = 2\pi d\Delta n/\lambda$) between the center and edge of the aperture as $f = \pi \cdot r^2/(\lambda \cdot \Delta\delta)$.

Figure 12.14(a) shows the calculated and measured voltage-dependent focal length of two cylindrical LC lens: configurations A and B. The agreement between the simulated and measured results is reasonably good. From the simulation results, we find that configuration B has the best positive refractive index profile and the shortest focal length. Configuration A has a severe image aberration problem and longer focal length (>15 cm) due to the broader and shallower refractive index profile as compared to B. For configuration C (Figure 12.14(b)), the simulated minimum focal length occurs at V ~ 60 V_{rms}, which is consistent with the experimental results reported by Ren et al [49]. For configuration D, if the aperture size is maintained at 2 mm then the fringing field-induced refractive index profile is far from the ideal parabolic shape so that the image quality is poor. In order to maintain a parabolic refractive index profile, the aperture size needs to be reduced to 1.2 mm. In this case, the simulation results indicate that both positive and negative lenses can be obtained depending on the applied voltage. When $V < 100\ V_{rms}$, the lens has a positive focal length but turns to negative as $V > 100\ V_{rms}$. The minimum focal length for the positive lens is ~3 cm and the maximum focal length for the negative lens is ~ –5 cm.

12.4.3 Switchable positive and negative microlens

The microlens array is a useful component for optical interconnections, optical fiber switches, shutters of optical super-resolution devices, light deflection devices, and image processing. In 3-D display systems based on integral photography, the microlens array with dynamically

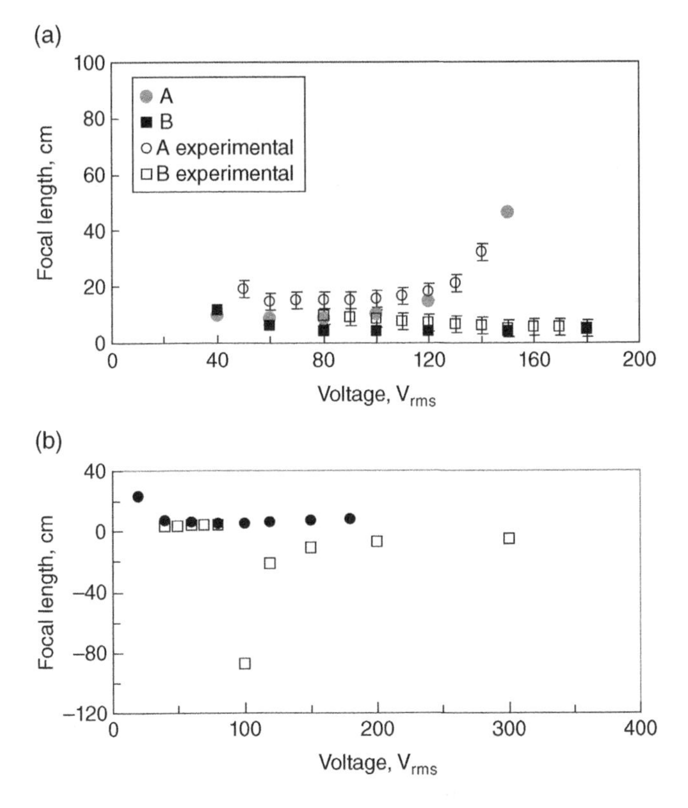

Figure 12.14 The voltage-dependent focal length of cylindrical LC lenses: (a) simulation and experimental results for configurations A and B, (b) simulation results for configurations C and D. The filled circles in (b) are for configuration C and the open squares are for configuration D with reduced aperture size $a = 1.2$ mm.

variable image planes is useful for enhancing the resolution of real and virtual images and for increasing the depth perception of images [51]. To display 3-D images, a lens with variable focal length is highly desirable. In conventional devices, this is commonly done by moving the lens array along the optical axis. This kind of system leads to a bulky device and requires extra mechanical elements for movement. A variable-focus lens can be obtained by changing the shape of the lens [52] or by creating a gradient refractive index profile in the materials with uniform thickness, as discussed in this chapter. For an LC microlens, spatial distribution of refractive index can be induced and varied continuously by the applied voltage instead of mechanical movement. Thus, the switching time is greatly reduced.

As shown in Equation (12.5), the focal length of an LC lens depends on the lens radius, LC layer thickness, and its gradient refractive index. Normally, a lens can either have a positive or negative focal length. In this section, an LC microlens array whose focal length can be switched from positive to negative or vice versa by the applied voltage is introduced.

Figure 12.15 illustrates the structure of a microlens array. To make the focal length electrically tunable, the convex surface of the top BK-7 glass substrate is coated with a thin

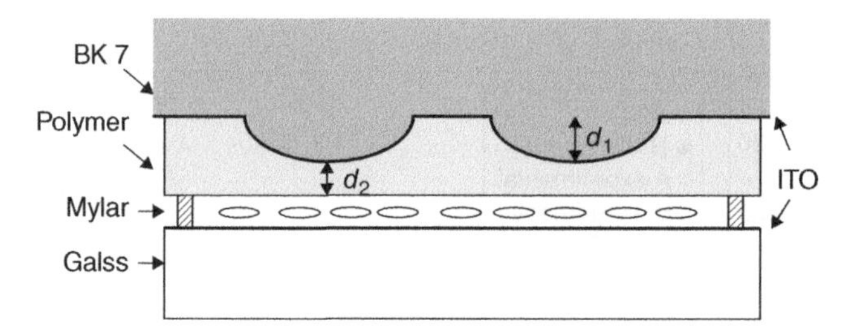

Figure 12.15 Device structure of a DFLC microlens array. Fan 2005. Reproduced with permission from IEEE.

ITO electrode and then flattened by a polymer film. For phase-only modulation, homogeneous LC alignment should be used. Thus, the polymer surface and the bottom ITO glass substrate with a thin polyimide alignment layer are rubbed in anti-parallel directions. The cell gap is controlled at 18 μm by Mylar spacers and hermetically sealed using ultraviolet-curable glue. To achieve a fast response time, a dual-frequency liquid crystal (DFLC) is used. The physical properties of the DFLC mixture are summarized as follows: crossover frequency f_c ~5 kHz, $\Delta n \sim 0.285$ (at $\lambda = 633$ nm and $T = 22\,°C$), and dielectric anisotropy $\Delta\varepsilon = 4.73$ at $f = 1$ kHz, and $\Delta\varepsilon = -3.93$ at 50 kHz.

As depicted in Figure 12.15, the top flattened substrate and the LC layer work together as a zoom lens. The refractive index of the filled polymer NOA-76 ($n_p \sim 1.51$, $\varepsilon = 3.33$) plays an important role in affecting the initial focal length of the microlens. If the filled polymer has a similar refractive index to that of the top glass substrate (BK7, $n_g = 1.517$), i.e. $n_p \approx n_g$, then the whole structure would have a uniform refractive index and does not focus light in the voltage-off state. If the polymer has a different refractive index from glass, i.e. $n_p \neq n_g$, then the microlens would have an initial focus f_s at $V = 0$. In Figure 12.15, each single convex glass microlens has radius $R = 450$ μm, aperture $D = 500$ μm, and height $d_1 = 76$ μm. The extra polymer layer has thickness $d_2 = 81$ μm. As a result, the microlens has a positive focal length ($f_s = 4.26$ cm) at $V = 0$. As the applied voltage increases, the LC directors are reoriented by the electric field. The voltage-induced refractive index change within the LC layer leads to a tunable-focus microlens array. The resultant focal length of the microlens array can be positive or negative depending on the applied voltage.

Figure 12.16 shows the CCD images of the microlens at $V = 0$, 50, 150, and 200 V_{rms}. At $V = 0$, the focusing effect is caused by the top substrate only, as shown in Figure 12.16(a). When the applied voltage exceeds a threshold, the LC directors are reoriented. The gradient refractive index is formed because of the inhomogeneous electric field, as Figure 12.15 depicts. The LC layer adds a diverging effect to the whole rooming lens system. At $V > 30\,V_{rms}$, the focal length of the whole system becomes negative. At 50 V_{rms}, the beam is diverged to the edges of each single microlens, as shown in Figure 12.16(b). In the high-voltage regime, the LC directors are all reoriented perpendicular to the substrates. The gradient refractive index profile is gradually flattened and erased. The diverging effect of the LC lens is weakening and the microlens becomes a converging lens again, as shown in Figures. 12.16(c) and (d).

To explain this focal length transition phenomenon quantitatively, we need to calculate the voltage-induced LC director reorientation numerically. First, we calculate the voltage-dependent refractive index change for a symmetric, uniform LC layer and use that to predict

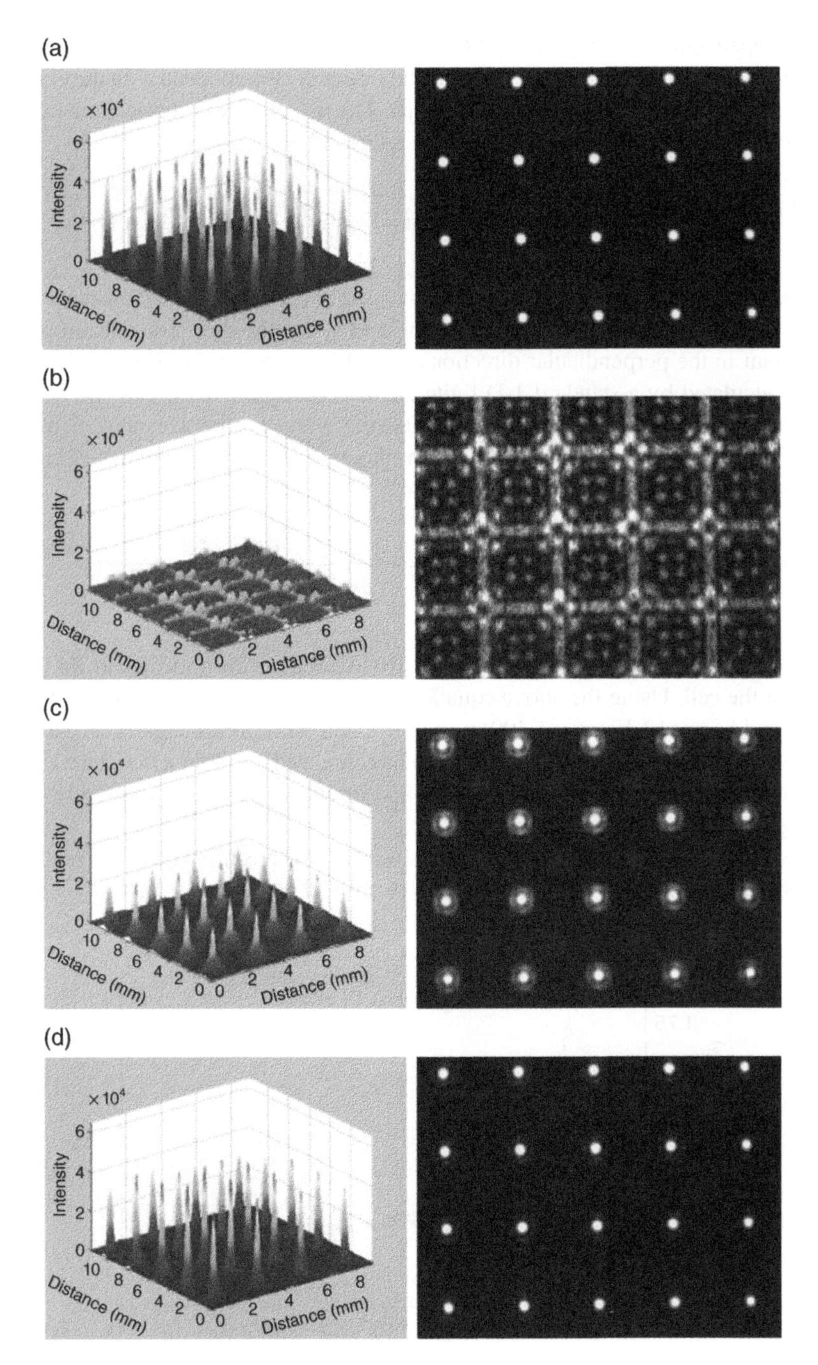

Figure 12.16 CCD images of the DFLC microlens array (right) and the corresponding 3-D light intensity profiles (left) at (a) 0, (b) 50 V_{rms}, (c) 150 V_{rms}, and (d) 200 V_{rms}. Fan 2005. Reproduced with permission from IEEE.

the voltage-dependent focal length. The LC directors are reoriented along the electric field (***E***) direction in order to minimize the free energy. The free energy associated with the elastic forces can be described in terms of three elastic constants. The free energy of the nematic LC directors in the static electric field ***E*** is generally expressed as [53]

$$\mathbf{F}=\frac{1}{2}k_{11}(\nabla\cdot\mathbf{n})^2+\frac{1}{2}k_{22}(\mathbf{n}\cdot\nabla\times\mathbf{n})^2+\frac{1}{2}k_{33}(\mathbf{n}\times\nabla\times\mathbf{n})^2-\frac{1}{2}\left[\varepsilon_{\perp}\mathbf{E}^2+\Delta\varepsilon(\mathbf{n}\cdot\mathbf{E})^2\right], \tag{12.9}$$

where $\mathbf{n}=(n_x, n_y, n_z)$ is the LC director vector, k_{11}, k_{22} and k_{33} are elastic constants associated with splay, twist, and bend deformations, and $\Delta\varepsilon$ and $\varepsilon_{\perp}$ are the dielectric anisotropy and dielectric constant in the perpendicular direction of the LC, respectively. The LC director reorientation is calculated by a standard 1-D finite element method.

The tilt angle profile $\theta(z)$ can be used to calculate the average refractive index, $\tilde{n}$, of the liquid crystal at different applied voltages. For a given tilt angle, the refractive index is given by the index ellipsoid equation [54]:

$$n(z)=\frac{n_e n_o}{\sqrt{n_o{}^2\cos^2\theta(z)+n_e{}^2\sin^2\theta(z)}}=\frac{n_e n_o}{\sqrt{n_o{}^2+(n_e{}^2-n_o{}^2)\sin^2\theta(z)}} \tag{12.10}$$

The tilt angle profile $\theta(z)$ is used in Equation (12.10) to calculate the optical path, $\tilde{n}d$, when light goes through the cell. Using the above equations and DFLC parameters ($k_{11}=33.5$ pN, $k_{33}=35$ pN, $\varepsilon_{\perp}=7.44$, $\varepsilon_{//}=12.17$, $n_o=1.490$, $n_e=1.775$, and with pretilt angle = 2°), the voltage-dependent average effective refractive index can be calculated, as plotted in Figure 12.17. Solid line and open circles represent the experimental and simulation results, respectively, at $\lambda=633$ nm. The agreement is very good.

When a voltage V is applied to the LC microlens array, the electric field at the center (E_{center}) and at the border (E_{border}) of each microlens is expressed as [55]

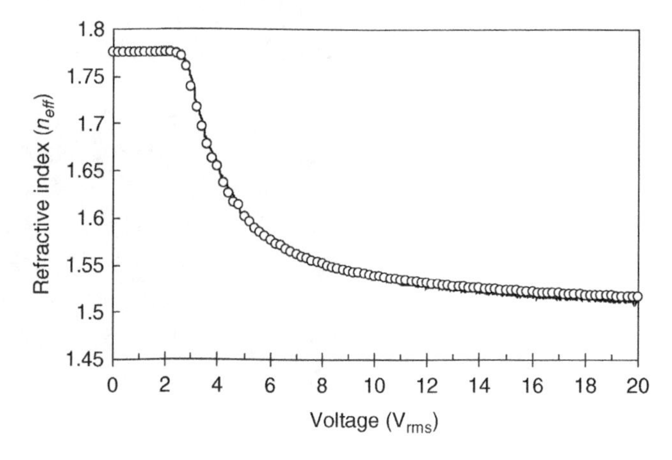

Figure 12.17 Voltage-dependent average refractive index of dual-frequency liquid crystal used in the experiment. Solid line and open circles represent the experimental and simulation results, respectively, at $\lambda=633$ nm. Fan 2005. Reproduced with permission from IEEE.

$$E_{center} = \frac{V/\varepsilon_{LC}}{\frac{d_{LC}}{\varepsilon_{LC}} + \frac{d_2}{\varepsilon_p}}, \tag{12.11}$$

$$E_{border} = \frac{V/\varepsilon_{LC}}{\frac{d_{LC}}{\varepsilon_{LC}} + \frac{d_1 + d_2}{\varepsilon_p}}, \tag{12.12}$$

where d_1 represents the microlens thickness and d_2 the thickness of the polymer layer shown in Figure 12.15, d_{LC} is the LC layer thickness, and ε_{LC} and ε_p represent the dielectric constant of the LC and polymer, respectively. At $V = 0$, $\varepsilon_{LC} = \varepsilon_{\perp}$. As $V > V_{th}$, the effective ε_{LC} will change and affect the electric field distribution through the cell. Finally, the equilibrium state is achieved. By using the final voltage distribution, the LC tilt angle, average effective birefringence, and the phase difference between the center and borders of each microlens can be calculated.

From Fresnel's approximation, the focal length of an LC lens is related to the lens radius r and $\delta n = \tilde{n}_{center} - \tilde{n}_{border}$ which is the refractive index difference between the lens center and borders, as described in Equation (12.5).

Figure 12.18 plots the measured (dots) and simulated (lines) voltage-dependent focal length of the microlens [56]. At $V = 0$, LC directors are aligned homogeneously, and no focusing effect occurs in the LC layer. The system shows the initial focus contributed solely by the top glass microlens array. As the voltage increases, the LC layer behaves like a diverging lens, so that the combined focal length increases accordingly. At $V \sim 30\ V_{rms}$, the microlens begins to behave like a diverging lens. At ~40 V_{rms}, the microlens reaches its shortest negative focal length. Further increasing the voltage would reorient all the LC directors perpendicular to the substrates and reduce the phase difference. The microlens becomes a converging lens again and gradually approaches the initial focal length but at a different rate. The simulation results

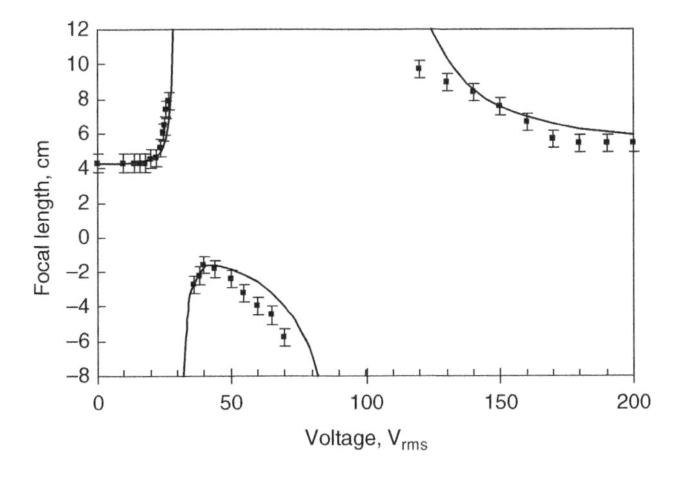

Figure 12.18 Voltage-dependent focal length of the DFLC microlens array. LC cell gap $d = 18\ \mu m$, the diameter of microlens $D = 500\ \mu m$, $\lambda = 633$ nm. Fan 2005. Reproduced with permission from IEEE.

agree with experiment quite well in the lower voltage regime (<40 V_{rms}) but in the higher voltage regime (40–130 V_{rms}) the fitting is somewhat deviated. This is because, in the simulations, the anchoring energy at the boundaries is assumed to be infinity, which means that the LC directors near the substrates will not be reoriented at all by the electric field. But in reality the LC directors near the substrates would still be reoriented slightly by the strong electric field at a high voltage. Therefore, in the high voltage regime the measured focus change is faster than that simulated. Above 140 V_{rms}, the LC directors are reoriented nearly perpendicular to the substrates in both experimental data and simulation results. Therefore, the measured focal length agrees well with the simulated values in the high voltage regime.

A key consideration for using DFLC material is to obtain a fast response time. Using 1 kHz and 50 kHz driving frequencies to switch the microlens array between 0 and 200 V_{rms}, the rise time is 3.9 ms and decay time is 5.4 ms for the 18 μm cell gap. Compared to a nominal nematic, the response time would be hundreds of millisecond. The high driving voltage (200 V_{rms}) results from the small dielectric anisotropy of DFLC at low and high frequencies. To lower the operating voltage, we can either reduce the d_2 shown in Figure 12.15 or increase the dielectric constant of the filled polymer.

12.4.4 Hermaphroditic LC microlens

Most LC lenses are polarization sensitive. When the incoming light polarization is parallel to the LC alignment direction (i.e. extraordinary ray) the focal length of the LC lens can be tuned continuously within a finite range which depends on the LC birefringence. However, for the ordinary ray (i.e. the incident light polarization is perpendicular to the LC directors), the focal length of the LC lens does not change with voltage. Both positive and negative lenses can be designed according to need, but once the lens is designed it exhibits as either a positive or a negative lens. Although under some special operation conditions the central part of a positive LC lens could exhibit negative focusing property [57], the surrounding part remains positive. This volcano type of LC lens has severe index distortion.

Figure 12.19 shows the side view of a hermaphroditic microlens which could exhibit either a positive or a negative focal length depending on the input light polarization [58]. Unlike a conventional LC lens whose focal length is tunable by the applied voltage, the hermaphroditic LC microlens changes focal length according to the angle between the polarization axis and the LC directors. For the extraordinary ray, the focal length is positive while for the ordinary ray the focal length becomes negative. By changing the relative angle between the incident light polarization and the LC directors, the focal length of the LC lens can be varied. This polarization rotation can be achieved manually or by an electrically controlled 90° twist nematic (TN) cell. The switching time is about 10–20 ms, depending on the LC cell gap and material employed.

In Figure 12.19 the flat lens is composed of a plano-convex LC lens and a plano-concave molded polymeric lens (shaded areas). The LC directors in the plano-convex lens are aligned along x axis. The ordinary and extraordinary refractive indices (n_o and n_e) are along the y and x axis, respectively. On the other hand, the plano-concave lens is made of UV-cured polymer/LC composite on a polyimide surface whose rubbing direction is along y axis. Thus, its refractive indices are also anisotropic: $n_1 > n_2$. The LC material chosen for this lens satisfies the following relationship: $n_e \sim n_1 > n_2 \sim n_o$. When the incident light passes through the convex and concave lenses from *the* z axis with its polarization at an angle θ with respect to x axis, the focal length of the microlens can be expressed as:

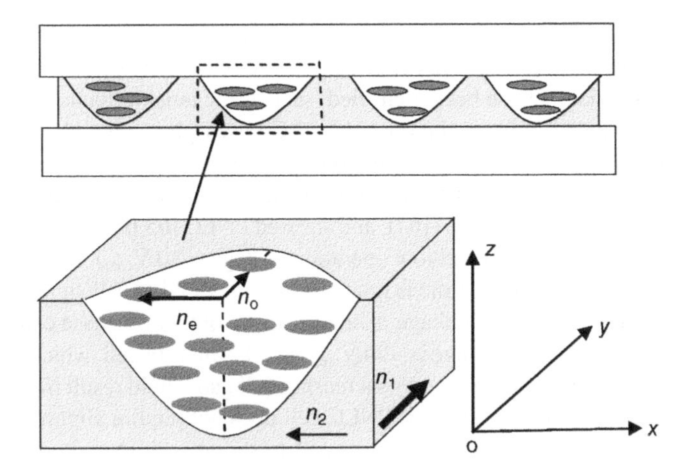

Figure 12.19 Side view of the hermaphroditic LC microlens arrays. n_1 and n_2 are the refractive indices of the molded microlens along y and x axes, respectively. n_o and n_e are the ordinary and extraordinary refractive indices of the LC material. Ren et al 2005. Reproduced with permission from the Optical Society of America.

$$f = R/(n_{LC} - n_{mold}). \tag{12.13}$$

Here, R is the radius of curvature of the lens surface and n_{LC} and n_{mold} denote the effective refractive indices of the LC and the molded polymeric lens, respectively. Both n_{LC} and n_{mold} are dependent on θ as:

$$n_{LC} = \frac{n_o \cdot n_e}{\sqrt{\left(n_o^2 \cos^2\theta + n_e^2 \sin^2\theta\right)}}, \tag{12.14}$$

$$n_{mold} = \frac{n_1 \cdot n_2}{\sqrt{\left(n_1^2 \cos^2\theta + n_2^2 \sin^2\theta\right)}}. \tag{12.15}$$

From above equations, when $\theta = 0$, the focal length of the lens is $f_1 = R/(n_e - n_2)$. In this case, the focal length f_1 is positive. If $\theta = 90°$, then the focal length of the lens is $f_2 = R/(n_o - n_1)$. Because $n_o < n_1$, the focal length f_2 is negative. When $n_{LC} \sim n_{mold}$, the focal length of the lens approaches infinity. By tuning the incident light polarization axis gradually from 0 to 90°, the focal length changes from positive to negative.

12.5 Polarization-Independent LC Devices

Most of the LC devices operate under a linearly polarized light in order to achieve high contrast ratio. The use of a polarizer reduces the optical efficiency dramatically. The maximum transmittance of a pair of polarizers is only about 38%. Polarization-independent LC devices for phase or amplitude modulation are highly desirable.

Phase-only modulation [59] plays an important role in adaptive optics, optical cross-connect switching, laser beam steering, and low-cost electro-optic sensors. Several interesting applications using phase modulators have been identified, such as the tunable-focus lens [60], grating, and prism [61], and spatial light modulators [62]. LC-based phase modulators offer several advantages: low cost, light weight, low power consumption, and no mechanical moving parts. Several LC-based phase modulators have been developed, such as homogeneous LC [63], polymer network liquid crystal (PNLC) [64], and sheared PNLC [65,66]. The homogenous cell is attractive for its large phase shift and low operation voltage (<10 V_{rms}). However, it is polarization dependent, and the response time is relatively slow. A PNLC cell significantly reduces the response time, but its operating voltage is increased. To obtain 2π phase change in a transmissive PNLC cell, the required voltage is ~90 V_{rms} for a 12 μm E44 cell, which corresponds to ~7 V/μm. To achieve more phase change by increasing cell gap would result in substantial light scattering and higher voltage. The sheared PNLC cell does not require alignment layers but it needs a shearing force to stress the LC directors and to suppress light scattering. Its response time is also in the submillisecond regime but its operating voltage is also relatively high. A common drawback of these three approaches is that they are polarization sensitive. For laser applications, the incident light polarization may not be always parallel to the LC directors to ensure a phase-only modulation. Thus, it is highly desirable to develop polarization-independent phase modulators.

Several approaches for obtaining polarization-independent LC phase modulation have been developed, such as the 90° twisted nematic cell operated at a voltage about three times higher than the threshold voltage [67,68], nano-scale polymer-dispersed liquid crystal (nano-PDLC) [69], voltage-biased PDLC [70], and voltage-biased polymer-stabilized cholesteric texture (PSCT) [71]. A common problem for these approaches is that their phase change is relatively small and the operating voltage is quite high. For instance, the nano-PDLC is scattering-free, polarization independent, and has a submillisecond response time. However, its phase shift is small and its operating voltage is around 15–20 V_{rms}/μm. Increasing the cell gap would enhance the phase change, but the operating voltage will also further increase. The voltage-biased micro-sized PDLC and PSCT are also polarization independent, but their residual phase is still small (~0.1π at $\lambda = 633$ nm) so that their applications are limited to micro-photonic devices, such as tunable-focus microlens array.

In the following sections, we introduce two polarization-independent LC phase modulators: (1) a double-layered structure with two ultra-thin anisotropic polymer films as cell separators, and (2) a double-layered LC gel without any separator.

12.5.1 Double-layered homogeneous LC cells

The double-layered structure has been proposed for guest–host LC displays [72,73]. The conventional approach uses a thin glass (~0.3 mm) or Mylar film (~0.1 mm) to separate the two orthogonal LC layers. In the former case, an ITO glass substrate is used as a middle substrate. To overcome the electric field shielding effect, both sides of the ITO layers should be pixelated and connected via feed-through holes, and then overcoated with a thin polyimide layer, which is rubbed in the orthogonal directions to match the LC alignment. This approach is difficult for high resolution devices because of the complicated pixel structures and precision registration between the passive ITO pixels in the middle substrate and the active elements.

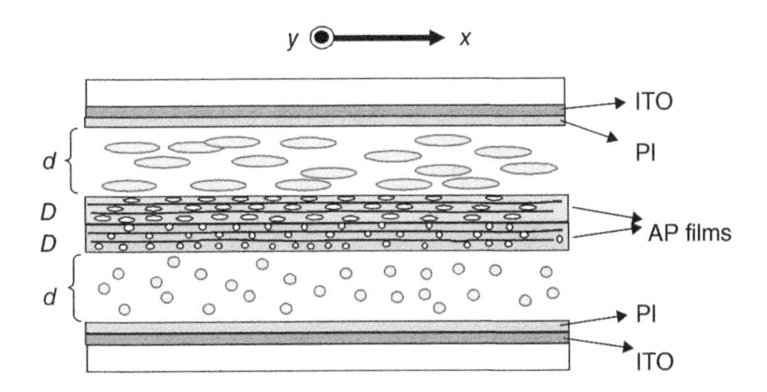

Figure 12.20 The structure of a polarization-independent phase modulator. AP = anisotropic polymer films. Lin 2005. Reproduced with permission from the Optical Society of America.

To reduce the parallax incurred by the middle glass substrate and to enable high resolution, a thin Mylar film has been considered. However, the Mylar film cannot align the LC molecules because the post baking temperature (~250 °C) of polyimide is higher than the glass transition temperature of the Mylar film.

In this section, we introduce an anisotropic polymer film [74] which is thin, optically anisotropic, and possesses alignment capability. Figure 12.20 shows the schematic design of the double-layered polarization-independent phase modulator. The cell consists of two glass substrates which are overcoated with thin (~80 nm), mechanically buffed polyimide layers, two anisotropic polymer films, and two LC layers. The top and bottom LC directors are oriented orthogonally. To achieve orthogonal homogeneous LC layers, the principal axes of these two anisotropic polymer films are also arranged to be orthogonal.

The anisotropic film is made of Merck E7 nematic LC mixture, photo-initiator IRG184, and an LC monomer RM-257 (4-(3-Acryloyloxypropyloxy)-benzoic acid 2-methyl-1,4-phenylene ester) at 19:1:80 wt% ratios. The LC/monomer mixture was injected into a homogeneous cell with 23 μm cell gap, which was controlled by the Mylar stripes and then the cell was exposed to a UV light with intensity $I = 10$ mW/cm [2] for ~30 min at 90°C. After UV exposure, the two substrates of the homogeneous cell were peeled off and a solidified anisotropic film with 23 μm thickness was obtained. The anisotropic polymer film is fully transparent. A large film can be sliced into two identical films. These two films are then stacked together in orthogonal directions. The LC mixture employed is also E7. The LC was filled to the empty cell by the one-drop-fill method. The cell gap of each LC layer was controlled by a Mylar film to be d ~ 12 μm. The total dimension of the cell is around 25 × 25 mm.

To characterize the phase shift of the double-layered LC cell, a Mach–Zehnder interferometer and an unpolarized He–Ne laser ($\lambda = 633$ nm) were used. Figure 12.21 plots the measured voltage-dependent phase shift of the double-layered E7 LC cell (filled circles). The threshold voltage is ~5 V_{rms}. For reference, the threshold voltage of the single E7 cell without any middle substrate is ~0.95 V_{rms}. The increased threshold voltage originates from the dielectric shielding effect of the two middle polymeric layers. In the interferometer, the measured phase shift is referenced to that at $V = 0$. The total phase shift reaches ~8.1π at V = 40 V_{rms}. This total phase shift is independent of the incident light polarization. Also included in Figure 12.21 are the simulated results (open circles) of a similar double-layered structure using a glass separator

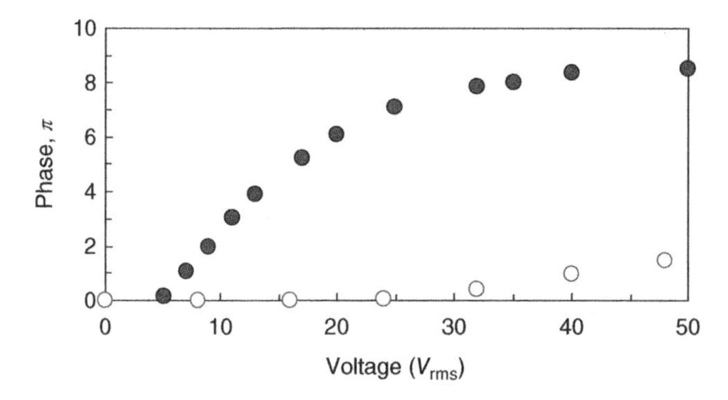

Figure 12.21 Voltage-dependent phase shift of the polarization-independent LC phase modulator at $\lambda = 633$ nm. Filled circles represent the measured data using two anisotropic polymeric films while open circles are the simulated results of the double-layered structure using a glass separator 0.3 mm thick. Lin 2005. Reproduced with permission from the Optical Society of America.

0.3 mm thick. Because of the electric field screening effect, the required voltage to reach a 2π phase change is beyond 50 V_{rms}.

The obtainable phase shift of the described double-layered structure is much larger, and the operating voltage is much lower than those of nano-PDLC, PDLC, and PSCT. To further lower the operating voltage of the double-layered structure, we can reduce the thickness of the anisotropic polymer films, but the trade-off is that a thinner polymer film may degrade the uniformity of the cell.

The response time of the double-layered LC cell is ~300 ms at $T \sim 23\,°C$. The slow response time originates from the thick LC layers ($d \sim 12\,\mu m$) and high viscosity of the E7 LC employed. To reduce response time, a high Δn and low viscosity LC should be used [75]. A high Δn LC enables a thinner cell gap to be used, which is helpful for reducing response time.

The thickness of polymer film is 23 μm which is more than the LC layer (12 μm). The flexibility and hardness can be controlled by the fabrication process, such as the UV curing condition and LC concentration. Although the anisotropic films are thin, their deformation during operation should not be a problem because the films are still sandwiched by two glass substrates. Therefore, the mechanical stability of the system is not a concern.

12.5.2 Double-layered LC gels

To achieve a fast response time, another polarization-independent phase modulator using two thin stratified LC gels has been developed [76]. The two homogeneously aligned gel films are identical, but stacked in orthogonal directions. Because of the high LC concentration and uniform molecular alignment, the LC gel possesses a large phase change ($>1\pi$). Meanwhile, because of the relatively high monomer concentration (28 wt%) the formed LC domains are in the submicron range. Therefore, the response time of the LC gel is around 0.5 ms.

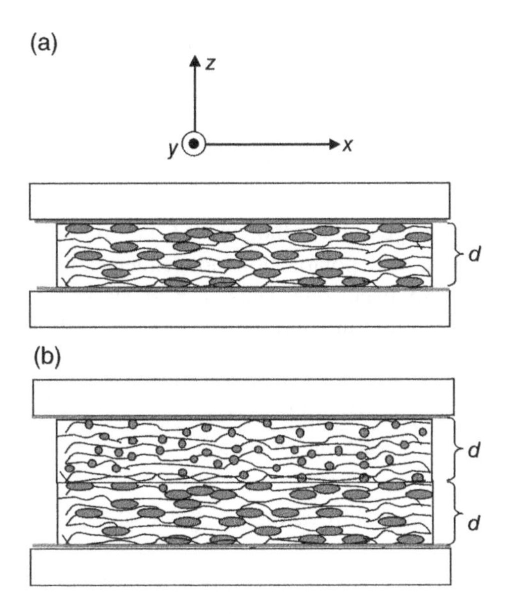

Figure 12.22 A homogeneous LC gel: (a) single layer and (b) two orthogonal layers. Ren 2006. Reproduced with permission from the American Institute of Physics.

In an LC gel, the homogeneously aligned LC is stabilized by dense polymer networks, as shown in Figure 12.22(a). The phase shift along z axis can be expressed as

$$\Delta\delta_{Gel}(V)=\frac{2\pi dc\left[n_e-n_{eff}(V)\right]}{\lambda}, \tag{12.16}$$

where d is the cell gap, c is the LC concentration, λ is the incident wavelength, n_e and $n_{eff}(V)$ are the extraordinary and effective refractive index of the LC, respectively. At $V\rightarrow\infty$, $n_{eff}\rightarrow n_o$, where n_o is the ordinary refractive index of LC. From Figure 12.22(a), the homogeneous LC gel is polarization dependent. To make it polarization independent, two identical homogeneous LC gels are stacked in the orthogonal directions, as shown in Figure 12.22(b).

As the voltage increases, the phase change occurs because of the electric field-induced LC director reorientation. At a very high voltage, the voltage-induced phase shift is reduced to:

$$\Delta\delta_{Gel}(V\rightarrow\infty)=\frac{2\pi dc\Delta n}{\lambda}, \tag{12.17}$$

where $\Delta n=n_e-n_o$ is the LC birefringence. In comparison, the LC droplets in a nano- or voltage-biased PDLC cell are almost randomly orientated. Thus, the phase shift is

$$\Delta\delta_{PDLC}(V)=\frac{2\pi d'c'\left[\bar{n}-n_{eff}(V)\right]}{\lambda}, \tag{12.18}$$

where $\bar{n} = (2n_o + n_e)/3$ is the average refractive index of the LC at $V = 0$, $\boldsymbol{d'}$ and $\boldsymbol{c'}$ are the cell gap and LC concentration, respectively. As $V \rightarrow \infty$, $n_{eff} \rightarrow n_o$, and the phase shift is reduced to

$$\Delta\delta_{PDLC}(V \rightarrow \infty) = \frac{2\pi d' c' \Delta n}{3\lambda}. \tag{12.19}$$

To fairly compare the phase change of the orthogonal LC gel films with the nano-PDLC, let us use the same LC material. To achieve polarization independence, the LC gel needs two orthogonal layers, but nano-PDLC only needs one. Thus, $d' = 2d$. However, the LC concentration in the gel is two times higher than that in nano-PDLC, i.e. $c = 2c'$. From Equation (12.17) and Equation (12.19), we find

$$\frac{\Delta\delta_{Gel}(V \rightarrow \infty)}{\Delta\delta_{PDLC}(V \rightarrow \infty)} = 3. \tag{12.20}$$

From Equation (12.20), the phase shift of the LC gel is three times higher than that of a nano-PDLC.

The LC gel is made by mixing 28 wt% of photocurable rod-like LC diacrylate monomer (RM257) in a nematic LC (E48: $n_o = 1.523$, $\Delta n = 0.231$ at $\lambda = 589$ nm). The mixture was injected into an empty cell in the nematic state. The inner surfaces of the ITO-glass substrates were coated with a thin polyimide layer and then rubbed in anti-parallel directions. The filled cell was exposed to UV ($\lambda \sim 365$ nm, $I \sim 10$ mW/cm [2]) for 30 min. The cell gap was controlled at 8 μm by spacer balls.

After UV exposure, the cell is highly transparent. To peel off the gel, the top glass substrate is cleaved off. The stratified gel remains on the bottom substrate surface without LC leakage. From microscope inspection, the LC gel is indeed aligned homogeneous without being damaged during cell cleaving. To assemble a double-layered structure, the LC gel was cut in half, stacked together at orthogonal direction, and then covered with another top ITO substrate, as Figure 12.22(b) shows. Similarly, the phase change is monitored by a Mach–Zehnder interferometer using an unpolarized He–Ne laser beam. When an AC voltage ($f = 1$ kHz) was applied to the LC gel, the interference fringes moved as recorded by a digital CCD camera.

Figure 12.23 shows the voltage-dependent phase shift of a 16 μm double-layered LC gel at $\lambda = 633$ nm. The threshold voltage is ~30 V_{rms}. This high threshold originates from the dense polymer networks. Beyond this threshold, the phase change increases almost linearly with the applied voltage. The estimated total phase change from an 8 μm LC gel which contains ~80 wt% E48 should be ~2π for a linearly polarized He–Ne laser ($\lambda = 633$ nm). Therefore, the applied voltage has not reached the saturation regime.

The rise time of the LC gel is ~200 μs and decay time is ~500 μs at room temperature (~22°C). Such a fast response time results from the small LC domain sizes and polymer stabilization. Due to the relatively high monomer concentration (28 wt%), the formed polymer networks are quite dense so that the formed LC domains are in submicron size. Similar to a nano-PDLC, the contact interfaces between the polymer networks and the LC molecules are large. As a result, the anchoring force of polymer networks exerted on the LC is very strong. This is the primary reason for the observed fast response time and high threshold voltage.

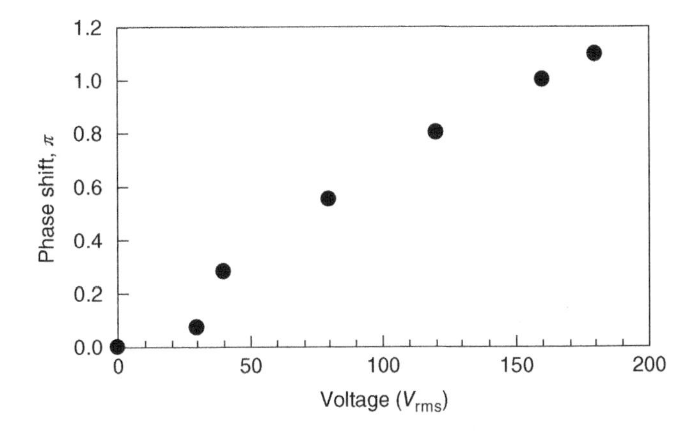

Figure 12.23 Measured phase shift of a 16-μm double-layered LC gel at different voltages. $\lambda = 633$ nm. Ren 2006. Reproduced with permission from the American Institute of Physics.

To get a 2π phase change for laser beam steering and other photonic applications, the LC gel can be operated in reflective mode without increasing the operating voltage. For practical applications, the operating voltage of the LC gel is still too high (11 V_{rms}/μm). To increase the phase change and reduce the operation voltage, an LC material with high Δn and high $\Delta\varepsilon$ should be considered, while optimizing the LC and monomer concentration. A high Δn LC also enables a thinner gel to be used which, in turn, helps reduce the operating voltage. A high $\Delta\varepsilon$ LC lowers the threshold and the operating voltages simultaneously. Increasing the LC concentration would boost the phase change and reduce the operating voltage. However, the gel may become too soft to stand alone. Its response time will also increase slightly.

Homework Problems

12.1 Using Figure 12.6 to derive Equation (12.4) and explain how to obtain a large steering angle.

12.2 A student wants to design a polarization-independent tunable-focus microlens using a 90° TN cell. The LC mixture employed has following properties: $\Delta\varepsilon = 12$, $\Delta n = 0.5$ at $\lambda = 550$ nm, $K_{11} = 10$ pN, $K_{22} = 6$ pN, $K_{33} = 20$ pN, and $\gamma_1 = 0.2$ Pa · s. If the microlens diameter is 200 μm, what is the maximum tunable range of the focal length at $\lambda = 550$ nm?

12.3 Nano-sized polymer-dispersed liquid crystal (nano-PDLC), voltage-biased PDLC, double-layered homogeneous LC, and double-layered LC gels are all polarization independent. Compare their pros and cons.

12.4 A 90° TN cell is filled with an LC whose physical properties are as follows: $\Delta\varepsilon = 10$, $\Delta n = 0.1$ at $\lambda = 550$ nm, $K_{11} = 14$ pN, $K_{22} = 7$ pN, $K_{33} = 18$ pN, and $\gamma_1 = 0.2$ Pas. (a) Under what conditions can the TN cell be used as a polarization-independent phase modulator? (b) Explain why this phase modulator is polarization independent?

References

1. T. T. Alkeskjold, J. Lagsgaard, A. Bjarklev, et al., All-optical modulation in dye-doped nematic liquid crystal photonic bandgap fibers, *Opt. Express* **12**, 5857 (2004).
2. F. Du, Y. Q. Lu, and S. T. Wu, Electrically tunable liquid crystal photonic crystal fiber, *Appl. Phys. Lett.* **85**, 2181 (2004).
3. J. Borel, J. C. Deutsch, G. Labrunie, and J. Robert, Liquid crystal diffraction grating, US patent **3**,843,231 (1974).
4. W. Kulchke, K. Kosanke, E. Max, et al., Digital light deflector, *Appl. Opt.* **5**, 1657 (1966).
5. H. Meyer, D. Riekmann, K. P. Schmidt, et al., Design and performance of a 20-stage digital light beam deflector, *Appl. Opt.* **11**, 1732 (1972).
6. T. K. Gaylord and M. G. Moharam, Planar dielectric grating diffraction theories, *Applied Physics B – Photophysics and Laser Chemistry* **28**, 1 (1982).
7. X. Wang, D. Wilson, R. Muller, et al., Liquid-crystal blazed-grating beam deflector, *Appl. Opt.* **39**, 6545 (2000).
8. D. P. Resler, D. S. Hobbs, R. C. Sharp, et al., High-efficiency liquid-crystal optical phased-array beam steering, *Opt. Lett.* **21**, 689 (1996).
9. C. M. Titus, J. R. Kelly, E. C. Gartland, et al., Asymmetric transmissive behavior of liquid-crystal diffraction gratings, *Opt. Lett.* **26**, 1188 (2001).
10. W. Klaus, M. Ide, S. Morokawa, et al., Angle-independent beam steering using a liquid crystal grating with multi-resistive electrodes, *Opt. Commun.* **138**, 151 (1997).
11. P. F. McManamon, T. A. Dorschner, D. L. Corkum, et al., Optical phased array technology, *Prof. IEEE*, **84**, 268 (1996).
12. S. T. Wu, U. Efron, and L. D. Hess, Birefringence measurement of liquid crystals, *Appl. Opt.* **23**, 3911 (1984).
13. K. Rastani, A. Marrakchi, S. F. Habiby,et al., Binary phase Fresnel lenses for generation of two-dimensional beam arrays, *Appl. Opt.* **30**, 1347 (1991).
14. L. Beiser, Laser beam information scanning and recording in *Laser Scanning and Recording. Bellingham, WA: SPIE*, **378**, 3 (1985).
15. L. B. Glebov, Volume hologram recording in inorganic glasses, *Glass science and technology, Suppl.* C1, **75**, 73 (2002).
16. O. M. Efimov, L. B. Glebov, L. N. Glebova, two, High-efficiency Bragg gratings in photothermorefractive glass, *Appl. Opt.* **38**, 619 (1999).
17. C. M. Titus, P. J. Bos, and O. D. Lavrentovich, Efficient accurate liquid crystal digital light deflector, *Proc. SPIE* **3633**, 244 (1999).
18. O. Pishnyak, L. Kreminska, O. D. Lavrentovich, et al., Liquid crystal digital beam steering device based on decoupled birefringent deflector and polarization rotator, *Mol. Cryst. Liq. Cryst.* **433**, 279 (2005).
19. C. Mao, M. Xu, W. Feng, et al., Liquid crystal applications in telecommunication, *Proc. SPIE* **5003**, 121 (2003).
20. L. Eldada, Optical communication components, *Review of Scientific Instruments* **75**, 575 (2004).
21. J. J. Pan, H. Wu, W. Wang, et al., *Proc. National Fiber Optics Engineers Conference*, Telcordia, Orlando, p. 943 (2003).
22. N. A. Riza and S. F. Yuan, Reconfigurable wavelength add-drop filtering based on a Banyan network topology and ferroelectric liquid crystal fiber-optic switches, *J. Lightwave Technol.* **17**, 1575 (1999).
23. S. T. Wu and U. Efron, Optical properties of thin nematic liquid crystal cells, *Appl. Phys. Lett.* **48**, 624 (1986).
24. V. V. Belyaev, S. Ivanov, and M. F. Grebenkin, *Sov. Phys. Crystallogr.* **30**, 674 (1985).
25. S. T. Wu, U. Efron, and A. M. Lackner, Optimal operating temperature of liquid crystal modulators, *Appl. Opt.* **26**, 3411 (1987).
26. H. K. Bucher, R. T. Klingbiel, and J. P. VanMeter, Frequency-addressed liquid crystal field effect, *Appl. Phys. Lett.* **25**, 186 (1974).
27. M. Schadt, Low-frequency dielectric relaxation in nematics and dual-frequency addressing of field effects, *Mol. Cryst. Liq. Cryst.* **89**, 77 (1982).
28. M. Xu and D. K. Yang, Dual frequency cholesteric light shutters, *Appl. Phys. Lett.* **70**, 720 (1997).

29. S. T. Wu, A nematic liquid crystal modulator with response time less than 100 μs at room temperature, *Appl. Phys. Lett.* **57**, 986 (1990).
30. Y. Q. Lu, F. Du, Y. H. Lin and S. T. Wu, Variable optical attenuator based on polymer stabilized twisted nematic liquid crystal, *Opt. Express* **12**, 1221 (2004).
31. S. T. Wu, Dual parallel-aligned cells for high speed liquid crystal displays, *J. Appl. Phys.* **73**, 2080 (1993).
32. S. T. Wu and D. K. Yang, *Reflective Liquid Crystal Displays* (Wiley, New York, 2001).
33. X. Liang, Y. Q. Lu, Y. H. Wu, et al., Dual-frequency addressed variable optical attenuator with submillisecond response time, *Jpn. J. Appl. Phys.* **44**, 1292 (2005).
34. S. Sato, Liquid-crystal lens-cells with variable focal length, *Jpn. J. Appl. Phys.* **18**, 1679 (1979)
35. B. Wang, M. Ye, M. Honma, et al., Liquid crystal lens with spherical electrode, *Jpn. J. Appl. Phys.* **41**, L1232 (2002).
36. H. S. Ji, J. H. Kim, and S. Kumar, Electrically controllable microlens array fabricated by anisotropic phase separation from liquid-crystal and polymer composite materials, *Opt. Lett.* **28**, 1147 (2003).
37. T. Nose and S. Sato, A liquid-crystal microlens obtained with a non-uniform electric-field, *Liq. Cryst.* **5**, 1425 (1989).
38. M. Ye and S. Sato, Optical properties of liquid crystal lens of any size, *Jpn. J. Appl. Phys.* **41**, L571 (2002).
39. N. A. Riza and M. C. DeJule, Three-terminal adaptive nematic liquid-crystal lens device, *Opt. Lett.* **19**, 1013 (1994).
40. W. W. Chan and S. T. Kowel, Imaging performance of the liquid-crystal-adaptive lens with conductive ladder meshing, *Appl. Opt.* **36**, 8958 (1997).
41. J. S. Patel and K. Rastani, Electrically controlled polarization-independent liquid-crystal fresnel lens arrays, *Opt. Lett.* **16**, 532 (1991).
42. H. Ren, Y. H. Fan, and S. T. Wu, Tunable Fresnel lens using nanoscale polymer-dispersed liquid crystals, *Appl. Phys. Lett.* **83**, 1515 (2003).
43. A. F. Naumov, M. Yu. Loktev, I. R. Guralnik, and G. Vdovin, Liquid-crystal adaptive lenses with modal control, *Opt. Lett.* **23**, 992 (1998).
44. H. Ren and S. T. Wu, Tunable electronic lens using a gradient polymer network liquid crystal, *Appl. Phys. Lett.* **82**, 22 (2003).
45. V. V. Presnyakov, K. E. Asatryan, and T. V. Galstian, Polymer-stabilized liquid crystal for tunable microlens applications, *Opt. Express* **10**, 865 (2002).
46. S. Gauza, H. Wang, C. H. Wen, et al., High birefringence isothiocyanato tolane liquid crystals *Jpn. J. Appl. Phys.* **42**, 3463 (2003).
47. S. T. Kowel, D. S. Cleverly, and P. G. Kornreich, Focusing by electrical modulation of refraction in a liquid-crystal cell, *Appl. Opt.* **23**, 278 (1984).
48. T. Nose, Y. Yamada, and S. Sato, Improvement of optical properties and beam steering functions in a liquid crystal microlens with an extra controlling electrode by a planar structure, *Jpn. J. Appl. Phys.* **39**, 6383 (2000).
49. H. Ren, Y. H. Fan, S. Gauza, and S. T. Wu, Tunable-focus cylindrical liquid crystal lens, *Jpn. J. Appl. Phys.* **43**, 652 (2004).
50. Y. H. Lin, et al, Tunable-focus cylindrical liquid crystal lenses, *Jpn. J. Appl. Phys.* **44**, 243 (2005).
51. B. Lee, S. Jung, S. W. Min, and J. H. Park, Three-dimensional display by use of integral photography with dynamically variable image planes, *Opt. Lett.* **26**, 1481 (2001).
52. H. Ren and S. T. Wu, Variable-focus liquid lens by changing aperture, *Appl. Phys. Lett.* **86**, 211107 (2005).
53. P. G. de Gennes and J. Prost, *The Physics of Liquid Crystals* (Clarendon Press, Oxford, 1993).
54. M. Born and E. Wolf, *Principle of Optics* (Pergamon Press, Oxford, 1993).
55. H. Ren, Y. H. Fan, S. Gauza, and S. T. Wu, Tunable-focus flat liquid crystal spherical lens, *Appl. Phys. Lett.* **84**, 4789 (2004).
56. Y. H. Fan, H. Ren, X. Liang, et al., Liquid crystal microlens arrays with switchable positive and negative focal lengths, *J. Display Technology*, **1**, 151 (2005).
57. S. Yanase, K. Ouchi, and S. Sato, Molecular orientation analysis of a design concept for optical properties of liquid crystal microlenses, *Jpn. J. Appl. Phys.* **40**, 6514 (2001).
58. H. Ren, J. R. Wu, Y. H. Fan, et al., Hermaphroditic liquid-crystal microlens, *Opt. Lett.* **30**, 376 (2005).
59. P. F. McManamon, T. A. Dorschner, D. L. Corkum, et al., Optical phased arrays technology, *Proc. IEEE* **84**, 268 (1996).

60. H. Ren, Y. H. Fan, S. Gauza, and S. T. Wu, Tunable-focus flat liquid crystal spherical lens, *Appl. Phys. Lett.* **84**, 4789 (2004).
61. H. Ren, Y. H. Fan, and S. T. Wu, Prism grating using polymer stabilized nematic liquid crystal, *Appl. Phys. Lett.* **82**, 3168 (2003).
62. U. Efron, *Spatial Light Modulators* (Marcel Dekker, New York, 1994).
63. V. Freedericksz and V. Zolina, Forces causing the orientation of an anisotropic liquid, *Trans. Faraday Soc.* **29**, 919-930 (1933).
64. Y. H. Fan, Y. H. Lin, H. Ren, et al., Fast-response and scattering-free polymer network liquid crystals, *Appl. Phys. Lett.* **84**, 1233 (2004).
65. Y. H. Wu, Y. H. Lin, Y. Q. Lu, et al., Submillisecond response variable optical attenuator based on sheared polymer network liquid crystal, *Opt. Express* **12**, 6377 (2004).
66. J. L. West, G. Zhang, and A. Glushchenko, Fast birefringent mode stressed liquid crystal, *Appl. Phys. Lett.* **86**, 031111 (2005).
67. J. S. Patel, Polarization insensitive tunable liquid-crystal etalon filter, *Appl. Phys. Lett.* **59**, 1314 (1991).
68. Y. Huang, T. X. Wu, and S. T. Wu, Simulations of liquid-crystal Fabry–Perot etalons by an improved 4 × 4 matrix method, *J. Appl. Phys.* **93**, 2490-2495 (2003).
69. R. L. Sutherland, V. P. Tondiglia, L. V. Natarajan, et al., Electrically switchable volume gratings in polymer-dispersed liquid crystals, *Appl. Phys. Lett.* **64**, 1074 (1994).
70. H. Ren, Y. H. Lin, Y. H. Fan, and S. T. Wu, Polarization-independent phase modulation using a polymer-dispersed liquid crystal, *Appl. Phys. Lett.* **86**, 141110 (2005).
71. Y. H. Lin, H. Ren, Y. H. Fan, et al., Polarization-independent and fast-response phase modulation using a normal-mode polymer-stabilized cholesteric texture, *J. Appl. Phys.* **98**, 043112 (2005).
72. T. Uchida, H. Seki, C. Shishido, and M. Wada, Bright dichroic guest–host LCDs without a polarizer, *Proc. SID*, **22**, 41 (1981).
73. M. Hasegawa,C. Hellermark, A. Nishikai, et al., Reflective stacked crossed guest-host display with a planarized inner diffuser, *SID Tech. Digest*, **31**, 128 (2000).
74. Y. H. Lin, H. Ren, Y. H. Wu, et al., Polarization-independent liquid crystal phase modulator using a thin polymer-separated double-layered structure, *Opt. Express* **13**, 8746 (2005).
75. S. Gauza, H. Wang, C. H. Wen, et al., High birefringence isothiocyanato tolane liquid crystals, *Jpn. J. Appl. Phys.* **42**, 3463 (2003).
76. H. Ren, Y. H. Lin, and S. T. Wu, Polarization-independent and fast-response phase modulators using double-layered liquid crystal gels, *Appl. Phys. Lett.* **88**, 061123 (2006).

13

Blue Phases of Chiral Liquid Crystals

13.1 Introduction

The history of blue phases can be traced back more than a century. In 1888, Reinitzer [1], an Austrian botanist, examined cholesteryl benzoate (a chiral organic compound) under an optical microscope. He observed that this material exhibited a blue color just below the isotropic phase as it cooled, and the color disappeared almost immediately. It is known now that cholesteryl benzoate exhibits blue phases – and they got their name because this first observed such phase had a blue color. This blue phase is also the first liquid crystal phase reported in literature.

Non-chiral liquid crystal transforms directly from the isotropic phase into the nematic phase. If there is a spatial fluctuation in these phases (i.e. spatial non-uniformity) a physical property $f(\vec{r})$ can be expressed in terms of a set of Fourier components $\sum_{\vec{q}} f(\vec{q})\exp(-i\vec{q}\cdot\vec{r})$, where $\vec{q}$ is the wave vector given by $q = 2\pi/L$ and L is the wavelength (spatial period) of the fluctuation. In both the isotropic and nematic phases, any variation in space costs energy and is energetically unfavorable. Therefore both phases are uniform in space in the ground state, and the transition from the isotropic phase to the nematic phase takes place at the origin of the wave vector space, namely, $q = 0$. Fluctuations do not play a role in this transition. However, for a chiral liquid crystal, fluctuations may play a very important role in the isotropic–cholesteric transition. In the low-temperature cholesteric phase, the liquid crystal has a helical periodic structure. In the wave vector space, the phase transition takes place on a spherical shell in q -space with the radius $q = 2\pi/\text{pitch}$ [2–4]. Any fluctuation with the wave vector on the surface of the shell does not cost energy. Therefore other phases may exist between the isotropic and cholesteric phases. Indeed, for chiral liquid crystals with short pitches, blue phases exist in narrow temperature regions (about 1 degree wide) between the isotropic and cholesteric phases. There are three blue phases named BPI, BPII, and BPIII. BPI and BPII have cubic structures and

Fundamentals of Liquid Crystal Devices, Second Edition. Deng-Ke Yang and Shin-Tson Wu.

exhibit Bragg reflections [2,4–6]. The color of the reflected light can be any color in the visible and UV light region. BPIII has an amorphous structure.

There was little investigation of blue phases for a long time after their discovery, partially because they only exist in a very narrow temperature region. Research interest on blue phases was renewed twice. The first intensive research occurred in the 1980s. The renewal in this period was due to two reasons. The first was the great success of theories based on Landau phase transition theory and disclination theory [7–10]. The second was the advance in experimental techniques that made it possible to control temperature within 1 mK, thus allowing the study of blue phases. In this period, most of the physics associated with blue phases was understood. Interests in blue phases were, however, mainly scientific curiosity. Beginning in the early 2000s, interest piqued again. First it was discovered that one could achieve mirrorless lasing in blue phases [11]. In addition, it was found that the blue phase temperature region can be dramatically increased by polymer stabilization, allowing the possibility of blue phase displays [12–14]. Blue phases gained great practical importance because they promise large viewing angles and ultrafast switching time.

13.2 Phase Diagram of Blue Phases

Blue phases exist in a narrow temperature region between the isotropic and cholesteric phases. As temperature is decreased, the order of appearance of the blue phases is BPIII, BPII, and BPI [15–17]. Whether a chiral liquid crystal has a blue phase depends on its molecular structure and chirality. The blue phases can be identified by an optical microscope under reflection mode. BPI and BPII have bright and colorful multi-domain crystal plate textures, while BPIII has a dim uniform foggy texture [5,18]. Therefore, BPIII is also called the fog phase. As will be discussed later, BPI and BPII have cubic crystal structures while BPIII has an amorphous structure.

In order to see the effect of chirality, let us consider the phase diagram of blue phases as a function of temperature and chirality. The liquid crystal material is CE2 (from Merck) whose chemical structure is shown in Figure 13.1. It is a chiral molecule with two chiral centers and itself exhibits liquid crystal phases. The helical twist power of CE2 is $HTP = 9.2\ \mu m^{-1}$. The phase diagram is shown in Figure 13.2 [17]. In the measurement of the phase diagram, m_L mole of left-handed CE2 and m_R mole right-handed CE2 are mixed. The pitch P of the mixture is determined by the excessive mole fraction of right-handed CE2: $m_e = (m_R - m_L)/(m_R + m_L)$ and the HTP: $P = 1/(m_e \cdot HTP)$. The chirality of the mixture is given by $q = 2\pi/P = 2\pi(HTP \cdot m_e)$, which is linearly proportional to the excessive mole fraction of right-handed CE2. In Figure 13.2, the horizontal axis is m_e and the vertical axis is temperature. When the chirality is low, the material transforms directly from the isotropic phase to the cholesteric phase. When the excessive mole fraction is increased to 0.1 (the corresponding pitch is 1.1 μm and the chirality is $0.9 \cdot 2\pi\ \mu m^{-1}$), BPI appears. The temperature range of BPI increases with the chirality.

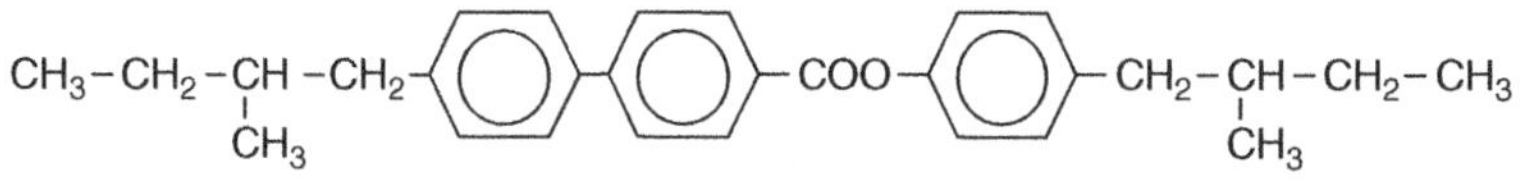

$CH_3-CH_2-\underset{\displaystyle CH_3}{CH}-CH_2-C_6H_4-C_6H_4-COO-C_6H_4-CH_2-\underset{\displaystyle CH_3}{CH}-CH_2-CH_3$

Figure 13.1 Chemical structure of CE2.

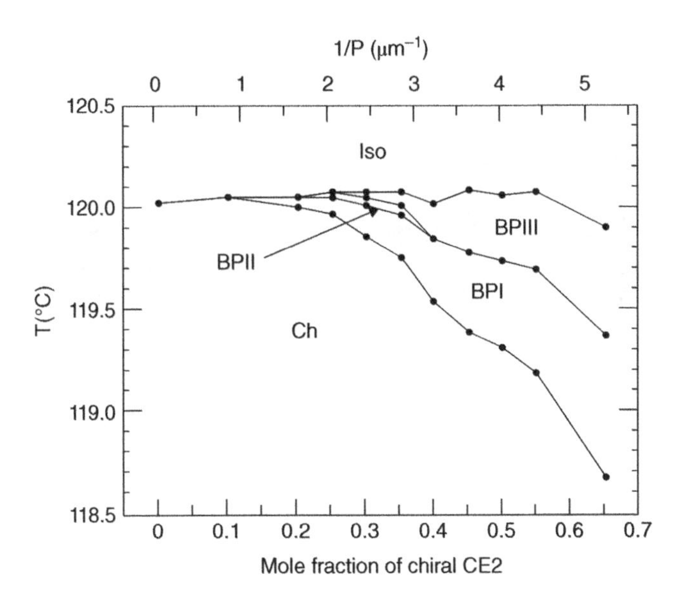

Figure 13.2 Blue phase diagram of CE2. Reproduced with permission from the American Physical Society.

When the excessive mole fraction is increased to 0.19 (the corresponding pitch is 0.57 μm and the chirality is $1.75 \cdot 2\pi\ \mu m^{-1}$), BPII appears. The temperature range of BPII first increases with the chirality and then decreases. It disappears when the excessive mole fraction is increased to 0.4. Then BPIII appears at the excessive mole fraction of 0.22 (the corresponding pitch is 0.50 μm and the chirality is $2.0 \cdot 2\pi\ \mu m^{-1}$). The temperature range of BPIII increases with the chirality. When the excessive mole fraction is increased beyond 0.6, the reflected light moves into the UV light region and it becomes difficult to visually observe the blue phases. The phase behavior of the materials with higher chiralities is unknown. The author conjectures that the liquid crystal phases will all disappear at a sufficiently high chirality, because the energy cost of defects decreases with increasing chirality. When the energy cost becomes comparable with thermal energy, the liquid crystal phases become unstable. The critical chiralities at which the blue phases appear vary from material to material. In many materials, blue phases appear with chirality around $4\pi\ \mu m^{-1}$, which corresponds to the pitch length of 0.5 μm.

13.3 Reflection of Blue Phases

13.3.1 Basics of crystal structure and X-ray diffraction

X-ray diffraction is a powerful experimental technique for studying the structure of crystals, because its wavelength is in the nanometer region, comparable to the periodicity of crystal structures. Let us consider the crystal structure schematically shown in Figure 13.3(a). The primitive vectors are $\vec{a}$, $\vec{b}$ and $\vec{c}$. The lattice is defined by the translations of the primitive vectors:

$$\vec{r} = u\vec{a} + v\vec{b} + w\vec{c} \tag{13.1}$$

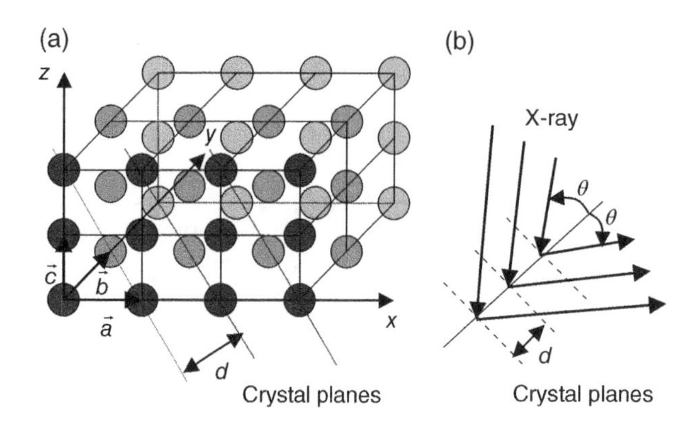

Figure 13.3 (a) Lattice of the crystal, (b) X-ray diffraction by a set of crystal planes.

where u, v, and w are integers. The constituent elements (atoms or molecules) are located at the lattice sites.

When an X-ray beam is shone on a crystal, it will be diffracted if the Bragg condition is satisfied.

$$m\lambda = 2d\cos\theta \tag{13.2}$$

where λ is the wavelength of the X-ray, m is a integer, d is the distance between the parallel crystal planes, and θ is the incident angle. For a crystal, there are many different sets of parallel crystal planes. A set of a crystal planes is specified by the index (j, k, l) of the planes, which is obtained by the following process: (1) find the intercepts of the plane on the axes along the primitive lattice vectors $\vec{a}$, $\vec{b}$, and $\vec{c}$, and (2) take the reciprocals of these intercepts and change them into integers by multiplying their least common multiplier. As an example, let us consider the set of crystal planes shown by the dashed lines in Figure 13.3(a). The intercepts on the x, y, and z axes are $(1, \infty, 2)$. Their reciprocals are $(1, 0, 1/2)$, and the index of the plane is $(2, 0, 1)$. The distance between the crystal planes the index (j, k, l) is

$$d = 1 \Big/ \sqrt{\frac{j^2}{a^2} + \frac{k^2}{b^2} + \frac{l^2}{c^2}}. \tag{13.3}$$

Crystals have periodic structures, and their physical properties vary periodically in space. A physical parameter can be expanded in Fourier components:

$$f(\vec{r}) = \sum_{\vec{G}} f\left(\vec{K}\right) e^{i\vec{G}\cdot\vec{r}} \tag{13.4}$$

The wave vectors $\vec{G}$ have the magnitude $2\pi/d_{jkl}$, along the normal direction of the crystal plane specified by the plane index (j, k, l). The right side of Equation (13.4) sums over all the plane indices. It is very convenient to introduce the reciprocal lattice vectors in identification of the wave vectors. The primitive reciprocal lattice vectors are defined by

$$\vec{a}_r = 2\pi \frac{\vec{b} \times \vec{c}}{\vec{a} \cdot \left(\vec{b} \times \vec{c}\right)}, \vec{b}_r = 2\pi \frac{\vec{c} \times \vec{a}}{\vec{a} \cdot \left(\vec{b} \times \vec{c}\right)}, \vec{c}_r = 2\pi \frac{\vec{a} \times \vec{b}}{\vec{a} \cdot \left(\vec{b} \times \vec{c}\right)} \tag{13.5}$$

where $\vec{a}_r$, $\vec{b}_r$, and $\vec{c}_r$ are orthogonal to $\vec{a}$, $\vec{b}$, and $\vec{c}$, respectively. Also,

$$\vec{\mu} \cdot \vec{\eta}_r = 2\pi \delta_{\mu\eta}, \ (\mu, \eta = a, \ b, \ c). \tag{13.6}$$

The possible wave vectors for the Fourier expansion are given by

$$\vec{G} = u\vec{a}_r + v\vec{b}_r + w\vec{c}_r, \tag{13.7}$$

where u, v and w are integers.

The intercepts of the crystal plane (j, k, l) are m/j, m/k, and m/l on the crystal axes $\vec{a}$, $\vec{b}$, and $\vec{c}$, respectively, where m is the least common multiplier of j, k, and l. Any vector lying in the crystal plane can be expressed as $\vec{e} = s\left[(m/j)\,\vec{a} - (m/k)\,\vec{b}\right] + t\left[(m/k)\,\vec{b} - (m/l)\,\vec{c}\right]$. The reciprocal vector $j\vec{a}_r + k\vec{b}_r + l\vec{c}_r$ is perpendicular to $\vec{e}$, because

$$\vec{e} \cdot \left(j\vec{a}_r + k\vec{b}_r + l\vec{c}_r\right) = 2\pi[sj(m/j) - sk(m/k) + tk(m/k) - tl(m/l)] = 0. \tag{13.8}$$

Therefore the reciprocal vector $j\vec{a}_r + k\vec{b}_r + l\vec{c}_r$ can also be chosen to represent the crystal plane. The index (j, k, l) is also called the Miller index.

13.3.2 Bragg reflection of blue phases

Since the periodicity of blue phases is in the submicron region, visible light spectroscopy is suitable for studying the structure of the blue phases. Furthermore polarization of the optical wave can provide additional information on the structure of the blue phases. The reflection spectrum of BPII of a multi-domain CE2 sample is shown in Figure 13.4(a), where the excessive mole fraction of right-handed CE2 is 0.25 [19,20]. There are three important features. First, the reflected light is circularly polarized with the same handedness as the helical structure of the material in the cholesteric phase [6]. Second, the reflection peaks are narrow, quite different from that of the cholesteric phase. The width of the reflection peaks of blue phases are less than 10 nm, which is determined only by the size of the blue phase crystal, while the width $\Delta\lambda$ of the reflection peak of the cholesteric phase is about 50 nm, which is governed by $\Delta\lambda = \Delta n P$, where Δn is the birefringence and P is the pitch. Third, there are multiple reflection speaks, corresponding to different crystal planes. The ratio between the peak wavelengths are $\lambda_1 : \lambda_2 : \lambda_3 \approx 1 : (1/\sqrt{2}) : (1/\sqrt{3})$, indicating that the first peak is produced by the $(1, 0,$

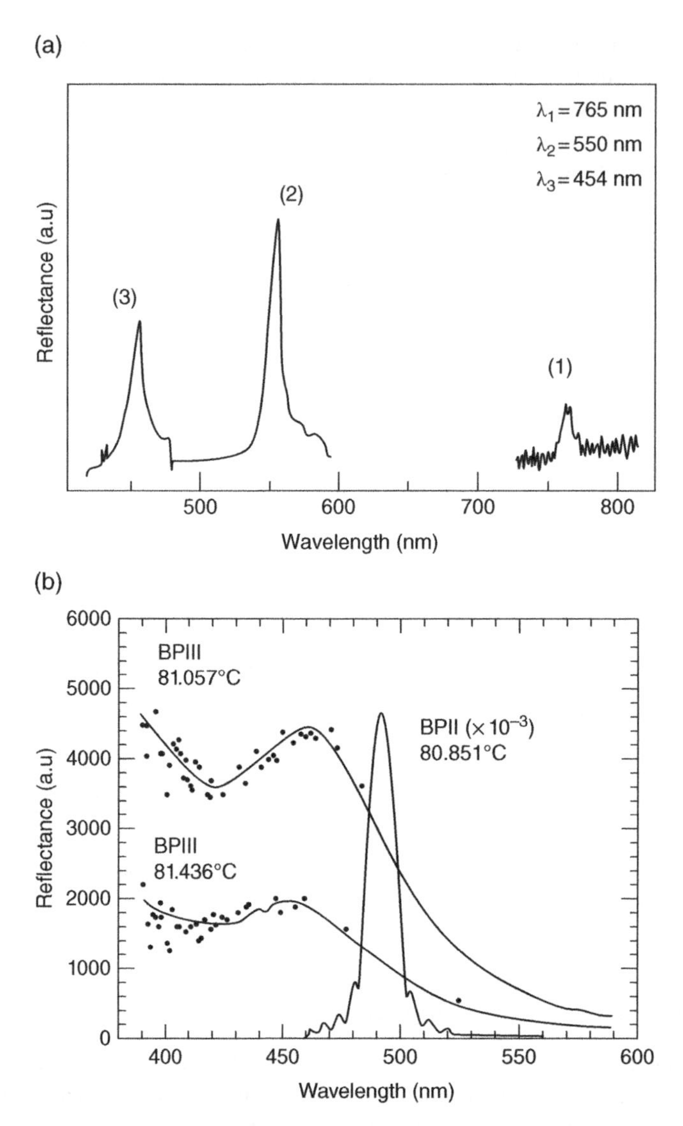

Figure 13.4 (a) Reflection spectrum of BPII of the CE2 sample. (b) reflection spectrum of BPII and BPIII of the 5CB/CE1/CE2 sample. Reproduced with permission from the American Physical Society.

0) plane, the second peak is produced by the (1, 1, 0) plane, and the third peak is produced by the (1, 1, 1) plane. In the experiment of Figure 13.4, the liquid crystal was sandwiched between two parallel glass substrates, and studied under an optical microscope with a spectrometer connected to it. Sometimes it is necessary to thermally recycle the sample in order to see the reflection from different crystal planes.

The reflection spectrum of BPII and BPIII of another sample is shown in Figure 13.4(b). The sample consists of 54.3% nematic liquid crystal 5CB, 17.6% chiral liquid crystal CE1, and 28.1% chiral liquid crystal CE2. The reflectance of BPIII is lower than that of BPII by three

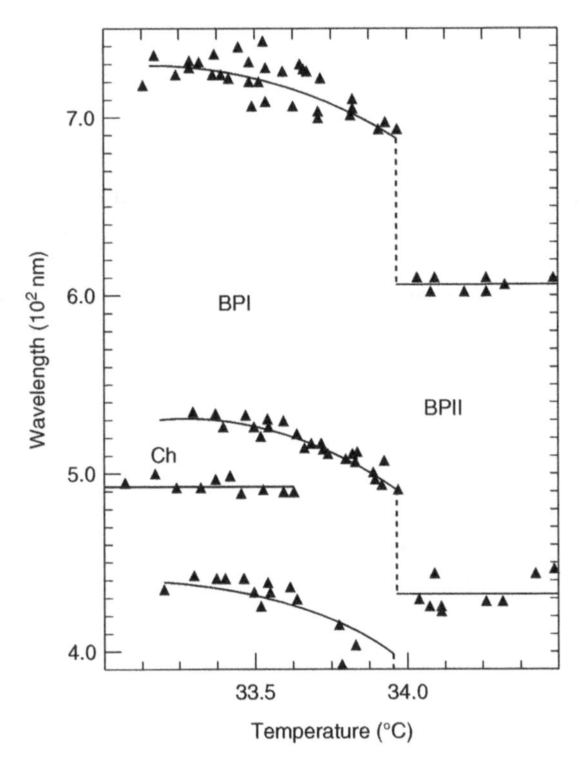

Figure 13.5 Selective reflection wavelengths of the cholesteric and blue phases [21].

orders of magnitude; therefore it is difficult to see BPIII under the microscope. Note that the peak width of BPIII is much wider than that of BPII. These two features indicate that there is no long-range order in BPIII. The reflection increases as the temperature is decreased in BPIII, indicating that the order increases with decreasing temperature in this phase. The reflection around 400 nm is probably produced by the ITO coating on the cell substrates.

The wavelengths of the reflection peaks in the blue phase are different in general from the central wavelength of the reflection band of the cholesteric phase. The selective reflection wavelengths of a chiral nematic liquid are shown in Figure 13.5 [21]. The longest selective reflection wavelengths of BPI and BPII are longer than that of the cholesteric phase. Therefore the periodicities of the blue phases are different from the periodicity of the cholesteric phase, which is equal to $P/2$. When the temperature is decreased toward the blue phase to cholesteric phase transition temperature, the selective wavelength increases.

13.4 Structure of Blue Phase

From the experimental results discussed in the above section, it is known that BPI and BPII have cubic structures. There are two theories that have successfully explained the existence of the blue phases and predict their symmetry and physical properties. One is known as the defect theory, in which the blue phases consist of packed double-twist cylinders and there are defects in the regions between the cylinders [4,22–27]. In the double-twist cylinder, the liquid crystal twists around any radius of the cylinder. The free energy of the blue phases is expressed in terms of the free energies of the double-twist cylinder and defect. In a narrow temperature region between the isotropic phase and cholesteric phase, the free energy is found to be lower than those of the isotropic and cholesteric phases. In this theory, it is easy to see the impact of material parameters on the blue phase behavior. The other theory is based on Landau theory, in which the free energy of the blue phases is expressed in terms of a tensor order parameter [7–10]. With certain cubic symmetries, the free energy is also found to be lower than those of the isotropic and cholesteric phases. Because the anisotropic part of the dielectric tensor is chosen to be the tensor order parameter, it is straightforward to see the optical properties of the blue phase in this theory.

13.4.1 Defect theory

A defect theory for blue phases was introduced by Meiboom, Sethna, Anderson, et al. [22–27]. In this theory, the liquid crystal is assumed to form double-twist cylinders where the liquid crystal molecules twist about any radius of the cylinder, as shown in Figure 13.6. The cylinder cannot, however, cover the whole 3-D space without topological defects. Instead of a single cylinder, the blue phases consist of packed double-twist cylinders. There are defects in the regions not occupied by the cylinders.

13.4.1.1 Double-twist cylinder

When the chiral liquid crystal transforms from the unordered isotropic phase to the ordered helical phase, the liquid crystal molecules start to twist with respect to one another. The

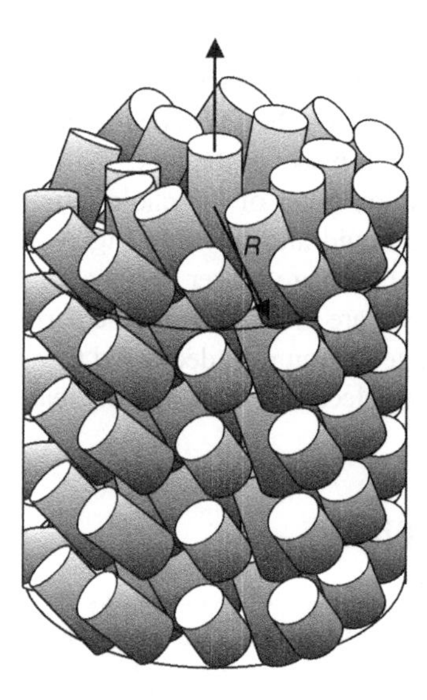

Figure 13.6 Schematic diagram of the double twist cylinder.

cholesteric planes (within which the liquid crystal molecules are parallel to each other) have, however, not yet formed in the early stage of the transition. The liquid crystal molecules twist along any radial direction with respect to the molecule in the center and form the double-twist cylinder as shown in Figure 13.6. In cylindrical coordinate, the director is given by

$$\vec{n} = -\sin(q_o r)\hat{\phi} + \cos(q_o r)\hat{z}. \tag{13.9}$$

The bulk elastic energy density is

$$f_{e/B} = \frac{1}{2}K_{11}(\nabla\cdot\vec{n})^2 + \frac{1}{2}K_{22}(\vec{n}\cdot\nabla\times\vec{n} + q_o)^2 + \frac{1}{2}K_{33}(\vec{n}\times\nabla\times\vec{n})^2. \tag{13.10}$$

From Equation (13.9), we can get

$$\nabla\cdot\vec{n} = \frac{1}{r}\frac{\partial}{\partial r}(rn_r) + \frac{1}{r}\frac{\partial n_\phi}{\partial\phi} + \frac{1}{r}\frac{\partial}{\partial z}(rn_z) = 0, \tag{13.11}$$

$$\nabla\times\vec{n} = \begin{vmatrix} \frac{\hat{r}}{r} & \hat{\phi} & \frac{\hat{z}}{r} \\ \frac{\partial}{\partial r} & \frac{\partial}{\partial\phi} & \frac{\partial}{\partial z} \\ n_r & rn_\phi & n_z \end{vmatrix} = q_o\sin(q_o r)\hat{\phi} - \left[q_o\cos(q_o r) + \frac{1}{r}\sin(q_o r)\right]\hat{z}. \tag{13.12}$$

Equation (13.10) becomes

$$f_{e/B} = \frac{1}{2}K_{22}\frac{1}{r^2}\sin^2(q_o r)\cos^2(q_o r) + \frac{1}{2}K_{33}\frac{1}{r^2}\sin^4(q_o r) \approx K_{22}\frac{1}{r^2}\sin^2(q_o r), \tag{13.13}$$

where the approximation $K_{22} = K_{33}$ is used. If the double-twist cylinder has a finite radius R (which is smaller than the pitch $P = 2\pi/q_o$), there are many double-twist cylinders in a macroscopic sample. The surface-to-volume ratio will be large, and thus the surface elastic energy is important. In the double-twist cylinder configuration, the negative K_{24} surface elastic energy term is crucial and is given by

$$f_{e/S} = -K_{24}\nabla\cdot(\vec{n}\,\nabla\cdot\vec{n} + \vec{n}\times\nabla\times\vec{n}) = -K_{24}\nabla\cdot\left[\frac{1}{r}\sin^2(q_o r)\hat{r}\right], \tag{13.14}$$

which can be converted into a surface integration. The total elastic energy (per unit length) is

$$F_e = 2\pi\int_0^R (f_{e/B} + f_{e/S})\,rdr = \pi K_{22}\int_0^{q_o R}\frac{1}{q_o r}\sin^2(q_o r)\,d(q_o r) - 2\pi K_{24}\sin^2(q_o R). \tag{13.15}$$

The surface elastic constant K_{24} is usually smaller than the bulk twist elastic constant K_{22}. As an example, we assume $K_{24} = 0.5K_{22}$ [28]. The total elastic energy is plotted as a function of the

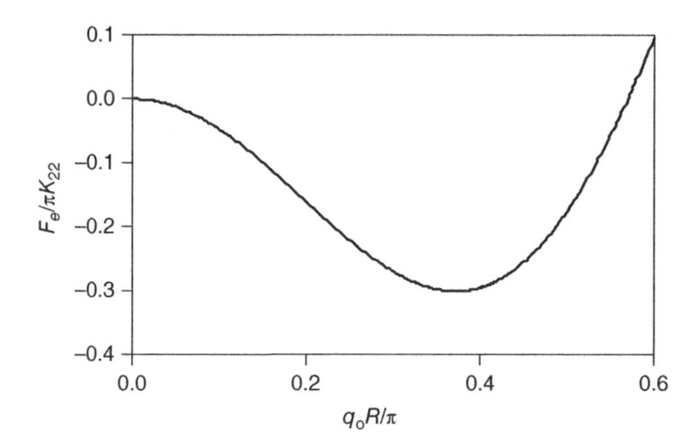

Figure 13.7 The elastic energy as a function of radius of the double-twist cylinder.

radius of the double-twist cylinder in Figure 13.7. When the radius is small, as the radius increases, the total elastic energy decreases because the rapid decrease of the surface elastic energy. When the radius is large, the bend elastic energy becomes large. As the radius increases, the bend elastic energy increases rapidly and the total elastic energy increases. Therefore the total elastic energy decreases first with the radius and then increases. The minimum elastic energy occurs at $q_oR/\pi = 0.36$. This shows that the double-twist cylinder has finite size with the radius of $R = 0.36\pi/q_o = 0.18P$. The corresponding twist angle on the surface of the cylinder is 65°. If $K_{33} = 2K_{22}$ is used, the minimum elastic energy occurs at $q_oR/\pi = 0.27$, and then the radius is $R = 0.135P$ and the twist angle on the surface of the cylinder is 49°. The actual radius of the cylinder may also depend on the packing of the cylinders in space. For smaller size, the surface-to-volume ratio is higher, which will also help to reduce the total free energy.

13.4.1.2 Packing of double-twist cylinders

As shown in the above section, the double-twist cylinder has finite size. The cylinders must pack to fill 3-D space. Now let us consider the packing of the (infinitely long) double-twist cylinders in 3-D space. The simplest structure, the hexagonal packing of the cylinders as shown in Figure 13.8(a), turns out to have the highest packing density (percentage of occupied volume) of 0.9069 [29]. The liquid crystal director on the cross-sections of nearest neighboring cylinders is shown in Figure 13.8(d). Let us consider the two double-twist cylinders on the top. At the right side boundary of the left cylinder, the twist angle is about 45°, while at the left side boundary of the right cylinder, the twist angle is about −45°. The liquid crystal director changes discontinuously when moving from one cylinder to the next. This discontinuity costs too much energy and thus the hexagonal packing is energetically unfavored. Any other structures, in which some nearest neighboring cylinders are parallel, will all have the same problem of discontinuity change of liquid crystal director, and therefore cannot be realized.

It is easy to see that a structure without parallel neighboring cylinders is the simple cubic structure shown in Figure 13.8(b). The neighboring cylinders are orthogonal to each other, and the packing density is 0.5890 [29]. The liquid crystal director on the cross sections of nearest neighboring cylinders is shown in Figure 13.8(e). If the twist angle at the surface of the

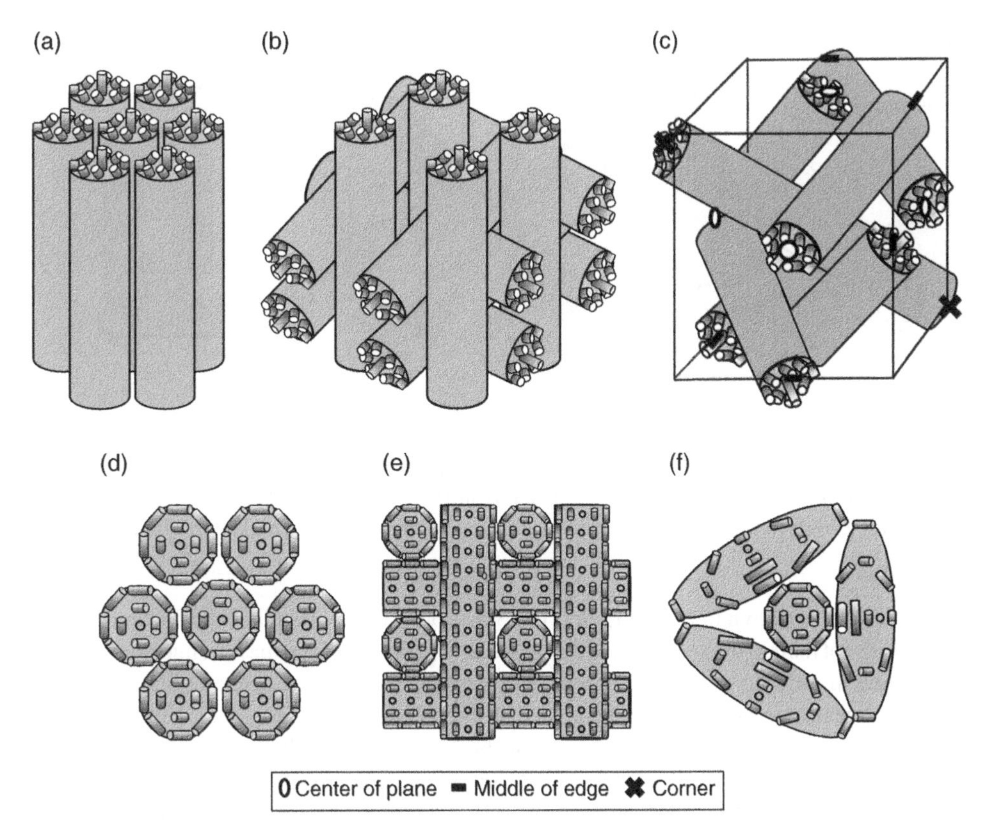

Figure 13.8 Schematic diagrams of the packing of the double-twist cylinders in 3-D space: (a) hexagonal packing. (b) simple cubic packing, (c) body-centered cubic packing, and (d), (e), and (f) show the liquid crystal director configurations on the corresponding cross sections.

cylinder is 45°, the director changes continuously when moving from one cylinder to the next. Therefore this cubic structure could be stable. Note that there are also liquid crystals in the unoccupied regions, whose structure will be discussed in next section.

Another possible structure is the body-centered cubic packing of the cylinders, as shown in Figure 13.8(c). One of the cylinders is along the diagonal of the cube and the other three cylinders are along the line connecting the center of one plane to the middle of one edge. The packing density is 0.6802 [29]. The liquid crystal director on the cross sections of nearest neighboring cylinders is shown in Figure 13.8(f). If the twist angle at the surface of the cylinder is 45°, the director changes continuously when moving from one cylinder to the next. Therefore this cubic structure also could be stable.

13.4.1.3 Disclination

As pointed out in the previous section, there are voids between the double-twist cylinders when they are packed in 3-D space. Liquid crystal must fill the void. Because of the boundary condition imposed by the cylinders, the liquid crystal director is not uniform in this space and forms a defect

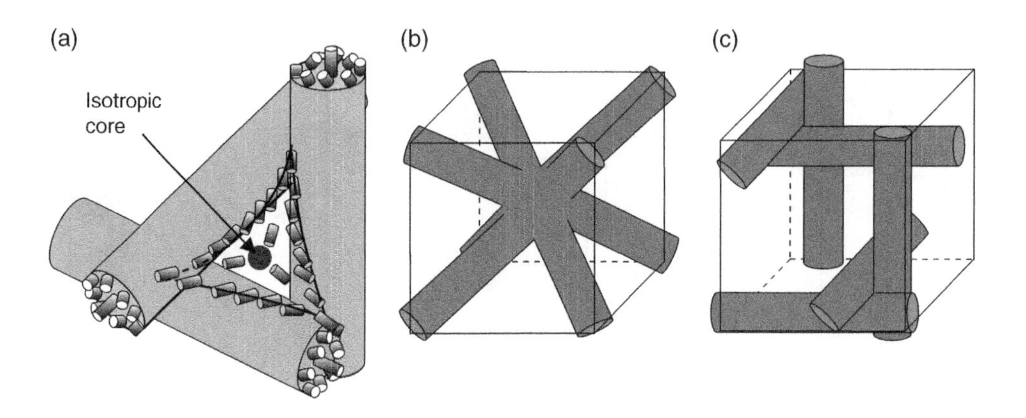

Figure 13.9 (a) The liquid crystal director configuration of the disclination formed between the double-twist cylinders. (b) The structure of the disclinations in the simple cubic packing of the double-twist cylinders. (c) The structure of the disclinations in the body-centered cubic packing of the double-twist cylinders.

of strength −1/2 as shown in Figure 13.9(a) [25,30]. Note that the boundaries of the double-twist cylinders are not hard in the sense that at the boundary the liquid crystal may change orientation to some degree in order to reduce the energy of the disclination. In the defect region, the elastic energy diverges when approaching the center of the defect, and an isotropic core will form to reduce the elastic energy. The linear size R_o of the defect is about the same as the radius R of the double-twist cylinder. The radius R_c of the isotropic core is governed by the elastic constant and by the free energy difference between the isotropic phase and the chiral nematic phase.

In 3-D space, the defects formed between the double-twist cylinders are line defects and are called disclinations. The organization of the disclinations in the 3-D space has the same symmetry as the structure of the packed the double-twist cylinders. The disclinations in the simple cubic packing and body-centered cubic packing are shown in Figure 13.9(b) and (c), respectively.

13.4.1.4 Free energy

The free energy associated with the blue phase structure is contributed by the free energy of the double-twist cylinder, the free energy of the disclination, the free energy of the isotropic core, and the free energy of the surface of the isotropic core. The free energy of the surface of the isotropic core is small and probably plays a minor role and thus is neglected here. The total free energy of one unit cell of the blue phase is given by [27]

$$F = F_{\mathrm{DTC}} + F_D + F_{IC}, \tag{13.16}$$

where F_{DTC} is the total free energy of the double-twist cylinder. The free energy per unit length of the cylinder is given by Equation (13.15). F_{DTC} also depends on the total length of the double-twist cylinders in one unit cell, which in turn depends on the packing of the cylinders:

$$F_{\mathrm{DTC}} = \pi K_{22} A \int_0^{q_o R} \frac{1}{q_o r} \sin^2(q_o r) d(q_o r) - 2\pi K_{24} A \sin^2(q_o R), \tag{13.17}$$

where A is a total length of the cylinders in the unit cell. F_D is the elastic energy of the disclination. As an approximation, we assume that the cross section of the disclination is a circle:

$$F_D = \alpha A\left\{\int_{R_c}^{R}\frac{1}{2}K_{33}\left(\nabla \vec{n}\right)^2 2\pi r dr\right\} = \alpha A\left\{\int_{R_c}^{R_o}\frac{1}{2}K_{33}\left(\frac{1}{r}\right)^2 2\pi r dr\right\} = \alpha A\pi K_{33}\ln\left(\frac{R_o}{R_c}\right), \quad (13.18)$$

where α is a constant and αA is the total length of the disclinations in one unit cell, R_o is the radius of the disclination (approximately the same as the radius R of the double-twist cylinder), and R_c is the radius of the isotropic core. The value of α depends on the structure of the packed double-twist cylinders. F_{iso} is the total free energy of the isotropic core in one unit cell and is given by

$$F_{iso} = \alpha A\pi R_c^2\left(f_{iso} - f_{lc}\right), \quad (13.19)$$

where f_{iso} is the free energy of the isotropic phase without orientational order, and f_{lc} is the free energy of the liquid crystal with orientational order. At a temperature T slightly below the transition temperature T_{iso} to the isotropic phase,

$$\left(f_{iso} - f_{lc}\right) \approx \left(\left.\frac{\partial f_{iso}}{\partial T}\right|_{T_{iso}} - \left.\frac{\partial f_N}{\partial T}\right|_{T_{iso}}\right)\left(T - T_{Iso}\right) = \left(-\frac{S_{iso}}{T_{iso}} + \frac{S_{lc}}{T_{iso}}\right)\left(T_{iso} - T\right) = \frac{L}{T_{iso}}\left(T - T_{iso}\right),$$

where S_{iso} is the entropy of the isotropic phase, S_{lc} is the entropy of the liquid crystal phase, and L is the latent heat of the transition. We can rewrite Equation (13.19) as

$$F_{iso} = \alpha A\pi R_c^2\beta\left(T_{iso} - T\right), \quad (13.20)$$

where $\beta = L/T_{iso}$ is a constant. The radius of the isotropic core depends on the free energy difference between the liquid crystal phase and the isotropic phase. At the outer boundary of the isotropic core, the elastic energy density becomes the same as the free energy density difference between the isotropic phase and the liquid crystal phase, namely,

$$\frac{1}{2}K_{33}\left(\frac{1}{R_c}\right)^2 = f_{iso} - f_{lc} = \beta\left(T_{iso} - T\right),$$

and thus

$$R_c = \left[\frac{K_{33}}{2\beta\left(T_{iso} - T\right)}\right]^{1/2}. \quad (13.21)$$

R_c is about a few tens of nanometers. From Equations (13.20) and (13.21) we get

$$F_{iso} = \frac{1}{2}\alpha\beta A\pi K_{33}. \quad (13.22)$$

The total free energy per unit cell becomes

$$F/\pi A = K_{22}\int_0^{q_oR}\frac{1}{q_o r}\sin^2(q_o r)d(q_o r) - 2K_{24}\sin^2(q_oR) + \alpha K_{33}\ln\left(\frac{q_oR}{q_oR_c}\right) + \frac{1}{2}\alpha\beta K_{33}. \quad (13.23)$$

In the above equation, the second term is negative and tends to stabilize the blue phase. The other terms are positive and tend to destabilize the blue phase. Introducing $\varphi = q_o r$ and $\rho = q_oR$, the above equation becomes

$$F/\pi A = K_{22}\int_0^{\rho}\frac{1}{\varphi}\sin^2\varphi d\varphi - 2K_{24}\sin^2\rho + \alpha K_{33}\ln\left(\frac{\rho}{q_oR_c}\right) + \frac{1}{2}\alpha\beta K_{33}. \quad (13.24)$$

R_c is given by Equation (13.21), independent of the chirality. Only the third term depends on the chirality q_o. For a given R_c, when the chirality is large, the third term is smaller. Physically that means that when the chirality is high, the radius of the double-twist cylinder is small and so is the radius of the disclination, and the free energy of the disclination is low when the radius R_c of the isotropic core is fixed. The total free energy as a function of ρ ($=q_oR$) is plotted in Figure 13.10, where $\alpha = 0.2$ and the last term of Equation (13.23) is left out, because it is independent of ρ. When $q_oR_c = 0.5$, the total free energy has a negative minimum and thus the blue phase is stable. When the chirality q_o is decreased such that $q_oR_c = 0.2$, the total free energy is always positive, and thus the blue phase is unstable. This explains why a blue phase only exist in liquid crystals with high chiralities.

The major cost of free energy is the elastic energy of the disclination, which tends to destabilize the blue phase. At a temperature slightly below the transition temperature to the isotropic phase, the free energy of the isotropic core is small and the liquid crystal in the disclination

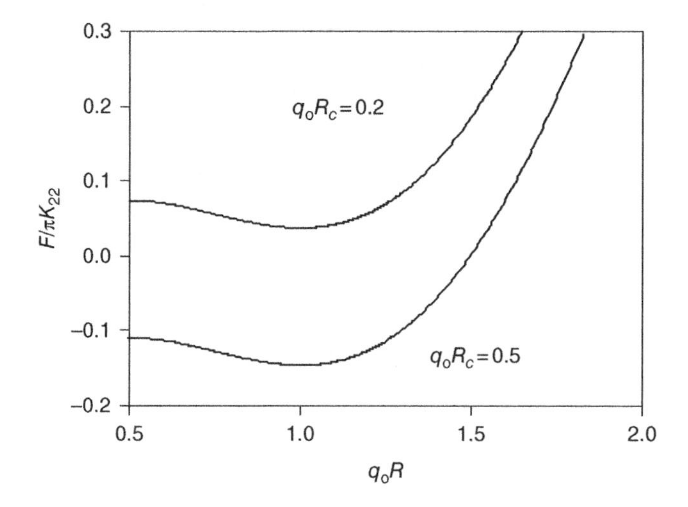

Figure 13.10 The total free energy of one unit cell as a function of radius of the double-twist cylinder.

center escapes from the nematic phase to isotropic phase to reduce the elastic energy. Therefore the blue phase is stable. However, at a temperature significantly below the transition temperature to the isotropic phase, the free energy of the isotropic core is high and the liquid crystal in the disclination center cannot escape to the isotropic phase. The elastic energy of the disclination is therefore high, which destabilizes the blue phase. Mathematically this effect is shown by the third term of the right of Equation (13.23), where the radius of the isotropic core decreases with decreasing temperature.

We estimate the chirality needed to allow the blue phase, and also the temperature range of the blue phase. When the temperature is close the isotropic transition temperature, the isotropic core does not cost much energy and can fill the void between the packed double-twist cylinders. The radius R_c of the isotropic core is approximately the same as radius R of the double-twist cylinder. From Figure 13.7, we know that $q_oR \sim 0.36\pi$. From Equation (13.21) we have

$$q_oR_c = q_o\left[\frac{K_{33}}{2\beta(T_{\text{iso}}-T)}\right]^{1/2} = q_o\left[\frac{K_{33}T_{iso}}{2L(T_{\text{iso}}-T)}\right]^{1/2} \sim q_oR \sim 0.36\pi. \tag{13.25}$$

The typical liquid crystal to isotropic transition latent heat is 100 cal/mole $\approx 10^6$J/m^3. Taking $K_{33} \sim 2\times10^{-11}$N, from Equation (13.25) we have $q_o^2 = 1.2\times10^{17}\text{m}^{-2}(T_{iso}-T)/T_{Iso}$. For $T_{iso} =$ 300K and for a 1 K wide blue phase, $(T_{iso}-T) = 1$ K, $q_o = 2\times10^7\text{m}^{-1}$, and the corresponding pitch is $2\pi/q_o = 0.31\times10^{-7}\text{m} = 310$ nm. This is the pitch needed in order to have 1 K wide blue phase.

Experiments have shown that BPI has the body-centered cubic structure and BII has the simple cubic structure. The latter has a lower packing density than the body-centered cubic structure. Therefore the total length of the disclinations in the simple cubic packing is longer. From Equations (13.21) and (13.23), we know the elastic energy of the disclination is $\alpha K_{33}\ln\left\{\rho\left[2\beta(T_{iso}-T)/K_{33}q_o^2\right]^{1/2}\right\}$. This energy is lower either when $(T_{iso}-T)$ is smaller or when q_o is larger. Therefore BPII exists either in the temperature region higher than that of BPI, or in the region with higher chirality than that of BPI, agreeing well with the experimental measured phase diagram shown in Figure 13.2.

13.4.2 Landau theory

A Landau theory for blue phase was proposed by Brazovskii, Dmitriev, Hornreich, and Shtrikman [7–10]. In this theory, the free energy of the blue phase is expressed in terms of a tensor order parameter which is expanded in Fourier components. The free energy is then minimized with respect to the order parameter with the wave vector in various cubic symmetries. In a narrow temperature region below the isotropic transition temperature, the structures with certain cubic symmetries have free energy lower than both the isotroic and cholesteric phases.

De Gennes used Landau theory to describe the isotropic–nematic transition. In his theory, he used a scalar order parameter S defined by

$$S = <\frac{1}{2}(3\cos^2\theta - 1)>, \tag{13.26}$$

where θ is the instantaneous angle between the long molecular axis and the average direction, and $< >$ indicates the average. The order parameter is 0 in the isotropic phase and 1 for a perfectly oriented nematic phase. Below the isotropic–nematic phase transition temperature, S has a value between 0 and 1. The average direction is represented by a unit vector $\vec{n}$, called the liquid crystal director. Sometimes, it is more convenient to use traceless tensor order parameters. For a uniaxial nematic liquid crystal, the tensor order parameter is defined by

$$\overleftrightarrow{Q} = S\left(\vec{n}\vec{n} - \frac{1}{3}\overleftrightarrow{I}\right), \tag{13.27}$$

where $\overleftrightarrow{I}$ is the 3×3 identity tensor. The components of the tensor order parameter are

$$Q_{ij} = S\left(n_i n_j - \frac{1}{3}\delta_{ij}\right)\ i,\ j = 1,\ 2,\ 3. \tag{13.28}$$

The trace of the tensor is

$$Tr\left(\overleftrightarrow{Q}\right) = \sum_{i=3}^{3} Q_{ii} = S\left[\left(n_1^2 - \frac{1}{3}\right) + \left(n_2^2 - \frac{1}{3}\right) + \left(n_3^2 - \frac{1}{3}\right)\right] = 0. \tag{13.29}$$

This tensor order parameter is traceless and symmetric and vanishes in the isotropic phase. The anisotropic physical properties of the liquid crystal are closely related to the tensor order parameter. For example, the dielectric tensor of the liquid crystal is

$$\overleftrightarrow{\varepsilon} = \begin{pmatrix} \varepsilon_{11} & \varepsilon_{12} & \varepsilon_{13} \\ \varepsilon_{21} & \varepsilon_{22} & \varepsilon_{23} \\ \varepsilon_{31} & \varepsilon_{32} & \varepsilon_{33} \end{pmatrix} = \frac{1}{3}\left(\varepsilon_{//} + 2\varepsilon_{\perp}\right)\overleftrightarrow{I} + \left[\left(\varepsilon_{//} - \varepsilon_{\perp}\right)/S\right]\overleftrightarrow{Q}, \tag{13.30}$$

where $\varepsilon_{//}$ and $\varepsilon_{\perp}$ are dielectric constants parallel and perpendicular to the liquid crystal director, respectively. If the dielectric tensor is known, the tensor order parameter can be calculated by

$$\overleftrightarrow{Q} = \frac{S}{\varepsilon_{//} - \varepsilon_{\perp}}\left[\overleftrightarrow{\varepsilon} - \frac{1}{3}Tr\left(\overleftrightarrow{\varepsilon}\right)\right]. \tag{13.31}$$

Therefore $\overleftrightarrow{\varepsilon} - (1/3)Tr\left(\overleftrightarrow{\varepsilon}\right)$ is sometimes used as the order parameter.

For a cholesteric liquid crystal with the chirality q and helical axis along the z direction, the tensor order parameter is

$$\overleftrightarrow{Q} = S\begin{pmatrix} \cos^2(qz) - 1/3 & \sin(qz)\cos(qz) & 0 \\ \sin(qz)\cos(qz) & \sin^2(qz) - 1/3 & 0 \\ 0 & 0 & -1/3 \end{pmatrix}$$

$$= \frac{S}{6}\begin{pmatrix} 1 & 0 & 0 \\ 0 & 1 & 0 \\ 0 & 0 & -2 \end{pmatrix} + \frac{S}{2}\begin{pmatrix} \cos(2qz) & \sin(2qz) & 0 \\ \sin(2qz) & -\cos(2qz) & 0 \\ 0 & 0 & 0 \end{pmatrix}. \tag{13.32}$$

Because the optical properties of blue phases are of great importance, we choose the traceless part of the dielectric tensor (at optical frequency) $\overleftrightarrow{Q}(\vec{r}) = \overleftrightarrow{\varepsilon}(\vec{r}) - (1/3)Tr\left[\overleftrightarrow{\varepsilon}(\vec{r})\right]$ to be the tensor parameter in future discussions. Similar to the Landau–de Gennes theory, the free energy density of the system is expressed in terms of the order parameter [9,10]:

$$f = \frac{1}{2}\alpha(T - T^*)\overleftrightarrow{Q}^2 - d\left(\nabla \times \overleftrightarrow{Q}\right)\cdot\overleftrightarrow{Q} - \beta\overleftrightarrow{Q}^3 + \gamma\overleftrightarrow{Q}^4 + c\left(\nabla\overleftrightarrow{Q}\right)^2, \tag{13.33}$$

where $\overleftrightarrow{Q}^n$ $(n=2,\ 3,\ 4)$ is the rotational invariant that can be made from the nth power of $\overleftrightarrow{Q}$ and α, β, γ, and c are constants. The second term of Equation (13.33) is the chiral term and d is proportional to the chirality. In the blue phases, the order parameter is not uniform but varies in space. The last term of Equation (13.33) is the elastic energy that describes the free energy caused by the spatial variation of the order parameter. In terms of the components of the order parameter, the free energy density is

$$f = \frac{1}{2}aQ_{ij}Q_{ji} - de_{ijl}Q_{in}Q_{jn,l} - \beta Q_{ij}Q_{jl}Q_{li} + \gamma\left(Q_{ij}Q_{ij}\right)^2 + \frac{1}{2}\left(c_1 Q_{ij,l}Q_{ij,l} + c_2 Q_{ij},iQ_{lj,l}\right), \tag{13.34}$$

where $a = \alpha(T - T^*)$ and $Q_{ij,l} = \partial Q_{ij}/\partial x_l$ is the derivative of Q_{ij} with respect to the coordinate x_l. The notation of sum over repeating subscript is used, such as $Q_{ij}Q_{ji} = \sum_{i=1}^{3}\sum_{j=1}^{3} Q_{ij}Q_{ji}$. Note that the terms in Equation (13.34) are the traces, which are invariant under any rotation of coordinates, of the powers of the tensor $\overleftrightarrow{Q}$, *that is,* $Tr\left(\overleftrightarrow{Q}^2\right) = Q_{ij}Q_{ji}$ and $Tr\left(\overleftrightarrow{Q}^3\right) = Q_{ij}Q_{jl}Q_{li}$. Note also that there are two elastic constants c_1 and c_2 in Equation (13.34), which is more accurate than Equation (13.33) where there is only one elastic constant c.

It is almost impossible to find an order parameter that gives the global minimum free energy. However, because experiments have shown that the blue phases have cubic symmetries, Hornreich and Shtrikman tried to minimize the free energy with respect to an order parameter that possesses cubic symmetry. Because the order parameter changes periodically in space, it can be expanded in Fourier components:

$$Q_{ij}(\vec{r}) = \sum_{h,k,l}\frac{1}{\sqrt{N_{hkl}}}Q_{ij}(h,k,l)e^{iq}(hx + ky + lz), \tag{13.35}$$

where (h, k, l) are Miller indices whose values are integers from $-\infty$ to ∞ (including 0), and $q = 2\pi/P$ is the primitive wave vector (where P is the lattice constant). Note that here the cubic symmetry is used such that the lattice constants in the x, y, and z directions are the same. The Fourier coefficient $Q(h, k, l) = Q(\sigma)$, where $\sigma = \sqrt{h^2 + k^2 + l^2}$, depends only on the magnitude σq of the reciprocal vector $q(h\hat{x} + k\hat{y} + l\hat{z})$ but not on its direction. $N_{hkl} = N(\sigma)$ is the number of reciprocal vectors that have the same amplitude σq. For example, when $\sigma = \sqrt{2}$, the possible Miller indices are (1, 1, 0), (1, 0, 1), (1, −1, 0), (1, 0, −1), (0, 1, 1), (0, −1, 1), (0, 1, −1), (0, −1, −1) (−1, 1, 0), (−1, 0, 1), (−1, −1, 0), and (−1, 0, −1), and therefore $N\left(\sqrt{2}\right) = 12$. The normalization coefficient is $1/\sqrt{N_{hkl}}$.

As defined, $\overleftrightarrow{Q}(\vec{r})$ is a 3×3 symmetric traceless tensor, and therefore $\overleftrightarrow{Q}(\sigma)$ is a 3×3 symmetric traceless tensor. $\overleftrightarrow{Q}(\sigma)$ has five independent parameters. It is convenient to expand the tensor order parameter in terms of the second-order spherical harmonics:

$$\overleftrightarrow{Q}(\sigma)=\sum_{m=-2}^{2}\varepsilon_m(\sigma)e^{i\Psi_m(\sigma)}\overleftrightarrow{M}_m \tag{13.36}$$

$$\overleftrightarrow{Q}(\sigma)=\varepsilon_2(\sigma)e^{i\Psi_2(\sigma)}\frac{1}{2}\begin{pmatrix}1 & i & 0\\ i & -1 & 0\\ 0 & 0 & 0\end{pmatrix}+\varepsilon_1(\sigma)e^{i\Psi_1(\sigma)}\frac{1}{2}\begin{pmatrix}0 & 0 & 1\\ 0 & 0 & i\\ 1 & i & 0\end{pmatrix}+\varepsilon_0(\sigma)e^{i\Psi_0(\sigma)}\frac{1}{\sqrt{6}}\begin{pmatrix}-1 & 0 & 0\\ 0 & -1 & 0\\ 0 & 0 & 2\end{pmatrix}$$

$$+\varepsilon_{-1}(\sigma)e^{i\Psi_{-1}(\sigma)}\frac{1}{2}\begin{pmatrix}0 & 0 & -1\\ 0 & 0 & i\\ -1 & i & 0\end{pmatrix}+\varepsilon_{-2}(\sigma)e^{i\Psi_{-2}(\sigma)}\frac{1}{2}\begin{pmatrix}1 & -i & 0\\ -i & -1 & 0\\ 0 & 0 & 0\end{pmatrix}, \tag{13.37}$$

where $\varepsilon_m(\sigma)\geq 0$ is the amplitude and $\Psi_m(\sigma)$ is the phase $[\Psi_m(h,k,l)=-\Psi_m(-h,-k,-l)]$. The spherical harmonic tensors are defined in the local right-handed coordinate system for each reciprocal vector $\vec{G}=q(h\hat{x}+k\hat{y}+l\hat{z})$, with the polar axis parallel to the reciprocal vector. The reason for using the spherical harmonics tensors is their following properties:

$$M_m(i,j)M_n^*(j,i)=M_m(i,j)M_{-n}(j,i)=\sum_i\sum_j M_m(i,j)M_n^*(j,i)=\delta_{mn}. \tag{13.38}$$

We first try to determine the primitive wave vector q. The part of the free energy density depending on the spatial variation of the order parameter is

$$f_2=-de_{ijl}Q_{in}Q_{jn,l}+\frac{1}{2}\left(c_1Q_{ij,l}Q_{ij,l}+c_2Q_{ij,i}Q_{lj,l}\right). \tag{13.39}$$

The average free energy density (integrated over the unit cell and divided by the unit cell volume) is

$$F_2=\frac{1}{2}\sum_{h,k,l}\sum_m\frac{1}{N}\left\{-mdq\left(h^2+k^2+l^2\right)^{1/2}+\left[c_1+\frac{1}{6}c_2\left(4-m^2\right)\right]q^2\left(h^2+k^2+l^2\right)\right\}\varepsilon_m^2(\sigma). \tag{13.40}$$

Because F_2 only depends on the magnitudes of the reciprocal wave vectors, but not their direction,

$$F_2=\frac{1}{2}\sum_\sigma\sum_m\left\{-mdq\sigma+\left[c_1+\frac{1}{6}c_2\left(4-m^2\right)\right]q^2\sigma^2\right\}\varepsilon_m^2(\sigma). \tag{13.41}$$

We minimize the average free energy density with respect to the primitive wave vector by using

$$\frac{\partial F}{\partial q}=\frac{\partial F_2}{\partial q}=\frac{1}{2}\sum_{\sigma}\sum_{m}\left\{-md\sigma+2\left[c_1+\frac{1}{6}c_2(4-m^2)\right]q\sigma^2\right\}\varepsilon_m^2(\sigma)\overset{\text{let}}{=}0.$$

We get

$$q=\frac{q_o\sum_{\sigma}\sum_{m}m\sigma\varepsilon_m^2(\sigma)}{\sqrt{2}\sum_{\sigma}\sum_{m}\left[1+\frac{1}{6}(c_2/c_1)(4-m^2)\right]\sigma\varepsilon_m^2(\sigma)}, \tag{13.42}$$

where $q_o=d/\sqrt{2}c_1$. Note that q_o is different from $2\pi/P_c$ (here P_c is the pitch in the cholesteric phase).

13.4.2.1 Cholesteric phase

Before further discussion of the cubic structured blue phase, let us consider a right-handed cholesteric phase with the helical axis in z direction. From Equation (13.32), we have the dielectric tensor

$$\overleftrightarrow{Q}_{Ch}(\vec{r})=-\varepsilon_0(0)\overleftrightarrow{M}_o+\frac{1}{2\sqrt{2}}\varepsilon_2(2)\left[e^{i2q_cz+i\Psi_2(2)}\overleftrightarrow{M}_2+c.c.\right], \tag{13.43}$$

where $q_c=2\pi/P_c$ and $c.\,c.$ denotes the complex conjugate. Note that Equation (13.32) is only for uniaxial cholesteric liquid crystals while Equation (13.43) is more general and includes biaxial cholesteric liquid crystals. Because

$$\overleftrightarrow{Q}_{Ch}(\vec{r})=-\varepsilon_0(0)\frac{1}{\sqrt{6}}\begin{pmatrix}-1&0&0\\0&-1&0\\0&0&2\end{pmatrix}+\varepsilon_2(2)\frac{1}{\sqrt{2}}\begin{pmatrix}\cos\delta&\sin\delta&0\\\sin\delta&-\cos\delta&0\\0&0&0\end{pmatrix}\overset{\text{let}}{=}\overleftrightarrow{R}_0+\overleftrightarrow{R}_2, \tag{13.44}$$

where $\delta=2q_cz+\Psi_2$. Because

$$\overleftrightarrow{Q}^2_{Ch}(\vec{r})=\overleftrightarrow{R}^2_0+\overleftrightarrow{R}_0\overleftrightarrow{R}_2+\overleftrightarrow{R}_2\overleftrightarrow{R}_0+\overleftrightarrow{R}^2_2,$$

$$\overleftrightarrow{Q}^3_{Ch}(\vec{r})=\overleftrightarrow{R}^3_0+\overleftrightarrow{R}^2_0\overleftrightarrow{R}_2+\overleftrightarrow{R}_0\overleftrightarrow{R}_2\overleftrightarrow{R}_0+\overleftrightarrow{R}_0\overleftrightarrow{R}^2_2+\overleftrightarrow{R}_2\overleftrightarrow{R}^2_0+\overleftrightarrow{R}_2\overleftrightarrow{R}_0\overleftrightarrow{R}_2+\overleftrightarrow{R}^2_2\overleftrightarrow{R}_0+\overleftrightarrow{R}^3_2,$$

$$\overleftrightarrow{R}_0\cdot\overleftrightarrow{R}_0=[\varepsilon_0(0)]^2\frac{1}{6}\begin{pmatrix}1&0&0\\0&1&0\\0&0&4\end{pmatrix},\overleftrightarrow{R}_2\cdot\overleftrightarrow{R}_2=[\varepsilon_2(2)]^2\frac{1}{2}\begin{pmatrix}1&0&0\\0&1&0\\0&0&0\end{pmatrix}.$$

$$\overleftrightarrow{R}_0\cdot\overleftrightarrow{R}_2=\overleftrightarrow{R}_2\cdot\overleftrightarrow{R}_0=\varepsilon_0(0)\varepsilon_2(2)\frac{1}{\sqrt{12}}\begin{pmatrix}\cos\delta&\sin\delta&0\\\sin\delta&-\cos\delta&0\\0&0&0\end{pmatrix}.$$

The trace of the terms with $\overleftrightarrow{R}_2$ to odd power is 0. Therefore

$$Tr\left[\overleftrightarrow{Q}^2_{Ch}(\vec{r})\right] = [\varepsilon_0(0)]^2 + [\varepsilon_2(2)]^2$$

$$Tr\left[\overleftrightarrow{Q}^3_{Ch}(\vec{r})\right] = \left[\overleftrightarrow{R}^3_0 + \overleftrightarrow{R}_0\overleftrightarrow{R}^2_2 + \overleftrightarrow{R}_2\overleftrightarrow{R}_0\overleftrightarrow{R}_2 + \overleftrightarrow{R}^2_2\overleftrightarrow{R}_0\right] = \frac{1}{\sqrt{6}}\varepsilon_0(0)\left\{[\varepsilon_0(0)]^2 - 3[\varepsilon_2(2)]^2\right\}.$$

From Equation (13.42) we have

$$q_c = \frac{q_o\left[0\cdot\varepsilon_0^2(0) + 2\cdot 2\cdot\varepsilon_2^2(2)\right]}{\sqrt{2}\left[1 + \frac{1}{6}(c_2/c_1)(4-2^2)\right]\cdot 2\varepsilon_2^2(2)} = \frac{2q_o}{\sqrt{2}}, \tag{13.45}$$

so that $q_o = q_c/\sqrt{2} = (2\pi/P_c)/\sqrt{2}$. The average free energy density of the cholesteric phase is

$$F_{Ch} = \frac{1}{2}\left\{a[\varepsilon_0(0)]^2 + (a - d^2/c_1)[\varepsilon_2(2)]^2\right\}$$
$$+ \frac{1}{\sqrt{6}}\beta\varepsilon_0(0)\left\{[\varepsilon_0(0)]^2 - 3[\varepsilon_2(2)]^2\right\} + \gamma\left\{[\varepsilon_0(0)]^2 + [\varepsilon_2(2)]^2\right\}^2. \tag{13.46}$$

Introduce reduced parameters:

$$\mu = \left(\sqrt{6}\gamma/\beta\right)\varepsilon, f = F/(\beta^4/36\gamma^3), t = 4(3\gamma/\beta^2)a, \xi^2 = 4(3\gamma/\beta^2)c_1 \text{ and } \kappa = q_c\xi,$$

where t is the reduced temperature and κ is the reduced chirality. Equation (13.46) changes to

$$f_{Ch} = \frac{1}{4}t[\mu_0(0)]^2 + \frac{1}{4}(t-\kappa)[\mu_2(2)]^2 + \mu_0(0)\left\{[\mu_0(0)]^2 - 3[\mu_2(2)]^2\right\} + \left\{[\mu_0(0)]^2 + [\mu_2(2)]^2\right\}^2. \tag{13.47}$$

Let us consider a uniaxial nematic phase (racemic mixture), where there is no twist. Then $d = 0$ and the reduced free energy density becomes

$$f_{Ch} = \frac{1}{4}t\left\{[\mu_0(0)]^2 + [\mu_2(2)]^2\right\} + \mu_0(0)\left\{[\mu_0(0)]^2 - 3[\mu_2(2)]^2\right\} + \left\{[\mu_0(0)]^2 + [\mu_2(2)]^2\right\}^2. \tag{13.48}$$

Comparing Equation (13.32) with Equation (13.44), we get $\varepsilon_2(2)/\varepsilon_0(0) = \mu_2(2)/\mu_0(0) = \sqrt{3}$. Defining

$$\mu^2 = [\mu_o(0)]^2 + [\mu_2(2)]^2, \tag{13.49}$$

Equation (13.48) becomes

$$f_N = \frac{1}{4}t\mu^2 - \mu^3 + \mu^4. \tag{13.50}$$

In the isotropic phase, $\mu = 0$ and $f_{Iso} = 0$. In the nematic phase, the order parameter can be found by minimizing the free energy with respect to the order parameter, namely $\partial f_N/\partial\mu = 0$. Also, at the isotropic–nematic phase transition temperature the minimized free energy of the nematic phase equals that of the isotropic phase: $f_N = f_{iso} = 0$. From these two conditions, we can get the reduced iso-nematic phase transition temperature $t_{NI} = 1$.

Now we consider a uniaxial cholesteric liquid crystal where $\mu_2(2)/\mu_0(0) = \sqrt{3}$ and $d \neq 0$. The average free energy density is

$$f_{Ch/U} = \frac{1}{4}\left(t - \frac{3}{4}\kappa\right)\mu^2 - \mu^3 + \mu^4, \tag{13.51}$$

where μ is defined in Equation (13.49). We can follow the same process as in the nematic case and get the reduced Iso–Ch phase transition temperature $t_{ChI/U} = 1 + 3/4\kappa$, which is higher than the Iso–N phase transition temperature.

We can also consider the general case of a biaxial cholesteric liquid crystal where $\mu_0(0)$ and $\mu_2(2)$ are independent. Introducing two new parameters μ and θ defined by

$$\mu_0(0) = \mu\sin\theta, \quad \mu_2(2) = \mu\cos\theta, \tag{13.52}$$

the free energy density becomes

$$f_{Ch/B} = \frac{1}{4}(t - \kappa\cos^2\theta)\mu^2 + \sin\theta(1 - 4\cos^2\theta)\mu^3 + \mu^4. \tag{13.53}$$

Minimizing the free energy with respect to μ we get

$$\partial f_{Ch/B}/\partial\mu = \frac{1}{2}(t - \kappa\cos^2\theta)\mu + 3\sin\theta(1 - 4\cos^2\theta)\mu^2 + 4\mu^3 = 0. \tag{13.54}$$

Minimizing the free energy with respect to θ we get

$$\partial f_{Ch/B}/\partial\theta = \frac{1}{2}\kappa\sin\theta\cos\theta\mu^2 + 3\cos\theta(4\sin^2\theta - 1)]\mu^3 = 0. \tag{13.55}$$

At the Iso–Ch phase transition, the free energy equals zero (the free energy of the isotropic phase), so for the biaxial case

$$f_{Ch/B} = \frac{1}{4}(t - \kappa\cos^2\theta)\mu^2 + \sin\theta(1 - 4\cos^2\theta)\mu^3 + \mu^4 = 0. \tag{13.56}$$

Solving the above three equations, we get the parameters at the Iso–Ch phase transition temperature. There are two cases:
(a) For $\kappa < 3$,

$$\sin^2\theta = \frac{1}{2} - \frac{1}{4}\left(1 + \frac{1}{3}k^2\right)^{1/2}, \tag{13.57}$$

$$\mu = \frac{1}{2}\sin\theta(3 - 4\sin^2\theta), \tag{13.58}$$

and the transition temperature

$$t_{ICh/B}=\frac{1}{2}\left[1+\kappa^2+\left(1+\frac{1}{3}\kappa^2\right)^{3/2}\right]. \tag{13.59}$$

As the chirality κ increases, both the ratio $\mu_0(0)/\mu_2(2)$ and the parameter μ of the cholesteric phase at the transition temperature decrease. The difference between the transition temperatures of the biaxial and uniaxial cholesteric liquid crystal is

$$\Delta t_{B/U}=t_{ICh/B}-t_{ICh/U}=\frac{1}{2}\left[-1-\frac{3}{2}\kappa+\kappa^2+\left(1+\frac{1}{3}\kappa^2\right)^{3/2}\right]. \tag{13.60}$$

When $\kappa>0.97$, $\Delta t_{B/U}>0$, the biaxial cholesteric phase has a higher transition temperature than the uniaxial cholesteric phase. When $\kappa=3$, $\mu_0(0)/\mu_2(2)=0$ and $\mu=0$, indicating the transition becomes second-order.

(b) For $\kappa>3$,

$$\sin^2\theta=0,\ \mu=0,\ t_{ICh/B}=\kappa^2,\ \Delta t_{B/U}=t_{ICh/B}-t_{ICh/U}=\kappa^2-1-\frac{3}{4}\kappa. \tag{13.61}$$

It is a second-order phase transition. It is interesting to note that when $\kappa\geq 3$, the free energy becomes

$$f_{Ch/B}=\frac{1}{4}(t-3)\mu^2+\mu^4. \tag{13.62}$$

The cubic term vanishes and therefore the transition becomes a second-order transition. This suggests the possibility that if the cubic term is not zero, a first-order transition may occur at a temperature higher than that of the second-order phase transition.

In the biaxial cholesteric phase, choosing the local coordinate with the x axis parallel to the long molecular axis, the tensor order parameter is

$$\overleftrightarrow{Q}_{Ch}=\varepsilon_0(0)\frac{1}{\sqrt{6}}\begin{pmatrix}1+\sqrt{3}\cot\theta & 0 & 0\\ 0 & 1-\sqrt{3}\cot\theta & 0\\ 0 & 0 & -2\end{pmatrix}$$

$$=\frac{\varepsilon_0(0)}{\sqrt{6}(1+\sqrt{3}\cot\theta)}\begin{pmatrix}1 & 0 & 0\\ 0 & \dfrac{1-\sqrt{3}\cot\theta}{1+\sqrt{3}vot\theta} & 0\\ 0 & 0 & \dfrac{-2}{1+\sqrt{3}\cot\theta}\end{pmatrix}. \tag{13.63}$$

The standard form of biaxial order parameter in terms of the asymmetry parameter η is

$$\overset{\leftrightarrow}{Q}_{Ch} = \frac{\varepsilon_0(0)}{\sqrt{6}(1+\sqrt{3}\cot\theta)} \begin{pmatrix} 1 & 0 & 0 \\ 0 & (-1+\eta)/2 & 0 \\ 0 & 0 & (-1-\mu)/2 \end{pmatrix}. \tag{13.64}$$

The asymmetry parameter is

$$\eta = \left(\frac{1-\sqrt{3}\cot\theta}{1+\sqrt{3}\cot\theta}\right) - \left(\frac{-2}{1+\sqrt{3}\cot\theta}\right) = \frac{3-\sqrt{3}\cot\theta}{1+\sqrt{3}\cot\theta}, \tag{13.65}$$

where $\cot\theta = \mu_2(2)/\mu_0(0) = [2 + (1+\kappa^2/3)^{1/2}]/[2 - (1+\kappa^2/3)^{1/2}]$.

13.4.2.2 Cubic phase

Now we consider chiral phases with cubic symmetries in the high-chirality limit. As we have shown in the above section, when the chirality of the liquid crystal is sufficiently high, $\varepsilon_0(0)$ becomes small and is negligible; the Iso–Ch transition becomes a second-order phase transition. Below the phase transition temperature, the isotropic phase is absolutely unstable with respect to the cholesteric phase. As we know of the Iso–N transition of a non-chiral liquid crystal, there is a virtual second-order phase transition temperature T^*, below which the isotropic phase is absolutely unstable. There is, however, a first-order transition at a temperature T_{NI}, which is higher than the second-order phase transition temperature T^*. In the Landau–de Gennes expansion of the free energy in terms of the orientational order parameter there are three terms: quadratic, cubic, and quartic. The cubic term is allowed, which is negative and decreases the free energy and is responsible for the first-order transition, because a state with a positive order parameter S_o is different from the state with a negative order parameter $-S_o$, and the free energy of the state with the positive order parameter is different from that of the state with the negative order parameter. At T_{NI}, the free energy of the nematic phase becomes equal to that of the isotropic phase. There is an energy barrier between the two phases and the material can transform from the isotropic phase to the nematic phase through a nucleation processes. Thus for a chiral liquid crystal with high chirality, there may be a first-order phase transition in which the material transforms from the isotropic phase to the unknown X phase. The phase transition temperature is T_{XI}, which is higher than the second-order Iso–Ch phase transition temperature T_{ChI}. For the physics point of view, the cubic term is negative and will decrease the free energy. Also the mathematical requirement for a first-order phase transition is a non-vanishing cubic term in the Landau free energy expansion. So let us examine the cubic term. From Equations (13.34) and (13.35), we have the cubic term

$$\begin{aligned} f_3 &\propto \left[Q_{ij}(h_1,k_1,l_1)e^{iq(h_1x+k_1y+l_1z)} + c.c.\right]\cdot\left[Q_{jm}(h_2,k_2,l_2)e^{iq(h_2x+k_2y+l_2z)} + c.c.\right] \\ &\quad \cdot\left[Q_{mi}(h_3,k_3,l_3)e^{iq(h_3x+k_3y+l_3z)} + c.c.\right], \\ &\propto Q_{ij}(\vec{\sigma}_1)Q_{jm}(\vec{\sigma}_2)Q_{mi}(\vec{\sigma}_3)e^{iq\vec{r}\cdot(\vec{\sigma}_1+\vec{\sigma}_3-\vec{\sigma}_3)} \end{aligned} \tag{13.66}$$

where $\vec{\sigma}_i = h_i\hat{x} + k_i\hat{y} + l_i\hat{z}$ $i = 1,2,3$. The corresponding average free energy density F_3 is obtained by integrating f_3 over one unit cell and dividing by the unit cell volume. For a non-vanishing F_3, it is required that [9,10]

$$\vec{\sigma}_1 + \vec{\sigma}_2 - \vec{\sigma}_3 = 0. \tag{13.67}$$

Before we go further, let us go back to consider the case of the cholesteric phase. As shown in Equation (13.46), the cubic term consists of $[\varepsilon_0(0)]^3$ and $3\varepsilon_0(0)[\varepsilon_2(2)]^2$. The wave vectors of the term $[\varepsilon_0(0)]^3$ are all 0 and their sum is 0. The wave vectors of the term $3\varepsilon_0(0)[\varepsilon_2(2)]^2$ are 0, $q_c\hat{z}$, and $-q_c\hat{z}$, respectively, and their sum is also 0. The condition given by Equation (13.67) is satisfied in the cholesteric phase.

Now we come back to our consideration of cubic phases. When the chirality is very high, namely, d (and κ) is very large, it can be seen from Equation (13.41) that the state with minimum free energy will have $m = 2$. When we put this value back into Equation (13.41), we have

$$F_2 = \frac{1}{2}\sum_{\sigma}\left[-2dq\sigma + c_1 q^2\sigma^2\right]\varepsilon_2^2(\sigma) = \frac{1}{2}c_1\sum_{\sigma}\left[-2\sqrt{2}q_o q\sigma + q^2\sigma^2\right]\varepsilon_2^2(\sigma). \tag{13.68}$$

F_2 is minimized when

$$q\sigma = q\sqrt{h^2 + k^2 + l^2} = \sqrt{2}q_o = q_c. \tag{13.69}$$

This is the magnitude of the wave vectors. The coordinate is chosen in such a way that the wave vector lies along the (1, 1, 0) directions. All the possible wave vectors are

$$\vec{q}_{21} = q_o(\hat{x} + \hat{y}),$$
$$\vec{q}_{22} = q_o(\hat{y} + \hat{z}),$$
$$\vec{q}_{23} = q_o(\hat{z} + \hat{x}),$$
$$\vec{q}_{24} = q_o(-\hat{x} + \hat{y}),$$
$$\vec{q}_{25} = q_o(-\hat{y} + \hat{z}),$$
$$\vec{q}_{26} = q_o(-\hat{z} + \hat{x}),$$

as shown in Figure 13.11. These vectors generate the face-centered cubic (fcc) structure in reciprocal space, which corresponds to the body-centered cubic (bcc) structure in real space.

In this high chirality limit, $\varepsilon_0(0) \approx 0$ and the order parameter is

$$\overleftrightarrow{Q}_c(\vec{r}) = \frac{1}{\sqrt{12}}\sum_{n=1}^{6}\varepsilon_2(2n)\left\{\overleftrightarrow{M}_2 e^{i\left[\vec{q}_{2n}\cdot\vec{r} + \Psi_2(2n)\right]} + c.c.\right\}. \tag{13.70}$$

Note that the matrix $\overleftrightarrow{M}_2$ is the $m = 2$ spherical harmonic tensor defined in the local coordinate, whose z axis is along $\vec{q}_{2n}$. Also note that the wave vector of $\overleftrightarrow{M}_2 e^{i\left[\vec{q}_{2n}\cdot\vec{r} + \Psi_2(2n)\right]}$ is $\vec{q}_{2n}$, and the

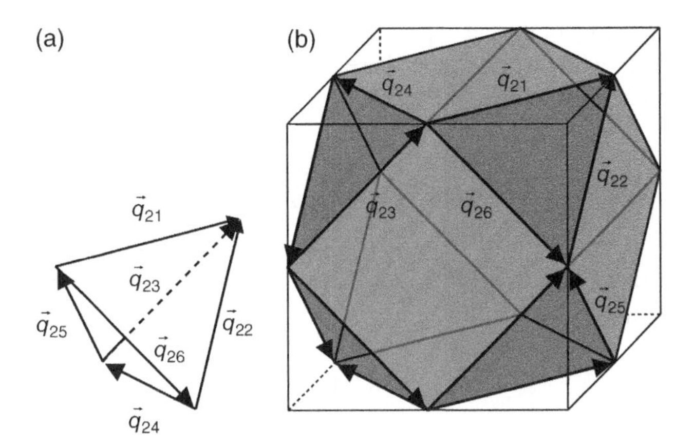

Figure 13.11 (a) Tetrahedron formed by the wave vectors. (b) fcc structure formed by the wave vectors.

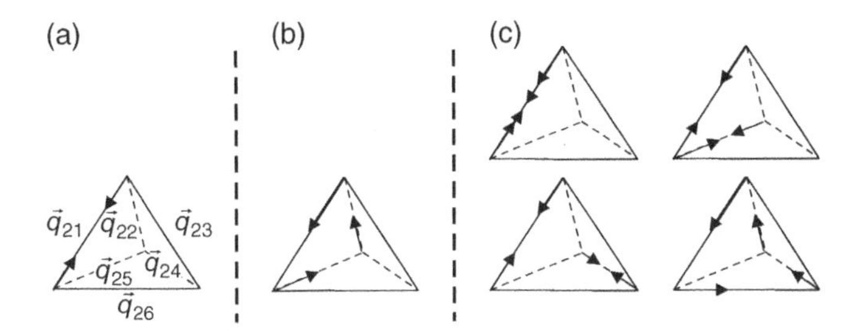

Figure 13.12 Diagrams of the wave vectors along the edges of the tetrahedron used in the calculation of the average free energy. (a) quadratic term, (b) cubic term, (c) quartic term.

wave vector of its complex conjugate is in the opposite direction. Now we consider the calculation of the average free energy of the cubic phase. In the calculation of the quadratic term of the free energy, we only have to consider the product of each component of the order parameter with its complex conjugate, where their wave vectors cancel each other, as shown in Figure 13.12(a). In the calculation of the cubic term of the free energy, we have to consider the products of three components of the order parameter whose wave vectors form a triangle as shown in Figure 13.12(b). The quartic term of the free energy is $\left[Tr\left(\overleftrightarrow{Q}^2\right)\right]^2 = \left[Tr\left(\overleftrightarrow{Q}^2\right)\right]\left[Tr\left(\overleftrightarrow{Q}^2\right)\right]$. In calculating the average value, we need to multiply $\left[Tr\left(\overleftrightarrow{Q}^2\right)\right]$ by $\left[Tr\left(\overleftrightarrow{Q}^2\right)\right]$ and then integrate the product over the unit cell. Therefore $<\left[Tr\left(\overleftrightarrow{Q}^2\right)\right]^2> \neq \left[<Tr\left(\overleftrightarrow{Q}^2\right)>\right]^2$. $\left[Tr\left(\overleftrightarrow{Q}^2\right)\right]^2$ produces terms consisting of the products of four components of the order parameter. In the calculation, we have to consider the products of four components of the order parameter whose wave vectors form one of the four patterns shown in Figure 13.12(c).

The average free energy of the cubic phase depends on the phase angles $\Psi_2(2n)$. After minimizing the free energy with respect to the phase angles under the constraint of bcc O^5 symmetry, we can obtain the reduced free energy [9]

$$f_{c/O^5} = \frac{1}{4}(t-\kappa^2)\mu_2^2(2) - \frac{23\sqrt{2}}{32}\mu_2^3(2) + \frac{499}{384}\mu_2^4(2). \tag{13.71}$$

As shown before, using $f_{c/O^5} = 0$ and $\partial f_{c/O^5}/\partial\mu_2(2) = 0$, gives the reduced isotropic–cubic phase transition temperature:

$$t_{O^5 I} = 1587/1996 + \kappa^2 \tag{13.72}$$

The transition temperature from the isotropic phase to phases with various structures are plotted as a function of the reduced chirality κ, as shown in Fig 13.13. It can be seen that when the chirality is sufficiently high, the isotropic–cubic phase transition temperature is higher than that of the isotropic–cholesteric phase transition. This explains why the blue phase exists in a temperature region below the isotropic phase for liquid crystals with high chiralities. At temperature far below the isotropic–cubic phase transition temperature, the order parameter becomes large, and the quartic term of the average free energy of the cubic phase, as shown in Equation (13.71), becomes larger than that of the cholesteric phase, as shown in Equation (13.51). Therefore at lower temperature, the cholesteric phase is the stable phase. Note that the derivation is correct when the chirality $\kappa \gg 1$, and is an approximation when $\kappa \sim 1$. Also note that here we only consider the bcc structure with group symmetry O^5. There are also other cubic structures whose free energies are lower than that of the cholesteric phase.

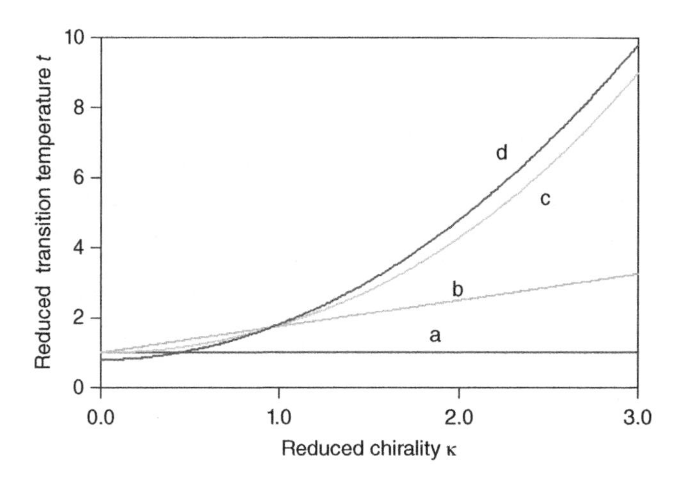

Figure 13.13 Phase transition temperatures from isotropic phase to various structures as a function of chirality. (a) isotropic–uniaxial nematic transition, (b) isotropic–uniaxial cholesteric transition, (c) isotropic–biaxial cholesteric transition, (d) isotropic–cubic O^5 transition.

13.5 Optical Properties of Blue Phase

The optical properties of the blue phase that interest us here are the transmittance and reflectance of the material and its effect on the polarization of light. These properties depend on the dielectric tensor (at optical frequencies) of the material. The dielectric tensor $\overleftrightarrow{\varepsilon}$ is related to the order parameter tensor $\overleftrightarrow{Q}$ shown in Equation (13.30). The first term on the right side of Equation (13.30) is the average dielectric constant and does not affect the anisotropic optical properties. The second term is the anisotropic part, which is linearly proportional to the tensor order parameter, and determines the optical properties. From Equations (13.35) and (13.36), we have

$$\overleftrightarrow{Q}(\vec{r}) = \sum_{h,k,l} \sum_{m=-2}^{2} \varepsilon_m(\sigma) \frac{1}{\sqrt{N_{hkl}}} e^{i\Psi_m(\sigma)} \overleftrightarrow{M}_m e^{iq}(hx + ky + lz), \tag{13.73}$$

where $\overleftrightarrow{M}_m$ is the spherical harmonic tensor defined with respect to the local coordinate whose z axis is parallel to the wave vector $q(h\hat{x} + k\hat{y} + l\hat{z})$. The components of $\overleftrightarrow{M}_m$ are the anisotropic part of the dielectric constants along the three axes of the local coordinates.

13.5.1 Reflection

The reflection from cholesteric and blue phases is Bragg-type scattering, similar to the diffraction of X-rays by crystals. The wave vector of the incident light $\vec{K}_o$, the wave vector of the scattered light $\vec{K}_s$, and the wave vector of the dielectric constant component $\vec{q}$ must satisfy the Bragg condition:

$$\vec{q} = \vec{K}_s - \vec{K}_o. \tag{13.74}$$

The polarization of the Bragg scattered light depends on the spherical harmonic tensor $\overleftrightarrow{M}_m$. The intensity I_s of the scattered light is proportional to the structure factor [31]

$$I_s(\vec{q}) \propto \left[\vec{P}_{in} \cdot \overleftrightarrow{Q}(\vec{q}) \cdot \vec{P}_s\right]^2 \propto \varepsilon_m^2(\sigma) \left[\vec{P}_{in} \cdot \overleftrightarrow{M}_m \cdot \vec{P}_s\right]^2, \tag{13.75}$$

where $\vec{P}_{in}$ and $\vec{P}_s$ are the polarization vectors of the incident and scattered light, respectively. The dielectric tensor $\overleftrightarrow{Q}$ is defined in the local frame whose z axis is parallel to the wave vector $\vec{q}$. When $\vec{K}_o$ and $\vec{K}_s$ are parallel (or anti-parallel), the wave vector $\vec{q}$ is parallel to them, and the light propagation direction is parallel to the z axis of the local frame. The polarization vectors only have non-zero components along the x and y axes:

$$\vec{P}_{in} = \begin{pmatrix} p_{x/in} \\ p_{y/in} \end{pmatrix}, \vec{P}_s = \begin{pmatrix} p_{x/s} \\ p_{y/s} \end{pmatrix} \tag{13.76}$$

Also, only the x and y components of $\overleftrightarrow{M}_m$ will affect the scattered light intensity.

For $m = 0$

$$I_s(\vec{q}) \propto \left[\left(p_{x/in} p_{y/in} \right) \cdot \begin{pmatrix} -1 & 0 \\ 0 & -1 \end{pmatrix} \cdot \begin{pmatrix} p_{x/s} \\ p_{y/s} \end{pmatrix} \right]^2 = \left(p_{x/in} p_{x/s} + p_{y/in} p_{y/s} \right)^2. \tag{13.77}$$

The x component of the incident light only produces the x component of the scattered light, and the y component of the incident light only produces the y component of the scattered light, thus the scattered light intensity is independent of the polarization and the polarization of the scattered light is the same as that of the incident light.
For $m = \pm 1$,

$$I_s(\vec{q}) \propto \left[\left(p_{x/in} p_{y/in} \right) \cdot \begin{pmatrix} 0 & 0 \\ 0 & 0 \end{pmatrix} \cdot \begin{pmatrix} p_{x/s} \\ p_{y/s} \end{pmatrix} \right]^2 = 0. \tag{13.78}$$

Therefore this harmonic does not produce scattered light.
For $m = \pm 2$,

$$I_s(\vec{q}) \propto \left[\left(p_{x/in} p_{y/in} \right) \cdot \begin{pmatrix} 1 & \pm i \\ \pm i & -1 \end{pmatrix} \cdot \begin{pmatrix} p_{x/s} \\ p_{y/s} \end{pmatrix} \right]^2 = \left[\left(p_{x/in} \pm i p_{y/in} \right) \left(p_{x/s} \pm i p_{y/s} \right) \right]^2. \tag{13.79}$$

The intensity is maximized when $p_{x/in}/p_{y/in} = \mp i$ and $p_{x/s}/p_{y/s} = \mp i$, namely, this harmonic produces right- or left-handed circular polarized light.

It can be seen from Equation (13.41) that the Fourier components with $m = 2$ (for right-handed liquid crystals) have lower free energy than the components with other m values, and therefore the blue phase only has the components with $m = 2$, as shown in Equation (13.70). The reflected light is circularly polarized with the same handedness as the chirality of the liquid crystal.

The reflection bandwidth of blue phases is different from that of the cholesteric phase. In the cholesteric phase, when the light propagates along the helical axis and the wavelength is within the reflection band, the eigenmode polarization is linear polarization. For a linearly polarized incident light, the angle between the polarization and the liquid crystal director remains unchanged, so the polarization rotates with the liquid crystal director, when it propagates through the sample. The low edge wavelength of the reflection band is $n_o P_c$ for light linearly polarized perpendicular to the liquid crystal director, and the high edge wavelength of the reflection band is $n_e P_c$ for the light linearly polarized parallel to the liquid crystal director. The bandwidth is given by $(n_e - n_o)P_c$ and is typically a few tens of nanometers.

In the blue phase, the reflection wavelength is governed by the period of the crystal planes, which is different from the helical pitch. At the reflection wavelength, for a linearly polarized light, the polarization does not rotate with the liquid crystal director in space. Therefore the reflection bandwidth is narrow, typically a few nanometers.

13.5.2 Transmission

As discussed in the above section, blue phases allow Bragg reflection of visible light. Therefore the transmittance is less than 100%. The precise treatments of the optics of blue phases are complex and are given by Belyakov, Dmitrienko, Hornreich, et al. [32–35]. Here we only give some qualitative discussions, mainly for the purpose of understanding blue phase display based on electric field induced birefringence.

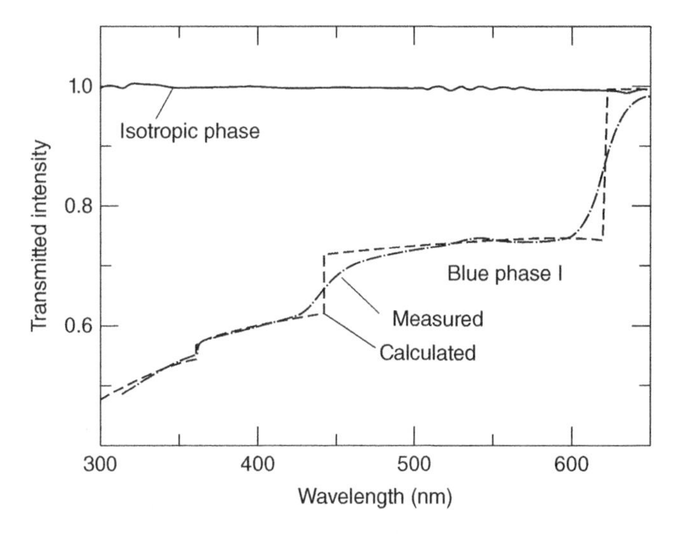

Figure 13.14 Transmission spectrum of blue phase I. The incident light is unpolarized and there is no polarizer in the transmission measurement [34,37].

First we consider the effect of light scattering (reflection) on the transmittance. The reflection of the blue phase deflects light from the incident light and thus results in a decrease of the transmittance. For the blue phase with face-centered cubic symmetry, the crystal plane with the longest periodicity is (1,1,0). The corresponding reflection has the wavelength $\lambda_{(100)} = 2\cos\theta P/\sqrt{2}$, where P is the lattice length of the cubic structure and θ is the incident light angle. As the incident angle θ increases, the reflection wavelength decreases. All other crystal planes have higher indices and have shorter periodicities, as shown by Equation (13.3), and thus will reflect light at shorter wavelengths. For a polycrystalline blue phase sample, due to the incident angle effect and more crystal planes with high indices, there are more reflections at shorter wavelengths. Therefore the reflection increases with decreasing wavelength and thus the transmittance decreases with decreasing wavelength. The transmission spectrum of a polycrystalline BPI sample is shown in Figure 13.14 [35–37]. At the high wavelength 650 nm, the transmission is close to 1. The transmission decreases with decreasing wavelength and becomes about 0.5 at 300 nm. Because the incident light is unpolarized and the BPI only reflects circular polarized light with one handedness, the minimum transmission is 0.5.

Second, we consider the optical anisotropy of the blue phases. Generally speaking, the refractive indices of a crystal form an ellipsoid, as discussed in Chapter 2. Now the blue phases have cubic symmetries. On a macroscopic scale, the refractive index ellipsoid must have the same cubic symmetries. Cubic symmetries contain four-fold rotational symmetry around three orthogonal axes. Therefore the refractive index ellipsoid must be a sphere, that is, the refractive index in any direction is the same at macroscopic scale. Due to this optical isotropy, when a blue phase sample is sandwiched between two crossed polarizers, the transmittance is zero. This is the dark state of the blue phase display based on field induced birefringence.

Third, we consider the effect of blue phases on the polarization of light. Locally the dielectric tensor (and refractive index) is not required to satisfy the symmetries of the blue phase cubic structures. Therefore it is possible that locally the material is optically birefringent. This is

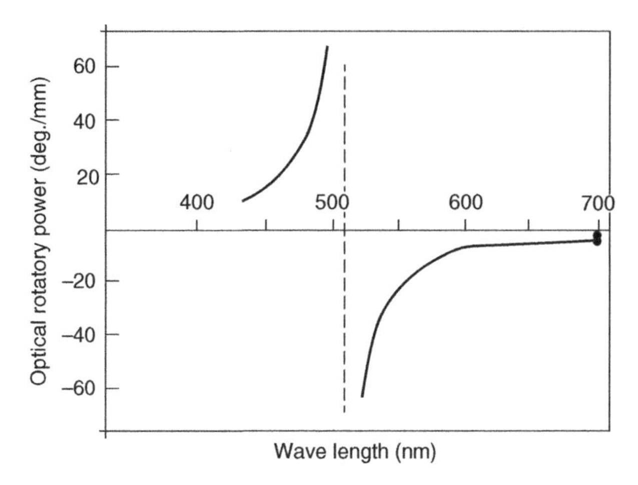

Figure 13.15 Optical rotatory power of blue phases. (1) single crystalline BPI, (2) single crystalline BPII, (3) polycrystalline BPI, (4) polycrystalline BPII, (5) BPIII, (6) isotropic phase. © Sov. Phys. JETP.

supported by experimental evidence that blue phases have non-zero optical rotatory powers (the rotation angle of linear polarization per unit length by the material), as shown in Figure 13.15 [33], due to the twisting of the liquid crystal director in space. The optical rotatory power is small at long wavelength, and first increases with decreasing wavelength. It diverges at the reflection wavelength and then changes sign. For polycrystalline samples, the typical optical rotatory power is less than 10°/mm, which is at least one order of magnitude smaller than that of cholesteric phase. For a blue phase display with the thickness of 10 μm, when a linear polarized beam propagates through the display, the polarization will be rotated by the angle $\Delta\alpha = 10\ \mu\text{m} \cdot 10°/\text{mm} = 0.1°$. Under crossed polarizers, the rotation of the polarization will result in a light leakage of $\sin^2(\Delta\alpha) = 3 \times 10^{-6}$, which is negligible. Nevertheless, the blue phases are not strictly optically isotropic for visible light.

We present here an approximate calculation of the optical rotatory power of the blue phase. As shown in the discussion of the cholesteric phase in Chapter 3, for a light beam propagating along the helical axis with $\Delta nP \ll \lambda$, the polarization vector $\vec{E}_o$ of the outgoing light is related to the polarization vector $\vec{E}_i$ of the incident light by (Equation (13.37))

$$E_o = \begin{pmatrix} \cos\Phi & -\sin\Phi \\ \sin\Phi & \cos\Phi \end{pmatrix} \begin{pmatrix} \cos\Theta & \sin\Theta \\ -\sin\Theta & \cos\Theta \end{pmatrix} \vec{E}_i. \tag{13.80}$$

For a thin slab of material with thickness Δd, the twist angle $\Phi = 2\pi\Delta d/P \ll 1$, $\Theta = \Phi + \pi P(\Delta n)^2\Delta d/4\lambda^2$. For a linear polarized light along the x axis (also parallel to the liquid crystal on the entrance plane), keeping up to first-order terms, the polarization of the outgoing light is

$$\vec{E}_o = \begin{pmatrix} 1 & -\Phi \\ \Phi & 1 \end{pmatrix} \begin{pmatrix} 1 & \Theta \\ -\Theta & 1 \end{pmatrix} \begin{pmatrix} 1 \\ 0 \end{pmatrix} = \begin{pmatrix} 1 \\ \Phi - \Theta \end{pmatrix} = \begin{pmatrix} 1 \\ -\pi P(\Delta n)^2\Delta d/4\lambda^2 \end{pmatrix}. \tag{13.81}$$

The polarization is rotated by the angle $\Delta\alpha = -\pi P(\Delta n)^2 \Delta d/4\lambda^2$. The optical rotatory power is

$$\Delta\alpha/\Delta d = -\pi P(\Delta n)^2/4\lambda^2. \tag{13.82}$$

For a material with $P = 300$ nm, $\Delta n = 0.1$, the optical rotatory power for light with wavelength 500 nm is

$$\pi \times 300 \text{ nm} \times (0.1)^2/\left[4 \times (500 \text{ nm})^2\right] = 3\pi \times 10^{-6} \text{ rad/nm} = 4.8 \times 10^2 \text{ deg/mm}.$$

This is the typical optical rotatory power of cholesteric liquid crystals with pitch shorter than the light wavelength. In the blue phases, the double-twist cylinders orient along many different directions, and therefore the optical rotatory power is smaller than that of cholesteric phase. The thickness of blue phases displays are typically a few microns, and over this distance the helical structure in the blue phases does not change much the polarization state of light.

Homework Problems

13.1 Calculate the free energy per unit length of the double-twist cylinder as a function of qR from 0 to π, where q is the chirality of the liquid crystal and R is the radius of the double-twist cylinder. Use the following elastic constants: K_{22}, $K_{33} = 2K_{22}$, and $K_{24} = 0.5K_{22}$.

13.2 For a pure biaxial cholesteric liquid crystal, (1) calculate the isotropic–cholesteric phase transition temperature, (2) calculate the order parameter as a function of reduced temperature.

References

1. F. Reinitzer, Beiträge zur Kenntniss des Cholestherins, *Monatsh Chem.*, **9**, 421 (1888).
2. P. P. Crooker, Blue phase in *Chirality in liquid crystals*, ed. H.-S. Kitzerow and C. Bahr (Springer, New York, 2001).
3. P. E. Cladis, A review of cholesteric blue phases in *Theory and applications of liquid crystals* ed. J. L. Ericksen and D. Kinderlehrer, p. 73 (Springer-Verlag, New York, 1987).
4. D. C. Wright and N. D. Mermin, Crystalline liquid: the blue phases, *Rev. Mod. Phys.*, **61**, 385 (1989).
5. P. P. Crooker, The blue phases; a review of experiments, *Liq. Cryst.*, **5**, 751 (1989).
6. H. Stegemeyer and K. Bergmann, Experimental results and problem concerning blue phases in the book *Liquid Crystals* of *One- and Two-Dimensional,* ed. W. Helfrich and A. Heppke (Springer-Verlag, Berlin, 1980).
7. S. A. Brazovskii and S. G. Dmitriev, *Zh. Eksp. Teor. Fiz.* 69, 979 (1975) (1976, *Soviet. Phys.-JETP,* 42, 497).
8. S. A. Brazovskii and V. M. Filev, *Zh. Eksp. Teor. Fiz.* 75, 1140 (1978) (1978, *Soviet. Phys.-JETP* 48, 573)
9. H. Grebel, R. M. Hornreich, and S. Shtrikman, Landau theory of cholesteric blue phases, *Phys. Rev. A*, **28,** 1114 (1983).
10. H. Grebel, R. M. Hornreich, and S. Shtrickman, Landau theory of cholesteric blue phases: the role of higher harmonics, *Phys. Rev. A*, **30,** 3264 (1984).
11. W. Cao, A. Munoz, P. Palffy-Muhoray, and B. Taheri, Lasing in a three-dimensional photonic crystal of the liquid crystal blue phase II, *Nature materials*, **1**, 111 (2002).
12. H. Kikuchi, M. Yokota, Y. Hisakado, et al., Polymer-stabilized liquid crystal blue phases, *Nat. Mater.* **1**(1), 64 (2002).

13. M. Lee, S.-T. Hur, H. Higuchi, et al., Liquid crystalline blue phase I observed for a bent-core molecule and its electro-optical performance, *J. Mat. Chem.* 20, 5765 (2010).
14. Jin Yan and Shin-Tson Wu, Polymer-stabilized blue phase liquid crystals: a tutorial, *Opt. Materials Express*, **1**, 1527 (2011).
15. M. Marcus, and J. W. Goodby, *Mol. Crystals Liq. Crystals Lett.*, **72**, 297 (1982).
16. P. J. Collings., *Phys. Rev. A*, 33, 2153 (1986).
17. D.K. Yang and P.P. Crooker, Chiral-racemic diagrams of blue phase liquid crystal, *Phys. Rev. A*, **35**, 4419 (1987).
18. H. Stegemeyer, T. H. Blümel, K. Hiltrop, et al., Thermodynamic, structural and morphological studies on liquid-crystalline blue phases, *Liquid Crystals*, **1**, 3 (1986).
19. D.-K. Yang, Optical studies of blue phase III of chiral liquid crystals, dissertation, University of Hawaii, 1989.
20. D. K. Yang and P. P. Crooker, Blue phase III of chiral liquid crystal in an electric field, *Phys. Rev. A*, **37**, 4001–4005 (1988).
21. D. L. Johnson, J. H, Flack, and P. P. Crooker, *Phys. Rev. Lett.*, **45**, 641 (1980).
22. J. P. Sethna, *Phys. Rev. Lett.*, **51**, 2198 (1983).
23. J. P. Sethna, D. C. Wright, and N. D. Mermin, *Phys. Rev. Lett.*, **51**, 467 (1983).
24. S. Meiboon, J. P. Sethna, P. W. Anderson, and W. F. Brinkman, *Phys. Rev. Lett.*, **46**, 467 (1981).
25. J. P. Sethna, Frustration, curvature, and defect lines in metallic glasses and the cholesteric blue phase, *Phys. Rev. B*, **31**, 6278 (1985).
26. S. Meiboom, M. Sammon and D. Berreman, *Phys. Rev. A*, **28**, 3553 (1983).
27. S. Meiboom, M. Sammon and W. F. Brinkman, *Phys. Rev. A*, **27**, 438 (1983).
28. M. Kleman and O. D. Laverentovich, *Soft matter physics: an introduction*, (Springer-Verlag, New York, 2003).
29. M. O'Keeffe and S. Anderson, Rod packing and crystal chemistry, *Acta Cryst.* **A33**, 914 (1977).
30. J. P. Sethna, in *Theory and applications of liquid crystals*, ed. J. L. Ericksen and D. Kinderlehrer (Springer-Verlag, New York, 1987).
31. P. G. de Gennes and J. Prost, *The physics of liquid crystals*, p. 264 (Oxford University Press, New York, 1993).
32. V. A. Belyakov, V. E. Dmitrienko, and S. M. Osadchii, Optics of the blue phase of cholesteric liquid crystals, *Sov. Phys. JETP*, **56** (2), 322 (1982).
33. V. A. Belyakov, E. I. Demikhov, V. E. Dmitrienko, and V. K. Dolganov, Optical activity, transmission spectra, and structure of blue phases of liquid crystals, *Sov. Phys. JETP*, **62**, 1173 (1985).
34. R. M. Hornreich and S. Shtrikman, Theory of light scattering in cholesteric blue phases, *Phys. Rev. A*, **28**, 1791 (1983).
35. D. Bensimon, E. Domany, and S. Shtrikman, Optical activity of cholesteric liquid crystals in the pretransitional regime and in the blue phase, *Phys. Rev. A*, **28**, 427 (1983).
36. S. Meiboom and M. Sammon, *Phys. Rev. Lett.*, **44**, 882 (1980).
37. S. Meiboom and M. Sammon, *Phys. Rev. A.*, **24**, 648 (1981).

14

Polymer-Stabilized Blue Phase Liquid Crystals

14.1 Introduction

An optically isotropic liquid crystal (LC) refers to a composite material system whose refractive index is isotropic macroscopically, yet its dielectric constant remains anisotropic microscopically [1]. When such a material is subject to an external electric field, induced birefringence takes place along the electric field direction if the employed LC host has a positive dielectric anisotropy ($\Delta\varepsilon$). This optically isotropic medium is different from a polar liquid crystal in an isotropic state, such as 5CB (clearing point = 35.4°C) at 50°C. The latter is not switchable because its dielectric anisotropy and optical anisotropy (birefringence) both vanish in the isotropic phase. Blue phase, which exists between cholesteric and isotropic phases, is an example of optically isotropic media.

Chiral nematic phase is usually referred to as cholesteric phase. Cholesteric liquid crystals have helical structure, in which the direction of the long molecular axes of each layer form an angle with that of the successive layer. In the vicinity of phase transition to isotropic phase, optically isotropic uniform textures are often observed. This texture is called blue phase (BP).When the first such compound was discovered, it happened to appear in a blue color because of Bragg reflection. Therefore, it has been termed blue phase liquid crystal (BPLC) ever since. An obvious feature of blue phases is the selective reflection of incident light due to Bragg reflection from its periodic structure. However, blue phases are not always blue; they may reflect other colors, depending on the pitch length of the periodic structure. For many applications, such as transmissive displays, Bragg reflection is intentionally shifted to the UV region so that the BPLC is actually transparent in the visible region. Another important characteristic of blue phases is frustration. In early days, blue phases existed in a fairly narrow temperature range (0.5–2 K). As a result, further investigation of blue phases was not pursued for several decades after its discovery by Reinitzer in 1888 [2]. Not until 1970s, did the study of

Fundamentals of Liquid Crystal Devices, Second Edition. Deng-Ke Yang and Shin-Tson Wu.

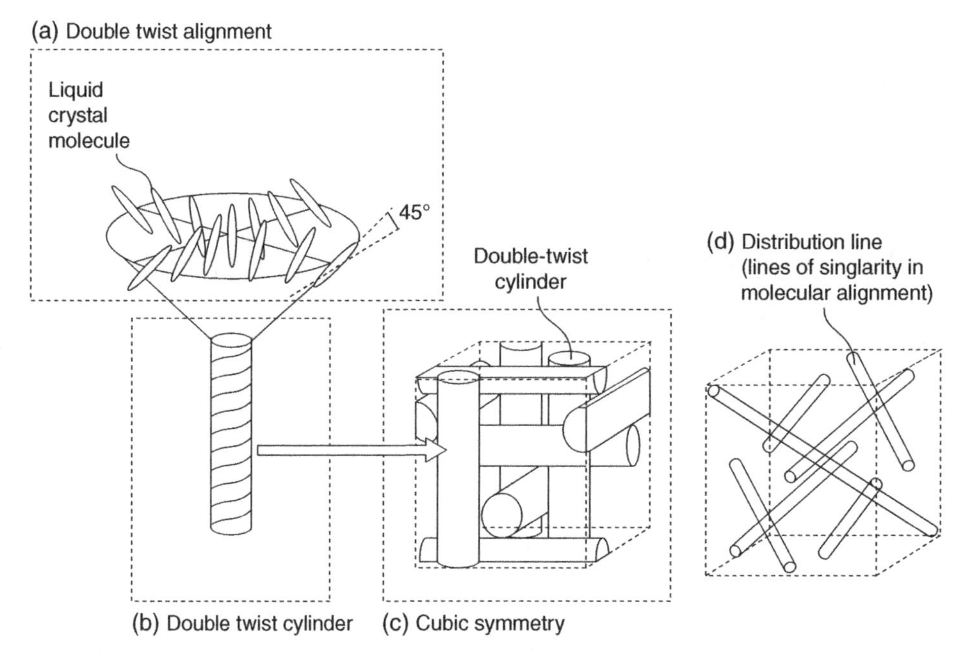

Figure 14.1 Blue phase LC structure at the microscopic level: (a) double-twist alignment of LC molecules, (b) double-twist cylinder, (c) lattice cubic formed by double-twist cylinders, and (d) disclination lines.

blue phases became popular, and tremendous progress has been made since then [3]. It was found that blue phase was optically isotropic while exhibiting unusually strong optical activity. Based on this phenomenon, Saupe proposed that blue phases have cubic superstructures [4]. After that, much effort was devoted to exploring the blue phase structures both experimentally and theoretically [5–11].

Figure 14.1 illustrates how the LC molecules are arranged in a blue phase [12,13]. As Figure 14.1(a) shows, the LC directors form a double-twist alignment in a cylinder. Similar to a chiral nematic, the LC directors are rotated in a helical structure. However, instead of a single helical axis there are many helical axes formed by the rotated LC directors and all of these axes are perpendicular to the center line. Although in reality an unlimited number of such helical axes could be present, we just use two of them to illustrate the molecular orientation and name it the double-twist structure. This double-twist structure is only extended over a small distance, with the boundary molecules aligned at 45° to the middle line. As Figure 14.1(b) shows, the double-twist structure extends and forms a cylinder. The helical lines drawn on the surface of the cylinder represent the LCs on boundary. The diameter of such a double-twist cylinder, which is related to the pitch length of the twisted LC structure, is usually ~100 nm. As Figure 14.1(c) shows, these double-twist cylinders are arranged in three directions perpendicular to each other and form a symmetric cubic structure, like a lattice structure. Here we describe the LC structure in a microscopic view. The LC directors are aligned in various directions in a lattice structure, so from a macroscopic viewpoint blue phase LC is optically isotropic. Defects occur at the contact

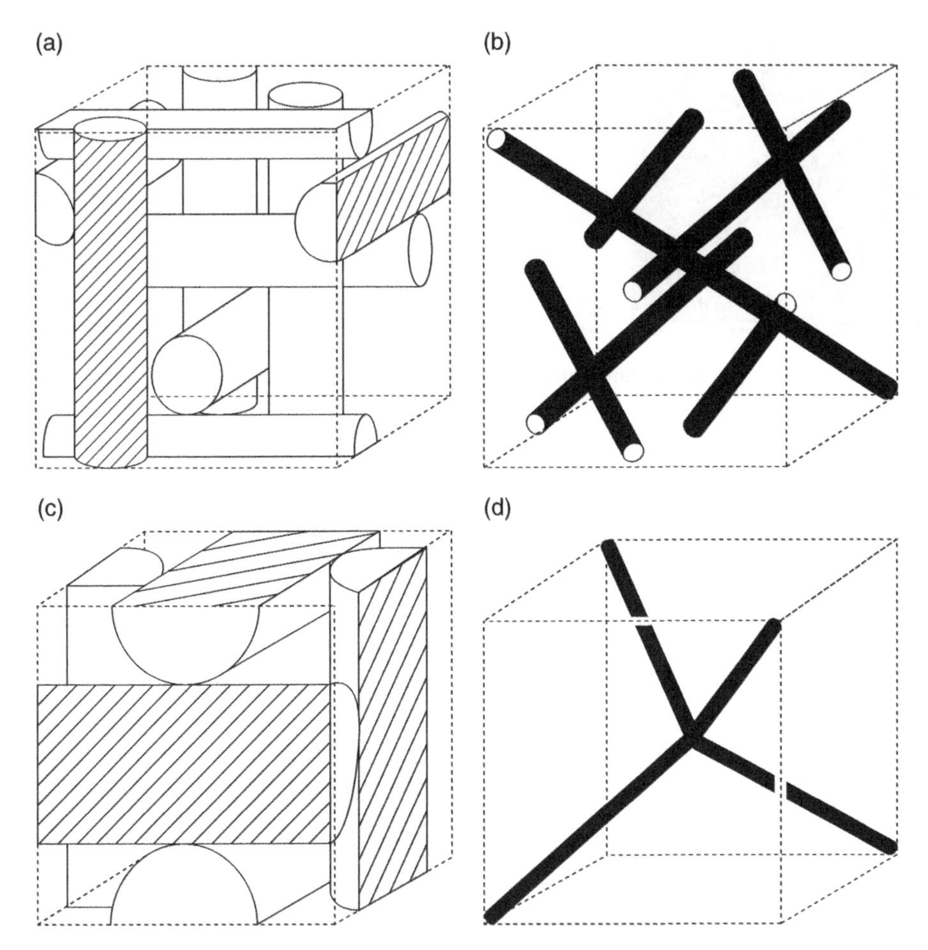

Figure 14.2 (a) and (b) Body-center cubic structure of BPI,(c) and (d) simple cubic structure of BPII filled with double-twist cylinders. The black lines in (b) and (d) represent the defect lines. Reproduced with permission from the Optical Society of America.

areas of the three perpendicular cylinders. In these areas, LC directors are randomly distributed. Disclination lines formed by such defects also form a lattice structure – Figure 14.1(d). The lattice dimension depends on the pitch length and arrangement order of the double-twist cylinders, usually around several hundred nanometers. Bragg reflection occurs over a certain wavelength range due to the periodic lattice structure of defects, and this is why BPLC appears colored.

As the temperature increases, up to three types of blue phases: BPI, BPII, and BPIII may exist [14]. BPIII is believed to possess amorphous structure. BPI (Figure 14.2(a)) and BPII (Figure 14.2(c)) are composed of double-twist cylinders arranged in cubic lattices. Inside each cylinder, the LC director rotates spatially about any radius of the cylinder. These double-twist cylinders are then fitted into a three-dimensional structure. However, they cannot fill the full space without defects. Therefore, blue phase is a coexistence of double-twist cylinders and disclinations. Defects occur at the points where the cylinders are in contact (Figures. 14.2(b) and 14.2(d)). BPI is known to have body-center cubic structure and BPII simple cubic structure.

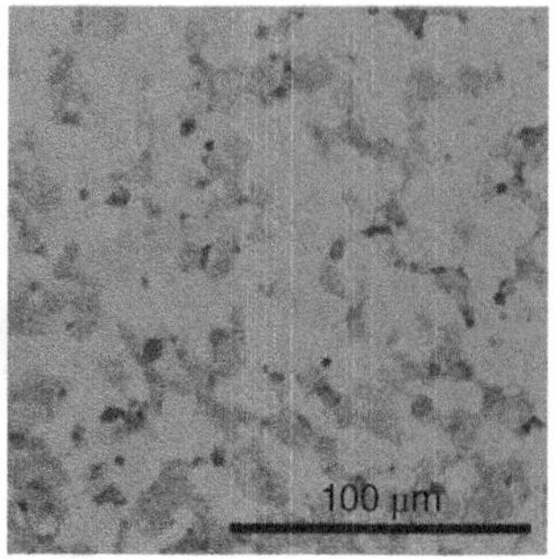

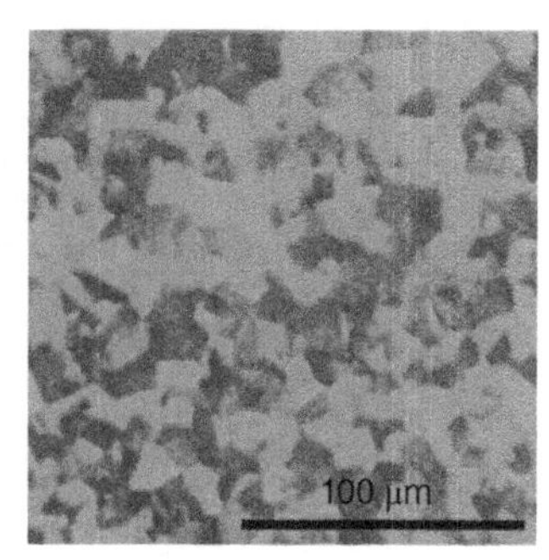

Figure 14.3 BPLC platelet textures under a polarizing optical microscope with different chiral concentrations.

In a chiral nematic liquid crystal, the selective reflection wavelength $\lambda = n{\cdot}P$, where n is the average refractive index and P is the pitch length. The reflection band is relatively broad with $\Delta\lambda = \Delta n{\cdot}P$, where Δn is the birefringence of the LC host. Unlike the chiral nematic phase, blue phases do not need any alignment layer and have several reflection wavelengths, corresponding to various crystal planes. The Bragg reflection wavelength can be expressed as:

$$\lambda = \frac{2na}{\sqrt{h^2 + k^2 + l^2}}, \tag{14.1}$$

where n and a denote average refractive index and lattice constant of blue phases, and h, k, and l are the Miller indices. In BPI, the lattice constant corresponds to one pitch length and diffraction peaks appear at (1,1,0), (2,0,0), (2,1,1), etc. The summation of Miller indices $h + k + l$ is an *even* number. In BPII, the lattice constant corresponds to one half of a pitch length and diffraction peaks appear at (1,0,0), (1,1,0), etc [11]. The pitch length of a BPLC is slightly different from that of the chiral nematic phase. The reflection bandwidth is also much narrower than that of the chiral nematic phase. Figure 14.3 shows the platelet textures of three BPLCs under crossed polarizers. The three photos exhibit different colors because of their different pitch lengths. The multiple colors in each photo correspond to different crystal planes.

14.2 Polymer-Stabilized Blue Phases

To widen BP temperature range, several approaches have been proposed [15–18] Here, we focus on the blue phases induced by incorporating chiral dopants into a nematic LC host. To make a polymer-stabilized blue phase liquid crystal, a small fraction of monomers (~8%) and photo-initiator (~0.5%) is added to the blue phase system. Figure 14.4 shows some exemplary nematic LC compounds, chiral dopants, and monomers [19]. Then we control the temperature within the narrow blue phase range to conduct UV curing. After UV irradiation, monomers are polymerized to form a polymer network, which stabilizes the blue phase lattice structures.

In a polymer-stabilized self-assembled blue phase system, each material component plays an important role while interacting with the others. In the following, we will discuss the optimization of materials in terms of nematic LC host, chiral dopant, and monomers, respectively.

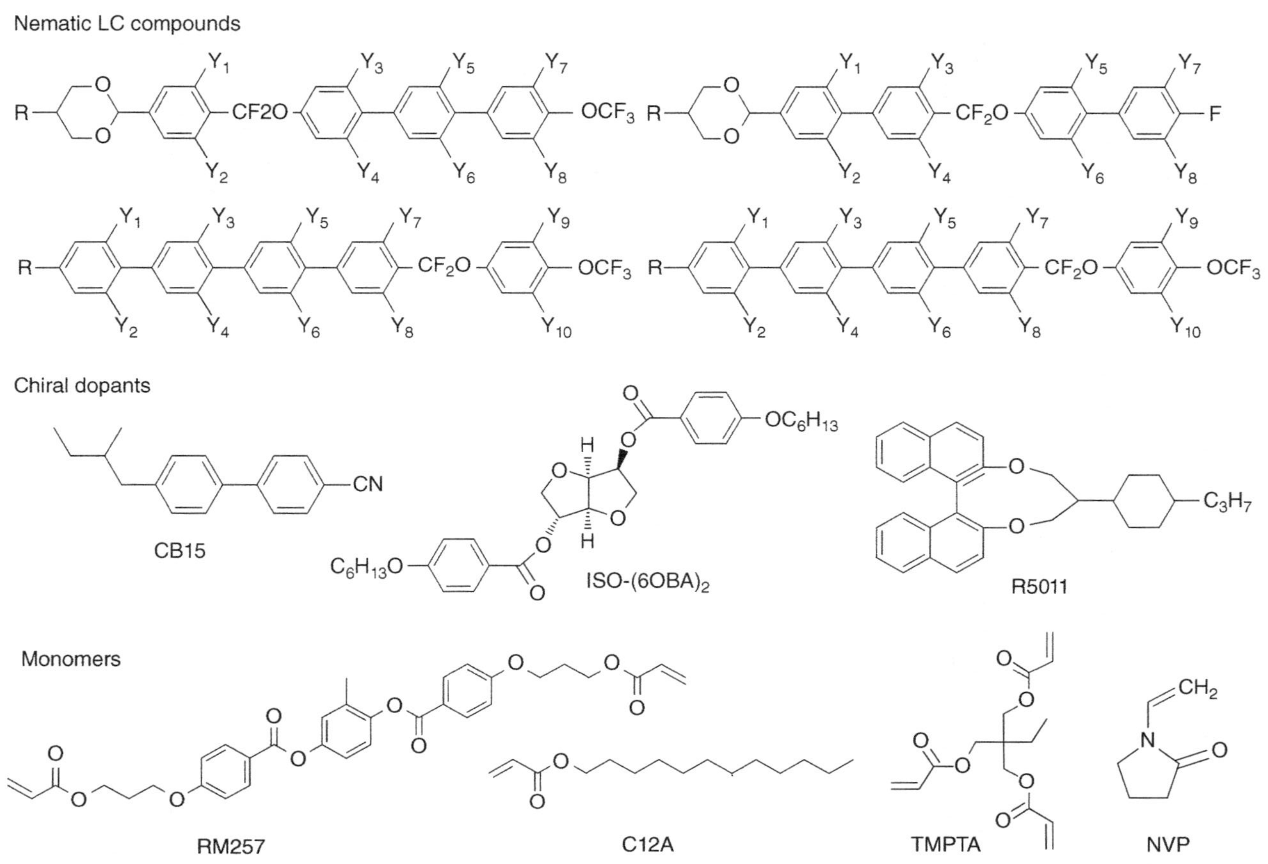

Figure 14.4 Examples of nematic liquid crystals, chiral dopants, and monomers.

14.2.1 Nematic LC host

In a polymer-stabilized BPLC material system, nematic LC host occupies the highest concentration. It serves as switching medium, and therefore it plays a major role in determining the performance of the system, such as temperature range, driving voltage, and response time. For display and photonics applications, the blue phase temperature range should ideally cover from −40°C to 80°C or higher. To obtain a wide temperature polymer-stabilized BPLC, a nematic LC host should first have a wide nematic range. Since other components in the polymer-stabilized BPLC system tend to lower the clearing temperature, it is desirable to choose a nematic LC host with clearing temperature $T_c > 80°C$.

The driving voltage of a BPLC device depends on the device structure and the Kerr constant (K) of the material employed. The on-state voltage (V_{on}) is inversely proportional to the square-root of Kerr constant [20]. For example, if K increases by four times then V_{on} would decrease by two times. The development goals for BPLC materials are threefold: (1) to increase the Kerr constant for lowering the driving voltage ($V_{on} < 10$ V), (2) to eliminate hysteresis for accurate gray-scale control and to minimize residual birefringence for high contrast ratio, and (3) to form a sturdy BPLC composite with long-term stability.

From Gerber's model [21], the Kerr constant of a BPLC is determined by following LC parameters as:

$$K \approx \frac{\Delta n \cdot \Delta \varepsilon}{k} \frac{\varepsilon_o P^2}{\lambda (2\pi)^2}, \tag{14.2}$$

where Δn, $\Delta \varepsilon$, and k are the intrinsic birefringence, dielectric anisotropy, and average elastic constant of the host LC, respectively, and P is the pitch length. From Equation (14.2), to enhance the Kerr constant, a liquid crystal with high Δn and large $\Delta \varepsilon$ is highly desirable [22]. Birefringence of an LC is determined by the conjugation length and dielectric anisotropy by the dipole moment. As Figure 14.4 shows, the exemplary compounds have 3–4 phenyl rings in order to obtain a high birefringence. They also have several polar groups for achieving a large $\Delta \varepsilon$. Moreover, to keep a high voltage holding ratio fluoro compounds are preferred [23]. An obvious drawback for having so many polar groups is the increased viscosity.

From Equation (14.2), pitch length also plays a crucial role affecting the Kerr constant. For display and photonic applications, the LC device should be clear in the visible spectral region. Thus, the Bragg reflection is usually hidden in the UV region, say $\lambda_B \sim 350$ nm, so that the BPLC is optically isotropic in the visible region. If the average refractive index of the BPLC is $n = 1.6$, then the pitch length should be $P \sim 220$ nm. Such a short pitch would lead to increased voltage. One strategy to enhance Kerr constant is to shift the Bragg reflection to a longer wavelength [24]. Although the long pitch approach helps to decrease the operating voltage, it requires two broadband circular polarizers in order to maintain a high contrast ratio.

The response time of a polymer-stabilized BPLC material is related to the LC parameters as [25]:

$$\tau \approx \frac{\gamma_1 P^2}{k (2\pi)^2}, \tag{14.3}$$

where γ_1 is the rotational viscosity of the BPLC system, which is closely related to the viscosity of the LC host and the chiral dopant. A low viscosity LC host is always favorable from the response time viewpoint. However, there are compromises between the large Kerr constant and fast response time. For example, a large $\Delta\varepsilon$ (>100) is favorable for enhancing the Kerr constant, but its viscosity is increased dramatically because of the multiple dipole groups involved. Similarly, increasing pitch length helps to enhance the Kerr constant, but it will slow down the response time as Equation (14.3) shows. All these factors have to be taken into consideration in order to develop a large Kerr constant BPLC while retaining a fast response time. For example, JNC JC-BP01M has a reasonably large Kerr constant ~13.7 nm/V^2 because its host LC has Δn ~ 0.17 and $\Delta\varepsilon$ ~ 94 [26]. Meanwhile, the clearing temperature of JC-BP01M is ~70°C and the response time in an IPS cell is ~1 ms at ~30°C. The key reason that such a high viscosity BPLC system can still maintain a relatively fast response time is because of the nano-structured (~100 nm) domain size.

14.2.2 Chiral dopants

The functionality of chiral dopants is to induce twist in blue phases. Blue phases only appear as the chirality ($q_0 = 2\pi/P$) exceeds a certain value. To increase chirality, we can either increase the chiral dopant concentration or employ a chiral dopant with a high helical twisting power (HTP). Since the solubility of a chiral dopant limits its maximum concentration, it is preferred to use a high HTP chiral dopant. For example, Merck developed a chiral dopant with HTP ~170/μm so that the required concentration is less than 5% [27].

Another important factor for selecting a chiral dopant is its melting point. For example, CB15 shown in Figure 14.4 has a very low melting point (~4°C). Therefore, after mixing with the LC host, the clearing temperature of the mixture decreases substantially. To make a wide temperature BPLC, the chiral dopant is preferred to have a reasonably high melting point while keeping good solubility. The solubility of chiral dopant in a nematic host depends on its melting point and heat fusion enthalpy. High melting point and large heat fusion enthalpy are two key factors limiting the solubility. As shown in Figure 14.4, the chiral dopant ISO-$(6OBA)_2$has a melting point of 90°C, solubility ~10 wt%, and HTP ~35/μm, which depends on the LC host. ISO-$(6OBA)_2$ has a similar HTP to ZLI-4572, but its solubility is better because of its lower melting point.

14.2.3 Monomers

Monomers are essential in determining the stability of BPLC. Typically, a polymer-stabilized BPLC requires two types of monomers: mono-functional (e.g. EHA or C12A) and di-functional (e.g. RM257). The overall monomer concentration is around 10 wt%, and the ratio between the two monomers, say RM257 to C12A is about 1:1. Although higher monomer concentration makes polymer network sturdier – which is helpful for reducing response time and suppressing hysteresis – it increases the operating voltage substantially [28]. As shown in Figure 14.4, TMPTA is a tri-functional monomer, while the normally used EHA or C12A are mono-functional monomers. The crosslink between TMPTA and RM257 is much stronger, resulting in an increased elastic constant (which is favorable for response time) and decreased hysteresis [29]. However, the trade-off is the increased operating voltage. Therefore, a delicate balance between operating voltage and response time should be carefully considered.

Besides material optimizations, photo-initiator and UV irradiation conditions also affect the performance of a polymer-stabilized BPLC composite, such as photo-initiator concentration,

UV exposure wavelength and intensity, exposure time, and curing temperature. For example, a typical UV dosage for preparing polymer-stabilized BPLC composite is ~3.6 J/cm^2 and the intensity varies from 1 mW/cm^2 to 20 mW/cm^2. Two types of UV light sources are commonly used: mercury lamp ($\lambda \sim 365$ nm) and LED lamp ($\lambda \sim 385$ nm) [30]. Finally, the curing temperature is near the chiral nematic to blue phase transition temperature.

14.3 Kerr Effect

Unlike nematic LC, whose LC directors are reoriented by an external electric field, the underlying physical mechanism of blue phase LC is electric-field-induced birefringence, known as Kerr effect [31]. The Kerr effect is a type of quadratic electro-optic effect caused by an electric-field-induced ordering of polar molecules in an optically isotropic medium. It usually exists in crystals with centro-symmetric point groups. Macroscopically, the induced birefringence of blue phase LC follows the Kerr effect, while microscopically the birefringence is still realized through LC molecular redistribution by an external electric field. Therefore, the maximum induced birefringence of a blue phase LC cannot exceed the birefringence of the LC composite.

According to the Kerr effect, the induced birefringence is proportional to the quadratic electric field as described by

$$\Delta n_{ind} = \lambda K E^2, \tag{14.4}$$

where K is the Kerr constant, λ is the wavelength, and E is the applied electric field. The following equation is commonly used to represent the refractive index ellipsoid of a medium in the presence of an electric field [32]:

$$\left(\frac{1}{n^2}\right)_1 x^2 + \left(\frac{1}{n^2}\right)_2 y^2 + \left(\frac{1}{n^2}\right)_3 z^2 + 2\left(\frac{1}{n^2}\right)_4 yz + 2\left(\frac{1}{n^2}\right)_5 xz + 2\left(\frac{1}{n^2}\right)_6 xy = 1, \tag{14.5}$$

where

$$\begin{bmatrix} \left(\frac{1}{n^2}\right)_1 \\ \left(\frac{1}{n^2}\right)_2 \\ \left(\frac{1}{n^2}\right)_3 \\ \left(\frac{1}{n^2}\right)_4 \\ \left(\frac{1}{n^2}\right)_5 \\ \left(\frac{1}{n^2}\right)_6 \end{bmatrix} = \begin{bmatrix} \left(\frac{1}{n_x^2}\right) \\ \left(\frac{1}{n_y^2}\right) \\ \left(\frac{1}{n_z^2}\right) \\ 0 \\ 0 \\ 0 \end{bmatrix} + \begin{bmatrix} r_{11} & r_{12} & r_{13} \\ r_{21} & r_{22} & r_{23} \\ r_{31} & r_{32} & r_{33} \\ r_{41} & r_{42} & r_{43} \\ r_{51} & r_{52} & r_{53} \\ r_{61} & r_{62} & r_{63} \end{bmatrix} \begin{bmatrix} E_x \\ E_y \\ E_z \end{bmatrix} + \begin{bmatrix} s_{11} & s_{12} & s_{13} & s_{14} & s_{15} & s_{16} \\ s_{21} & s_{22} & s_{23} & s_{24} & s_{25} & s_{26} \\ s_{31} & s_{32} & s_{33} & s_{34} & s_{35} & s_{36} \\ s_{41} & s_{42} & s_{43} & s_{44} & s_{45} & s_{46} \\ s_{51} & s_{52} & s_{53} & s_{54} & s_{55} & s_{56} \\ s_{61} & s_{62} & s_{63} & s_{64} & s_{65} & s_{66} \end{bmatrix} \begin{bmatrix} E_x^2 \\ E_y^2 \\ E_z^2 \\ E_yE_z \\ E_zE_x \\ E_xE_y \end{bmatrix}. \tag{14.6}$$

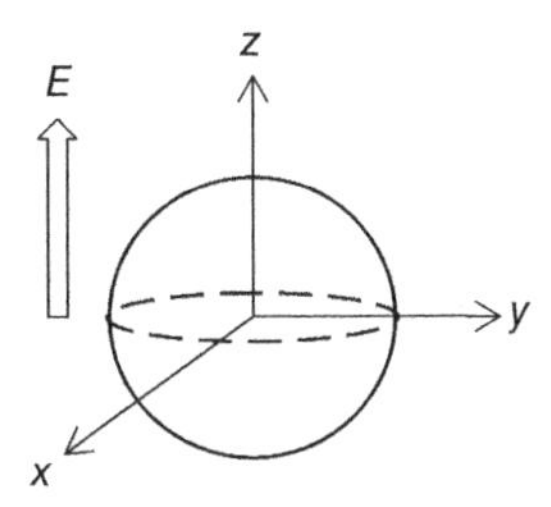

Figure 14.5 A coordinate showing the refractive index ellipsoid and the direction of applied electric field.

The first term in Equation (14.6) is related to initial refractive indices of the medium at three primary directions, n_x, n_y, n_z. The second term refers to the linear electro-optic effect, which is known as the Pockels effect, and the third term refers to the quadratic electro-optic effect, known as the Kerr effect. Here, r_{ij} and s_{ij} are electro-optic tensors for the linear and quadratic electro-optic effects, respectively. The second-order Kerr effect is small as compared to the first-order linear effect, so it is usually neglected in the presence of linear effect. However, in crystals with centro-symmetric point groups, the linear effect vanishes and then the Kerr effect becomes dominant.

Blue phase is optically isotropic when the external electric field is absent, therefore the refractive index in any direction is equal, say n_i. Therefore, we assume that the applied electric field is along the z direction in a Cartesian coordinate system, as shown in Figure 14.5.

Under these circumstances, the electric field components in x and y directions are both zero:

$$\begin{bmatrix} E_x^2 \\ E_y^2 \\ E_z^2 \\ E_yE_z \\ E_zE_x \\ E_xE_y \end{bmatrix} \Rightarrow \begin{bmatrix} 0 \\ 0 \\ E_z^2 \\ 0 \\ 0 \\ 0 \end{bmatrix} \tag{14.7}$$

In Equation (14.7), the linear term vanishes and only the Kerr effect term survives. The electro-optic tensor for the Kerr effect varies with different molecular structure. For an isotropic liquid, its quadratic electro-optic effect coefficients can be represented by the following matrix:

$$\begin{bmatrix} s_{11} & s_{12} & s_{12} & 0 & 0 & 0 \\ s_{12} & s_{11} & s_{12} & 0 & 0 & 0 \\ s_{12} & s_{12} & s_{11} & 0 & 0 & 0 \\ 0 & 0 & 0 & \frac{1}{2}(s_{11}-s_{12}) & 0 & 0 \\ 0 & 0 & 0 & 0 & \frac{1}{2}(s_{11}-s_{12}) & 0 \\ 0 & 0 & 0 & 0 & 0 & \frac{1}{2}(s_{11}-s_{12}) \end{bmatrix} \tag{14.8}$$

Substituting matrices (14.7) and (14.8) into Equation (14.6), we obtain the refractive index ellipsoid of blue phase LC under electric field as

$$\left(\left(\frac{1}{n_i^2}\right)+s_{12}E_z^2\right)x^2+\left(\left(\frac{1}{n_i^2}\right)+s_{12}E_z^2\right)y^2+\left(\left(\frac{1}{n_i^2}\right)+s_{11}E_z^2\right)z^2=1. \tag{14.9}$$

From Equation (14.9), we can tell that the ordinary refractive index is along the x and y directions and the extraordinary refractive index is along the z direction:

$$n_o=\left(\left(\frac{1}{n_i^2}\right)+s_{12}E_z^2\right)^{-1/2}\approx n_i-\frac{1}{2}n_i^3s_{12}E_z^2 \tag{14.10}$$

$$n_e=\left(\left(\frac{1}{n_i^2}\right)+s_{11}E_z^2\right)^{-1/2}\approx n_i-\frac{1}{2}n_i^3s_{11}E_z^2 \tag{14.11}$$

$$\Delta n_{ind}=n_e-n_o\approx\frac{1}{2}n_i^3(s_{12}-s_{11})E_z^2 \tag{14.12}$$

Both Equations (14.4) and (14.12) can represent the induced birefringence of a blue phase LC, the former using Kerr constant and the latter using the quadratic electro-optic coefficients. Comparing these two equations, we can see that the Kerr constant K is also dependent on the wavelength. It is a constant only at a given wavelength and temperature.

14.3.1 *Extended Kerr effect*

In the off-resonance region, blue phases are optically isotropic. When an electric field (E) is applied, liquid crystal molecules tend to align with the electric field if the dielectric anisotropy is positive ($\Delta\varepsilon > 0$) (or perpendicular to the electric field if $\Delta\varepsilon < 0$). As a result, birefringence is induced. In the low-field region, the induced birefringence is described by Equation (14.4). The Kerr effect exhibits a fast response time (<1 ms) because of the short coherent length of BPLC, which is quite attractive for both display and photonics applications. A higher electric field could lead to lattice distortion (the electrostriction effect), which results in a shift in the Bragg reflection wavelength [33]. For a sufficiently high electric field, blue phase may transform to new phases, to chiral nematic phases, and ultimately to nematic phases [34]. This transition is usually the slowest process (in the order of a few seconds) and is irreversible which causes undesirable hysteresis, residual birefringence, or permanent structural damage.

In a polymer-stabilized blue phase liquid crystal, the polymer network restricts the lattice structure so that color switching behavior is hardly observed. The response time is very fast (<1 ms), which originates from the Kerr effect. However, the Kerr effect is only valid in the low-field region because Equation (14.4) would lead to divergence if the electric field keeps on increasing. For a finite material system, the induced birefringence should gradually saturate in the high-field region once all the molecules have been reoriented. To verify this assumption, Yan et al.measured the ordinary refractive index change and developed an extended Kerr effect model to explain the saturation phenomenon.

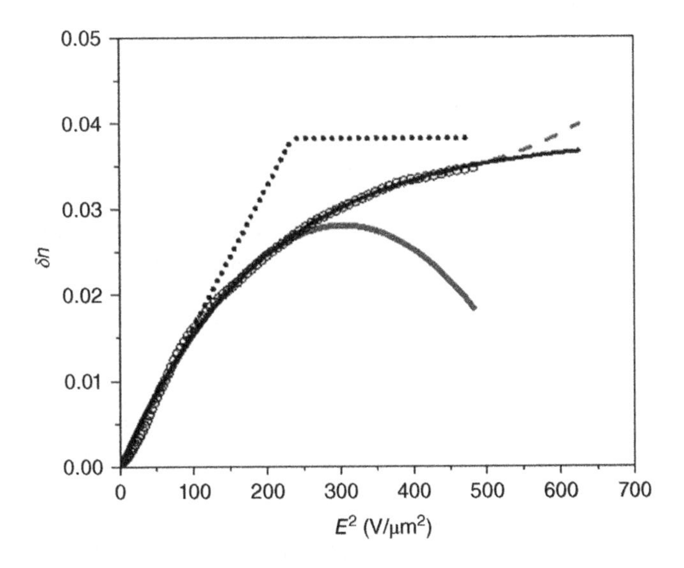

Figure 14.6 Measured refractive index change (open circles) and fittings with truncation model (Equation (14.13); dotted lines), Equation (14.18) (gray line), the model including second-, fourth-, and sixth-order terms (dashed lines), and the extended Kerr effect Equation (14.19) (black line). $\lambda = 633$ nm. Reproduced with permission from the American Physical Society.

Figure 14.6 depicts the measured refractive index change (dots) and fitting curves. The solid lines represent a truncation model [35, 36]:

$$\Delta n_{ind} = (\Delta n)_o (E/E_s)^2, \tag{14.13}$$

where $(\Delta n)_o$ denotes the maximum induced birefringence of the LC composite and E_s is the saturation field. In the truncation model, the induced birefringence increases linearly with $(E/E_s)^2$when $E < E_s$ and then saturates at $(\Delta n)_o$ when $E \geq E_s$ in order to prevent divergence. In reality, as the electric field increases, the induced birefringence will gradually saturate because all the LC directors will eventually be reoriented by the electric field.

Figure 14.6 depicts the measured refractive index change of a polymer-stabilized blue phase liquid crystal at $\lambda = 633$ nm. The PS-BPLC employed in this study is a mixture consisting of nematic LC (49 wt% Merck BL038, chiral dopants (21% Merck CB15 and 6% ZLI-4572), and monomers (9% EHA and 15% RM257). The BPLC was sandwiched between two indium-tin-oxide (ITO) glass substrates with a cell gap of 8 μm. The applied electric field is in the longitudinal direction. To measure the phase change, a Michelson interferometer was employed.

In the voltage-off state, the BPLC is optically isotropic. Its refractive index can be described through Maxwell relation ($\varepsilon = n^2$) as

$$n_i = \sqrt{(n_e^2 + 2n_o^2)/3}, \tag{14.14}$$

where n_e and n_o are the extraordinary and ordinary refractive indices of the LC composite, respectively. When the birefringence is small, Equation (14.14) can be approximated as

$$n_i \approx (n_e + 2n_o)/3. \tag{14.15}$$

To validate this approximation, let us assume that the BPLC composite has $n_e = 1.70$ and $n_o = 1.50$. The difference between Equation (14.14) and Equation (14.15) is only ~0.2%. When an electric field is applied, the refractive index is changed from n_i to $n_o(E)$:

$$\delta n = n_i - n_o. \tag{14.16}$$

The optic axis of the induced refractive-index ellipsoid is along the electric field direction. From Equation (14.15) and (14.16), we can rewrite the induced birefringence as [37]

$$\Delta n_{ind} = n_e(E) - n_o(E) = 3\delta n. \tag{14.17}$$

The data shown in Figure 14.6 is δn. It is indeed linearly proportional to E^2 as expected from the Kerr effect in the weak-field region. As the electric field increases, the induced refractive index change gradually saturates. The dotted lines represent the truncation model (Equation14.13), which is correct only in the low-field region. To explain the saturation phenomenon in the high-field region, higher-order electro-optical effects have been considered [38]. For a centro-symmetric crystal, the odd-order terms vanish due to the inversion symmetry, and only the even-order terms remain. Including the fourth-order term, the refractive index change can be written as

$$\delta n = \frac{\lambda(K_1 E^2 + K_2 E^4)}{3}, \tag{14.18}$$

where K_1 and K_2 are the Kerr constant and the fourth-order coefficient, respectively. Equation (14.18) is used to fit the experimental data with K_1 and K_2 as adjustable parameters. As depicted in Figure 14.6, good fitting (gray line) is found when E^2 is below 200 $V^2/\mu m^2$. However, above this field the fourth-order term begins to dominate and the curve eventually bends down because Equation (14.18) is a downward parabola if K_2 is negative. To avoid this bending-down phenomenon, one may further include the sixth-order term. Indeed, the fitting (using K_1, K_2 and K_3 as adjustable parameters) is very good in the entire region (dashed lines). Nevertheless, this curve predicts a fast divergent trend in the high-field region, as Figure 14.6 shows.

To explain the saturation phenomenon shown in Figure 14.6, Yan et al. proposed the convergence model [39]

$$\delta n = \delta n_s \left(1 - \exp\left[-\left(\frac{E}{E_s}\right)^2\right]\right), \tag{14.19}$$

where δn_s is the saturated refractive index change and E_s the saturation field. The fitting with the experimental data is quite good in the entire region and, more importantly, it shows the anticipated saturation trend in the high-field region. For convenience, this convergence model is called the extended Kerr effect.

It is interesting to note that if we expand Equation (14.19) into a power series, we can obtain the E^2 term (Kerr effect) under the weak field approximation, and the Kerr constant can be written as

$$K = 3\delta n_s / \left(\lambda E_s^2\right). \tag{14.20}$$

The higher-order terms become increasingly important in the high-field region. However, the inclusion up to the sixth-order term still does not lead to saturation in the high fields (Figure 14.6), despite this model involving three fitting parameters. The extended Kerr effect correctly predicts the saturation behavior of the induced birefringence and fits well with experimental data with just two parameters. This leads to a more accurate simulation of the electro-optic properties of polymer stabilized optically isotropic LCs.

14.3.2 Wavelength effect

As shown in Equation (14.4), Kerr-effect-induced birefringence seems to increase linearly with wavelength. However, such a relationship is not explicit because the Kerr constant K is also wavelength-dependent. Jiao, et al. conducted a nice experiment and proved that Kerr constant actually decreases with wavelength as [40]

$$K=\frac{G}{E_s^2}\frac{\lambda\lambda^{*2}}{\left(\lambda^2-\lambda^{*2}\right)}, \tag{14.21}$$

where G is a proportionality constant and λ^* is the resonance wavelength of the LC host, according to the single-band model [41].

Figure 14.7 shows the wavelength-dependent Kerr constant of a polymer-stabilized blue phase LC composite studied by Jiao, et al. Dots are the measured data and the solid line is the fitting result with Equation (14.21) using two adjustable parameters: $\lambda^* \sim 216$ nm and proportionality constant $G/E_s^2 \sim 2.62 \times 10^{-2}$ nm^{-1}. This $\lambda^* \sim 216$ nm agrees with that obtained from the employed LC host very well.

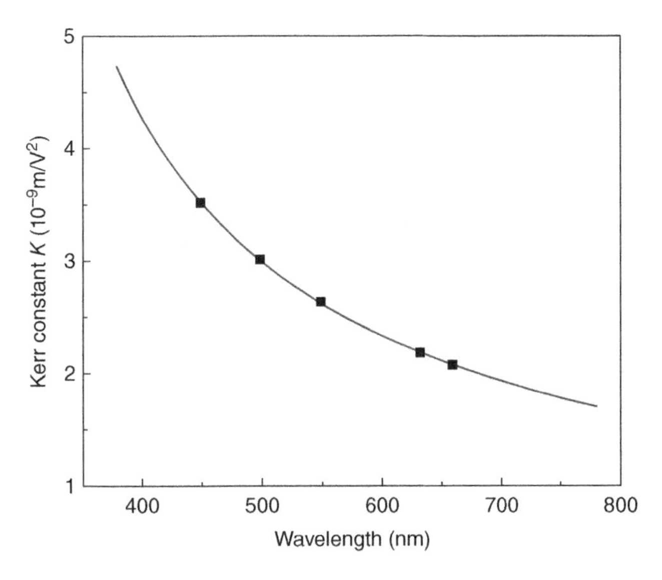

Figure 14.7 Measured wavelength dependent Kerr constant (dots) of a polymer-stabilized blue phase liquid crystal and fitting results with Equation (14.21). Reproduced with permission from the American Physical Society.

In the original Kerr effect, λK is used as a coefficient to calculate the induced birefringence under certain electric field. The term λK follows the same dispersion trend as Δn_{sat} when $\lambda \gg \lambda^*$. Correspondingly, the Kerr-effect-induced birefringence also decreases with wavelength and gradually saturates in the long wavelength (near infrared) region. Finally, the wavelength and electric field dependency of induced birefringence of a blue phase LC can be explicitly expressed as

$$\Delta n_{ind} = \frac{G\lambda^2\lambda^{*2}}{\lambda^2 - \lambda^{*2}}\left(1 - \exp\left[-\left(\frac{E}{E_s}\right)^2\right]\right). \tag{14.22}$$

In this equation, the saturation electric field E_s governs the electric behavior of blue phase LC, while saturated induced birefringence determines the optical behavior and thus the wavelength dispersion of blue phase. It still follows the normal dispersion trend and agrees very well with the single-band model.

14.3.3 Frequency effect

Polymer-stabilized BPLC is a self-assembled nano-structured soft matter. Owing to the small diameter of the double-twist cylinders, the response time of a BPLC is in microseconds to the submilliseconds range despite its high viscosity. Therefore, it enables color sequential displays using RGB LEDs without spatial color filters. As a result, both optical efficiency and resolution density are tripled. However, the operating electric field frequency is also tripled since three RGB subframes replace one color frame. Most of the current nematic LCD TVs are operated at 120 Hz frame rate. For color sequential, the operation frequency should be at least 360 Hz, in order to avoid color breakup. For low molecular weight nematics, dielectric relaxation usually occurs in the 100 kHz region. However, BPLC requires an extremely large dielectric anisotropy ($\Delta\varepsilon > 100$) in order to lower the operating voltage. Such a bulky and high viscosity compound cannot follow the high frequency electric field too well. Consequently, its relaxation frequency could drop to the hundreds of hertz region, which might impact the operation for color sequential displays. Therefore, the frequency effect is an important subject for BPLC materials.

From Equation (14.2), the only frequency-dependent part of Kerr constant is the dielectric anisotropy. For an LC mixture, the frequency dependent dielectric constants $\varepsilon_{//}$ and $\varepsilon_{\perp}$ can be described by the Cole–Cole equation [42]:

$$\varepsilon^*(f) = \varepsilon_\infty + \frac{\varepsilon_s - \varepsilon_\infty}{1 + \left(i\frac{f}{f_r}\right)^{1-\alpha}}, \tag{14.23}$$

where $\varepsilon* = \varepsilon' + i\varepsilon'$ is the complex dielectric constant at frequency f, ε_s and ε_∞ are the dielectric constants at static and high frequencies, respectively, f_r is the relaxation frequency, i is the imaginary unit, and α is a value between 0 and 1, which allows us to describe different spectral shapes. The real part of the complex dielectric constant (ε') is the one we commonly measure ($\varepsilon_{//}$ and $\varepsilon_{\perp}$), and the imaginary part (ε') is responsible for the dielectric heating.

For a rod-like compound, $\varepsilon_{\perp}$ has a much higher relaxation frequency than $\varepsilon_{//}$ due to its shorter dipole length in the direction perpendicular to molecular axis [43]. Thus in the low frequency region, the relaxation of $\varepsilon_{//}$, as described by the real part of Cole–Cole equation, is noticeable, while $\varepsilon_{\perp}$ remains unchanged ($\varepsilon_{\perp} = \varepsilon_{\perp s}$). As a result, dielectric anisotropy ($\Delta\varepsilon = \varepsilon_{//} - \varepsilon_{\perp}$) also follows the real part of Cole–Cole equation, i.e. it has the same relaxation frequency as $\varepsilon_{//}$:

$$\Delta\varepsilon(f) = \Delta\varepsilon_{\infty} + (\Delta\varepsilon_s - \Delta\varepsilon_{\infty})\frac{1 + \left(\frac{f}{f_r}\right)^{1-\alpha}\sin\frac{1}{2}\alpha\pi}{1 + 2\left(\frac{f}{f_r}\right)^{1-\alpha}\sin\frac{1}{2}\alpha\pi + \left(\frac{f}{f_r}\right)^{2(1-\alpha)}}, \tag{14.24}$$

where $f_r = f_{r//}$, $\Delta\varepsilon_{\infty} = \varepsilon_{//\infty} - \varepsilon_{\perp s}$, $\Delta\varepsilon_s = \varepsilon_{//s} - \varepsilon_{\perp s}$, $\Delta\varepsilon_s$ is the static dielectric anisotropy, and $\Delta\varepsilon_{\infty}$ is the dielectric anisotropy in the high-frequency region. Similarly, the average dielectric constant $< (2\varepsilon_{\perp} + \varepsilon_{//})/3 >$ should have the same relaxation frequency as well.

Since the Kerr constant is linearly proportional to $\Delta\varepsilon$, Li et al. modify Equation (14.2) to describe the frequency-dependent Kerr constant as [44]

$$K(f) = K_{\infty} + (K_s - K_{\infty})\frac{1 + \left(\frac{f}{f_r}\right)^{1-\alpha}\sin\frac{1}{2}\alpha\pi}{1 + 2\left(\frac{f}{f_r}\right)^{1-\alpha}\sin\frac{1}{2}\alpha\pi + \left(\frac{f}{f_r}\right)^{2(1-\alpha)}}, \tag{14.25}$$

where K_s and K_{∞} are the Kerr constant at static and high frequency, respectively. For convenience, Equation (14.25) is called *extended Cole–Cole equation.*

To validate the extended Cole–Cole equation, Li et al. measured the frequency effect using a JNC JC-BP01M polymer-stabilized BPLC. The employed IPS cell has 10 μm electrode width and 10 μm electrode gap, and 7.5 μm cell gap. Results are plotted in Figure 14.8, where the *VT* curves gradually shift to the right side and V_{on} increases as frequency increases. At 5 kHz, the transmittance at 60 V_{rms} is only ~10% of that of the peak transmittance. These results indicate that frequency has a tremendous impact on the electro-optic properties of this BPLC cell.

Equation (14.25) has four unknowns: K_{∞}, K_s, f_r, and α. However, in the low-frequency region, K_s is insensitive to frequency and can be treated as a constant. For JC-BP01M BPLC, its saturation birefringence $\Delta n_s = 0.15$ and $K_s = 10.4\ \text{nm/V}^2$. Through fitting the measured *VT* curves (Figure 14.8) with extended Kerr model, we can obtain the Kerr constant at those specified frequencies. Figure 14.9 shows the frequency-dependent Kerr constant of a polymer-stabilized BPLC using JNC JC-BP01M host; the squares are experimental data and the solid line denotes fitting using Equation (14.25) with $K_{\infty} = 0$, $f_r = 1300$ Hz, and $\alpha = 0.13$. The agreement is quite good. From Figure 14.9, the Kerr constant decreases rapidly as frequency increases. This trend is less pronounced for the BPLC hosts with a smaller dielectric anisotropy. However, the low-*K* materials would lead to a high operating voltage. Thus for practical devices, a delicate balance between Kerr constant and operation frequency should be taken into consideration. One way to overcome this dilemma is to improve device structure so that the demand on high-*K* materials can be greatly relaxed.

14.3.4 Temperature effects

For a thermotropic liquid crystal, its physical properties, such as birefringence, viscosity, dielectric anisotropy, and elastic constant, are all dependent on the operation temperature – except at different rates. Polymer-stabilized BPLC is no exception [45]. Figure 14.10 shows

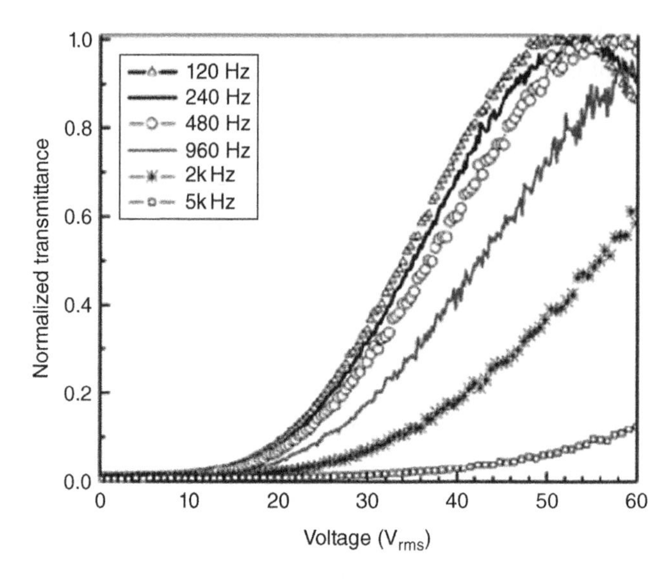

Figure 14.8 Measured *VT* curves of a polymer-stabilized BPLC (JNC JC-BP01M) at the specified frequencies. IPS cell: electrode width 10 μm, electrode gap 10 μm and cell gap 7.5 μm. $\lambda = 633$ nm. Reproduced with permission from the American Physical Society.

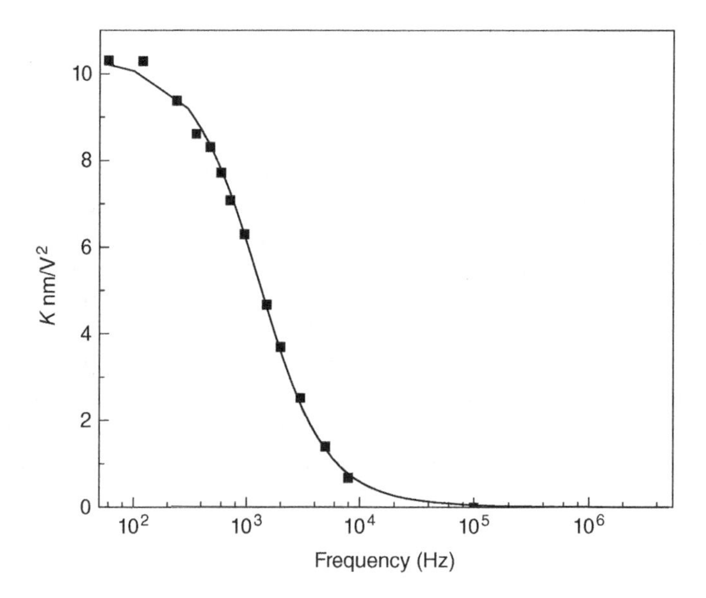

Figure 14.9 Frequency-dependent Kerr constant for JC-BP01M BPLC. Squares are Kerr constant obtained from *VT* curves, and solid line is the fitting of Kerr constant Equation (14.25) with $K_\infty = 0$, $f_r = 1.3$ kHz, and $\alpha = 0.13$. Reproduced with permission from the American Physical Society.

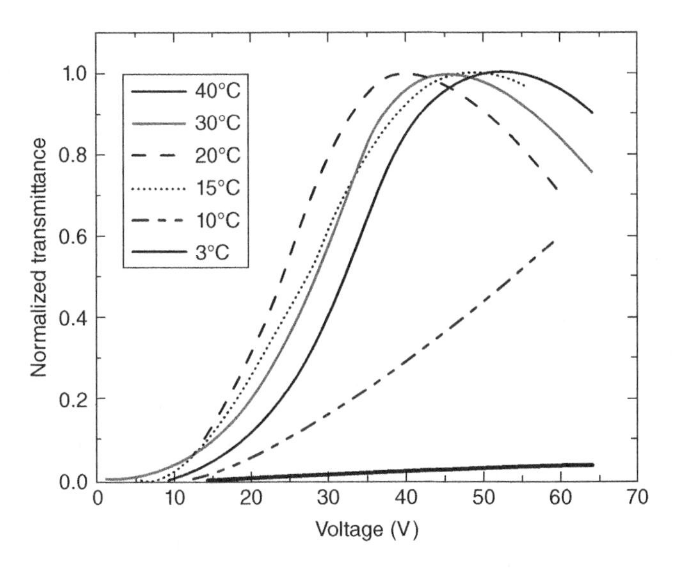

Figure 14.10 Measured *VT* curves of PSBP-06 at the specified temperatures; $f = 480$ Hz and $\lambda = 633$ nm. Reproduced with permission from the Royal Society.

the measured *VT* curves of PSBP-06 from 40°C down to 3°C at $f = 480$ Hz. As the temperature decreases, the *VT* curves shift leftward first and then rightward, indicating the on-state voltage (V_{on}) bounces back at low temperatures. The lowest V_{on} occurs at ~20°C. As the temperature continues to decrease to 3°C, which is still above the melting point of the PSBP-06 ($T_{mp} = -2$°C), the transmittance stays below 5% (normalized to the peak transmittance at 60 Hz under the same temperature) even the applied voltage is as high as 65 V_{rms}. Therefore, this imposes a practical low temperature operation limit for PSBP-06.

To investigate the temperature effects quantitatively, here we analyze the temperature and frequency effects on Kerr constant. Based on Gerber's model (Equation (14.2)), Kerr constant is governed by the birefringence (Δn), average elastic constant (k), dielectric anisotropy ($\Delta\varepsilon$) and pitch length (P) of the chiral LC host. The temperature effects on Δn, k, and $\Delta\varepsilon$ are described by the following relations: [46]

$$\Delta n = \Delta n_o S, \tag{14.26}$$

$$\Delta\varepsilon \sim S \cdot \exp(E_1/k_B T), \tag{14.27}$$

$$k \sim S^2, \tag{14.28}$$

$$S = (1 - T/T_c)^\beta, \tag{14.29}$$

where S denotes the order parameter, Δn_o is the extrapolated birefringence at $T = 0$ K, E_1 is a parameter related to dipole moment, k_B is the Boltzmann constant, T_c is the clearing point of the nematic host, and β is a material constant. On the other hand, pitch length (P) is not sensitive to the temperature. Substituting Equations (14.26), (14.27), and (14.28) into Equation (14.2), we find that $K \sim \exp(E_1/K_{BE}T)$. Generally speaking, as the temperature increases, K decreases.

The frequency effect of K mainly originates from $\Delta\varepsilon$ because the remaining parameters are all independent of frequency in the low frequency region. Based on Debye relaxation model, $\Delta\varepsilon$ has following form:

$$\Delta\varepsilon = \Delta\varepsilon_\infty + \frac{\Delta\varepsilon_0 - \Delta\varepsilon_\infty}{1 + (f/f_r)^2} \tag{14.30}$$

where $\Delta\varepsilon_\infty$ and $\Delta\varepsilon_o$ are the dielectric anisotropy at high and low frequency limits respectively, f is the operation frequency, and f_r is the relaxation frequency. For a low viscosity nematic LC host, its f_r is usually over 100 kHz, which is much higher than the intended operation frequency (e.g. 120 – 960 Hz) of the LC device. As a result, the f/f_r term in Equation (14.30) can be neglected and $\Delta\varepsilon \approx \Delta\varepsilon_o$, which is insensitive to the frequency. However for a large $\Delta\varepsilon$ BPLC, the bulky molecules cannot follow the electric field in the high frequency region. The Debye relaxation frequency is usually in the 1–2 kHz region. Thus, the f/f_r term in Equation (14.30) becomes significant. As a result, $\Delta\varepsilon$ (or Kerr constant) is strongly dependent on the frequency.

Figure 14.11 depicts the measured $\Delta\varepsilon$ (dots) and fitting curve (solid lines) with Equation (14.30) for JC-BP06N. Through fittings, f_r at each temperature is obtained. Results indicate that f_r decreases exponentially with T as:

$$f_r = f_0 \cdot \exp(-E_2/k_B T) \tag{14.31}$$

Here, E_2 is the activation energy of molecular rotation and f_0 is the proportionality constant. From Equation (14.31), as T decreases, f_r decreases exponentially so that the ratio of f/f_r increases, which in turn leads to a decreased $\Delta\varepsilon$ (Equation (14.30)]. As Figure 14.11 shows, at 480 Hz $\Delta\varepsilon$ decreases by about two times as T decreases from 20°C to 15°C. Thus, the on-state voltage 'bounces back' as the temperature decreases (Figure (14.10)). Substituting Equations (14.26), (14.27), and (14.28) into Equation (14.2), the Kerr constant can be expressed as:

$$K \sim A \cdot \frac{\exp(E_1/k_B T)}{1 + (f/f_r)^2} = A \cdot \frac{\exp(E_1/k_B T)}{1 + [(f/f_0) \cdot \exp(E_2/k_B T)]^2}, \tag{14.32}$$

where A is a proportionality constant. However, when the temperature approaches the clearing point (T_c) both Δn and $\Delta\varepsilon$ vanish, so does the Kerr constant (at least, it dramatically decreases). To satisfy this boundary condition, we modify Equation (14.32) to [47]

$$K = A \cdot \frac{\exp\left[\frac{E_1}{k_B}\left(\frac{1}{T} - \frac{1}{T_C}\right)\right] - 1}{1 + [(f/f_0) \cdot \exp(E_2/k_B T)]^2}. \tag{14.33}$$

From Equation (14.33), an optimal operation temperature (T_{op}) exists, at which the Kerr constant has a maximum value. Generally speaking, T_{op} is governed by several parameters, such as frequency, relaxation frequency, and temperature.

Figure 14.12 depicts the temperature-dependent Kerr constant at different frequencies for two PSBP composites: PSBP-06 and BPLC-R1. The latter has a smaller Kerr constant because its LC host has a smaller $\Delta\varepsilon$ (~50), but its viscosity is also much lower than that of JC-BP06N.

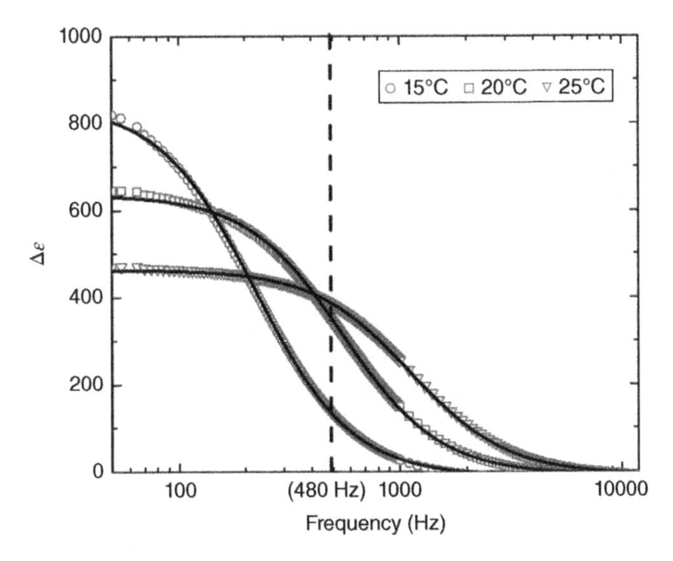

Figure 14.11 Frequency dependent $\Delta\varepsilon$ of JC-BP06N at the three specified temperatures. Dots are measured data and lines are fittings with Equation (14.30). Reproduced with permission from the Royal Society.

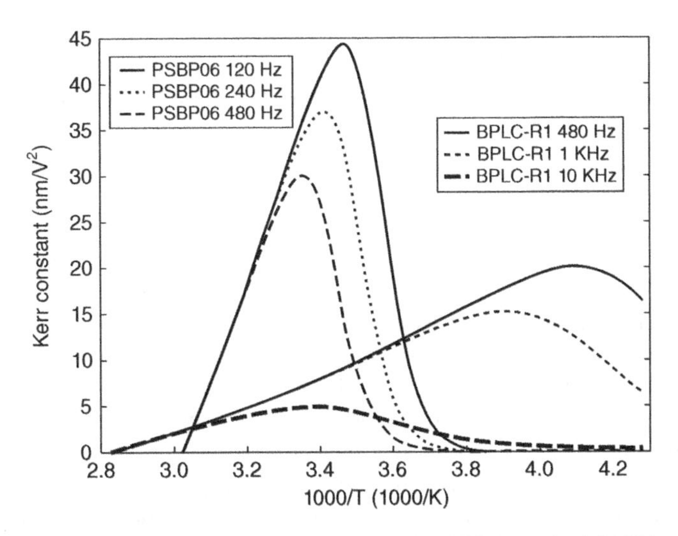

Figure 14.12 Temperature-dependent Kerr constant of PSBP-06 ($f_r \sim 1.2\,\text{kHz}$) and BPLC-R1 ($f_r \sim 15\,\text{kHz}$) at the specified frequencies, $\lambda = 633\,\text{nm}$. Reproduced with permission from the Royal Society.

The Debye relaxation frequency of the BPLC-R1 host is ~15 kHz at 25°C, which is about ten times higher than that of JC-BP06N. For the two BPLC samples shown in Figure 14.12, the three curves (representing low, medium, and high frequencies for each sample) overlap in the high temperature region, which means that their Kerr constant is proportional to $1/T$, but is quite insensitive to the frequency. This can be explained as follows. In the high temperature region, the BPLC has a lower viscosity so that its relaxation frequency is higher (Equation (14.30)). From Equation (14.32), when $f_r \gg f$ the frequency part can be ignored and K is inert to the frequency. As the temperature decreases (or $1/T$ increases), f_r decreases exponentially (Equation (14.30)). The high frequency curve (whose f is closer to f_r) bends down first due to the dramatically reduced $\Delta\varepsilon$ (Figure 14.11). As a result, its maximum Kerr constant is smaller, which leads to a higher operating voltage. For the lower frequency curves, their peak Kerr constant and bending-over phenomenon occur at a lower temperature, as Figure 14.12 shows.

As shown in Figure 14.12, for a given BPLC material its T_{op} increases as the frequency increases. Let us illustrate this concept using PSBP-06 as an example. In Figure 14.12, as the frequency increases from 120 Hz to 480 Hz, the T_{op} increases gradually from 15°C to 22.5°C, but in the meantime the Kerr constant decreases from 44.4 nm/V^2 to 30.1 nm/V^2. If the relaxation frequency of a BPLC is too high, then its T_{op} might shift outside the intended operation temperature range. Let us take BPLC-R1 as an example: at 480 Hz its T_{op} occurs at −30°C, as Figure 14.12 shows. At such a low operation temperature, the viscosity of BPLC would increase dramatically. If we want to shift T_{op} to room temperature, then the operation frequency should be increased to ~10 kHz, which would increase the power consumption dramatically. An optimal relaxation frequency for a blue phase LCD should be in the 2–3 kHz range. For high speed spatial light modulator applications (e.g. 1 kHz operation rate), then the relaxation frequency of the employed BPLC should be higher, say 5–10 kHz.

14.4 Device Configurations

From application viewpoint, polymer-stabilized BPLC exhibits three attractive features: (1) self-assembly process, so that no surface alignment layer is needed for producing uniform molecular arrangement, which is generally required in most nematic devices, (2) nano-sized (~100 nm) double-twist cylinder diameter and short coherence length, which leads to submillisecond response time, and (3) three-dimensional lattice structure resulting in Bragg reflections. When the Bragg reflection wavelength is controlled to occur in the ultraviolet region, BPLC appears optically isotropic in the voltage-off state. That means that if such a BPLC is sandwiched between two crossed linear polarizers, no light is transmitted.

The alignment-layer-free feature greatly simplifies device fabrication. Fast response time not only produces crisp pictures without image blur but also enables color sequential displays. By eliminating spatial color filters, both optical efficiency and resolution density are tripled, although the required frame rate is also tripled. This feature is particularly important for reducing the power consumption of a high resolution display. Moreover, short coherence length significantly improves the diffraction efficiency of BPLC gratings. Finally, the optically isotropic state makes adaptive BPLC lenses polarization-independent. Therefore, BPLC has great potential for display and photonics applications.

However, a relatively high operating voltage, noticeable hysteresis [48], and slow charging time due to large capacitance [49] still hinder the widespread applications of BPLC. Among these three technical barriers, high voltage is the main problem. If voltage is sufficiently low (<10 V) so that the peak electric field is lower than the critical field of the BPLC composite, then the electrostriction effect which causes lattice distortion would be minimized, and hysteresis would be negligible (<1%). Another critical issue of BPLC is that its dielectric anisotropy is usually quite large ($\Delta\varepsilon > 100$) in order to boost the Kerr constant and lower the operating voltage. Such a huge $\Delta\varepsilon$ leads to three potential problems: (1) high viscosity, which increases response time, (2) low Debye relaxation frequency, which limits the operation frequency, and (3) large capacitance, which requires a much longer charging time for each pixel, which in turn reduces the frame rate. A higher frame rate, say 240 Hz, helps to reduce image blurs. However, the electric power consumption increases accordingly. For color sequential displays, we need a frame rate higher than 360 Hz in order to minimize the annoying color breakup. To boost frame rate, we can either use oxide semiconductor, e.g. indium-gallium-zinc-oxide (IGZO) whose electron mobility is about twenty times higher than that of amorphous silicon [50], or use bootstrapping circuitry.Therefore, the most fundamental issue for BPLC devices is to lower the operating voltage to below 10 V, without sacrificing other desirable properties, such as high transmittance, submillisecond response time, high contrast ratio, and wide viewing angle.

To achieve this goal, both device structures and BPLC materials have been investigated extensively. From a device structure viewpoint, two major approaches have been developed: (1) implementing protrusion electrodes [51] so that the fringing field can penetrate deeply into the LC bulk, and (2) using vertical field switching (VFS) [52] to generate uniform longitudinal field across the entire BPLC layer. From a material aspect, developing BPLC materials with a large Kerr constant helps to reduce operating voltage because the on-state voltage is inversely proportional to $\sqrt{K}$ [53].

14.4.1 In-plane-switching BPLCD

For display applications, BPLCs are usually driven by in-plane switching (IPS) electrodes, where the electric fields are primarily in the lateral direction. Figure 14.13 shows a planar IPS-based blue phase LCD, in which w stands for the electrode width and l the electrode gap. The substantial lateral electric fields would induce birefringence along the direction of electric fields, provided that the employed LC has a positive dielectric anisotropy. Macroscopically, the BPLC whose Bragg reflection does not occur in the visible region can be treated as an isotropic medium when no external field (E) is present. Under crossed polarizers, no light is transmitted and the BPLC cell appears dark. As E increases, the induced birefringence increases and the refractive index sphere turns to ellipsoid with its optic axis along the electric field

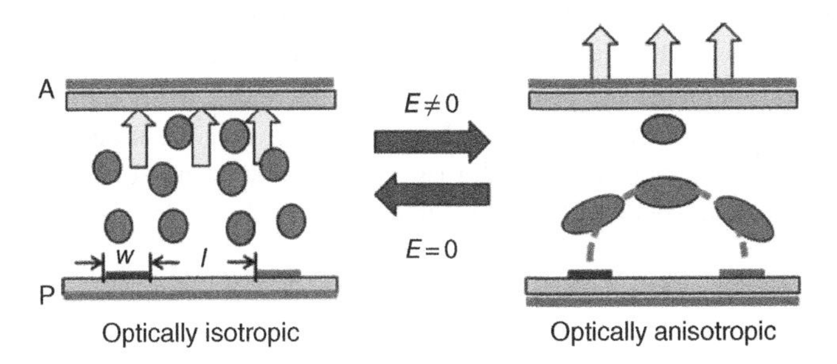

Figure 14.13 Operation principle of a planar IPS BPLCD between crossed polarizers: (a) $E = 0$ and (b) $E > 0$.

vector. Therefore, the incident linearly polarized light would experience phase retardation. As a result, the light would gradually transmit through the crossed polarizers.

In the following sections, we will look at how electrode dimension, cell gap, and Kerr constant affect the electro-optic behaviors of a BP LCD.

14.4.1.1 Electrode dimension effect

Electrode dimension affects the electro-optics of IPS-BPLC significantly. Figure 14.13 shows the simulated voltage-dependent transmittance (*VT*) curves of JC-BP01 in an IPS structure with electrode dimensions $l/w = 1$ and $l/w = 2$. The material parameters for JC-BP01 at $\lambda = 550$ nm are saturated induced birefringence $\Delta n_s = 0.154$ and saturation electric field $E_s = 4.05$ V/μm. As Figure 14.14(a) shows, the on-state voltage (V_{on}) decreases as electrode dimension decreases from IPS-10/10 to IPS-5/5 and then IPS-3/3. However, for IPS-2/2 this trend is reversed. To explain this, we need to consider two determining factors for V_{on}: (1) penetration depth of electric field, and (2) induced birefringence Δn. A smaller electrode dimension allows a lower voltage to achieve comparable induced birefringence. This is why IPS-3/3 shows a lower V_{on} than IPS-5/5 and IPS-10/10. But a smaller electrode dimension also leads to a shallower penetration depth, which in turn demands a larger induced birefringence in order to accumulate sufficient phase retardation for high transmittance. These two factors compete with each other and result in a higher V_{on} for IPS-2/2 because the deficiency of shallow penetration depth outweighs the merit of strong electric field. However, for IPS-3/3 and IPS-5/5, although their electric field intensities are weaker than that of IPS-2/2, their deeper penetration helps to lower V_{on}. Therefore, from material viewpoint we cannot only emphasize large Kerr constant, at the same time we need to pay attention to the individual Δn and $\Delta\varepsilon$ values according to the device structures employed. For small electrode gap, high birefringence is preferred because of its shallow electric field penetration depth. For large electrode gap, large dielectric anisotropy is preferred because of its weaker electric field.

Another clear trend shown in Figure 14.14(a) is that the peak transmittance decreases as the electrode dimension increases. IPS-BPLCs with different electrode dimensions have different on-state voltages and electric field intensities. According to the extended Kerr model, stronger electric field would result in a higher induced birefringence. As a result, the refraction effect in the BPLC medium is larger, which contributes to larger aperture ratio for higher transmittance. Especially on the top of electrodes, the larger angle with respect to the optical axis and higher induced birefringence would result in a higher transmittance and reduced dead zones. This is why IPS-5/5 exhibits a larger effective aperture ratio than IPS-10/10 [54].

Figure 14.14(b) shows the simulated *VT* curves with $l/w = 2$. Similar to $l/w = 1$, as the electrode dimension increases the peak transmittance decreases. Although a larger l leads to a deeper penetration depth, the electric field is weaker so that the induced birefringence is smaller. Moreover, a larger electrode width (w) causes more dead-zone area, which in turn lowers the transmittance. This trend is also found for $l/w = 3$ and 4.

14.4.1.2 Cell gap effect

In a nematic IPS cell, both transmittance and response time are affected by the cell gap. However, the transmittance of IPS-BPLC is insensitive to the cell gap as long as it exceeds the

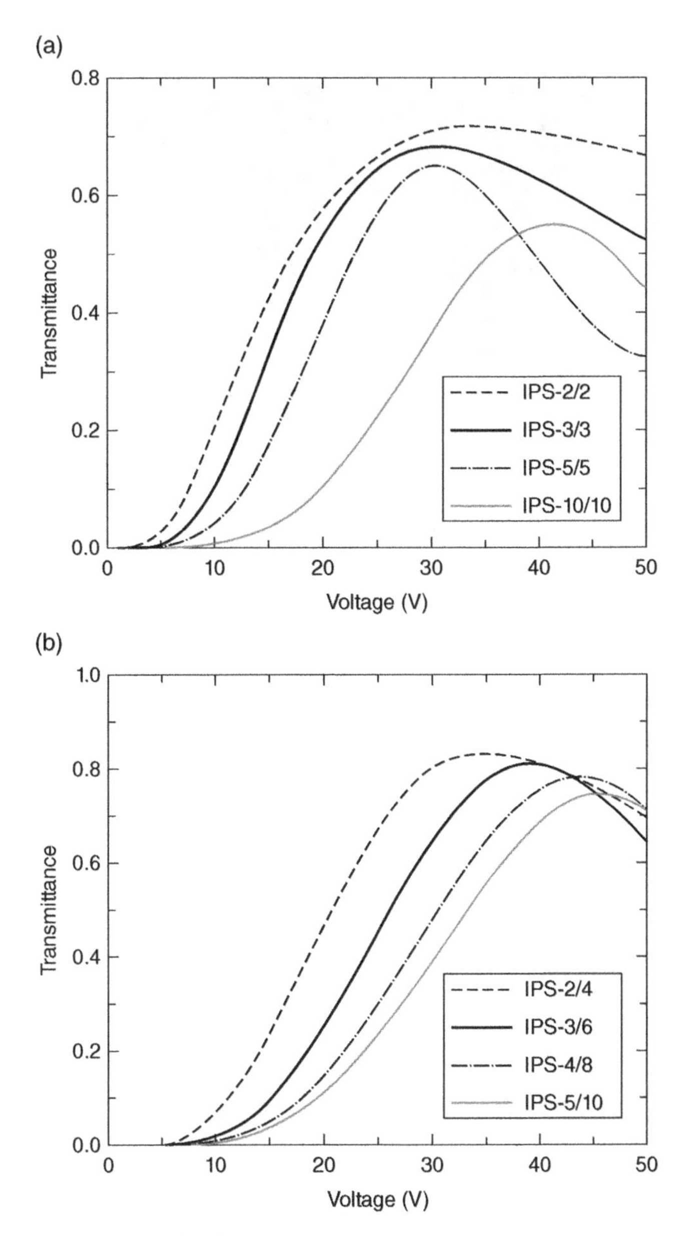

Figure 14.14 Simulated *VT* curves of IPS cells with (a) $l/w = 1$ and (b) $l/w = 2$ using JC-BP01 at 25°C and $\lambda = 550$ nm. Reproduced with permission from the Optical Society of America.

penetration depth of the electric field. This feature is particularly desirable for fabricating large LCD panels, in which uniform cell gap control is a big concern.

Figure 14.15(a) depicts the induced birefringence distribution of a 10 μm thick IPS-5/5 cell and Figure 14.15(b) shows the *VT* curves of the IPS-5/5 cell with different cell gaps. The material employed here is JC-BP01. From Figure 14.15(a), the induced birefringence is largest near

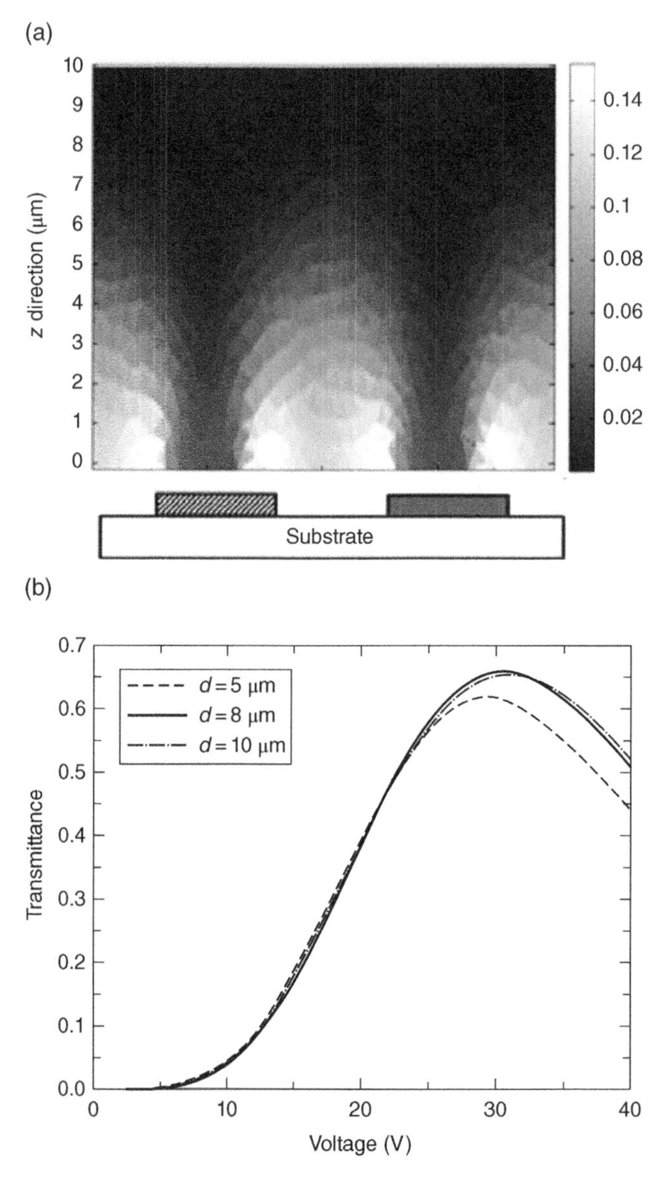

Figure 14.15 (a) Induced Δn profile of a 10 μm thick IPS-5/5 cell at 31 V and (b) simulated *VT* curves of IPS-5/5 with different cell gaps using JC-BP01 (25°C and $\lambda = 550$ nm). Reproduced with permission from the Optical Society of America.

the electrode surface and it gradually decreases as the distance increases. For JC-BP01, as the depth increases to 5.5 μm (from the bottom electrode) the induced birefringence decreases to 0.05. Hence, for IPS-BPLC the *VT* curve is insensitive to cell gap as long as the cell gap is larger than the field's penetration depth, which is governed by the electrode dimension through the Poisson equation. This is verified in Figure 14.15(b). Therefore, the *VT* curve of an IPS BPLC cell is insensitive to the cell gap variation, as long as the cell gap is above the field's penetration depth.

14.4.1.3 Saturated birefringence and saturation field effects

Ideally, for an IPS-BPLC cell to achieve 100% transmittance the required phase retardation $\delta = 2\pi d\Delta n_{ind}/\lambda$ should be equal to 1π. However, as Figure 14.15(a) shows, the electric field distribution in an IPS cell is not uniform in either horizontal or longitudinal direction. Therefore, we can only take spatial averaging to get an averaged transmittance. Moreover, the electric field penetration depth is limited, depending on the electrode dimension. A large induced birefringence, that is, large saturated birefringence Δn_s becomes critical. Meanwhile, to lower the operating voltage to below 10 V, a large dielectric anisotropy, or small saturation field E_s plays an equally important role.

Based on JC-BP01, Figure 14.16(a) shows the simulated *VT* curves of IPS-2/4 with different Δn_s values while keeping E_s unchanged ($E_s = 4.05$ V/μm). The dashed lines represent BPLC using JC-BP01. As Δn_s increases, phase retardation increases, resulting in a higher transmittance. Figure 14.16(b) shows how E_s affects the *VT* curves of IPS-2/4 based on JC-BP01. As E_s increases, the on-state voltage also increases while peak transmittance remains unchanged. This indicates that E_s only determines the voltage where the BPLC reaches its Δn_s, but the maximum induced Δn still remains unchanged at the on-state voltage. Therefore, in the IPS BPLC structure, a larger Δn_s helps reduce the operating voltage and enhance the transmittance, not only from increasing the Kerr constant on the material side but also from a stronger refraction effect in the cell. A smaller E_s is effective for lowering operating voltage but does not affect the peak transmittance. In order to achieve low voltage while keeping high transmittance, we should boost Δn_s while keeping E_s as low as possible.

14.4.2 *Protruded electrodes*

Based on the discussion above, we can enhance the transmittance by increasing the *l*/*w* ratio of an IPS structure. This is because a larger *l*/*w* ratio leads to a smaller dead zone area, and meanwhile it increases the electric field penetration depth. However, the major trade-off is increased voltage because of the wider electrode gap. An effective way to overcome this problem is to employ protrusion electrodes, which enable the horizontal electric fields to penetrate more deeply into the bulk LC layer. The detailed performance depends on the protrusion height and the *l*/*w* ratio.

Figure 14.7(a) depicts protruded rectangular electrodes with a height (h') from the bottom surface. Figures 14.7(b) and 14.7(c) show the simulated *VT* curves of planar and protruded IPS cells with protrusion height $h' = 1$ μm and 2 μm for IPS-2/4 and IPS-3/6 employing two different BPLC materials: JC-BP01 ($\Delta n_s = 0.154$, $E_s = 4.05$ V/μm) and JC-BP06 ($\Delta n_s = 0.09$, $E_s = 2.2$ V/μm). Compared to a planar IPS, the protruded IPS shows about the same transmittance because the protruded electrodes mainly generate electric fields in the electrode gaps and do not change the field distribution above the electrodes. But due to the dimension effect described above, IPS-3/6 exhibits a slightly lower transmittance than IPS-2/4 with the same protrusion height. By comparing these curves, we find that enhancing the protrusion height is an effective way to reduce the operating voltage. However, high-protrusion electrodes are more difficult to fabricate. On the other hand, JC-BP06 shows a higher voltage than JC-BP01 for planar IPS-2/4 structure (40 V cf. 34 V). This is because JC-BP06 has a relatively small induced birefringence ($\Delta n_s = 0.09$), and IPS-2/4 has a relatively shallow

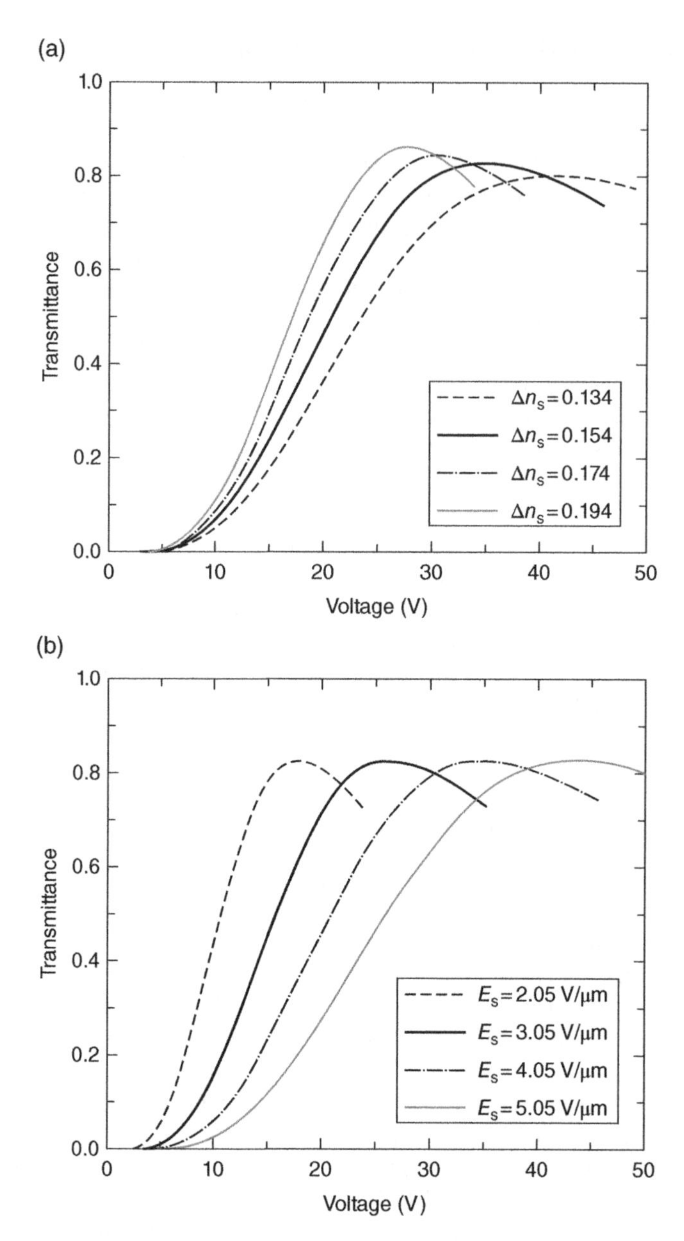

Figure 14.16 Simulated *VT* curves of IPS-2/4 using BPLCs with (a) different Δn_s but the same $E_s = 4.05$ V/μm, and (b) different E_s but the same $\Delta n_s = 0.154$ at 25°C and $\lambda = 550$ nm. Dashed lines: JC-BP01. Reproduced with permission from the Optical Society of America.

penetration depth. Thus, it requires a higher operating voltage to reach the peak transmittance. However, in the protrusion electrode configuration the relatively small Δn_s of JC-BP06 can be compensated by the increased penetration depth. Therefore, with the same device structure JC-BP06 exhibits a lower operating voltage than JC-BP01 because JC-BP06 has a larger Kerr

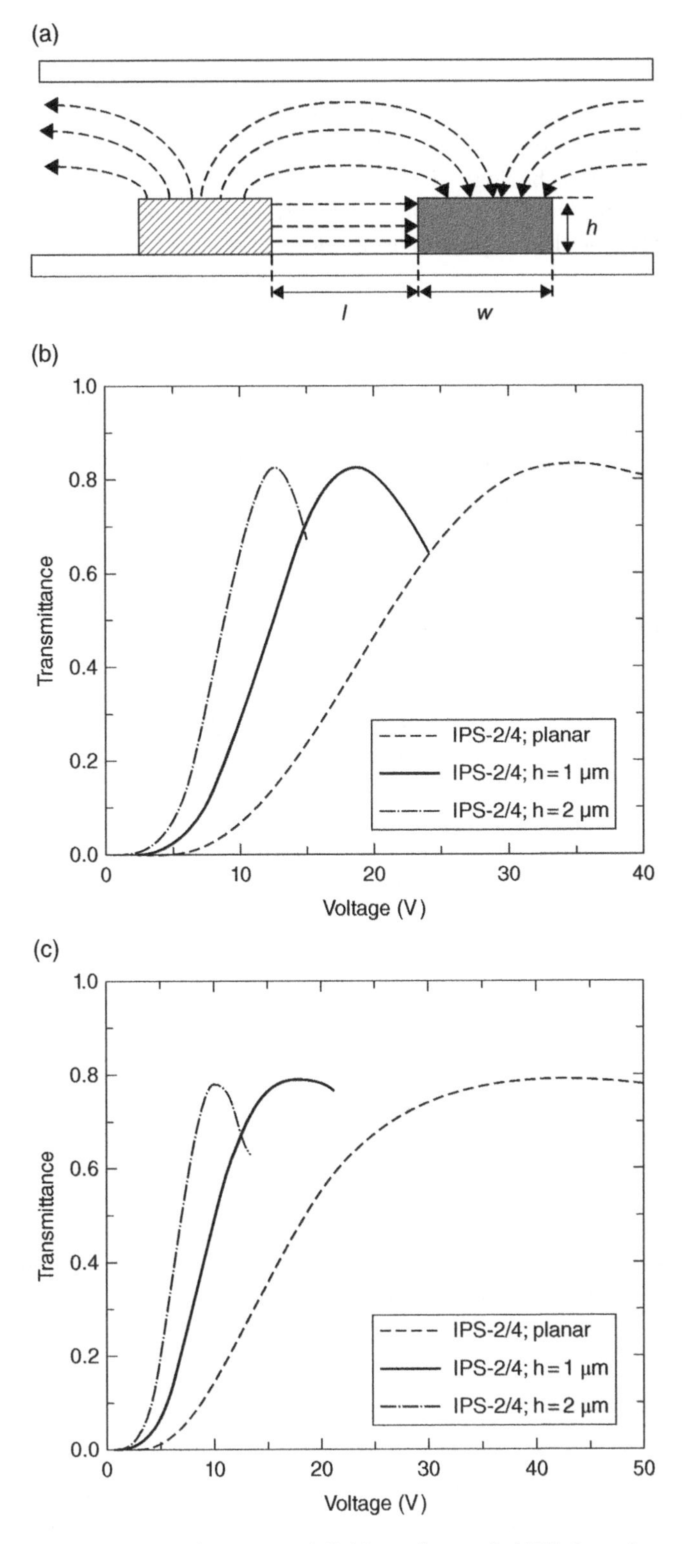

Figure 14.17 (a) Cell structure and parameter definitions of protruded IPS electrodes, and simulated *VT* curves of the protruded IPS-BPLC with different electrode dimensions employing (b) JC-BP01 and (c) JC-BP06 (25°C and $\lambda = 550$ nm). Reproduced with permission from the Optical Society of America.

constant. As Figure 14.17(c) shows, for IPS-2/4 with $h' = 2\ \mu m$ its operating voltage is reduced to 10 V while keeping 80% transmittance. This is an important milestone to enable BPLC to be addressed by a-Si TFT technology. To boost transmittance further, a higher birefringence BPLC host can be considered. For example, as Figure 14.16(b) shows, JC-BP01 has a higher induced birefringence so that its peak transmittance can reach 83%.

14.4.3 Etched electrodes

In contrast to protruded electrodes, etching is another way to create deeper penetrating field for lowering the operating voltage [55]. Figure 14.18(a) shows an IPS cell with an etched depth (h'). Taking IPS-2/4 as an example, the etching takes place along the 4 μm electrode gaps. As a result, the fringe fields occur both above and under the 2 μm ITO electrodes. These doubled penetrating fringe fields help to reduce the operating voltage. Similar to a protruded IPS, etched IPS using JC-BP06 also shows lower operating voltage than that using JC-BP01, since the bottom fringe fields provide an extra phase retardation to compensate for the relatively small Δn_s of JC-BP06. Therefore, let us use JC-BP06 as an example to demonstrate the effectiveness of this etched electrode approach.

Figure 14.18(b) compares the simulated *VT* curves of IPS-2/4 structure with h' increased from 0, 1, 2, to 4 μm. As the etching depth increases, the operating voltage decreases rapidly and then saturates gradually. With $h' \sim 2\ \mu m$, the operating voltage is reduced to ~10 V_{rms}. This saturation phenomenon originates from the finite penetration depth of the electric field. Hence, when the etching depth is larger than the penetration depth, the operating voltage does not continue to decrease anymore. This feature makes the etched-electrode IPS easy to fabricate because the etching depth does not need to be controlled precisely, as long as it is larger than the penetration depth. Similar to the protruded IPS, the etched electrodes have a minor influence on the transmittance, since the electric field distribution on the top of the electrode is not changed.

To improve transmittance, we can increase the *l*/*w* ratio, such as IPS-2/6 or IPS-2/8, while still using the etched electrodes. Similar to etched IPS-2/4, both etched IPS-2/6 and IPS-2/8 also exhibit similar saturation effect, but at a deeper etching depth (4 μm for IPS-2/6 and 6 μm for IPS-2/8) due to their larger penetration depths. Figure 14.18(c) depicts the simulated *VT* curves of etched IPS-2/6 and IPS-2/8 cells with 4 μm and 6 μm etching depths, respectively. Compared to IPS-2/4, their transmittance can exceed 80% due to their larger *l*/*w* ratios, but the required operating voltage is somewhat higher. As nanotechnology continues to advance, smaller electrode dimension will one day be fabricated with high yield. If the electrode width can be reduced to 1 μm, then the operating voltage as low as 8 V and transmittance higher than 80% will be achieved, as Figure 14.18(c) shows.

14.4.4 Single gamma curve

In an IPS-BPLC cell, the phase retardation depends on the wavelength and induced birefringence. The latter in turn depends on the Kerr constant, which decreases as the wavelength increases. Therefore, as plotted in Figure 14.19(a) the *VT* curves of an IPS-BPLC depend on the wavelength, and three gamma curves are required to drive the red (R = 650 nm), green

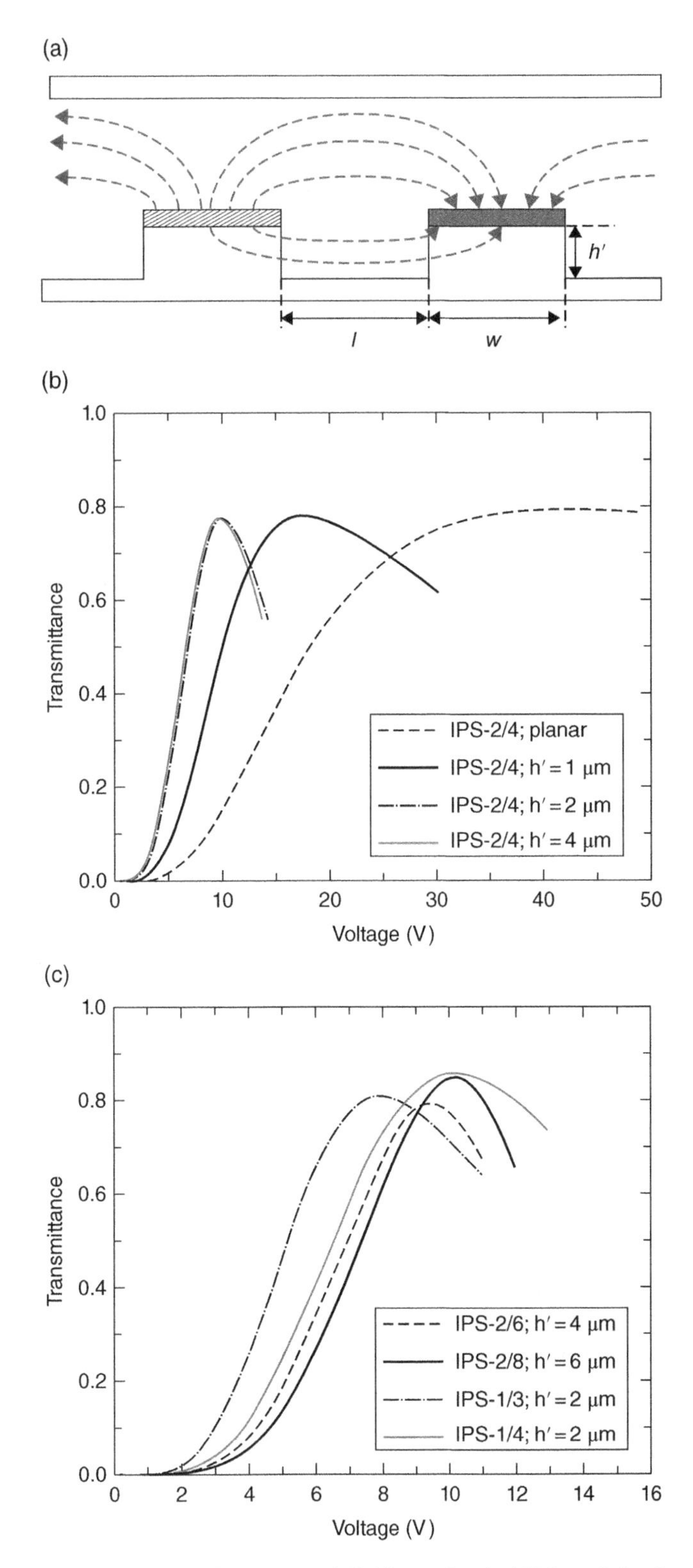

Figure 14.18 (a) Cell structure and parameter definitions of etched-IPS, and simulated VT curves of etched-IPS cells with different electrode dimensions: (b) $l/w = 2$ and (c) $l/w = 3$ and 4, using JC-BP06 (25°C and $\lambda = 550$ nm). Reproduced with permission from the Optical Society of America.

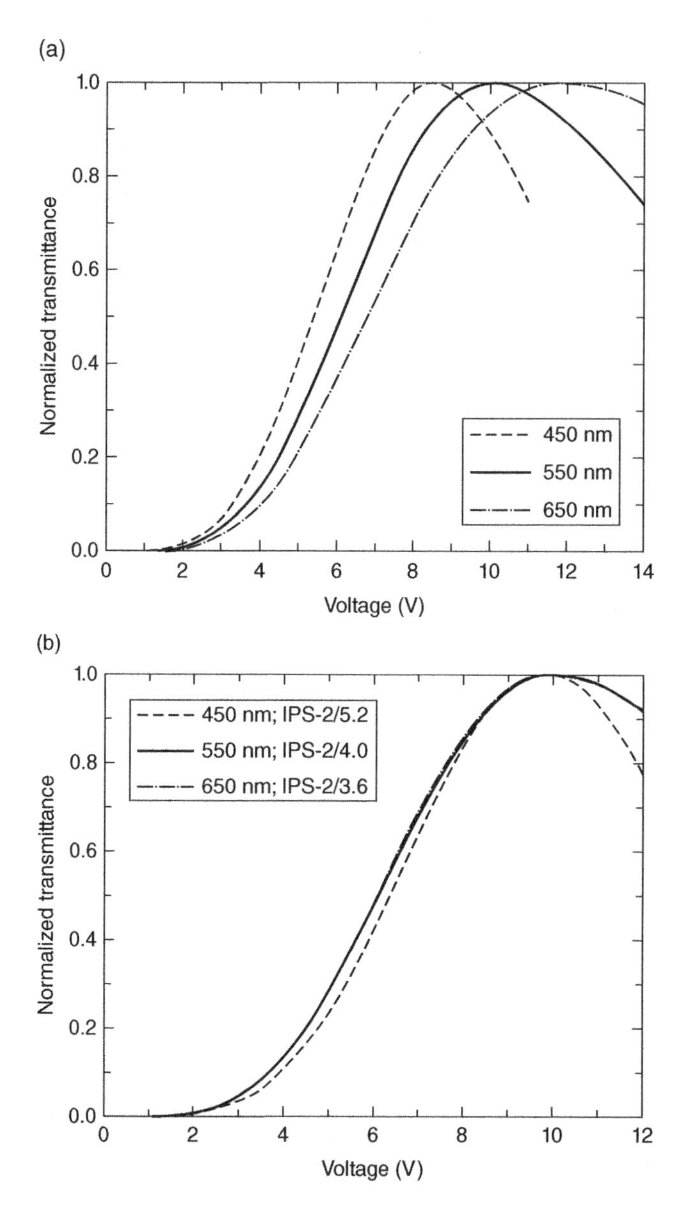

Figure 14.19 (a) Simulated *VT* curves for the specified RGB wavelengths. BPLC cell: etched IPS-2/4 with $h = 2.5$ μm and JC-BP06, and (b) simulated VT curves for IPS-2/3.6 (R), IPS-2/4 (G), IPS-2/5.2 (B) with $h = 2.5$ μm and $d = 7.5$ μm. Reproduced with permission from the Optical Society of America.

(G = 550 nm) and blue (B = 450 nm) subpixels, which increases the complexity of driving electronics.

To obtain a single gamma curve, Xu et al. proposed an interesting approach by varying the electrode gap (l) for RGB subpixels while keeping the same electrode width (w) and etching depth (h) [56]. Let us continue to use IPS-2/4 with $h = 2.5$ μm and $d = 7.5$ μm for the green wavelength to illustrate the design principles. From Figure 14.19(a), we need to lower the on-state voltage for the red color. Thus, we choose a smaller electrode gap, say $l = 3.6\ \mu m$, while still keeping $h = 2.5\ \mu m$ and $d = 7.5\ \mu m$. On the other hand, we increase l to $5.2\ \mu m$ for the blue wavelength. Figure 14.19(b) shows the normalized *VT* curves for the unequally spaced IPS cell. Indeed, the RGB gamma curves overlap fairly well.

14.5 Vertical Field Switching

In an IPS cell, the electric field is mainly in the lateral direction but it is quite non-uniform spatially. Near the electrode edges the field intensity is particularly strong, which could deform the lattice structure and cause hysteresis. To suppress hysteresis, Cheng et al. proposed a vertical field switching (VFS) mode [57], in which the electric field is in the longitudinal direction and is uniform. By using a thin cell gap and a large oblique incident angle (70°), the operating voltage can be reduced to below 10 V. The uniform electric field also helps to suppress hysteresis and shorten response time.

14.5.1 Device structure

Figure 14.20 depicts a device configuration of VFS BPLC. Unlike with an IPS cell, the electric field inside the VFS cell is in the *longitudinal* direction, and only the incident light at an oblique angle ($0 < \theta < 90°$) can experience the phase retardation effect. For a given BPLC layer thickness, a larger incident angle results in a larger phase retardation, which is helpful for lowering the operating voltage. However, as will be discussed later, a larger incident angle requires a more sophisticated top coupling film for achieving wide viewing angle.

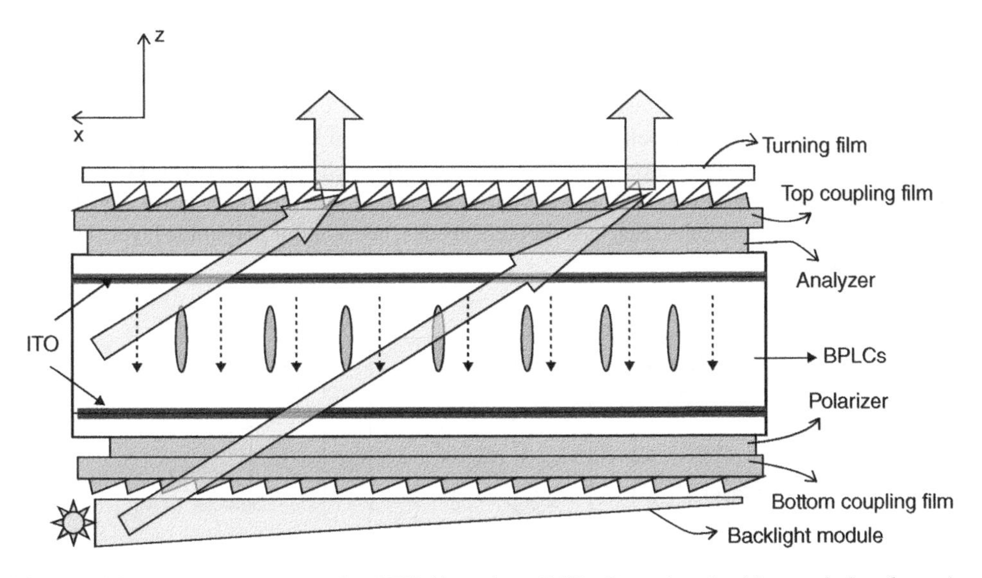

Figure 14.20 Device structure of a VFS blue phase LCD. Reproduced with permission from the American Physical Society.

Since a larger incident angle in the BPLC layer results in a larger phase retardation – which helps to reduce the operating voltage – we can choose a well-collimated directional backlight with 70° incident angle and ±5° divergence angle [58]. The bottom coupling film is designed to substantially couple the oblique incident light to the bottom substrate and the BPLC layer. The prismatic structure not only couples the oblique input light to the cell but also keeps a large incident angle in the BPLC layer. Without the bottom coupling film, according to Snell's law the refraction angle in the BPLC layer will be reduced dramatically. Subsequently, the phase retardation will be smaller and the operating voltage will be higher. Therefore, the design optimization of the prism structure greatly affects the performance of VFS mode. The prism pitch should be smaller than the pixel size of the LCD; it could range from ~5 μm to ~50 μm. The turning film steers the output light to viewer's direction. Without this top coupling film, the oblique light will be trapped in the cell module because of total internal reflection (TIR). Therefore, the major function of the top coupling film is to couple the oblique light to the air while keeping a wide viewing angle and uniform brightness [59].

14.5.2 *Experiments and simulations*

To simulate the large incident angle, Figure 14.21 depicts the experimental setup for exploring the electro-optic properties of a VFS device. The VFS BPLC cell is immersed in a transparent container filled with an index-matching fluid glycerol (n = 1.47@ $\lambda = 633$ nm) and it can be rotated freely. Because of matched index between glass and glycerol, the light can pass through the BPLC without refraction at a very large angle.To compare the performance of a VFS cell with an IPS cell, a polymer-stabilized BPLC material using JC-BP01M host was employed. The LC host has a dielectric anisotropy $\Delta\varepsilon \sim 94$ and birefringence $\Delta n \sim 0.17$. The phase transition temperature of the precursor is BP 42.4°C N* during the cooling process and N* 44.5°C BP during the heating process, where N* denotes chiral nematic phase. The phase transition temperature between isotropic phase and blue phase was not easy to determine precisely because Bragg reflection occurred at ~350 nm. UV stabilization curing process was performed at 44°C for 30 min. with an intensity of 2 mW/cm^2. After UV curing, the blue-phase temperature range was widened from below 0°C to ~70°C. To make a fair comparison, both IPS and VFS cells are filled with the same BPLC material. The IPS cell has patterned ITO electrodes with 10 μm electrode width and 10 μm electrode gap, and 7.5 μm cell gap. For the VFS cell, both top and bottom glass substrates have ITO electrodes, but without polyimide layer. The cell gap is $d \sim 5.74$ μm.

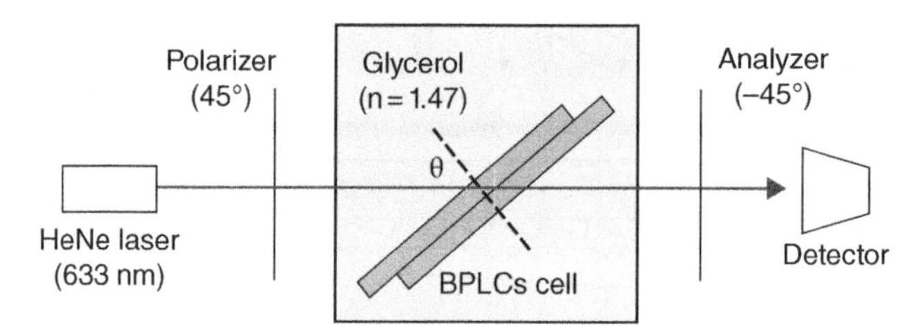

Figure 14.21 Experimental setup for characterizing the VFS cell. Reproduced with permission from the American Physical Society.

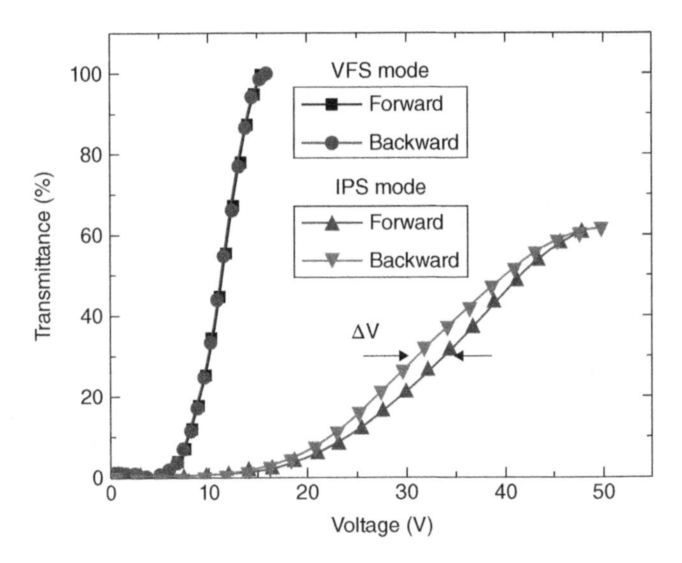

Figure 14.22 Measured *VT* curves and hysteresis of IPS and VFS cells. $\lambda = 633$ nm. Reproduced with permission from IEEE.

Figure 14.22 compares the measured voltage-dependent transmittance (*VT*) curves (at $\lambda = 633$ nm and $T \sim 23°C$) of the IPS cell at normal incidence and VFS cell at $\theta = 70°$. For the IPS cell, the peak voltage occurs at $V_p \sim 50\ V_{rms}$. For the VFS cell at $\theta = 70°$, its $V_p \sim 16\ V_{rms}$ which is about 3.2 times lower than that of the IPS structure.For a smaller θ, the gain factor will decrease accordingly.

Hysteresis is a common phenomenon for polymer-stabilized LCDs. For display applications, hysteresis affects the accuracy of gray-scale control and should be minimized. Hysteresis is defined by the voltage difference (ΔV) at half-maximum transmittance between forward and backward scans. From Figure 14.22, the measured $\Delta V/V_p$ is ~5.8% for the IPS cell, but it is nearly free for the VFS cell. The reason is that the VFS cell has a much lower operating voltage, which corresponds to an electric field $E \sim 2.8$ V/μm so that the electrostriction effect does not occur. On the contrary, in an IPS cell the generated electric fields are not spatially uniform. The electric fields are much stronger near the electrode edges than those in the electrode gap [60]. The peak electric fields could cause lattice deformation locally, which results in a noticeable hysteresis.

Residual birefringence is another serious problem for BPLC as it affects the long-term device operation reliability. When this occurs, the dark state light leakage accumulates with time, which deteriorates the contrast ratio. In an IPS cell, the residual birefringence arises in the region with strong field. In a VFS cell, the required field intensity for producing a π-phase retardation is relatively weak. Thus, the resultant residual birefringence is negligible.

Unlike an IPS cell, whose *VT* curve is insensitive to the cell gap, the peak transmittance voltage of our VFS cell is dependent on the cell gap. Here, two factors have to be considered: induced birefringence and effective cell gap. At a given voltage, as the cell gap decreases the electric field gets stronger because $E = V/d$. According to the Kerr effect, the induced birefringence is proportional to E^2. Thus, the induced birefringence is larger as the cell gap decreases. On the other hand, for a given incident angle, the beam path in the BPLC cell for accumulating phase retardation decreases as the cell gap decreases. Therefore, there ought to be an optimal cell gap for achieving the desired phase retardation. If the cell gap is too thin, although the induced birefringence is large, the optical beam path is decreased so that the accumulated phase may not be adequate for achieving maximum transmittance. Moreover, the saturation phenomenon in the high-field region should also be considered. In a real situation, the induced birefringence will gradually saturate as the electric field increases. Therefore, the driving voltage could not decrease indefinitely for an ultra-thin cell gap, as described by the extended Kerr effect equation.

14.6 Phase Modulation

In addition to amplitude modulation (displays), blue phase can also be used as a phase-only modulator. In particular, the isotropic-to-anisotropic transition of BPLC is insensitive to the incident light polarization. Based on this principle, some polarization-insensitive tunable-focus lenses have been demonstrated [61–63]. However, the induced refractive index change $(n_i - n_o)$ is three times smaller than that of a corresponding nematic LC $(n_e - n_o)$ so that the resultant focal length is relatively large.

Another unique property of BPLC is its short coherence length, in the order of 100 nm. In an IPS-based BPLC cell, the electric field-induced phase profile is sharp, while the electric field above the electrodes is fairly weak. So a rectangular-like phase grating is formed [64]. The first-order diffraction efficiency reaches ~40%, which is approaching to the 41% theoretical limit. By stacking two IPS electrodes in orthogonal direction, a two-dimensional polarization-independent phase grating has been demonstrated [65].

References

1. J. Yan, L. Rao, M. Jiao, et al., Polymer-stabilized optically isotropic liquid crystals for next-generation display and photonic applications, *J. Mater. Chem.* **21**, 7870–7877 (2011).
2. F. Reinitzer, Beiträge zur Kenntniss des Cholestherins, *Monatsh Chem.* **9**, 421–441 (1888).
3. P. P. Crooker, *Chirality in Liquid Crystals,* ch. 7 (Springer, New York, 2001).
4. A. Saupe, On molecular structure and physical properties of thermotropic liquid crystals, *Mol. Cryst. Liq. Cryst.* **7**, 59–74 (1969).
5. S. A. Brazovskii and S. G. Dmitriev, Phase transitions in cholesteric liquid crystals, *Zh. Eksp. Teor. Fiz.* **69**, 979–989 (1975).
6. R. M. Hornreich and S. Shtrikman, *Liquid Crystals of One- and Two- Dimensional Order* (Springer-Verlag, Berlin, 1980).
7. S. Meiboom, J. P. Sethna, W. P. Anderson, and W. F. Brinkman, Theory of the blue phase cholesteric liquid crystals, *Phys. Rev. Lett.* **46**, 1216–1219 (1981).
8. E. Dubois-Violette and B. Pansu, Frustration and related topology of blue phases, *Mol. Cryst. Liq. Cryst.* **165**, 151–182 (1988).
9. D. L. Johnson, J. H. Flack, and P. P. Crooker, Structure and properties of the cholesteric blue phases, *Phys. Rev. Lett.* **45**, 641–644 (1980).
10. P. E. Cladis, T. Garel, and P. Pieranski, Kossel diagrams show electric-field-induced cubic-tetragonal structural transition in frustrated liquid-crystal blue phases, *Phys. Rev. Lett.* **57**, 2841–2844 (1986).
11. R. J. Miller and H. F. Gleeson,Order parameter measurements from the Kossel diagrams of the liquid-crystal blue phases, *Phys. Rev.* **E52**, 5011–5016 (1995).
12. H. Kikuchi, *Liquid Crystalline Blue Phases* pp. 99–117 (Springer Berlin, Heidelberg, 2008).
13. P. G. De Gennes, and J. Prost, *The Physics of Liquid Crystals, 2nd edn* (Clarendon, Oxford, 1993).

14. For a review, see A. Yoshizawa, *RSC Adv.* **3**, 25475–25497 (2013).
15. H. S. Kitzerow, H. Schmid, A. Ranft, et al.,Observation of blue phases in polymer networks, *Liq. Cryst.* **14**, 911–916 (1993).
16. H. Kikuchi, M. Yokota, Y. Hisakado, et al., Polymer-stabilized liquid crystal blue phases, *Nat. Mater.* **1**, 64–68 (2002).
17. Y. Haseba, H. Kikuchi, T. Nagamura, and T. Kajiyama, Large electro-optic Kerr effect in nano-structured chiral liquid-crystal composites over a wide temperature range, *Adv. Mater.* **17**, 2311 (2005).
18. H. J. Coles and M. N. Pivnenko, Liquid crystal blue phases with a wide temperature range, *Nature* **436**, 997–1000 (2005).
19. J. Yan and S. T. Wu, Polymer-stabilized blue phase liquid crystals: A tutorial *Opt. Materials Express* **1**, 1527–1535 (2011).
20. L. Rao, J. Yan, and S. T. Wu, Prospects of emerging polymer-stabilized blue-phase liquid crystal displays, *J. Soc. Inf. Display* **18**, 954–959 (2010).
21. P. R. Gerber, Electro-optical effects of a small-pitch blue-phase system, *Mol. Cryst. Liq. Cryst.* **116**, 197–206 (1985).
22. S. W. Choi, S. Yamamoto, Y. Haseba, et al., *Appl. Phys. Lett.* **92**, 043119 (2008).
23. M. Hird, Fluorinated liquid crystals – properties and applications, *Chem. Soc. Rev.* **36**, 2070–2095 (2007).
24. J. Yan, Z. Luo, S.T. Wu, et al., Low voltage and high contrast blue phase liquid crystal with red-shifted Bragg reflection, *Appl. Phys. Lett.* **102**, 011113 (2013).
25. H. F. Gleeson and H. J. Coles, Dynamic properties of blue-phase mixtures, *Liq. Cryst.* **5**, 917–926 (1989).
26. L. Rao, J. Yan, S. T. Wu, et al., A large Kerr constant polymer-stabilized blue phase liquid crystal, *Appl. Phys. Lett.* **98**, 081109 (2011).
27. M. Wittek, N. Tanaka, M. Bremer, et al., New materials for polymer-stabilized blue phase, *SID Int. Symp. Digest Tech. Papers* **42**, 292–293 (2011).
28. J. Yan and S. T. Wu, Effect of polymer concentration and composition on polymer-stabilized blue-phase liquid crystals, *J. Display Technol.* **7**, 490–493 (2011).
29. Y. Chen, J. Yan, J. Sun, et al., A microsecond-response polymer-stabilized blue phase liquid crystals, *Appl. Phys. Lett.* **99**, 201105 (2011).
30. T. N. Oo, T. Mizunuma, Y. Nagano, et al., *Opt. Mater. Express* **1**, 1502–1510 (2011).
31. J. Kerr, A new relation between electricity and light: Dielectrified media birefringent, *Philos. Mag.* **50**, 337 (1875).
32. A. Yariv and P. Yeh, *Optical Waves in Crystal: Propagation and Control of Laser Retardation* (Wiley, Hoboken, 2002).
33. G. Heppke, B. Jerome, H. S. Kitzerow and P. Pieranski, Electrostriction of the cholesteric blue phases BPI and BPII in mixtures with positive dielectric anisotropy, *J. Phys. France* **50**, 2991–2998 (1989).
34. H. Stegemeyer and F. Porsch, Electric field effect on phase transitions in liquid-crystalline blue-phase systems, *Phys. Rev. A* **30**, 3369–3371 (1984).
35. Z. Ge, S. Gauza, M. Jiao, et al., Electro-optics of polymer-stabilized blue phase liquid crystal displays, *Appl. Phys. Lett.* **94**, 101104 (2009).
36. Z. Ge, L. Rao, S. Gauza and S.-T. Wu, Modeling of blue phase liquid crystal displays, *J. Disp. Technol.* **5**, 250–256 (2009).
37. J. Yan, M. Jiao, L. Rao, and S. T. Wu, Direct measurement of electric-field-induced birefringence in a polymer-stabilized blue-phase liquid crystal composite, *Opt. Express* **18**, 11450–5 (2010).
38. W. Jamroz, J. Karniewicz, and W. Kucharczyk, The Electro-Optical Effect of the Fourth Order, *J. Phys. D: Appl. Phys.* **11**, 2625 (1978).
39. J. Yan, H. C. Cheng, S. Gauza, et al., Extended Kerr effect of polymer-stabilized blue-phase liquid crystals, *Appl. Phys. Lett.* **96**, 071105 (2010).
40. M. Jiao, J. Yan, and S. T. Wu, Dispersion relation on the Kerr constant of a polymer-dispersed optically isotropic liquid crystal, *Phys. Rev.E*, **83**, 041706 (2011).
41. S. T. Wu, Birefringence dispersions of liquid-crystals, *Phys. Rev. A* **33**, 1270–1274 (1986).
42. K. S. Cole and R. H. Cole, Dispersion and absorption in dielectrics. I. Alternating current characteristics, *J. Chem. Phys.* **9**, 341–351 (1941).
43. T. K. Bose, B. Campbell, S. Yagihara, and J. Thoen, *Phys. Rev. A*, **36**, 5767 (1987).
44. Y. Li, Y. Chen, J. Sun, et al., Dielectric dispersion on the Kerr constant of blue phase liquid crystals, *Appl. Phys. Lett.* **99**, 181126 (2011).

45. Y. Chen, and S.T. Wu, "Recent advances on polymer-stabilized blue phase liquid crystal materials and devices", *J. Appl. Polym. Sci.*, **131**, 40556 (2014).
46. J. Yan, Y. Chen, S. T. Wu, and X. Song, Figure of merit of polymer-stabilized blue phase liquid crystals, *J. Display Technol.* **9**, 24–29 (2013).
47. F. Peng, Y. Chen, J. Yuan, et al., Low temperature and high frequency effects of polymer-stabilized blue phase liquid crystals with a large dielectric anisotropy, *J. Mater. Chem.* C. **2**, 3597–3601 (2014). DOI: 10.1039/c4tc00115j
48. K. M. Chen, S. Gauza, H. Xianyu, and S. T. Wu, Hysteresis effects in blue-phase liquid crystals, *J. Display Technol.* **6**, 318–322 (2010).
49. C. D. Tu, C. L. Lin, J. Yan, et al., Driving scheme using bootstrapping method for blue-phase LCDs, *J. Display Technol.* **9**, 3–6 (2013).
50. J. F. Wager, D. A. Keszler, and R. E. Presley, *Transparent Electronics* (Springer, 2008).
51. L. Rao, Z. Ge, S. T. Wu, and S. H. Lee, Low voltage blue-phase liquid crystal displays, *Appl. Phys. Lett.* **95**, 231101 (2009).
52. H. C. Cheng, J. Yan, T. Ishinabe, and S. T. Wu, Vertical field switching for blue-phase liquid crystal devices, *Appl. Phys. Lett.* **98**, 261102 (2011).
53. Y. Chen, D. Xu, S.T. Wu, et al., A low voltage and submillisecond-response polymer-stabilized blue phase liquid crystal, *Appl. Phys. Lett.* **102**, 141116 (2013).
54. K. M. Chen, J. Yan, S. T. Wu, et al., Electrode dimension effects on blue-phase liquid crystal displays, *J. Display Technol.* **7**, 362–364 (2011).
55. L. Rao, H. C. Cheng, and S. T. Wu, Low voltage blue-phase LCDs with double-penetrating fringe fields, *J. Display Technology* **6**, 287–289 (2010).
56. D. Xu, Y. Chen, Y. Liu, and S.T. Wu, Refraction effect in an in-plane-switching blue phase liquid crystal cell, *Opt. Express* **21**, 24721–24735 (2013).
57. H. C. Cheng, J. Yan, T. Ishinabe, et al., Blue-phase liquid crystal displays with vertical field switching, *J. Display Technol.* **8**, 98–103 (2012).
58. K. Käläntär, A monolithic segmented functional lightguide for 2-D dimming LCD backlight, *J. Soc. Inf. Display*, **19**, 37–47 (2011).
59. J. Yan, D. Xu, H. C. Cheng, et al.,Turning film for widening the viewing angle of a blue phase liquid crystal display, *Appl. Opt.* **52**, 8840–8844 (2013).
60. L. Rao, J. Yan, S. T. Wu, et al., Critical field for a hysteresis-free blue-phase liquid crystal device, *J. Display Technol.* **7**, 627–629 (2011).
61. Y. H. Lin, H. S. Chen, H. C. Lin, et al., Polarizer-free and fast response microlens arrays using polymer-stabilized blue phase liquid crystals, *Appl. Phys. Lett.* **96**, 113505 (2010).
62. Y. Li and S. T. Wu, Polarization independent adaptive microlens with a blue-phase liquid crystal, *Opt. Express* **19**, 8045–8050 (2011).
63. Y. Li, Y. Liu, Q. Li, and S. T. Wu, Polarization independent blue-phase liquid crystal cylindrical lens with a resistive film, *Appl. Opt.* **51**, 2568–2572 (2012).
64. J. Yan, Y. Li, and S. T. Wu, High-efficiency and fast-response tunable phase grating using a blue phase liquid crystal, *Opt. Lett.* **36**, 1404–1406 (2011).
65. G. Zhu, J. N. Li, X. W. Lin, et al., Polarization-independent blue-phase liquid crystal gratings driven by vertical electric field, *J. Soc. Inf. Disp.* **20**, 341–346 (2012).

15

Liquid Crystal Display Components

15.1 Introduction

Liquid crystal displays (LCDs) have found enormous success in the past couple of decades. They are used everywhere, from cellular phones, ebooks, GPS devices, computer monitors, and automotive displays to projectors and TVs to name a few. They play a critical role in the information age and are import elements of our daily life. In LCDs, besides liquid crystal, there are other important components. A typical LCD system is schematically shown in Figure 15.1. We will discuss some of the components in this chapter.

15.2 Light Source

Liquid crystals do not emit light. Their function is to modify the state of light produced by a light source in order to display images. The light is produced by either a direct backlight, which is placed directly beneath the liquid crystal panel, or edge light which is placed at the edge of a waveguide sheet [1]. Backlight is more suitable for large-size LCDs, because it can provide high light intensities, but it is bulky. Edge light is more suitable for small-size handheld LCDs, because it is compact, but its light output is limited. The common light sources for LCD lighting are cold cathode fluorescentlamps (CCFL), light emitting diodes (LED), external electrode fluorescent lamps (EEFL), and flat fluorescent lamps (FFL).

CCFL consists of a glass tube with a cathode and an anode at the ends [2,3]. The tube is filled with mercury gas. The inner surface of the tube is coated with a fluorescent (phosphor) material. When a voltage is applied across the two electrodes, some (primary) electrons are emitted by thermal motion in the cathode and accelerated toward the anode. In the path from the cathode to

Fundamentals of Liquid Crystal Devices, Second Edition. Deng-Ke Yang and Shin-Tson Wu.

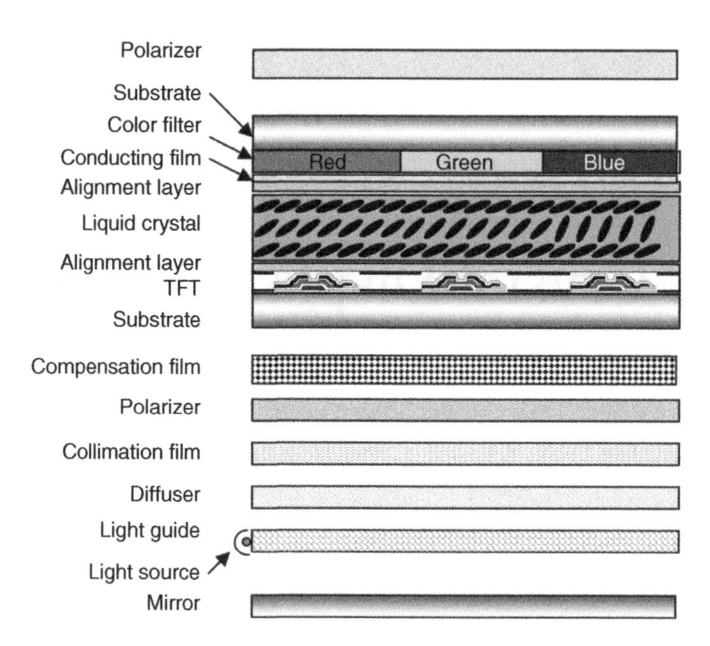

Figure 15.1 Schematic diagram of a active matrix liquid crystal display.

the anode, the electrons collide with the vapor molecules and may generate electrons and positively charged atoms. Under the applied electric field, the atoms fly to the cathode. When they collide with the cathode, more (secondary) electrons are produced. These electrons are accelerated toward the anode and cause more collisions. The collision between an electron and a mercury atom may excite the atom. When the excited atom relaxes back to lower energy states, UV light is emitted. The UV light hits the fluorescent material to generate white light. A typical spectrum of CCFL, shown in Figure 15.2 [4,5], has three primary intensity peaks located at red, green, and blue light wavelength, and can match well with the spectrum of color filters. CCFL has a long tube shape. Multiple CCFL tubes are used in one LCD. The spatial light intensity distribution is not uniform and thus a diffuser is usually used to make the light intensity uniform. EEFL is similar in its light emission mechanism to CCFL, except that its electrodes are outside of the glass tube.

Flat fluorescent lamp (FFL) is more suitable for backlight of large size LCD TVs [6]. It has a flat rectangular shape instead of a tube shape. It consists of two parallel substrates with mercury gas (or xenon gas) sandwiched between them. The inner surfaces are coated with fluorescent (phosphor) material. The operating mechanism is similar to that of CCFL. The electrodes are at the edges of the rectangle. One of the inner surfaces is flat while the other inner surface is grooved to maintain the continuous discharge of the excited mercury atoms.

Light emitting diode (LED) is made from semiconductor material doped with impurities to create a p-n junction [7]. When a voltage is applied cross the p-n junction, electrons and holes are generated. When an electron and a hole recombine, the energy is released as a photon, producing light. The LED has many advantages from high electric energy to light conversion efficiency, fast switching, and long lifetime.

The light emitted by an LED is usually colored with a bandwidth of a few tens of nanometers. The color of a LED depends on the semiconductor material. For example, gallium arsenide

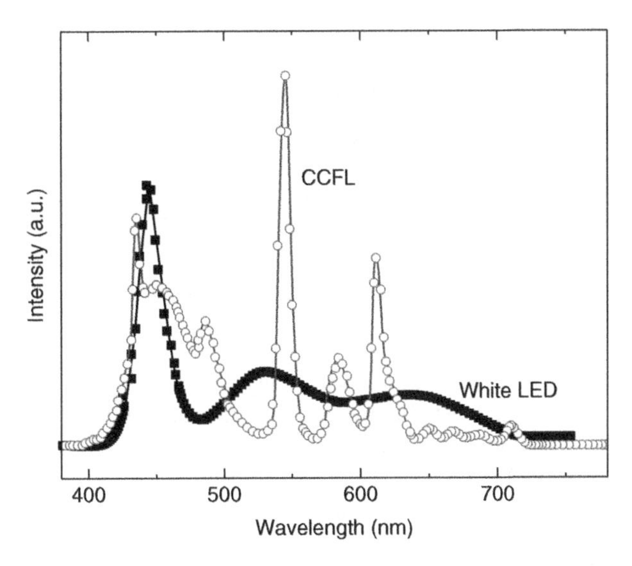

Figure 15.2 Spectrum of CCFL and LED © J. of Korean Inst. of Illum. and Elec. Install. Engineers.

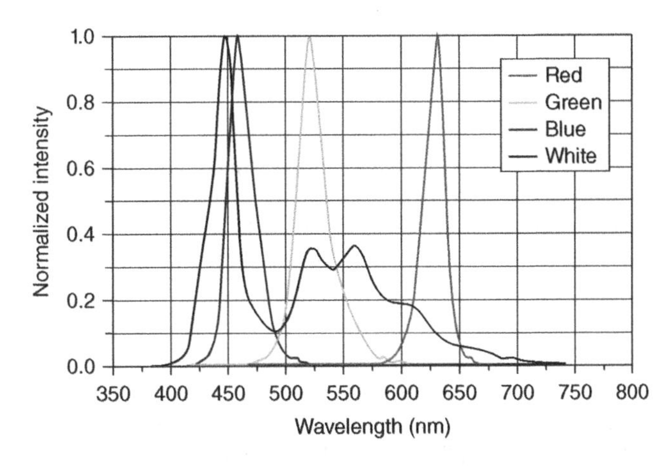

Figure 15.3 Spectra of red, green, blue, and white LEDs © www.growblu.com.

phosphide emits red light, gallium phosphide emits green light, and indium gallium nitride emits blue light. White light is produced when a material, which emits UV (or blue) light, is doped with yellow phosphor. The spectra of red, green, blue, and white LEDs are shown in Figure 15.3 [8]. When LED is used for the backlight unit in LCDs, many separated LEDs are mounted beneath the LCD panel [9–11]. A diffuser sheet is placed on top of the LEDs to make the light intensity uniform.

The switching time of LEDs is very fast, usually about 1 μs or less. This fast switching makes it possible for LEDs to be used in color sequential display and adaptive dimming [12]. In color sequential display, RGB colors in time domain are used to generate full color images [13,14], in contrast to regular displays where color filters in spatial domain are used. Each frame is divided

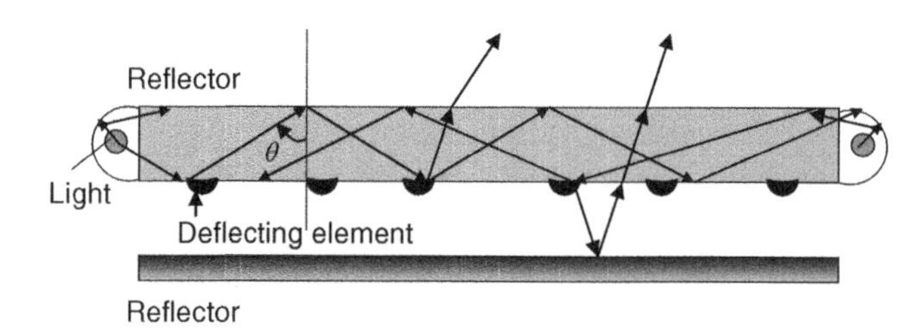

Figure 15.4 Schematic diagram of light-guide in LCD.

into three subframes for R, G, and B color, respectively. During one subframe, only LEDs with one of the three colors are turned on to illuminate the LCD panel. The energy efficiency of color sequential displays is at least three times higher than displays with color filters. In the adaptive dimming technology, the LED backlight is divided into multiple blocks which can be turned on or off independently. When an image has a spatial part in (near-) dark state, then less power is supplied to the LEDs behind that part to dim the light. This dimming not only increases the contrast ratio of the image, but also reduces the consumed electric energy.

15.3 Light-guide

For edge lighting, a light-guide plate is needed as shown in Figure 15.4 [15–17]. The light produced by the edge light is coupled into the light-guide plate. When a light ray hits the surface of the plate with an incident angle $\theta > \theta_c = \arcsin(n_a/n)$, where $n_a(=1)$ is the refractive index of air and n (≈ 1.5) is the refractive index of the plate, it will be total internally reflected back to the plate. This is the mechanism of light guiding. In the light-guide plate (or on the surface) there are deflecting (scattering) elements. When a light ray hits the deflecting element, it will be scattered in all directions. If the scattered light hits the surface of the plate with an incident angle $\theta < \theta_c$, it will come out of the plate. Some light is scattered upward to the LCD panel and some light is scattered downward. There is a reflector below the light-guide plate which reflects downward scattered light toward the LCD panel. The distribution of the deflecting elements must be controlled to achieve spatially uniform light intensity.

15.4 Diffuser

Diffusers are usually used in LCDs to achieve uniform light intensity distribution. There are two types of diffusers: bulk and surface [18–21]. The bulk diffuser consists of a transparent polymer film with dispersed inorganic particles, as shown in Figure 15.5(a). The size of the particles is usually 2–8 μm, slightly larger than the wavelength of visible light. The refractive index of the particles is different from that of the polymer film. When light encounters the particle, it will be scattered in directions away from the original direction. The surface diffuser has a rough topography, as shown in Figure 15.5(b). The characteristic size of the surface bumps is comparable to the wavelength of light. The surface diffuser can be manufactured

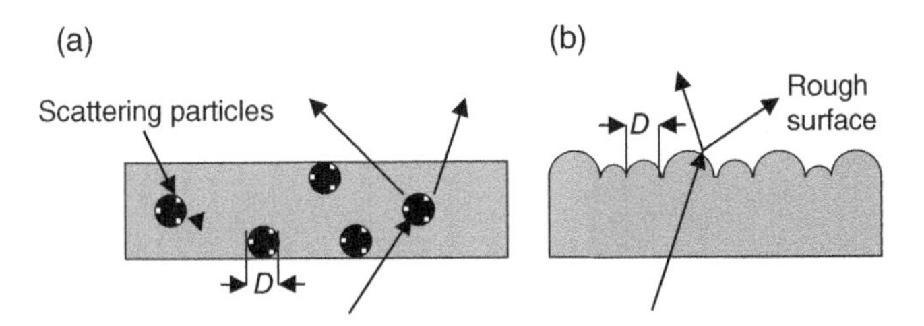

Figure 15.5 Schematic diagram of diffuser (a) bulk diffuser with embedded scattering particles, (b) surface diffuser with rough surface.

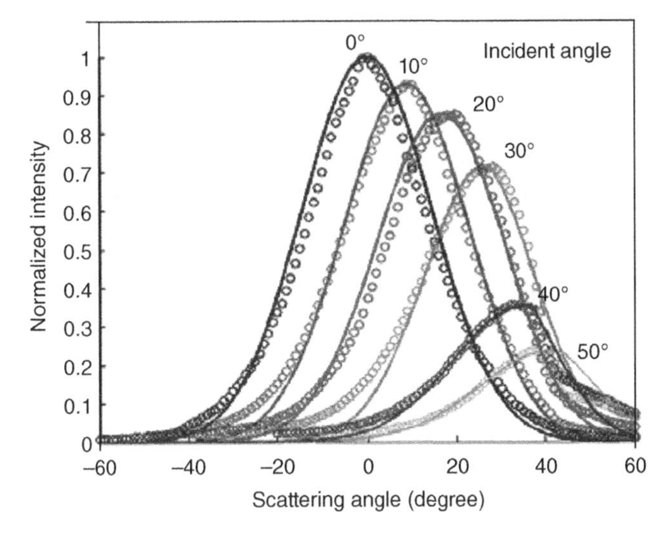

Figure 15.6 Scattered light intensity profile of the diffuser for incident light with various incident angles. Reproduced with permission from Wiley.

either by stamping or by dispersing transparent polymer particles on the surface of a transparent polymer sheet.

The scattered light intensity profile of a bulk diffuser is shown in Figure 15.6, where the scattering angle is defined with respect to the diffuser film normal [22]. For normal incident light (with 0° incident angle), the scattering profile peak has a width of about 40°. Light is mainly scattered in the forward direction. As the incident angle θ_i increases, the peak intensity decreases, because the optical path length inside the diffuser increases and so light encounters more scattering particles. At small incident angles, the scattering profile peak is symmetric about the incident angle. At large incident angles, the peak becomes asymmetric so that the scattered light intensity is higher at scattering angles closer to the normal of the film. This indicates that there is multiple scattering. When light is scattered for the first time, the probability for the light being scattered at the scattering angle $(\theta_i - \alpha)$ is the same as the probability of the light scattered at the scattering angle $(\theta_i + \alpha)$. The scattered light will be scattered again. For the light scattered at the angle $(\theta_i - \alpha)$, the optical path length inside the diffuser is shorter and it is less likely that it will be scattered again than the light scattered at the angle $(\theta_i + \alpha)$. Therefore for an isotropic incident light (light incident at all angles), the scattered light intensity is no longer isotropic but is higher at angles closer to the film normal. This is called the focusing effect of the diffuser.

15.5 Collimation Film

In LCDs, the light produced by the backlight is more or less isotropic, with light incident on the LCD panel at all directions. It is desirable to have the light incident at directions close to the normal of the LCD panel for two reasons. First, the LCD panel does not work well for light with large incident angle. Second, viewers usually look at the display in the normal direction and thus light coming out at large angles is wasted. Collimation films are used to convert isotropic incident light into collimated light [23–26]. Collimation films are also called brightness enhancement films (BEF), and their structure is schematically shown in Figure 15.7(a). The bottom surface of the BEF is flat and the top surface has a one-dimensional saw-tooth shape. As an example, say the refractive index of the polymer is $n = \sin(90°)/\sin(45°) = \sqrt{2} = 1.41$ and the angle of the saw-tooth is 90°. The critical angle θ_c for total internal reflection is 45°. The light incident on the film is isotropic, and the incident angle on the bottom surface (outside the film) is the region from −90° to 90°. The incident angle on the top surface is θ_i and the exit angle is θ_o, as defined in Figure 15.7(b). Inside the film, the incident angle θ_i of the light is in the

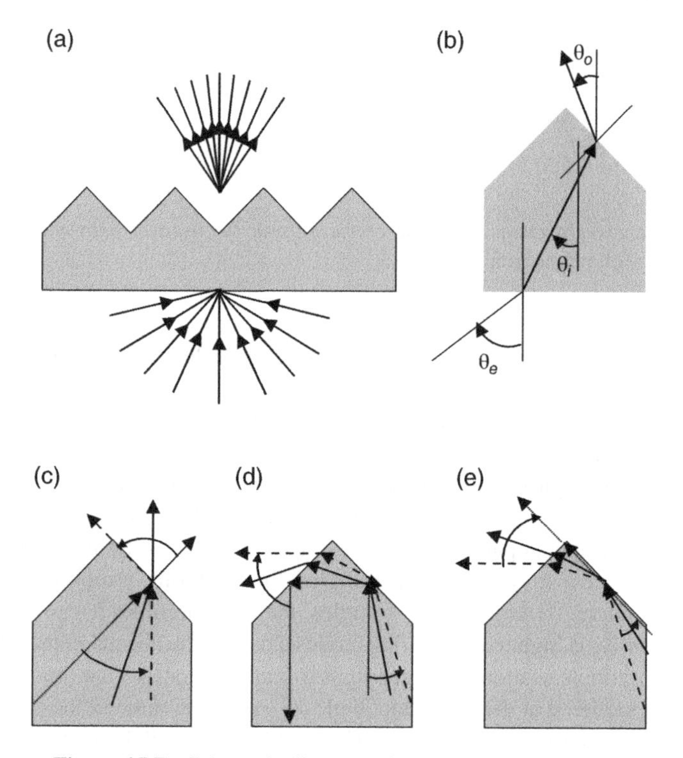

Figure 15.7 Schematic diagram of the structure of the BEF.

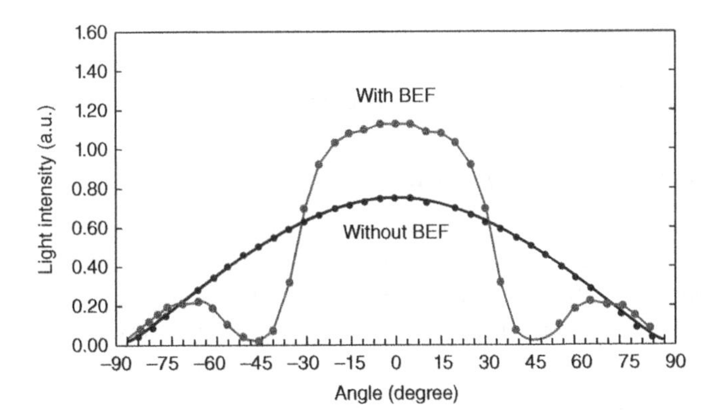

Figure 15.8 Angular distribution of light intensity produced by the collimation film. Reproduced with permission from Wiley.

region from −45° to 45°. When the light hits the saw-tooth surface, there are two possibilities: on the left side or on the right side of the saw-tooth. The refractions occurring on the two sides, however, are the same. Therefore we only have to consider the case of the left side of the saw-tooth.

1. When $45° \le \theta_i < 0°$, the incident light is refracted once and the exit angle is given by

$$n\cdot \sin(45° - \theta_i) = \sin(45° - \theta_o) \tag{15.1}$$

   Therefore the exit angle θ_o is in the region 45° to −45°, as shown in Figure 15.17(c).
2. When $-15° \le \theta_i \le 0°$, the incident light is either totally reflected back or refracted into exit angle smaller than −90°, which will hit the neighboring saw-tooth and be refracted backward, as shown in Figure 15.17(d). Therefore the exit angle θ_o is in the region from −180° to −90°. The light reflected or refractive backward is recycled.
3. When $-45° \le \theta_i < -15°$, the incident light is first totally reflected and then refracted,

$$n\cdot \sin(45° + \theta_i) = \sin(-45° - \theta_o) \tag{15.2}$$

   Therefore the exit angle θ_o is in the region from −90° to −45°, as shown in Figure 15.17(e).

As a rough approximation, the percentage of light is reflected or refracted backward is $[0 - (15°)]/90° = 16.6\ \%$. Therefore it can be seen that 83.3% of the incident light in the region from −90° to 90° before BEF is collimated into the outgoing light in the angular region from −90° to 45°. The simulated light intensity profile as a function of angle of a BEF with the refractive index 1.6 is shown in Figure 15.8 [22]. When two BEFs with orthogonal saw-tooth stripes are stacked together, the light will be collimated into smaller angles.

15.6 Polarizer

In most LCDs, the liquid crystal only modifies the polarization state of polarized incident light. Because light generated by a backlight is unpolarized, a polarizer must be used ahead of the liquid crystal panel to produce polarized incident light. Also in order to block light a polarizer must be used after the liquid crystal panel. Because LCDs are flat and compact, the polarizers must be thin films. Also good polarizers must have high dichroic ratio, high transmittance, and be broadband in order to cover the entire visible light region.

15.6.1 Dichroic absorbing polarizer

The most common polarizer for LCDs is a sheet polarizer made from iodine crystal and stretched polymers [27]. This type of polarizer is also called Polaroid polarizer because it was first developed by Edwin H. Land at Polaroid Corporation [28–30]. The chemical in the iodine crystal is iodoquinine sulfate. It is also called herapathite crystal, because it was first discovered by William Bird Herapath in 1852 [31]. The crystal exhibits dichroic absorption: it absorbs light polarized along one crystalline axis but does not absorb light polarized along the other crystalline axis. Figure 15.9(a) shows an electron microscope photograph of iodine crystal needles in a sheet polarizer after some chemical treatment [28]. The needles must be sufficiently small to minimize light scattering.

The sheet polarizer is fabricated by roll-to-roll process in the following way: (1) polyvinyl alcohol (PVA) film is coated with iodine crystalline needles by sending it through a solution dyeing bath. After this stage the orientation of the iodine needles is random. (2) The PVA film is stretched at an elevated temperature in a thermal oven. The stretching unidirectionally aligns the linear polymer chains. After this stage, the iodine needles are aligned parallel to the polymer chains, as shown in Figure 15.9(b). (3) A tri-acetyl cellulose (TAC) film is laminated on top of the PVA film to sandwich the iodine needles between them. It is highly desirable that the birefringence of the polymer film is minimized to reduce light leakage. Nevertheless the polymer film does have non-zero birefringence, which must be taken into account in modeling LCDs in order to achieve accurate results.

The transmission spectrum of a sheet polarizer is shown in Figure 15.10 [32]. For an incident light polarized parallel to the transmission axis of the polarizer, the transmittance of the polarizer is about 90%, because of the non-perfect alignment of the iodine needles. For an incident light polarized perpendicular to the transmission axis of the polarizer, the transmittance of the polarizer is less than 0.1%. The absorption of the polarizer is wavelength dependent. For light with wavelength higher than 700 nm or shorter than 400 nm, the absorption is much weaker.

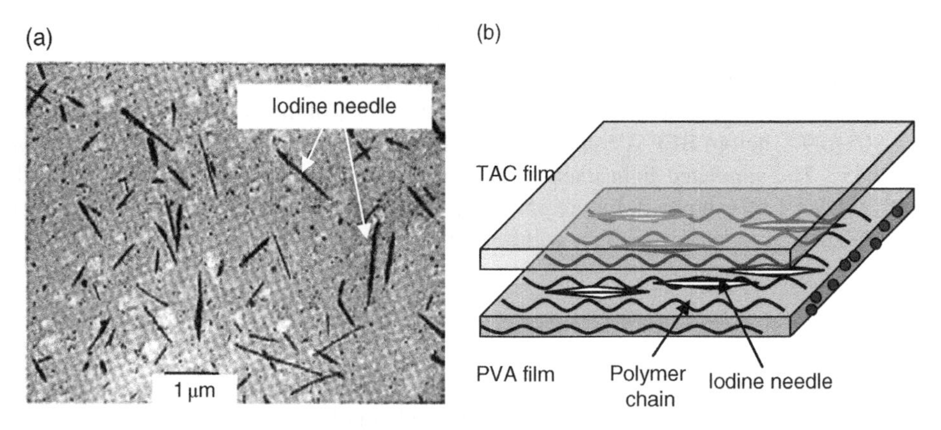

Figure 15.9 (a) Electron micrograph of iodine crystal needles in sheet polarizer by C. E. Hall, Massachusetts Institute of Technology, 1949 [28]. (b) Schematic diagram of sheet polarizer. Reproduced with permission from the Optical Society of America.

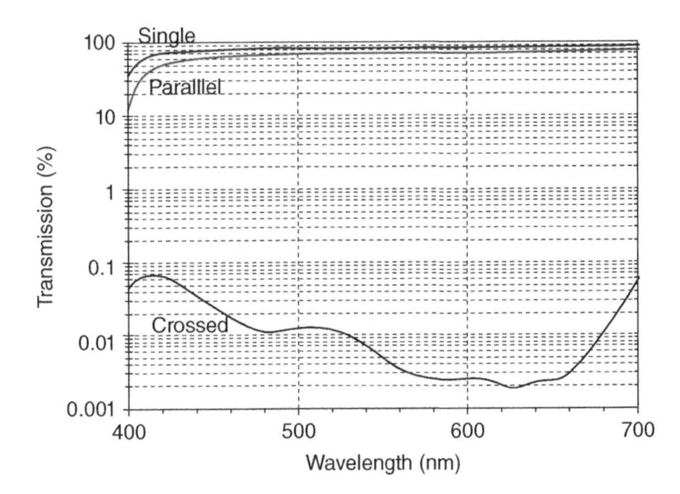

Figure 15.10 Transmission spectrum of sheet polarizer. Single: one polarizer; Parallel: two polarizers with parallel transmission axes; Cross: two polarizers with orthogonal transmission axes [32]. Reproduced with permission from Meadowlark Optics.

There are other ways to make dichroic absorption polarizers. One example is the dichroic dye polarizer [33,34]. Dye molecules are first mixed with a polymer solution, which is cast into a film. When the film is stretched, the linear polymer chains reorient to one direction and the dye molecules are aligned unidirectionally along the linear polymer chain. The film exhibits dichroic absorption. Another example is the lyotropic liquid crystal polarizer [35], which uses chromonic dye molecules self-assembled to form cylindrical super-molecular complexes with aspect ratio (the ratio between the length and the diameter of the cylinder) larger than 100:1. The super-molecular complexes are water-soluble. The solution of the complexes is cast on a substrate. If shear force is applied during casting, the cylinders are aligned along the direction of the force. Then the water is allowed to evaporate and the cylinders crystallize to form a submicron thick film which exhibits dichroic absorption. Furthermore, the film can also act as a homogeneous alignment layer.

15.6.2 Dichroic reflective polarizer

There are also dichroic reflective polarizers, which have the advantage of high light efficiency. They pass incident light polarized in one direction and reflect incident light polarized in the orthogonal direction. The reflected light can be recycled by rotating its polarization into the direction of the transmission axis of the polarizers. The rotation of the polarization can be achieved either by a half waveplate or by a scattering medium.

15.6.2.1 Alternating stacking of polymer layers

3 M developed a reflective polarizer by alternately stacking two different polymer layers as shown in Figure 15.11 [36]. One of the polymer layers is isotropic with the refractive index n_i. The other polymer layer is anisotropic with the ordinary and extraordinary refractive indices n_o and n_e, respectively. Also, n_i is matched to n_o. For light polarized parallel to the optical axis

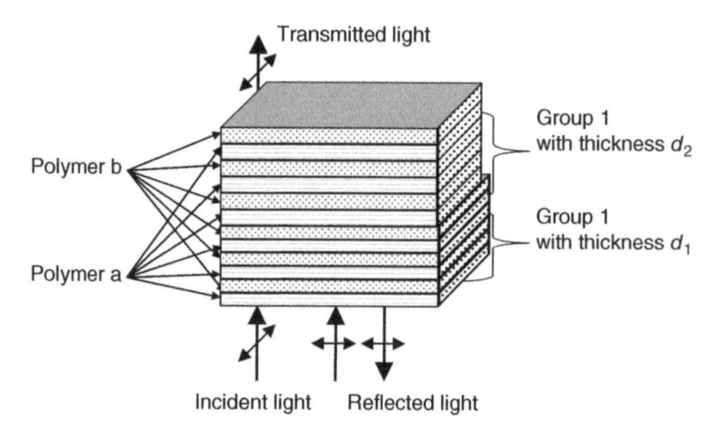

Figure 15.11 Schematic diagram of the reflective polarizer formed by alternating polymer layers.

of the ordinary refractive index, the stack is a uniform optical medium and the light passes through it. For light polarized parallel to the optical axis of the extraordinary refractive index, the stack is a periodically varying optical medium and the light is reflected. Unpolarized incident light can be decomposed into two linearly polarized lights with polarization parallel and perpendicular, respectively, to the optical axis of the ordinary refractive index. The parallel component is transmitted while the perpendicular component is reflected. The transmitted light is linearly polarized. The reflected light can be recycled by using a combination of a bottom reflector and a quarter waveplate. The polarization of the reflected light is converted into left-handed circular polarization by the quarter waveplate. The light is then reflected by the bottom mirror and its polarization is changed into right-handed circular polarization. When the light passes the quarter waveplate again, the polarization is converted into linear polarization orthogonal to the initial polarization.

Now we quantitatively consider the reflection of a dichroic mirror consisting of alternating stacks of two dielectric materials with refractive indices n_1 and n_2, respectively, as shown in Figure 15.12. When light in the layer with refractive index n_2 is incident to the layer with refractive index n_1, some light is transmitted, and the rest of the light is reflected. The transmission and reflection coefficients are respectively given by $2n_2/(n_1+n_2)$ and $(n_1-n_2)/(n_1+n_2)$ [37]. In layer $(i+1)$, the electric field of the up-going light is contributed by the transmission of the up-going light in layer i and the reflection of the down-going light in layer $(i+1)$,

$$E_{u(i+1)} = \frac{2n_1}{(n_1+n_2)} e^{i2\pi n_1 d/\lambda} E_{ui} + \frac{(n_2-n_1)}{(n_1+n_2)} E_{d(i+1)}. \tag{15.3}$$

In layer i, the electric field of the down-going light is contributed by the transmission of the down-going light in layer $(i+1)$ and the reflection of the up-going light in later i,

$$E_{di} = \frac{(n_1-n_2)}{(n_1+n_2)} e^{i4\pi n_1 d/\lambda} E_{ui} + \frac{2n_2}{(n_1+n_2)} e^{i2\pi n_1 d/\lambda} E_{d(i+1)}. \tag{15.4}$$

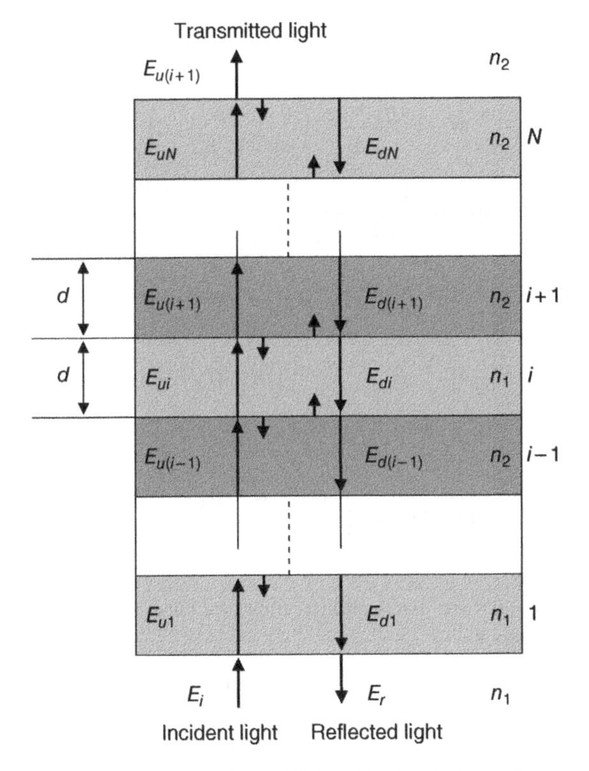

Figure 15.12 Schematic diagram showing the multiple reflection in the reflective polarizer. The electric fields are the ones at the bottom surface of each.

Solving the above two equations we get

$$E_{ui} = \frac{(n_1 + n_2)}{2n_1} e^{-i2\pi n_1 d/\lambda} E_{u(i+1)} + \frac{(n_1 - n_2)}{2n_1} e^{-i2\pi n_1 d/\lambda} E_{d(i+1)}, \tag{15.5}$$

$$E_{di} = \frac{(n_1 - n_2)}{2n_1} e^{i2\pi n_1 d/\lambda} E_{u(i+1)} + \frac{(n_1 + n_2)}{2n_1} e^{i2\pi n_1 d/\lambda} E_{d(i+1)}. \tag{15.6}$$

We put them into matrix form

$$\vec{E}_i = \begin{pmatrix} E_{ui} \\ E_{di} \end{pmatrix} = \frac{1}{2n_1} \begin{pmatrix} (n_1 + n_2)e^{-i2\pi n_1 d/\lambda} & (n_1 - n_2)e^{-i2\pi n_1 d/\lambda} \\ (n_1 - n_2)e^{i2\pi n_1 d/\lambda} & (n_1 + n_2)e^{i2\pi n_1 d/\lambda} \end{pmatrix} \begin{pmatrix} E_{u(i+1)} \\ E_{d(i+1)} \end{pmatrix} = \overleftrightarrow{M}_1 \cdot \vec{E}_i. \tag{15.7}$$

In the same way we can get the relation between the electric fields of the up-going and down-going light in layer i and the electric fields of the up-going and down-going light in layer $(i-1)$,

$$\vec{E}_{(i-1)} = \begin{pmatrix} E_{u(i-1)} \\ E_{d(i-1)} \end{pmatrix} = \frac{1}{2n_2} \begin{pmatrix} (n_1 + n_2)e^{-i2\pi n_2 d/\lambda} & -(n_2 - n_1)e^{-i2\pi n_2 d/\lambda} \\ -(n_2 - n_1)e^{i2\pi n_2 d/\lambda} & (n_1 + n_2)e^{i2\pi n_2 d/\lambda} \end{pmatrix} \begin{pmatrix} E_{ui} \\ E_{di} \end{pmatrix} = \overleftrightarrow{M}_2 \cdot \vec{E}_i. \tag{15.8}$$

Define $a = (n_2 - n_1)/(n_1 + n_2)$, $\alpha_1 = 2\pi n_1 d/\lambda$, and $\alpha_2 = 2\pi n_2 d/\lambda$. Then $\overleftrightarrow{M}_1$ and $\overleftrightarrow{M}_2$ become

$$\overleftrightarrow{M}_1 = \frac{(n_1 + n_2)}{2n_1} \begin{pmatrix} e^{-i\alpha_1} & ae^{-i\alpha_1} \\ ae^{i\alpha_1} & e^{i\alpha_1} \end{pmatrix}, \tag{15.9}$$

$$\overleftrightarrow{M}_2 = \frac{(n_1 + n_2)}{2n_2} \begin{pmatrix} e^{-i\alpha_2} & -ae^{-i\alpha_2} \\ -ae^{i\alpha_2} & e^{i\alpha_2} \end{pmatrix}. \tag{15.10}$$

For a stack of N layers of polymer 1 and N layers of polymer 2,

$$\vec{E}_0 = \left(\overleftrightarrow{M}_1 \cdot \overleftrightarrow{M}_2\right) \cdot \vec{E}_1 = \left(\overleftrightarrow{M}_1 \cdot \overleftrightarrow{M}_2\right)^2 \cdot \vec{E}_2 = \ldots\ldots = \left(\overleftrightarrow{M}_1 \cdot \overleftrightarrow{M}_2\right)^N \cdot \vec{E}_{(N+1)} = \overleftrightarrow{M}^N \cdot \vec{E}_{(N+1)}, \tag{15.11}$$

where

$$\overleftrightarrow{M} = \overleftrightarrow{M}_1 \cdot \overleftrightarrow{M}_2 = b \begin{pmatrix} e^{-i\beta} - a^2 & a\left(1 - e^{-i\beta}\right) \\ a\left(1 - e^{i\beta}\right) & e^{i\beta} - a^2 \end{pmatrix}, \tag{15.12}$$

where $b = (n_1 + n_2)^2/4n_1 n_2$, $\beta = (\alpha_1 + \alpha_2)$ and $(\alpha_1 - \alpha_2) = 2\pi d(n_1 - n_2)/\lambda \approx 0$, because usually n_1 and n_2 are close to each other.

Below the stack, there is incident light and reflected light: $\vec{E}_0 = \begin{pmatrix} E_i \\ E_r \end{pmatrix}$. Above the stack, there is only transmitted light: $\vec{E}_{(N+1)} = \begin{pmatrix} E_t \\ 0 \end{pmatrix}$. Note that for the purpose of simplicity, the reflection and refraction between the dielectric layer and air are omitted. This gives

$$\vec{E}_0 = \begin{pmatrix} E_i \\ E_r \end{pmatrix} = \left(\overleftrightarrow{M}\right)^N \cdot \vec{E}_{(N+1)} = \left(\overleftrightarrow{M}\right)^N \cdot \begin{pmatrix} E_t \\ 0 \end{pmatrix}. \tag{15.13}$$

From the above equation, we can calculate E_i and E_r as functions of E_t. Using Cayley–Hamilton theory, we have

$$\left(\overleftrightarrow{M}\right)^N = \lambda_1 \overleftrightarrow{I} + \lambda_2 \overleftrightarrow{M}, \tag{15.14}$$

where λ_1 and λ_2 are the solution of the following equations:

$$q_1^N = \lambda_1 + \lambda_2 q_1 \tag{15.15}$$

$$q_2^N = \lambda_1 + \lambda_2 q_2 \tag{15.16}$$

q_1 and q_2 are the eigenvalues of $\overleftrightarrow{M}$:

$$\begin{vmatrix} b\left(e^{-i\beta} - a^2\right) - q & ba\left(e^{-i\beta} - 1\right) \\ ba\left(1 - e^{i\beta}\right) & b\left(e^{i\beta} - a^2\right) - q \end{vmatrix} = 0 \tag{15.17}$$

The solutions are

$$q_{1,2}/b = \left(\cos\beta - a^2\right) \pm \left[\left(\cos\beta - a^2\right)^2 - \left(1-a^2\right)^2\right]^{1/2} = u \pm v, \tag{15.18}$$

where $u = (\cos\beta - a^2)$, $v = [(\cos\beta - a^2)^2 - (1 - a^2)^2]^{1/2}$.
From Equations (15.15) and (15.16), we can get

$$\lambda_1 = -\frac{q_1 q_2\left(q_1^{N-1} - q_2^{N-1}\right)}{(q_1 - q_2)} = -b^{N-1}\left(1-a^2\right)^2\left[(u+v)^{N-1} - (u-v)^{N-1}\right]/2v, \tag{15.19}$$

$$\lambda_2 = \frac{\left(q_1^N - q_2^N\right)}{(q_1 - q_2)} = b^{N-1}\left[(u+v)^N - (u-v)^N\right]/2v. \tag{15.20}$$

From Equation (15.14), we have

$$\left(\overleftrightarrow{M}\right)^N = -b^{N-1}(1-a^2)^2\left\{[(u+v)^{N-1} - (u-v)^{N-1}]/2v\right\}\begin{pmatrix} 1 & 0 \\ 0 & 1 \end{pmatrix}$$

$$+ b^{N-1}\left\{\left[(u+v)^N - (u-v)^N\right]/2v\right\}\begin{pmatrix} e^{-i\beta} - a^2 & a\left(1-e^{-i\beta}\right) \\ a\left(1-e^{i\beta}\right) & e^{i\beta} - a^2 \end{pmatrix}. \tag{15.21}$$

From Equation (15.13), we get the reflectance

$$R = \frac{E_r \cdot E_r^*}{E_i \cdot E_i^*} = \frac{\left[\left(\overleftrightarrow{M}\right)^N\right]_{21}\left[\left(\overleftrightarrow{M}\right)^N\right]_{21}^*}{\left[\left(\overleftrightarrow{M}\right)^N\right]_{11}\left[\left(\overleftrightarrow{M}\right)^N\right]_{11}^*},$$

$$R = \frac{\left|a\left[(u+v)^N - (u-v)^N\right]\left(1-e^{i\beta}\right)\right|^2}{\left|-(1-a^2)^2\left[(u+v)^{N-1} - (u-v)^{N-1}\right] + \left[(u+v)^N - (u-v)^N\right](e^{-i\beta} - a^2)\right|^2}. \tag{15.22}$$

The reflectance is maximized when phase retardation angle $\beta = \beta_{max} = (2m+1)\pi$. The corresponding wavelength is given by

$$\beta_{max} = 2\pi(n_1 + n_2)d/\lambda_{max} = (2m+1)\pi,$$

$$\lambda_{max} = \frac{2(n_1 + n_2)d}{2m+1}. \tag{15.23}$$

At this wavelength, $\cos\beta = -1$, $e^{\pm i\beta} = -1$, $u = -1 - a^2$, $v = 2a$, and therefore $u + v = -(1-a)^2$ and $u - v = -(1+a)^2$. The maximum reflectance is given by

$$R_{\max} = \frac{\left[(1+a)^{2N} - (1-a)^{2N}\right]^2}{\left[(1+a)^{2N} + (1-a)^{2N}\right]^2}. \tag{15.24}$$

When N is large enough,

$$R_{\max} \approx 1 - 4\left(\frac{1-a^2}{1+a^2}\right)^{2N}. \tag{15.25}$$

As an example, $n_1 = 1.7$ and $n_2 = 1.5$, in order to get the reflection $R_{\max} = 0.9$, $N = 15$, so 15 layers are needed.

Now let us consider the width of the reflection band. It can be seen from Equation (15.22) that the reflection is a minimum when

$$\left[(u+v)^N - (u-v)^N\right] = 0, \tag{15.26}$$

namely,

$$v = \left[\left(\cos\beta - a^2\right)^2 - \left(1-a^2\right)^2\right]^{1/2} = 0,$$

$$\cos\beta - a^2 = \pm\left(1-a^2\right).$$

The solution for wavelength near the principal reflection peak is

$$\cos\beta_{\min} - a^2 = -\left(1-a^2\right), \qquad \cos\beta_{\min} = -1 + 2a^2. \tag{15.27}$$

If $n_1 = 1.7$ and $n_2 = 1.5$, $\beta_{\min} = \beta_{\max} + \Delta\beta$ and $\Delta\beta \ll 1$

$$\cos\beta_{\min} = \cos\left(\beta_{\max} + \Delta\beta\right) = \cos\left[(2m+1)\pi + \Delta\beta\right] \approx -1\cdot\left[1 - (\Delta\beta)^2/2\right] = -1 + 2a^2,$$

$$\Delta\beta = \frac{2\pi(n_1+n_2)d}{\lambda_{\max}} - \frac{2\pi(n_1+n_2)d}{\lambda_{\min}} = 2a = 2\frac{(n_1-n_2)}{(n_1+n_2)},$$

$$\Delta\lambda = \lambda_{\min} - \lambda_{max} \approx \frac{(n_1-n_2)}{\pi(n_1+n_2)d}\cdot\lambda_{\max}^2 = \frac{(n_1-n_2)}{\pi(n_1+n_2)d}\left[\frac{2(n_1+n_2)d}{2m+1}\right]^2 = \frac{4\left(n_1^2-n_2^2\right)}{\pi}d\left(\frac{1}{2m+1}\right)^2.$$

For the first principal reflection peak, $m = 0$ and thus

$$\Delta\lambda = \frac{4(n_1+n_2)}{\pi}(n_1-n_2)d. \tag{15.28}$$

If $n_1 = 1.7$ and $n_2 = 1.5$, the first principal reflection band width is $\Delta\lambda = 0.8d$.

A calculated reflection spectrum of a dichroic mirror is shown in Figure 15.13, where the following parameters are used: $n_1 = 1.7$, $n_2 = 1.5$, and $d = 80$ nm. The reflection peak is located at a wavelength of $[2(n_1+n_2)]d = 512$ nm. The reflection bandwidth is about 60 nm.

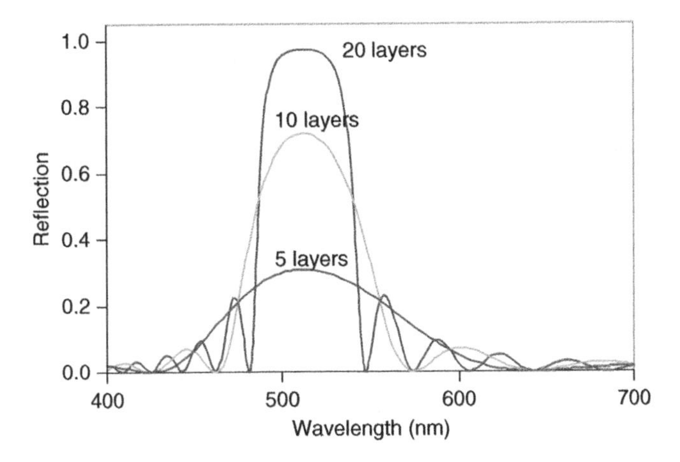

Figure 15.13 The reflection spectrum of the dichroic mirror.

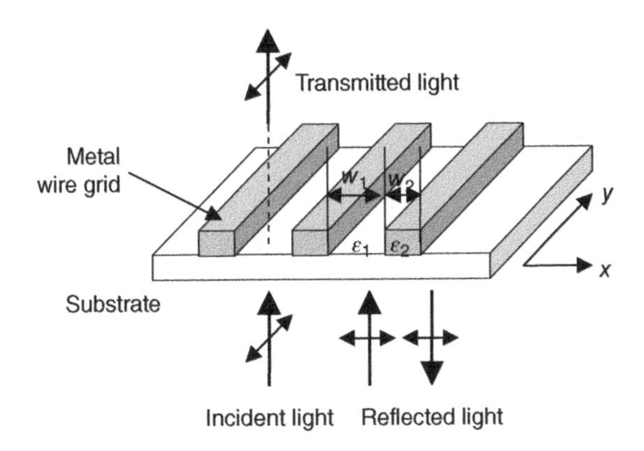

Figure 15.14 Schematic diagram of the reflective polarizer formed by metal wire grid.

A couple of issues should be mentioned for the reflective polarizer. First, different layer thicknesses must be used to reflect different wavelengths of light. From Equation (15.28) and Figure 15.13, it can be seen that in order to cover the entire visible spectrum, a few groups of alternating stacked polymer layers with different layer thicknesses must be used. The second issue is that there is some light leakage for light polarized parallel to the optical axis of the extraordinary refractive index. In order for the liquid crystal display to achieve high contrast ratio, an absorbing polarizer must be used after the reflective polarizer.

15.6.2.2 Wire grid reflective polarizer

There is another type of reflective polarizer made from wire grid, as shown in Figure 15.14 [38–41]. Metal stripes are fabricated on the surface of a transparent substrate. The width w_2 of the stripe and the gap w_1 between stripes are much smaller than the wavelength of the incident light. The dielectric constant of the metal stripe is ε_2 and the refractive index of the material between the metal stripes is ε_1. At the vertical surface of the metal stripe, the boundary conditions are that the tangential component of the electric field $\vec{E}$ is continuous and the normal component of the electric displacement $\vec{D}$ is continuous [42]. For light polarized parallel to the metal stripe, the boundary condition for the electric field of the light is

$$E_{1y} = E_{2y} = E_y. \tag{15.29}$$

The average electric displacement can be calculated by

$$\bar{D}_y = \frac{D_{1y}w_1 + D_{2y}w_2}{w_1 + w_2} = \frac{(\varepsilon_1 E_{1y})w_1 + (\varepsilon_2 E_{2y})w_2}{w_1 + w_2} = \left(\frac{\varepsilon_1 w_1 + \varepsilon_2 w_2}{w_1 + w_2}\right)E_y = \bar{\varepsilon}E_y. \tag{15.30}$$

The corresponding refractive index is

$$n_{//} = \sqrt{\bar{\varepsilon}} = \sqrt{\frac{\varepsilon_1 w_1 + \varepsilon_2 w_2}{w_1 + w_2}}. \tag{15.31}$$

The dielectric constant of the metal is complex and its absolute value is much larger than that of the material (air) between the metal stripes. $|w_2\varepsilon_2| \gg |w_1\varepsilon_1|$. Therefore approximately we have [40]

$$n_{//} = \sqrt{\frac{w_2}{w_1 + w_2}\varepsilon_2} = \sqrt{\frac{w_2}{w_1 + w_2}}n_2. \tag{15.32}$$

The resulting refractive index is complex, and thus the light is reflected.

For light polarized perpendicular to the metal stripe, the boundary condition for the electric displacement of the light is

$$D_{1x} = D_{2x} = D_x. \tag{15.33}$$

The average electric field can be calculated as

$$\bar{E}_x = \frac{E_{1x}w_1 + E_{2x}w_2}{w_1 + w_2} = \frac{(D_{1x}/\varepsilon_1)w_1 + (D_{2x}/\varepsilon_2)w_2}{w_1 + w_2} = \left(\frac{w_1/\varepsilon_1 + w_2/\varepsilon_2}{w_1 + w_2}\right)D_x = D_x/\bar{\varepsilon}. \tag{15.34}$$

The corresponding refractive index is

$$n_\perp = \sqrt{\bar{\varepsilon}} = \sqrt{\frac{1}{\frac{w_1/\varepsilon_1 + w_2/\varepsilon_2}{w_1 + w_2}}} = \sqrt{\frac{w_1 + w_2}{w_1/\varepsilon_1 + w_2/\varepsilon_2}}. \tag{15.35}$$

The dielectric constant of the metal is complex, and its absolute value is much larger than that of the material (air) between the metal stripes, $|w_1/\varepsilon_1| \gg |w_2/\varepsilon_2|$. Therefore approximately we have

$$n_\perp = \sqrt{\frac{w_1 + w_2}{w_1/\varepsilon_1}} = \sqrt{\frac{w_1 + w_2}{w_1}}n_1. \tag{15.36}$$

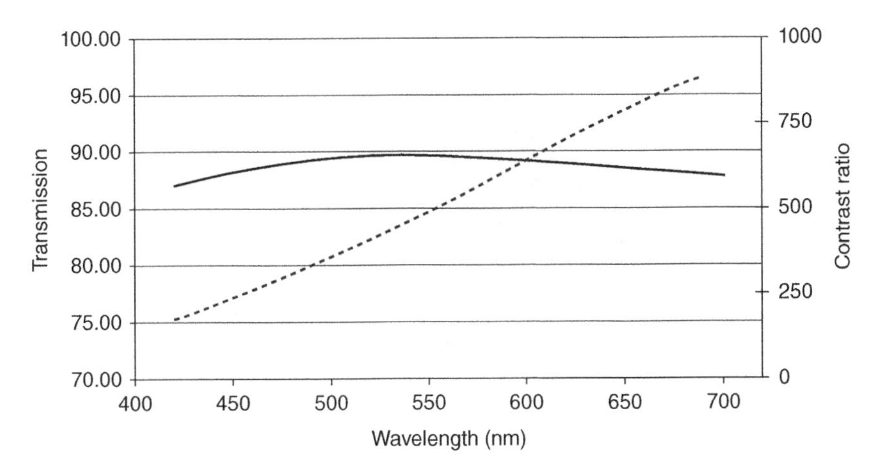

Figure 15.15 The transmission spectrum of the wire grid reflective polarizer. Reproduced with permission from Wiley.

The resulting refractive index is real, and thus the light is transmitted.

The transmission spectrum of a wire grid reflective polarizer made by Moxtel Inc. is shown in Figure 15.15 [39]. The shown transmission is the transmittance of incident light polarized parallel to the transmission axis (which is perpendicular to the metal wire). The contrast ratio (dichroic ratio) is the ratio between the transmittances of incident light polarized parallel and perpendicular to the transmission axis, respectively. The contrast ratio of metal wire grid reflective polarizer is higher than that of the birefringent polymer reflective polarizer, but the transmittance of the metal wire grid reflective polarizer is lower.

15.6.2.3 Cholesteric liquid crystal reflective polarizer

Cholesteric liquid crystals (CLCs) can also be used to make reflective polarizers. CLCs reflect circularly polarized light with the same handedness as the helical structure of the liquid crystal. An unpolarized incident light can be decomposed into a left-handed circular polarized light and a right-handed circular polarized light. One component is reflected and the other component is transmitted. The reflected light is reflected toward the CLC polarizer by a back mirror and its handedness is converted to the opposite handedness and thus it passes the CLC polarizer. The transmitted circular polarized light is converted into linear polarization by a quarter waveplate.

The reflection band of a CLC is located at the wavelength given by $\lambda = [(n_o + n_e)/2]P$, and the bandwidth is given by $\Delta\lambda = (n_e - n_o)P$, where P is the pitch of the liquid crystal, and n_e and n_o are the extraordinary and ordinary refractive indices of the liquid crystal, respectively. If the birefringence Δn $(= n_e - n_o)$ is 0.2 and the pitch P is 300 nm, the bandwidth is about 60 nm, which is not able to cover the visible light region. Broadband CLCs are produced by using CLC/polymer composites [43,44]. A CLC is mixed with a mono-functional chiral monomer, a multifunctional monomer, and a photo-initiator. The mixture is sandwiched between two parallel substrates. The cell is irradiated by UV light to polymerize the monomers. During the polymerization, a UV intensity gradient is created across the cell, which in turn produces a free radical gradient. When the chiral monomers diffuse to the high free radical density region, they are polymerized and stay there. Thus the formed chiral polymer density

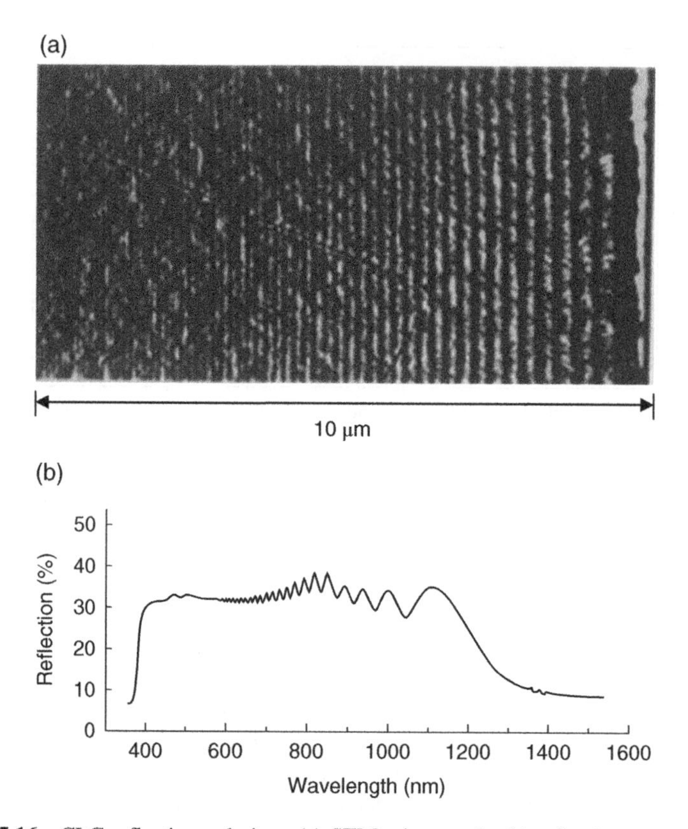

Figure 15.16 CLC reflective polarizer. (a) SEM micrograph, (b) reflection spectrum. [44].

varies across the cell and induces a variation of helical pitch. The cross-linked multi-functional monomers freeze the non-uniform helical structure. Thus the polymer stabilized CLC has a pitch gradient, as shown by the SEM micrograph in Figure 15.16(a), and exhibits broadband reflection. The broadband reflection of a polymer stabilized CLC is shown in Figure 15.16(b) [44].

15.7 Compensation Film

LCDs are designed in such a way that in the dark state the liquid crystals exhibit certain optical retardation for normal incident light. Undesirably the optical retardation changes when light incidents at oblique angles. This change of optical retardation causes light leakage, and thus results in a bad viewing angle. Furthermore, crossed polarizers also leak light at oblique incident angles. The common way to improve the viewing angle is to use compensation films. As viewing angle is increased from 0° (cell normal direction), the optical retardation of the liquid crystal increases (or decreases). The optical retardation of the compensation film must change in the opposite way to that of the liquid crystal. It is highly desirable that the variations of the liquid crystal and compensation film compensate each other, and the net change of optical retardation is zero as the viewing angle is changed.

The liquid crystals used in displays are usually calamitic (rod-like). Their refractive index ellipsoid has prolate shape: the refractive index is the largest in the direction of the uniaxial axis. The compensation films should have oblate-shaped refractive index ellipsoids: the refractive index should be the smallest in the direction of the uniaxial axis. According to the orientation of the uniaxial axis with respect to the film, compensation films are categorized into three types: *c* plate whose uniaxial axis is perpendicular to the film, *a* plate whose uniaxial axis is parallel to the film, and *o* plate whose uniaxial axis makes an oblique angle to the film. If the index along the uniaxial axis of the refractive index ellipsoid is smaller than that along the orthogonal axis, the compensation is called negative plate. In the opposite, if the index along the uniaxial axis of the refractive index ellipsoid is larger than that along the orthogonal axis, the compensation is called positive plate. Besides uniaxial compensation films, there are biaxial compensation films, whose refractive indices along the three principal axes of the ellipsoid are all different.

15.7.1 Form birefringence compensation film

Negative *c* plate can be made from 'form' birefringence [42]. It consists of alternating stacks of two dielectric layers with different refractive indices n_1 and n_2, and layer thicknesses w_1 and w_2, respectively, as shown in Figure 15.12. The layer thicknesses are much smaller than the wavelength of light. For normal incident light, the electric field of the light is parallel to the layers, and the effective refractive index can be derived in the same way as Equation (15.31):

$$n_o = \sqrt{\frac{\varepsilon_1 w_1 + \varepsilon_2 w_2}{w_1 + w_2}} = \sqrt{x n_1^2 + (1-x) n_2^2}, \tag{15.37}$$

where $x = w_1/(w_1 + w_2)$. For incident light propagating parallel to the layers, if the electric field of the light is parallel to the layer normal, the effective refractive index can be derived in the same way as Equation (15.35):

$$n_e = \sqrt{\frac{w_1 + w_2}{w_1/\varepsilon_1 + w_2/\varepsilon_2}} = \sqrt{\frac{n_1^2 n_2^2}{x n_2^2 + (1-x) n_1^2}} \tag{15.38}$$

Because

$$n_o^2 - n_e^2 = \left[x n_1^2 + (1-x) n_2^2\right] - \frac{n_1^2 n_2^2}{x n_2^2 + (1-x) n_1^2} = \frac{x(1-x)\left(n_1^2 - n_2^2\right)^2}{x n_2^2 + (1-x) n_1^2} > 0, \tag{15.39}$$

then it is a negative *c* plate. This is a relatively old technology, and it is not suitable for large-size displays.

15.7.2 Discotic liquid crystal compensation film

Discotic liquid crystals (DLC) consist of molecules with a rigid core from aromatic rings and flexible branches from hydrocarbon chains as discussed in Chapter 1. The molecules can be regarded as discs [45]. On average, the normal axes of the discs are aligned along a common direction, the liquid crystal director, in the nematic phase. The physical properties of discotic nematic liquid crystals are invariant under rotation around the liquid crystal director. Thus the uniaxial axis is parallel to the liquid crystal director. The refractive index for light polarized parallel to the disc is larger than that for light polarized perpendicular to the disc. Therefore the refractive index ellipsoids of DLCs are oblate. They are good candidates for compensation films for calamitic LCDs. If the flexible branches contain an acrylic reactive group, the liquid crystals can be polymerized to form cross-linked polymeric liquid crystal films [46].

When a discotic liquid crystal is sandwiched between two substrates (or exposed to air), the direction of the uniaxial axis can be controlled by alignment layers, external electric fields, and chiral dopants [46,47]. It is therefore possible to develop discotic compensation films with spatially varied uniaxial axis orientations. For example, Fuji Photo Film Co. developed discotic compensation films for TN LCDs. In both the TN display and discotic compensation film, the liquid crystal directors vary in the vertical direction. Each layer of nematic liquid crystal with a certain director orientation is compensated by a layer of discotic liquid crystal with the same director orientation.

15.7.3 Compensation film from rigid polymer chains

Long rigid polymer chains may be aligned parallel to one another in the global minimum energy state. In reality, they cannot reach this state, because of limited space. Instead, they have the tendency to from in-plane orientation [48]. When the rigid polymer chains are dissolved in a solvent with a low chain number density, the orientation of the chains are random in three dimensions, as shown in Figure 15.17(a). The orientational order parameter of the chains is 0. As the solvent evaporates, the thickness of the solution shrinks. Effectively the polymer chains are compressed. When the solvent completely evaporates, the chains are randomly oriented in the xy plane. The orientational order parameter becomes –0.5. The uniaxial axis is parallel to the z axis. The refractive indices are $n_x = x_y > n_z$, resulting in a negative $\boldsymbol{c}$ plate.

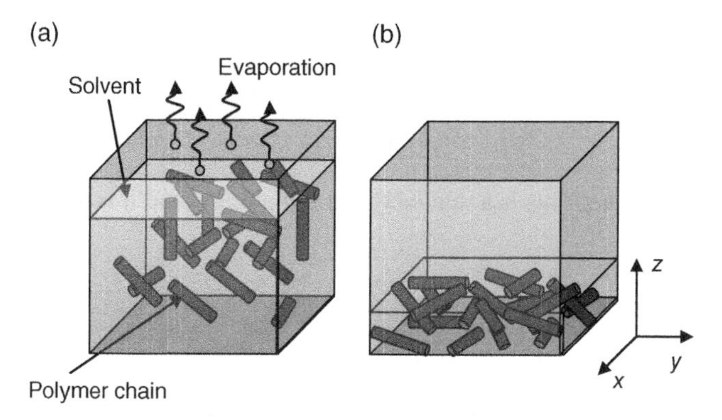

Figure 15.17 Schematic diagram showing how the negative c plate is formed from long rigid polymer chains: (a) random orientation in 3-D, (b) random orientation in 2-D.

15.7.4 Drawn polymer compensation film

Polymers with flexible chains are isotropic because of the random orientation of the chains. They become, however, anisotropic when they are drawn (stretched) because the chains tend to align parallel to the drawing [49]. Now we consider the orientational order and birefringence induced by drawing.

A flexible polymer chain can be modeled by self-avoiding random walking: the orientation of each monomer of the polymer chain is random except that two (or more) monomers cannot occupy the same position. When the chain is stretched, the number of possible configurations of the monomers is decreased, which results in an entropic force. Furthermore, there is interaction between monomers, which also results in a resistant force. For a cylindrical fiber consisting of linear polymer chains, when a tensile stress (force per unit area) is applied, the produced strain (change of length per unit length) is given by [50]

$$e=\sigma/E, \tag{15.40}$$

where E is the Young's modulus of the polymer. For most linear polymers, E is on the order of $1\times10^9 \mathrm{N/m^2}$. Orientational order of the chains is produced when a stress is applied. Consider a polymer chain consisting of N monomers, and assume that each monomer is a straight rigid segment with a flexible linkage. The orientation of the ith monomer is described by the polar angle θ_i, defined with respect to the drawing direction. The projection of the monomer in the drawing direction is $a\cos\theta_i$, where a is the monomer length. In the unstretched state, the orientation of the monomers is random, and the average shape of the chain is a sphere with the radius given by

$$R=a(N/3)\left[\frac{1}{N}\sum_{i=1}^{N}\cos^2\theta_i\right]^{1/2}=a(N/3)\left[\frac{\int_0^{\pi/2}\cos^2\theta\sin\theta d\theta}{\int_0^{\pi}\sin\theta d\theta}\right]^{1/2}=aN/3\sqrt{3}, \tag{15.41}$$

because θ_i is randomly distributed between 0 and $\pi/2$. When the fiber is stretched, the monomer is tilted toward the drawing direction. The shape of the polymer chain becomes an ellipsoid with the long axis given by $R(1+e)$ and short axis given by $R(1-e/2)$. Then θ_i is no longer randomly distributed but follows a distribution function $f(\theta)$:

$$R(1+e)=a(N/3)\left[\frac{1}{N}\sum_{i=1}^{N}\cos^2\theta_i\right]^{1/2}=a(N/3)\left[\frac{\int_0^{\pi/2}f(\theta)\cos^2\theta\sin\theta d\theta}{\int_0^{\pi}f(\theta)\sin\theta d\theta}\right]^{1/2} \tag{15.42}$$

From the above two equations, we get

$$1+e=\sqrt{3}\left[\frac{\int_0^{\pi/2} f(\theta)\cos^2\theta\sin\theta d\theta}{\int_0^{\pi} f(\theta)\sin\theta d\theta}\right]^{1/2}. \tag{15.43}$$

The orientational order parameter is defined by

$$S=\frac{1}{N}\sum_{i=1}^{N}\frac{1}{2}\left(3\cos^2\theta_i-1\right)=\frac{\int_0^{\pi/2} f(\theta)\frac{1}{2}\left(3\cos^2\theta-1\right)\sin\theta d\theta}{\int_0^{\pi} f(\theta)\sin\theta d\theta}=\frac{\frac{3}{2}\int_0^{\pi/2} f(\theta)\cos^2\theta\sin\theta d\theta}{\int_0^{\pi} f(\theta)\sin\theta d\theta}-\frac{1}{2}$$

$$S=\frac{3}{2}\left[\frac{1}{\sqrt{3}}(1+e)\right]^2-\frac{1}{2}=\frac{1}{2}\left[(1+e)^2-1\right]=\frac{1}{2}\left(2e+e^2\right)\approx e. \tag{15.44}$$

The birefringence $\Delta n \propto S$. From Equation (15.41) we have

$$\Delta n=(\Delta n)_o S=\frac{(\Delta n)_o}{E}\sigma=C\sigma, \tag{15.45}$$

where $(\Delta\lambda)_o$ is the birefringence of perfectly ordered polymer chains, and C is called the stress optical constant [49,51]. For example, for polyvinyl chloride (PVC), when the stress is 1, the induced birefringence is about 2×10^{-4} [50].

When the applied stress is removed, the polymer chains can relax back to the original random orientation state and the induced birefringence will disappear. There are two methods to retain the induced birefringence. The first method is to draw with a sufficiently large stress, such that the polymer yields, then the polymer chains will no longer be able to relax back to the random state and thus the induced birefringence becomes permanent. The second method is to draw the polymer at a temperature higher than the glass transition temperature of the polymer and then cool it down quickly to freeze the deformation of the polymer chains. The induced birefringence becomes permanent.

When a polymer film is drawn uniaxially, the induced birefringence is usually uniaxial. The produced compensation film is a positive *a* plate. The drawing can be carried out by the machine in a roll-to-roll process, as shown in Figure 15.18 [50]. The stretching rate can be adjusted by changing the relative speed between the feed rolls and the stretching rolls.

In some liquid crystal displays, however, biaxial compensation films are needed. These can be made by drawing polymer films in the two directions in the film plane [50]. They can also be made by drawing uniaxial liquid crystal polymer films in the direction perpendicular to the uniaxial axis [52,53].

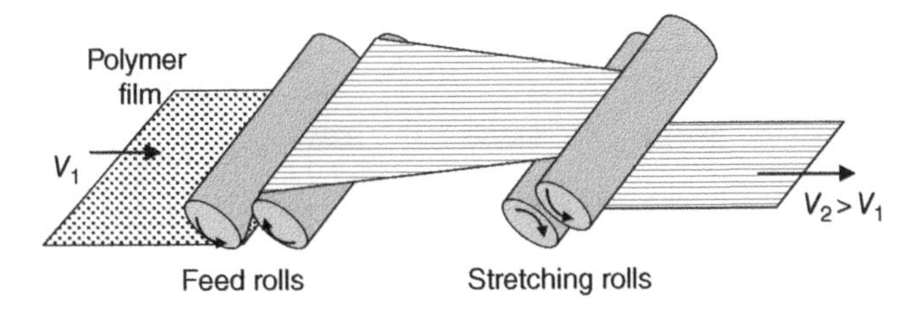

Figure 15.18 Schematic diagram of the drawing machine. The stretching rolls rotate at a higher speed than the feed rolls.

15.8 Color Filter

In LCDs, colors are produced by color filters [54,55]. The structure of a color filter is schematically shown in Figure15.19 [56]. It consists of three types of color pixels: red (R), green (G), and blue (B), because they are sufficient to generate any other colors. In some new designs, there is one more color pixel, white, in order to increase light efficiency of normal white displays such as computer monitors. Under each color pixel, there is a TFT element to control the light intensity. A black matrix is located in the area between the color pixels to prevent light leakage and to provide light shield for the TFT driving element. On top of the color pixels there is an overcoat layer whose function is to reduce the thickness variation of the color pixels and protect the color pixels from chemicals used in processing. On top of the overcoat layer is the indium-tin-oxide (ITO) transparent electrode. The color pixels are on the inner side of the substrate of the LCD in order to avoid the image parallax problem. If the color pixels are on the outer side, light coming from one liquid crystal pixel may reach multiple color pixels, because the out-coming light is not collimated and the substrate is thick.

Color filters can be fabricated by methods such as pigment dispersing, dyeing, printing, and electrodepositing [56,57]. The pigment dispersing method is the most commonly used because of its low manufacturing cost and high thermal and light resistance. Pigment is made from selective color light-absorbing materials which can be either inorganic or organic, natural or synthesized [58]. The raw materials are ground into fine powers to make the pigments, which are also used in painting. The pigment particle size is usually in the range 50–500 nm. Small pigment size is preferred, because if the size is comparable to the wavelength of visible light, the pigments will scatter light and thus depolarize the polarized incident light and decrease the contrast ratio of displays.

The pigment color filters can be manufactured either by an etching method or a polymerization method [56]. In the etching method, the pigments are mixed with a polyimide and coated on the substrate. A photoresist is then coated on top of the polyimide. The system is irradiated by UV light under a photomask. In the region exposed to UV light, the photoresist is decomposed and then removed. The unprotected polyimide is then etched off. In the region not exposed to UV light, the photoresist remains and prevents the polyimide from etching. The polyimide with the dispersed pigments remains on the substrate. The process is repeated to fabricate the three types of color pixels. In the polymerization method, the pigments are mixed with reactive monomers and photo-initiators. The system is then irradiated by UV light to photopolymerize the monomers under a photomask. In the region unexposed to UV light, the monomers are not polymerized and afterward are washed off. In the region exposed to UV light, the

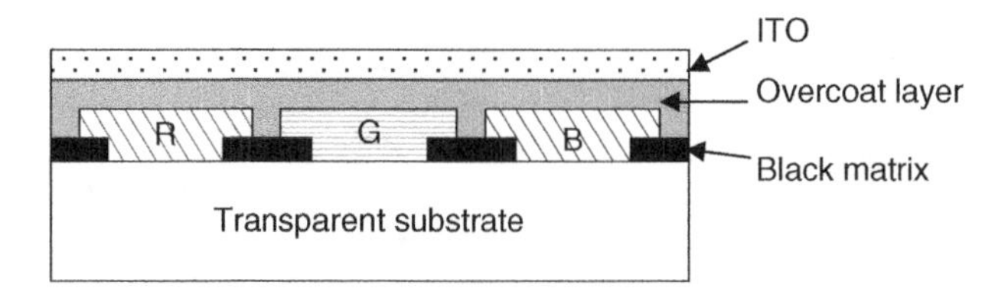

Figure 15.19 Schematic diagram of the structure of color filter.

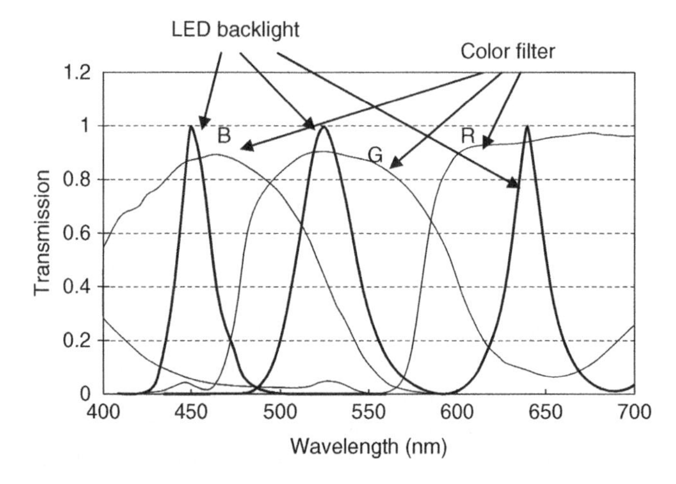

Figure 15.20 Spectra of color filter and LED backlight. Reproduced with permission from Wiley.

monomers are polymerized and stick to the substrate. The process is repeated to fabricate the three types of color pixels. The polymerization method is simpler than the etching method and therefore is more commonly used.

A typical spectrum of RGB color filter is shown in Figure 15.20 [59]. The spectrum of a LED backlight is also shown for reference. In choosing a color filter, the following properties should be considered: high transmittance, high color purity and contrast, and large chromaticity area.

References

1. S. Kobayashi, S. Mikoshiba, and S. Lim, *LCD Backlights* (John Wiley and Sons, 2009).
2. S. K. Lim, LCD Backlights and Light Sources, *Proceeding of Asia Display*, 160 (2006).
3. J.-H. Ko, Recent Research Trends in the Development of New Light Sources for the Backlight Unit of Liquid Crystal Display, *ResearchGate (*http://www.researchgate.net*)*.
4. J.-H. Ko, J.-S. Ryu, M.-Y. Yu, et al., Initial photometric and spectroscopic characteristics of 55-inch CCFL and LED backlights for LCD-TV applications, *J. of Korean Inst. of Illum. and Elec. Install. Engineers*, **24**, p. 8 (2010).
5. S. S. Kim, B. H. Berkeley, and T. Kim, Advancements for highest-performance LCD-TV, *SID Tech. Digest*, **37**, 1939 (2006).
6. J.-H. Park and J.-H. Ko, Optimization of the emitting structure of flat fluorescent lamps for LCD backlight applications, *J. of Opt. Soc. of Korea*, **11**, 118 (2207).

7. http://en.wikipedia.org/wiki/Light-emitting_diode
8. http://growblu.com/
9. W. Schwedler and F. Nguyen, LED Backlighting for LCD TVs, *SID Tech. Digest*, **41**, 1092 (2010).
10. M. Anandan, LED backlight for LCD/TV monitor: issues that remain, *SID Tech. Digest*, **37**, 1510 (2006).
11. K. Kakinuma, M. Shinoda, T. Arai, et al., Technology of wide color gamut backlight with RGB light-emitting diode for liquid crystal display television, *SID Tech. Digest*, **38**, 1232 (2007).
12. T. Shirai, S. Shimizukawa, T. Shiga, and S. Mikoshiba, RGB-LED backlights for LCD-TVs with 0D, 1D, and 2D adaptive dimming, *SID Tech. Digest*, **37**, 1520 (2006).
13. F. Yamada, H. Nakamura, Y. Sakaguchi, and Y. Taira, Color sequential LCD based on OCB with an LED backlight, *SID Tech. Digest*, **31**, 1180 (2000).
14. F. Yamada, H. Nakamura, Y. Sakaguchi, and Y. Taira, Sequential-color LCD based on OCB with an LED backlight, *J. of the SID*, **10**, 81 (2002).
15. Y. Ishiwatari, Light guide plates, in *LCD Backlights*, ed. S. Kobayashi, S. Mikoshiba and S. Lim, (John Wiley and Sons, 2009).
16. K. Käläntär, S. Matsumoto, T. Onishi, and K. Takizawa, Optical micro deflector based functional light guide plate for backlight unit, *SID Tech. Digest*, **31**, 1029 (2000).
17. C.-J. Li, Y.-C. Fang, W.-T. Chu, and M.-C. Cheng, Design of a prism light guide plate for an LCD backlight module, *Journal of the SID*, **16**, 545 (2008).
18. M. Tjahjadi, G. Hay, D. J. Coyle, and E. G. Olczak, Advances in LCD backlight film and plate technology, *Information Display*, **10**, 22 (2006).
19. G. H. Kim, W. J. Kim, S. M. Kim, and J. G. Son, Analysis of thermo-physical and optical properties of a diffuser using PET/PC/PBT copolymer in LCD backlight units, *Displays*, **26**, 37 (2005).
20. G. Park, T. S. Aum and J. H. Kwon, Characterization and modeling light scattering in diffuser sheets, *J. of Korean Phys. Soc.*, **54**, p. 44 (2009).
21. C.-H. Hung and C.-H. Tien, Modeling diffuse components by bidirectional scatter distribution function for LCD applications, *SID Tech. Digest*, **30**, 518 (2009).
22. Y. Cui, D.-K. Yang, R. Ma, J. J. Brown, Characterization and modeling of light diffusing sheet, *SID Tech. Digest Tech.*, **44**, 291 (2013).
23. M.-W. Wang and Ch.-Ch. Tseng, Analysis and fabrication of a prism film with roll-to-roll fabrication process, *Opt. Express*, **17**, 4718 (2009).
24. http://solutions.3m.com/wps/portal/3M/en_US/IndustrialFilms/Home/Products/DisplayEnhancementFilms/
25. J. Lee, S. C. Meissner, and R. J. Sudol, Optical film to enhance cosmetic appearance and brightness in liquid crystal displays, *Opt. Express*, **15**, 8609 (2007).
26. T. Okumura, A. Tagaya, and Y. Koike, Highly-efficient backlight for liquid crystal display having no optical films, *Appl. Phys. Lett.* **83,** 2515 (2003).
27. J. Ma, X. Ye, and B. Jin, Structure and application of polarizer film for thin-film-transistor liquid crystal displays, *Displays*, **32**, 49 (2011).
28. E. H. Land, Some aspects of the development of sheet polarizers, *J. of Opt. Soc. of Am.*, **41**, 957 (1951).
29. W. J. Gunning and J. Foschaar, Improvement in the transmission of iodine-polyvinyl alcohol polarizers, *Appl. Optics*, **22**, 3229 (1983).
30. M. E. Denker, A. T. Ruff, K. Derks, et al., Advanced polarizer film for improved performance of liquid crystal displays, *SID Tech. Digest*, **37**, 1528 (2006).
31. B. Kahr, J. Freudenthal, S. Phillips, et al., *Science*, **324**, 1407 (2009).
32. http://www.meadowlark.com/store/PDFs/Polarizers.pdf
33. E. Beekman, C. Kocher, A. Kokil, et al., UV polarizers based on oriented poly(vinyl alcohol)–chrysophenine–Congo red blend films, *Journal of Applied Polymer Science*, **86**, 1235 (2002).
34. D. H. Song*, H. Y. Yoo, and J. P. Kim, Synthesis of stilbene-based azo dyes and application for dichroic materials in poly(vinyl alcohol) polarizing films, *Dyes and Pigments*, **75**, p. 727 (2007).
35. Y. Bobrov, C. Cobb, P. Lazarev, et al., Lyotropic thin film polarizers, *SID Tech. Digest*, **31**, 1102 (2000).
36. M. F. Weber, C. A. Stover, L. R. Gilbert, et al., Giant birefringent optics in multilayer polymer mirrors, *Science*, **287**, 2451 (2000).
37. R. D. Guenther, *Modern optics*, (John Wiley and Sons, New York 1990).
38. S. H. Kim, J.-D. Park, and K.-D. Lee, Fabrication of a nano-wire grid polarizer for brightness enhancement in liquid crystal display, *Nanotechnology*, **17**, 4436 (2006).

39. D. Hansen, E. Gardner, and R. Perkins, The display applications and physics of the ProFlux™ wire grid polarizer, *SID Tech. Digest*, **33**, 730 (2002).
40. P. Yeh, A new optical modeling for wire grid polarizer, *Opt. Communications,* **26**, 289 (1978).
41. X. J. Yu and H. S. Kwok, Optical wire-grid polarizers at oblique angles of incidence, *J. of Appl. Phys.* **93**, 4407 (2003).
42. M. Born and E. Wolf, p. 705, *Principles of optics*, (Pergamon Press, Oxford, 1980).
43. D. J. Broer, J. Lub and G. N. Mol, Wide-band reflective polarizers from cholesteric polymer networks with a pitch gradient, *Nature,* **378**, 467 (1995).
44. L. Li and S. M. Faris, A single-layer super broadband reflective polarizer, *SID Tech. Digest*, **27**, 111 (1996).
45. R. J. Bushby and O. R. Lozman, Discotic liquid crystals 25 years, *Current Opinion in Colloid & Interface Science*, **7**, 343 (2002).
46. H. Mori, M. Nagai, H. Nakayama, et al., Novel optical compensation method based upon a discotic optical compensation film for wide-viewing-angle LCDs, *Technical Digest of SID,* **34**, 1058 (2003).
47. H. Mori, Y. Itoh, Y. Nishiura, et al., Performance of a novel optical compensation film based on negative birefringence of discotic compound for wide-view-angle twisted-nematic liquid crystal displays, *Jpn. J of Appl. Phys.*, **36**, 143 (1997).
48. J. J. Ge, B. F. Li, F. W. Harris, and S. Z. D. Cheng, Novel polymer wide view angle compensation films for liquid crystal displays (LCDs), *Chinese J of Polymer Science,* **21**, 223 (2003).
49. H. Janeschitz-Kriegl, p. 522 in *Polymer Melt Rheology and Flow Birefringence* (Springer-Verlag: Berlin, 1983).
50. D. I. Bower, *An introduction to polymer physics*, (Cambridge University Press, 2002).
51. J. Mulligan and M. Cakmak, Nonlinear mechanooptical behavior of uniaxially stretched poly(lactic acid): dynamic phase behavior, *Macromolecules,* **38**, 2333 (2005).
52. S. Okude, Retardation film made from negative intrinsic birefringent material, *SID Tech. Digest*, **40**, 888 (2009).
53. Y.-C. Yang and D.-K. Yang, Drawing-induced biaxiality change from a positive C to a negative A plate and its application in wide viewing angle IPS LCDs, *SID Tech. Digest*, **41**, 495–498 (2010)
54. E. Chino, K. Tajiri, H. Kawakami, et al., Development of wide-color-gamut mobile displays with four-primary-color LCDs, *SID Tech. Digest*, **37,** 1221 (2006).
55. T. Sugiura, EBU color filter for LCDs, *SID Tech. Digest*, **32**, 146 (2001).
56. R. W. Sabnis, Color filter technology for liquid crystal displays, *Displays*, **20**, 119 (1999).
57. H.-S. Koo, M. Chen, and P.-C. Pan, LCD-based color filter films fabricated by a pigment-based colorant photo resist inks and printing technology, *Thin Solid Films*, **515**, 896 (2006).
58. http://en.wikipedia.org/wiki/Pigment
59. K. Kakinuma, M. Shinoda, T. Arai, et al., Technology of wide color gamut backlight with RGB light-emitting diode for liquid crystal display television, *SID Tech. Digest*, **1232** (2007).

16

Three-Dimensional Displays

16.1 Introduction

Three-dimensional (3-D) display is an ultimate display technology. From the theaters to TVs at home, to naked eye mobile devices, 3-D displays [1–3] have been gaining popularity in our daily lives. In this chapter, we will outline the basic operation principles for generating depth perception, in order to realize 3-D displays. Several types of 3-D display devices: stereoscopic displays, autostereoscopic displays, integral imaging, holography, and volumetric displays are discussed.

16.2 Depth Cues

Depth cues enable us to perceive the world in three dimensions and estimate the distance of an object [4,5]. Complex, natural objects contain a variety of depth cues. Most of them are available in 2-D images, such as occlusion (opaque closer object partially covering distant objects), perspective (different points of view resulting in different scenes), and size (closer objects seeming to be larger than distant objects). As a result, even looking at 2-D media we could get some reasonably good sense of the depths. But there are several depth cues that are missing from 2-D media.

16.2.1 Binocular disparity

Since human eye pupils are horizontally separated with a distance called inter-pupil distance (IPD), each eye has a slightly different viewing point and obtains a different observation of an object. The closer the object is, the more different it appears in the two eyes. Therefore, depth is estimated by the human brain based on experience. An example is illustrated in Figure 16.1,

Fundamentals of Liquid Crystal Devices, Second Edition. Deng-Ke Yang and Shin-Tson Wu.

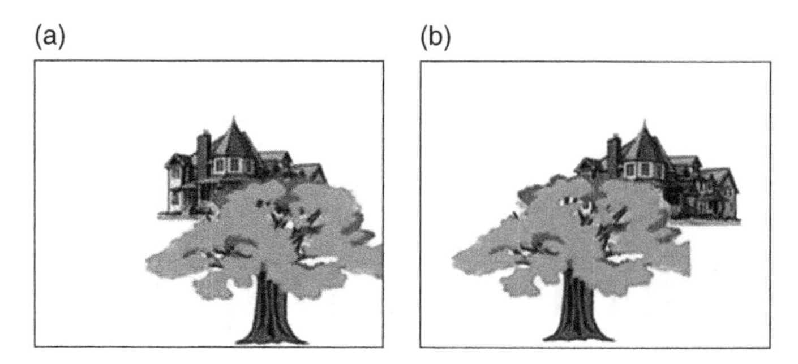

Figure 16.1 Illustration of binocular disparity: an image observed from (a) left eye, and (b) right eye.

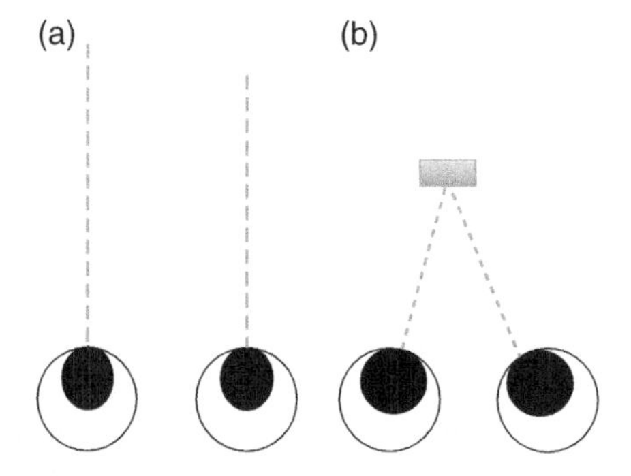

Figure 16.2 Illustration of convergence: (a) parallel visual axes for a distant object, and (b) visual axes at a large convergence angle for a nearby object.

where the distant house is positioned similarly from both eyes' fields of view, while the closer tree has a distinct horizontal shift. Disparity effect is dependent on the IPD value. The average IPD for adults is ~63 mm, and almost all adults' IPD is within 45–80 mm [6]. A smaller IPD makes binocular disparity less sensitive.

16.2.2 Convergence

When we look at an object, the eyeballs will rotate and the visual axes converge at the object. This effect is called convergence. Depending on how far away the object is, the convergence angle (the angle between the two visual axes) varies. As Figure 16.2 shows, when the object is very far, the visual axes of the two eyes are basically parallel to each other. But when the object is close, the eyeballs rotate and a larger convergence angle is subtended.

16.2.3 Motion parallax

When a human head moves sidewise, the pupil location changes so that the viewing direction changes accordingly. The closer objects appear to move faster across the field of view than those further away. This effect is, in principle, similar to binocular disparity. The former is a result of the temporal change of viewing points, and the latter is a result of spatially separated

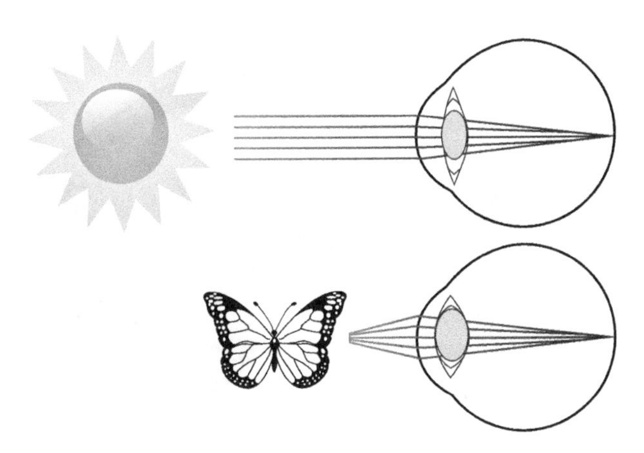

Figure 16.3 Illustration of eye accommodation due to different strains.

viewing points. Motion parallax could be observed by using either binocular or monocular. For 3-D displays that do not inherently include a motion parallax cue, head tracking is a common technique to add this cue artificially [7–10].

16.2.4 Accommodation

Accommodation is the optical power change of the human eye lens to focus on objects at different distances. As shown in Figure 16.3, if the eye is looking at a distant object such as the Sun, the eye muscles are relaxed, and a less curved lens shape is formed so that the image of the Sun is sharply formed on the retina. If the eye is looking at a nearby object, the muscles are strained to curve the lens so that the near object is focused. Due to the limits of the human eye, only within certain depth range would the objects be in focus at a specific lens power. The objects that are either too far or too close will be blurry because they are out of focus.

Binocular disparity and convergence require the involvement of both eyes, while motion parallax and accommodation could be observed even with a single eye. For natural 3-D objects, all the four major depth cues mentioned above should be present at the same time. Various 3-D display technologies employ at least one of the four major depth cues to generate 3-D depth sensation. The more consistent the depth cues are, the more realistic and natural a 3-D image appears.

16.3 Stereoscopic Displays

A stereoscopic display requires viewers to wear special glasses in order to see two slightly different 2D images in two different eyes. The 2D images are integrated by human brain to generate 3D depth perception. Apparently, the primary depth cue of stereoscopic displays is binocular disparity. Several types of stereoscopic displays have been developed, as discussed below.

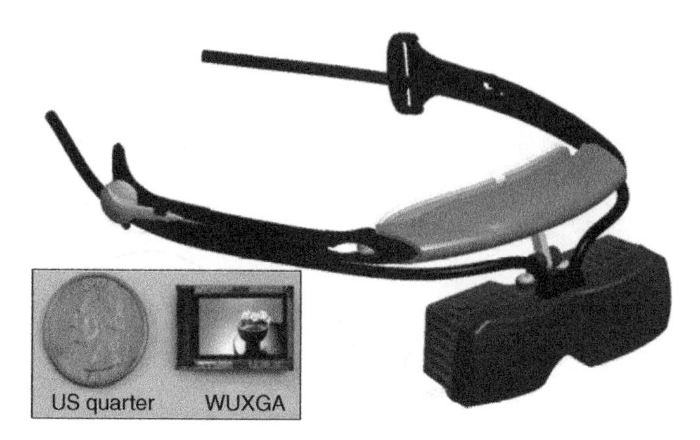

Figure 16.4 Binocular 3-D stereoscopic HMD using two full-color WUXGA resolution (1920 × 1200) AMOLED microdisplays. Reproduced with permission from Wiley.

16.3.1 Head-mounted displays

The simplest concept for realizing stereoscopy is to use a head-mounted display (HMD), which employs two small displays to present left and right offset images to each eye [11–15]. HMD used to be heavy, bulky, and uncomfortable to wear, but recently, lightweight, high-resolution and high-brightness microdisplays, such as active matrix organic light emitting diode (AMOLED) microdisplays and microLCDs [16] have been developed, so head mounted 3-D displays have gained renewed interest. Figure 16.4 shows a binocular 3-D stereoscopic HMD using full color 1920 × 1200 resolution AMOLED microdisplays [11].

16.3.2 Anaglyph

Anaglyph [17,18] gives stereoscopic 3-D effect by encoding each eye's image using complementary color, such as red and cyan as shown in Figure 16.5. The viewer wears a pair of corresponding color filters to separate the stereoscopic images. The human brain would blend the color and extract the depth information. The color filter glasses are cheap and easily available, but the major disadvantage of this approach is the absence of color information.

16.3.3 Time sequential stereoscopic displays with shutter glasses

Recently, stereoscopic displays that provide full-color high-resolution 3-D experience have been developed. One approach combines the uses of a fast-response flat panel display and active shutter glasses [19–22]. The working principle is demonstrated in Figure 16.6. The flat panel delivers left-eye and right-eye images alternately in the time sequence. During odd frames, when the left-eye image is displayed, the left-eye shutter is synchronized to be transparent while the right-eye shutter is at a dark state. During even frames, the right-eye shutter is transparent and the left-eye shutter blocks the light. As a result, the viewer sees left- and right-eye images separately in the time sequence and the brain integrates them into 3-D images.

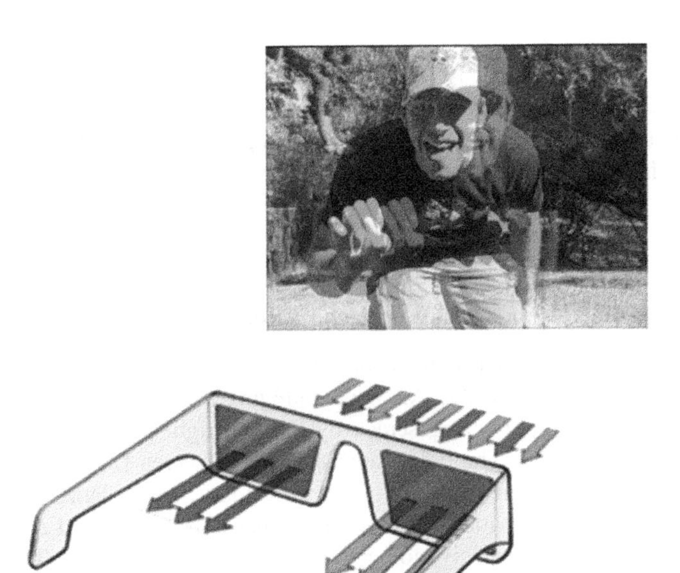

Figure 16.5 Working principle of Anaglyph 3-D using a two-color eye glasses: Left eye is cyan color filter which transmits red, and right eye is red color filter which transmits cyan.

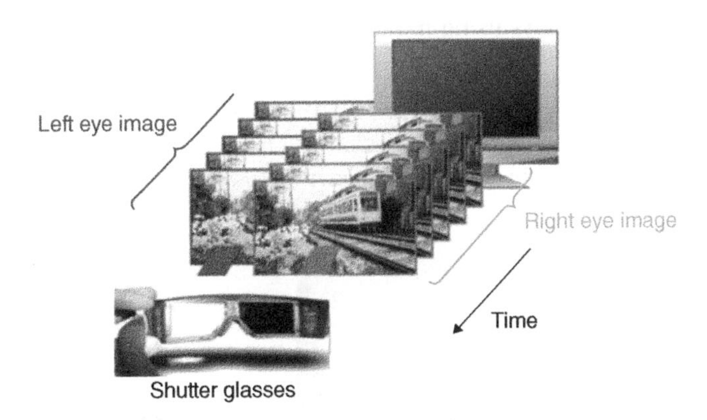

Figure 16.6 Working principle of a stereoscopic display using LC shutter glasses.

A similar concept is employed in systems where the flat panel display is replaced with a fast-response projector [23].

In a time-sequential stereoscopic display system, the flat panel display preserves full spatial resolution, but needs to be operated at a doubled frame rate. The shutter glasses need to be synchronized with the display, and also be operated at a high speed. Most shutter glasses are made of optically compensated bend (OCB) LC cells because of their fast response time [24–26]. In addition, the shutter glasses require power connection or batteries, and therefore are relatively heavy and expensive.

16.3.4 Stereoscopic displays with polarizing glasses

An alternative to shutter glasses in stereoscopic displays is passive polarizing glasses. Due to the low cost, light weight, and convenience of polarizing glasses, these displays are becoming more popular. There are a number of technologies to realize 3-D stereoscopy based on polarizing glasses.

16.3.4.1 Stereoscopic displays using projectors and polarizing glasses

A common 3-D system in cinemas consists of polarizing projectors, a polarization-preserving screen and polarizing glasses [27–29]. As shown in Figure 16.7, left- and right-eye images are projected to a polarization-preserving screen by left and right projectors respectively with orthogonal polarization states. The polarization-preserving screen [30], as its name implies, preserves the polarization of incident light while diffusing it. The polarization of light could be linear (with orthogonal transmission axes) or circular (left-handed and right-handed), the latter of which would be less sensitive to head tilting, but at a higher cost. The viewer wears a pair of orthogonal polarizers, which filter out the unwanted polarization for each eye. As a result, the left eye only sees the image from the left projector, and the right eye only sees the image from the right projector. In this approach, the left and right images are present to the viewer simultaneously. There are also systems using a single projector with active retarder to switch between the orthogonal polarizations in the time sequence [29].

16.3.4.2 Stereoscopic displays using patterned retarders and polarizing glasses

To bring the fantastic cinematic 3-D experience to ordinary family homes, 3-D TVs with patterned retarders have been developed [31,32]. The concept is shown in Figure 16.8. The odd-row pixels of the LCD are designated for left-eye images and even-row pixels for right-eye images. Correspondingly, the strips of the patterned $\lambda/4$ retarder, which overlays odd-row

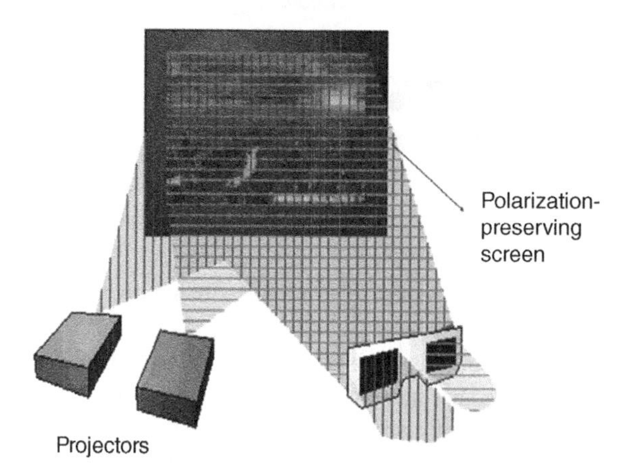

Figure 16.7 Working principle of stereoscopic display using two projectors and polarizing glasses.

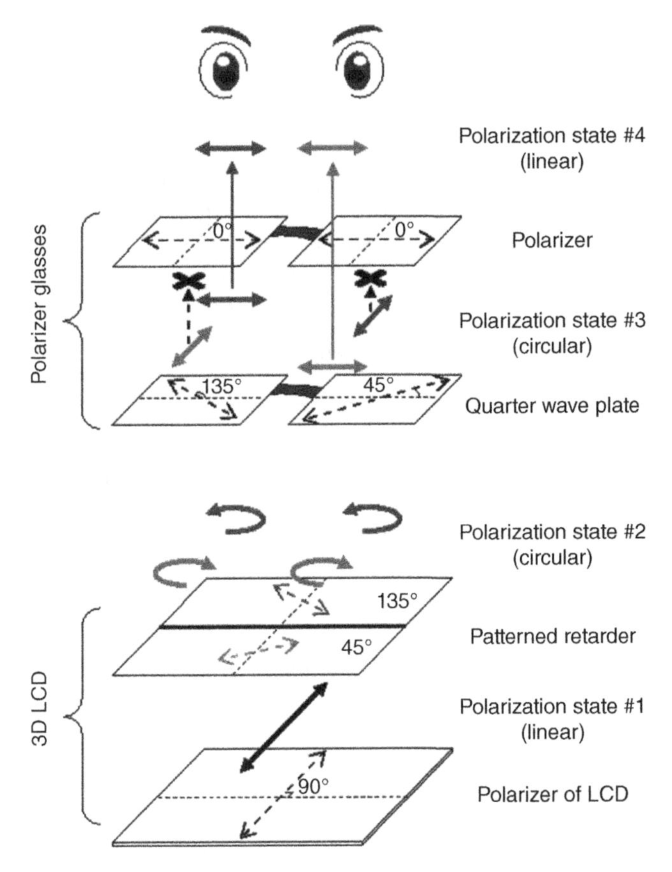

Figure 16.8 Working principle of a 3-D TV with a patterned phase retarder. Reproduced with permission from Wiley.

(even-row) pixels, have the optical axis 45° (–45°) with outer-polarizer transmission axis, and thus convert the linear polarization to left-handed (right-handed) circular polarization. The viewer wears a pair of glasses made of left- and right-handed circular polarizers that separate the left and right images. In this method, however, the spatial resolution of the display is halved.

16.3.4.3 Stereoscopic displays using active retarders and polarizing glasses

In order to maintain high spatial resolution, while using inexpensive polarizing glasses, 3-D displays using active retarders have been developed [33–35]. As Figure 16.9 shows, during odd frames, the left-eye image is shown on the whole screen, while the active half-wave retarder is switched off by applying a voltage across it. Therefore, the vertically polarized light exiting the LCD panel only encounters a quarter-wave plate whose optical axis is 45° and is converted to left-handed circular polarization. During even frames, the right image is shown on the whole screen and the active half-wave retarder is switched on with no voltage applied. The vertically polarized light is first converted into horizontally polarized by the half-wave retarder, and later encounters the quarter-wave plate, resulting in a right-handed circular polarization.

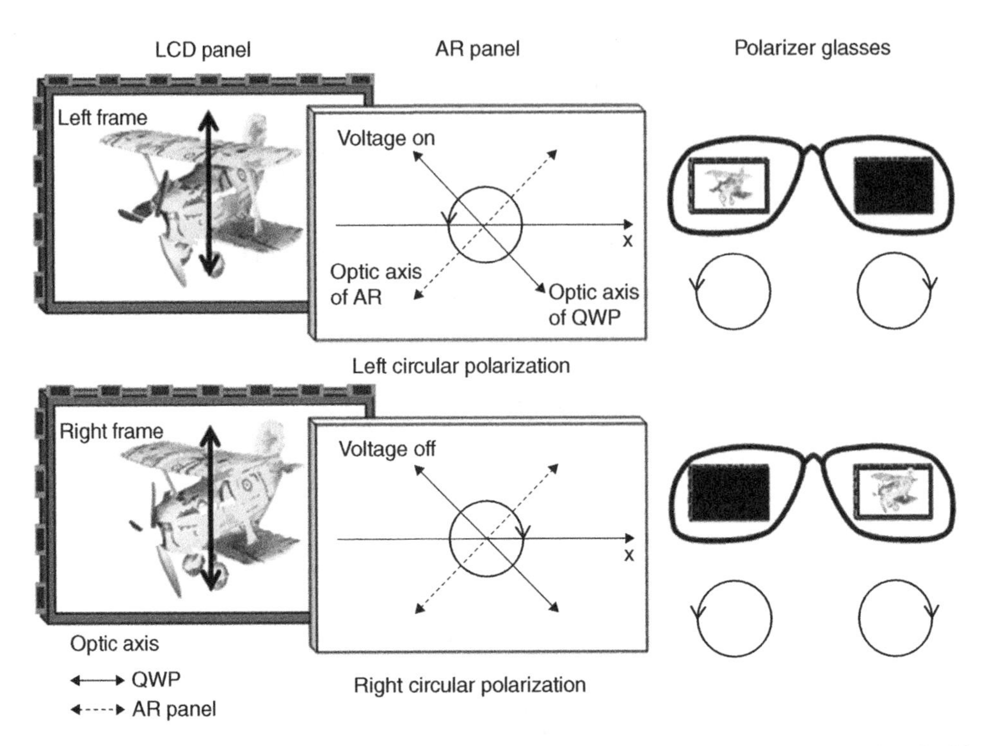

Figure 16.9 Working principle of a stereoscopic display using an active retarder. Reproduced with permission from Wiley.

As the viewer wears a pair of glasses made of left- and right-handed circular polarizers respectively, the left and right eye images are directed to the corresponding eye in the time sequence. Using this method, the full resolution of the screen is preserved, the glasses are inexpensive passive polarizers, but the flat panel frame rate has to be doubled.

16.4 Autostereoscopic Displays

Autostereoscopic displays [36] can generate stereoscopic images and form 3-D perception without the viewer wearing special glasses.

16.4.1 Autostereoscopic displays based on parallax barriers

16.4.1.1 Fixed parallax barriers

The autostereoscopic technique employing parallax barriers is one of the most common and earliest [37]. The working principle for a 2-view display based on a parallax burner is shown in Fig. 16.10. The left and right images are displayed on alternative columns of the flat panel display. An absorptive barrier is placed in front of the display so that left and right images are directed to the corresponding eye. When standing at the ideal distance and in the correct position, the viewer will perceive a stereoscopic image. However, the spatial resolution is halved.

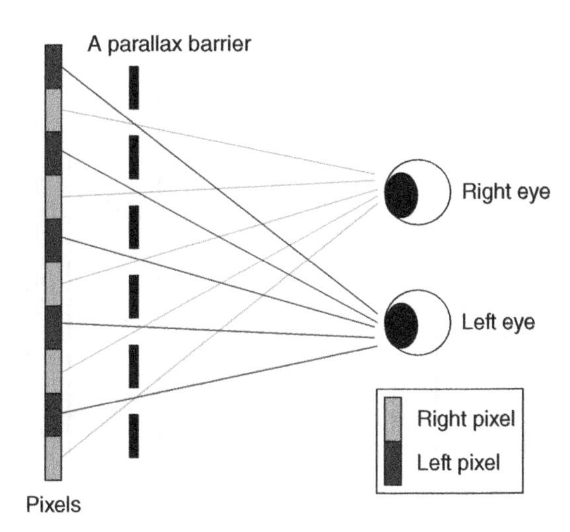

Figure 16.10 Working principle of a two-view autostereoscopic display based on a parallax barrier (top view).

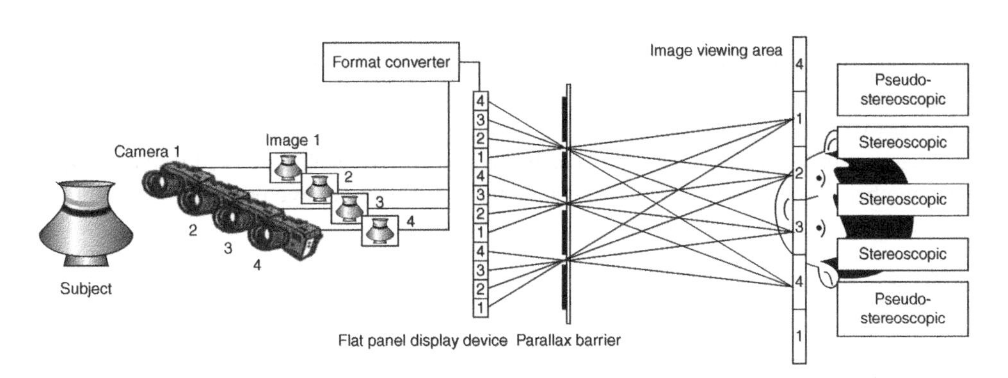

Figure 16.11 Working principle of a four-view camera and display system based on a parallax barriers (top view). Reproduced with permission from the Optical Society of America.

Multi-view autostereoscopic displays have been developed recently, allowing more than one viewer to view the 3-D scenes, from different points of view [38,39]. A four-view camera and display system is shown in Figure 16.11: cameras are horizontally separated and capture the 3-D scene from slightly different viewing angles. The images captured by the four cameras are interlaced by columns and displayed on the flat panel. The number 1, 2, 3, and 4 designated on the pixel columns indicates that they are showing images captured by camera 1,2, 3, and 4 respectively. And the barrier directs four-view images to different viewing zones 1, 2, 3, and 4. When the viewer is appropriately located in two of the four viewing zones – e.g. viewing zones 2 and 3, as shown in the figure – they could see the stereoscopic image. As a viewer moves to zones 1 and 2, or zones 3 and 4, they would be able to see another stereoscopic image but from a different perspective. In this case, to some degree, motion parallax can be observed in addition

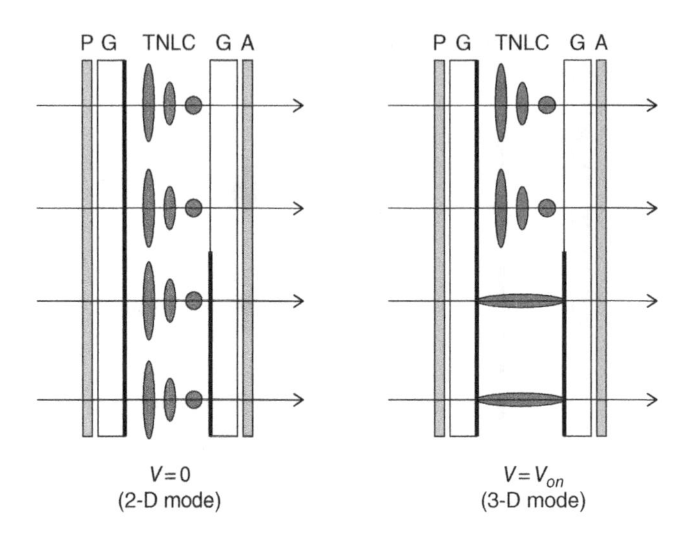

Figure 16.12 Working principle of a switchable TNLC barrier: P = polarizer, G = glass substrate, and A = crossed analyzer. Reproduced with permission from Wiley.

to binocular disparity. The more views it has, the more natural and smooth the transition between different views is. However, that is at the cost of reduced spatial resolution.

Similar to a two-view display, there is an optimal viewing distance and in some regions it only gives a pseudo-stereoscopic view. The horizontal spatial resolution is reduced by a factor of the number of views. Moreover, the opaque barriers absorb a large portion of the light and significantly reduce the brightness.

16.4.1.2 Switchable parallax barriers

2-D/3-D switchable displays employing liquid crystal barriers have been developed so that viewers can switch between full-resolution, high-brightness 2-D mode and low resolution, low brightness 3-D mode [40]. Figure 16.12 depicts the working principle of a switchable barrier. The barrier is made of a 90° TN liquid crystal cell with patterned ITO electrode on one substrate and planar ITO electrode on the other. At voltage-off state, the TN cell is normally white between crossed polarizers, as if the parallax barrier was switched off – the 2-D mode is thus realized. When a voltage is applied across the cell, in some regions the liquid crystal directors reorient vertically in response to the electric field, resulting in a dark state. In other regions, liquid crystal directors remain undisturbed and a white state is maintained. The periodical interlace of black and white regions forms the barrier, and a 3-D mode is realized.

16.4.1.3 Time-division parallax barriers

In order to maintain full resolution in 3-D mode, a time division barrier method is proposed as shown in Figure 16.13 [41]. During odd frames, odd-column pixels display the left-eye image and even-column pixels display the right-eye image. The LC barrier is operated in barrier-B

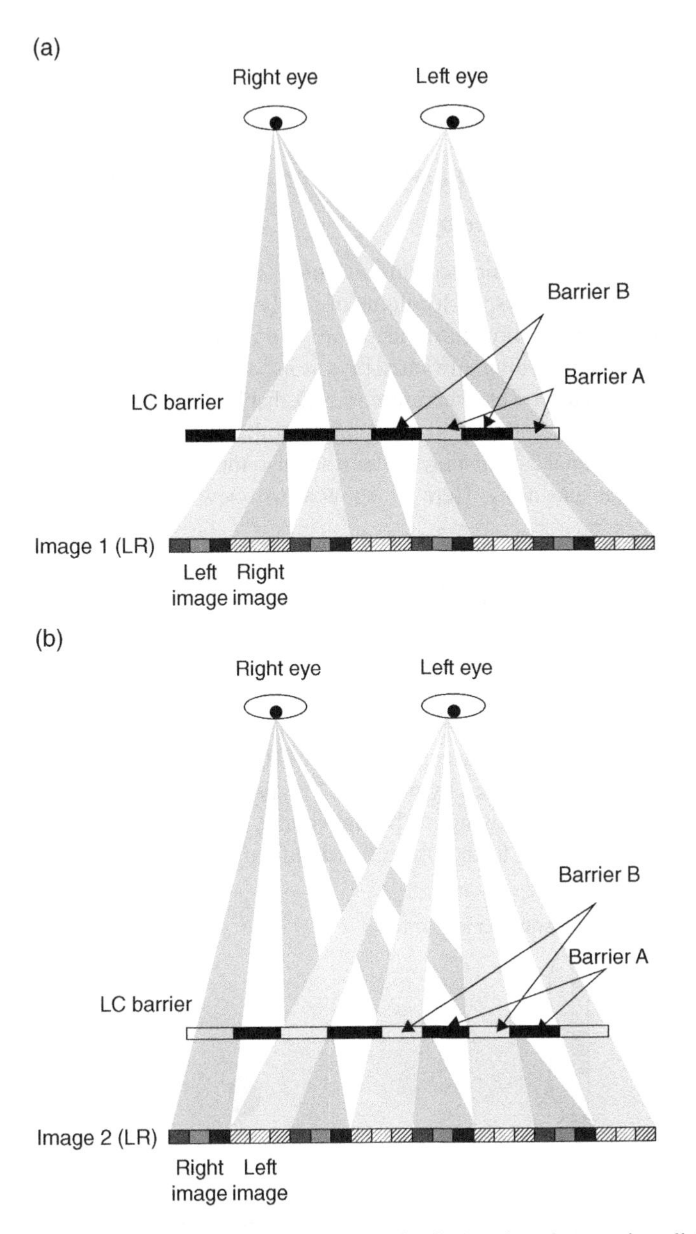

Figure 16.13 Working principle of an autostereoscopic display based on a time-division parallax barrier: (a) odd frame, and (b) even frame. Reproduced with permission from Wiley.

mode, which directs odd-column pixel images (left-eye image) to the left eye, and even-column pixel images (right-eye image) to the right eye. During even frames, on the other hand, odd-column pixels display the right-eye image and even-column pixels display the left-eye image. The LC barrier in operated in barrier A mode which directs odd-column pixel images (right-eye image) to the right eye, and even-column pixel images (left-eye image) to the left eye.

By this method, spatial multiplexing and time multiplexing are used simultaneously. Each eye can see the image from both odd- and even-column pixels in the time sequence, and the full resolution is preserved.

16.4.2 Autostereoscopic displays based on lenticular lens array

16.4.2.1 Fixed lenticular lens array

Lenticular lens array [42–47] is another common micro-optics method for autostereoscopic displays. Similar to parallax barriers, each column of pixels is visible only in one particular zone in space, thus the spatial resolution is divided into a number of distinct views as shown in Figure 16.14. Instead of absorbing unwanted rays as in parallax barriers, lenticular lenses direct the light using the optical power. Therefore, a 3-D display using a lenticular lens array has a major advantage over its counterpart of using a parallax barrier: high brightness, which is comparable to conventional 2-D displays. There are also multi-view autostereoscopic displays based on the lenticular lens array. Here we show a two-view display for simplicity.

16.4.2.2 Switchable lenticular lens array

Similar to the switchable parallax barrier, a switchable liquid crystal lenticular lens array could be used for 2-D/3-D switchable displays.

Figure 16.15 shows the basic operating principle of an autostereoscopic display using a switchable liquid crystal lens array proposed by Philips [42].

The lens array is sandwiched between two planar ITO glass substrates. On the top substrate there is a fixed negative lens formed by polymer. The liquid crystal is homogeneously aligned (perpendicular to the plane of drawing). When no voltage is applied, if the light exiting the LCD has polarization perpendicular to the plane, it would experience the extraordinary refractive

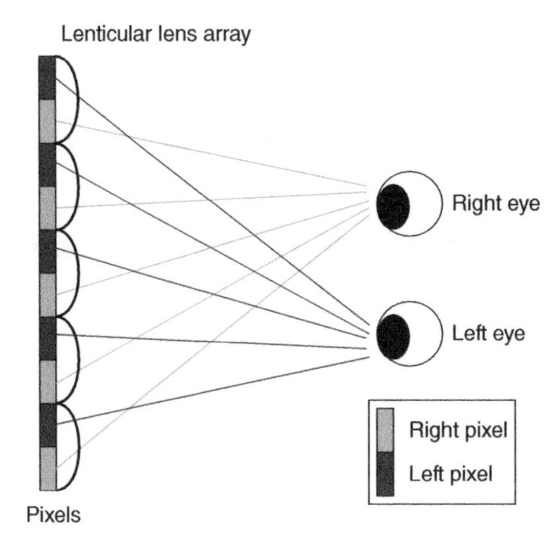

Figure 16.14 Working principle of a two-view autostereoscopic display based on a fixed lenticular lens array (top view).

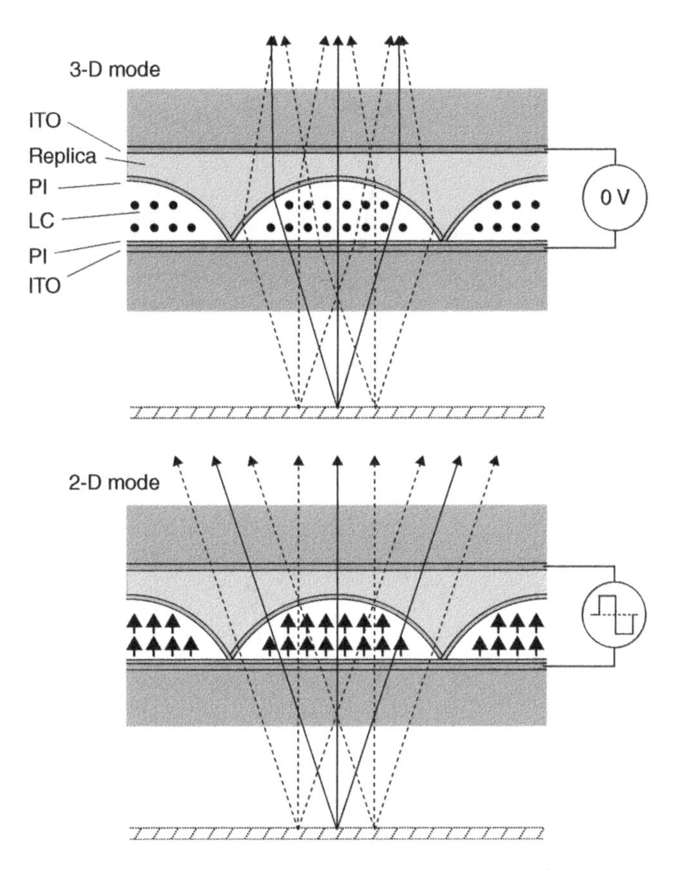

Figure 16.15 Working principle of an autostereoscopic display using a LC lens array. Reproduced with permission from Wiley.

index (n_e) of the liquid crystal, which is larger than the refractive index of the polymer. Therefore, a positive optical power is achieved, and autostereoscopic 3-D mode is realized.

On the other hand, when voltage is applied to the lens cell, the LC directors would be reoriented vertically following the electric field. Thus, the light would encounter ordinary refractive index, which is matched with the polymer index. As a result, the lenses are effectively switched off, and the displayed image is in 2-D mode.

This display could be switched between 2-D and 3-D modes electronically. But the LC alignment on the curved surface is difficult. Moreover, the switching time is slow because a relatively thick liquid crystal layer is needed to have a sufficient phase difference.

Recently, Ren et al. proposed a 2-D/3-D switchable display using a polarization rotator [43]. As depicted in Figure 16.16, the structure consists of a 2-D LCD panel, an outer polarizer, a switchable TN polarization rotator, and a polymeric liquid crystal microlens array.

In the voltage-off state, the light polarization is rotated 90° by the TN rotator, and becomes perpendicular to the drawing plane. Thus, it encounters n_o of the LC in microlens array. Since the refractive index of isotropic polymer n_p matches the ordinary refractive n_o of the homogeneously aligned liquid crystal, there is no focusing effect on the light, and the microlens array is effectively switched off, resulting in a 2-D mode. In a high voltage state, the liquid crystal

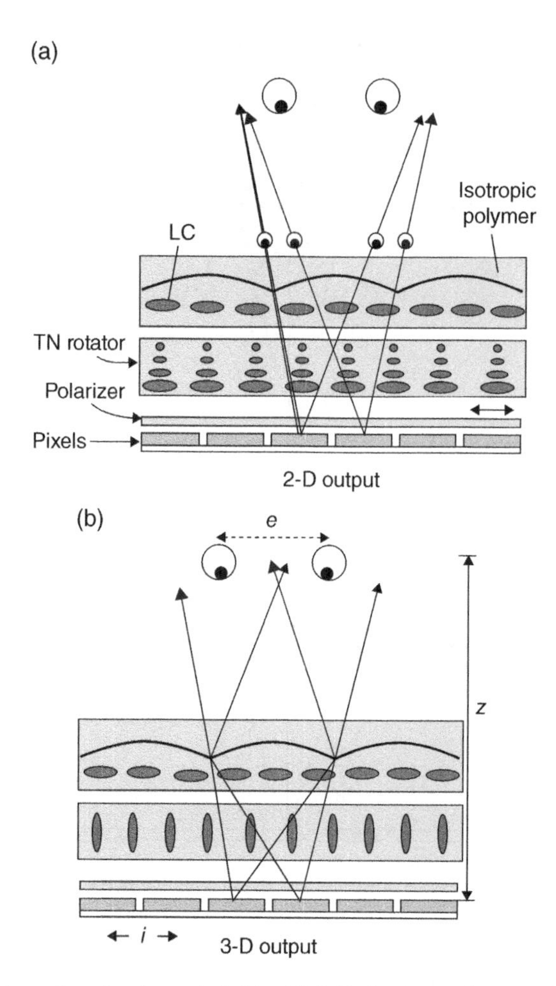

Figure 16.16 Working principle of a switchable 2-D/3-D display with a broadband TN polarization rotator and a polymeric microlens array: (a) $V = 0$ (2-D mode), and (b) $V = Von$ (3-D mode).

directors tilt up to align with the vertical electric field. The incident light polarization remains parallel to the drawing plane and encounters n_e of the LC in microlens array. Since $n_e > n_p$, the microlens array is switched on, and 3-D mode is realized. In this design, because the voltage is applied on a thin planar TN polarization rotator, the voltage is low (<5 V) and response time is fast (~1 ms, if a 1.5 μm TN cell is employed).

16.4.3 Directional backlight

3 M has demonstrated a autostereoscopic 3-D display based on a directional backlight unit [48–50].

As shown in Figure 16.17, the directional backlight unit consists of two LED light sources, a polymeric light-guide plate, and a 3-D film. During odd frames, the left LED is on, while the right one is off. The 3-D film directs the light from the left LED to the left eye, while

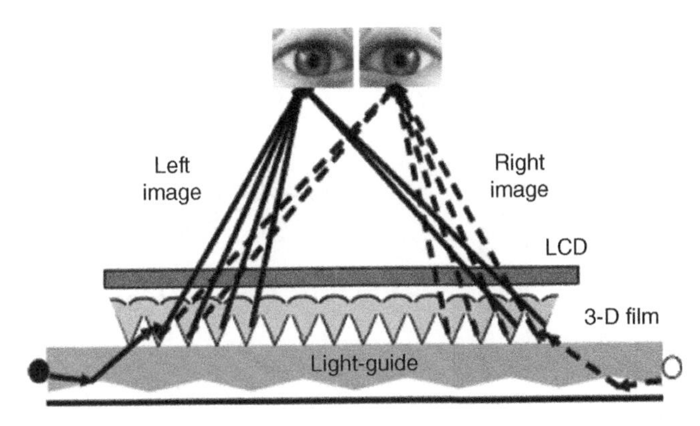

Figure 16.17 Working principle of an autostereoscopic display with a directional backlight unit. Reproduced with permission from Wiley.

the fast-response LCD is showing the left-eye image. During even frames, the film directs light from the right LED to the right eye while the LCD is showing the right-eye image. Therefore, in the time sequence, both eyes are able to get full-resolution stereoscopic images to form 3-D perception. In these two-view 3-D displays, the refresh rate of the display device is doubled, and the sophisticated light-guide and film designs play an important role affecting the display performance.

16.5 Integral imaging

The integral imaging method [51–53] was first proposed by Lippmann in 1908. The basic concept is to record 2-D projections of a 3-D object from many different perspectives. Its working principle is shown in Figure 16.18. Figure 16.18(a) depicts the image recording process, which is also called 'pick-up'. A microlens array is placed in front of the recording media. Each image formed by a single microlens is an elemental image. Because each microlens has a slightly different position in respect to the object, each has an elemental image from a unique perspective of view. The more microlenses used, the more continuous perspectives of views of the object could be obtained. Figure 16.18(b) shows the image reconstruction principle. The images captured on the recording media are displayed on a high resolution display device. Each elemental image is projected back to the 3-D space, via the microlens array, from different perspectives of view. The ray bundles that reach the eye are very much the same as those emitted by a real object.

Also, because different microlens reproduces the 3-D object from a different perspective, as the viewer move around, motion parallax could be observed.

Integral imaging is not only picked up or reconstructed optically as depicted in Figure 16.18, but can also be realized digitally, based on computer simulation [54–56]. During the pick-up process, since there is only one recording plane, only one plane in the object space could be imaged sharply by the microlens array. That plane is called the object reference plane (ORP). The farther away the object deviates from the ORP, more blurry the image on the recording media would be, as shown in Figure 16.18(a). Similarly, during reconstruction, only

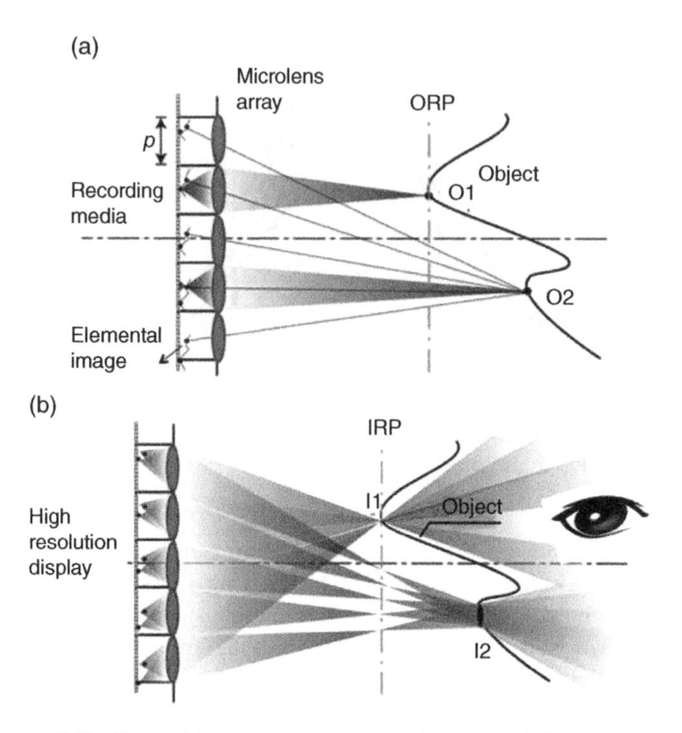

Figure 16.18 Integral imaging processes: (a) pick-up, and (b) reconstruction.

one plane in the image space, the image reference plane (IRP), could be reconstructed sharply. For the objects that are not on the IRP, although the chief rays still converge to the original image points, the ray bundles get fatter and fatter as the image deviates from IRP, as shown in Figure 16.18(b). The reconstructed 3-D image has reversed depth information to the viewer. As shown in Figure 16.18(a) during the recording process, O1 appears closer than O2. However, when the image is reconstructed, I1 appears farther from the viewer than I2. There have been various efforts to reverse the depth, both optically and digitally [57–60]. In integral imaging, very high resolution recording media and displays are needed to provide sufficient information on each elemental image. Therefore, it is relatively difficult to realize dynamic imaging [61,62].

16.6 Holography

Holography was first proposed by Dennis Gabor in 1948 [63]. The principles of recording and reconstructing a hologram are demonstrated in Figure 16.19. A coherent beam is split into two beams by a beam splitter. One serves as a reference, and the other illuminates 3-D objects, gets scattered and becomes the object beam. The object beam carries the information about the objects. The recording medium is placed where the reference beam and object beam interfere. To reconstruct the 3-D image, a reconstruction beam, which is identical to the reference beam, is incident to the hologram, and the reconstructed wavefront is generated in the opposite direction to the object beam. As a result, a virtual 3-D image is formed.

With advances in computer technologies and electronic devices, it is possible to reconstruct 3D scenes without the conventional optical interference recording and film developing process. The hologram, which includes the wavefront information of object light, could be either virtually generated by a computer or acquired from digital imaging devices such as charge coupled devices (CCDs) or cameras. Then, the hologram is loaded to a display device like a spatial light modulator (SLM), which spatially modulates the either amplitude or phase, or both amplitude and phase of the incident light. When illuminated with an appropriate reference wave, the SLM physically reconstructs the object wavefront and displays the 3D scene. The updatability of SLMs makes dynamic holographic displays feasible. However, achieving video-rate real-time 3D holographic displays remains challenging mainly due to the huge amount of data and the limitations of SLM devices.

To precisely reconstruct the 3D scene wavefronts in real time, huge amount of data is involved in the calculation, sampling or communication process, far beyond the capability of the state-of-art computers. To address this issue, various methods from both software aspect and hardware aspect have been proposed, such as the look-up table algorithm [64], graphics processing unit (GPU) [65] devices and field-programmable gate array (FPGA) devices [66]. Integral holography [67], which, instead of reproducing the wavefronts as in the conventional way, records and reconstructs 2D hogels (note: hogel is a compound word of *holographic*

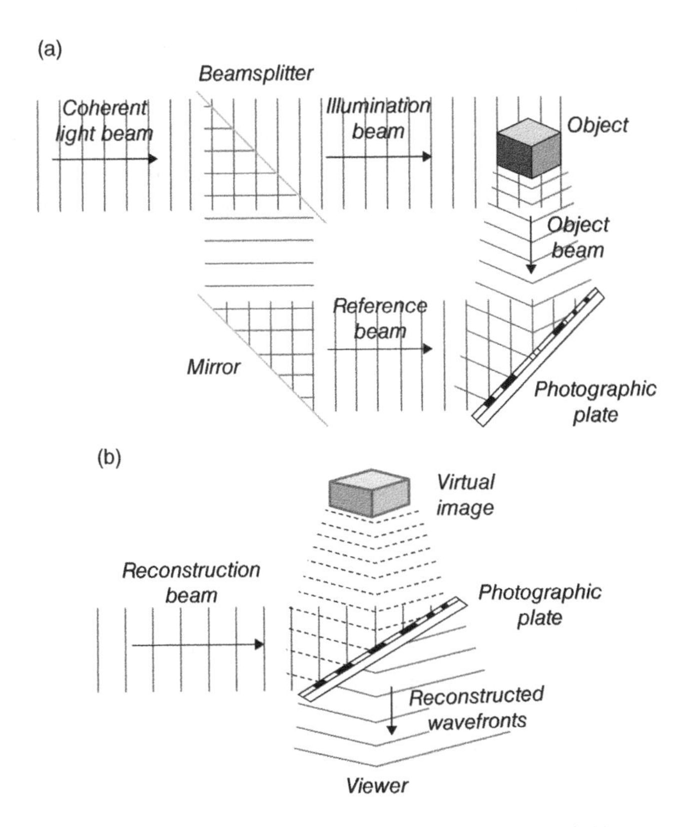

Figure 16.19 Working principle of a hologram: (a) recording and (b) reconstruction.

and element) from different perspectives, have been adopted in 3D holographic displays to effectively reduce the computational burden while providing a fascinating 3D sensation.

As for the physical display devices, the most commonly used SLMs in holographic displays are: LCD-SLM [68], LCOS-SLM [69], and DMD-SLM [70]. Relatively low visualization quality [71] could be achieved in such systems: the projection angle is limited to less than 10° due to the relatively large pixel size (larger than 2μm); the small SLM display area (about several square centimeters) limits the size of display content; the presence of quantization noise, unwanted diffractive orders and zero-order light are also obstacles to be overcome. Over the decades, great efforts have been endeavored to developing suitable SLM devices for holographic displays.

Since 1990s, MIT Media Lab has been building dynamic holographic display systems based on acousto-optic modulator (AOM). This type of device changes the refractive index in an acousto-optic medium through acoustic wave, thus modulating the light illuminating on the medium. Several prototypes have been built with larger viewing angles [71]: 15° for Mark I, 30° for Mark II and 24° for Mark III. Recently, they have developed a new holographic scanner design [72,73], based on the principle of anisotropic leaky-mode coupling. It supports a bandwidth of more than 50 billion pixels per second, a 10X improvement to the current state of the art, and can be constructed at a fairly low cost.

In 2011, QinetiQ used active tiling method, in combination of a fast electronic addressed SLM, and a large-area optically addressed SLM, and achieved a pixel density of over 2.2 million pixels per square centimeter [74]. Tay et al. from University of Arizona developed an updatable photorefractive polymer material [75,76] which could be made larger size and refreshed at a rate 0.5 s^{-1}. Other promising display devices include magneto-optical SLM [77], polymer-dispersed liquid crystal (PDLC) [78], photochromic and photodichroic materials [79] and so on. Nevertheless, the current display devices are not ready for practical applications, and the realization of full parallax, real-time, full-color, low cost 3D holographic displays remains an open challenge.

16.7 Volumetric displays

Volumetric displays [80,81] use voxels, pixels that use three-dimensional coordinates (x, y, z), to present image in space. Each voxel scatters or emits light as if the light were coming from a real 3-D object. The key concept of a volumetric 3-D display system is to have the entire display volume filled with voxels that can be selectively excited at any desired locations. For most volumetric displays, it is difficult to have shadows, because they cannot generate black pixels.

16.7.1 Swept volumetric displays

One of the earliest volumetric 3-D displays is the swept volumetric display [82,83]. The key component is a fast-switching light emitter array panel that rotates at high speed. The *xy* addressing of the light emitter array shows the 2-D slide at one time, and 3-D images can be formed within the volume swept by the rotating panel. A recent demonstration of a swept volumetric display using RGB LED arrays is shown in Figure 16.20 [84].

Another swept approach is shown in Figure 16.21 [85]. In the center, a semi-transparent bidirectional Lambertian scattering projection screen rotates around its center axis at

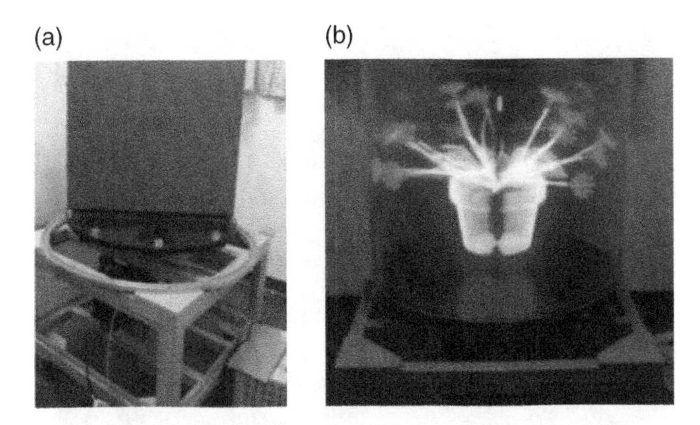

Figure 16.20 (a) A volumetric display system based on rotating LED array; (b) The displayed 3-D flower. Reproduced with permission from Wiley.

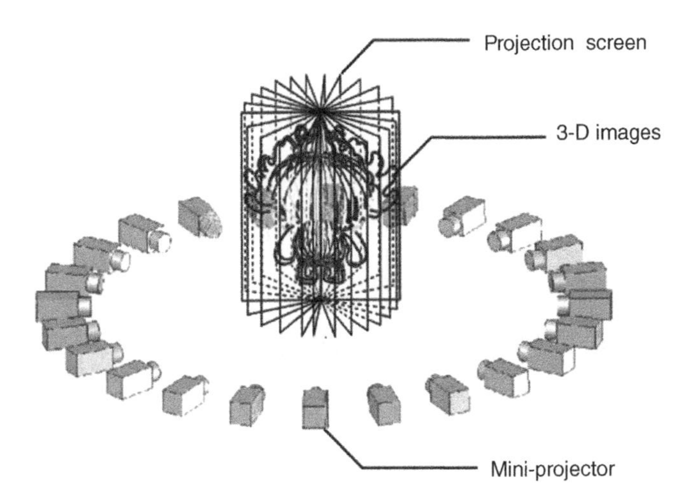

Figure 16.21 Schematic diagram of a volumetric display using multiple mini-projectors and a rotating screen. Reproduced with permission from Wiley.

high speed. One hundred and twenty mini-projectors are placed in a circle around the rotating screen. At any instant during the screen rotation, only one mini-projector whose optical axis is perpendicular to the screen is turned on, to project a 2-D image slide to the screen. As the screen rotates, different projectors are turned on in time sequence, and all the voxels have been accessed. Due to the persistence of the human vision, the 3-D volumetric image is perceived.

16.7.2 Multi-planar volumetric displays

A multi-planar volumetric display presents the volume by a stack of 2-D images at different depths. An example is shown in Figure 16.22 employing 20 polymer-dispersed liquid crystal

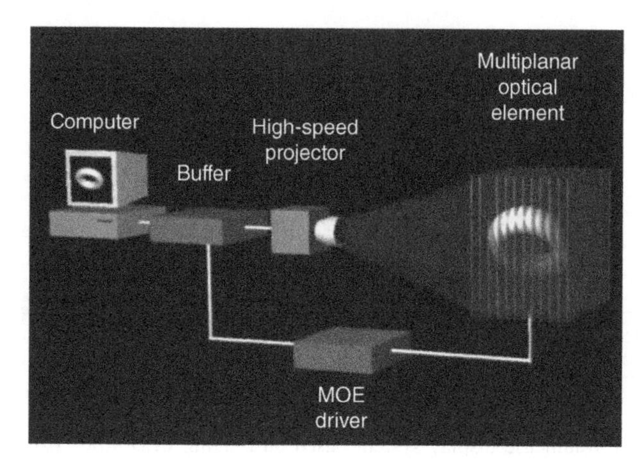

Figure 16.22 A multi-planar volumetric display system using PDLC screens. Reproduced with permission from Wiley.

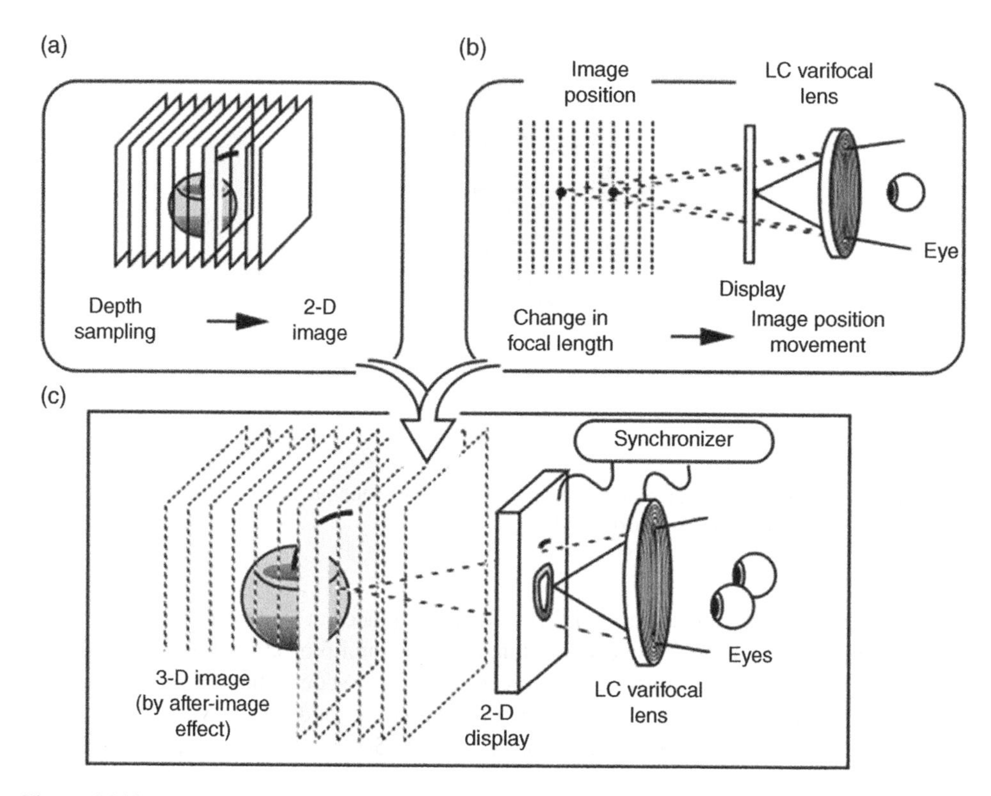

Figure 16.23 A multi-planar volumetric display using high-speed varifocal lens. (a) 3-D object divided into 2-D image slides. (b) The virtual images of the 2-D display can be shifted by changing the focal length of the LC varifocal lens. (c) A 3-D image is reconstructed volumetrically by displaying all the depth-sampled images at corresponding depth positions. Reproduced with permission from the Japan Society of Applied Physics. Copyright © 2000.

screens, each of which could be electronically switched between scattering (voltage off) and transparent (voltage on) states at high speed [86]. One 3-D frame is divided into 20 subframes. During each subframe, only one of the liquid crystal screens is at scattering state while all the others remain transparent. A high-speed DLP projector projects a corresponding 2-D image slide to that scattering liquid crystal screen, as if the image is emitted from that plane. As all the liquid crystal screens have been switched into scattering state one by one, 2-D image slides at different depths are shown in the time sequence. Both the projector and liquid-crystal screens are switched so fast that it appears to human eyes that all the 2-D image slides are presented simultaneously, and the 3-D effect is perceived.

Another multi-planar volumetric display based on high-speed, varifocal lens is shown in Figure 16.23 [87]. Figure 16.23(a) shows that a 3-D object could be divided into several 2-D slides, 16.23(b) shows that with a variable-focal-length lens, the virtual image of the 2-D display could be shifted to different depths, and 16.23(c) shows that 3-D image could be reconstructed volumetrically by displaying all depth-sampled images at corresponding depth positions within the after-image time (<1/60 s). The system requires high speed operation of a 2-D display (e.g. CRT, ferroelectric display, or micro-mirror array display), and a varifocus lens.

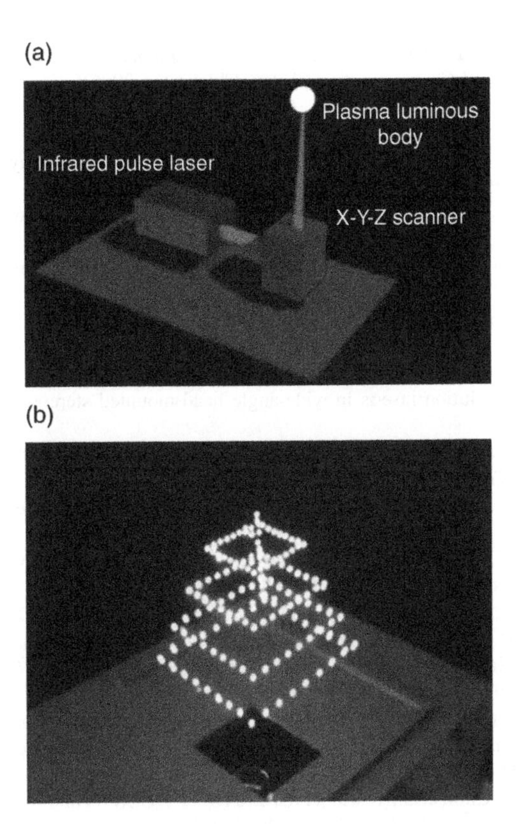

Figure 16.24 (a) Free space 3-D display using an infrared pulsed laser, and (b) a 3-D object displayed by the glowing plasma excited by the pulsed infrared laser. Reproduced with permission from SPIE.

The response time of a conventional varifocal lens is relatively slow [88]. To improve the response time, a Fresnel lens with a dual frequency liquid crystal [89,90] can be used. Thus, the response time of the LC lens is sufficient for 60 Hz operation to avoid flicker.

16.7.3 Points volumetric displays

Another volumetric display technique involves the excitation of voxel in air by lasers [91]. Figure 16.24(a) shows a 3-D display using a pulsed infrared laser. The laser creates balls of glowing plasma at the focal point in the air. The glowing ball is directed by an *xyz* scanner, and can excite any voxel within the volume cube in the air. Each voxel glows for only about a nanosecond but it appears to last longer due to the persistence of human vision. A continuous image is perceived by refreshing the glowing points. A generated 3-D image is shown in Figure 16.24(b).

References

1. E. Lueder, *3-D Displays* (Wiley, Chichester, 2011).
2. B. Lee, Three-dimensional displays, past and present, *Physics Today* **66**, 36 (2013).
3. T. Okoshi, *Three-Dimensional Imaging Techniques* (Academic Press, New York, 1976).
4. O. Schreer, P. Kauff, and T. Sikora, *3-D Video communications: Algorithms, Concepts and Real-Time Systems in Human Centred Communication* (Wiley, Chichester, 2006).
5. R. Patterson, Human factors of 3-D displays, *J. SID*, **15**, 861 (2007).
6. N. A. Dodgson, Variation and extrema of human interpupillary distance, *Proc. SPIE*, **5291**, 36 (2004).
7. Y. Yang, A. Higashi, T. Uehara, et al., A wide-view high resolution 3D display using real-time rendering relating to viewer position, *SID Tech. Digest*, **44**, 78 (2013).
8. L. McMillan and G. Bishop, Head-tracked stereoscopic display using image warping, *Proc. SPIE*, **2409**, 21 (1995).
9. Y. Kitamura, T. Konishi, S. Yamamoto, and F. Kishino, Interactive stereoscopic display for three or more users, *Proc. of the 28th Annual Conference on Computer Graphics and Interactive Techniques*, 231 (2001).
10. S. Fisher, Viewpoint dependent imaging: An interactive stereoscopic display, *Proc. SPIE*, **0367**, 41(1983).
11. I. I. Khayrullin, I. Wacyk, T. A. Ali, et al., WUXGA resolution 3D stereoscopic head mounted full color AMOLED microdisplay, *SID Tech. Digest*, **43**, 244 (2012).
12. E. M. Howlett, High-resolution inserts in wide-angle head-mounted stereoscopic displays, *Proc. SPIE*, **1669**, 193 (1992).
13. E. Kambe, M. Nakamura, J. Yamada, et al., Stable white OLED device structure for 3D-compatible head mounted display, *SID Tech. Digest* **43**, 363 (2012).
14. M. B. Spitzer, P. M. Zavracky, G. Hunter, and N. Rensing, Wearable, stereo eyewear display, *SID Tech. Digest*, **34**, 261 (2003).
15. R. Zhang and H. Hua, Design of a polarized head-mounted projection display using ferroelectric liquid-crystal-on-silicon microdisplays, *Appl. Opt.* **47**, 2888 (2008).
16. H. L. Ong, Low voltage, high contrast ratio, and wide viewing angle TN LCDs for microdisplay, mobile phone and PDA mobile video applications, *SID Tech. Digest*, **37**, 748 (2006).
17. A. J. Woods and C. R. Harris, Comparing levels of crosstalk with red/cyan, blue/yellow, and green/magenta anaglyph 3D glasses, *Proc. SPIE*, **7524**, 75240Q (2010).
18. A. J. Woods and T. Rourke, Ghosting in anaglyphic stereoscopic images, *Proc. SPIE*, **5291**, 354 (2004).
19. S. Shestak and D. Kim, Application of pi-cells in time-multiplexed stereoscopic and autostereoscopic displays based on LCD panels, *Proc. SPIE* **6490**, 64900Q (2007).
20. S. S. Kim, B. H. You, H. Choi, et al., World's First 240 Hz TFT-LCD technology for full-HD LCD-TV and its application to 3D Display, *SID Tech. Digest*, **40**, 424 (2009).
21. H. M. Zhan, Z. Xu, Y. C. Wang, et al., Fast response fringe-field switching mode liquid crystal development for shutter glass 3D, *J. SID*, **21**, 137 (2013).

22. A. Srivastava, J. de Bougrenet de la Tocnaye, and L. Dupont, Liquid crystal active glasses for 3D cinema, *J. Disp. Technol.* **6**, 522 (2010).
23. S. M. Faris, Novel 3D stereoscopic imaging technology, *Proc. SPIE*, **2177**, 780 (1994).
24. P. J. Bos and K. Koehler, The pi-cell: A fast liquid crystal optical switching device, *Mol. Cryst. Liq. Cryst.* **133**, 329 (1984).
25. D. Suzuki, T. Fukami, E. Higano, et al., Crosstalk-free 3D display with time-sequential OCB LCD, *SID Tech. Digest*, **40**, 428 (2009).
26. T. Ishinabe, K. Wako, and T. Uchida, A fast-switching OCB-mode LCD for high-quality display applications, *J. SID*, **18**, 968 (2010).
27. V. Walworth, S. Bennett, and G. Trapani, Three-dimensional projection with circular polarizers, *Proc. SPIE*, **0462**, 64 (1984).
28. A. J. Woods, Optimal usage of LCD projectors for polarized stereoscopic projection, *Proc. SPIE*, **4297**, 5 (2001).
29. M. Schuck and G. Sharp, 3D digital cinema technologies, *SID Tech. Digest*, **43**, 629 (2012).
30. D. Coleman and G. Sharp, High efficiency polarization preserving cinema projection screens, *SID Tech. Digest*, **44**, 748 (2013).
31. H. Kang, S. D. Roh, I. S. Baik, et al., A novel polarizer glasses type 3D displays with a patterned retarder, *SID Tech. Digest*, **41**, 1(2010).
32. Y. J. Wu, Y. S. Jeng, P. C. Yeh, et al., Stereoscopic 3D display using patterned retarder, *SID Tech. Digest*, **39**, 260 (2008).
33. S. M. Jung, Y. B. Lee, H. J. Park, et al., Polarizer glasses type 3D TVs having high image quality with active retarder 3D technology, *SID Tech. Digest*, **42**, 168 (2011).
34. S. M. Jung, J. U. Park, S. C. Lee, et al., A novel polarizer glasses-type 3D displays with an active retarder, *SID Tech. Digest* **40**, 348 (2009).
35. C. W. Su, M. S. Shih, and J. T. Lien, A novel polyimide-free patterned retarder 3D display, *SID Tech. Digest* **42**, 1590 (2011).
36. A. R. Travis, "Autostereoscopic 3-D display," *Appl. Opt.* **29**, 4341–4342 (1990).
37. F. E. Ives, A novel stereogram, *J. Franklin Inst.* **153**, 51(1902).
38. N. A. Dodgson, Analysis of the viewing zone of the Cambridge autostereoscopic display, *Appl. Opt.* **35**, 1705 (1996).
39. K. Mashitani, G. Hamagishi, M. Higashino, et al., Step barrier system multiview glassless 3D display, *Proc. SPIE*, **5291**, 265 (2004).
40. H. Nam, J. Lee, H. Jang, et al., Auto-stereoscopic swing 3D display, *SID Tech. Digest*, **36**, 94 (2005).
41. H. J. Lee, H. Nam, J. D. Lee, et al., A high resolution autostereoscopic display employing a time division parallax barrier, *SID Tech. Digest*, **37**, 81(2006)
42. M. G. H. Hiddink, S. T. de Zwart, O. H. Willemsen, and T. Dekker, Locally switchable 3D displays, *SID Tech. Digest*, **37**, 1142 (2006).
43. H. Ren, S. Xu, Y. Liu, and S. T. Wu, Switchable focus using a polymeric lenticular microlens array and a polarization rotator, *Opt. Express* **21**, 7916 (2013).
44. Y. Li and S. T. Wu, Polarization independent adaptive microlens with a blue-phase liquid crystal, *Opt. Express* **19**, 8045 (2011).
45. C. T. Lee, Y. Li, H. Y. Lin, and S. T. Wu, Design of polarization-insensitive multi-electrode GRIN lens with a blue-phase liquid crystal, *Opt. Express* **19**, 17402 (2011).
46. J. Sun, S. Xu, H. Ren, and S. T. Wu, Reconfigurable fabrication of scattering-free polymer network liquid crystal prism/grating/lens, *Appl. Phys. Lett.* **102**, 161106 (2013).
47. C. van Berkel and J. A. Clarke, Characterisation and optimisation of 3D-LCD module design, *Proc. SPIE*, **3012**, 179 (1997).
48. T. Sasagawa, A. Yuuki, S. Tahata, et al., Dual directional backlight for stereoscopic LCD, *SID Tech. Digest* **34**, 399 (2003).
49. R. Brott and J. Schultz, Directional backlight lightguide considerations for full resolution autostereoscopic 3D displays, *SID Tech. Digest* **41**, 218 (2010).
50. S. E. Brigham and J. Schultz, Directional backlight timing requirements for full resolution autostereoscopic 3D displays, *SID Tech. Digest* **41**, 226 (2010).
51. M. G. Lippmann, Epreuves reversibles donnant la sensation du relief, *J. Phys.*, **7**, 821 (1908).

52. B. Lee, S. Min, and B. Javidi, Theoretical analysis for three-dimensional integral imaging systems with double devices, *Appl. Opt.* **41**, 4856 (2002).
53. S. H. Hong, J. S. Jang, and B. Javidi, Three-dimensional volumetric object reconstruction using computational integral imaging, *Opt. Express* **12**, 483 (2004).
54. H. Arimoto and B. Javidi, Integral three-dimensional imaging with digital reconstruction, *Opt. Lett.* **26**, 157 (2001).
55. Y. Frauel and B. Javidi, Digital three-dimensional image correlation by use of computer-reconstructed integral imaging, *Appl. Opt.* **41**, 5488 (2002).
56. H. Kakeya and T. Kurokawa, Energy-efficient integral imaging with suppression of pseudo images, *Opt. Lett.* **38**, 3227 (2013).
57. H. Navarro, R. Martínez-Cuenca, G. Saavedra, et al., 3D integral imaging display by smart pseudoscopic-to-orthoscopic conversion (SPOC), *Opt. Express* **18**, 25573 (2010).
58. H. Hoshino, F. Okano, H. Isono, and I. Yuyama, Analysis of resolution limitation of integral photography, *J. Opt. Soc. Am. A*, **15**, 2059 (1998).
59. J. Arai, F. Okano, M. Kawakita, et al., Integral three-dimensional television using a 33-megapixel imaging system, *J. Disp. Technol.* **6**, 422 (2010).
60. M. Oikawa, M. Kobayashi, T. Koike, K. et al., Sample applications suitable for features of integral videography, *SID Tech. Digest*, **39**, 748 (2008).
61. Y. Kim, B. Lee, S. W. Min, et al., Projection-type integral imaging system using convex mirror array, *SID Tech. Digest* **39**, 752 (2008).
62. Y. Liu, H. Ren, S. Xu, et al., Adaptive focus integral image system design based on fast-response liquid crystal microlens, *J. Display Technol.* **7**, 674 (2011).
63. D. Gabor, A new microscopic principle, *Nature*, **161**, 777 (1948).
64. M. E. Lucente, Interactive computation of holograms using a look-up table, *J. Electron. Imaging* **2**, 1 (1993).
65. T. Shimobaba, T. Ito, N. Masuda, et al., Fast calculation of computer-generated-hologram on AMD HD5000 series GPU and OpenCL, *Opt. Express* **18**, 10 (2010).
66. Y. Ichihashi, H. Nakayama, T. Ito, et al., HORN-6 special-purpose clustered computing system for electroholography, *Opt. Express* **17**, 16 (2009).
67. N. T. Shaked, J. Rosen, and A. Stern, Integral holography: white-light single-shot hologram acquisition, *Opt. Express* **15**, 9 (2007).
68. T. Ito, T. Shimobaba, H. Godo, and M. Horiuchi, Holographic reconstruction with a 10-μm pixel-pitch reflective liquid-crystal display by use of a light-emitting diode reference light, *Opt. Lett.* **27**, 16 (2002).
69. A. Michalkiewicz, M. Kujawinska, J. Krezel, et al., Phase manipulation and optoelectronic reconstruction of digital holograms by means of LCOS spatial light modulator, *Proc. SPIE*, **5776**, (2005).
70. M. Huebschman, B. Munjuluri, and H. Garner, Dynamic holographic 3-D image projection, *Opt. Express* **11**, 5 (2003).
71. J. Geng, Three-dimensional display technologies, *Advances in Optics and Photonics* **5**, 4 (2013).
72. D. Smalley, Q. Smithwick, V. Bove, et al., Anisotropic leaky-mode modulator for holographic video displays, *Nature* **498**, 7454 (2013).
73. D. E. Smalley, Q. Y. Smithwick, and V. M. Bove Jr, Holographic video display based on guided-wave acousto-optic devices, *Proc. SPIE*, **6488,** (2007).
74. M. Stanley, M. A. Smith, A. P. Smith, et al., 3D electronic holography display system using a 100-megapixel spatial light modulator, *Proc. SPIE,* **5249**, (2004).
75. S. Tay, P. A. Blanche, R. Voorakaranam, et al., An updatable holographic three-dimensional display, *Nature* **451**, 7179 (2008).
76. P. A. Blanche, A. Bablumian, R. Voorakaranam, et al., Holographic three-dimensional telepresence using large-area photorefractive polymer, *Nature* **468**, 7320 (2010).
77. H. Takagi, K. Nakamura, T. Goto, et al., Magneto-optic spatial light modulator with submicron-size magnetic pixels for wide-viewing-angle holographic displays, *Opt. Lett.* **39**, 11 (2014).
78. D. Coates, Polymer-dispersed liquid crystals, *J. Mater. Chem.* **5**, 12 (1995).
79. X. Li, C. P. Chen, H. Y. Gao, et al., Video-Rate Holographic Display Using Azo-Dye-Doped Liquid Crystal, *J. Disp. Technol.* **10**, 6 (2014).
80. G. E. Favalora, Volumetric 3D displays and application infrastructure, *Computer*, **38**, 37 (2005).

81. B. G. Blundell, A. J. Schwarz, and D. K. Horrell, Volumetric three-dimensional display systems: their past, present and future, *IEE Sci. Ed. J.* **2**, 196 (1993).
82. R. J. Schipper, Three-dimensional display, US Patent **3**,097,261 (July 9, 1963).
83. E.P. Berlin, Three-dimensional display, US Patent 4,160,973, (July 10, 1979).
84. J. Wu, C. Yan, X. Xia, et al., An analysis of image uniformity of three-dimensional image based on rotating LED array volumetric display system, *SID Tech. Digest*, **41**, 657 (2010).
85. W. Song, Q. Zhu, T. Huang, et al., Volumetric display system using multiple mini-projectors, *SID Tech. Digest*, **44**, 318 (2013).
86. A. Sullivan, A solid-state multi-planar volumetric display, *SID Tech. Digest*, **34**, 1531 (2003).
87. S. Suyama, M. Date, and H. Takada, Three-dimensional display system with dual-frequency liquid-crystal varifocal lens, *Jpn. J. Appl. Phys.*, **39**, 480 (2000).
88. H. Ren and S. T. Wu, *Introduction to Adaptive Lenses* (Wiley, Hoboken, 2012).
89. H. Xianyu, S. T. Wu, and C. L. Lin, Dual frequency liquid crystals: A review, *Liq. Cryst.*, **36**, 717 (2009).
90. H. Xianyu, Y. Zhao, S. Gauza, et al., High performance dual frequency liquid crystal compounds, *Liq. Cryst.*, **35**, 1129 (2008).
91. H. Saito, H. Kimura, S. Shimada, et al., Laser-plasma scanning 3D display for putting digital contents in free space, *Proc. SPIE*, **6803**, 680309 (2008).

Index

Note: Page numbers in *italics* refer to Figures; those in **bold** to Tables

Fundamentals of Liquid Crystal Devices, Second Edition. Deng-Ke Yang and Shin-Tson Wu.

Printed and bound by CPI Group (UK) Ltd, Croydon, CR0 4YY

16/04/2025

14658381-0005